WAR MADE NEW

WEAPONS, WARRIORS, AND THE MAKING OF THE MODERN WORLD

MAX BOOT

GOTHAM BOOKS

GOTHAM BOOKS
Published by Penguin Group (USA) Inc.
375 Hudson Street, New York, New York 10014, U.S.A.

Penguin Group (Canada), 90 Eglinton Avenue East, Suite 700, Toronto, Ontario, Canada M4P 2Y3 (a division of Pearson Penguin Canada Inc.); Penguin Books Ltd, 80 Strand, London WC2R 0RL, England; Penguin Ireland, 25 St Stephen's Green, Dublin 2, Ireland (a division of Penguin Books Ltd); Penguin Group (Australia), 250 Camberwell Road, Camberwell, Victoria 3124, Australia (a division of Pearson Australia Group Pty Ltd); Penguin Books India Pvt Ltd, 11 Community Centre, Panchsheel Park, New Delhi–110 017, India; Penguin Group (NZ), 67 Apollo Drive, Rosedale, North Shore 0745, Auckland, New Zealand (a division of Pearson New Zealand Ltd); Penguin Books (South Africa) (Pty) Ltd, 24 Sturdee Avenue, Rosebank, Johannesburg 2196, South Africa

Penguin Books Ltd, Registered Offices: 80 Strand, London WC2R 0RL, England

Published by Gotham Books, a division of Penguin Group (USA) Inc.

Previously published as a Gotham Books hardcover edition, October 2006
First trade paperback printing, August 2007

1 3 5 7 9 10 8 6 4 2

A Council on Foreign Relations Book

Founded in 1921, the Council on Foreign Relations is an independent, national membership organization and a nonpartisan center for scholars dedicated to producing and disseminating ideas so that individual and corporate members, as well as policymakers, journalists, students, and interested citizens in the United States and other countries, can better understand the world and the foreign policy choices facing the United States and other governments. The Council does this by convening meetings; conducting a wide-ranging Studies program; publishing *Foreign Affairs*, the preeminent journal covering international affairs and U.S. foreign policy; maintaining a diverse membership; sponsoring Independent Task Forces; and providing up-to-date information about the world and U.S. foreign policy on the Council's Web site, www.cfr.org.

Maps by David Lindroth

Gotham Books and the skyscraper logo are trademarks of Penguin Group (USA) Inc.

The Library of Congress has catalogued the hardcover edition of this book as follows:
Boot, Max, 1969–
War made new : technology, warfare, and the course of history / Max Boot.
p. cm.
Includes bibliographical references.
ISBN 1-592-40222-4 (hardcover) 978-1-592-40315-8 (paperback)
1. Military history, Modern. 2. Military art and science—Technological innovations. I. Title.
D214.B67 2006
355.0209'03—dc22 2006015518

Printed in the United States of America · Set in Minion with Electra · Designed by Sabrina Bowers

Praise for *War Made New*:

"A timely and important work." —*The Christian Science Monitor*

"Magisterial." —*The Weekly Standard*

"A terrific read; it's a cheap education on how, for half a millennium, machines made war and war made machines. The research is impressive, the judgments are sound—and Max Boot's a strong, clear writer. . . . Riveting. . . . This is a book for both the general reader and reading generals." —Ralph Peters, *New York Post*

"Never does [Boot] bog down in detail—and never does he lose sight of the fact that without good people, good weapons are useless. Boot has bitten off a big chunk of history. But thanks to his knowledge of the facts and his skill in setting them down, he has served up a first-class book." —*St. Louis Post-Dispatch*

"[Boot] is, all hyperbole aside, a modern-day Thucydides, telling the story of war and why it matters." —*The Philadelphia Inquirer*

"Superb . . . utterly fascinating." —*The Washington Times*

"A fascinating look at the complicated relationship between warfare and technological development by a master historian." —Barry Gewen, nytimes.com

"Mr. Boot is ably filling the role occupied for many years by John Keegan, the famed British author of classics like *The Face of War* and *The Mask of Command*. Both use a similar approach: illustrate broad military trends with specific examples, and embed the analysis in an entertaining historical narrative accompanied by commentary. Fans of Mr. Keegan's will enjoy Mr. Boot."
—Bruce Berkowitz, *New York Sun*

"*War Made New* is a tour-de-force of warfare over the past half-millennium. . . . It is fast-paced and reads like a novel. Boot grabs the readers and causes him or her to turn the page to find out what happened next. This is not only essential reading for anyone who is a serious student of warfare, technology, and the like, but for anyone wanting to know what has made the world unfold the way it has over the past five centuries." —*Naval Institute Proceedings*

"Excellent. . . . Cogent and compelling, *War Made New* should be required reading for anyone dealing with military issues." —Military.com

"A vivid and engaging mix of historical narrative and analysis, showing the bloody real-world results of abstract decision-making."
—Victor Davis Hanson, *Commentary*

"Max Boot has the intellectual audacity and meticulous scholarship to rearrange the kaleidoscope of military history. *War Made New* is a classic that must be savored. A wonderful book, combining impressive scholarship and keen insights. It is not possible to read this book without stopping every twenty or so pages to say, 'I didn't know that,' and without frequently pausing to reflect on the future."

—Bing West, *Marine Corps Gazette*

"Readable and informative, this book provides a valuable overview of how military innovations can abruptly affect the course of history. Highly recommended."

—*Library Journal*

"Engaging. . . . Boot distills five hundred years of military history into a well-paced, insightful narrative." —*Publishers Weekly*

"Boot's detail-packed discussion of the impact of military revolutions on the course of modern history makes *War Made New* one of the most provocative, thought-stimulating books in recent memory."

—The Editors, Barnesandnoble.com

"Boot's magisterial grasp of the long trend lines of history is impressive and compelling. . . . *War Made New* is an ambitious effort that ultimately succeeds in capturing the general sweep of history." —Frank Hoffman, *Armed Forces Journal*

"A dazzling history of war." —*Arkansas Democrat Gazette*

"Max Boot has produced another American classic with *War Made New*. . . . Fast-paced. . . . Very readable. . . . Take the time to read this one. It will provide new perspectives." —Col. Walt Ford, *Leatherneck* magazine

"I really enjoyed reading it. Boot is a fantastic writer." —Philip Carter, *Slate*

"The subject of military transformation is one that is difficult to make interesting—some think it impossible—but the book is not just interesting, it is compelling."

—Powerlineblog.com

"Max Boot traces the impact of military revolutions on the course of politics and history over the past 500 years. In doing so, he shows that changes in military technology are limited not to war-fighting alone, but play a decisive role in shaping our world. Sweeping and erudite, while entirely accessible to the lay reader, this work is key for anyone interested in where military revolutions have taken us—and where they might lead in the future." —U.S. Senator John McCain

"[Max Boot] not only tells a remarkable tale, but he compels us all, even those obsessed solely with contemporary military affairs, to ask the right questions and to distinguish what is truly new and revolutionary from what is merely ephemeral. He has rendered a valuable service, and given us a fascinating read at the same time, so we are doubly in his debt."

—Paul Kennedy, professor of History at Yale University and author of *The Rise and Fall of the Great Powers*

"*War Made New* is impressive in scope. What is equally impressive is its unique interpretation of the causal relationship between technology, warfare and the contemporary social milieu. This is a superb thinking person's book which scrutinizes conventional historical wisdom through a new lens."

—Lt. Gen. Bernard E. Trainor, USMC (ret.), co-author of *Cobra II: The Inside Story of the Invasion and Occupation of Iraq*

"Max Boot's book takes hundred of years of tactical battle history and reduces it to an incisive narrative of how war has changed. . . . What is doubly impressive is how he draws surprising, fresh lessons from wars we thought we knew so much about."

—Robert D. Kaplan, author of *Imperial Grunts*

Courtesy of the *Los Angeles Times*

Max Boot is a historian and the author of the award-winning *The Savage Wars of Peace: Small Wars and the Rise of American Power*. A senior fellow in national security studies at the Council on Foreign Relations and a regular contributor to many magazines and newspapers, he also lectures at numerous military schools and advises the Department of Defense on transformation issues.

To Olga and
Yan Kagan

CONTENTS

PART III: THE SECOND INDUSTRIAL REVOLUTION

PART IV: THE INFORMATION REVOLUTION

PART V: REVOLUTIONS PAST, PRESENT, FUTURE

LIST OF MAPS

AUTHOR'S NOTE

All quotations are rendered in modern English, with spelling and capitalization regularized where necessary. Archaic usage makes no sense when translating foreign-language documents; it is also, in my view, an unnecessary distraction when quoting English-language documents of centuries past.

Chinese and Japanese names are rendered Western-style: given name followed by family name (e.g., Heihachiro Togo, not Togo Heihachiro).

Casualties, unless otherwise specified, refers to all those killed, wounded, missing, and taken prisoner.

The structure of armed forces has varied widely over the years, but readers unfamiliar with the military may find it useful to keep in mind the organization of the modern U.S. Army, which is roughly similar to that of most other state-run forces since the Napoleonic era. The smallest standing unit is the squad, which typically consists of ten soldiers led by a staff sergeant. Next comes the platoon—three to four squads (30–40 soldiers) led by a lieutenant; followed by the company, artillery battery, or cavalry troop—three to four platoons (100–200 soldiers) led by a captain; the battalion or armored cavalry squadron—three to seven companies (500–900 soldiers) led by a lieutenant colonel; the brigade or cavalry regiment—three or more maneuver battalions (3,000–5,000 soldiers) led by a colonel; the division—three or more brigades (10,000–18,000 soldiers) led by a major general; the corps—two to five divisions (20,000 to 40,000 soldiers) led by a lieutenant general; the field army—two to five corps (50,000 or more soldiers) led by a lieutenant general or general; and finally, a formation not employed since World War II—the army group: two or more field armies led by a full general.

War is both king of all and father of all, and it has revealed some as gods, others as men; some it has made slaves, others free.

Heraclitus (c. 500 B.C.)

In war, moral considerations account for three-quarters, the actual balance of forces only for the other quarter.

Napoleon Bonaparte (1808)

Tools, or weapons, if only the right ones can be discovered, form 99 percent of victory. . . . Strategy, command, leadership, courage, discipline, supply, organization and all the moral and physical paraphernalia of war are nothing to a high superiority of weapons—at most they go to form the one percent which makes the whole possible.

J. F. C. Fuller (1919)

PROLOGUE:

The Blitzkrieg of 1494

One of the most technologically forward-looking military expeditions in history was launched for the most retrograde of reasons. Fourteen ninety-four was a time of momentous change. Less than half a century had passed since the first printed book had appeared. The Moors and Jews had been expelled from Spain just two years before. Only a year before, an Italian sailor in Spanish employ named Columbus had returned from an overseas voyage claiming to have discovered a new route to the Indies. Yet none of these events loomed as large at the time as the loss of Constantinople forty-one years earlier to the Turks. Byzantium, the bastion of Christianity in the East, had fallen to the Mohammedans! Any self-respecting Christian monarch felt a duty to take up arms. King Charles VIII of France had the means to act and the inclination to do so.

His kingdom had been greatly enlarged and substantially strengthened over the past half century. The English had been kicked out of Normandy and Guienne in 1453 at the end of the Hundred Years' War. In subsequent decades, Armagnac, Burgundy, Provence, Anjou, and Brittany had been wrested from their feudal rulers and added to the crown domains. With France almost at its modern boundaries, Charles VIII presided over the most powerful nation in Europe at a time when the very concept of a "state" was just taking shape.

With his small stature, skinny body, large head, and massive nose, the twenty-four-year-old king hardly looked the part of a mighty monarch. His hands twitched nervously and he stuttered when he spoke. "In body as in mind," wrote a contemporary chronicler, "he is of no great value." He had so little education that he was unable to speak Latin, the language of civilized discourse, and he was able to sign his own name only with great difficulty. But he loved tales of chivalrous derring-do like Thomas Malory's evocation of the Camelot legend, *Le Morte d'Arthur*, published in 1485. Charles longed to be another El Cid, Roland, or Charlemagne—a great hero who would smite the infidels and reclaim the Holy Land. He wanted to go on a Crusade.

The opportunity to act out his anachronistic dream presented itself early in 1494 when the king of Naples died. Charles VIII felt he had a dynastic claim to the throne and he decided to make good on it in order to seize an operating base for use against the Ottoman Empire. Italy was then at the height of the Renaissance. Riches from agriculture, manufacture, and trade—Italy was Europe's gateway, via the Mediterranean, to Africa, the Middle East, and Asia—created a class of wealthy businessmen and nobles who financed an outpouring of art the likes of which the world had never seen before or since. Botticelli, Leonardo, Michelangelo, Raphael, Titian: All were alive in the 1490s. They benefited from a loosening of religious restrictions that was savagely denounced by moralists such as the Florentine friar Girolamo Savonarola, who complained that Pope Alexander VI (one of the Borgia popes) was guilty of incest, murder, and corruption. Wallowing in decadent luxury, Italy let its defenses molder. This problem was compounded by the peninsula's lack of unity. It was split into small states poised in carefully balanced equilibrium. The most powerful were the Papal States, the Republic of Venice, the Duchy of Milan, the Kingdom of Naples, and the Republic of Florence. In all of them, leading families such as the Borgias, Medicis, and Sforzas schemed for power and position with a bag of gold ducats in one hand and a dagger in the other. Seeking an ally against his rivals, the Duke of Milan invited Charles VIII into Italy, and the French king eagerly seized the opportunity.

In the fall of 1494, Charles led some 27,000 men across the Alps, bringing with him a new way of war. Since most Italians disdained military service, they had turned over the protection of their city-states to mercenary captains, the *condottieri* (contractors), whose paid followers, many of them foreigners, were grouped into *compagnie* (companies). They had some primitive cannons and arquebuses—a handheld firearm ignited with a burning match—but mostly they were mounted men-at-arms. Lacking much loyalty to their paymasters, and always acutely conscious that today's enemies could be tomorrow's employers, the *condottieri* evolved a style of warfare that was highly stylized and strikingly ineffectual. Machiavelli told the story, perhaps apocryphal, of a four-hour battle between two mercenary armies that resulted in only one man dying—and that happened when he fell off his horse and suffocated in the mud.

Charles VIII's forces fought with a brutality and determination altogether alien to Italy. Most of his soldiers were French, and they were part of one of the first national armies in Europe since the days of the Roman legions. Slightly fewer than half were traditional cavalrymen carrying lances and swords and protected by bulky suits of plate armor. The rest were a combination of French crossbowmen and archers along with Swiss pikemen,

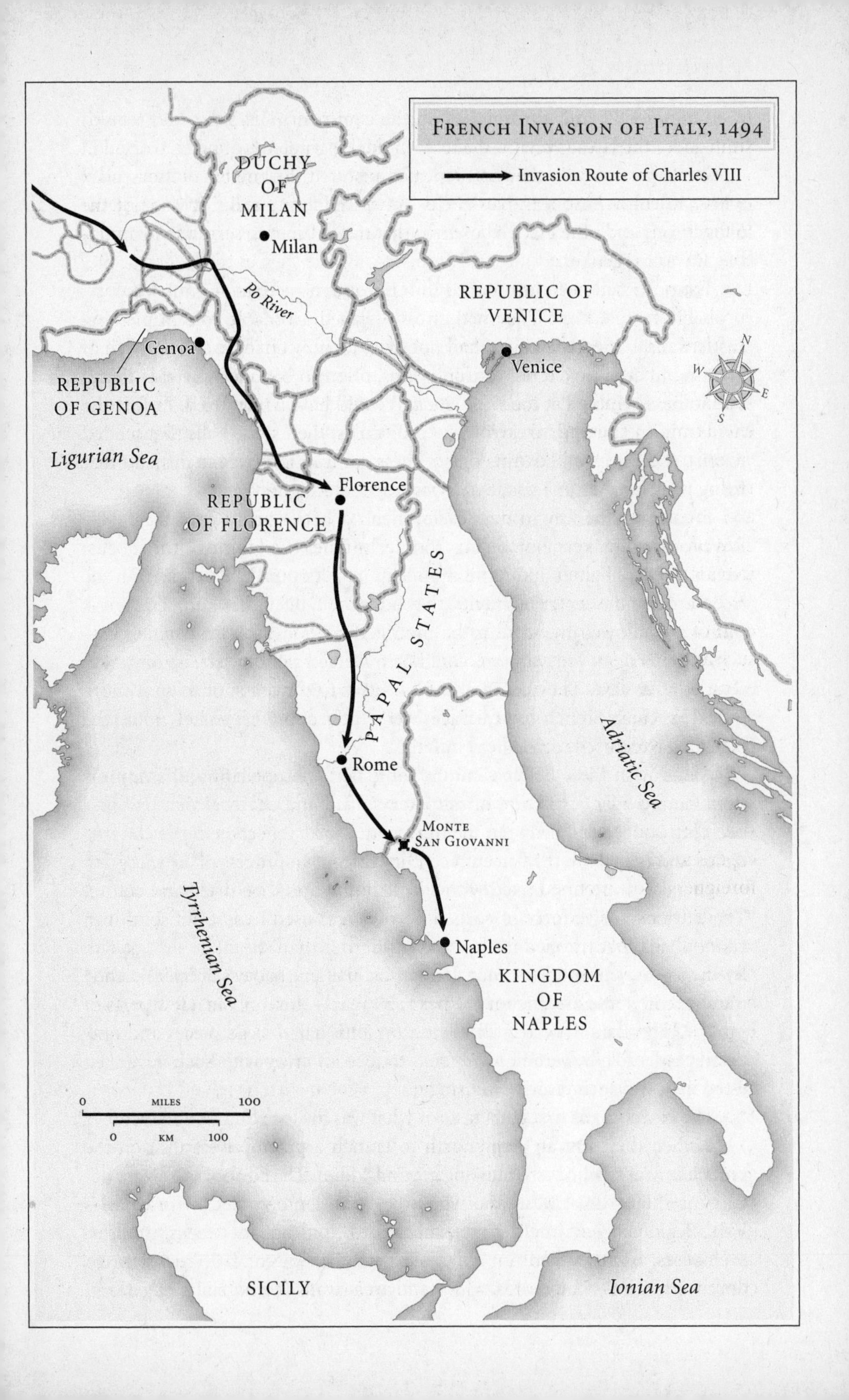

French Invasion of Italy, 1494
Invasion Route of Charles VIII
DUCHY OF MILAN
Milan
Po River
REPUBLIC OF VENICE
Venice
Genoa
REPUBLIC OF GENOA
Ligurian Sea
Florence
REPUBLIC OF FLORENCE
PAPAL STATES
Rome
Monte San Giovanni
Adriatic Sea
Tyrrhenian Sea
Naples
KINGDOM OF NAPLES
0 MILES 100
0 KM 100
SICILY
Ionian Sea
N
W
E
S

then the most feared mercenaries on the continent. The Swiss had revived Alexander the Great's phalanx, and with their eighteen-foot pikes arrayed in a hedgehog formation they had repeatedly bested the knights on horseback who had ruled European battlefields for a millennium. But it was not the Swiss pikemen or the French bowmen that made Charles's army so formidable. It was his artillery.

Cannons had been introduced into Europe more than a century before, probably from China. They had already played a notable role in piercing castle walls, but early artillery had not been terribly effective. It consisted of giant bombards woven out of iron hoops, often so big and unwieldy that it had to be assembled at the siege site and could barely be moved. At first the ammunition consisted of arrowlike projectiles, then stone balls that tended to shatter on impact. Loading one of these contraptions was so difficult that firing three rounds in a single day was considered quite a feat.

France led the way in the development of better artillery in the 1400s. Royal cannon makers, borrowing from techniques used to cast church bells, began to make lighter and more mobile guns out of molded bronze. On the sides, around the center of gravity, they added small handles known as trunnions that allowed the guns to be mounted on two-wheeled wooden carriages. Thus deployed, cannons could be traversed right or left, up or down, with relative ease. They could also be transported from spot to spot more quickly because French gun carriages were hitched to swift horses, not to the plodding oxen used for artillery in Italy.

Along with these better cannons came better propellant and ammunition. Gunpowder, a mixture of saltpeter, sulfur, and charcoal that had first been introduced to Europe in the thirteenth century, became more concentrated and reliable in the fifteenth century through a process of "corning," in which flakes were first mixed with a little liquid and then dried and cut up. The extra explosive force of corned powder was used to spit out solid iron cannonballs that traveled farther and hit harder than the stone shot of old. By the 1490s, smoothbore, muzzle-loading artillery had essentially reached the shape it would assume for the next 350 years—but only in a few parts of Europe. The Italians were still relying on antiquated siege pieces and outdated castles. They were not prepared to face an army with such advanced weaponry and such ferocity in using it.

The Neapolitans first got a taste of what was in store for them in October 1494 when they sent an army north to launch a preemptive attack on the French invaders who were massing around Milan. The Neapolitans were occupying the castle at Mordano when they came into contact with Charles VIII's legions. They might have expected to hold out for weeks, months, even years, but they had not reckoned with the power of Charles's three-dozen-odd guns, which breached the fortress walls (or possibly its gate) in three hours. The Frenchmen rushed inside and slaughtered all the occupants. Then they moved into Tuscany, capturing a series of frontier fortresses with shocking ease. Piero de Medici, ruler of Florence, was so terrified that he gave up power and allowed Charles VIII to enter his city uncontested. The French then walked into Rome before proceeding on to the largest Italian state, the Kingdom of Naples. Their way was barred by the fortress at Monte San Giovanni, which had once withstood a siege of seven years' duration. Charles's cannons breached its walls within eight hours, allowing his troops to slay everyone inside. More than seven hundred Neapolitans died, including women and children, compared to only ten Frenchmen. A contemporary marveled at the impact of French cannon: "They were planted against the walls of a town with such speed, the space between each shot was so little, and the balls flew so quickly, and were impelled with such force, that as much execution was done in a few hours, as formerly, in Italy, in the like number of days."

King Alfonso II of Naples, knowing he could not withstand this onslaught, abdicated his throne. On February 22, 1495, Charles entered Naples—then one of the biggest cities in Europe, with a population of about 150,000—beneath a canopy of gold cloth borne by four Neapolitan noblemen.

Within less than six months, Charles had marched the length and breadth of Italy, brushing aside resistance wherever he went. The speed and power of his artillery, the discipline of his troops, and the savagery with which they were employed—all left the Italians awestruck. Their days of sheltering behind old-fashioned castle walls had ended in a crash of shattered masonry. Contemporaries saw 1494 as a momentous year, much as subsequent generations were to regard 1789, 1914, or 1939.

Not even the setbacks subsequently suffered by *li franzisi* (the French) could dispel the lasting impression they had made. The French army was chased out of Italy in the summer of 1495 by a coalition of Italian states buttressed by the might (and military technology) of newly unified Spain. Charles never got a chance to fight for the Holy Land. Within three years, he was dead—killed, stupidly enough, when he cracked his outsize cranium against a low doorjamb in one of his palaces.

But Charles's invasion had left a lasting legacy—and not only by spreading all over Italy the malady that would become known as syphilis (the French called it "the Italian disease" or "the Neapolitan disease," while to the Italians it was "the French disease"). The French incursion triggered a sixty-year contest for hegemony in Italy pitting France against Spain and the Holy Roman Empire, the House of Valois against the House of Habsburg. The Habsburgs eventually prevailed in the first major coalition war in modern European history, thanks to the efficient Spanish army organized in massive formations of musketeers and pikemen known as *tercios*. The French suffered for not having adopted handheld firearms as readily as they did cannons.

The events of 1494–95 may not have made France a long-term winner, but they definitely left Italy a long-term loser. City-states could flourish in the days of edged weapons but not in the Gunpowder Age. Having failed to develop polities big enough, complex enough, and rich enough to deploy advanced gunpowder armies, the Italians lost control of their own destiny and would not become a unified nation for almost four centuries. In the meantime, their countryside was ravaged by foreign armies, their cities sacked by drunken soldiers. (Italians' persistent, if perhaps unfair, reputation for being poor soldiers—so at odds with the glory of the Roman legions—dates back to these dark days.) Within a few years, Italians like Niccolò Machiavelli were speaking nostalgically about the pre-1494 world as a lost golden age.

Writing in the 1520s, Francesco Guicciardini, a Florentine politician and historian who was a friend of Machiavelli's, recalled how ineffectual combat had been before the French came: "When war broke out, the sides were so evenly balanced, the military methods so slow and the artillery so primitive, that the capture of a castle took up almost a whole campaign. Wars lasted a very long time, and battles ended with very few or no deaths."

All that changed with the arrival of Charles VIII. "The French came upon all this like a sudden tempest which turns everything upside down. . . ." Guicciardini continued. "Wars became sudden and violent, conquering and capturing a state in less time than it used to take to occupy a village; cities were reduced with great speed, in a matter of days and hours rather than months; battles became savage and bloody in the extreme. In fact states now began to be saved or ruined, lost and captured, not according to the plans made in a study as formerly but by feat of arms in the field."

INTRODUCTION:

Revolutions in Military Affairs

A sudden tempest which turns everything upside down.

Those plaintive words, uttered by an Italian of the early sixteenth century, could just as easily have come from a Sudanese warrior in 1898 confronting British machine guns, a Frenchman in 1940 experiencing the German *blitzkrieg*, or an Iraqi in 1991 facing American smart bombs. All found themselves in the midst of a military revolution they did not understand—and all paid a heavy price for their backwardness.

The French invasion of Italy in 1494 inaugurated the modern age in which warfare, which had been relatively static for a thousand years, was to change with bewildering and accelerating rapidity. This process would lead some states to domination, others to oblivion. It would profoundly disturb the balance of power first within Europe and then in the rest of the world, giving rise to the Western hegemony that has not been eclipsed even to this day. It would change the very nature of the state itself, providing a powerful impetus for the rise of modern governments and their inexorable expansion into the leviathans of the twentieth century. Its impact is still being felt, as armed forces around the world grapple with the consequences of information technology that is reshaping war in ways that are as profound as they are unpredictable.

From one perspective, it might seem that in warfare, as in many other realms, change is slow and gradual: that it is characterized, in other words, by a process of continual evolution, not by a few wrenching revolutions. This is to some extent true; we must not pretend that change was more sudden or sweeping than it actually was, or unduly emphasize novelty at the expense of continuity, which is always considerable. But in the military sphere, just as in science, economics, art, or culture, change is not evenly distributed across space and time. Sometimes innovations cluster together to produce a major change in the way people live—or, in the case of the military, the way they die. Obvious examples of transformational technologies include the steam engine in the late eighteenth century and the computer in the late twentieth century, both of which spread from one area to another, transforming everything from production to transportation. When this happens, a *revolution* is said to have occurred.

This book examines four such instances of great change in warfare over the past five hundred years: the Gunpowder Revolution, the Industrial Revolution, the Second Industrial Revolution, and the Information Revolution. Scholars may quibble about exactly how many revolutions have occurred or when they started and ended. But few would deny that at least the first three were periods when new technology combined with new tactics to reshape the face of battle. The verdict on the distinctive nature of the ongoing Information Revolution is not yet in, and some caution is necessary in reaching any preliminary conclusions. But even if the current transformation does not prove as radical in its implications as the preceding revolutionary periods, it has already exerted a substantial impact and one that is likely to grow over time. Indeed, advances in computing, which have made possible precision-guided munitions, stealth aircraft, and stellar surveillance systems, have sparked much of the current interest in what are known as *Revolutions in Military Affairs.*

This term derives from the old Soviet general staff, which, as early as the 1960s, began to speak of a "military technical revolution"—a description first applied to nuclear missiles and then to information technology that multiplied the power of conventional (i.e., nonnuclear) weaponry. The Soviet debate over the impact of these technologies spurred a similar discussion in the United States. By the early 1990s, following the impressive U.S.-led victory in the Gulf War, which served as an advertisement for the power of precision weaponry, it became commonplace to hear heated debate in the hallways of the Pentagon about the "the RMA." (The switch from "military technical revolution" to the "revolution in military affairs" was made to signify that the changes in question involved more than simply new technology.) Radicals suggested that all the old verities of warfare had been outmoded and that sweeping changes were necessary to compete in this brave new world. Reactionaries scoffed that nothing significant had changed and that no revolution had occurred. The debate rages to the present day, not only in the U.S., but also in countries as diverse as Britain, France, China, Russia, and Israel that seek to keep pace with the American armed forces.

Intense as this controversy is, it has been curiously unsatisfying for a historically minded observer. Most books and articles on the current RMA are concerned only with the present and future; most writing on historical RMAs is concerned only with the long-ago past; and studies that take a more comparative approach tend to address only a small audience of specialists.

Moreover, writing on the current RMA is dominated by partisans who seek to either inflate or deflate the importance of information technology, while ignoring the broader political implications of these military transformations.

This book is intended to present a fair-minded account that brings together past and present to offer fresh insights about the future. It does not attempt to construct a definitive model of how military innovations occur; that is a job for social scientists. Nor is it a manifesto calling for a radical overhaul of the world's militaries; that is a job for seers and strategists. This is a work of history that attempts a less ambitious but no less challenging task: trying to tell the story of the past five hundred years of war through the prism of four great revolutions that changed the nature of politics and society as much as they changed combat.

THE ROLE OF TECHNOLOGY

While this book is organized around technical innovations, it is by no means a brief for technological determinism—the theory, held by almost no one in its pure form, that all significant historical developments flow inevitably from the development of certain machines, tools, and weapons. This argument is no more tenable than any attempt to explain all of human history with reference to just one factor, whether that factor is economics, culture, politics, religion, geography, climate, birth order, or the alignment of the stars. Technology itself, it may plausibly be argued, is often not a prime mover but a second-order consequence of food surpluses, urbanization, secularism, political stability, strong patent laws, and other conditions that create a fertile climate for innovation. Moreover, the history of warfare has been profoundly altered by forces, such as the rise of nationalism and democracy, that have little to do with new tools. But just as there is a danger of technological determinism, so there is an equal danger of ignoring the effects of technology and thinking that willing bodies and stout hearts can overcome anything. This was the view of many military strategists prior to World War I. A few years of trench warfare revealed that no amount of élan could allow a soldier to outrun a machine-gun bullet or stop an artillery shell. The tools of war do matter.

This book seeks to avoid the Scylla of overvaluing technology and the Charybdis of undervaluing it—the extreme views represented in the opening epigrams from J. F. C. Fuller and Napoleon Bonaparte. I agree with another historian who wrote "that we must steer between two dangerous determinisms: the technological—'what can be done will be done'—and the psychological—'where there's a will there's a way.'" My view is that technology sets the parameters of the possible; it creates the *potential* for a military revolution. The extent to which various societies and their armies exploit the possibilities inherent in new tools of war and thereby create an *actual* military revolution depends on organization, strategy, tactics, leadership, training, morale, and other human factors. Ultimately, a military revolution, like a scientific revolution, demands a "paradigm shift" from one set of assumptions to another. For this reason, this book will highlight the personalities of important commanders—the likes of Gustavus Adolphus, Helmuth von Moltke, Isoroku Yamamoto, and Curtis LeMay—who struggled to take advantage of new weapons. No technical advance by itself made a revolution; it was how people responded to technology that produced seismic shifts in warfare.

Readers should not be misled by the section and chapter headings which highlight innovations such as "gunpowder," "rifles," and "tanks." The scope of this work is much broader than that. The weapons mentioned are only a shorthand way of referring to more sweeping changes that occurred at the organizational and doctrinal level.

Nor do I mean to emphasize military revolutions to the exclusion of everything else. I will not attempt to challenge most of the theses put forward in a number of prominent recent works that have sought to explain the course of human development. Among the more celebrated are Jared Diamond's *Guns, Germs, and Steel*, which emphasizes the importance of geography, demography, and environment; Paul Kennedy's *The Rise and Fall of the Great Powers*, which stresses economics; and David Landes's *The Wealth and Poverty of Nations* and Victor Davis Hanson's *Carnage and Culture*, both of which focus, in different ways, on cultural factors. In my view, all of these books have a large measure of validity and are not, for the most part, mutually contradictory. Rather than attempting to supplant them, this book will supplement them by highlighting the importance of certain vital military developments in the making of the modern world.

Each section begins with a brief introduction to make clear that what was happening on the battlefield was only part of far-reaching changes occurring in the wider society. Cannons and muskets were only two of many innovations that swept Europe during the Renaissance; machine guns, rifles, and gunboats were only some of the products mass-produced in factories during the Industrial Revolution; the bombers and tanks of World War II were not too dissimilar from civilian airplanes and tractors; and, in our own day, computing power has not only made it possible for the U.S. armed forces to track vast numbers of targets but also made it possible for Wal-Mart to track vast numbers of detergents. Inevitably, there was a lag, ranging from a few decades to a few centuries, between the initial development of a technology and the moment when it transformed the battlefield.

Following each scene-setter, we will examine key battles that show how the revolution in question played out. In describing these developments, I will seek to avoid the anodyne language typically employed by soldiers and academics. It is all too easy to speak of flanks turned, positions entrenched, battles won and lost, and to lose sight of limbs shattered, heads blown off, bodies reduced to bloody pulp. It is typical to speak of weapons in terms of range and rate of fire, and it is important to do so. But it is also important to investigate the impact of grapeshot on a skirmish line, of a machine gun on a formation of spear-toting warriors, of an incendiary bomb on a neighborhood packed with civilians. Much of the debate over "military revolutions" today has a sterile quality to it, as if the subject under discussion were a better way to manufacture microchips or to deploy a new type of broadband fiber. My intention is to return the focus where it belongs—to the soldier struggling to kill or avoid being killed, and to his commander struggling to master the remorseless logic of carnage.

In the business world, the usefulness of technical innovations can be checked easily by glancing at the stock tables. In the military world, the only test worth anything is the test of battle, and it is here that we must look for the impact of technology upon military affairs.

WHICH BATTLES, AND WHY

There is a long tradition of "battle histories," stretching all the way back to Sir Edward Creasy's 1850 classic, *Fifteen Decisive Battles of the World*. Unlike Sir Edward's selections, or those of J. F. C. Fuller in a later work in a similar vein, the battles described here were not chosen because they were historical turning points, though some were. Nor were they chosen simply because they are interesting and dramatic, though all are. They were chosen with three primary considerations in mind:

- First, to illustrate different aspects of the various revolutions—battles pitting Europeans against one another and against non-Europeans; battles on land, sea, and, in the later stages, in the air.

- Second, to demonstrate the development of the four revolutions over time. Needless to say, technological revolutions, unlike some political ones, do not occur in a day. They may stretch out over decades, even centuries, and change shape as they go. The Gunpowder Revolution started with clumsy arquebuses and wound up with more deadly flintlocks; the First Industrial Revolution began with single-shot rifles and wound up with machine guns; the Second Industrial Revolution began with propeller-driven biplanes and wound up with jets firing missiles; the Information Revolution began with laser-guided smart bombs and where it will end no one yet knows. The battles described in this book are meant to offer snapshots of these developments at various stages.

- Finally, to show how one side can gain a decisive edge over an enemy—though not necessarily keep it. What role does technology play, as opposed to training, strategy, leadership, logistics, and other factors?

Readers might wonder why there is no separate heading for the atomic revolution and the concomitant rocketry revolution. Books written soon after 1945 on this subject, the history of warfare, inevitably treated nuclear weapons as a transformative event and speculated feverishly about their impact on the battlefield. The U.S. Army even acted on this advice in the 1950s by creating the Pentomic Army, which fielded atomic artillery shells. As it happens, while the atomic bomb helped to deter some wars from happening (for instance, direct combat between the U.S. and USSR), and helped to limit the reach of some wars that did occur (for instance, the Korean War), its impact on the operational level of war was . . . close to nil. At least so far. Because atomic weapons have not had much direct effect on warfare, they will not receive a separate chapter in this volume, although their implications for the future will be discussed in the concluding pages.

Readers also may be surprised to see many prominent conflicts slighted here. Why isn't there more on the Seven Years' War? The Napoleonic Wars? The American Civil War? The First World War? The Vietnam War? The Six-Day War? It can rightfully be argued that many battles that are not described—whether Lepanto (1571), Rocroi (1643), Yorktown (1781), Trafalgar (1805), Waterloo (1815), Gettysburg (1863), the Marne (1914), Stalingrad (1942–43), or Tet (1968)—had profound political repercussions, often greater than those of the battles that *are* described. I can only reply that this book is not intended to be a comprehensive history of warfare, nor even a comprehensive history of decisive battles. Its focus is strictly on the impact of military revolutions on the modern world. What gets left out are wars and battles that were largely static from the standpoint of tactics and technology, a description that applies to most conflicts through the ages. The focus here is not on grinding, attritional struggles but on technological and tactical breakthroughs that illustrate the transformation of the military art.

Other examples could just as easily have been chosen. I could have chronicled the Battle of the Downs (1639) instead of the Battle of the Spanish Armada; the conquest of northern Nigeria (1902–3) instead of the conquest of the Sudan; the invasion of Poland (1939) instead of the invasion of France. Obviously any selection of a dozen or so battles intended to highlight military developments over the past five hundred years will leave out a good deal. But to cover everything would be, in effect, to cover nothing.

FOUR REVOLUTIONS

For the Gunpowder Revolution (a shorthand way of describing what other historians call "the military revolution of early modern Europe"), I have chosen to look at the Battles of the Spanish Armada (1588), Breitenfeld and Lützen (1631, 1632), and Assaye (1803). The defeat of the Armada heralded the rise of England and the eclipse of Spain, but how were Elizabeth I's commanders able to achieve this feat? By mastery of the emerging technologies of oceangoing sailing ships and heavy cannon. Breitenfeld and Lützen, turning points of the Thirty Years' War, were the greatest victories won by the Swedish King Gustavus Adolphus. His forces, armed with matchlock muskets, pikes, and cannons, introduced many of the organizational techniques that still define barracks life today. Assaye is less known but worth studying for a number of reasons: It was a major early success for Arthur Wellesley, not yet the Duke of Wellington; it was a significant step in the British conquest of India; and, most important for our purposes, it was a demonstration of how European soldiers, armed with flintlock muskets and bayonets and schooled in a "battle culture of forbearance," could defeat Asian armies that had a major edge in manpower and artillery but a crippling deficit in discipline and tactics.

For the Industrial Revolution (which includes the democratization of warfare associated with the French Revolution), I have chosen to look at the Battles of Königgrätz (1866), Omdurman (1898), and Tsushima (1905). At Königgrätz (also known as the Battle of Sadowa), one of the most important stepping-stones toward the creation of the German Reich, the Prussians dealt the Austrian Empire a decisive defeat, primarily because their general staff had figured out how to make effective use of new technologies such as railroads and breech-loading rifles. These technologies, along with others, such as machine guns, quick-firing artillery, and field telephones, were soon incorporated by all the armies of Europe, producing a bloody stalemate on the Western Front between 1914 and 1918.

If the Industrial Revolution did not give one European power a clear advantage over another, at least not for long, it vastly increased the gap between the West and the Rest, making it relatively easy for a handful of Europeans to conquer much of Asia and Africa. To illustrate this aspect of the story, I examine the Battle of Omdurman. It later became famous because of the participation of a young lieutenant named Winston Churchill, but Omdurman was really notable for General Horatio Herbert Kitchener's industrial techniques, utilizing railroads, gunboats, and machine guns to make the reconquest of the Sudan a foregone conclusion.

One of the few non-Western states able to emulate the industrial warfare of the Europeans in this period was Japan. Its achievement was particularly impressive in the naval realm. In the late nineteenth century, the line-of-battle ship, which had changed little since the days of the Spanish Armada, was transformed by the replacement of sails with steam engines, wooden hulls with iron and steel, muzzle-loading smoothbore cannons with breech-loading rifled guns, and solid shot with explosive shells. Japan, which had only recently ended a long period of self-imposed isolation, took full advantage of these inventions to build a navy that demolished the less capable Russian fleet at the Battle of Tsushima, altering the balance of power in the Pacific for decades.

For the Second Industrial Revolution, which transformed warfare in the 1920s and 1930s and whose repercussions were felt in the 1940s, I will examine the fall of France (1940), the Japanese attack on Pearl Harbor (1941), and the U.S. firebombing of Tokyo (1945). The blitzkrieg is one of the best-known examples of a "military technical revolution"—and one of the most misunderstood by the general public. It is commonly assumed, based on the ease with which German armies overran Poland, Norway, Denmark, the Low Countries, and France, that they possessed a big technological and numerical edge over their adversaries. Nothing could be further from the truth; Hitler actually fielded fewer tanks and aircraft than the British and French, and the quality of the Allied weapons was in many cases higher than the Germans'. The German edge lay in their superior ability to coordinate their forces, and in their high quality of leadership, training, and morale. They figured out how to make the best use of the technology of the day; the Allies did not.

The Japanese, at least initially, also had an edge in morale and training. Moreover, by the time of the Pearl Harbor raid, Japan was a world leader in aircraft carriers, naval aviation, torpedoes, and night fighting. Tokyo fully exploited these advantages to destroy much of the U.S. Pacific Fleet at its anchorage and, during the following six months, to overrun all of the principal Western bases between Hawaii and Australia. Eventually, the German and Japanese advantages were negated, in part by the sheer weight of materiel on the Allied side. But the Allies also possessed some superior weapons, such as the American long-range bombers, which General Curtis LeMay employed with brutal efficiency to bring Japan to its knees even before the two atomic bombs were dropped. All of these weapons systems were enhanced by the use of radio, radar, and other inventions of the Second Industrial Age.

The fourth and final revolution to be examined is the ongoing Information Revolution. To explore its impact, I look at the Gulf War (1991),

Afghanistan (2001), and the Iraq War (2003–5). There is some debate over whether the Gulf War was the last of the industrial wars or the first of the Information Age wars. There is merit in both views. The clash of tank armies in the desert is reminiscent of the great battles of World War II. But the disproportionate impact of "smart" bombs, cruise missiles, satellite navigation systems, and stealth planes suggests that the 1991 conflict rightly belongs to the new era, even if only to its early stages.

The success of precision-guided munitions in Desert Storm convinced many of America's enemies that they could not hope to win a conventional struggle. Al Qaeda tried a different tack with its September 11, 2001, attacks which killed almost three thousand people. This was a spectacularly evil—and spectacularly effective—example of guerrilla and terrorist tactics that have always given conventional armies fits. But the U.S. military showed in Afghanistan that a combination of highly trained commandos and precision weaponry—a high-low mix of technologies—could subdue, at least temporarily, even a land renowned as the graveyard of empires.

The U.S. armed forces produced equally spectacular results in the early stages of the Iraq War, concluding with the fall of Baghdad on April 9, 2003. But the aftermath, with its messy guerrilla war and demanding nation-building, proved a challenge for the high-tech American military. Stealth bombers and nuclear-powered aircraft carriers are not much use for keeping order in Fallujah or Baghdad, leading many to question whether technological supremacy is all that it was cracked up to be. Had a few spectacular victories led America into the kind of strategic overstretch that brought down the German and Japanese empires after their initial successes in World War II? Chapter Twelve will examine the first two years of a conflict that, as of this writing, remains far from over.

The summary of the Information Revolution and Chapter Thirteen move beyond contemporary events and into the realm of the immediate future. I examine where the current military revolution is going and what other revolutions are on the horizon in such fields as robotics, cyberwar, biological and chemical weapons, space systems, and nanotechnology. The Epilogue will survey the lessons of the last five hundred years for clues about what lies ahead.

FIVE THEMES

As this book explores four military revolutions, five major themes will emerge:

First, that technology alone rarely confers an insurmountable military edge; tactics, organization, training, leadership, and other products of an effective bureaucracy are necessary to realize the full potential of new inventions.

For this reason, ever since the rise of modern nation-states in the sixteenth and seventeenth centuries, shifts in military power have been closely associated with shifts in governance.

Second, that countries able to take advantage of these shifts have been history's winners while those that have fallen behind in harnessing military innovations have usually been consigned to irrelevance or oblivion. Thus, each revolution has been accompanied by a shift in the international balance of power.

Third, that even if a country figures out how to harness military power, it still needs the wisdom to know the capabilities and limitations of its war machine and thereby avoid squandering it on impossible projects, as have too many successful innovators from Charles VIII to Adolf Hitler.

Fourth, that even with the best strategy, tactics, and technology in the world, no military revolution has ever conferred an indefinite advantage upon its early innovators. Rivals inevitably copy what they can and come up with tactics or technologies to blunt the effectiveness of what they cannot produce or acquire.

And, fifth, that innovation has been speeding up. It took at least two hundred years for the Gunpowder Revolution to come to fruition (c. 1500–1700); one hundred and fifty years for the First Industrial Revolution (c. 1750–1900); forty years for the Second Industrial Revolution (c. 1900–1940); and just thirty years for the Information Revolution (c. 1970–2000). That means that keeping up with the pace of change is getting harder than ever, and the risks of getting left behind are rising. Today, there is no room for error.

America's early lead in the Information Revolution can easily be lost—it may be being lost already—if it does not stay at the forefront of military developments. Other countries and even subnational entities such as al Qaeda have an opportunity to exert power that would have been unthinkable before the spread of personal computers, cell phones, satellite navigation devices, and other Information Age technologies.

There is no magic formula for coping with the complexities of changing warfare. The course of future developments can be glimpsed only fleetingly and indistinctly through the fog of contemporary confusion. As any reader of H. G. Wells, Jules Verne, or Isaac Asimov must know, few prognostications about the future have ever come true. The best way to figure out the path ahead is to examine how we arrived at this point.

In the case of modern warfare, that means casting our eyes back five hundred years to a time when Europe was just emerging from the Middle Ages and cannons and muskets were as new and unsettling as stealth bombers and smart bombs are today.

PART I

THE GUNPOWDER REVOLUTION

We all know, that what now makes a nation formidable, is not the number nor riches of its inhabitants, but the number of ships of war provided with able seamen, and the number of regular and well disciplined troops they have at their command.

English parliamentarians (1734)

The Rise of the Gunpowder Age

Sometime between 1504 and 1515—a decade or more after the French invasion of Italy, which is often cited as the dividing line between modern and pre-modern times—Desiderius Erasmus and John Colet paid a visit to the famous shrine of Thomas à Becket in Canterbury, England. According to Erasmus's account, the two writers were revolted by the opulence of the shrine and the garish collection of religious relics on display. When a beggar offered them the supposed shoes of St. Thomas to be kissed, Colet said scornfully to Erasmus, "What, do these brutes imagine that we must kiss every good man's shoes? Why, by the same rule, they might offer his spittle to be kissed, or what else."

It is hard to imagine such a scene transpiring a century earlier—or if it had and their comments were overhead, that the two great humanists would have escaped with their heads still on their shoulders. But this was no longer the Europe of the Middle Ages. Between approximately 1450 and 1650—the period dubbed the Renaissance, Reformation, and Counter-Reformation by historians of a later epoch—the continent changed more than it had in a millennium. The transformation was incremental and incomplete, more advanced in some regions than in others, and it grew out of, rather than marking a sharp break with, medieval Christian civilization, but it was nevertheless hugely significant. Any brief list of the most important changes would have to include the following:

Once-weak kings became powerful, even absolute, monarchs. The autonomy of nobles correspondingly declined and feudal levies were largely replaced by standing armies in royal employ. Once-all-powerful guilds saw their grip on manufacturing and trade weaken as new market structures developed. Economic exchange came to be based on cash, not on barter and obligation. Wealth began to be more glorified than poverty. Towns grew in size and importance. Science began to challenge religion. Vernacular tongues began to displace Latin among the educated classes. Schools and universities added the study of man ("humanities") to the study of God. Artists turned to contemporary and classical themes, not just to those of Christianity. And, as the example of Erasmus and Colet would indicate, certitude—some would say credulity—gave way in some quarters to skepticism as the defining mode of thought, while for many others one orthodoxy (Protestantism) was simply substituted for another (Catholicism).

Europe had changed, and before long the rest of the world would, too. Before this period—before 1500 or so—Europe had been an isolated, splintered, and relatively impoverished region, trailing badly in many areas of science and culture behind the cosmopolitan polities of Turkey, Persia, Arabia, India, China, and Japan. The mightiest military forces in the world until at least the 1400s belonged to the Mongols and Turks, whose mounted archers terrorized much of Asia and penetrated into southern and eastern Europe. On those infrequent occasions when Europeans fought warriors of other cultures—for instance, the Second and Third Crusades in the twelfth century—they usually lost. In 1450, at the dawn of the Gunpowder Age, Europeans controlled just 15 percent of the earth's landmass. But by the early sixteenth century, Europe would emerge as the richest, most dynamic, and most powerful region on the planet. In the years ahead, its explorers, merchants, preachers, settlers, sailors, and soldiers would subdue much of the rest of mankind, controlling 35 percent of the earth's landmass by 1800 and a staggering 84 percent by 1914.

THE IMPACT OF FIREARMS

Technology played a crucial role in these momentous developments. The increase in iron production, the substitution of all-iron for wooden plows, the introduction of carts with swiveling front axles, the development of three-field crop rotation, the spread of watermills, windmills and sail-driven cargo ships—all these improvements led to a marked increase in agricultural productivity, trade, and hence prosperity in the late Middle Ages. This, in turn, made possible the rise of magnificent cities like Florence, Venice, Bruges, and Antwerp, where the culture of the Renaissance flourished.

The most important invention of all was movable type, which, when combined with cheap paper (both Chinese innovations introduced to Europe in the thirteenth century), helped to disseminate knowledge more widely than ever before. The first printed book was the Gutenberg Bible. It came out in 1455, at a time when the total number of tomes in Europe was no more than a hundred thousand. Within five years, an estimated six million volumes were crowding the bookshelves of Europe—a figure that, fifty years later, had climbed to between fifteen million and twenty million. Printing proved to be a powerful force for advancement but also a subversive one, because many of the new books spread doctrines such as Copernicus's heliocentric hypothesis (published in 1543) and Martin Luther's denunciations of the Roman Catholic Church (starting in 1517). The Popes reacted by introducing censorship, but the relentless march of modernity could not be reversed.

Military technology, too, played a vital role in the waning of the Middle Ages. The feudal order had been based on the battlefield superiority of heavy cavalry. The vast estates held by noblemen were designed to produce and support mounted warriors, while the courtly culture of chivalry glamorized their accomplishments. Historian Lynn White argued that the dominance of the armored knights was made possible by advances in horse breeding and the introduction of the stirrup in early eighth-century Europe. There had been cavalry in the days of antiquity but it had been light cavalry, utilizing bows and lacking armor, because absent a stirrup it is hard to fight effectively from the saddle. The stirrup made possible a melding of man and horse, leading to the creation of virtual centaurs whose shock effect in battle proved irresistible. At least at first.

The dominance of the man on horseback was challenged during the fourteenth century in what some historians call the infantry revolution. English longbowmen and Swiss pikeman proved to be more than a match for cumbersome heavy cavalry, the pikemen winning their first notable victory at Laupen in 1339, the longbowmen at Crécy seven years later. Firing longbows or holding pikes were not activities deemed worthy of a gentleman in medieval Europe, but warfare was beginning to be democratized. Politics and society would soon follow.

The coup de grâce to the old order came from the introduction of gunpowder. Like the stirrup, compass, paper, and much else, this was a Chinese invention that made its way west. Gunpowder is thought to have appeared in China in the ninth century or even earlier. Contrary to the old misconception that the Chinese employed it only for fireworks, gunpowder was used from the start to create weapons such as rockets, grenades, flamethrowers, bombs, and mines. By the twelfth century, after considerable experimentation designed to concentrate its explosive effect, there is evidence of gunpowder being used to propel projectiles such as flaming arrows, stones, and bombs out of a tube. Eventually, the gun replaced the crossbow as the primary missile weapon of Chinese armies. By that time, however, gunpowder had already spread westward.

The first record of gunpowder in Europe dates from 1267, the first record of guns from 1326. By the end of the 1400s, thanks to advances in iron and copper mining, metallurgy, and gunpowder manufacture, Europeans were making firearms in great quantity and great variety, from enormous cannons to handheld arquebuses. If one of the essential characteristics of modernity is the substitution of chemical for muscle power, then firearms may be regarded as the first modern invention.

The spread of gunpowder was not welcomed by everyone. The nobility, in particular, did not like weapons that rendered obsolete old notions of chivalry and allowed, as one contemporary complained, "so many brave and valiant men" to be killed by "cowards and shirkers who would not dare to look in the face the men they bring down from a distance with their wretched bullets." Some cavaliers were so outraged by this "instrument sent from hell" (to quote the fourteenth-century poet Petrarch) that they cut off the hands or pierced out the eyes of captured arquebusiers. Such rearguard actions notwithstanding, no European captain seeking competitive advantage against his rivals could possibly afford to boycott firearms, the impact of which was so powerfully demonstrated in the French invasion of the Italian peninsula in 1494. In Japan, the shogun eventually unified the country and created a long era of peace (c. 1615–1853) when guns fell into disuse; no European nation experienced a comparable period of peace after the fall of the Roman Empire.

Firearms were by no means a European monopoly—Mogul India, Ming China, Safavid Persia, Choson Korea, Ottoman Turkey, and Tokugawa Japan also made effective use of them—but Europe early on became the worldwide leader in their production and development. The reasons why this was so continue to be hotly debated. Some historians point to Europe's political and geographical fragmentation, which spurred an arms race; others to European culture, with its rationalism, liberalism, and dynamism, which spurred scientific advances in all areas; still others to European markets, which were freer and less regulated than elsewhere and hence provided greater incentives for innovation. No doubt all of these factors helped Europe to pull ahead of its competitors, including the originators of firearms, the Chinese. China and many other non-Western countries continued producing guns, but by 1500 their weapons were markedly inferior to those being crafted in the workshops of Europe. Even the siege guns used by the Ottomans to conquer Constantinople in 1453, including two titanic cannons firing stone projectiles weighing more than eight hundred pounds, were created not by a Turk but by a Hungarian in Sultan Mehmed II's employ.

The most immediate impact of firearms, once they became more reliable, was to end the military ascendancy of the horse archers of central Asia. Cavalrymen equipped with bows and arrows were no match for infantrymen armed with guns, and since the Mongolian nomads could not manufacture firearms of their own, their reign of terror came to an end. The impact of firearms in altering the intra-European balance of power was equally profound, if more complex.

In the popular imagination, the spread of artillery ended the feudal order by knocking down the walls of the castles that were nobles' power centers. There is much truth to this account, but it is incomplete. Artillery did make quick work in the fifteenth century of many of the old stone castles scattered around Europe. As vividly demonstrated by the French invasion of Italy, fortresses that had once held out for years now fell in hours. Using cannons, the French were able to expel the English from France, the Spaniards to kick the Moors out of Spain, the Ottomans to take Constantinople. In 1519, Machiavelli wrote, "No walls exist, however thick, that artillery cannot destroy in a few days."

This triumph of offensive technology was, however, transitory. Charles VIII's invasion spurred the best minds in Italy, including Michelangelo and Leonardo da Vinci, to come up with an answer to the formidable French artillery. The Italians built thick, low-slung earthwork fortresses bristling with artillery ingeniously positioned in angled bastions to command interlocking fields of fire. The lower profile of the ramparts, protected by loose-packed dirt, made them harder for cannonballs to penetrate; the bastions in front of the castle, packed full of artillery, made it difficult for enemy soldiers to get close enough to even try. The *trace italienne*, as this Italian-style fortress was known, was developed too late to save Italy from being ravaged by foreign armies. But once this fortress design spread through Italy and then the rest of Europe beginning in the 1530s, the advantage enjoyed by attacking armies was largely negated. (The edge would swing back to the offense in the early seventeenth century, thanks in large measure to the Dutch and Swedish innovations that are the subject of Chapter Two.) These new fortresses could be taken only by a very large army and then only at great length. Their prevalence meant that no conqueror could hope to unify all of Europe.

The cost of both a state-of-the-art fortress and the forces needed to besiege it properly was steep. When Charles VIII's successor, King Louis XII of France, asked what would be necessary to carry out his planned invasion of Milan in 1499, one of his advisers replied bluntly, "Money, more money, and again more money." The petty lords of Europe did not have enough money. To compete in the Gunpowder Age required the resources of a super-lord, a king, ruling over a large kingdom providing substantial revenues. Thus the dictates of the battlefield—or the siege site—gave a powerful impetus to the development of sovereign states.

A circular effect developed: Richer monarchs could afford larger armies; the possession of these armies in turn allowed those monarchs, at least in theory, to become richer still by expanding their sphere of control. It did not always work out that way in practice (military expeditions often cost more than they made), but there is a fair amount of truth to social scientist Charles Tilly's aphorism: "War made the state, and the state made war."

THE HABSBURG BID FOR MASTERY

France, as we have seen, was an early leader in applying the new military technology in Europe. Spain and Portugal led the way in exporting it abroad. By the mid-sixteenth century the Portuguese had established forts and trading stations along the coasts of Africa, India, China, the East Indies, and South America, and the Spanish had carved out a large empire in the New World. These outposts, which returned great wealth to the Iberian peninsula, would not have been possible without some vital military innovations. The Portuguese and Spanish reached the Americas, Asia, and Africa by utilizing oceangoing, sail-driven galleons; they never could have made it in the oar-driven galleys of old. Once they arrived, steel blades, crossbows, cannons, and arquebuses allowed the Iberians to make short work of native opposition.

Even after half a millennium, the feats of the conquistadors defy belief. Hernán Cortés conquered the Aztec empire (estimated population: eight million to ten million) with eight hundred men and sixteen horses. Francisco Pizarro defeated the Inca empire (estimated population: nine million to sixteen million) with just 180 men and eighty horses. Native allies and the spread of European diseases helped a great deal, but it was the Spaniards' military technology that made the difference—otherwise the Aztec and Inca empires would have been overthrown long before by their native enemies. The Spanish infantry proved almost as effective on the battlefields of Europe. In 1525 they inflicted a crushing defeat on the French at the Battle of Pavia (near Milan), ensuring that the Habsburgs, not the Valois, would dominate Italy.

The head of the House of Habsburg was Emperor Charles V. Unlike Louis XIV, Napoleon, Wilhelm II, Hitler, or Stalin, he had no conscious design to dominate Europe, but his web of inheritances and acquisitions was pushing him in the same direction as those latter-day conquerors. His domains encompassed Spain, Germany, Austria, Alsace, Franche-Comté, Bohemia, Hungary, the Netherlands, much of Italy, and the Americas. Upon his abdication and retirement to a monastery in 1555–1556, Charles V split his realm. His younger brother, Ferdinand, was given the Holy Roman Empire and the Habsburg lands in Austria, Bohemia, and Hungary. To his son, Philip, went everything else. Philip added Portugal and its colonies to his domain by right of dynastic succession in 1580, his mother being a Portuguese princess.

Like his father, King Philip II saw himself as "God's standard bearer," defender of Christendom against "infidels" without and "heretics" within. He once proclaimed, "If my own son was a heretic, I would carry wood to burn him myself." Although the Muslim threat had receded somewhat (the Ottomans were repulsed at Vienna in 1529 and defeated at Lepanto in 1571), the number of heretics had been growing steadily since that fateful day in October 1517 when Martin Luther had nailed his 95 theses protesting the sale of indulgences and other papal abuses to the door of the church in Wittenberg.

The battle pitting Reformation vs. Counter-Reformation was only one of many that convulsed Europe during the early modern era. Religious rivalries melded with dynastic ones to produce a particularly bellicose age. "Between 1480 and 1700," writes historian Frank Tallett, "England was involved in 29 wars, France in 34, Spain in 36 and the [Holy Roman] Empire in 25." As their names imply—the Eighty Years' War (also known as the Dutch War of Independence), the Thirty Years' War—many of these struggles were quite protracted. This nearly constant strife proved a powerful impetus for the development of nation-states with their attendant ideology of nationalism. (The very term *state* entered the French vernacular during the Italian Wars of the early sixteenth century.) All this fighting also led to the development and deployment of new tools of war, especially cannons, muskets, and galleons. By the mid-sixteenth century, the Habsburgs had seized from the French the early advantage in utilizing these innovations, but their lead—and with it their empire—would be challenged by increasingly assertive competitors such as England, Holland, and Sweden. The ensuing battle for control of Europe and eventually the entire world would be decided in substantial measure by mastery of the tactics and technology of the Gunpowder Age.

CHAPTER 1

SAIL AND SHOT:

The Spanish Armada, July 31–September 21, 1588

They first glimpsed the southern coastline of England through the afternoon mist and haze on Friday, July 29, 1588. The Spanish did not realize that they, in turn, had been spotted by an English patrol boat, the *Golden Hind*. Its captain, Thomas Fleming, raced back to Plymouth Sound, where the bulk of the Royal Navy was gathering. Legend has it that he found the English vice admiral, Sir Francis Drake, playing bowls—akin to the modern Italian game of bocce—on Plymouth Hoe. The old privateer was said to have drawled insouciantly in his West Country accent, "We have time enough to finish the game and beat the Spaniards too." Scholarly research has cast doubt on the veracity of this tale, but it has a plausible ring because Drake would have realized that with the tide and wind against him there was no chance of leaving the anchorage immediately. That night, the English began warping out of the harbor, a laborious process that involved towing their warships with rowboats. By the afternoon of Saturday, July 30, the English fleet was fully under way. They caught a brief glimpse of the Spanish Armada in the distance as it began its advance up the English Channel, but the arrival of storm clouds and nightfall prevented any chance of a clash.

The first fight between these mighty fleets would not occur until the morning of Sunday, July 31. The day dawned clear and bright, with a light wind blowing from west-northwest. Somehow during the night the English fleet, commanded by Lord Howard of Effingham, had managed to slip around to the rear of their Spanish foes. The English now had the wind gauge, meaning the breeze was at their backs—a crucial advantage in the age of sail, because it gave them the initiative. But the Royal Navy was badly outnumbered. With only part of his force present, Lord Howard had just sixty-four ships. The Spanish commander, the Duke of Medina Sidonia, had at least 125.

As the fleets maneuvered within sight of shore, they must have presented quite a spectacle: Tarred black hulls glistening in the sun, canvas sails and colorful banners rippling in the breeze, massive bronze and iron guns bristling from every porthole. The tens of thousands of men packed aboard these warships were no less impressive in appearance. Uniforms had not yet become standard in Western militaries, while suits of armor had gone out of style. Many of the combatants, especially on the Spanish side, wore brightly colored finery—plumed hats, doublets, cloaks, ruffle collars, sheer hose, velvet breeches, buckled shoes—that would have made them look to the modern eye like fashionable dandies rather than the ruthless killers they were.

Upon seeing the English on the horizon, the Spanish ships had shifted from "line of march"—they employed army terminology even at sea—into line of battle. Medina Sidonia, a landlubber who had been barely four months in command, arranged his fleet in the classic crescent formation that had characterized naval warfare since the battle of Salamis in 480 B.C. From the days of antiquity until the fifteenth century, warfare on the water had been dominated by rowed galleys. The development of powerful, three-masted sailing rigs meant that by 1588 almost all warships were driven by the wind, not by straining rows of oarsmen. But the Spanish made few tactical adjustments for this new age. They arrayed their vessels in a close-packed mass, just as the ancient Greeks had. The bulk of the armada was in the center; jutting out on either end were two "horns," ready to close and crush any foe unlucky enough to fall within their grasp.

The English sailors marveled at the ability of the Spanish captains to hold this tight formation. However anachronistic, it was quite a feat of seamanship. They were also awed by the sheer size of the enemy armada, which spanned two miles end to end, "the ocean groaning under their weight." The Spanish, in turn, were impressed by the deft maneuvering of the swift English galleons in grabbing the wind gauge.

The combat began with a gesture straight out of the annals of chivalry. At 9 A.M. on July 31, Lord Howard sent his personal pinnace, appropriately named the *Disdain*, to challenge the Spanish admiral. The small ship raced out from the English formation, fired a single shot across the bow of what it mistakenly took to be the Spanish flagship, then scooted back to its own lines while a Spanish salvo in reply fell short. This ceremony out of the way, Howard led his ships toward the Spanish fleet in a straight line, with his flagship, the *Ark Royal*, in front. It was the first time that this simple yet flexible formation, which by the eighteenth century would become standard practice in the Royal Navy, had been employed. The English division commanders—Drake aboard the *Revenge*, John Hawkins aboard the *Victory*, Martin Frobisher aboard the *Triumph*—concentrated their fire on the *San Juan de Portugal*, a giant galleon on the right wing of the Spanish formation. The thunder of their guns shattered the morning stillness and the resulting smoke wreathed the cream-colored sails in a grayish haze. The English, with their bigger guns and better crews, were estimated to have fired two thousand cannonballs, to only seven hundred fifty from the Spanish. Yet they caused little damage. The *San Juan* had some spars and rigging shot away but suffered hardly any killed or wounded. The problem was that the English ships were firing from a range of three hundred yards—too far away to do much damage. But they were afraid to get too close because they knew this was what their enemies wanted.

In another throwback to galley warfare, the entire Spanish strategy depended on closing with and boarding their foes. For this reason the Spanish fleet carried far more soldiers than sailors. There were nineteen thousand troops in all, most armed with shoulder-fired arquebuses or muskets. "We durst not adventure to put in amongst them, their fleet being so strong," Lord Howard wrote. Knowing they did not have nearly as many men, the English captains preferred to batter the Spanish from long range with their heavy artillery. But they had no clear idea of how to employ their guns effectively in a major fleet action. No one did. This was the first large battle in history between heavily gunned sailing fleets. With the age of sail-and-shot still in its infancy, tactics had to be improvised on the spot.

The first day's fighting petered out inconclusively in the early afternoon. The Royal Navy had landed some blows, but it had barely delayed the progress of *La Armada Invencible*—the Invincible Armada. The Duke of Medina Sidonia directed his ships to continue sailing northeast through the English Channel, heading for a fateful rendezvous in the Netherlands, where they were supposed to pick up the Spanish Army of Flanders and transport it across the channel to invade England. Queen Elizabeth I had almost no army of her own, just a pitiful militia that could not possibly stand for long against these battle-hardened Spanish veterans. England's only hope lay in preventing the Spanish Armada from landing. And that would require the Royal Navy to master an entirely new style of warfare at sea.

They had about a week to figure out how.

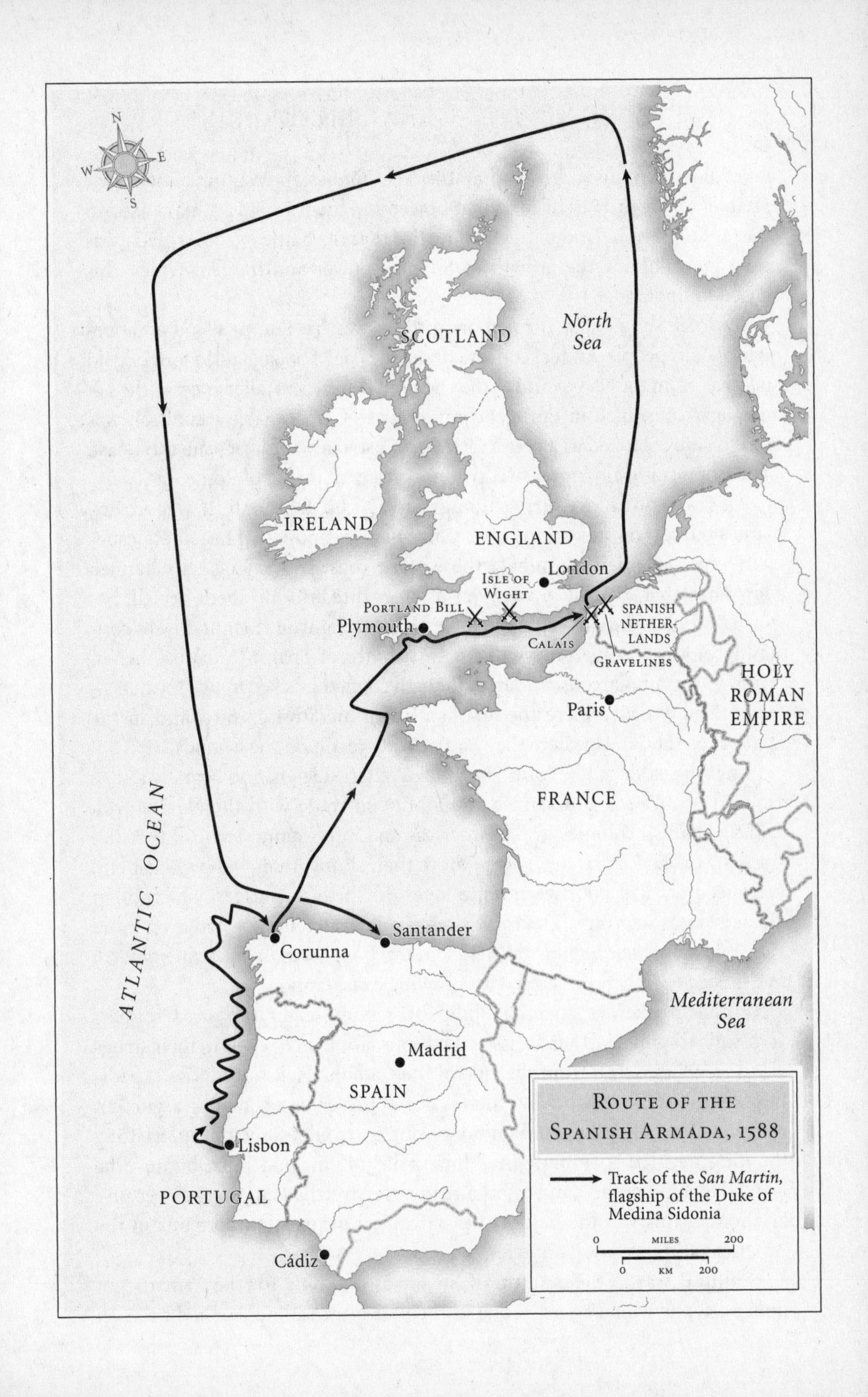
N
E
S
W
North Sea
SCOTLAND
IRELAND
ENGLAND
London
Isle of Wight
Portland Bill
Plymouth
Calais
Spanish Nether-lands
Gravelines
Holy Roman Empire
Paris
FRANCE
ATLANTIC OCEAN
Corunna
Santander
Mediterranean Sea
Madrid
SPAIN
Lisbon
PORTUGAL
Cádiz
Route of the Spanish Armada, 1588
Track of the San Martin, flagship of the Duke of Medina Sidonia
0
MILES
200
0
KM
200

PLANNING THE GREAT ENTERPRISE

Thucydides famously wrote that the Peloponnesian War was made inevitable by "the growth of Athenian power and the fear which this caused in Sparta." It might be said with equal justice that the battle of the Armada was made inevitable by the growth of Spanish power and the fear which this caused in England.

By 1588, King Philip II ruled one-fifth of western Europe's land and one-quarter of its people. Annual convoys under the flag of Spain hauled tons of gold and silver from the New World to the Old. Philip even controlled most of the salt and sugar consumed in Europe, commodities of incalculable value (salt was used to cure meat and as a preservative). A Castilian writer had cause to boast: "The empire of Spain is over twenty times greater than that of Rome was."

This was not a reassuring thought to Queen Elizabeth. It was in England's interest to prevent the emergence of any continental hegemon, especially one that might dominate the eastern coast of the English Channel. More than that, "Papist" power was a direct threat to Elizabeth herself, because Catholics did not regard her as England's rightful ruler and were constantly plotting, sometimes with the assistance of Spain, to depose her in favor of her Catholic cousin, Mary. Elizabeth counterattacked in two theaters—in the New World, where the Spanish had a lucrative empire, and in the Spanish Netherlands, where they had a costly revolt on their hands.

English sea dogs like John Hawkins, Walter Raleigh, and Francis Drake violated Spain's self-proclaimed monopoly on trade with the New World, raiding Spanish colonies in the Americas and intercepting Spanish treasure ships at sea. These privateers were more than simply tolerated by Elizabeth; the queen and her counselors were investors in their voyages, which often yielded fantastic profits. Their raids cost Philip much money and even more face. *Who were these impudent heretics to steal from the most powerful monarch in Christendom?* he wondered with growing exasperation.

Elizabeth likewise angered Philip by sending eight thousand soldiers across the Channel in 1585 to assist the Protestant Dutch rebels in their struggle to free themselves from the rule of the Catholic Habsburgs. The expedition was not successful, but in Philip's court it presented a clinching argument in favor of invading England and deposing its troublesome queen. In 1585 the king decided to mount the "Enterprise of England," not because he wanted to add another land to his already overstretched empire, but because he thought this was the best way to safeguard his existing domains in the Americas and the Low Countries.

Philip II was in the autumn of his life as planning for the Armada got under way; he would be sixty-one in the year of its sailing. He had been on the throne for more than forty years, during which he had known but six months of peace. He would have little more than ten years to live. His blond hair and blue eyes were fading with age. His teeth were almost gone, he was hobbled by gout, and his stomach gave him constant trouble despite his precisely regulated diet. He had seen eight children (out of eleven) and four wives die. Gradually this ailing widower had withdrawn from life, from the feasts, jousts, and masquerades he had enjoyed as a young man. He spent more and more time in seclusion at El Escorial, a sprawling monastic palace in the mountains thirty miles from Madrid. He attended mass daily, read almost nothing but devotional tracts for pleasure, and collected a staggering array of religious relics—more than seven thousand of them, including, a biographer writes, "ten whole bodies, 144 heads, 306 arms and legs, [and] thousands of bones of various parts of holy bodies . . ." His intense religious faith convinced the king not only that he was on God's side, but, more important, that God was on *his* side. "You are engaged in God's service and in mine," he wrote to one nobleman, "which is the same thing."

The soft-spoken monarch hated meeting officials face-to-face and trusted few of his subordinates to make decisions in his name. He preferred to run his empire's affairs personally and through the written word—a management style made possible by the spread of cheap paper. He even personally reviewed applications for jobs as cooks and porters in the royal household. Thus, like any overworked executive, he was drowning in a sea of paper. He toiled long and hard every day in his modest rooms, at least as hard as his fragile health would permit, but he could never get ahead of the paper flow. It sometimes seemed as if "the Bureaucrat King" was also his own chief clerk. Though he lacked any formal military or naval training, he carefully scrutinized letters from his chief military commanders proposing different plans for the invasion of England.

Spain's greatest admiral, the Marquis of Santa Cruz, wrote that this would require a fleet of more than five hundred ships and 94,000 men. The cost would be a staggering 3.8 million ducats. It was money that the heavily indebted Philip simply did not have; his obligations far exceeded the value of bullion shipped from the New World. Spain's greatest general, the Duke of Parma, suggested instead that the invasion could be accomplished by just 34,000 men of his Army of Flanders. They could steal across the Channel in barges, he suggested, escorted by twenty-five great ships, and catch the English by surprise. But where would they get the seven to eight hundred barges required, and how could they cross the Channel without being intercepted by English or Dutch warships? Philip decided to combine the two plans. Santa Cruz would assemble a fleet in Spain of about 130 ships, then sail to Flanders, where he would rendezvous with Parma and ferry his veterans across the Channel.

It was a fine plan on paper, but it did not answer a crucial question: How would the naval and land forces link up? It would be no mean achievement to synchronize such far-flung operations when communication between Spain and the Netherlands could take weeks under the best of conditions. The task was complicated by the fact that Spanish warships would be too heavy to navigate the shallow waters along the Flemish coast. His Most Catholic Majesty was aware of these obstacles but assumed that because the expedition was being undertaken in God's name, He would somehow deliver a solution when the time came.

England did not wait to be invaded. The spy networks of the Dutch rebels and the English government provided plenty of early warning about Philip II's plans, and in March 1587 Queen Elizabeth instructed Sir Francis Drake to launch a preemptive strike. The forty-seven-year-old Drake was one of the world's most experienced navigators and naval fighters, and his short, stout frame showed the signs of long years at sea. His reddish-blond beard was turning gray, his hair thinning, his body thickening. His right leg was scarred with a bullet wound, his face with an arrow wound—mementos of numerous forays to the New World.

Drake undertook his voyages for private gain, but he was happy to wrap his thievery in the respectable cloaks of patriotism and religion. Anything, in his mind, was fair game against the "Antichrist" who called himself the King of Spain, and his men did not hesitate to destroy or deface "Popish" churches. Drake was ruthless, vain, unscrupulous, and absolutely fearless. He dealt harshly with his own men and even more harshly with the enemy.

His crowning moment had come seven years earlier when, in the course of one of his raids on the New World, he had almost casually made the decision to take the long way home, thereby leading only the second expedition (1577–80) to circle the globe. This feat had won him a knighthood. More recently, in 1585, he had terrorized the Caribbean, sacking the Spanish towns of Cartagena and Santo Domingo. Such audacity had made this son of a Devonshire parson one of the most feared men in Spain and one of the richest men in England. He bought a large house in London and several country estates, and showered the Virgin Queen with expensive baubles, including a New Year's gift in 1581 of "an emerald-studded crown and a diamond cross, together worth more than 50,000 ducats."

However much he might masquerade as a rich courtier in London, Drake was still a sailor at heart, and he wasted no time in putting out to sea after he received the queen's orders. He led his squadron of twenty-four ships—four of them belonging to the crown, the others to private investors including himself—out of Plymouth harbor on April 12, 1587. Drake had no firm destination in mind. Finding Lisbon too heavily fortified to be attacked, he headed for the port of Cádiz.

He arrived there on the afternoon of Wednesday, April 29, and instantly decided to attack, notwithstanding the objections of some of his more cautious captains. It turned out to be the right decision. El Draque (The Dragon) caught the Spanish by surprise just before sunset, easily brushing aside a dozen galleys that came out to challenge him; their battering rams were no match for his heavily gunned galleons. He anchored in a bustling roadstead crowded with some sixty merchantmen, many destined for the Armada. Drake set his men to looting and burning as many vessels as possible, "to the great terror of our enemies and honour to ourselves." Though they were prevented from sacking the town by the arrival of a large force of Spanish infantry and cavalry under the Duke of Medina Sidonia, the English sailors managed to destroy at least two dozen ships and large quantities of provisions intended for the Armada. Having "singed the King of Spain's beard," Drake then trawled the waters off Spain for more prizes, not quitting until he had bagged a lucrative Portuguese treasure ship.

ARMING THE ARMADA

A Spanish squadron chased fruitlessly after the "English corsair." By the time they returned to port, it was no longer practical to mount the Enterprise of England in 1587. This massive undertaking would have to be pushed back into the following year. In the meantime the assembled ships lay moldering in Lisbon, the men suffering from all manner of disease. Like many of his subordinates, the armada's veteran commander, the Marquis of Santa Cruz, was felled by typhus. To replace him Philip II settled on a wealthy nobleman with vast landholdings but no experience at sea: Don Alonso Pérez de Guzmán, the seventh Duke of Medina Sidonia. The thirty-seven-year-old grandee had no desire for the assignment; sailing would only aggravate his rheumatism. He wrote Philip a long letter pleading to be excused and citing his own limitations: "I should not give a good account; for I do not understand it, know nothing about it, have no health for the sea, and have no money to spend on it." This was not exactly the reckless spirit of a Cortés or Pizarro, but the king took this to be false modesty and waved it aside.

However little he wanted it, Medina Sidonia now had a commission as Captain General of the Ocean Sea, and he was determined to make the best of it. The industrious nobleman managed to whip the Armada into tolerable shape over the course of the next few months. Ships were repaired and restocked with provisions. Sick men were mustered out and healthy ones recruited. The duke's top priority was to round up as many guns and cannonballs as he could, for he had seen with his own eyes in Cádiz how heavily armed the English ships were.

It was not easy for Medina Sidonia to find the armaments he needed. Spain did not have much of an arms industry. Most of what it needed had to be imported. In the best of circumstances this took time; this was far from the best of times, however, because Spain was at war with England, one of its traditional munitions suppliers. Spain's few arsenals went to work, but they lacked master gunsmiths and the cannons they produced turned out to be poorly made and unreliable.

Medina Sidonia was forced to requisition cannons from foreign ships in Spanish ports. This helped somewhat, but it produced a fleet armed with a motley array of weapons of varying calibers, many lacking the right supply of shot. Most of the guns were small in size; of 2,411 artillery pieces, only 138 were big enough to shoot a ball of twenty-five pounds or more, the size needed to really damage a ship. (The English navy had 251 guns of such heavy caliber.) The bulk of the Armada's offensive punch came from the soldiers crammed onto its decks, most of them armed with arquebuses capable of firing a half-ounce lead ball. There was also an elite corps of musketeers whose heavier weapons, which discharged a lead ball of 1½ ounces, had to be fired from a forked stand.

This fit the traditional Spanish style of naval warfare, which called for closing with and boarding the enemy. This was how battles were decided in the Mediterranean, but it was an utterly inappropriate tactic for northern waters, where a new style of war was being developed, one in which victory would go to the side most proficient in the use of artillery rather than in hand-to-hand combat.

Spain was not yet ready for such warfare—and many of its most experienced naval officers knew it. As one of them told a papal envoy before the Armada departed: "If we come to close quarters, Spanish valour and Spanish steel (and the great masses of soldiers we shall have on board) will make our victory certain. But unless God helps us by a miracle the English, who have faster and handier ships than ours, and many more long-range guns, and who know their advantages as well as we do, will never close with us at all, but stand aloof and knock us to pieces with their culverins [a type of cannon] without our being able to do them any serious harm. So we are sailing against England in the confident hope of a miracle."

To ensure that it would meet with divine favor, the Armada carried 180 priests aboard when it sailed from Lisbon on May 30, 1588. The Spanish fleet had 130 ships, but only twenty of them were powerful galleons, the frontline fighting ships of the day. Another four were galleases from Naples which combined oars with sails, a hybrid technology typical of times of transition.

Four more were pure galleys, propelled entirely by straining ranks of rowers. Both galleys and galleases would prove almost useless in the choppy waters of the English Channel and the North Sea. The second line of the Armada was made up of forty-seven armed merchantmen, many of them huge but not especially weatherly or deft in maneuvering. The rest of the Armada was composed of transport hulks to haul the invasion force and small craft used for scouting and communications. The bulk of the fleet was intended neither for serious artillery combat nor for heavy seas.

Its deficiencies became obvious at the start. The Armada crawled along the Spanish coast, limited by the speed of "the scurviest ship in the fleet," in their commander's words. ("Verily," Medina Sidonia complained, "some of them are dreadfully slow.") Twenty days out, on June 19, 1588, a storm scattered part of the fleet. Medina Sidonia was forced to put into the port of Corunna, on the northwest tip of the Iberian peninsula, where he spent a month repairing ships that had been battered by the summer squall and replenishing their dwindling supply of provisions. Medina Sidonia wrote to Philip urging him to abandon the expedition altogether and make "some honourable terms with the enemy," but the king brusquely refused. On July 21, 1588, the Spanish ships once again headed for the English Channel under a fresh northeasterly breeze.

ALBION'S ADVANTAGES

The Royal Navy was ready to meet them. It had ancient roots, but as an official body it had existed for only half a century. In medieval times English kings would raise fleets from among the merchant marine when necessary in time of war, and few if any ships were built expressly for fighting. Even the king's personal ships, when not needed in a campaign, would be used to transport Bordeaux wine or other goods for the royal household. Henry VIII (r. 1509–47), Elizabeth's father, had pioneered among European monarchs a standing fleet belonging to the crown, and, as important, a standing department to administer it. This was part of what is sometimes called the Tudor Revolution, which gave England the prototype of a modern bureaucracy long before Spain possessed one. Under Henry VIII and his energetic ministers, Thomas Wolsey and Thomas Cromwell, the center of English administration shifted away from the royal household and toward new governmental departments.

To manage the navy, officials were appointed with such titles as Master of Naval Ordnance, Lieutenant of the Admiralty, Treasurer, Controller, Clerk of the King's Ships, and General Surveyor of the Victuals for the Seas. Beginning in 1546, many of these senior managers sat together on the Council of the Marine, popularly known as the Navy Board, direct ancestor of the modern Admiralty. The slightly older Ordnance Board was responsible for procuring weapons and everything needed to operate them. Together, these two organizations provided England with more efficient naval administration than that of any contemporary state, with the possible exceptions of Portugal and Venice. The Spanish navy was a virtual one-man operation by comparison, and that man was the overworked King Philip II, isolated in his gloomy cell at the Escorial.

Spain had nothing like the royal dockyards and storehouses that had sprung up around southern English ports like Portsmouth, Woolwich, and Deptford. Nor did it have officials, as England had, who carefully drew up mobilization plans to make full use of its maritime might. England, not yet possessing lucrative colonies, was much poorer than Spain (Elizabeth's ordinary revenues were a tenth of Philip's) and could not keep a large fleet mobilized for long periods. It needed accurate intelligence and ready contingency plans to defend itself when danger materialized. There was no margin of safety. As part of this planning, the Elizabethan navy launched an ambitious program of construction in the 1560s and 1570s to take advantage of a (so to speak) sea change in warship design.

Most ships at the time could be divided into two classes: those propelled by oars and those propelled by sails. (There were also some hybrids such as the Spanish galleases mentioned above.) Until the fifteenth century it was by no means clear which type would become dominant. Both forms of propulsion went back to ancient times. In the Mediterranean, where European naval warfare first developed, a division of labor emerged: "roundships" with sails were used for trade because they had greater cargo capacity; "longships" with oars were used for fighting because they were more maneuverable and capable of faster speeds for short bursts. Oar-driven galleys had obvious limitations, however. They lacked staying power because of the inherent limits of human strength. They could be used only for short passages along coastlines; they could not cross oceans. Nor could they handle well in the rougher seas of northern Europe. Thus, sails had always been more important for northern seafarers. Longships that combined a single sail with oars had enabled the Vikings to terrorize Europe during the Dark Ages. Later, from approximately 1100 to 1400, single-masted cogs plied the waters of northern Europe for both commerce and conquest.

It took until the fifteenth century for triple-masted, square-rigged vessels known as caravels and carracks to emerge. With these extra sails, Europeans were able to sail faster and farther than ever before. They were even able to discover new worlds: Columbus's ships in 1492 were essentially caravels. The caravels that ventured overseas carried sufficient artillery to blast foreign opposition—whether Arab dhows or Chinese junks—out of the way. At first they were outgunned in European waters by galley fleets. By the mid-1500s, however, the balance of power had swung decisively toward sailing ships as caravels and carracks gave way to bigger and sturdier galleons. With their distinctive high stern, low forecastle, and taller masts that permitted more topsails to be put out, galleons handled better than caravels in rough seas even when loaded down with heavy artillery.

Galleons were by no means an English invention; they were employed by many nations, including Spain, Portugal, Denmark, Sweden, and France. But the English added a crucial modification.

John Hawkins, a leading merchant, occasional pirate, unabashed slave trader (his coat of arms included a picture of a chained African), and a cousin of Sir Francis Drake, became Queen Elizabeth's Treasurer of the Navy in 1577. Having sailed all over the world, he had picked up some ideas on galleon design. He has traditionally been credited with persuading the queen's shipwrights to reduce the size of the fore and aft castles, to increase the size of the gun decks, and to make the hulls longer and slimmer. It may not have all been Hawkins's doing; other officials were certainly involved. Whatever their origins, the resulting "race-built" galleons (the first was the *Dreadnought*, launched in 1573) were marvels of marine design. And no wonder: They were the first ships designed using blueprints. They looked less imposing than their Spanish or Portuguese counterparts, but they were faster and better able to handle rough weather. They lacked the cargo capacity of Spanish ships, but then England, which did most of its trade closer to home, did not need a lot of storage space. What it needed was speed and firepower, and that is what the race-built galleons delivered. They could literally run circles around the clumsier Spanish competition.

By 1588 England had eighteen frontline warships that had superior handling and artillery to anything they might meet at sea; sixteen other royal galleons were not far behind. This core would be supplemented by thirty large private warships provided by their owners as part of a Volunteer Reserve. Aboard these ships was the most formidable array of artillery ever assembled.

Two inventions made this possible: the lidded gunport and the truck carriage. No one knows who designed either one. Portholes began to be cut into the sides of ships around 1501, allowing guns to be mounted in the hull, not just on the top deck. Hatches were used to keep out the sea when the guns were not in use. This made possible a vast increase in a ship's armaments without imperiling its stability. The truck carriage emerged slightly later. It was an attempt to answer one of the perennial problems of gunnery at sea: how to manhandle a gun and discharge it through a small port in the confined space of a ship. Large guns mounted on field carriages of the type used in land warfare, with their large twin-spoked wheels and long trails, were too unwieldy to be practical belowdecks. The more compact truck carriage, with four small, solid wooden wheels, allowed a gun to be run out easily and fired, then brought back inside to be reloaded. A gun on a truck carriage could be reloaded twice as fast as one mounted on a field carriage. While the lidded gunport had spread to all the major navies of Europe by the mid-sixteenth century, the Spanish had few if any truck carriages in 1588. This proved to be a crucial English advantage; perhaps *the* crucial one.

With these innovations, the way was now clear to put many more heavy guns aboard ships. The English were ready to seize the opportunity, thanks to the work of Henry VIII. At the beginning of his reign, England had to import almost all of its guns from abroad; by the time of his death, England's cannon industry was among the finest in the world. Under the supervision of the Ordnance Board, which carefully parceled out contracts to a small group of private firms, English foundries developed the first cannons made of cast iron. This was cheaper than using bronze, allowing English gunsmiths to produce a vast array of ordnance that ranged in size from 300-pound robinets firing half-pound balls to 7,000-pound cannons firing 66-pound balls. Such heavy guns could not easily be maneuvered on land but they were perfect for the "mobile forts" of the sea. Potent as they were, however, these cannons had severe limitations. Artillery could generally smash wooden hulls at no more than two hundred to three hundred yards, and aiming from a rolling deck was no easy feat. Because the advent of sail-and-shot was so recent, few officers had a good idea of the potential and limitations of this new weapon system.

Both the English and Spanish navies had a number of captains familiar with artillery. Some of Philip II's skippers were veterans of the Battle of Lepanto (1571), where they had used their guns to perforate the Ottoman fleet, but the English had a major advantage in their greater knowledge of sailing ships. Two of England's most fearsome privateers, the cousins John Hawkins and Francis Drake, were given commissions as vice-admirals in order to fight the Spanish. Their experience more than compensated for their commander's inexperience: Lord Howard, the Lord Admiral of England and a cousin of the queen, had no more direct knowledge of fighting at sea than had his Spanish counterpart, the Duke of Medina Sidonia. But Howard had the advantage of capable subordinates who would have the leeway to exercise their own initiative. Philip II was a micromanager; Elizabeth I was not. She instructed her commanders to use "your own judgment and discretion, to do that thing you may think may best tend to the advancement of our service." The best judgment of Drake, Hawkins, and the other English captains turned out to be very good indeed.

The English edge extended to ordinary sailors, mostly seamen impressed from privateers and merchant ships. The English fleet carried one sailor for every two tons of shipping; the Spanish had one sailor for every seven tons. Moreover, the English trained their naval crews to a high state of proficiency in gunnery and ship handling. The infantrymen aboard Spanish ships were tyros by comparison. To make matters worse, while the English ships were manned for the most part by Englishmen, the Spanish decks were loaded with men speaking a dozen different languages—a polyglot crew (including two hundred Englishmen) who were suspicious of each other and of their Spanish masters.

All this meant that English commanders could be justified in their confidence as they sailed off to confront the supposedly invincible Armada. "I think there were never in any place in the world worthier ships than these are," Lord Howard wrote of the Royal Navy. "And few as we are, if the King of Spain's force be not hundreds, we will make good sport with them."

THE BROKEN CRESCENT

Yet, in the beginning, the English had trouble coming to grips with their adversaries, much less making sport with them. On the first day of combat, Sunday, July 31, 1588, the Spanish lost only two ships, both by accident. The *San Salvador*, one of the best-armed Spanish ships, was rocked by an explosion around 5 P.M. No one knows what caused it, though rumor had it that a disgruntled gunner—like many in the Armada, a German—had set fire to the powder barrels. Some two hundred men were either killed or badly burned, and its steering was disabled. The ship had to be abandoned the next day.

While the *San Salvador* was still burning, the *Nuestra Señora del Rosario*, a converted merchantman that was one of the newest ships in the Armada, collided with another vessel. It then underwent a series of accidents that left it dead in the water. On Monday morning, Drake appeared as if by magic in front of the stricken vessel aboard his own ship, the five-hundred-ton *Revenge*. "I am Francis Drake and my matches are burning!" he announced melodramatically. The Spanish captain was so awed by El Draque's reputation that he agreed to surrender on the spot, turning over part of the Armada's payroll, amounting to fifty thousand gold ducats.

What was Drake doing there? Apparently he had been drawn by the scent of treasure, and never mind the consequences. The previous night, Lord Howard had ordered Drake to lead the English fleet in pursuit of the Spanish by lighting a lantern in the *Revenge*'s stern for the rest of the ships to follow. Instead of doing his duty, Drake went off treasure-hunting, explaining his conduct with the unlikely story that he had seen some mysterious sails in the night and had gone off to investigate. In today's Royal Navy this would have earned him a court-martial, but in the age of privateering, when the border between public and private military action was as thin as lace, few thought there was anything amiss about his conduct, even though it had almost led to the destruction of the English flagship. Throughout the night the *Ark Royal* had followed a lantern that Howard thought belonged to *Revenge*; in the morning he realized that it had actually belonged to the enemy's flagship. The *Ark Royal* was almost surrounded by Spanish ships and barely managed to escape.

The two fleets next clashed on Tuesday, August 2, off Portland Bill, not far from the southern English coast. This time the Spanish had the weather gauge and used it to try to close with and board the enemy. But the English ships proved too nimble. Every time a Spanish vessel got close, the English would stand off and pound it with artillery. "It may well be said," wrote Lord Howard, "that for the time there was never seen a more terrible value of great shot, nor more hot fight than this was." All that shooting did not achieve much, however. The English did not sink a single Spanish ship; the Spanish did not board a single English ship. Both sides called off the fighting after a few hours, and the Armada continued its on its way up the channel, heading for a rendezvous with the Duke of Parma's invasion army at Calais.

The third battle of the campaign occurred the next day, Wednesday, August 3, off the Isle of Wight. A Spanish hulk, the 650-ton *Gran Grifon*, fell out of formation and lagged behind the rest of the Armada. An English ship—almost certainly Drake's *Revenge*—pounced at once. The English vessel gave the *Gran Grifon* a broadside, came about, delivered another broadside, then crossed her stern and raked her a third time at close range. It was a dazzling example of the new gunnery techniques, but, as on the previous day, the battle proved indecisive. The *Gran Grifon* was towed back to the Spanish lines, and the Armada continued sailing north. The fourth battle, on Thursday, August 4, also near the Isle of Wight and also triggered by some stragglers from the Spanish ranks, was equally inconclusive.

In the first four battles, the Spanish had fewer than a thousand men killed, captured, or wounded; the English losses were perhaps half that. Not a single ship had been sunk so far by gunnery. The English, starting to run low on powder and shot, decided to conserve their resources until they could make their weapons count. Medina Sidonia was perfectly happy to avoid further battles; his goal was to pick up Parma, not to grapple with the English fleet.

In the early evening hours of Saturday, August 6, the Armada anchored off Calais. The decisive phase of the campaign was about to begin.

During his journey up the Channel, Medina Sidonia had been dispatching messengers to let the Duke of Parma know of his progress and coordinate the linkup between their forces. But communicating in the sixteenth century was not as easy as it is today. Medina Sidonia's messengers, sent in small oared vessels, had to cope with enemy ships, foul weather, bad roads, and numerous other obstacles. Even when letters reached Parma, it was hard for him to send a reply to a fleet whose movements were unpredictable. The upshot of all this was that Medina Sidonia did not receive his first word back from Parma until August 6, and it was not encouraging: The Army of Flanders "had not embarked a beer barrel, still less a soldier," and would not be ready to set off for six days. *Six days?* Medina Sidonia had expected that Parma would be ready the instant the Armada docked. He could not anchor off Calais for a whole week, not with the undefeated English fleet hovering dangerously close offshore.

The Royal Navy had swelled to 140 ships with the arrival of another squadron. On Sunday morning, August 7, a council of war convened aboard the English flagship, the *Ark Royal*. The commanders assessed the situation: The Armada was clustered in an unprotected anchorage; the English had the windward gauge; a strong tide was running toward shore. This was an obvious situation in which to utilize fire ships, the sixteenth-century version of torpedoes. Eight English ships were packed full of combustibles such as hemp and tar, and their guns were double-shotted so that as soon as fire reached their matches they would go off.

Around midnight on Sunday, anxious Spanish sailors could see several lights twinkling near the English fleet anchored offshore. The lights became brighter and more numerous as they came closer. Eight flaming ships were headed straight for the Armada. Each had a skeleton crew aboard to steer, lash the helm, and then make their escape aboard a dinghy. The current was moving so fast that there was little time to react. Small boats managed to intercept two of the fireships and steer them safely toward the beach. But the other six drifted on, their guns releasing a terrifying salvo, their flames sending sparks into the night sky.

Panic swept the anchorage. Captains cut their cables, sending their anchors to the bottom of the sea, and scattered to the winds. When dawn broke on Monday, August 8, and Medina Sidonia, aboard the *San Martin*, looked around for his mighty fleet, all he saw were four other ships. The rest of the Armada had drifted off toward Dunkirk, farther up the coast of France to the northeast. The formidable crescent formation, so laboriously maintained until now, had been broken at last. The English might have finished off the Armada then and there, but Lord Howard was inexplicably distracted by looting the *San Lorenzo*, a galleass that had grounded the night before. This gave Medina Sidonia time to regroup off Gravelines, about ten miles north of Calais. Here, Drake began the attack in the *Revenge*, closing to less than one hundred yards of the *San Martin* before opening a devastating fire. Martin Frobisher aboard the *Triumph* and John

Hawkins aboard the *Victory* joined in, turning the Spanish flagship into the focus of English attacks.

The bombardment lasted from morning until late afternoon. At some points it seemed as if the *San Martin* were fighting the entire English fleet by itself. "So tremendous was the fire," wrote one Spanish officer, "that over 200 balls struck the sails and hull of the flagship on the starboard side, killing and wounding many men, disabling and dismounting three guns, and destroying much rigging." Only heroic work by two divers who applied an underwater patch prevented the flagship from sinking. Other Spanish vessels took a similar pounding.

For much of the day, the two sides were "within speech of one another," in the words of an English officer. At such close range the English guns splintered oak, shredded sails and spars, and turned the decks of Spanish ships into blood-soaked killing zones. Don Francisco de Toledo, captain of the eight-hundred-ton Portuguese galleon *San Felipe*, saw "five of her starboard guns dismounted. . . . his upper deck was destroyed, both his pumps broken, his rigging in shreds and his ship almost a wreck." More than two hundred of his men lay wounded or dying after this withering assault. So what did he do? According to a Spanish officer's account, he "ordered the grappling hooks to be got out, and shouted to the enemy to come to close quarters. They replied, summoning him to surrender in fair fight; and one Englishman, standing in the maintop with his sword and buckler, called out, 'Good soldiers that ye are, surrender to the fair terms that we offer ye.' But the only answer he got was a gunshot, which brought him down in sight of everyone. . . . The enemy thereupon retired, whilst our men shouted out to them that they were cowards, and with opprobrious words reproached them for their want of spirit, calling them Lutheran hens, and daring them to return to the fight."

The courage of the Spanish was impressive, but daring and pluck proved no match for shot and powder. The English had no desire to board a Spanish vessel when they could stand off and pound it to pieces. The *San Felipe* grounded the next day and was captured by Dutch rebels allied with the English. By Monday evening, after nine hours of battle and with their shot lockers nearly exhausted, the English commanders decided to call it a day. They knew that they had damaged the enemy, but they did not know how much. They assumed that they would have to fight the Armada again. "Their force is wonderful great and strong," Lord Howard wrote that night, "and yet we pluck their feathers by little and little."

Actually, no more feather plucking would be necessary. The fight off Gravelines marked the last battle of the Armada. The Spanish had suffered grievously that day, with more than a thousand killed and eight hundred wounded. (Only about a hundred Englishmen were killed, and not a single

English ship was lost.) Although only one Spanish ship actually sank during the battle—the *Maria Juan*—others were badly damaged and would not long survive in the hostile waters of the North Sea. The next day, the entire Armada almost grounded upon the Zeeland Sands. A last-minute change of winds allowed the Armada to escape into deeper water. But this was only a temporary reprieve.

Medina Sidonia was noted for his bravery, and he was willing to turn around and tangle again with the formidable English fleet which continued to menace his rear. But the duke noted that the ship's pilots told him that this was impossible, "as both winds and tide were against us. They said that he would be forced either to run up into the North Sea, or wreck all the Armada on the shoals." After consulting his captains, the duke decided that unless the winds shifted they had no alternative but to turn tail and run.

Their attempt to save the Armada would lead to its greatest peril. Ahead lay the uncharted waters of the North Sea—uncharted by the Spaniards, that is, who had no pilots familiar with this area and no reliable maps. The fleet would have to sail around Scotland and Ireland before it could return home to Spain. Late summer was giving way to early fall. The winds were kicking up and temperatures were dropping. The Spanish had brought no cold-weather clothing. Nor did they have provisions enough for a voyage that was sure to last more than a month. Medina Sidonia ordered rations to be cut back for all officers and men, including himself, to half a pound of biscuit, half a pint of wine, and a pint of water a day—"just enough being served out to keep them alive," he told Philip II.

To lighten the load and save precious water, the Armada's horses and mules, hauled along for use once they had landed in England, were tossed overboard. Their thrashing and neighing as they drowned must have been dispiriting to behold. Blasphemy and swearing were officially forbidden aboard these ships that sailed on behalf of "God's obvious design," but no doubt more than one man uttered oaths under his breath as the Armada continued on its frightening course.

As they struggled along the coasts of Scotland and Ireland, the weaker ships—many of them built for the placid Mediterranean, not the tempestuous North Sea—fell out of formation. Some tried to seek shelter along the coast, but most could not make anchorage because they had cut their anchor lines on August 8. They were at the mercy of the elements, and the elements that stormy fall were not merciful. Savage storms dashed ship after ship onto razor-sharp rocks. Stout wooden vessels disintegrated in the pounding surf as if they were mere sand castles. Sailors drowned by the thousands, pulled down into the dark, icy depths by Neptune's unrelenting grip.

Those Spaniards who washed up in Scotland, England's traditional enemy, usually found refuge. Most who made it ashore in English-occupied Ireland did not. Sir William Fitzwilliam, the Lord Deputy sent from London to preside over the rebellious Irish lands, had a scant 750 troops at his disposal, and he was petrified that Spaniards would land and make common cause with their fellow Catholics ashore. Therefore he cruelly commanded his men "to apprehend and execute all Spaniards found, of what quality soever. Torture may be used in prosecuting this enquiry."

The English were merciless in carrying out their orders, as the survivors of *La Trinidad Valencera*, a bulky Venetian grain ship, found out. After their vessel grounded in Kinnagoe Bay on September 1, 1588, the crew landed and skirmished with the local English garrison. Cold, hungry, and outgunned, the Spanish agreed to surrender in return for a promise from the English commander, Major John Kelly, to treat them fairly. It turned out that Major Kelly was not a man of his word. His troops first stripped the defenseless crew of everything they owned, then separated them into two groups, officers over here, men over there. More than three hundred of the Spanish soldiers were butchered with lance and bullet. Their officers were marched off to a jail more than a hundred miles away, with little food or clothing along the way. Those who survived this ordeal were ransomed back to Spain.

It was later calculated that some six thousand Spanish sailors shipwrecked along the Irish coast. Only 750—one out of eight—survived. One of the few who made it back to Spain later wrote of "the terrible hardships and misfortunes," the "cruel deaths and tortures," endured by his countrymen.

Those who did not wreck along the coast were scarcely better off. Some of the surviving ships ran out of water during the final weeks of their voyage. Provisions also were lacking and what remained was frequently putrid. Dysentery, typhus, pox, and other illnesses ravaged the crews. Spanish ships were never particularly clean—English officers had been disgusted by "the foul and beastly" conditions on captured vessels—but now they must have become nearly unbearable, as men fouled themselves, screamed in misery, hallucinated feverishly, bled copiously, and suffered unspeakable agonies of dehydration and dysentery.

The Armada's flagship, the *San Martin*, finally reached Santander, a small port in northern Spain, on September 21, 1588, accompanied by only 8 other ships. In all, just 60 out of 130 ships that had set out with the armada—fewer than half—made it back to Spain. More than 12,000 men were lost out of 26,000. "The troubles and miseries we have suffered cannot be described to your Majesty," wrote Medina Sidonia, who was himself prostrated with fever. "On the flagship 180 men died of sickness, three out of the four pilots on board have succumbed, and all the rest of the people on the ship are ill, many of typhus and other contagious maladies." The maladies extended to

Medina Sidonia's own household. Of sixty-two retainers, ranging from barbers to musicians, he lamented to the king, "Only two have remained able to serve me."

Legend has it that upon hearing of this tragedy, the deeply pious Philip exclaimed, "I sent my ships to fight against men, and not against the winds and waves of God." It is not hard to see why the king would want to ascribe his men's failure—and by extension his own—to divine will. Yet even those who shared his faith could see that he was mistaken. The destruction of the Spanish Armada may have been the work of winds and waves, but its defeat was very much the work of men: Englishmen who had managed to best the mightiest fleet of the greatest empire in the world through their superior training, morale, organization, strategy, seamanship—and technology.

HOW BRITANNIA CAME TO RULE THE WAVES

The Spanish lost for a simple reason: They were outgunned. The English fleet had one-third more firepower, and the English were better able to utilize what they had because they had better-trained crews and more efficient gun carriages. Spanish crews did not even stay at their guns after discharging an initial blast; because of their old-fashioned ideas about naval warfare, they went topside to rake the enemy decks with small-arms fire and serve in boarding parties. The English gunners, by contrast, continued firing round after round. They were said to be three or four times more proficient in firing rate and accuracy.

England's superiority in armaments was symptomatic of the more fundamental underlying disparity stemming from its more effective naval administration. But even though the Navy and Ordnance Boards provided the English fleet with superior weapons and ships, its commanders still had to figure out how to make the best use of them. This took time.

The early engagements produced scant results because the effective range of English artillery was much shorter than their gunners realized. "At ranges of three to seven hundred yards a sixteenth-century culverin [firing eighteen-pound balls] or demi-culverin [firing nine-pound balls] might fail altogether to pierce the thick hull of a galleon or stout greatship," writes naval historian Garrett Mattingly, "and when it did would only make a small hole quickly caulked by an alert crew." It seems likely that Drake first realized the English guns' ineffectiveness when he captured the *Nuestra Señora del Rosaria* on August 1 and saw how little she had suffered. The first close-in attack on a Spanish ship took place two days later when an English vessel pounded the *Gran Grifon* from point-blank range. It is no coincidence that the ship in question was probably Drake's *Revenge*. This experience showed the English commanders how to fight the Spanish, and Howard acted on the results. He decided to conserve English shot until they could get close enough to make it count—"to go within musket shot of the enemy before they should discharge any one piece of ordnance." ("Within musket shot" meant under a hundred yards.) That opportunity finally arrived on August 9 after the fireships had broken the Armada's formidable array.

Thus a crucial element of English success was their commanders' ability to learn on the fly, make adjustments, and attempt new tactics. The Spanish paid a heavy price for their lack of equal flexibility.

Victorian historians liked to suggest rather grandiloquently that the defeat of the Spanish Armada was a "decisive moment in the history of Europe" with "more important and far-reaching consequences upon the destinies of nations than any other since the Norman Conquest." Some modern iconoclasts, by contrast, argue the outcome was not really an English victory. This is going too far. It is true, however, that while the Armada's defeat was a way station on the road to England's rise and Spain's decline, it did not mark a hard boundary between the two imperial epochs.

When England sent a counter-armada to invade Portugal the following year it fared almost as badly as Medina Sedonia had done. A fleet led by Francis Drake failed to take Lisbon and failed to place the pro-English pretender Dom António on the Portuguese throne. The war between England and Spain dragged on for fourteen more years, until the death of Elizabeth in 1603. For all that, it seems fair to say that the Armada did mark the peak of Spanish power. Although Spain would continue as a great power for more than a century, it would never achieve the same level of influence in Europe or the world.

The significance of the Armada becomes more evident the more one considers what would have happened had Spain succeeded: England might have been converted back to Roman Catholicism, the Reformation across Europe dealt a heavy blow, Dutch independence delayed or denied, and Habsburg power vastly enhanced. Because the Armada failed, the prospects of the Dutch revolt against Spanish rule appreciably improved.

More important for our purposes, the Armada inaugurated a style of fighting—pitting oceangoing, gun-toting sailing ships against one another—that would dominate naval warfare for almost three centuries to come. It is remarkable, looking back, how little naval warfare changed from the late sixteenth century to the early nineteenth century. The galleons that defeated Philip II, while smaller and less powerful, were not very different from the ships of the line which defeated Napoleon.

This is not to suggest that English dominance at sea was inevitable. In the first half of the seventeenth century, England was exhausted by the aftermath of the Spanish war and paralyzed by conflicts between crown and Parliament, leaving an opening for the Dutch Republic to become the first true maritime superpower. The United Provinces of the Netherlands depended on fishing and seaborne commerce. To safeguard their livelihood, they developed a formidable navy based, like that of England, on broadside gunnery tactics and led by brilliant admirals such as Maarten Tromp and Michiel de Ruyter, who were every bit the equals of Hawkins and Drake. With their merchant and naval vessels, the Dutch carved out an impressive empire centered on the Indonesian archipelago. Overseas, powerful and nimble Dutch East Indiamen bombarded clumsier Portuguese warships into submission. Closer to home, the Dutch repeatedly bested the Spanish navy. In both cases they used ships and tactics similar to those employed by the English in 1588; the Iberians still had not come up with an effective answer. The Battle of the Downs (1639) was a resounding Dutch victory that signaled the final eclipse of Spanish power at sea.

Once they had vanquished the Spanish, the Dutch turned on the English, who were competing for the same overseas markets. The two naval powers fought three major, inconclusive trade wars between 1652 and 1674. Their rivalry finally ended in 1688 when the Protestant Dutch ruler William of Orange and his English wife, Mary, staged what amounted to a friendly takeover of England—the "Glorious Revolution," which overthrew Mary's father, the Catholic King James II. Thereafter, the Netherlands became in effect a junior partner to Britain in a larger struggle to prevent France from dominating Europe.

King Louis XIV of France was able to pose a serious challenge to Anglo-Dutch fleets because he developed an effective naval administration under his energetic Minister of Marine, Jean-Baptiste Colbert (1619–83). The French navy performed capably in a series of epic battles against the Royal Navy, but Britain won more than it lost and ultimately emerged victorious at the Battle of Trafalgar in 1805. For the next 137 years, Britannia ruled the waves. (How Britain lost her lead in 1942 will be considered in Chapter Eight.)

That the Netherlands and Britain were able to vanquish Spain and France in the battle for supremacy at sea appears, in retrospect, to be nothing short of remarkable. Keep in mind that the population of England and Wales in 1600 was 4.2 million; the Netherlands had 1.5 million. By contrast, Spain had 8.1 million people and France 20 million. Spain and France were also much wealthier than Britain and Holland, at least at first.

Why did the smaller, poorer powers prevail? This is a question that has puzzled historians and strategists for centuries. Part of the answer may be found in geography. France and Spain were quintessential "hybrid" states that had to split their budgets between armies and navies. England was able to concentrate a higher percentage of its resources on the sea, because, following its dynastic union with Scotland in 1603, it no longer had hostile land frontiers to defend. The Dutch Republic was somewhere in between: It faced dangers on its borders, but it was relatively secure behind its swamps, dikes, and artillery citadels. Thus it could devote a higher percentage of its resources to its navy than France or Spain but a lower percentage than Britain. But geography is not destiny; otherwise, Japan, a nation as advantageously positioned off the coast of Asia as Britain is off the coast of Europe, would have been a naval power for more than a few decades of its long history. Australia, Indonesia, Madagascar, and Taiwan are other examples of large islands that have never wielded much if any naval power.

Unlike Japan or other maritime states, Britain and the United Provinces benefited greatly from the fact that both were leaders in the creation of market economies and representative governments. Many historians have detected a connection between these developments and the fact that both were seafaring powers. In naval historian Peter Padfield's summation: "[S]eafaring and trade begets merchants; merchants accumulate wealth and bring the pressure of money to bear on hereditary monarchies and landowning aristocracies, usually poor by comparison; and sooner or later merchant values prevail in government." In continental states, the army might block the expansion of liberty, but navies could not be used for internal repression. Thus, in states with powerful navies but weak armies, those "merchant values"—the sanctity of contracts, private property rights, freedom of political and religious expression, representative government, the rule of law—could become firmly established. Once liberal institutions took hold, the economy grew rapidly, flooding the treasuries of Britain and Holland with revenues that could be used to outfit large fleets that safeguarded colonies and commerce. In a virtuous circle, this created still more wealth that could pay for even greater naval power.

It was not always the case, of course, that navies paid for themselves. In wartime, costs often exceeded revenues, and those deficits grew over time as fleets and armies got bigger. But this was hardly an insurmountable obstacle for the most dynamic economies in the world. The United Provinces and England were able to borrow all they needed to underwrite their defense budgets. The pressures of war gave a powerful impetus to the growth of stocks, bonds, loans, and paper currencies during the late seventeenth and early eighteenth centuries and helped to turn Amsterdam and then London into international financial centers. To take one example, the Bank of England was established in 1694 to raise funds to allow England to wage war against France.

Despotic continental states, with their cash-poor agrarian economies, found themselves unable to keep up. They could mobilize manpower for armies by diktat but not the money and technical expertise necessary to sustain superior naval establishments. Skilled shipwrights who could build warships, wealthy merchants who could provision them, experienced gunsmiths who could arm them, and international bankers who could finance them could not be rounded up by press gangs. These skilled men did not stay rooted to one spot like the peasants who formed the cannon fodder of the continental armies. They went where opportunities beckoned, and the bustling freedom of London and Amsterdam was more alluring to them than the stifling oppression of Paris, Lisbon, Madrid, and St. Petersburg.

Even on those infrequent occasions when the great land powers of Europe suddenly decided to pour a lot of money into their navies (as Spain did in the late sixteenth century, France did in the late seventeenth and eighteenth centuries, Germany did in the late nineteenth century, and the Soviet Union did in the late twentieth century), they usually found themselves unable to match the traditional maritime powers: the Netherlands, Britain, and, later, the United States. Take the case of the Armada. Spain was much wealthier than Britain in 1588 but, as historian N. A. M. Rodger suggests, "no amount of money could have created an efficient Spanish fleet in two or three years. . . . Navies demanded long-term planning and commitment."

Britain made that long-term commitment, backed up the wholehearted support of its populace, and the result, after a few twists and turns of history, was its domination of the seas. Still, there is no predestination. Even eighteenth-century Britain was capable of losing its naval dominance temporarily. Had the French fleet not defeated the Royal Navy at the Virginia Capes outside Chesapeake Bay in 1781, the American colonies might never have won their independence.

Controlling the seas was a much more uncertain business for sixteenth-century England. The whole course of history might have been different if the Royal Navy had failed to prevent the Armada from landing. But it did, because as early as 1588 the English had already mastered the emerging style of naval warfare, and the Spanish never would.

CHAPTER 2

MISSILE AND MUSCLE:

Breitenfeld and Lützen, September 17, 1631–November 16, 1632

They were not much to look at. The Swedish army, composed for the most part of German, English, and Scottish mercenaries, had slept in the fields the previous two nights, and now, a Scottish officer named Robert Monro observed, "they were so dusty, they looked . . . like kitchen servants with their uncleanly rags." Because uniforms were not yet universal among European armies, the bedraggled Swedes and their better coifed Saxon allies stuck green twigs or ribbons in their hats to identify themselves. Their enemies, the armies of the Holy Roman Empire and the Catholic League, wore white ribbons or cords.

At dawn on Wednesday, September 17, 1631, as the larks began to peep and the trumpets and drums began to sound, the Swedish and Saxon armies started their day with a prayer. Having "begged for reconciliation in Christ," Lieutenant Colonel Monro recorded, they marched out "in God's name," colors flying, to the field where the Imperial forces were waiting for them, between the villages of Breitenfeld and Podelwitz, about five miles north of Leipzig. At 9 A.M., the Protestant forces halted half a mile from the enemy and began the elaborate process of setting up in their battle array, the Swedes on the right and center, the Saxons on the left. There was little to indicate that these "kitchen servants" led by an upstart young king from an impoverished and inconsequential country on the northern periphery of the continent would pose the most serious challenge yet made to the Habsburgs' mastery of central Europe.

For the past thirteen years, the German states had been convulsed by religious and dynastic war, and the Habsburg forces had won all the major battles so far. The conflict that would become known as the Thirty Years' War (1618–48) was fought to determine the future of the Holy Roman Empire of the German Nation, a loose agglomeration of states presided over by an emperor. Since 1438, that august position had been held by Habsburgs, cousins of those who ruled Spain, and they were determined to turn this medieval entity into something more closely resembling a modern state. That ambition was fiercely resisted by many of the princes, dukes, knights, and other noblemen who ruled over more than a thousand individual polities in the empire. There was an important religious dimension to the conflict, for the Habsburgs were ardent champions of Catholicism, while many Germans had converted to either Lutheranism or Calvinism. This colossal struggle would draw in most of the major states of Europe: Spain on the side of its fellow Habsburgs; the Dutch Republic, England, Denmark, Sweden, and even Catholic France in an alliance to prevent Habsburg domination of the continent. The succeeding battles would turn much of Germany into a ravaged wasteland, causing more destruction in Mitteleuropa than any conflict until World War II.

Open warfare had first erupted in the Habsburg province of Bohemia with the famous "defenestration of Prague." On May 23, 1618, a group of Protestant noblemen tossed two Catholic regents and a secretary out the window of Hradschin Castle. All three men landed on a dung pile and survived the fifty-foot fall. Protestants called it an accident, Catholics a miracle. Whatever the case, the Bohemian revolt was crushed at the Battle of White Mountain two years later. By 1630 forces loyal to Emperor Ferdinand II had advanced deep into northern Germany, endangering Protestantism's strongholds and threatening to turn the Baltic Sea into a Habsburg lake.

That was not an outcome the Lutheran kingdom of Sweden could contemplate with equanimity, any more than Protestant England could look with favor upon Habsburg attempts to crush Dutch rebels. Seeking to protect German Protestants and Sweden's own political interests, King Gustavus II Adolphus landed with a small army in northern Germany on July 4, 1630. He spent the next year skirmishing against Imperial armies, taking fortified towns and establishing his control over the southern shore of the Baltic but trying to avoid pitched battles while he built up his meager forces. Most German Protestant princes were afraid to openly support their self-proclaimed champion, for they knew that any battles he fought would be on their soil. But much as they wanted to avoid choosing sides in this epic struggle between king and emperor, Protestant and Catholic, the brutal nature of this conflict made neutrality increasingly difficult to sustain.

On May 20, 1631, Imperial armies conquered the Protestant town of Magdeburg. "Here commenced a scene of horrors," the dramatist Friedrich von Schiller later wrote, "for which history has no language, poetry no pencil." The city was reduced to ashes and at least twenty thousand inhabitants were slaughtered. Women were beheaded in churches, infants thrown into flames or stabbed at their mother's breast. The river Elbe, which ran through the town, was choked with corpses. When the Imperial army moved into Saxony and threatened to give Leipzig, its capital, the same treatment, the elector of Saxony, John George, felt he had no choice but to fight alongside Gustavus. On Monday, September 15, 1631, John George's 18,000 men rendezvoused with Gustavus Adolphus's 23,000 soldiers. Together they were stronger than the nearby Imperial army of 35,000 men, and ready to do battle.

THE MONK'S MEN VS. THE KITCHEN SERVANTS

Even though he was now slightly outnumbered, the seventy-two-year-old Imperial commander, Count Johann Tilly, could not have been very concerned about the prospect of tangling with this force commanded by a Swedish king half his age who had yet to fight a single major battle on German soil. Tilly, a short, thin Fleming known as the "monk in armor" for his abstemious habits (his only vice, it was said, was an insatiable appetite for sweetmeats), had spent more than fifty years on active service, the last twenty-seven as a general. He had commanded Imperial forces from the very beginning of this already protracted conflict. He had defeated every general the "heretics" (his derisive term for the Protestants) had put forward and destroyed every army. He had never lost a major battle.

On September 17, 1631, Tilly chose a favorable position for his men on a slight rise outside Breitenfeld. They had the sun and the wind at their backs, the sun shining strongly on this warm fall day, the wind kicking up dust devils from the parched, dry earth. "Father Tilly," as his men affectionately called him, arrayed his forces in tercios, the fearsome infantry formation that had enabled the Habsburgs to dominate European warfare for the last century. Tercios were large combinations of pikemen and musketeers (or, in earlier days, arquebusiers) standing in a giant square. These had begun as formations three thousand strong in the sixteenth century; by 1631 they were down to about 1,500 men. Tilly lined up seventeen tercios side by side, with the men standing thirty ranks deep and fifty across. This unwieldy formation could roll forward inexorably, crushing everything in its path, but it

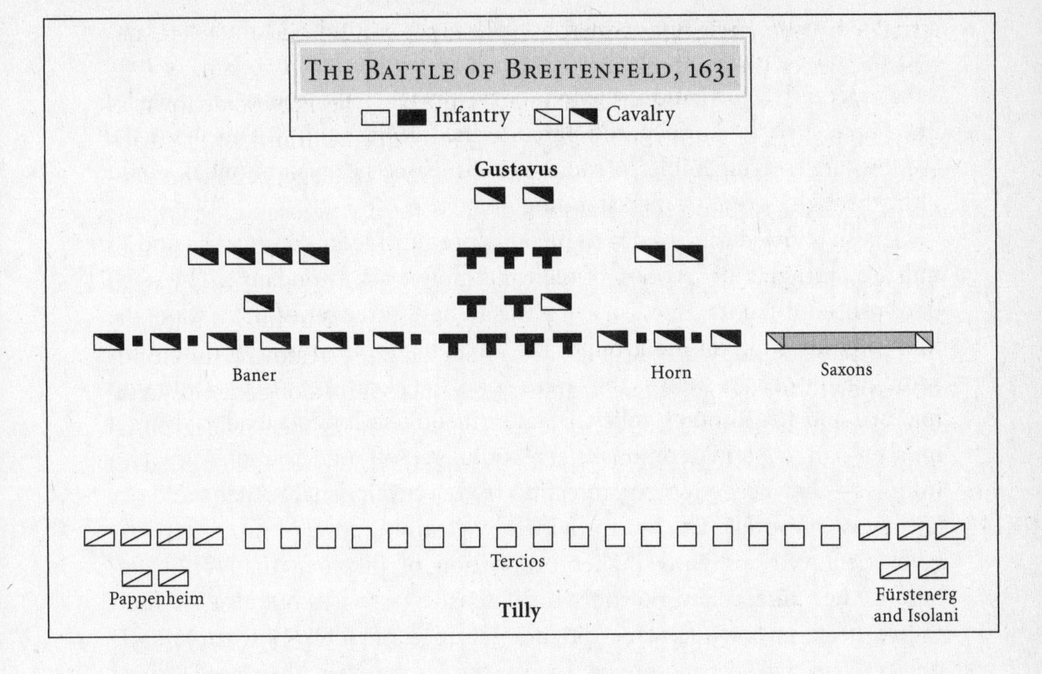

could not maneuver very well. Nor could it make efficient use of most of its men. Those who were not at the front of the square—the vast majority—could not bring their weapons to bear; all they could do was push those in front.

With his tercios deployed in the center, about twenty-five thousand men in all, Tilly sent his cavalry, ten thousand strong, to the wings. The horsemen's style of fighting was no more efficient than that of the infantry. They would trot up to the enemy and discharge their wheel lock pistols, turn around and then repeat this maneuver until they were out of ammunition—or, more accurately, until they were out of pistols, for, lacking time to reload during battle, cavalrymen would attack with two or three pistols loaded with a single shot. These weapons were not very reliable and not very potent beyond point-blank range; they could not penetrate armor beyond five paces. Even if a horseman managed to discharge his pistols at the enemy—no easy feat when firing with one hand from a moving mount—they were not likely to cause much damage. The caracole, as this maneuver was known, was pretty to behold but hardly the most effective use of cavalry.

Notwithstanding their severe limitations, Spanish-style armies had proven effective against all manner of foes for decades. Yet now they found themselves facing an enemy that fought in an entirely new style.

Gustavus did not employ the great clumps favored by the Imperial army. His center was held by four brigades of foot, 7,200 men armed with muskets and pikes, with two more brigades and a cavalry regiment in a second line, and three brigades and another cavalry regiment in a third, reserve line. What made this formation revolutionary was its shallowness. The Swedes lined up just six ranks deep, compared to the Empire's thirty. This made the Swedes far less vulnerable to enemy gunfire (one heavy cannonball could plow through dozens of men standing in a tercio).

It also allowed the Swedes to be far more flexible in their tactics and to utilize a far higher proportion of their manpower at any one time. The standard procedure in the Swedish army was to have three ranks fire a salvo, the first rank kneeling on the ground, the second stooping over, and the third standing upright. As soon as they were done, the next three ranks would step forward and fire another volley. Because it took so long to load and fire a musket—an experienced musketeer could get off one round every two minutes—they needed some protection from a cavalry charge that could occur between rounds. This was provided by gleaming rows of pikemen. Gustavus deployed a slightly higher proportion of pikemen than Tilly did, because he valued them not only for defensive purposes but also for their shock effect in battle. But he did not sacrifice missile power for muscle power: The Swedes outgunned their enemies because their well-drilled troops fired faster and deployed far more cannons.

Gustavus was a pioneer in the development of field artillery, further refining the techniques the French had pioneered in their invasion of Italy in 1494. He standardized his artillery into three weights and slimmed them down to the point where troops could more easily haul them along, even move them around as the fighting changed direction. Every Swedish squadron of four hundred to five hundred men deployed at least two "regimental pieces" for fire support in much the way that a modern infantry company might field heavy machine guns. These regimental pieces were known as "three-pounders," because they fired a three-pound ball or, more often, grape or canister shot composed of some two dozen tiny balls that proved especially effective against massed infantry ranks. The range of these cannon was three hundred yards, and they weighed six hundred pounds, which sounds like a lot, but they were light enough to be wheeled around by one horse or four men. The Empire's "24-pounders," by contrast, required twenty horses to move. As a consequence, at Breitenfeld, Gustavus enjoyed an almost three-to-one advantage in artillery, seventy-five guns to twenty-six.

The Swedish cavalry was designed to exploit the openings created by these gunpowder weapons. No elaborate caracoles for them. They had pistols, but mostly they relied on sword and saber to slash their way through enemy lines. Gustavus arrayed the horsemen on his wings with platoons of musketeers interspersed among them. It was this deftly arrayed combination of shot and shock that would make the Swedish army so formidable.

The Habsburg army did not let their enemies set up unmolested. Monro noted the "noise, and roaring whistling and flying of cannon bullets," but cannonading did not stop the steady work of these professional soldiers. When holes appeared in their lines, they calmly called for a surgeon to take away the wounded and marshaled fresh men to fill the gaps. By noon, the Protestants' preparations were complete and their artillery began to return fire, "paying the enemy with the like coin," in Monro's words. The thunderous roar of cannons continued for two and a half hours.

Count Gottfried Pappenheim, the impetuous cavalry commander on the Imperial left wing, eventually tired of the unrelenting artillery barrage. At 2:30 P.M. he ordered his five thousand horsemen to trot forward. He hoped to turn the Swedish right flank, but his effort failed. The Swedish cavalry, artillery, and musketeers skillfully maneuvered to keep the Imperial charge from breaching their ranks. The caracoles of the horsemen proved no match for the withering fire of the musketeers interspersed among the Swedish cavalry, "the musket ball carrying and piercing farther than the pistolet," as Monro noted. Pappenheim was repulsed time after time—seven charges in all—then sent reeling back by a counterattack from the Swedish reserve cavalry.

Tilly's men were more successful elsewhere. On the other side of the field, their cavalry charged the Saxon positions, mistakenly interpreting Pappenheim's impetuous assault as a signal for a general advance. (Such mix-ups were common in this age when communication was extremely difficult amid the noise and smoke of battle.) However unintended, this bold plunge paid off. The Saxons were reluctant warriors to begin with, and the Imperial cavalry's concerted attack sent them fleeing from the field, with Elector John George leading the headlong retreat. They paused in their panic only long enough to loot the Swedish baggage wagons. In an instant, Gustavus had lost 18,000 men and much materiel. He was left with just 24,000 soldiers to face 35,000 Imperial troops.

Tilly took advantage of this unexpected opportunity by ordering the full fury of his tercios—25,000 infantrymen—to fall on the now-exposed Swedish left flank, held by just 4,000 men under Marshal Gustav Horn. The Imperial veterans advanced amid shouts of "Victoria," confident that victory was already theirs. And so it would have been against any other army. But the Swedes, organized in independent squadrons, had a tactical flexibility unmatched by any other military of the day. Gustavus, surveying the battle from horseback, ordered Horn to wheel left to face the Imperial advance, and he sent forward two brigades from the center as reinforcements. Not waiting for the Empire to strike, the Swedish troops attacked while Tilly's men were still milling around chaotically. The Swedish cannons and muskets took a terrible toll on the close-packed enemy formations. The pikemen and horsemen charged to finish off the remnants of Tilly's tercios with their steel blades, hacking away with fierce abandon.

This was warfare in its rawest, bloodiest form, the kind of fighting that had been perfected by the Greek hoplites and the Roman legions long ago. "Our brigade," wrote Monro, "advanc[ed] unto them with push of pike, putting one of their [battalions] in disorder . . . so that they were put to the route."

Seeing what was happening, Gustavus Adolphus ordered the rest of his army to counterattack. He was no armchair strategist, this king of Sweden. In his day a general could command only as far as his eyes could see and his voice could carry. Gustavus always commanded from the front, even if he could not see very far because of his short-sightedness. With his large, bulging blue eyes, pale gold hair, delicate pink skin, and Vandyke beard, the stout, slightly paunchy king looked more like a scholar than a warrior. But in his thick elk-skin jacket—worn instead of armor because of an old bullet wound that made his right shoulder ache—he was an inspirational figure to his men. Even an enemy officer wrote that there was "something so majestic in his bearing, and something so tender in his gaze, that there was not a man who saw him was not filled with reverence, wonder and love."

Sweating under the hot sun, squinting through clouds of dust, Gustavus led four squadrons of cavalry in a sweeping right hook that took them to the center of the field where the heavy, immobile Imperial artillery presented an alluring target. The Imperial gunners fled or died on the spot, leaving the Swedes in possession of their cannons, which were then turned on the dense Catholic ranks. The Swedish reserve artillery under the master gunner Lennart Torstensson joined the unrelenting barrage.

The Imperial soldiers, proud veterans who had never been defeated, did not fold right away. They continued to stand and fight until they could fight no more. After five grueling hours, the tercios finally buckled and broke. The Imperial army fled the field, with the Swedish cavalry in hard pursuit. They did not stop running until they had reached Halle, more than fifteen miles away. They left behind 7,600 dead, 9,000 prisoners, and all their artillery. Many more were wounded, including Father Tilly himself, who was hit three times. The Swedish dead and wounded were no more than 2,000.

Across Europe, Protestants celebrated their first triumph in Germany. Gustavus Adolphus was acclaimed as the "Lion of the North," the savior of the faith. Habsburg hopes of hegemony had been broken. A new and unlikely Great Power—Sweden—had risen in the heart of Europe.

ANCIENT METHODS, MODERN WEAPONS

The roots of Gustavus Adolphus's success at Breitenfeld may be found four decades earlier in another country resisting Habsburg rule: the Netherlands. The Dutch had been fighting to gain independence from Spain since 1566. From 1585 to 1625 their struggle was led by Maurice of Nassau, Prince of Orange, who served as admiral-general of the United Netherlands and held various offices in Holland, Zeeland, and other individual provinces. Maurice, along with his cousins and fellow commanders, William Louis and John of Nassau, were among the first military intellectuals in the modern West. Both Maurice and William had studied at the University of Leiden with the noted classical scholar Justus Lipsius, who looked to ancient authorities for insights into contemporary politics and warfare.

The challenge these Dutch princes faced was how to turn their army into an instrument capable of standing up to the mighty Spanish Army of Flanders, which had already reconquered the southern Netherlands (roughly present-day Belgium) and now bid fair to do the same for the rebellious northern provinces. To start with, they built on the foundations laid by their enemies. In the early sixteenth century, Spain had been one of a handful of European states that pioneered a standing army that was kept together from one campaigning season to the next, as opposed to the old feudal levies that disbanded every winter. Some lasting terms of military organization owe their origin to Spain: *colonel* comes from *cabo de colunela*, or head of a column; *infantry* most likely comes from *infante*, the name for a Spanish prince, who often led these formations of foot soldiers. The Spanish army was made up of a core of native Castilians supplemented by a larger number of foreign mercenaries. Likewise, Maurice designed his army with a core of native Dutch augmented by mercenaries. The weapons at Maurice's disposal were also similar to those of the Spanish: cannons, muskets, pistols, and pikes.

There the similarities ended. Maurice, John, and William Louis, following the teachings of Justus Lipsius, had concluded that the tercio was not the best use of manpower because its massed ranks were too immobile and too vulnerable to enemy firepower. They devised a host of smaller units, starting with squads, platoons, and companies and ending in battalions—all formations that survive to the present day. Because they divided their men into so many tactical units, the Dutch had to deploy a higher percentage of officers and noncommissioned officers than was standard in the Spanish service. They also deployed their men in lines, not in Spanish-style squares. And they reversed the relationship between pikes and muskets. In the tercio, the pikemen reigned supreme; the musketeers were there largely to protect them from cavalry charges. The princes of Orange decided to make the musket the dominant weapon and thereby reduce the pike to a supporting role. But they faced formidable obstacles in getting much firepower out of the clumsy small arms of their day.

Since the early sixteenth century, infantrymen had relied on the arquebus, which typically weighed ten pounds and fired a .66 caliber ball. It was accurate to no more than fifty yards and took several minutes to reload. It was, in fact, markedly inferior in range and rate of fire to longbows and crossbows, which is why it took a long time to displace those older missile weapons. The arquebus's principal advantages were two: First, it was easier to use than a longbow, which could require a lifetime of training. Second, it was deadlier. "A slug of some weight hitting a human torso at near supersonic speeds not only pierced any body armor available but shattered bone, induced trauma, and inflicted a large jagged wound," writes military analyst Robert O'Connell. "Unlike an arrow or [crossbow] quarrel, which, barring a lucky shot, required some time to induce sufficient internal bleeding to debilitate, a single shot by a gun generally put a combatant down to stay."

By the end of the sixteenth century, the arquebus was being superseded by the musket. It was more accurate and more powerful—its two-ounce lead ball could penetrate plate armor at a hundred yards—but also more cumbersome. It was four to six feet long, weighed fifteen to twenty pounds, and required a forked rest to support its muzzle. Both arquebuses and muskets generally operated through a matchlock mechanism. As historian William H. McNeill notes: "Service of a matchlock was indeed complex, requiring each soldier to insert powder, wad, bullet, and wad successively into the muzzle, tamp each down with a ramrod, then level the piece and pour a different sort of powder into the firing pan before attaching a smoldering match (held all the while in the left hand) to the trigger mechanism; and then aiming and, finally, pulling the trigger. Failure to perform any of these acts in the right sequence meant a misfire."

An alternative was provided by the wheel lock mechanism developed around 1510 in workshops in southern Germany famed for their clocks. Like a clock, the wheel lock was powered by a tightly wound steel spring. Once it was released with the pull of a trigger, the spring would cause a steel hammer to strike iron pyrite, which produced the spark necessary to ignite gunpowder. (Modern cigarette lighters operate on a similar principle.) Because they were easier to use one-handed, wheel lock pistols became a favorite of horsemen. They generally did not catch on with the infantry, however, because they were too expensive and too unreliable.

To get steady fire out of such inefficient weapons, William Louis suggested to Maurice that they steal a march from the Roman legions. Literally. After studying how the Romans had employed legionnaires armed with javelins and sling shots, the Dutch princes came up with the countermarch: A row of soldiers would discharge their muskets, then march to the rear to reload, allowing the next row to fire. In this way, continuous fire could be maintained by ten rows of soldiers. But to carry out this complicated maneuver in the heat of battle would require extensive training. This realization became the impetus for the creation of the first truly professional army in the modern world.

European warfare since the fall of Rome had been dominated by warriors, not soldiers. The two categories are not mutually exclusive, because even modern soldiers try to cultivate warrior virtues such as courage, honor, loyalty, and strength. The difference lies in discipline and cohesion: Soldiers have it, warriors do not. Warriors may fight as part of a large army but in essence they fight alone, seeking one-on-one duels with enemy warriors in order to prove their manhood. They disdain death administered by machines from afar as unchivalric and dishonorable. Soldiers, on the other hand, are supposed to stay in the ranks, follow orders, and do their assigned jobs, no matter how inglorious. They are concerned more with efficiency than with displaying personal prowess. They are in essence cogs in a machine, a concept that would have been utterly alien to the average European of the sixteenth century, whose ideas of military service still derived from the medieval cavaliers—warriors par excellence.

To learn how to make soldiers, the princes of Orange were forced once again to consult classical sources. After closely studying military textbooks written by Greek and Roman authors such as Aelianus and Vegetius, they updated the drill of the Roman legions for a modern gunpowder battlefield. John of Nassau counted forty-two distinct steps necessary to fire a musket. He assigned simple words of command for each one and drilled his men incessantly in their execution. To propagate his ideas, he commissioned the noted artist Jacob de Gheyn to make engravings of each step. These handsome pictures were collected in a sort of comic strip that took a musketeer from the first step ("March with the musket shouldered") to the last ("Guard your pan and be ready"). De Gheyn's book, *The Exercise of Arms* (1619), the first modern manual of arms, proved so popular that it was translated into many languages. John of Nassau also set up the first military academy in Europe, the *Schola Militaris*, in the town of Siegen to teach his methods to young Protestant gentlemen from across the continent.

In addition to having his soldiers load and fire their weapons in unison, Maurice taught them to march in step. This vastly improved their efficiency and responsiveness on the battlefield. An added bonus was that drilling took up most of the men's time while in garrison, thus decreasing drunkenness, violence, and some of the other evils traditionally associated with an idle soldiery. The mostly mercenary soldiers accepted this discipline and drill because the Dutch were rich enough to pay them regularly year-round.

The Dutch reforms had implications that not even its authors could foresee. Centuries of experience have revealed that drill does far more than teach people to perform tasks by rote. When groups engage in choreographed movements such as marching or chanting, a feeling of fellowship rapidly develops. William McNeill attributes this phenomenon to "muscular bonding" that taps into something deep in human biology. This is the reason why all modern armies continue to practice close-order drill using commands similar to those employed by the Roman legions, even though the skills being taught had lost their utility on the battlefield by the 1840s. Soldiers no longer fight in tight formation (against modern weapons, this would be tantamount to suicide), but they still train that way because their commanders know that it creates a powerful *esprit de corps*. This was a realization that Maurice stumbled onto almost by accident.

Japanese warlords had developed similar methods of volley fire a few years earlier to make effective use of firearms imported from the West, but their techniques were not exported abroad and were soon forgotten even in Japan. By contrast, the Dutch reforms quickly spread throughout the West, especially among Protestant nations. Dutch drill sergeants were soon in demand across Europe and even in the New World.

Yet while Maurice made his army the envy of Europe, he did not produce many great victories. Maurice was a cautious commander who did not want to risk losing his army in battle. His goal was not to destroy the Spanish tercios but to keep them from occupying the Dutch provinces. He largely succeeded through skillful maneuver and siege warfare, of which he was also a master. Until then, soldiers had regarded digging as peasant's work, a carryover from the warrior's mind-set. Maurice taught them better. His disciplined soldiers used pick and shovel to build fortifications, stymieing one Spanish advance after another. The geography of the Netherlands, with its broken ground, numerous dikes, and heavy concentration of artillery fortresses, was particularly suited to this low-risk, defensive approach.

It was left to Gustavus Adolphus to further refine the Maurician schemes and turn them into a recipe for audacious offensive warfare that would change the course of history.

THE HOT KING

Gustavus II Adolphus was born in Stockholm in 1594 to the Vasa dynasty, which had ruled Sweden for seventy-one years, and he was trained from boyhood to succeed his father, King Charles IX. His primary schooling came from a tutor named Johan Skytte, who educated him in the classics, theology, languages (he was said to speak eleven languages with some degree of fluency), law, and, most important of all, the military arts. Skytte's instruction in the Maurician methods was supplemented by Jakob de la Gardie, a Dutch soldier who served in the Swedish army. Gustavus took the throne following the death of his father in 1611, determined at age seventeen to implement the Dutch way of war in his native land.

There were few more unlikely candidates in 1611 for Great Power status than Sweden. Even with its satrapy of Finland, the total population numbered no more than 1.3 million. They lived on the northern edge of Europe in an impoverished, barren, half-frozen country with almost no industry to speak of. So poor was Sweden that by winter's end peasants were often reduced to eating tree bark to survive. It was a society, writes historian Michael Roberts, "which was half-isolated, culturally retarded, and still in all essentials mediaeval." Sweden lagged economically, culturally, and militarily far behind not only France, Spain, England, and the Holy Roman Empire, but also its regional rival, Denmark. The political and religious situation was also unsettled upon Gustavus's ascension. Sweden had been Lutheran since the early sixteenth century, but there were still many Catholics who did not accept the legitimacy of Gustavus's rule and considered his Catholic cousin, King Sigismund of Poland, to be the rightful heir. As in Elizabethan England a few years earlier, plots against the Protestant monarch were rife and many noblemen were implicated.

But, just like Elizabeth, young Gustavus was not fazed by these challenges. He responded with a burst of Tudor-style reforms designed to strengthen the state at the expense of the nobility. He was assisted in this endeavor by Count Axel Oxenstierna, whom he appointed as chancellor in 1612. As the years went by and their trust deepened, the two men came to rule almost as coequals. The gray and taciturn chancellor perfectly complemented the energetic and impetuous king. A famous story, perhaps apocryphal, has Gustavus telling Oxenstierna, "If we were all as cold as you we should freeze," to which the chancellor supposedly replied, "If we were all as hot as Your Majesty, we should burn."

Their first project together was creating for Sweden the rudiments of a modern government. They injected fresh vigor into the judiciary, the treasury, the chancery, local government, the Diet (or parliament), and the council of state. To finance these improvements they regularized and increased taxation. Under their leadership, for the first time the government came to be run by full-time bureaucrats trained in newly established schools and universities. Education began to supplant birth as the key to advancement. By the time Gustavus was done, Sweden was, in the judgment of a modern historian, "at least two generations ahead of other European countries in the efficient organization of the realm for carrying out royal policy."

Important as all this was, Gustavus's focus from the first was on naval and military affairs. Because Sweden required a strong navy to ensure the security of its Baltic coastline, Gustavus built a modern fleet of sail and shot that could more than hold its own against regional rivals like Denmark. Having this powerful navy allowed Gustavus to transport an army to Germany in 1630 and to keep its lines of communication secure. But while the Swedes would never wield global naval power to rival the Dutch or English, for a time their army emerged as the best in Europe.

Gustavus based his military on conscription of all men over fifteen, giving Sweden the first draftee army in modern Europe. He divided his kingdom into recruiting districts and assigned each to a specific regiment, which were among the earliest permanent military organizations. The impact of these reforms should not be overestimated. Many Swedes were exempt from service, and if they were drafted they were employed mainly for the navy or defense of the home front. In the army abroad, Swedes formed a distinct minority, largest in the artillery and cavalry, smallest in the infantry. By the end of his German campaign, nine-tenths of Gustavus's army was made up of foreign mercenaries.

The use of mercenaries was not unusual at the time. Armies, like navies, were still quasi-private. Just as sea captains would enter the royal service in times of crisis in return for prize money and other emoluments, so army captains would agree to raise regiments for a monarch in return for a set fee and the promise of plunder. Ordinary soldiers would enlist for a bounty, "the king's shilling." The reliability of these paid fighters was often open to question. Mercenaries became notorious for mutinying, deserting, and for switching sides if captured. But on the whole, soldiers of fortune fought with surprising resolve for causes not their own so long as they were paid on time. One seventeenth-century soldier pithily expressed their outlook: "We serve our master honestly, it is no matter what master we serve."

Gustavus's mercenary regiments were generally of high quality, because he insisted on drilling and disciplining them to the same high standards as his Swedish conscripts. Swearing, dueling, and blasphemy were among the long list of offenses forbidden in Gustavus's Articles of War, and he was not afraid to enforce harsh penalties for transgressions. One time, his brother-in-law, Elector George William of Brandenburg, asked him what should be done with some Swedish officers who had violated the Articles of War. Gustavus's laconic reply: "Has my brother-in-law no gallows in his country, or is he short of timber?"

Like many geniuses, Gustavus was more a synthesizer of others' ideas than a pure inventor. He took the Dutch military model and during the course of lengthy campaigns against Muscovy and Poland from 1617 to 1629 added considerable modifications of his own. He has traditionally been credited with six major tactical innovations. First, he reduced the number of ranks in a formation from Maurice of Nassau's ten to six. Second, he increased the percentage of pikemen. Third, he reduced the size of pikes from eighteen feet to eleven feet to make them easier to wield. Fourth, and for similar reasons, he reduced the size of the musket so that it could be fired without a fork rest. Fifth, he put more emphasis on artillery, which he standardized and lightened. He organized a plethora of cannons into three classes: 24-, 12-, and 3-pounders. And, sixth, he abandoned the caracole for the cavalry charge, a tactic borrowed directly from the Poles, who had used it to great effect against him.

His constant quest for improvement based on experimentation and observation marked Gustavus as a true child of the scientific age then dawning. He was as much an empiricist as Galileo Galilei, whose heretical defense of a sun-centered cosmos, *Dialogue on the Two Chief World Systems, Ptolemaic and Copernican*, was published five months after the Battle of Breitenfeld. Just as Galileo ground lenses and built telescopes to study the sky, so Gustavus personally fired guns, dug trenches, drilled troops, constructed fortifications, and reconnoitered the terrain with his men. As Lieutenant Colonel Robert Monro remarked, "He thought nothing well done which he did not [do] himself."

It was hard for one man to do everything, however, for the Swedish army was growing by leaps and bounds. When he first went on campaign in Livonia (a region encompassing present-day Latvia and Estonia) in 1621, Gustavus led fewer than 18,000 men. By the end of his German campaign in 1633 he had more than 149,000 under his command. And that was only in Germany; there were thousands more at home and in the navy. Similar increases were taking place in other European states, many of which saw the size of their armed forces grow tenfold between 1500 and 1700 because of a general increase in wealth and population, as well as a growing ability by governments to tap those resources through the kind of "nation-building" reforms launched by Gustavus and Oxenstierna. The spread of firearms also contributed to this trend: Unlike bows, which could be skillfully wielded only by recruits with years of practice, muskets could be used effectively by peasants with only a few weeks' training, vastly expanding the pool of potential recruits.

Such large forces made up primarily of mercenaries were expensive to maintain. The revenues of the Swedish state upon Gustavus's accession would have been wholly inadequate to the task, but by the end of his reign this vast expenditure was affordable, if just barely. What had changed?

For one thing, Gustavus encouraged foreign capital to flock to his lands by offering lucrative copper and iron concessions, the only mineral wealth that Sweden possessed in abundance. The wealthy Dutch financier and industrialist Louis de Geer was lured in 1617 into extending his operations to Sweden with the promise of a royal monopoly on armaments. Within a decade, Sweden, which hitherto had possessed almost no armaments industry, became completely self-sufficient and even began exporting cannons and muskets to turn a tidy profit. But tax revenues extracted from Sweden's growing business sector and its increasingly strapped peasantry could not keep up with the costs of Gustavus's foreign campaigns. To pay the bills, Gustavus turned to foreign sources of income.

He extracted subsidies from France, the United Provinces, Saxony, and other states which feared Habsburg power as much as Sweden did. His customs officers, backed by a formidable navy, extracted heavy tolls from shipping in the Baltic and along several important rivers in central Europe. As part of an agreement to suspend his war against Poland in 1629, he compelled his enemies to provide a continuing stream of customs revenue from Poland and Prussia. Most important, his troops lived off the land. The Swedish army, like every other military force up to that time, would pillage and plunder occupied territory. It scarcely mattered whether the people were Protestant or Catholic. The only way a town could avoid being sacked was to pay a large ransom, which often turned into a continuing stream of taxes. Even then, there was always a danger that ill-disciplined soldiers would get drunk and run amok. Gustavus kept a better rein on his men than most contemporary commanders, but he was not averse to their helping themselves to the fat of the land. "War must pay for itself," the king declared, and he largely succeeded in achieving this goal.

THE TYRANNY OF LOGISTICS

Following the victory at Breitenfeld, the king of Sweden was on top of the world. Half of Germany was now in his grasp and there was even talk of placing Gustavus on the throne of the Holy Roman Empire, although there is no clear evidence that he seriously entertained such grandiose designs.

The question was how to follow up his triumph. He has been criticized by some historians for not marching straight on to Vienna, the Habsburg capital, or alternatively for not pursuing Tilly's battered army. His failure to follow either course allowed the Habsburgs to regroup during the winter of 1631–32 and to field fresh armies in the spring. It should be remembered, however, that Gustavus's resources were severely limited, and he lacked the logistics necessary for a lengthy campaign. He could not simply march anywhere he desired. He needed to find a route through land that had not already been ravaged by previous armies and that was close to a major river,

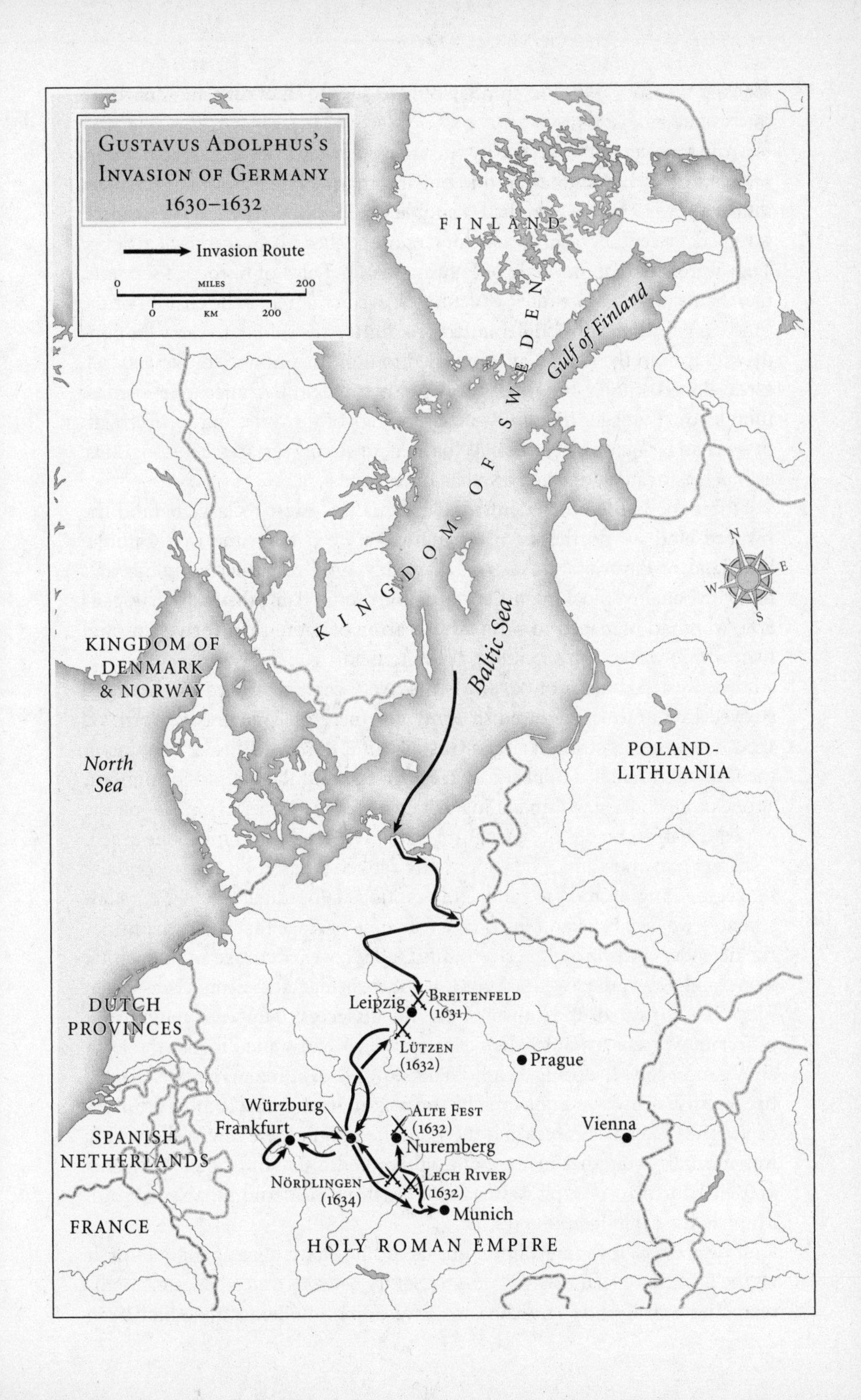
Gustavus Adolphus's
Invasion of Germany
1630–1632
Invasion Route
0
miles
200
0
km
200
FINLAND
KINGDOM OF SWEDEN
Gulf of Finland
Baltic Sea
N
E
W
S
KINGDOM OF
DENMARK
& NORWAY
North
Sea
POLAND-
LITHUANIA
Leipzig
Breitenfeld
(1631)
Lützen
(1632)
Prague
DUTCH
PROVINCES
Würzburg
Alte Fest
(1632)
Frankfurt
Nuremberg
Vienna
SPANISH
NETHERLANDS
Nördlingen
(1634)
Lech River
(1632)
Munich
FRANCE
HOLY ROMAN EMPIRE

for land transport was too cumbersome to supply all of an army's needs in the seventeenth century.

And those needs were vast. By Gustavus's time, armies contained not only many more men than they used to but also many more munitions; firearms, unlike swords or pikes, required a sizable ammunition train. In one respect, seventeenth-century armies had not entirely shed their medieval origins: They were made up not only of fighting men but also of hordes of servants, merchants, wives, prostitutes, concubines, even children. "The custom of the time," as Friedrich von Schiller noted, "permitted the soldier to carry his family with him in the field." Gustavus tried to limit the number of hangers-on, especially prostitutes, but he was only partly successful. Armies were in effect mobile towns, bigger than most urban centers of the time, and their growth in size had far outstripped the development of supply services. Thus logistics, as much as strategy, dictated Swedish movements.

After the Battle of Breitenfeld, Gustavus was eager to leave behind the ravaged lands of northeastern Germany. He headed for the rich Catholic heartland in Bavaria and Austria, where he could resupply his troops and harry the enemy. Blocking his way was old Count Tilly, the Habsburg general, who had managed to raise a fresh army of more than forty thousand men within weeks of his mauling at Breitenfeld.

The western border of Bavaria is the river Lech. Its bridges had been destroyed by the time the Swedish army got there on Wednesday, April 14, 1632. His generals strongly urged Gustavus not to hazard a river crossing in the face of entrenched opposition. He went to scout the ground for himself, at one point coming within hailing distance of some Imperial sentries on the opposite bank who did not recognize him. ("Where's your king?" one asked. "Nearer than you think," he supposedly replied.) Based on this reconnaissance, the king decided to ignore his cautious subordinates. On Thursday, seventy-two Swedish cannons fired a fearsome barrage that drove the Imperial defenders away from the riverbank. Under cover of smoke from burning wood and straw, the Swedes built a wooden bridge across the river. When Tilly's army rushed up to throw back the attackers, they were caught in a pincer movement by Gustavus's cavalry, which had waded across the Lech above and below the bridgehead. Thanks to this dazzling maneuver, the entire Swedish army was in Bavaria by the next morning. The combined forces of the Holy Roman Empire and the Catholic League were sent reeling. Tilly was mortally wounded in the fighting; he died a few days later. Gustavus marched forth to occupy Bavaria, culminating in a triumphal entry into Munich, the capital, on May 17, 1632.

It was Gustavus's finest hour, but he did not have long to enjoy himself on the Bavarian Elector's tennis courts before news arrived of disasters in his rear. The Saxons had launched an invasion of Bohemia that had been thrown back by Imperial forces led by Count Wallenstein, who was now threatening to occupy all of Saxony. To avert a total collapse among his feckless allies, Gustavus had to abandon his planned march on Vienna and head north. This began a months-long duel between two of the most celebrated commanders of the age.

WALLENSTEIN VS. GUSTAVUS

Albrecht von Waldstein, commonly known as Wallenstein, was one of the most enigmatic and controversial figures of the Thirty Years' War. Born in 1583, the son of a minor Czech nobleman, he was raised a Lutheran before converting to Catholicism and marrying a wealthy widow who died not long thereafter, leaving him a rich man. He used his newfound wealth to become the most successful military entrepreneur in Europe. Most of his competitors were small-timers; they raised companies or regiments. Wallenstein thought big. He raised whole armies out of the revenues generated by his vast estates. He led his forces on behalf of Holy Roman Emperor Ferdinand II, who would reward him with additional lands and titles seized from Protestant nobles. This, in turn, increased Wallenstein's wealth and allowed him to build a series of foundries and arsenals to equip his forces. In the 1620s, he enjoyed considerable success as a general; it was his victories that had brought Catholic power to the shores of the Baltic and precipitated Sweden's entry into the war. But he was relieved of command in 1630 by Ferdinand, who suspected that this scheming careerist did not always have the Empire's best interests at heart. This suspicion proved correct, for before long Wallenstein was secretly negotiating with Gustavus to join the Swedish cause, although these talks came to nothing. With the Empire in dire straits after Tilly's defeat, Ferdinand had no choice in 1632 but to give command back to Wallenstein.

By then Wallenstein had amassed riches to rival a king, and indeed the suspicion was prevalent that this wealthy upstart had his eye on a throne—possibly even the emperor's. He shuttled between his baroque palace in Prague and his many country houses accompanied by no fewer than fifty carriages for his court and one hundred wagons carrying his baggage. He was said to hate noise so much that upon his arrival in a town he would order all dogs and cats killed and all streets around his residence cordoned off. The wearing of spurs or loud talking was forbidden in his presence. Those who disobeyed his diktats were liable to get a call from his household executioner. One servant reportedly was hanged for waking him in the middle of the night. Such cruel and mercurial behavior did not make the "tall, thin, forbidding" Wallenstein very popular. But he was nevertheless an effective military leader, one who appointed officers based strictly on merit and was not averse to having Protestants in senior positions.

In mid-June 1632, Gustavus Adolphus reached the city of Nuremberg, where he fortified his army. Wallenstein decided to occupy his own fortifications nearby at Alte Feste and starve the Swedes out. The siege dragged on for two months before Gustavus felt compelled to risk an attack because he could no longer feed his troops within the city walls. His offensive on September 3–4 failed to dislodge Wallenstein. It was the first major Swedish setback after a long string of victories. Gustavus lost 1,000 troops. In the next two weeks, epidemics and desertions cost his dispirited army one-third its strength. As Monro dourly observed, "oftimes an army is lost sooner by hunger than by fighting."

In search of fresh supplies and fresh manpower, Gustavus marched out of Nuremberg on September 18, heading south along the Danube. The two armies spent the next two months shadow-boxing, never able to land a punch, seldom able even to figure out where the enemy was—a testament to the poor quality of communications and intelligence in the seventeenth century.

In late October 1632, Gustavus received word that Wallenstein had led his army northward where it had rendezvoused with another Imperial force under Count Pappenheim. Together the two Imperial commanders were once again threatening to seize Saxony and cut off Gustavus's rear. For a second and final time, Gustavus hurried north. His troops covered an impressive 380 miles in seventeen days. Unfortunately Gustavus could take only 19,000 men with him; the rest (more than 163,000) were scattered in garrisons or committed to other operations around northern Europe.

Reaching Naumburg on November 10, 1632, Gustavus made a fortified camp there. Wallenstein mistakenly believed that the Swedes were suspending campaigning for the year and going into winter quarters, so he decided to do the same. He began to scatter his forces, because his base near Leipzig could not provide enough sustenance for his large army. When he heard on Sunday, November 14, what Wallenstein was up to—dispersing his forces in the face of the Swedish army—Gustavus felt as if his prayers had been answered. "Now in very truth I believe that God has delivered him into my hands," he cackled.

Gustavus knew he had not a moment to lose. Pappenheim, with a force of 3,000 cavalry, had already left Wallenstein's encampment for the town of Halle, thirty-five miles away. As long as Pappenheim was absent, the Swedes had a slight numerical advantage, with 19,000 men to 16,000 Imperial soldiers. Pappenheim's arrival would even the odds. The Swedes needed to strike before Wallenstein became aware of their strategy and combined his forces again.

In the early morning hours of Monday, November 15, the Swedish troops

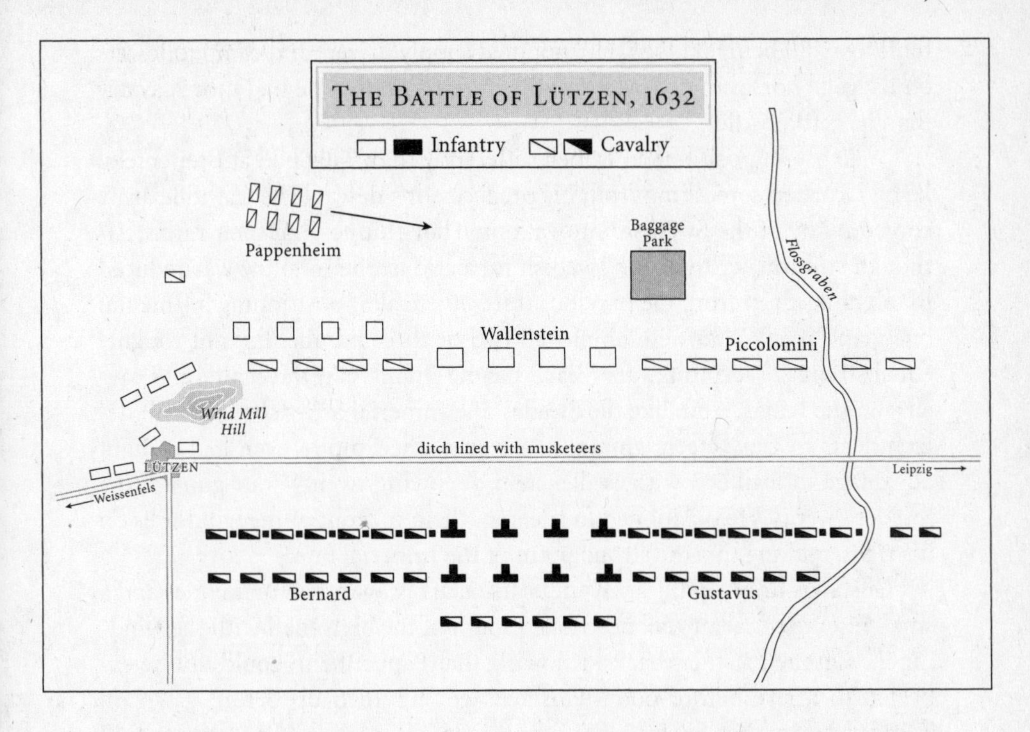

marched out of Naumburg in the direction of Wallenstein's camp located fifteen miles southwest of Leipzig near the village of Lützen. They might have caught the unsuspecting Imperial forces by surprise were it not for one of those strokes of ill luck that often frustrates the best-laid plans. In their advance, the Swedes stumbled upon a small Imperial reconnaissance party near a stream called the Rippach. The patrol put up an unexpectedly stiff resistance that delayed the Swedes for a few vital hours in the midafternoon. Not wanting to start his attack at dusk, Gustavus reluctantly decided to wait until the morning. Now amply forewarned, Wallenstein spent all night working feverishly by torchlight to mobilize his men and prepare a strong defensive position. He immediately sent an order recalling Pappenheim, which arrived around midnight. Pappenheim's cavalry set off at once, though his infantry waited another six hours to start back.

Wallenstein prepared a strong position running parallel to the high road from Leipzig to Lützen. In a deep ditch just north of the road, Wallenstein positioned musketeers to protect his front. Only the tops of their feathered hats and the muzzles of their muskets must have been visible to the approaching Swedes. Wallenstein anchored his position on the right along some windmills. On the left his line was less secure, not quite reaching a small canal called the Flossgraben. Wallenstein hoped that Pappenheim's men would fill in the rest of the space, but until their arrival he had to resort to bluff to avoid being flanked. He mounted supply drivers and camp followers on baggage horses in order to give the impression that he had more cavalry than he actually did.

Wallenstein positioned his men differently than Tilly had at Breitenfeld; he had learned something from his predecessor's defeat. The Catholic army copied many of the Swedish innovations. They thinned out their ranks: Although still thicker than the Swedish formations, the infantry was reduced to ten ranks deep from the previous thirty. Instead of positioning his men in one giant line, Wallenstein employed two or three parallel lines of roughly equal strength (accounts vary), each having infantry in the center and cavalry on the wings. And, like the Swedes, the Imperial army spread small detachments of musketeers among its cavalry. The Empire even had a slight advantage in artillery, with Wallenstein deploying twenty-four guns to the Swedes' twenty. He positioned fourteen of them in front of the windmills on his right, making this the strong point of the Imperial line.

Gustavus drew up his army in his usual array, with two main lines and a small reserve of cavalry, all positioned south of the high road with the winding Flossgraben at their backs. Knowing that Pappenheim could not be expected to arrive before noon, Gustavus got his men up before dawn on Tuesday, November 16. By 7 A.M., the Swedes were ready to go, only to find the whole battlefield covered in a thick, impenetrable fog, a hindrance that would beset the combatants on and off all day. While he waited for the fog to clear, Gustavus led the troops in a prayer and delivered a pre-battle oration in which he exhorted his "true and valiant brethren" to "see that you do valiantly carry yourselves this day, fighting bravely for God's Word and your King." He promised that if they stood with him, he would "hazard my body and blood with you."

He got his chance soon enough. The sun broke through the mist at 11 A.M., the signal for the Swedes to attack. Gustavus led his right wing into action against the enemy's weaker left wing, the Swedes crying, "God with us!", the Imperial soldiers countering with "Jesus Maria!" The Swedes took heavy losses from the fire of the Imperial musketeers but managed, in Monro's account, to chase "the enemy a little out of the ditch, and took seven of the Imperialist cannon, that were planted along the [ditch]." Gustavus's horsemen were on the verge of wheeling left and falling on the main body of Imperial soldiers when, as if out of a storybook, Pappenheim's cavalry appeared to restore the Imperial position at the last possible moment. The fortunes of battle instantly shifted against the Swedes.

A few minutes later they shifted back when Pappenheim was mortally wounded by two musket balls (or, according to some accounts, by a cannonball). Seeing this inspirational commander taken off the field in a cart unnerved the Imperial army. Panic spread in their ranks. Their left flank once again began to disintegrate, once again the Swedish cavalry started to attack the Imperial center, and once again victory appeared to be in Gustavus's grasp—only to have it snatched away once again by a caprice of nature. Another heavy mist rolled in, thicker than before, which concealed from the Swedes the extent of their enemies' disarray and ground their offensive to a halt.

On the other side of the field, the battle was not going the Swedes' way. The Imperial guns in front of the windmills were taking a heavy toll on the Swedish left wing led by Duke Bernard of Saxe-Weimar. Gustavus decided to lead a cavalry regiment over to help strengthen Bernard's attack. As usual, he led from the front, and now he paid a catastrophic price for his daring. Gustavus rode so fast that he left most of his escort behind and found himself in close proximity to the enemy lines. An Imperial musketeer took careful aim, fired, and shattered Gustavus's left arm. His horse galloped wildly out of control, leading him straight into a party of enemy cavalry, one of whom fired a pistol into his back. As usual, because his shoulder was sore from an old bullet wound, Gustavus was not wearing any armor, and the pistol ball wounded him grievously. He fell from the saddle. With a foot still stuck in his stirrup, he was dragged along the ground by his horse until he finally twisted free. Lying facedown in the mud, he was shot a third time, through the head. It was an inglorious end for such a historic military career, made all the more so by the fact that the Imperial soldiers plundered the king's body, stripping him down to his shirt before abandoning the corpse in the mud.

The sight of the king's wounded horse, Streiff, dashing madly around without a rider and covered with blood made clear to both sides what had happened. As word spread through the Swedish ranks that their sovereign was dead, panic momentarily set in, but the court chaplain rallied the men by singing, "Sustain Us by Thy Mighty Word." As soldiers joined in the psalm, they gained a fresh injection of courage and determination to avenge their fallen leader. The Imperial soldiers, for their part, were heartened by the news of their foe's demise. Pappenheim died a happy man, knowing that his arch-enemy had predeceased him, if only by a few hours.

The Imperial cavalry, now under Count Ottavio Piccolomini, surged forward and swept over the Swedish right wing, taking back the lost Imperial guns. By 2 P.M., the battle was going against the Swedes. Duke Bernard, left in charge of the Swedish army after Gustavus's death, decided to stage one last desperate lunge to turn things around. The Swedes, closely supported by their regimental artillery, charged straight into the heart of the Imperial position. In hard, bloody fighting, they crossed the high road and took the windmills one by one. By 5 P.M., as dusk was descending over the battlefield, every piece of Imperial artillery was in Swedish hands. To compound Wallenstein's problems, his powder wagons caught fire and exploded, leading many of his men to jump to the erroneous conclusion that they were being attacked from the rear.

After dark, Wallenstein took a survey of his forces and found they had suffered crippling losses. He had no choice but to retreat. He did not stop running until he had reached the safety of Bohemia. The battlefield was left to the Swedes. (Although Wallenstein had survived, he was disgraced and did not have long to live. He would be assassinated two years later at the instigation of the emperor, who suspected him of treason.)

It was a battle fought "with such fury as no man hath ever seen or heard," Wallenstein wrote the next day, and both sides were thoroughly drained by their exertions. Sweden lost an estimated five thousand to six thousand men (about a third of its force), the Empire even more. "The entire plain from Lützen to the [Flossgraben] Canal," wrote Friedrich von Schiller, "was strewed with the wounded, the dying and the dead." The outcome was hardly as one-sided as at Breitenfeld, but once again superior Swedish firepower and maneuverability had won the day. Monro was right to pay tribute to his fellow soldiers who "did crown the lamentable death of the King's majesty with a stately and heroical victory."

ABSOLUTISM ASCENDANT

Gustavus Adolphus earned lasting fame for the twenty-eight months he spent campaigning in Germany. Contemporaries, at least those of the Protestant faith, wept at the demise of the Hero of the North, who was just thirty-seven years old. "[W]hile the world stands," wrote Monro, in a typical effusion of grief, "our King, Captain and Master cannot be enough praised." Napoleon paid his own tribute more than 150 years later by putting Gustavus on his short list of the world's greatest captains, alongside Julius Caesar, Hannibal, and Alexander the Great. More recently, historians have acclaimed him for realizing the potential of the Gunpowder Revolution by creating the prototype of the modern army. He was in many ways the military equivalent of Drake, Hawkins, and the other Elizabethans who found a new way to fight at sea.

Naturally, this view has not won universal assent; no historical interpretation ever does. Skeptics suggest that Gustavus's tactics were not all that revolutionary or successful. As evidence they point to the battle of Nördlingen in 1634, pitting the Swedish army against a combined force of Spanish and Austrian troops. The Swedes were routed, losing more than half their force. But this defeat does not invalidate Gustavus's achievement. In the first place, the Swedish army at Nördlingen was heavily outnumbered and lacked unified leadership, with its command split between two jealous rivals, Field Marshal Gustav Horn and Duke Bernard of Saxe-Weimar. In the second place, Nördlingen was an anomaly; Breitenfeld was closer to the norm.

The Swedish military presence in Germany did not end with Gustavus's death. The Swedes kept fighting until the end of the Thirty Years' War sixteen years later—and they kept winning. The record of Swedish triumphs is a long one. In addition to Breitenfeld, the Lech, and Lützen, there were major victories at Wittstock (1636), Breitenfeld II (1642), and Jankau (1645). These battles were won by a number of different commanders trained in the Swedish system, which suggests that there was more to Swedish success at Breitenfeld and Lützen than the singular genius of Gustavus Adolphus, though without him the system could not have been developed.

Further confirmation of the superiority of the Dutch and Swedish military innovations comes from the Battle of Rocroi in 1643, when the French army used many of these same methods to destroy the cream of the Spanish tercios. Coupled with Dutch naval victories made possible by utilizing techniques that had worked so well for the English in 1588, Rocroi signaled the end of Spain as a Great Power.

When the Peace of Westphalia ending the Thirty Years' War was concluded in 1648, Sweden and France emerged as the big winners. France received part of the provinces of Alsace and Lorraine. Sweden received western Pomerania along the Baltic coast (part of modern Germany) as well as an indemnity of five million Imperial dollars. Habsburg ambitions to dominate the Holy Roman Empire had been crushed. Thereafter the Austrian Habsburgs would concentrate on their crown domains, principally Austria, Bohemia, Moravia, Hungary, and parts of Italy, which would be known as the Austrian Empire. Sweden would continue to be the dominant power in northern Europe until it was eclipsed by the rise of Russia and Prussia in the early eighteenth century. Its fall from preeminence was ratified by the 1721 Peace of Nystad, which concluded the Northern Wars and forced it to give up most of its Baltic possessions.

What seems remarkable in retrospect is not that Sweden eventually tumbled from the Great Power ranks but that it ever reached them in the first place. There was little in its geography, demography, or economy to explain such a rise, especially when pitted against a much larger and richer rival such as the Habsburg Empire. Much of the explanation must rest with the military genius of Gustavus Adolphus and his successors, culminating with King Charles XII (1682–1718), who would develop and apply a military system second to none. Sweden's standing reached its apex at the Battle of Narva in 1700, when eight thousand Swedes crushed a Russian army more than four times as large. Peter the Great was able to recover from this setback and eventually defeat the Swedes only by imitating their methods of military and social regimentation.

The military revolution of which Gustavus was such an integral part played a vital role in the emergence of modern Europe. There were in fact a host of mini-revolutions that combined to create the Gunpowder Age. Its immediate antecedent was the infantry revolution of the fourteenth century, when Swiss pikemen and English longbowmen eclipsed mounted knights as the dominant force on the battlefield. Then came the artillery revolution of the fifteenth century, when the spread of huge siege guns made old-fashioned fortresses obsolete. (It was this development that allowed the French to invade Italy so effectively in 1494.) This was followed in the sixteenth century by the artillery fortress revolution, the spread of the *trace italienne*, which restored the advantage to the defensive. At roughly the same time, the shot-and-sail revolution was giving unwieldy heavy cannons—similar to those utilized in the *trace italienne*—offensive utility at sea, as the Royal Navy proved in defeating the Spanish Armada. The final stage of the broad revolution occurred from 1594 to 1632 when Maurice of Nassau and Gustavus Adolphus restored offensive punch and maneuverability to armies that had become sluggish and unwieldy. This last development has been labeled a revolution in "field warfare" or "combined arms" to distinguish it from the siege warfare that had characterized the first part of the Gunpowder Age. These methods would be utilized with great success by Louis XIV's French army, Oliver Cromwell's New Model Army, the militias of England's North American colonies, and a host of other armed forces, many of which employed Dutch and Swedish veterans. "[T]he Swedish pattern," writes historian Gunther Rothenberg, "became the standard for all commanders of the next century."

This revolution not only transformed warfare. It also transformed politics and society. Only large states could afford large, well-trained, well-equipped, well-supplied armies, and they in turn used their growing resources to suppress all opposition. War steadily became nationalized, with the role of privateers and mercenaries reduced. Absolute monarchies sprang up, which could raise more revenue than previous regimes and spend almost all of it on their armed forces—75 percent in the case of Louis XIV, 85 percent for Peter the Great, 90 percent for Frederick the Great. Discipline (or, to put it less charitably, despotism) spread from the military to government in general, thus creating the forerunners of the modern centralized state.

Armies such as those led by Gustavus Adolphus were prototypes of the modern *bureaucracy*, a word first used in 1818 to describe an organization characterized (according to the *Merriam-Webster Dictionary*) "by specialization of functions, adherence to fixed rules, and a hierarchy of authority."

"The success of this type of hierarchical organization," according to political scientist Bruce Porter, "would make it an attractive template for organizing all kinds of large-scale human endeavors in the future, from civil service bureaucracies to industrial factories to totalitarian states."

Power and profit went to those states that mastered these techniques and their attendant technologies—in sea power, England and the Dutch Republic; in land power, Sweden and France, followed in the eighteenth century by Prussia and Russia. Prussia's performance was especially impressive. In a sense, it was the Sweden of the eighteenth century: another relatively small, poor country that managed to wield military power out of all proportion to its meager resources by creating an efficient state bureaucracy and a highly disciplined army led by a series of able rulers. The process began with the Great Elector, Frederick William (r. 1640–1688), and culminated with his great-grandson, King Frederick II (r. 1740–1786), who was able to annex the province of Silesia and repeatedly repulse the armies of his much larger neighbors, Russia, Austria, and France.

The losers in the European scramble for power included Spain, Portugal, Poland, Scotland, the Italian city-states, the Holy Roman Empire, the Ottoman Empire, and a host of lesser entities such as Navarre, Brittany, and Savoy, which never attained full sovereignty. None of them could keep pace with the latest military technology and techniques, and they fell by the wayside in relative terms as their rivals modernized. Poland, which never developed any effective central authority, paid the most catastrophic price for its failure to keep pace: It would be swallowed by its neighbors between 1768 and 1795 and cease to exist as an independent state for more than a century.

Spain's downfall, if less complete, was more dramatic, because in the sixteenth century it had been the most powerful state in Europe. The brilliance of Spanish armies and fleets had made possible the conquest and exploitation of much of Italy, the Netherlands, and the Americas, a veritable El Dorado filled with gold and silver. But the Habsburgs spent their escudos faster than they came in and never developed an administrative apparatus remotely capable of coping with the demands of the military revolution sweeping Europe. Before long, the military edge enjoyed by the Habsburgs in the early years of the Gunpowder Age was lost to its fast-rising rivals.

Sweden did not—could not—replace Spain as would-be hegemon of Europe. No matter how effective its army, no matter how brilliant its rulers, the land of reindeer and long winters was simply too small, too isolated, and too lacking in natural resources to play such a role. France, however, was big enough and ambitious enough. During the reign of Louis XIV (r. 1643–1715), a series of talented ministers led by Michel Le Tellier and his son, the Marquis de Louvois, overhauled the administrative machinery along the lines pioneered by Gustavus Adolphus and Axel Oxenstierna. They professionalized the army, created a war ministry, and set loose royal agents known as *intendants d'armée* to ensure that the logistical requirements of field armies were met. In many ways French bureaucrats went further than the Swedes by, for example, setting up a series of prestocked magazines for supplying armies on the march. These reforms gave France the most formidable army on the continent and allowed it to bid for European hegemony until Napoleon's defeat in 1815.

If French ambitions were ultimately frustrated, this can be ascribed in large measure to the fact that its rivals—Prussia, Austria, Russia, and England foremost among them—imitated the Swedish model as well. In a dynamic that would be repeated many times over in the centuries to come, innovations spread quickly across the armies of Europe, nullifying the advantage of the "early adopter." They would take much longer to be adopted overseas, giving Europeans a crucial edge in their wars of conquest abroad.

CHAPTER 3

FLINTLOCKS AND FORBEARANCE:

Assaye, September 23, 1803

Having covered fourteen miles since sunrise, the tired troopers finally ended their march in the Indian village of Naulniah. It was eleven o'clock in the morning on Friday, September 23, 1803. They were advancing deeper into the heart of Maharashtra, the powerful Hindu confederacy that dominated central India. The 3,300 infantrymen—a combination of 1,300 Scottish Highlanders serving King George III and 2,000 Indian sepoys serving the British East India Company—were no doubt delighted amid the stultifying midday heat to put down their long Brown Bess flintlocks and their bulky canvas knapsacks. They must have luxuriated in taking off the uncomfortable leather stocks from around their necks and unbuttoning their heavy scarlet wool coats better suited for a parade ground in England than for the scorching heat of the tropics. As soldiers pitched their tents, quartermasters bustled around, arranging to have bread baked and bullocks slaughtered to provide fresh meat. *Puckalees*, the native water carriers, filled their tanks. Lascars, the army servants, set out to gather forage for the horses. Local drivers looked after the specially bred white bullocks that pulled the thousands of carts containing the expedition's baggage and supplies.

Although the men knew that a fight was in the offing against the Marathas, as the people of Maharashtra were known, they expected to spend the day quietly in this dusty little village. Their commander, a thirty-four-year-old major general with a lean, muscular frame, aristocratic lineage, and curt manner, had not planned a battle until the following day. His best intelligence indicated that the enemy camp was still at least a dozen miles to the north and a supporting column of British troops was still a half day's march to the west.

Just as the troops were settling in, a patrol of dragoons led in some Indian *banjaras*, merchants who dealt with both sides even in wartime. The English commander questioned the men in Hindi and learned, much to his surprise, that the Maratha army was not more than twelve miles away but fewer than six. The Anglo-Indian column had inadvertently set up camp on the enemy's doorstep.

The general decided to see for himself the lay of the land. He galloped out of camp atop his fine gray stallion accompanied by only two staff officers, with some cavalrymen trailing far behind. About three miles away, the trio of British officers came to a crest that afforded a fine view of the plain below—and a disconcerting sight: Directly in front of them was a broad muddy river bounded by steep banks. On the other side, as far as the eye could see, stretched a horde of humanity. It was more than an army—it was a virtual nation in arms that sprawled across more than six miles of plains between the Kailna River and its tributary river farther north, the Juah.

The Maratha Confederacy was the last major power that could challenge the British for mastery of India. One of its chieftains was already a virtual vassal of the East India Company, but his powerful rivals remained unbowed. Here were the armies of just two chieftains—Daulat Rao Sindia, the maharaja of Gwalior, and Raghuji Bhonsle II, the raja of Berar—and together they must have totaled at least fifty thousand men. The actual number of people in camp was far higher, for Indian armies did not travel light. There were giant canvas tents lined with silk to provide suitable accommodations for the leading commanders (while the ordinary soldiers had to squat in the sun). Great teeming bazaars where all manner of merchants, from confectioners to goldsmiths, tailors to wine merchants, plied their trades. Women were everywhere: wives, concubines, servants, dancing girls. And, of course, the myriad animals: small wiry cavalry mounts, giant elephants with gem-encrusted capes and silver-sheathed tusks, camels carrying rockets on their backs, pack bullocks milling around. Overhead hung greasy clouds of smoke from camp fires fueled by dried cakes of bullock dung.

As the major general examined this scene through his telescope, he quickly realized that most of what he saw lacked military significance. The dancing girls would not fight. Nor, for that matter, would most of the swarms of cavalry spread out on the plain below. The Maratha horsemen were raiders par excellence, specialists at harassing enemy formations, looting peasants, and ambushing isolated outposts. But numerous as they were (around 30,000 to 40,000), they would never dare come to close quarters with a determined foe. Neither would the horde of Maratha irregular infantry, some 10,000 to 20,000 strong. Of greater concern was the smaller number of infantry arrayed on the left of the Maratha camp. These men hailed from the warlike tribes of northern India, but they were dressed in white coats like Europeans and they were officered by Europeans in the employ of the Marathas. One of their commanders had been a former sergeant in the East India Company, now promoted to colonel in the Marathas' service, and they were arraying themselves in ranks that would not put the Company's army to shame. In front of their formation was a vast array of field guns, gaudily painted, well maintained, and expertly manned by Portuguese artillerymen. These were positioned to command all the obvious fords across the river.

More than one hundred field guns for the Marathas; only twenty-two for the British. Almost 11,000 regular infantry for the Marathas; only 3,300 for the British. More than 30,000 cavalry for the Marathas; only 1,200 for the British. These were long odds, exceedingly long odds: a daunting three to one if only the Maratha regulars were counted; a ridiculous ten to one or more if the rest of the vast throng was included.

More experienced officers, confronted with such a daunting array of foes, might have beaten a hasty retreat and waited for reinforcements. Not this young major general. He had never before commanded a large force in battle, but he possessed the dauntless optimism of youth and the ingrained certainty of noble birth. He calculated that if he pulled back, the full weight of the Maratha cavalry might fall upon his camp and at the very least seize part of his supply train. As much as he was afraid of being attacked, he was even more afraid that the enemy would retreat and he would lose a chance to bring them to battle. He had been pursuing the Maratha army for more than a month, and he did not want them to escape.

As the general was pondering these matters, some enemy cavalry spotted him and set off in pursuit. Atop his fine Arab charger Diomed, he had no trouble outdistancing the slower enemy ponies. As he galloped back to camp, he was already barking orders. "I thought there was no time to be lost," Major General Arthur Wellesley would later explain. "Accordingly, I determined to march on to attack them."

That he would so eagerly face such overwhelming odds says much about the future victor of Waterloo, the man who would become the Duke of Wellington. It says even more about the balance of power between East and West more than three hundred years after the advent of the Gunpowder Age. Against the rich states of Asia, European soldiers did not yet enjoy a vast advantage in armaments—that would come with the Industrial Revolution—but their edge in tactics and discipline was already so great that Wellesley was willing, even eager, to charge into what against a European opponent would have been a suicidal confrontation, confident in the knowledge that no Indian army, no matter how large or well armed, could possibly stand for long against his better-drilled troops.

THE FIRST TRANSOCEANIC EMPIRES

The Portuguese had inaugurated the Age of Exploration around 1400. They wanted gold, spices, silk, slaves, and other precious goods, and, because of advances in shipbuilding, artillery, and navigation, they could seek them farther and farther afield. The compass, astrolabe, and cross-staff were especially important because they made possible the accurate measurement of direction and latitude.

A series of voyages edged ever farther down the west coast of Africa until in 1488 Bartolomeu Dias rounded the Cape of Good Hope. Nine years later Vasco da Gama sailed all the way to India. His voyage was not a commercial success because the Muslim merchants who dominated India's Malabar Coast had no intention of trading with infidel interlopers, and anyway da Gama had nothing they wanted. A second Portuguese expedition in 1500 resulted in a number of Portuguese being massacred and their fleet bombarding the port of Calicut in retaliation. In 1502 da Gama returned with a squadron of twenty heavily armed sailing ships known as carracks and caravels, determined to settle matters once and for all. He was met by a much larger fleet of Arab dhows and Indian prahus (two types of non-European sailing vessels) assembled by the Hindu ruler of Calicut and the Muslim ruler of Mameluke Egypt, whose merchants dominated the Indian Ocean trade. Da Gama's forces were outnumbered by at least ten to one, but with his superior sailing rigs, more seaworthy vessels, and heavier cannons, da Gama was able to stand off and batter the opposing vessels into submission, much as the English would later do to the Spanish Armada. He capped this victory by cutting off the hands, ears, and noses of captured crewmen and sending them to the king of Calicut with the facetious suggestion that he turn them into "a curry made to eat."

The Portuguese were lucky that their biggest potential rival—the Chinese Empire—had sidelined itself from maritime competition. In the early fifteenth century, China possessed the most powerful navy in the world. The eunuch admiral Cheng Ho, who explored the Indian Ocean from 1405 to 1431, had commanded hundreds of ships, some as large as fifteen hundred tons. (By comparison, da Gama's flagship displaced a mere three hundred tons.) But by the 1430s the Ming imperial court had forbidden all foreign trade and outlawed the construction of oceangoing ships. This isolationist policy inadvertently opened the door to the Portuguese, who were arriving in the Indian Ocean just as the Chinese were withdrawing.

Da Gama and his compatriots still faced some formidable foes, including many which could put large fleets to sea, such as the Ottoman Empire and the sultanate of Aceh (in present-day Indonesia). The Portuguese did not win every battle nor capture every city they set their sights on. But by the sixteenth century they had established a series of heavily fortified trading stations around the rim of Africa (Accra, Luanda, Mombasa), the Middle East (Hormuz), South Asia (Goa, Ceylon), Southeast Asia (Malacca), and East Asia (Macao). Portugal, a tiny nation whose population was no more than 1.5 million (comparable to present-day Estonia), had created the first transoceanic empire. It was an amazing achievement but a limited one. What the Portuguese could not do was to move very far inland. Their empire extended no farther than the reach of their naval guns.

The Spanish did better. At least in the New World. Following Columbus's pioneering journey in 1492, conquistadors eradicated the natives of the Caribbean islands, whose neolithic technology was thousands of years behind that of Europe. The West Indies became launching points for the conquest of Mexico in the 1520s and of Peru a decade later. That a few small bands of adventurers led by Cortés and Pizarro subdued roughly one-fifth of the world's population remains startling even after the passage of almost five hundred years. Their success has been ascribed to a variety of advantages, including the virulent germs they unknowingly brought with them (an early version of biological warfare) and the willingness of many Indians to help them against their cruel Aztec and Inca overlords. Their triumph owed even more to the inescapable fact that they used swords and guns against people who not only lacked firearms but even iron, steel, horses, oxen, and wheels. Spanish troops had similar success in conquering the Philippines against scarcely more sophisticated opposition. The Portuguese likewise prevailed, if more slowly, in Brazil. By 1600 the empire of the Spanish Habsburgs (who had taken over Portugal in 1580) spanned the globe.

They did not have long to enjoy the fruits of their overseas conquests before other Europeans arrived to share in the spoils. France, Britain and the Netherlands, which by the early 1600s had overtaken the Iberians militarily, were now determined to help themselves to their colonial possessions. Like their predecessors, the newcomers had no trouble establishing a substantial presence in the New World; the primary obstacle here was not the native people but competing Westerners. The French, Dutch, and English all seized Caribbean islands and planted colonies in eastern North America. Even though many Indians adopted horses and guns, they were inexorably pushed aside by the arriving Europeans. The same thing would happen to the native peoples of Siberia, who were conquered by the Russian Cossacks between the 1580s and the 1630s.

But, just like the Portuguese, the new imperial powers found it hard to move very far inland in much of Asia, Africa, and the Middle East. The newcomers generally stuck to islands or coastal cities that could be defended by naval gunfire. The chief of the Dutch factory at the Japanese port of Hirado wrote in 1623 that he had "barely sufficient force to set ashore, unless under the protection of the ship's cannon." As a result, the European presence in Asia had not expanded much by 1700; it had simply changed faces, with the Dutch seizing important bases such as Malacca and Ceylon from the Portuguese.

The limitations on European penetration were different in different places. In sub-Saharan Africa, the problem was not the people, who were technologically primitive, but the stifling weather, the inhospitable terrain, and a terrifying number of contagious diseases, all of which made it the "white man's grave." In the Middle East and Asia, the problem was less geographic than demographic. Even in their dotage, the great states of China, Japan, India, Persia, and Turkey remained stubbornly resistant to Western incursions. They all had large armies, heavily fortified cities, and plenty of firearms. In the 1590s the Mogul emperor Akbar, a contemporary of Queen Elizabeth I and King Philip II, commanded more than four million warriors—a force bigger than any army in Europe until the twentieth century. The Ottoman Empire actually encroached on Europe even as Europeans were invading the rest of the world. As late as 1683, the Turkish army besieged Vienna.

By that time, of course, the Ottomans' power was in steep decline. Their Janissary corps, made up of slave soldiers, was falling farther and farther behind the technological innovations of the West. So were the Manchu bannermen, the Mogul cavalrymen, and the Japanese samurai. But all of that became obvious only in retrospect. Until the mid-1700s, the notion of Europeans conquering these ancient civilizations was generally regarded as too fantastic to contemplate. When some Spanish conquistadors suggested to King Philip II in 1586 that they use the Philippines as a base to invade China, the most powerful monarch in Christendom was horrified. He directed "that this matter should be dropped and that on the contrary good friendship should be sought with the Chinese." This was good advice in dealing with the great powers of the East. A delegation of Portuguese would have done well to heed it in 1640 before complaining to the shogun about the expulsion of foreigners from Japan. He executed them for their effrontery. Dutch merchants had no choice but to meekly accept their confinement to a tiny island in Nagasaki harbor.

It would have been hard to predict in 1614 or even 1714 that by 1914 84 percent of the world's land surface would be dominated by Europeans and that both Africa and Asia would lie prostrate at the white man's feet. As we shall see in the succeeding chapters, most of those conquests can be ascribed to the military consequences of the Industrial Revolution. But not all. By 1800, Europe had already conquered 35 percent of the world, and industrialization had made almost no impact on warfare by that point. That first 35 percent was crucial, for it was the launching point for the rest. The British Empire reached its heights under Queen Victoria because it was able to exploit the manpower and financial resources of India—an entire subcontinent whose conquest was well under way by 1800 but had barely begun in 1700.

What happened in the eighteenth century that led the Europeans to begin the subjugation of Asia? A combination of changing attitudes and changing capabilities.

European colonial advancement was undertaken by such profit-seeking entities as the Dutch and British East India Companies, which had been loath initially to launch costly conquests. It was more lucrative simply to buy and sell goods. As the British East India Company put it in 1677: "Our business is trade, not war." But however much they sought to avoid the expense of warfare and colonial administration, the Europeans found it unavoidable. The scramble for trade routes inevitably brought them into contact with hostile native states, brigands, and, even more important, rival Europeans. The decay of established empires like those of the Moguls and Ottomans, which became apparent early in the eighteenth century, created both danger and opportunity: the danger that hated rivals might horn in or that chaos would develop; the opportunity to grab power for themselves. If they wanted to do business, Europeans decided, they had to take up sword and musket. In the succinct words of a Dutch proconsul: "Trade cannot be maintained without war, nor war without trade."

As it happened, Europeans had become steadily more proficient at making war. The continuing improvements associated with the Gunpowder Revolution had been expanding their military capabilities until, in the eighteenth century, the prospect of taking on such mighty monarchs as the Grand Mogul of India could be regarded with equanimity, not trepidation. The sine qua non for this expansion was dominance of the seas, which the West had enjoyed since the sixteenth century. But ships could not deliver victory on land. The more recent combined-arms revolution associated with Gustavus Adolphus could.

THE PROFESSIONALIZATION OF WAR

In the century and a half following the Thirty Years' War, the wild and unruly passions of religious conflict were tamed in Europe. War became more limited, fought by professional armies in the service of kings who did not want to upset the established order but merely wanted to seize a province or two from their neighbors or to prevent their neighbors from seizing a province or two from them. Civilians for the most part were spared the ravages of marauding armies. Hugo Grotius's *De Jure Belli ac Pacis* (*On the Law of War and Peace*), a seminal rulebook published in 1625, attempted to limit the savagery of combat.

But if war became less destructive for civilians, it did not become less frequent. Even before the lengthy wars triggered by the French Revolution, the eighteenth century saw virtually nonstop intra-European bloodshed, often waged on a global scale. Conflicts like the War of the Spanish Succession (1701–14) and the Seven Years' War (1756–63) were really world wars conducted from the West Indies to the East Indies. Out of this cauldron emerged some first-rate commanders, such as England's Duke of Marlborough, Prussia's King Frederick the Great, Savoy's Prince Eugene, and France's Marshal Maurice de Saxe. They did not radically change the tactics of Gustavus, but they did add some important refinements, with France usually in the lead.

Institutions were created to care for wounded, retired, and disabled soldiers, such as France's Hôtel des Invalides, which opened its doors in 1674. Uniforms became standardized, with the English choosing red, the French white, the Russians green, and the Prussians blue. Fortifications and siegeworks reached new heights of complexity thanks to the innovations of the French engineer Sébastien Le Prestre de Vauban. Armies grew in size, with King Louis XIV of France commanding four hundred thousand men in the early eighteenth century. This growth led to the creation in France in 1759 of the first division, a self-contained, combined-arms organization of about twelve thousand men; previously the biggest standing military unit had been the regiment. Logistical services were improved, and it became common for armies to be fed from well-stocked commissary carts rather than being forced to forage from the countryside. Permanent barracks were built so troops in peacetime no longer had to be quartered in civilian houses. Toward the end of the eighteenth century, light infantry and light cavalry units going by such names as chasseurs, Jägers, and hussars were introduced into most armed forces for scouting and skirmishing.

Artillery continued to be slimmed down, standardized, and improved, although its essence—smoothbore muzzle-loaders firing metal balls and pellets—had remained unchanged since the late 1400s. In the early eighteenth century an English mathematician and engineer named Benjamin Robins invented the modern science of ballistics. He devised practical rules for estimating the range and impact of artillery fire under different conditions. Applying his insights, Jean-Baptiste Vaquette de Gribeauval, the French inspector general of artillery, invented a new system of carefully calibrated sights that would allow a gunner to compensate for the dropping trajectory of a projectile. He also came up with a screw mechanism that allowed gunners to elevate and depress their guns precisely and easily. A Swiss engineer named Jean Maritz, meanwhile, developed better cannon-casting techniques. His more precisely cast guns reduced the amount of windage (the empty space between the barrel and the shot), decreasing the amount of gas that would escape when the gun was fired. This meant that smaller charges could be used and that gun barrels could be lightened and shortened. Between the 1750s and the 1760s the weight of a French four-pounder (a cannon firing a four-pound shot) dropped from 1,300 pounds to 600 pounds. Artillery thus became more mobile and more prevalent on the battlefield. Frederick the Great created horse-drawn artillery armed with six-pound guns, and most other European armies followed suit. Despite all these innovations, neither the range nor the rate of fire of artillery changed much; smoothbore guns generally fired two to three rounds a minute and were lethal at no more than eight hundred yards.

Handheld firearms were also becoming more efficient but not radically so. The matchlock of Gustavus's day was replaced by the flintlock starting in the 1690s. No longer was it necessary for a musketeer to hold a burning match in his hand in order to ignite the powder in his pan. Now the job would be performed by a mechanical device that struck a piece of flint against a rough steel surface when the trigger was pulled. Flints tended to wear out and had to be replaced after twenty or so firings, but they were much more reliable than matches; the rate of misfires dropped from 50 percent to 33 percent. The firing process was facilitated by the replacement of powder horns with paper cartridges containing a set amount of powder. "This," notes a modern British general, "greatly improved the accuracy of measurement of the gunpowder charge and, as a result, not only gave more uniform ballistic performance but also increased the rate of fire." Loading a flintlock was still a laborious twenty-six-step process that involved biting open the paper cartridge, dribbling some powder in the flash-pan, putting the remaining powder and ball down the muzzle, and ramming the paper cartridge into the barrel as wadding. But this was much faster than the forty-two steps needed to load a matchlock. A further advantage was that, because musketeers no longer had to hold lit matches in their hands, they could be grouped much closer together on the battlefield.

Armies also benefited from the invention of the socket bayonet. This long knife (whose name derived from a dagger made in Bayonne, France, in the fifteenth century) was attached by a sleeve underneath the flintlock muzzle. Although seemingly an obvious innovation, it took a number of years to perfect; previous bayonets either fitted into the muzzle itself, making firing impossible, or were attached by a loose ring that tended to slip. By 1700, the ubiquity of the socket bayonet meant that musketeers no longer needed pikemen to protect them while firing or to drive home cold steel to follow up a volley. They could do the job themselves. Pikemen, whose numbers had been shrinking for over a century, disappeared from the battlefield altogether.

With the increased firepower available to infantry, their ranks could be thinned out. Alignments dropped from six deep in Gustavus's day to only two or three deep a century later, while their firepower actually increased. Veterans utilizing the Brown Bess—a .76 caliber smoothbore, muzzle-loading musket that was the standard weapon of British infantry from the 1690s to the 1840s—could get off two or three volleys a minute. Thus a force of five thousand well-trained soldiers could, in theory, unleash a barrage of fifteen thousand lead balls in sixty seconds. (In actual battles the rate of fire was often less, due to smoke, fear, fatigue, and other factors.)

But could they hit what they were aiming at? The extreme range of flintlocks was about three hundred yards, but it was impossible to achieve reasonable accuracy at more than one hundred yards and often much less. (A modern M-16A2 rifle has an effective range of six hundred yards.) In order to make muskets easier to load, most of them used round bullets that fit very loosely into the barrel and came bouncing out at unpredictable angles. Most flintlocks, including the Brown Bess, did not even have rear sights. Nor did armies have any command for "aim." The typical instructions were simply: "Make Ready . . . Present . . . Fire!" Since accuracy was not possible, troops had to shoot from near point-blank range. Hence the famous order "Don't fire until you see the whites of their eyes."

Battles became a test of wills to see which side could stand more punishment and survive as a cohesive fighting force. The old knightly ideal of what Robert O'Connell calls "ferocious aggressiveness" was replaced by a new definition of courage: "passive disdain" of danger. It became a point of pride for professional soldiers not to flinch as bullets whizzed by.

When seen in a painting or movie, eighteenth-century battles can appear foppish and dainty compared to the carnage of twentieth-century warfare: A line of brightly clad toy soldiers brought to life marches toward the enemy in parade-ground formation accompanied by the thump of drums and the flutter of flags. How quaint! But from the perspective of the participants—who must have felt as if they were in a shooting gallery with bull's-eyes on their chests—the experience was horrific. With musket balls and artillery shot tearing great holes in their ranks and acrid powder smoke obscuring the field, infantrymen were expected to step over their fallen comrades, close up ranks, and keep walking at a measured pace into the face of death. Since metal breastplates and helmets had been largely discarded by the eighteenth century, the soldiers had no protection against all the metal whizzing around them. They must have wanted to hide or flee. Instead they had to stoically endure the enemy's fire until the command was finally given to pause and unleash a devastating volley or two in response. Then they were expected to charge, hack, and stab at the enemy soldiers with their bayonets.

In such battles, casualties of 20 percent or more were common even for the winning side. (By way of comparison, if the U.S. armed forces had suffered 20 percent casualties in the Gulf War of 1991, they would have had 110,000 killed and wounded. The actual figure was 614, or 0.1 percent of the 550,000 men and women who served in the war.) At one particularly bloody encounter, Zorndorf in 1758, the defeated Russian army lost 21,000 men, or 50 percent of its force. Frederick the Great won the battle but at a cost of 13,500 dead, wounded, or missing, or 37.5 percent of his army. No wonder only one in fifteen Prussian soldiers who started the Seven Years' War survived it.

Part of the reason for these high losses was the crude state of battlefield medicine, which meant that a far higher percentage of the wounded died than would be the case today. Field hospitals were usually overcrowded and unsanitary, medical services rudimentary, doctors badly trained. Bandages, operating tables, and surgical instruments were used on multiple patients with no attempt at sterilization, further spreading infection. When dealing with wounds in the legs or arms, surgeons often unnecessarily resorted to amputation with a dull saw and without any anesthesia beyond a slug of brandy. Amputation was designed to stop the spread of gangrene, but the patient often died anyway of shock or infection. One British lieutenant who visited an aid station in the Iberian Peninsula in 1812 was horrified by what he saw: "[T]o the right and left were arms and legs, flung here and there, without distinction, and the ground was dyed with blood."

Bad as this was, disease claimed more lives. In particularly unhealthy climes like the West Indies, entire regiments would be wiped out by yellow fever and malaria. Napoleon's attempt to quell a slave rebellion in Haiti in 1802 foundered when fifteen thousand men out of an army of twenty thousand were felled by yellow fever. At a time when there was no known prophylactic or treatment, such epidemics were a commonly accepted part of a soldier's miserable lot.

What would make men serve under such conditions? Most European states had voluntary enlistment, at least in theory, supplemented with press gangs (a form of selective conscription tantamount to kidnapping) and the hiring of foreign mercenaries such as the Hessians who served in the British Army. Those who took the king's shilling often did so because the alternatives were penury or prison, and the army at least offered steady pay, rations, and the prospect of a pension if you lived long enough. Once soldiers found what they had gotten themselves into, however, many deserted.

Turning "the scum of the earth" (as the Duke of Wellington affectionately described his men) into effective soldiers required iron discipline. Drill, which had been revived by Maurice and Gustavus, became a secular religion with most European armies. Soldiers who disobeyed orders were subject to flogging or worse. The name of Louis XIV's inspector general of infantry, Lieutenant Colonel Jean Martinet, became a byword for authoritarianism. Only if soldiers feared their officers and noncommissioned officers more than the enemy would they willingly face almost certain death. Or so the prevailing view had it in the *ancien régimes.* With the onset of the English Civil War in 1642 and the French Revolution in 1789, Europeans would discover that ideological or patriotic fervor could also induce men to risk their necks.

The discipline of Western armies, what historian John Lynn calls their "battle culture of forbearance," was, in many ways, their secret weapon. Non-Europeans were able to manufacture or purchase European "hardware" (their guns) with reasonable ease. They found it much more difficult to duplicate the "software" that made these weapons effective. Drill and discipline did not come easily to warrior societies that were used to one-on-one battles. The brutality of Western warfare, its single-mindedness, its imperative to kill or be killed, ran counter to most non-Western traditions.

This was not because most non-Westerners were "noble savages" who lived in a pacifist's paradise, as some Europeans once imagined. There was nothing edenic about the temples where Aztec priests ripped the still-beating hearts out of thousands of victims. But even for the Aztecs warfare was severely constrained by tradition, ritual, and the scarcity of lethal weapons; their "flower battles" were designed to capture, not kill, their neighbors. Neither they nor most other tribal societies were prepared for the organized slaughter of the Western battlefield. As Victor Davis Hanson has written, "The most gallant Apaches—murderously brave in raiding and skirmishing on the Great Plains—would have gone home after the first hour of Gettysburg." And no wonder. From a traditional warrior's standpoint, there was little sense in fighting anonymously in the ranks where no one could see your feats of courage. If you did fight, it should be to gain loot or to protect your clan. Why risk your neck for an abstract cause or a distant ruler? Especially if the odds of getting killed were so high.

A similar mind-set—which, it must be admitted, has a powerful logic of self-preservation behind it—prevailed in Europe during the Middle Ages. The nobility, which derived much of its status from chivalric jousting on horseback, ceded power slowly and grudgingly. The emergence of modern states, whose most important unit of military force was the humble infantryman, took hundreds of years. Most non-European states woke up to the need to make this transformation, if they ever awoke at all, only by the time Western armies were already on their doorsteps. It was in large part the failure to master the Western way of war that led to the fall of such once-mighty empires as those of the Ottomans and the Moguls. Marshal de Saxe's valedictory for the Turks applied equally well to other non-Western states: "It is not valour, numbers, or wealth that they lack; it is order, discipline and technique."

Sensing the approach of doomsday, at the eleventh hour a few non-Western states desperately tried to close the gap by importing Western soldiers to train and lead new armies infused with the "order, discipline and technique" that they so conspicuously lacked. The Marathas, who by the early eighteenth century controlled 75 percent of the Indian subcontinent, were a prominent example.

Their empire was established by the great Shivaji Bhonsle (1627–80), a master of fast-moving light cavalry whose forces had great success against the increasingly decrepit armies of the Mogul empire. After Shivaji's death, a variety of chieftains vied for supremacy. None possessed his military genius, and they allowed Maratha military capabilities to atrophy. In 1761 the Marathas were routed by Afghan invaders. Although they managed to hold their own in the First Anglo-Maratha War (1775–82), which was essentially a draw, the Marathas saw the urgent necessity to improve their military by imitating the example of the Europeans who were increasingly encroaching upon their domains.

The Marathas had already introduced firearms, uniforms, and the rudiments of drill to some of their forces, but their efforts had not been terribly successful. Their armies were still composed for the most part of feudal levies raised, like those of medieval Europe, for one campaign season and then disbanded—a badly supplied, ill-paid, undisciplined rabble lacking a formal command structure and loyal primarily to local landlords (the *jagirdars*). "They fought," John Lynn notes, "as collections of skilled individuals, not as cohesive units obedient to central command." In 1784, Mahadji Sindia hired a soldier named Benoît de Boigne (a native of Savoy, then part of the Kingdom of Sardinia, today part of France) to change that by fashioning his existing units into a true European-style army. Other Maratha chiefs followed his example.

De Boigne and his fellow mercenaries did their work as well as they could. Because most Marathas refused to learn a new way of fighting, they recruited sepoys from supposedly warlike tribes in northern India. (*Sepoy* comes from *sipahi*, the word for "soldier" in Persian, the official language of the Mogul empire.) This enabled them to create a growing number of *compoos* (brigades) that were equipped and drilled in the European manner and taught to seek pitched battle, eschewing the traditional hit-and-run tactics of the Maratha horsemen. These soldiers were equipped with locally produced muskets, artillery, and ammunition, many of which were as good as those employed by the British. Perhaps most important, de Boigne and other European officers saw to it that their men collected taxes from the peasantry, which allowed them to be paid regularly just like British sepoys. In turn, the European officers were showered with riches that far exceeded what their counterparts earned in the British East India Company or King George III's army.

By 1803, the regulars appeared formidable. Even though they were only a tiny fraction of the overall Maratha force, they were without a doubt the most impressive native military in South Asia—"veterans who had ever been victorious, and smiled in the face of danger," one of their European officers proudly wrote. But while they had defeated every Indian foe they had faced, they had not yet clashed with Europeans. It was their misfortune that their first test occurred against a force led by one of the greatest generals in history.

FIT ONLY FOR POWDER

No one would have expected early in his career that Arthur Wellesley would ever be so described. He was born in 1769 to a family of Anglo-Irish nobility. They did not have much of a martial tradition—his father, the Earl of Mornington, was a professor of music at Trinity College, Dublin; his mother a banker's daughter—but imperialism did run in their veins. His family had been a part of the English elite ruling over the Irish peasantry since the thirteenth century. Although born in Ireland, he always thought of himself as an Englishman; he was supposed to have quipped later in life, "To be born in a stable does not make a man a horse."

Upon the death of his father when Arthur was just twelve years old, his older brother Richard succeeded to the family title. The traditional options for a younger brother from such a family were the clergy, the military, or a government appointment. Arthur did not display much interest in matters theological, martial, or political; he was more interested in playing his violin. A shy, dreamy, indifferent student, he left Eton after only three years. He almost certainly never made the comment later attributed to him that "the Battle of Waterloo was won on the playing-fields of Eton," because, not being much of a sportsman, he did not spend much time on those fields himself. Nor did he particularly enjoy his time at this training ground of the English aristocracy. He completed his schooling in Brussels, where he learned to speak French with a Belgian accent. His mother did not know what to do with the unpromising lad, considered so much less intelligent than his brilliant older brother. She decided he was "fit only for powder."

As this remark suggests, the reputation of the British army at the time was not high. The satirical magazine *Spy* commented acerbically in 1700 that the soldier "is generally beloved by two sorts of Companion, in whores and lice." The magazine might have added rich gentlemen to its list, for they were still expected to raise and equip regiments. Traditionally, whoever paid for the regiment had become its colonel and owner, free to run things according to his own lights. This was gradually changing. Starting in the 1740s, the government issued regulations transferring some functions from the colonels to an army commander in chief, an adjutant-general, and two different cabinet officers. The Royal Military Academy at Woolwich was set up in 1741 to train engineering and artillery officers, whose professions required a high degree of technical competence. No such school would be set up for infantry and cavalry officers until 1799. (The Royal Military College opened its doors in 1800; its Junior Department relocated to Sandhurst, its present site, in 1812.) The only training Arthur received was a year at a privately run French "equitation academy" where the emphasis was on horse riding, swordplay, and aristocratic hauteur.

Most officers got their appointments by purchase once they met the minimal qualifications: One had to be certified as a "gentleman," recommended by another officer, and deemed acceptable by the secretary of state at war. The lowest officer's rank, ensign, cost four hundred pounds. Once appointed, a young officer would have to spend hundreds of pounds more for his uniform, equipment, and mess bills. Because these costs far exceeded the ninety-five pounds a year an ensign could earn, the officer corps was generally limited to men of means, although a few rankers were able to win promotion for outstanding feats of bravery. Promotion was by seniority, so if an officer lived long enough he could, in theory, expect to retire a general.

To the modern mind, this system, which was not abolished until 1871, seems hopelessly quaint and maddeningly inefficient. But while there were a fair number of abuses, such as noblemen buying commissions for their ten-year-old sons, there was also an informal weeding-out process designed to sideline the most incompetent officers and to give important commands to the most talented. The real saving grace of the British army was its enlisted men and NCOs. However humble their social origins, they attained a degree of professionalism unmatched by any other army in the world except the Prussian. The ordinary redcoats' competence made up for many faults of their dilettante officers.

Arthur Wellesley was one of those dilettantes when his older brother, at their mother's suggestion, first purchased a commission for him at eighteen.

During the next six years, while he advanced by purchase from ensign to lieutenant to captain to major, he did almost no soldiering. As an aide-de-camp to the Lord Lieutenant of Ireland, he spent most of his time running around Dublin with other young bucks, drinking, gambling, piling up debts, and wooing a pretty Anglo-Irish lass whose older brother ruled out marriage on the grounds that Arthur had neither money nor prospects. He also took the family seat in the Irish House of Commons. It was only when he became a major at twenty-four that Arthur began to take his duties seriously. That year, 1793, he burned his violin and vowed never to play again. He determined to see some action and won a transfer as a lieutenant-colonel to the 33rd Regiment of Foot, which in 1794 sailed to Flanders to fight the armies of Revolutionary France.

The expedition turned into a fiasco. For young Wellesley, it was an early primer on the importance of having a good commissariat, good planning, and good leadership from the front—all missing in that instance. "I learnt what one ought not to do, and that is always something," he observed many years later.

Wellesley returned home in 1795. The next year, as a twenty-seven-year-old colonel, he sailed with his regiment on a passage to India, where he would first earn the fame that would follow him for the rest of his life.

COMPANY BUSINESS

The British East India Company had been established in India since the early seventeenth century. Its three principal bases were Madras, Bombay, and Calcutta. Until the mid-eighteenth century it had been engaged primarily in trade. Then came the battle of Plassey: In 1757, three thousand English soldiers and sepoys under the command of Robert Clive routed fifty thousand men of the nawab of Bengal, largely by bribing the nawab's allies not to fight. This triumph helped make the Company master of Bengal, whose forty million people far outnumbered the ten and a half million inhabitants of the British Isles. The Company soon found that collecting taxes was even more lucrative than trading textiles. Those revenues made it possible to build a substantial Company army officered by Europeans and manned by Indians, supplemented by royal army regiments sent directly from home at Company expense. The French could not withstand the onslaught of these forces and in 1761 they lost their citadel at Pondicherry. The Dutch were evicted from Ceylon shortly afterward. Britain still had native competitors, but it was unquestionably the strongest power in South Asia.

The Company's increasing sway generated scandals in which its critics charged "nabobs," as the leading executives were known, with unfairly enriching themselves and oppressing their Indian subjects. Parliament's response was to bring the Company under political supervision by creating a Board of Control in England and appointing a governor-general in India. In 1797 Richard Wellesley, Arthur's older brother, was appointed as the fourth governor-general. He arrived in India the following year, shortly after Arthur. (A third brother, Henry, came along as Richard's private secretary.)

The Wellesley boys did not set out to be caretakers. Their goal was an audacious one. They wanted to extend British domination over the independent polities that had sprouted since the decline of Mogul authority. They aimed, in short, to lay the foundations of what came to be known as the Raj. The still-lingering threat of French power gave them the excuse they needed to act.

Their first target was Tipu Sultan, the Muslim ruler of the southern state of Mysore (now called Karnataka). The "Tiger Prince," as he styled himself, had no love for the English. (He had a mechanical toy fabricated showing a tiger mauling a Company officer: The tiger roared, the soldier shrieked.) He had already fought an English army to a draw, and his own force was full of French advisers. To court the favor of revolutionary France, the sultan even styled himself, absurdly enough, as "Citizen Tipu." In 1799 an expedition was organized to capture Tipu's capital, Seringapatam. Colonel Arthur Wellesley's role was to oversee an allied contingent of soldiers provided by the Nizam of Hyderabad. He skillfully deployed the Nizam's men and his own 33rd Regiment to beat off attacks by Tipu's cavalry and infantry. He was less skillful in the execution of a bungled night attack in a wooded area, one of the few blemishes on his sterling military career. Seringapatam was successfully stormed after British siege guns had punched a hole in its thick walls. Wellesley subsequently was made governor of Mysore and received four thousand pounds in prize money—quite a haul. (British privates got seven pounds, sepoys five pounds.) In his new position, he forcefully put down bandits and rebels. Within a year Mysore was fully secure, and the Wellesley brothers could turn their attention elsewhere.

The biggest state that still remained hostile to the British was the Maratha Confederacy, a loose grouping of five great Hindu families that ruled over some forty million people. Their feuding gave the Wellesleys the chance they were looking for. In 1802, the nominal leader of the Marathas, Baji Rao II, the Peshwa or prime minister, was chased out of his capital, Pune, by two rival chiefs, Dowlut Rao Sindia and Jaswant Rao Holkar. The Peshwa sought refuge near Bombay and appealed for British help to get back his throne. The governor-general was happy to help—as long as the Peshwa agreed to certain conditions. The Treaty of Bassein, signed in December 1802, made the Peshwa a British client. His affairs would be directed behind the scenes by a British resident and his domains would be garrisoned by

Company troops whose expenses he was expected to pay. This was typical of the way Britain extended its authority over nominally independent Indian princes.

To restore the Peshwa, an army was organized in 1803 under the command of Arthur Wellesley, recently promoted to major general. He dashed to Pune and occupied the city with no opposition on April 20, 1803. But the other Maratha chiefs refused to accept the Treaty of Bassein, and their armies continued to threaten British interests. Or so the Wellesleys thought. After several months of negotiations as arid as the Gobi desert, Arthur Wellesley declared war on August 6, 1803, and set off in pursuit of the Marathas.

The ostensible reasons for this conflict are as difficult to comprehend today as they were to many at the time. It is hard to avoid the conclusion that, however much Wellesley pretended to be the injured party, he was waging a war of aggression or at best a preventive war to bring a potential danger to heel. Most of his men neither understood why they were marching nor cared overmuch. A Company officer reflected the prevailing attitude: "Having never troubled my head with the intricacy of state affairs, I have, therefore, never learned the real cause of this war; but as an idle life in camp is always most irksome to a soldier, we hailed with delight the order for advancing, not much caring who the enemy might be, or what was the bone of contention."

ASSAYE: "THE NEAREST RUN THING"

The British advance took them on August 8 to the fortified city of Ahmadnagar, seventy-five miles from Pune. Wellesley was not fazed by the thick, twenty-foot-high stone walls or the thousands of defenders. His attitude was the same as that of his contemporary, Vice Admiral Lord Nelson, who instructed his captains, "Never mind maneuvers. Go straight at 'em." Instead of ordering a long siege, as would have been customary when facing a European fortress, Wellesley told his soldiers to assault the *pettah* (city) as soon as they arrived. The Highlanders of the 78th Regiment dutifully clambered up scaling ladders. It did not take them long to burst inside and cut down the defenders. The city fell in twenty minutes; the nearby fort followed a few days later. A Maratha chief allied with the British expressed his amazement in a letter to a friend: "These English are a strange people, and their General a wonderful man. They came here in the morning, looked at the *pettah* wall, walked over it, killed all the garrison and returned to breakfast! What can withstand them?"

The success of this audacious attack encouraged Wellesley to be equally bold when he caught up with the bulk of the Maratha force six weeks later,

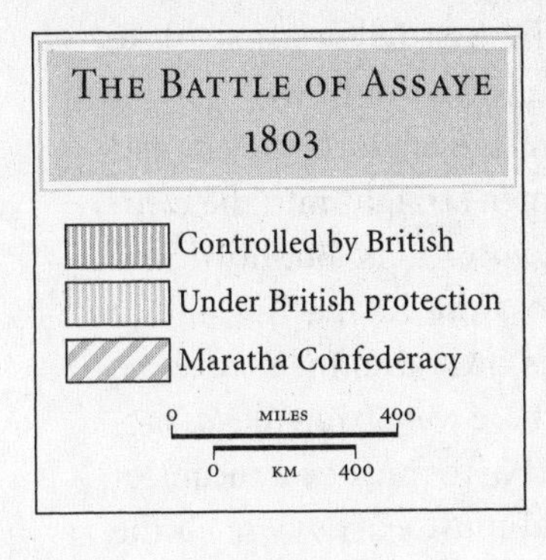

The Battle of Assaye
1803
Controlled by British
Under British protection
Maratha Confederacy
0 miles 400
0 km 400

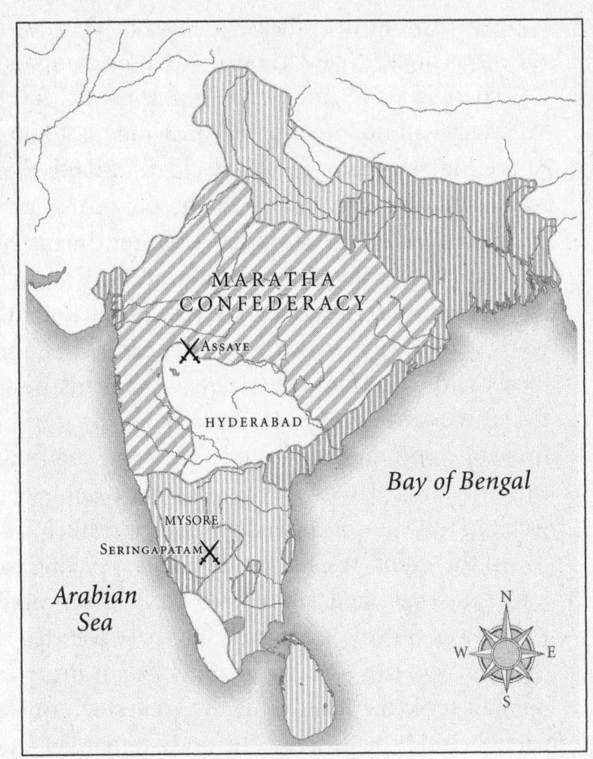

Maratha Confederacy
Assaye
Hyderabad
Bay of Bengal
Mysore
Seringapatam
Arabian Sea
N
W
E
S

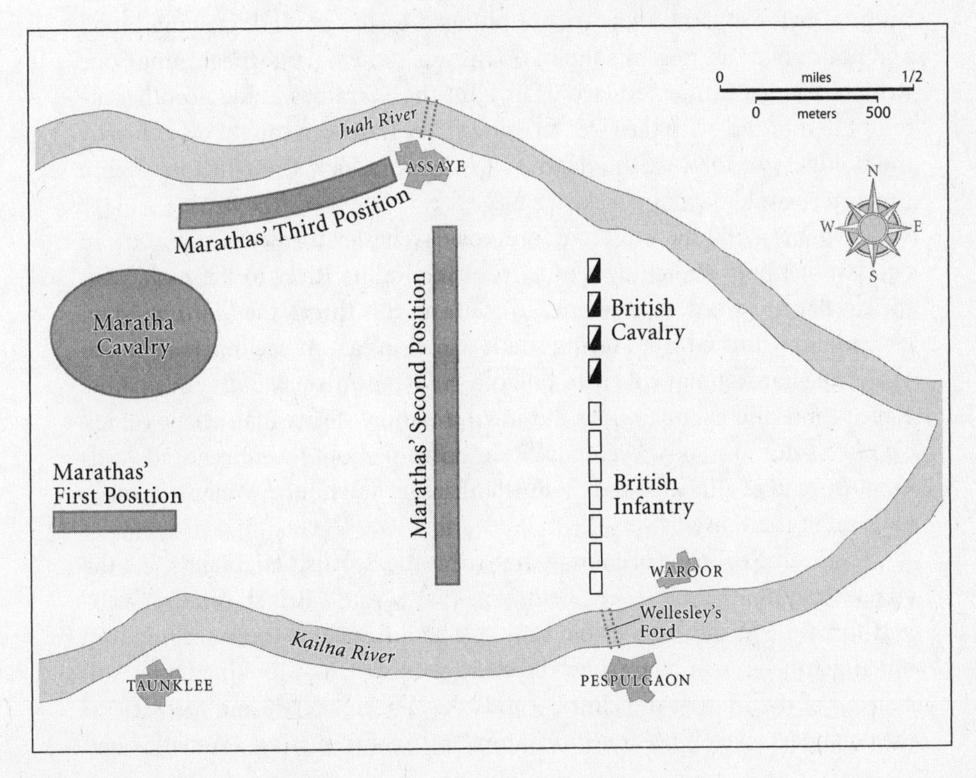

0 miles 1/2
0 meters 500
Juah River
Assaye
Marathas' Third Position
N
W
E
S
British Cavalry
Maratha Cavalry
Marathas' Second Position
British Infantry
Marathas' First Position
Waroor
Wellesley's Ford
Kailna River
Taunklee
Pespulgaon

camped around the village of Assaye. "Dash at the first fellows that make their appearance," he reckoned, "and the campaign will be our own."

There is a fine line between risk-taking and recklessness, however, and Wellesley had no intention of throwing his 4,500 men straight into the teeth of the Maratha defenses. He decided to flank their positions. But how? The local guides assured him that the only fords across the Kailna River were right in front of the Maratha guns. Yet during his own reconnaissance, he had seen that to the east of the Maratha positions there were two villages facing each other from opposite banks of the Kailna. There must be a means of communicating between them, he reasoned; a ford must exist where all the locals said it did not. Wellesley ordered his infantry to cross the river between the two villages. He later modestly ascribed his intuition about the ford to "common sense," but it was an uncommon risk. What if no ford had existed? What if the river were not crossable? His army would have been stuck in full view of the enemy and most likely decimated.

In the event, Wellesley's intuition was vindicated. He was the first man into the water, and from horseback he immediately discovered that the Kailna was shallow, no more than three feet deep. As the British were crossing the river, the Maratha artillery began firing from long range. One cannonball took the head right off Wellesley's orderly, who was riding a few paces behind the general. His body remained stuck in the saddle for a few minutes before the headless corpse plunged to the ground, gushing blood and brains over the nearby soldiers. "This was rather an ugly beginning," one British officer commented laconically, but the Marathas made no other attempt to interfere with the river crossing. They had been caught off guard by the boldness of the British advance. Who would have thought that such a tiny force would dare attack such a large foe?

Once his 4,500 men were safely across, Wellesley formed his infantry in two parallel lines stretching a mile from the Kailna River to the Juah, with the cavalry in a third, reserve line. To confront this threat, the Maratha regular battalions turned from facing south to facing east. Wheeling around was one of the hardest maneuvers to pull off, but to the dismay of the British the eleven thousand enemy sepoys did it "in the most steady manner possible." The two sides were now eyeing each other from about seven hundred yards away. Instead of attacking the enemy flank, as he had hoped, Wellesley would have to hit them head-on.

The 78th Regiment of Foot, a unit from the Scottish Highlands, led the attack straight at one hundred field guns that, wrote a British officer, "vomited forth death into our feeble ranks." The advance of these six hundred kilted giants—much taller than the average Indian, and looking larger still because of their bearskin helmets—must have been a terrifying spectacle to the defenders, even if the traditional pipe music was missing. (Wellesley had ordered the bagpipers to act as corpsmen.) Heavy shot would turn men into grisly puddles of blood and bone, but the thin red line did not falter. When they were just sixty yards from the enemy guns, the familiar orders boomed out—"Halt! . . . Make Ready! . . . Present! . . . Fire!"—followed by the crash of musketry.

The next command was "Charge . . . bayonets!" The Highlanders moved through the gray-white haze of gun smoke toward the Indian artillery. They overran the guns and barely paused as they reformed their line and kept marching at the prescribed seventy-five paces a minute toward the white-coated Maratha infantry waiting for them behind the cannons. The Indians fired first, a ragged blast that did little damage. The 78th did better. Their crisp volley scattered the Maratha regulars, already demoralized by the loss of their artillery. The Highlanders finished off the job in a few minutes of vicious hacking and thrusting with their already bloodstained bayonets. Next to them, Company sepoys from Madras were enjoying similar success.

The Marathas were folding at the first punch. Or so it appeared. In reality some of the Maratha artillerymen were only playing possum. "No sooner had our soldiers passed them," wrote a British officer, "than they suddenly arose, seized the cannon which had been left behind by the army, and began to reopen a fierce and destructive fire upon the rear of our troops." Wellesley had to go to the rescue, personally leading the 78th Regiment and a unit of native cavalry to finish off the Maratha gunners. In the melée that followed, the general had a horse shot out from under him, but he managed to crush the Maratha gunners.

Things were looking more grim on the other side of the field. The extreme right of the British line was entrusted to Lieutenant Colonel William Orrock at the head of the pickets, the advance guard made up of half-companies drawn from all the British battalions. Orrock had been given strict orders to avoid Assaye at all costs. But for some reason he marched straight at the fortified village. The fire from thousands of muskets and dozens of cannons virtually wiped out his column. The 74th Regiment of Foot, another Scottish unit, marched to their rescue, only to fall victim in turn to what Wellesley said was "a most terrible cannonade."

Seeing the Scotsmen fall and sensing their opportunity, the Maratha cavalry charged, thinking the battered survivors would be easy pickings. They were not. The Highlanders hastily formed a square, a formation directly descended from the Spanish tercio. By the eighteenth century, it had been hollowed out and used mainly in the face of cavalry attacks. Squares were extremely difficult to break, as the Marathas learned. Shielded behind a low rampart of enemy corpses and dead horses, the 74th fought off the cavalry attacks with their steady musketry. Still, they would not have lasted long had not British and sepoy cavalry been sent to their rescue. The British blades, made of finely honed Sheffield steel, cut deep into horses and humans alike, tearing through living flesh as if it were tender steak. The Marathas reeled back and the British horsemen "drove them with great slaughter into the river."

The remaining Maratha regular battalions managed to form a new line—their third of the day—but they did not last long. Although the 78th Regiment had been on the move since the early morning hours and had been fighting all afternoon in hot, muggy weather, the exhausted Highlanders formed up into ranks once more and advanced on the enemy position. The Maratha regulars panicked and fled across the River Juah.

With the regulars routed, the twenty thousand defenders of Assaye—Maratha irregular infantry—had no heart left for a fight. They were cleared out by the British with little trouble. Every piece of enemy artillery (102 guns) fell into British hands. The Maratha cavalry, which still numbered in the tens of thousands, did nothing to help. Not having been trained to support the Maratha sepoys and unable to operate effectively in the confined spaces between the Juah and Kailna Rivers, they had been bystanders almost the entire afternoon. "Had their numerous cavalry but supported them as they ought to have done," wrote a British officer, "God knows what would have been the issue."

By 6 P.M., night was falling and the fighting was ending. The British were too exhausted and too badly hurt to undertake any pursuit. The wounded were left to howl in anguish on the battlefield; some would receive no succor for two or three days. The survivors drank greedily from the blood-tinged river and lay down wherever they could find a spot to sleep. Wellesley slept fitfully in a farmyard, next to one officer who was dead and another who had had his leg shot off. He awakened often to think "that I had lost all my friends, so many had I lost in that battle."

Although a battle had been won at heavy cost, the war was not yet over. In the next three months, Wellesley harried the remnants of the Maratha forces, defeating their armies once again at Argaum and then taking the fortress at Gawilghur. As at Assaye, a considerable part of his success was due to his ability to carefully plan a campaign and organize the necessary logistics to keep his men in the field for extended periods. Another British army operating in the north under General Gerald Lake had similar success against the Maratha armies in that region. By the end of 1803 the Marathas had been forced to sign treaties that ceded vast tracts of land to the Company and recognized British stewardship of their affairs. When they unsuccessfully revolted in 1817, the Marathas lost the last remaining vestiges of their independence.

The Wellesleys had succeeded in quadrupling the size of the British empire in India, an achievement worthy of Napoleon. As his reward Arthur received a knighthood. Richard got a marquessate—and a recall home to face the wrath of the East India Company's directors. They were furious that he had abandoned trade for conquest and run up a budget deficit.

For the British army, there would be many other battles as they consolidated their control over the entire subcontinent. Few would be as costly or as hard-fought as Assaye. The Marathas lost an estimated 1,200 men killed and 4,800 wounded. The British suffered 428 killed and 1,156 wounded out of about 4,500 troops engaged. Their overall loss was 35 percent, mainly due to Orrock's blunder in attacking Assaye head-on. Every officer of the 74th Regiment was either killed or wounded. It was one of the two bloodiest engagements that the Duke of Wellington ever fought (Waterloo was the other), and the one of which he was proudest later in life. He should have been. As he later said of Waterloo, it had been "the nearest run thing in your life."

WHY THE MARATHAS COULD NOT WIN

The Battle of Assaye has important things to tell us about why the great empires of the East fell to a relative handful of Westerners. For one thing, the outcome should dispel any crude economic determinist theories about the rise and fall of nations. The Mogul emperor Aurungzebe (r. 1658–1701) had ten times the revenue of his contemporary Louis XIV, the richest monarch in Europe. Yet French and British soldiers were in India; Indian soldiers were not in Britain and France.

One common explanation is that India was undone by its disunity. If only Indian peasants had not been willing to serve in European armies . . . if only Indian princes had not been willing to call in European allies against their rivals . . . if only a spirit of nationalism had prevailed over narrow caste and clan loyalties . . . then the people of the subcontinent would have stood a better chance of repelling the Western invasion. So lament modern Indian nationalists.

All this is true, but its explanatory power is limited. Disunited as they were, the major states of India, such as Mysore, Bengal, and the Maratha Confederacy, were still much bigger than most European states, and their armies always vastly outnumbered the tiny European forces sent to subdue them. That the Indians lost anyway suggests that the Europeans had developed military techniques that allowed them to win almost regardless of how badly they were outnumbered. They had mastered the Gunpowder Revolution, and the Indians had not.

Although some Indian politicians tried, in most cases the experiments in Westernization were halfhearted and incomplete. Take the Marathas. True, they had plentiful artillery that Wellesley found after Assaye was "so good . . . that it answers for our service." But the bulk of their army remained resolutely traditional, and although some of their troops adopted firearms, others never gave up the bow and arrow. The Marathas did not develop a doctrine that could meld their traditional hit-and-run tactics (*Ganimi Kava*, or predatory warfare) with the frontal assaults of European-style warfare. As a result, at Assaye the bulk of the Maratha force—the light cavalry—made no attempt to influence the outcome. This was a reflection of the Marathas' dysfunctional command structure. Lacking a single commander like Wellesley whose orders were binding on all, the Maratha chieftains often would act at cross-purposes in battle, to the benefit of their enemies. Moreover, their supply service was virtually nonexistent and most of their soldiers were not paid regularly. The Marathas had nothing like the efficient British commissariat that contracted with Indian merchants (the *banjaras*) to supply food for Wellesley's expedition and hired ten thousand bullocks to carry his provisions.

Even the Marathas' most modern forces never quite attained the professionalism of their European adversaries. At Assaye, the Maratha regulars broke after their artillery was captured, while British soldiers fought on against far more desperate odds. The explanation cannot be that Indians were incapable of fighting in the new style, since East India Company sepoys performed well (though not quite as well as British regulars). Nor can the problem simply be that the Marathas' mercenaries were not very loyal; there was no intrinsic reason why the British army's mercenaries should have been any more steadfast.

The real problem was that the Marathas were handicapped by a lack of officers and NCOs schooled in the new way of fighting. At full strength, East India Company battalions had twenty-seven European officers. But even in the best of times a Maratha battalion had only two European or partly European officers and six Portuguese gunners to oversee more than seven hundred soldiers. At Assaye the number was lower still because the British authorities by threats and inducements had lured away many of the Marathas' European officers before the battle. Even the commander of Sindia's army, the French general Pierre Perron, had defected. There were almost no native Marathas capable of picking up the slack. It had never occurred to the Marathas to set up a war college to train a professional officer corps.

Even if the Marathas had had more officers schooled in Western tactics, however, overall control still would have been exercised by tribal chiefs who were more influenced by reading chicken entrails than by reading any treatise on strategy. Even if the Marathas had possessed a European officer who was as fine a strategist as Arthur Wellesley, he still would not have exercised as much sway as Wellesley did over his own force. Nor could it possibly have been otherwise. To turn over ultimate military power to Westerners or even to Western-educated Indians would have meant giving up their own absolute authority, and this the Maratha chiefs would never do.

As the Marathas discovered, it is no simple matter to truly modernize a traditional tribal force. Equipping and managing an army requires an administrative apparatus whose members are chosen at least partly on merit, who receive a secular technical education, and who perform their duties relatively honestly and efficiently and without excessive favoritism to family and friends. That would have been quite a change from the nepotism practiced by the Marathas and other tribal rulers, and it would have meant breaking the power of traditional military chieftains—the powerful *sirdars.* Moreover, staying abreast of the latest developments in military technology requires a certain amount of intellectual freedom and scientific inquiry, which would have been incompatible with the absolute rule of the Maratha nobility. Filling up the ranks requires arming and educating commoners, whose docility the nobles had always taken for granted. Above all, what was needed was an openness to new ideas and a willingness to evolve beyond old ways of doing things, something that would have been anathema to the Marathas, whose lives were governed by the ways of their ancestors. In sum, *to fight like Europeans you had to become "European."* You had to adopt at least some of the dynamism, intellectual curiosity, rationalism, and efficiency that has defined the West since the advent of the Gunpowder Age.

This was a realization reached by Peter the Great, the leaders of the Meiji Restoration, Kemal Mustafa Ataturk, and all the other great Westernizers of modern times. They knew that to generate military power comparable to the West they had to transform their societies, not just their armies. And they knew this would require a wrenching period of transition when they would have to wrest power from streltsy, samurai, janissaries, bannermen, and other powerful elites whose status was based on traditional ways of fighting.

Few if any of these reformers had any desire to introduce political liberalism. They wanted to borrow selectively, to use Western military methods to strengthen their own countries against the West. But they often found that Western political ideas wafted in along with Western military ideas. Thus, in many cases, a modernizing army became a tool of social change—or, depending on your perspective, of sedition. Two examples will suffice: The Ottoman sultan was toppled by dissatisfied officers known as the Young Turks in 1908. Four years later another military mutiny overthrew the emperor of China. In both cases the process of military transformation led, against the rulers' will, to political transformation. As historian David Ralston notes, "Rather than providing a defense against incursions from abroad, above all by the Europeans, the new-style army ended by serving as an agent, indeed a veritable channel, for the penetration of European ways."

It is hardly surprising that brittle polities like the Maratha Confederacy were not eager to undergo such traumatic changes. The Marathas must have realized implicitly that even if they won they would still lose, for to defeat the enemy they had to become the enemy. That was too high a price to seriously contemplate, and so they went down to defeat.

The Consequences of the Gunpowder Revolution

Historians differ over how long the Gunpowder Revolution lasted. A good case can be made that the major innovations occurred between 1500 and 1700, though it took time for their effects to spread across Europe and then across the world. To review, the most important developments were:

LAND WARFARE: The mounted knights of the Middle Ages were supplanted by infantrymen armed with missile weapons, first the longbow, then arquebuses and muskets. The spread of artillery made old-fashioned fortresses indefensible but gave rise to formidable redoubts known as the *trace italienne*. Cannons, which had started off as siege weapons in the 1400s, got lighter and more mobile over the centuries, making it possible to wheel them onto the battlefield. As the capabilities of both handheld firearms and cannons steadily improved, pikemen found their numbers reduced. By 1700, when the flintlock musket with a socket bayonet became the standard infantry weapon, they had vanished from the battlefield altogether. The saber remained the primary weapon of the cavalryman following a brief and unsuccessful flirtation with pistols. But cavalry's role was much diminished; horsemen became used primarily for scouting and harrying enemy forces once they were defeated. To take advantage of increasing firepower, Maurice of Nassau and Gustavus Adolphus developed a system of organization, tactics, and drill that harked back to the Roman legions. Disciplined soldiers replaced brave but undisciplined warriors in European armies. In a related development, native-born volunteers and draftees began to displace foreign mercenaries.

NAVAL WARFARE: Galleys powered by oars were supplanted by square-rigged galleons propelled by the winds. The old method of fighting at sea—closing with and boarding the enemy—gave way to artillery duels. Two key innovations were the development of lidded gunports and four-wheeled truck carriages that made it possible to run cannons in and out of the sides of ships. This made the broadside (all of a ship's guns on one side firing at once) the standard naval tactic. European ships were able to sail around the world as a result of advances in navigation that began with the compass and cross-staff in the 1400s and culminated with the invention in the 1760s of the chronometer to measure longitude. The combination of powerful sailing rigs and heavy artillery allowed European vessels to dominate almost every sea and coastline in the world.

The most important changes of all were common to sea and land warfare. Both realms saw the development of professional, standing armed forces that were larger and more lethal than the ad hoc feudal levies of old. These new armies and navies were directed and funded by centralized state bureaucracies that began to emerge in western Europe in the 1500s.

The rise and fall of nations between 1500 and 1800 is largely a tale of which ones took advantage of these military and naval revolutions and which ones did not. Sheer size or wealth was not a good predictor of military outcomes. In all three of the battles highlighted in this section, the richer and bigger polity lost: Spain (Spanish Armada), the Holy Roman Empire (Breitenfeld and Lützen), and the Maratha Confederacy (Assaye). (There are no reliable economics statistics for Maharashtra, the largest Indian state at the time, but the Indian subcontinent as a whole in 1800 had 19.7 percent of world manufacturing output, to only 4.3 percent for the United Kingdom.) Although it's not unusual to lose a battle and still win a war, in all three cases these defeats were not flukes; they signaled the long-term decline of once-mighty behemoths.

This is not to suggest that being rich or populous doesn't matter. Sustaining power in the long term ultimately requires both soldiers *and* silver, and there are many examples in this period where victory went to the side with—to use the wonderful neologism of Confederate General Nathan Bedford Forrest—the "mostest." France, for instance, was defeated under both Louis XIV and Napoleon by coalitions of other European states that were no more advanced militarily but could, in aggregate, bring more resources to bear. Sweden, despite its astounding triumphs under Gustavus Adolphus and his successors, faded from the Great Power ranks less than a century later because it was too small to compete against the big boys. The point isn't that population size and economic power are irrelevant. They simply aren't enough by themselves to generate effective military power.

Nor is it enough to manufacture or buy modern weapons. Winning generally requires having modern tactics and a modern organizational structure. This helps explain why Francis Drake and John Hawkins defeated the

Spanish Armada, Gustavus Adolphus defeated the Holy Roman Empire, and Arthur Wellesley defeated the Marathas. In all three instances, the winning side had developed an effective bureaucracy, with appropriate funding, staffing, and leadership, to manage its armed forces. The defeated side, by contrast, lacked an effective command structure and remained wedded to outdated military doctrines—the Spanish to boarding enemy ships, the Imperial armies to the tercio, the Marathas to using their cavalry hordes to harass the enemy. All three realized they were behind the times and had to change. The Spanish desperately requisitioned all the cannons they could find before their Armada sailed; the Imperial forces decreased the size of their tercio and, at Lützen, adopted the linear tactics of the Swedes; the Marathas formed regular infantry units run by European officers. But none adapted fast enough or completely enough to stave off catastrophic defeat.

Spain and Portugal, along with some non-European states (especially the Ming, Ottoman, and Mogul Empires), did well in the first stage of the Gunpowder Revolution. (Amerindians were the big losers during this period, which coincided with Columbus's discovery of the New World.) But in the seventeenth and eighteenth centuries, northern European nations—Britain, France, the Netherlands, Prussia, Austria, Russia, and, for a time, Sweden—rose to dominance.

A vast literature has been devoted to the subject of why some cultures proved more dynamic than others. Why did technological innovation accelerate in Europe after 1400 while stopping in the Islamic world by 1200, in China by 1450, and in Japan by 1600? And why did technological innovation slow down in southern Europe after 1600 while speeding up in the north? Those questions have been pondered by scholars as diverse and distinguished as Max Weber, Oswald Spengler, Arnold Toynbee, William McNeill, Thomas Sowell, and David Landes, without reaching any definitive conclusions.

For our purposes, the important point is simply that states that were once intellectually curious and technologically innovative—and hence militarily formidable—ossified into bastions of reaction. This was true across many fields, but it was in the military sphere that non-Western societies paid the highest price for their stagnation, because Europeans had now gained the means to project significant amounts of power to the far corners of the globe. The civilizations of East Asia, South Asia, and the Near East were rich enough, populous enough, and sophisticated enough that, unlike the tribes of Africa, Australia, and the Americas, they could have competed with the Europeans on the gunpowder battlefield. Many tried. Few succeeded. Their failure made possible the rise of the West.

PART II

THE FIRST INDUSTRIAL REVOLUTION

The gun can be discharged at the rate of two hundred shots per minute, and it bears the same relation to other firearms that the McCormack's [sic] *Reaper does to the sickle, or the sewing machine to the common needle. It will no doubt be the means of producing a great revolution in the art of warfare from the fact that a few men with it can perform the work of a regiment.*

Advertisement for the Gatling gun (1865)

The Rise of the Industrial Age

Britain was in a self-congratulatory mood in 1851, and the opening of the Great Exhibition in London—the first world's fair—showed that it had much to congratulate itself on. The exhibition was housed in a giant building in Hyde Park constructed entirely of glass and iron. The Crystal Palace, it was dubbed, and Queen Victoria was not alone in thinking it "one of the wonders of the world." Such a structure would not seem at all remarkable today, but this was a time when most construction was still done with brick or wood. No one was sure that this airy design—the largest enclosed space in the world—would survive the elements. Not only did it survive, it flourished. More than six million paying visitors came to marvel at the building and all the wonders contained within its nineteen acres of interior space.

There were one hundred thousand objects in all, ranging from the mundane (prizewinning vegetables, ornate umbrella stands, fancy toast racks) to the exotic (African loincloths, rhubarb champagne, a false nose made of silver) and the absurd (a carriage drawn by kites, a vase made of mutton fat, a silent alarm bed that tipped a sleeper onto the floor). The stars of the show were the mechanical displays—steam hammers, marine engines, textile looms, printing presses, a thirty-one-ton broad-gauge locomotive—many of them actually chugging and hissing away to the delight of awestruck spectators. Like many of her subjects, the queen found it all "so vast, so glorious, so touching" that she declared the Exhibition's opening "the *greatest* day in our history."

What they were really celebrating, these visitors to the Crystal Palace, was the transformation of their rainy isles during the past hundred years. In 1750 most Britons still lived in villages and worked the land with their hands, as their ancestors had always done. Machines were used in some industries, but they were small-scale and usually powered by muscle, wind, or water. To get anywhere on land required riding a horse or getting into a horse-drawn carriage and jostling over bad roads at a top speed of about six miles per hour. Water travel was a little faster and cheaper but even more uncertain, being subject to the vagaries of the weather. Long-distance communication was possible only by mail, and it could take two months for a letter to get from New York to London.

By 1851, all of these constraints had been cast off as completely as powdered wigs and silk breeches. Population had increased in general and urban population in particular. Manchester, "the metropolis of the commercial system," shot up from 25,000 souls in 1772 to 455,000 by 1850. London reached 2.3 million that year, making it the world's largest city. In 1851 for the first time the census found that the majority of the British population lived in cities and towns. While more Britons remained employed in agriculture and domestic service than in any other occupations, a growing number toiled over steam-powered machinery in dimly lit factories. Once upon a time, a visitor knew he was approaching a city by the appearance of church spires; now it was the sight of belching smokestacks. "Nothing can be conceived more grand or more terrific," wrote one observer in 1830, "than the yellow waves of fire that incessantly issue from the tops of these furnaces."

Few had such admiring words for the communities that grew up around the smokestacks. With urbanization had come greater attention to such perennial evils as child labor, prostitution, pollution, crime, and poverty, which were so eloquently exposed by writers like Dickens and Disraeli. Even in England, that placid "nation of shopkeepers," there had been outbreaks of class conflict such as the "Peterloo Massacre" of 1819. But by 1851 the rising middle class—a new term and a new concept—could assure themselves that the worst excesses had been addressed and that a new golden age was dawning. Statistics bore out their faith: Average wages in Britain, which had risen by 15 to 25 percent between 1815 and 1850, soared 80 percent in the next fifty years.

Not many Britons were aware of wage figures. All of them could see how transportation was changing, and it seemed nothing short of miraculous. Travelers could now zoom along at thirty-five miles per hour aboard a train—"swifter than a bird flies," marveled the actress Fanny Kemble, who, after taking her first train ride in 1830, found that "this sensation of flying was quite delightful and strange beyond description." Luckily, one learned professor's warning that "Rail travel at high speed is not possible because the passengers, unable to breathe, would die of asphyxia," was not borne out by experience. A railway boom in the 1840s crisscrossed the British Isles with five thousand miles of tracks, and it was now common to hear a locomotive's distinctive toot-toot puncturing the tranquility of the countryside.

Sea travel likewise had been altered beyond all recognition since Robert Fulton's first steamboat had been exhibited on New York's East River in 1807. Since 1838 vessels powered exclusively by steam had been crossing the Atlantic, dramatically reducing both the time and cost of long-distance travel and making possible the greatest mass migration in history as tens of millions of people left Europe for the Americas and Australia.

Communications had become even faster than a speeding train thanks to Samuel Morse's invention of the electric telegraph in 1837. In 1851 the first working telegraph line was laid under the English Channel. Within two decades, submarine cables would link Europe with America, Asia, Africa, and Australia. In 1750, using the mail, it had taken six months to communicate between Britain and India; by 1870, the telegraph had cut the time down to five hours. By 1886, with the completion of a cable line from Europe to West Africa, the entire world was wired together. (One of the unforeseen consequences of this development was to give a big boost to magazines and newspapers, which in turn subjected military actions to far greater public scrutiny than in the past. The Crimean War [1853–56] saw the advent of the first professional war correspondents.)

All of these events were part of what came to be known as the Industrial Revolution, and, in the view of many sober historians, they changed the world more than anything that had happened since the advent of farming (the Agricultural Revolution) some twelve thousand years before. Like the Gunpowder Revolution, the Industrial Revolution was a complex, long-term process that was the work of many minds in many countries over many years. It first developed in Britain, which had the advantages of peace (no major battles on English soil after 1651), stable government, freedom from arbitrary arrest or confiscation, secure property rights, well-established patents, courts that would enforce contracts, an entrepreneurial culture, and plenty of cheap capital.

The precursor to industrialization was a substantial improvement in agriculture. Better planting, harvesting, and ranching techniques increased the yield of English farmland early in the eighteenth century and made it possible to feed more people. Large landowners took advantage of these developments by enclosing fields that had once been farmed communally. The spread of private ownership and market techniques yielded further advances in productivity while pushing many small farmers off the land. Thus England had a growing and increasingly mobile labor force just as machines were being invented to harness their labor in unconventional ways.

The most important of these inventions was the steam engine. Thomas Newcomen produced the first practical model in 1705, but it was so inefficient that its usefulness was limited to coal fields where the fuel supply was essentially inexhaustible. A Scottish inventor named James Watt improved on Newcomen's design in the 1760s, making it much more fuel-efficient. Watt's steam engine ushered in a new age in which power would come primarily from fossil fuels, first coal and then oil, and machines would replace much of the labor of man and beast.

The steam engine had its biggest immediate impact on the textile industry. Previously most cloth-making machines had been run by hand in cottages or by falling water in riverside mills. Now these contraptions became steam-powered and it made sense to congregate them in factories which sprouted in industrial towns like Manchester and Birmingham. In 1813 Britain had 2,400 power looms; in 1820, 14,000; in 1833, 100,000.

No sooner had textiles been transformed than other industries followed suit. British iron production increased from 17,000 tons in 1740 to 700,000 tons in 1830 and 4 million tons in 1860. By fashioning a smaller and more powerful steam engine capable of hauling a load along a "railway," Richard Trevithick opened up new uses for this suddenly abundant metal. Utilizing his invention, George and Robert Stephenson designed the first really successful locomotive, the *Rocket*, which won a competition sponsored by the newly constructed Liverpool and Manchester Railway in 1829. From this humble beginning, the global rail network would grow to 220,000 miles in 1880, which not only made possible the rapid movement of goods and passengers but also provided a plentiful source of demand for iron producers, coal miners, and engine makers. The increasing use of steamboats, which began to be made of iron starting in the 1820s, also proved a boon for manufacturers and consumers alike.

Railroads and steamships demonstrated how the various innovations of the Industrial Age interacted with one another. Each new product seemed to create demand for other products, some not yet invented. Historian David Landes has argued that the distinctive feature of the age was the "routinization of discovery, the invention of invention." The number of patents issued in England soared from 7 in 1750 to 250 in 1825. Scientists in many different lands competed against one another and sometimes worked together to expand the frontiers of knowledge. Their work did not always lead to practical benefits, and in fact science lagged behind engineering throughout most of the nineteenth century. The principles of steam power, for instance, were not understood until decades after steam engines had become widespread. It was not until after 1850 that scientific research began to drive technological progress. But even in the eighteenth century, professional organizations like the British Royal Society and the French Académie des Sciences had created a fertile intellectual climate that encouraged the work of myriad inventors.

Their more practical inventions could be brought to market because of the availability of cheap capital, spread by institutions such as the London

Stock Exchange, which opened its doors in 1773, and a growing number of private banks that usurped some of the functions previously reserved for the Bank of England.

The result was continuous, self-sustaining economic growth that enabled industrialized nations to support ever-growing numbers of people in ever-greater comfort. This material progress rendered old methods of thought as obsolete as old methods of production. New ideas like socialism, nationalism, and Marxism would now jostle for attention. Influential as they became, the most important ideology of the early Industrial Age was laissez-faire capitalism. Its prophets, Adam Smith and David Ricardo, preached that limitless growth was possible as long as government did not interfere with the "invisible hand" of the market. By 1846 Britain had become such a firm convert to this creed that Parliament repealed the protectionist Corn Laws. Legislation that had prohibited the export of advanced technology and the emigration of skilled craftsmen was also lifted. The era of free trade had begun.

THE IMBALANCE OF POWER

Britain could be so generous because it had taken such a big lead over the rest of the world. In 1851 it owned half of the oceangoing ships and railway tracks in the world and produced more than half of the world's cotton cloth, coal, and iron. Such dominance by such a tiny island could not and did not last forever. A half century later Britain would be overtaken in manufacturing output by the United States, with Germany close behind. Other European states such as France, Russia, Italy, Spain, and Austria would lag behind, but they were still heading in the same general direction as America, Britain, and Germany. Not so the rest of the world. The Industrial Revolution opened up a yawning economic chasm between the West and the Rest that still has not been bridged. In 1750 Europe accounted for just 23 percent of world manufacturing output; what came to be known as the Third World had a hefty 73 percent. By 1900 the European share had risen to 62 percent (the United States added another 23.6 percent), while the Third World's portion had shrunk to a paltry 11 percent. There is good reason to think that leading Asian states like India and China actually got poorer during the nineteenth century, at least on a per-capita basis, even as Europe was witnessing the greatest explosion of wealth in its history.

The resulting imbalance in power would make it much easier for European empires to extend their dominions abroad. But it took a while for industrialization to affect warfare. Wellesley's army in 1803 fought much as the Duke of Marlborough's had 100 years before, and not all that differently from Gustavus's 70 years before that; its transport (bullock-drawn carts) was identical to that employed 2,100 years earlier by Alexander the Great. "The hardware of war," writes one historian, "was essentially the same in 1815 as in 1631."

The biggest change associated with the French Revolutionary and Napoleonic Wars (1792–1815) was universal conscription, the *levée en masse*, which made armies bigger, better motivated, and less reliant on mercenaries, though its impact was hardly as revolutionary as some historians claim. The principle of draftee armies had already been established by Gustavus Adolphus in the seventeenth century and expanded by Russia's Peter the Great in the early eighteenth century. (Prussia's Frederick the Great had managed to conscript a higher percentage of his population than the French revolutionaries ever would.)

In the late eighteenth and early nineteenth centuries, logistics and command-and-control systems could not keep pace with the growing size of armies. Supplies still had to be hauled by horses; muskets, cannons, ammunition, and uniforms still had to be handmade by skilled artisans; orders still had to be relayed by bugles, shouts, and messages carried on horseback; and commanders still had to depend on their spy glasses to figure out what was happening on a smoke-shrouded battlefield. All this severely limited the number of troops that could be deployed effectively. Napoleon commanded six hundred thousand men when he invaded Russia in 1812—50 percent more than had served Louis XIV a century earlier, although only half were Frenchmen—but he could not support that many in the field, and most of his Grande Armée did not make it home.

Other reforms associated with post-1789 France—selecting officers based on talent, not birth; organizing armies into combined-arms divisions and corps; arraying attackers into columns, not lines—were also, for the most part, incremental expansions of what the áncien regimes had already been doing. Napoleon brought the old style of war to its ruthless zenith; he did not invent a new style. At most, by stoking the fires of nationalism, the Napoleonic era blazed a path for a true revolution in warfare when mass production could be combined with mass mobilization.

After 1815 most armies shrank again. The period of relative tranquility that followed the end of the Napoleonic Wars did not provide strong incentives for military innovation. The Royal Navy patrolled the world's oceans virtually without challenge, while the Holy Alliance of Russia, Prussia, and Austria tried to preserve a reactionary status quo on the continent. Britain was the most industrialized nation in the world, but guided by its traditional antipathy toward standing armies and its equally passionate attachment to parsimonious government, it made little attempt to apply advanced technology to its armed forces.

It was not until the Crimean War that warfare really began to be transformed, with cataclysmic consequences. A century that had begun with swords and muzzle-loading, smoothbore muskets and cannons, would end with repeating rifles, quick-firing artillery, and machine guns. Sailing ships firing solid cannonballs would give way to turbine-powered dreadnoughts spewing high-explosive shells. The eventual result was a slaughter of such frightful proportions that it seemed to mock the very idea of progress that had been the centerpiece of the Great Exhibition of 1851.

CHAPTER 4

RIFLES AND RAILROADS:

Königgrätz, July 3, 1866

Where, oh where, was the Crown Prince?

That was the question on everyone's minds at the Prussian command post atop the heights at Roskosberg, two miles behind the front. King Wilhelm I and his large retinue of staff officers, diplomats, and hangers-on anxiously directed their telescopes to the left, looking for some sign that Crown Prince Friedrich Wilhelm's Second Army was approaching the battle raging in front of the fortress at Königgrätz, about fifty miles east of Prague in the Habsburg province of Bohemia. But little could be seen through the thick rain and even thicker fog, which combined with the white smoke billowing from hundreds of thousands of guns to almost completely obscure the entire area. As the fog of war settled lower and lower, the anxiety of the Prussian commanders crept higher and higher.

Wilhelm had allowed himself to be drawn into this *Brüderkrieg* (brothers' war) against Austria by his ambitious minister-president, Count Otto von Bismarck, to ensure that Berlin, not Vienna, would dominate the future course of Germany. Few would have bet on Prussia emerging victorious. The Habsburg Empire had 78 percent more people, 38 percent more soldiers, and a defense budget that was 54 percent larger. Its generals were also more experienced, having fought France just seven years before. Prussia had barely fought at all since the Napoleonic Wars, when it had suffered a catastrophic defeat that was only partly redeemed by Napoleon's final defeat at

Waterloo. Friedrich Engels, the journalist and co-author of *The Communist Manifesto*, who had served a year in a Prussian artillery regiment, spoke for received opinion when he wrote that "the odds are against the Prussians."

The odds tilted even further once Prussia's war plans became apparent. The Prussians had dared to violate the cardinal principle of war as laid down by the great Napoleon himself: Never divide your forces in the face of the enemy. The Prussian General Staff had broken up its troops into three field armies (First Army, Second Army, Elbe Army) which had crossed the mountains into Bohemia at three widely separated points. General Helmuth von Moltke, chief of the General Staff, hoped that one of his armies would fix the Austrians in place while the other two would circle around and annihilate them. But it was equally likely that the Austrian main force, larger than any of the three individual Prussian armies, would defeat each in turn before they could reinforce each other. Engels, in a newspaper column published on July 3, 1866, the very day of the battle, was withering in his assessment of Moltke's novel strategy: "an officer proposing such a plan of campaign was not fit to hold even a lieutenant's commission."

Indeed, the grand envelopment envisioned in Berlin was proving hard to implement. Lacking solid intelligence about the enemy, the Prussians had blundered along blindly until some scouts had run into the main Austrian encampment on Monday, July 2. The Austrians were spread out over nine miles of farmland, villages, and low, rolling hills in front of Königgrätz. Prince Friedrich Karl, the king's nephew, who was in command of both the First Army and the Elbe Army, resolved on an immediate attack, even though the Crown Prince's Second Army was still thirteen miles away, a whole day's march. Friedrich Karl thought he could take on the entire Austrian army by himself and earn all the glory. Moltke knew better. Upon learning of Friedrich Karl's plan in the early morning hours of July 3, the chief of staff immediately dispatched a messenger to the Crown Prince's headquarters with orders that the Second Army was to march at once. Unfortunately the order did not arrive until 4 A.M., and the troops did not start moving until three hours later.

The Second Army was still far away when the battle began around 7:00 A.M. on Tuesday in what a correspondent described as "a downpour worthy of a monsoon." Elbe and First Armies advanced through the "gray and cheerless" morning across Bistritz Creek, the northern boundary of the Habsburg position. Positioned on the other side was Austria's Northern Army, 206,000 strong, occupying what Moltke described as "an extremely strong position on the heights . . . behind the Bistritz." The Elbe and First Armies together deployed only 124,000 cold, tired, hungry men; they would be badly outnumbered until the Second Army's 97,000 reinforcements arrived.

Sure enough, the initial Prussian offensive quickly ground to a halt. The chief Prussian advantage lay in their needle gun, a breech-loading rifle that

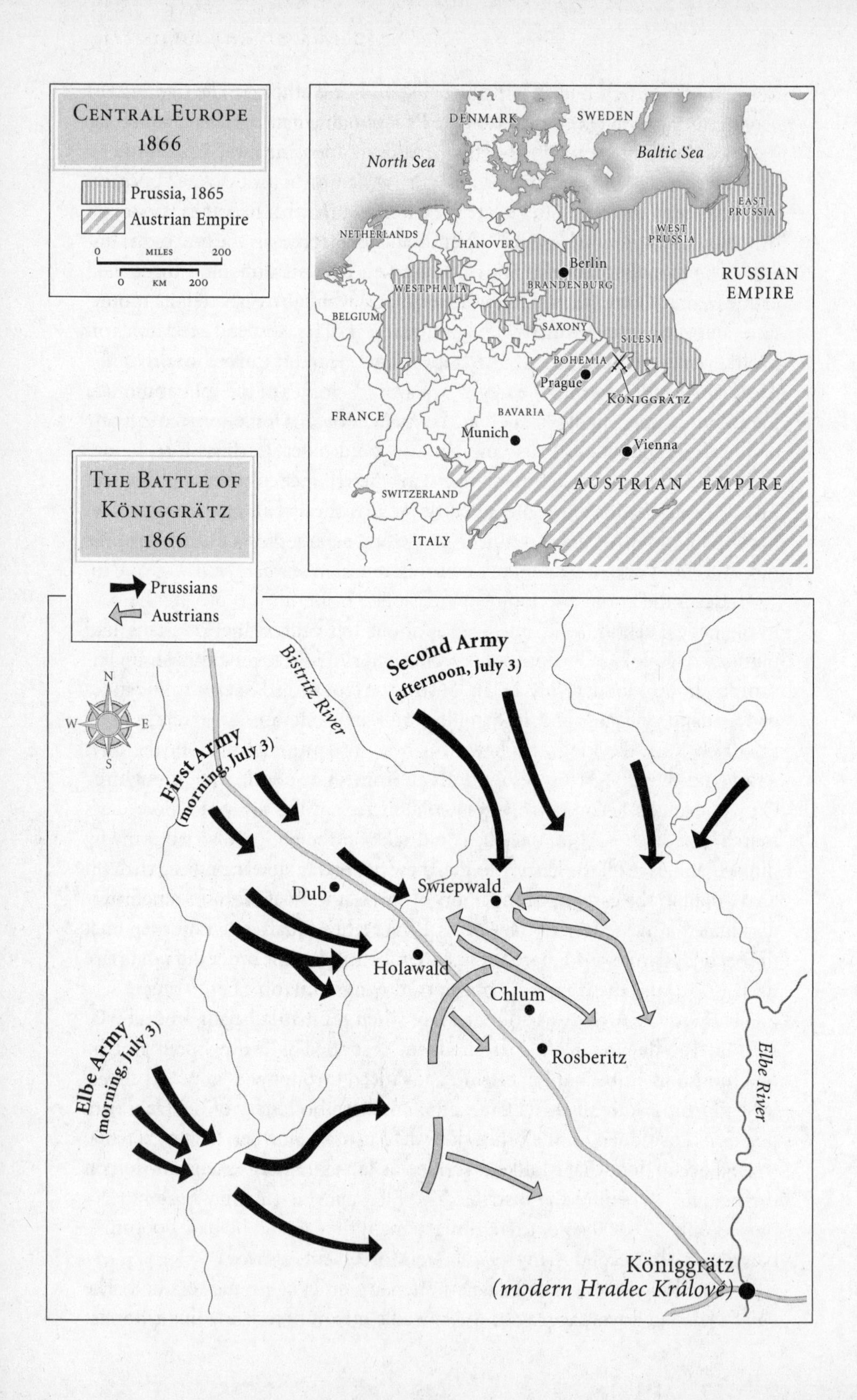

Central Europe
1866
Prussia, 1865
Austrian Empire
0 MILES 200
0 KM 200
DENMARK
SWEDEN
North Sea
Baltic Sea
EAST PRUSSIA
WEST PRUSSIA
NETHERLANDS
HANOVER
Berlin
BRANDENBURG
RUSSIAN EMPIRE
WESTPHALIA
BELGIUM
SAXONY
SILESIA
BOHEMIA
Prague
KÖNIGGRÄTZ
FRANCE
BAVARIA
Munich
Vienna
AUSTRIAN EMPIRE
SWITZERLAND
ITALY
The Battle of Königgrätz
1866
Prussians
Austrians
N
W
E
S
Bistritz River
Second Army
(afternoon, July 3)
First Army
(morning, July 3)
Dub
Swiepwald
Holawald
Chlum
Rosberitz
Elbe Army
(morning, July 3)
Elbe River
Königgrätz
(modern Hradec Králové)

was far superior to the muzzle-loading muskets still employed by the Austrians. But in the early going the Prussians had trouble getting close enough to make their rifle fire count. The Austrians were blowing away the attackers from long range with 250 field guns entrenched on a series of hills. These were not antiquated smoothbore cannons of the kind that had been the norm for 350 years, but modern rifled pieces firing explosive shells. The Austrians had had a couple of days to prepare the battlefield. Ranges and trajectories had been carefully calculated in advance, so that the Prussians felt as if they were advancing straight into a shooting gallery. "The Austrian artillery shot exceedingly well," Moltke had to concede later. "Scarcely did a column of infantry or cavalry make itself visible in one of the gorges of the valley before a shell came along and exploded in most unenjoyable proximity, and they withstood the fire of our own batteries with the utmost steadiness."

Less phlegmatic observers spoke of the "terrible . . . cannonade" which was sending "whizzing shells among the Prussian artillery, dismounting guns, killing men and horses, and splintering carriages in all directions." As shrapnel exploded around them, ordinary Prussian soldiers "felt we were in God's hands."

The worst slaughter was occurring on the left side of the Prussian lines, in a small forest known as the Swiepwald. The 7th Prussian Division under General Eduard Friedrich Karl von Fransecky had occupied this position by 9:30 A.M., and ever since then the Austrians had been mounting increasingly desperate offensives to dislodge them. One Austrian unit after another marched into the Swiepwald, drums beating and horns blaring, supported by the fire of dozens of field guns. The Prussians more than held their own; their riflemen, concealed behind thick firs and stout oaks, took a terrible toll on the attackers. But the sheer weight of numbers was slowly pushing the 7th Division back to the edge of the forest. By noon Fransecky had lost almost two thousand officers and men. Both the general and his aide-de-camp had their horses shot out from under them. As the Prussian army's official history noted, "the situation became every minute more critical." How much longer could nineteen Prussian battalions hold off fifty Austrian battalions?

The 7th Division was as important to the Prussian line at Königgrätz as the 20th Maine Regiment had been to the Union troops at Gettysburg three years before. If the men of Maine under Colonel Joshua Chamberlain had been routed, Little Round Top would have fallen and the Union defense might have collapsed. Likewise, if the men of Magdeburg under Fransecky were routed, the Austrians would be able to turn the left flank of the Prussian position. The battle might be lost.

At 11 A.M., the embattled Fransecky sent a messenger to the king's headquarters to beg for reinforcements. Moltke refused to send any. He wanted to keep every reserve to block a possible Austrian counteroffensive in the center.

Fransecky would just have to fend for himself until Crown Prince Friedrich Wilhelm arrived with his army. Moltke's comment—"I know General Fransecky and I know he will stand firm"—must have been cold comfort to a general who was telling his own men, "We've got to stand here or die!"

Other Prussians were showing less resolution. As the Austrian artillery continued to hammer away, the Prussian center began to buckle. Whole battalions streamed off the battlefield, desperate to escape the relentless rain of artillery shells. King Wilhelm himself tried to turn back the tide, shouting, "Let's see you fight like brave Prussians!" To little avail. Brave though they undoubtedly were, many Prussian soldiers had been deeply demoralized by hours of shelling.

Around noon, the king exclaimed in desperation, "Moltke, Moltke, we are losing this battle!" Bismarck, who was hovering nearby, wearing the uniform of a reserve major in the Landwehr (militia), complete with a *Pickelhaube* (spiked helmet), was equally concerned. The only battles he was used to fighting occurred in drawing rooms, the only weapons he regularly employed were barbed words and stiletto wit. He was not prepared for the ferocity of real warfare and he feared for the safety of his son, Herbert, who was wearing the blue uniform of the Prussian army. By early afternoon preparations were being made for a Prussian retreat.

Only one man kept his head while all about him were losing theirs and blaming it on him. Bald, emaciated, deeply wrinkled, and nursing a cold, Helmuth von Moltke, a sixty-six-year-old general who had never before commanded an army in battle, remained confident that his carefully prepared plans would yet bear fruit. Supposedly, when the situation looked grimmest, he reassured the king, "Your Majesty will win today not only the battle but the campaign."

That was more than wishful thinking. Behind that bold prediction lay years of meticulous, detailed preparation to ensure that Prussia would take advantage of the Industrial Revolution that was transforming warfare in the mid-nineteenth century. The world did not yet know how well the Prussian General Staff had done its work. Moltke did. He had taken a conscript army, organized it with his staff officers, trained it with his drill instructors, mobilized it with railroads and telegraphs, equipped it with breech-loading rifles, and now he had every reason to believe that it would prevail over the less advanced forces of the Austrian Empire.

THE GENERAL STAFF AND ITS CHIEF

As with so many military renaissances, Prussia's rise had its origins in a defeat. At the battles of Jena and Auerstädt in 1806, Napoleon shattered the

Prussian army and destroyed any mystique remaining from the days of Frederick the Great. The French army then entered Berlin and turned Prussia into a tributary state. The memory of this humiliation was only partly erased seven years later when Prussia joined with Russia, Austria, and Sweden to defeat Napoleon at the epic Battle of the Nations near Leipzig in 1813.

To a whole generation of Prussians, Jena had shown the rotten underpinnings of the old Prussian state. The years after 1806 saw a burst of reforms including the freeing of serfs, the emancipation of Jews, the strengthening of government bureaucracy, and the weakening of trade guilds. The changes were especially significant in the military realm. The overhaul of the army was led by two officers, General Gerhard von Scharnhorst and Count August von Gneisenau, who sought to replace Frederick the Great's force of aristocrats and mercenaries with a French-style nation in arms. They stopped recruiting foreigners and instituted a universal draft that did not allow the rich to buy an exemption. They also created a citizen militia called the Landwehr and a substantial force of reserves. After 1815, the army would conscript forty thousand men annually to serve for three years. Upon leaving active duty they would serve a further two years in the reserves and fourteen years in the Landwehr. In this way Prussia limited its military expenditures in peacetime while creating a vast pool of trained manpower that could be called to the colors in wartime. By 1850 Berlin had around a half million trained soldiers at its beck and call.

And increasingly these soldiers were not the ignorant peasants of old. Starting in 1809, under the direction of Baron Wilhelm von Humboldt, Prussia created one of the best systems of public education in the world, offering elementary schooling for all, secondary schools for some, and university education for the elite. Out of the classrooms came citizen-soldiers who were able to master the complexities of industrial warfare. Special schools were set up to train a corps of noncommissioned officers, the sergeants and corporals who would become the backbone of the Prussian army.

As important as Scharnhorst's and Gneisenau's reforms were for the rank and file, they were equally significant for the officer corps. Their goal, in which they were only partly successful, was to break the stranglehold of the Junker aristocracy ("heartless, wooden, half-educated men," one reformer called them) on the leadership ranks in the army. They wanted to make merit, not birth, the most important criterion for officer selection, so they put many old warhorses out to pasture and forced every officer seeking promotion to pass an exam. Military academies and staff colleges were set up to train officers, the first one being the Kriegsakademie (War Academy), whose most illustrious early director was Carl von Clausewitz, author of the classic exposition of military philosophy, *On War*. Under the guidance of Clausewitz and his colleagues, soldiering became a profession, not a pastime for the nobility.

At the center of these reforms was the creation between 1803 and 1809 of a unique institution called the *Grosser Generalstab* (Great General Staff), charged with war planning. Its ranks, which initially numbered fewer than fifty and never rose to more than one hundred, were composed of the best and brightest officers. Their relationship to the rest of the army was ambiguous. The chief of the General Staff did not exercise formal command, but he and his officers did exert a strong influence. Each army, corps, and division commander was assigned a General Staff officer to be his adjutant; Scharnhorst wanted them to "support incompetent generals, providing the talents that might otherwise be wanting among leaders and commanders." This system spread the tentacles of the General Staff throughout the army.

All previous generals had made plans, of course, but usually they were made in haste only when war was imminent or had already broken out. The Prussian General Staff inaugurated a new era, one in which, as historian John Keegan writes, "plans [were] conceived at leisure, pigeon-holed and pulled out when eventuality became actuality." To facilitate this task, the General Staff was organized into geographical sections that planned for hostilities against such likely enemies as France, Austria, and Russia. War plans were updated every year. Staff officers also oversaw vital functions such as mapping, intelligence, and logistics.

Although the General Staff was future-oriented, a good deal of its time was spent in the study of the past. Its dedication to learning the lessons of battles past has endeared it to historians ever since. Twice a year, officers undertook "staff rides" in which they studied old battles on the very ground where they had been fought. Their findings were published in a series of books and a monthly journal whose focus was not just on what had happened but why it had happened and whether the various participants had made the correct decisions.

This was only one of many practices started by the General Staff that has become routine in virtually all professional armies. Another was war gaming. Staff officers played out elaborate scenarios at a map or sand table. Metal symbols were used to denote the opposing armies—red for the enemy, blue for friendly forces, a color code that has persisted to the present day in the armed forces of the United States and other nations. A roll of the dice indicated the element of chance and an umpire scored the results. In the summer and fall, the General Staff took their war games outdoors, supervising large-scale maneuvers by the army. War gaming became a fashionable pursuit among Prussian officers, with various regiments pitted against each other.

The General Staff system created a group of military intellectuals without peer, an elite group instantly recognizable by their distinctive red-striped trousers. Other states had general staffs, but none was as brilliant or influential as Prussia's. Helmuth Carl Bernhard von Moltke was an outstanding, but not unrepresentative, product of this system.

Curiously enough, some of the greatest conquerors of modern times were not natives of the countries they led. Napoleon hailed from Corsica, Hitler from Austria, Stalin from Georgia, and Moltke from Denmark. His parents were Germans—his mother the daughter of a Lübeck merchant, his father a former Prussian officer turned unsuccessful farmer—who settled in the duchy of Holstein, which was under Danish rule. Helmuth, who was born in 1800, was educated at a military academy in Copenhagen, which he found "strict, even harsh," and then commissioned a lieutenant in the Danish army. Seeing scant opportunities in this tiny force, he applied to join the Prussian army, and after passing an officer's exam, he was accepted as a second lieutenant in 1822.

A decade later, after having graduated from the War Academy and served an apprenticeship as an army surveyor, Moltke joined the General Staff. He would serve in its ranks for the rest of his life. He never led so much as a company before becoming commander of the entire Prussian army, and he never served with troops in the field after his time as a lieutenant. His only experience of war prior to the 1860s occurred in 1836–39 when he served as an adviser to the Ottoman sultan who was fighting an Egyptian army in Mesopotamia. He was an intellectual whose abilities to apply his theories in practice would be revealed only when he was well into his sixties.

Moltke hardly fit the image of a Prussian militarist. He loved music, poetry, art, archaeology, and theater. He knew seven languages (German, Danish, English, French, Italian, Spanish, and Turkish). He was a prolific artist who filled sketchbooks with landscapes and portraits, as well as a popular author. His German translation of Edward Gibbon's *History of the Decline and Fall of the Roman Empire* was never published, and a novella he penned in the 1820s was not well received, but his account of his travels in Turkey, released after his return to Berlin in 1840 and illustrated with his own drawings, turned him into a literary celebrity, a role that he embraced by donning a Turkish fez and giving public lectures. Basking in the glow of fame, the forty-two-year-old officer married a sixteen-year-old English girl who was his sister's stepdaughter. Despite the disparity in their ages and their lack of children, they would live together happily until Marie's death in 1868. Out of devotion to her memory, he refused to remarry during the remaining twenty-three years of his life.

For all his catholicity of interests, Moltke was no closet liberal. He was a nationalist and a monarchist to the core who was appalled by the liberal revolutions that swept Europe in 1848. He placed his faith in the king and the forces of the old regime. Eventually the king returned his trust. In the 1840s and 1850s, while serving as a personal aide to several princes, Moltke made a favorable impression on the Hohenzollern court and got to know the future king Wilhelm, himself a professional officer who was known as the Soldier Prince. In 1857 Wilhelm's brother, the king, was incapacitated by illness, and Wilhelm was made regent. One of his first acts was to appoint Moltke chief of the General Staff after the death of the previous chief.

The fifty-seven-year-old major general immediately moved into the General Staff 's official mansion in Berlin. The first two floors served as offices; upstairs the general lived with his wife. In later years, after the German Wars of Unification, Moltke ascended to the status of a demigod and numerous memoirs appeared of his personal life. The picture that emerged was of a hardworking, modest man who dutifully took part in the whirl of court socializing but disliked dinner parties and had a horror of shaking hands. Instead of going out, he preferred a quiet dinner at home, followed by some music (Mozart was his favorite), a cigar, and a game of whist with a few close friends. His only vanity was to wear an ill-fitting wig to cover his completely bald head.

The chief of staff 's outstanding characteristic was his taciturnity. He became known as "The Great Mute" (*der grosse Schweiger*), or "the man who could be silent in seven languages." According to one of his relatives, "This silence was a mixture of reflection and shyness, as he himself has sometimes confessed. He did not feel that he possessed the talent of expressing himself easily on the idea of the moment, much less of making, as one says, 'Fine speeches,' and so his silence was often interpreted as pride."

Moltke preferred to express himself not in public flights of oratory but in top-secret war plans.

OVERCOMING GEOGRAPHY

The strategic challenges that Moltke inherited as chief of staff would have daunted a lesser man. The 1815 Congress of Vienna had added to Prussia's domains but also made them harder to defend. Prussia acquired much of the Rhineland and Westphalia in the west. Eventually, this would be a great boon, for the Rhineland would become one of the wealthiest and most industrialized regions in the world. But from a military viewpoint the added territories were quite a headache. They were not directly connected to the rest of Prussia; in between were the independent German states of Hanover and Hesse-Kassel. It was as if Prussia were made of two giant icebergs floating in treacherous, shark-infested waters, for all around were potential foes—France, Russia, Austria, Denmark, and the other German states.

Prussia was not considered one of the leading powers of Europe; it was not even the dominant power in Germany. That position was occupied by the Austrian Empire, leader of the German Confederation, which had replaced the defunct Holy Roman Empire. The thirty-nine-member Confederation did not have much power, but in its Diet in Frankfurt, Prussia was in a distinct minority. Prussia tried to counteract its isolation by creating a Zollverein, or customs union, which excluded Austria, but this could hardly disguise Vienna's preeminence in central Europe.

Along with the difficulties of geography, Moltke confronted the challenges of changing technology. It was a tribute to his peculiar genius that he figured out a way to harness three key innovations—the railroad, telegraph, and rifle—so as to compensate for Prussia's inherent vulnerability and turn it, improbably enough, into the strongest state in Europe, if not the world.

Railroads came to Germany a decade later than they did in Britain: The first steam-driven railway, covering less than four miles, was not opened until 1835, when Britain already had 340 miles of track. Railroad promoters had trouble raising enough capital and getting rights of way, so they turned to governments for help. Initially the Prussian army was not too impressed by the possibilities of railways whose carrying capacity, in the early days, was severely limited. One of the exceptions was Moltke, who as a major in 1841 joined the board of the Berlin Hamburg Railway and sank most of his savings into its stock. Moltke was fascinated by the possibilities of steam power once he discovered that railroads could transport troops ten times faster than Napoleonic armies had marched and that a single train could carry as much as one thousand horse-drawn carts. He produced a steady stream of memoranda and reports on the military uses of this new form of transportation.

As the size of the German railway system expanded to 3,638 miles by 1850, the rest of the Prussian army awakened to its possibilities. The Prussians made use of railroads in 1848 to rush troops from city to city to squelch the liberal uprisings. Further confirmation of railways' usefulness came from other conflicts that were carefully studied by the General Staff, such as the Italian War of 1859 (pitting Austria against France and her Italian allies) and the American Civil War of 1861–65.

With railways proving their worth, the state got more involved. Laws and conventions were signed to ensure the uniformity of rail lines in Prussia and other German states. A Prussian Railway Fund was set up to subsidize the construction and operation of some lines that were militarily valuable but not commercially profitable. By 1860 about half of Prussia's railway lines were state-owned or -administered. And all Prussian freight cars were required to have fittings that would allow them to be used in wartime to transport soldiers and horses. A special Railway Section was set up within the General

Staff to synchronize rail movements in wartime, and Field Railway Detachments were created within the army to repair damaged tracks and build new ones. The General Staff even pressed successfully for the adoption of standard time for the entire country to facilitate planning and execution.

By taking full advantage of railways, Moltke hoped that his soldiers could mobilize and deploy before their enemies did. They could then dash from one end of the kingdom to the other to deal with different threats. Railroads were particularly important because Prussia depended heavily on reserves and had a smaller standing army than neighboring states like France and Austria. Speed was of the essence if Prussia was to beat its enemies to the punch. All reservists were told exactly where they had to go when the mobilization order was given, and all regiments were provided with railheads at which they would assemble. Mobilization was facilitated by the fact that each regiment was based in the district from which it recruited. The General Staff knew that it would have to plan everything precisely, down to the minute, otherwise bottlenecks would develop and troops and supplies would not be able to move fast enough.

If hundreds of thousands of soldiers had to be mobilized within a matter of weeks, there was no time to send orders by horse courier. Luckily, the development of the electric telegraph allowed nearly instantaneous communications between headquarters and various divisions. In theory, that link could be maintained even in the field if soldiers strung copper wire as they advanced; in practice, armies usually outran their tether to headquarters. Nevertheless, improvements in communications allowed effective command and control to extend over a much wider area than in the past, and hence made possible another leap in army size. Napoleon had been able to control an army only as far as he could see with the aid of a telescope; Moltke hoped to control troops spread over hundreds, even thousands, of miles.

The chief of staff realized that while he could send general orders over long distances he could not manage a battle in detail from afar. He insisted that his subordinates digest the general principles of the war plan and then exercise their own initiative in carrying them out. "No plan of operations can look with any certainty beyond the first meeting with the major forces of the enemy," Moltke famously wrote. "All consecutive acts of war are, therefore, not executions of a premeditated plan, but spontaneous actions, directed by military tact."

To avoid the temptation to micromanage, Moltke instructed that "an order shall contain everything a commander cannot do by himself, but nothing else." These spare *Auftragstaktik*, or "mission-type orders," became a hallmark of the Prussian, and later the German, army. So ingrained did this system of delegation become that in 1864, when the Prussian commanding general ordered an invasion of the Danish duchy of Schleswig, his order simply stated: "On February 1st, I want to sleep in Schleswig."

Mobilization and deployment could get an army only so far. It still had to come face-to-face with the enemy forces and defeat them. Prussian soldiers were well equipped for this task because by the 1860s they possessed a better infantry weapon—the breech-loading rifle—than any other army in Europe.

The advantages of rifled barrels had been recognized early in the Gunpowder Age: The grooves imparted a spin to the projectile which made it possible to fire over longer ranges with greater accuracy. (The same principle explains why football quarterbacks, tennis players, and other athletes put spin on their balls.) But a rifle was hard to load, because a ball had to be laboriously forced down the barrel in order to grip the grooves. This limited its suitability for general military use. Most infantrymen were issued less accurate but faster-loading smoothbores. Rifles were generally reserved for hunters or specialized units of snipers. Yankee backwoodsmen made especially effective use of them against British redcoats during the American War of Independence.

Rifles began to spread in the nineteenth century following the invention of a cone-shaped bullet, hollow at the base and pointed at the nose, that was easy to drop down a muzzle and expanded when fired to grip the barrel's grooving. A French army captain named Claude-Étienne Minié perfected this design in 1843. His bullets were generally combined with a copper percussion cap that first appeared in the 1820s and was much more reliable than the old flintlock firing mechanism, especially in foul weather. The result was still a muzzle-loading musket but one that had been brought to new heights of deadly efficiency. Minié rifles could be fired with reasonable accuracy up to five hundred to eight hundred yards, or eight times the range of the old flintlocks, and their rate of misfires was twenty-five times lower. They proved their worth in the Crimean War, where the British and French troops, who had Miniés, mowed down the Russians, who did not.

Four years later, in the American Civil War, both sides started off largely with smoothbores but had completely changed over to percussion-cap Minié rifles by 1863. This contributed to the death of 620,000 soldiers—more than would be killed in all of America's other wars combined. Unlike in the past, most wounds were caused by small arms, not artillery. (Most deaths, however, were the result of disease, not gunshot.) This slaughter occurred in large measure because commanders on both sides had trouble coming to grips with the destructive potential of these new weapons. The leading officers of both the Union and the Confederacy had learned their trade in the Mexican War (1846–48), the last major conflict fought mainly with smoothbores. Having seen frontal assaults work against the Mexican army, they tried the same tactics against each other and turned farm fields into abattoirs. It took a few years of slaughter for both sides to start hiding their troops in trenches or dispersing them in order to mitigate the rifle's impact.

The Civil War brought muzzle-loaders about as far as they could go. Further progress would necessitate changing over to breech-loaders, which were faster and easier to load. A German craftsman named Johann Nikolaus von Dreyse led the way. Starting in the 1820s he began tinkering with new designs for what eventually became the *Zündnadelgewehr*, or needle gun, so called because the firing mechanism was a needle that drove through the paper casing of a cartridge to ignite its primer, thereby creating the mini-explosion needed to propel a lead ball out the barrel. This in itself was no improvement over the competition; it was in many respects less efficient than the percussion-cap musket. What made the needle gun special was the method of loading: the bullet went into the breech, not the muzzle. It was locked into place by pushing down a knob resembling the turn bolt of a door, which also automatically cocked the firing pin. Dreyse had made one of the first bolt-action rifles, a design still in common use to this day.

The disadvantages of the needle gun were many. The novel breech mechanism leaked gas, dissipating the force of the explosion. The needle gun's effective range was a bit less than the Minié rifle's (though still much greater than that of a smoothbore like the Brown Bess). The gun was also prone to malfunction. Jams caused by overheating were common, and soldiers were sometimes forced to use a rock to hammer the bolt open. The delicate needle was even more unreliable; it was so prone to break that riflemen had to carry spares.

But, flawed as it was, the needle gun represented a quantum advance over the Minié rifle. For one thing, it could be reloaded lying down or crouching, much to the relief of infantrymen who were exposed to enemy fire when they had to stand erect to load a traditional musket. Even more important, it could be reloaded much faster than any muzzle-loader—more than three times as fast as the Minié. This meant, as one historian notes, that "a ten-thousand-man unit armed with breechloaders was the equivalent of thirty thousand or forty thousand muzzleloaders or more."

The potential of breech-loading small arms first became apparent during the American Civil War. A regiment of Union "sharpshooters" armed with the Sharps single-shot breech-loading rifle (roughly comparable to the needle gun) shredded larger Confederate formations armed with muzzle-loaders at battles ranging from Antietam to Gettsyburg. The U.S. 1st Mounted Rifles, popularly known as Wilder's Lightning Brigade, had similar success in the Western theater of operations with their Spencer repeating rifles. (A breech-loader capable of firing seven shots from a magazine hidden in its wooden stock, the Spencer, patented in 1861, represented a considerable advance over the needle gun.) But the inflexible conservatism of the Union army and in particular of its mulish ordnance chief, General James W. Ripley, prevented these revolutionary rifles from becoming standard issue for all soldiers, in spite of the importuning of President Abraham Lincoln, who had personally fired, and been impressed by, the Spencer and Sharps guns. Ripley was convinced that these easy-to-fire rifles would simply waste ammunition, and he wanted nothing to do with them. Seldom has there been a more egregious example of an army failing to take advantage of readily available technology that might have shortened a war.

The Prussian military, despite its reputation as a den of reactionary aristocrats, showed itself more willing to innovate even without the spur of a major conflict. The Prussians had seen the possibilities in Dreyse's work early on and had been subsidizing him since 1833, though it took several years for him to improve his design sufficiently to win a large order. After extensive field trials and his personal test firing, King Friedrich Wilhelm IV ordered sixty thousand needle guns in 1840. The production of guns had been simplified by Eli Whitney and other American inventors who in the early nineteenth century had come up with a system of interchangeable parts. But it took a long time for Dreyse to ramp up his operation to meet the Prussian army's growing demand. At first only 10,000 needle guns a year were produced, a figure that eventually rose to 22,000. At that rate it took more than twenty years to reequip all of Prussia's 300,000 active-duty and reserve soldiers. By 1866 all Prussian infantrymen finally had the needle gun.

Dreyse, an independent businessman, would have been happy to sell his products to France, Austria, and other nations too, but they weren't interested; they remained devoted to the Minié rifle and the bayonet charge. The French beat the Austrians in 1859 despite having an inferior musket. They had won through élan, by pushing home infantry assault after infantry assault. The Austrians decided that if it had worked against them, it would work for them. Their field regulations of 1862 downplayed the importance of rifle fire in favor of mass bayonet charges. While Prussian infantrymen went to rifle practice, their Austrian counterparts engaged in wind sprints. Like the Union Ordnance Department, the Austrian high command thought that breech-loading rifles would simply waste valuable ammunition by allowing soldiers to fire too often. In 1866 their infantry would be equipped with the Lorenz-model Minié.

The Austrians were not totally oblivious to the march of technology. Following their defeat to the French in 1859, they had reequipped their artillery with rifled muzzle-loading bronze cannons. The advantages of grooving were as great for cannons as for small arms. An old smoothbore cannon was useless beyond 1,500 yards; new rifled guns were effective at 4,000 to 5,000 yards. Their ammunition changed also, from mainly solid cast-iron balls to explosive shells. The Prussians, too, were buying rifled cannons from a manufacturer named Alfred Krupp. But because the Prussians had spent so much money reequipping their infantry, they were unable to complete the process with their artillery by 1866. More than a third of Prussia's cannons were still antiquated bronze smoothbores, whereas Austria had gone over almost entirely to rifled pieces. Austria had 736 rifled cannons, Prussia only 492.

The Austrian edge in artillery was insufficient to cancel out the Prussian advantage in small arms, however, because the rifle had usurped the cannon's traditional place as the queen of the battlefield. In the American Civil War, rifles accounted for 86 percent of casualties, cannons only 9 percent. This was because the increased range of the rifle made artillerymen, who traditionally plied their trade from the front line, vulnerable to being picked off by infantrymen. The solution was to increase the range of artillery and move it into protected positions in the rear, allowing gunners to hit targets that they could not see by relying on forward spotters. But effective "indirect fire" did not become practical until the advent of field telephones and radios in the early twentieth century. In the meantime, as the Austrians were to learn to their sorrow, artillery's role on the battlefield had been much diminished.

BLOOD AND IRON: THE ROAD TO KÖNIGGRÄTZ

German unification had first become a real possibility in 1848. Liberals hoped to unite the various German states in a parliamentary democracy. But while granting a few of their demands, the king of Prussia, the emperor of Austria, and other German rulers had no intention of giving power to the people. After the liberal uprisings were crushed, the forces of conservatism schemed to snatch away the banner of nationalism. In 1862 Otto von Bismarck was appointed minister-president of Prussia, and the crafty politician maneuvered aggressively to unite all of Germany under the grip of King Wilhelm. He vowed, "The great questions of the day will not be decided by speeches and the resolutions of majorities—that was the great mistake from 1848 to 1849—but by iron and blood."

To ensure that Prussia would prevail in the coming struggle, King Wilhelm increased the military budget and the number of conscripts. The Prussian parliament, which had been created in 1849, refused to approve these measures, so the king simply ignored it. Military and foreign policy was to remain a royal prerogative as long as the Hohenzollerns ruled.

Prussia's revamped military met its first test in 1864 against a weak foe, Denmark. Popular sentiment in the German states had long been agitated by Danish rule over two duchies full of ethnic Germans, Schleswig and Holstein. Bismarck saw this as a convenient opportunity to enlarge Prussia's domains while converting Berlin from the enemy of German nationalism to its leading champion. Austria and the rest of the German Confederation went along, contributing a substantial force for the invasion of the duchies, which began on February 1, 1864.

The Danes had constructed some formidable defensive positions, but their weaponry was hopelessly outdated; they had neither breech-loading muskets nor rifled cannons. The Prussians, on the other hand, had not fought a real war in more than fifty years, and their rustiness showed. The Danes were not decisively defeated until the original Prussian commander, a relic of the Napoleonic Wars, was replaced by Prince Friedrich Karl, who brought Helmuth von Moltke with him as chief of staff. Moltke designed a daring amphibious operation that seized the fortified Danish island of Alsen and persuaded the king of Denmark to sue for peace.

Once Denmark was out of the way, the next target for Prussia was Austria, for, as Bismarck had once observed, "Germany is too small for us both." This was a much more formidable adversary than Denmark. Moltke, who predicted, "The struggle will be terrific," had been planning for this conflict since 1860, but he revised his plans in light of the lessons of the war with Denmark. He had seen the high casualties sustained both by German troops attacking Danish fortifications and by Danish troops attacking Prussians armed with the needle gun. With weapons becoming deadlier, Moltke realized that a massed frontal assault would be suicidal. But trying a flanking maneuver on the battlefield was also extremely hazardous because of the greater range of guns—artillery shells could now cover three miles, rifle bullets more than half a mile.

The answer, he decided, was an envelopment on a grand scale. His flanking maneuver would begin from the first day of mobilization when three Prussian armies would take trains to three different locations. From there, they would march toward the Austrians, hoping to arrive before the other side had completed its mobilization and catching the enemy in a vise. Instead of concentrating their forces before facing the enemy—the usual practice of Napoleon and countless other generals—the Prussians would unite only after the battle had already been joined. Moltke thought this risky scheme was justified because if all his armies started off from one location, moving and provisioning so many men would take too long. To seize and keep the initiative, Moltke proposed to take full advantage of railroads and telegraphs. He knew that Prussia had the edge here. It had five railroads running to the Habsburg frontier in Bohemia, while the Austrians had only one railroad running from Vienna. Prussia could complete its mobilization and deployment in three weeks, twice as fast as the Austrians. That opened a window of opportunity when the Prussians would have "numerical superiority, and this," Moltke calculated, "is the period which we must use with all speed and energy to bring about great decisive battles."

The author of this daring plan did not anticipate all the difficulties of implementing it, not the least of which would be persuading King Wilhelm I to go along with it. Moltke had not reckoned with the king's hesitancy to attack his fellow German monarch, Emperor Franz Josef of Austria. In early 1866, Bismarck and Moltke urged Wilhelm to strike, warning him that "every day's delay may mean an incalculable loss." Strengthening the case for immediate action, Bismarck concluded an alliance with the new Kingdom of Italy in April to open a second front against Austria if war broke out within ninety days. But even after Austria ordered a full call-up on April 21, Prussia waited to follow suit until May 12. The Prussians thus lost an opportunity to strike before the enemy was ready, showing how political calculations can upset the best-laid military plans. But Austria was not able to take advantage of its three-week head start. Prussia used its superior mobilization capabilities to catch up fast.

Once their forces were assembled, the Prussians' immediate priority was to deal with the other German states, most of which were friendly to Austria. If 150,000 German troops joined with the Austrian Northern Army, 240,000 strong, the Prussians would face an insurmountable disadvantage. The Prussians solved this problem by issuing ultimatums on June 15, 1866, to the other German states to disarm immediately. When they did not respond, Prussian armies invaded the next day. Hanover, Hesse, and Saxony fell in short order. Only the Saxon army of thirty-two thousand men managed to escape the fast-moving Prussians and link up with the Austrians. But to do so they left their homeland undefended, giving the Prussians another invasion route into Bohemia. Bavaria, Austria's major German ally, was taken aback by this rapid Prussian offensive and decided to use its army for defensive purposes only, leaving the Austrian and Saxon armies on their own. There was still the danger of French intervention (France would not want Prussia to strengthen itself by defeating Austria), but Moltke, gambling that Emperor Louis Napoleon would not move fast enough, denuded western Prussia of troops, sending every man he could find to the east.

By June 21, 1866, the day that King Wilhelm declared war on Austria, the bulk of Prussia's army was deployed in a 250-mile arc around the borders of Bohemia from the Elbe River to the Oder. The Prussians had been holding their forces back, not certain what the Austrians would do. They wanted to be sure they had enough troops to block a sudden Habsburg thrust toward Berlin or into the coal-rich Prussian province of Silesia. Within a few days, however, it became apparent that the Austrians had no intention of taking the offensive; they were too disorganized. So on June 22, the order went out from headquarters in Berlin to the First Army in Lusatia and the Second Army in Silesia: "His Majesty commands that both armies march into Bohemia and seek a juncture in the direction of Gitschin." (These orders also affected the Elbe Army in Saxony, which was considered subordinate to the First Army.)

The man giving the orders was Helmuth von Moltke. That was by no means foreordained, since the chief of the General Staff had not, until this point, been in operational control of field armies. Traditionally the General Staff had been subordinate to the minister of war, who in turn reported to the king. But Moltke's stature had grown since his successful role in the Danish War. He had won the right to report directly to the king, bypassing War Minister Albrecht von Roon. On June 2, 1866, Wilhelm had gone further and given Moltke the right to issue orders in the king's name. Since this command structure was new, the little-known chief of staff was not always able to get his way with the strong-willed generals in the field. One division commander, upon receiving an order, was said to have exclaimed, "This is all quite correct, but who is General *Moltke*?"

The slight confusion caused by the Prussian change in command was as nothing compared to the chaos that gripped the Austrian high command. The commander of the Austrian Northern Army was Lieutenant General Ludwig von Benedek. A Hungarian Protestant and the son of a doctor, Benedek had established his reputation by fighting gallantly in a losing cause against France in 1859. He had spent virtually his entire career in Italy and knew nothing of Bohemia. When Franz Josef tried to offer him the job, he declined, finally agreeing only after the emperor insisted. Benedek was no military intellectual like Moltke; he was a "muddy boots" general who mocked "professors in shoulder straps" and enjoyed drinking with enlisted men. In his view, "the only talents required in a staff chief are a strong stomach and a good digestion."

Nor did Benedek have a competent staff to make up for his deficiencies. His chief advisers were cautious and pessimistic. Austria had its own general staff, but its work was shoddy and slipshod. One staff officer who was sent on a mapping expedition to central Germany in the early 1860s made a beeline for the casinos at Bad Ems. He wrote back to Vienna that his work wasn't very important anyway; the staff could learn everything they needed to know by perusing Baedeker's guidebook.

Compounding the Habsburg army's problems was its polyglot nature. Nine languages of instruction were used in peacetime, ranging from Serbo-Croatian to Rumanian. In battle, all orders were issued in German, which many soldiers did not understand. Since they were part of a multiethnic empire, Habsburg soldiers lacked the kind of national cohesion that bound together their Prussian foes, whose battle cry was "With God for King and Fatherland!"

Austria, unlike Prussia, did not undertake wide-ranging reforms to improve its governance after suffering defeat at Napoleon's hands. The state structure remained rickety, its administration corrupt and inefficient, its finances uncertain, its ideology reactionary. The chief focus of the monarchy was repressing its own population, not mobilizing it to defeat outside threats. Austrian soldiers were brave and well motivated but poorly led. Benedek and the other top generals displayed a curious passivity even as the bulk of the Prussian army descended upon them. They placed their faith not in rapid movement but in antiquated fortresses such as those at Josephstadt and Königgrätz. Instead of moving decisively against any one of Prussia's three armies, they hesitated everywhere, thus ceding the initiative to the invaders.

The Prussian forces crossed the Bohemian frontier on June 23. Three days later, as they were emerging from the Giant Mountains, they made their first sustained contact with Austrian defenders. The results did not bode well for the Habsburg army. In the village of Hühnerwasser, one Prussian company used its needle guns to annihilate two Austrian battalions that were foolish enough to attempt a bayonet charge over open ground. The Prussians lost 50 men, the Austrians 277. A Prussian witness wrote that "the wood and the road were plastered with dead and wounded. There were Austrian bodies and backpacks as far as the eye could see. Trees had been stripped of their bark by our fire and the cries of the wounded were heart-rending."

The outcome was equally one-sided in the village of Podol later that same day (June 26). In a murderous nighttime fight to control one of the crossings over the Iser River, the Prussians cleared out the Austrians house by house. When the Austrians tried a frontal attack, the needle gun again proved an efficient killer. A force of two thousand Austrians suffered 50 percent casualties.

Those engagements set the pattern for the next three days. As one Prussian force after another popped out of the mountain passes, the Austrians tried to block their way, only to be mowed down by the needle gun. On the one occasion when the Austrians prevailed, the victory was a pyrrhic one. At Trautenau on June 27, the Austrians managed to repel the vanguard of the Second Army, but in the process they lost 5,780 casualties to the Prussians' 1,280. And the Austrians had to pull back anyway when the bulk of Prussia's Guard Corps arrived. The outcome might have been very different if the Austrians had gambled by throwing a large force against one of the Prussian armies when it was most vulnerable, while making its way through the narrow mountain passes. But the irresolute Benedek hesitated to commit himself.

The opening rounds of the war cost the Austrian army thirty-one thousand casualties, including a substantial number of prisoners, and left its commander deeply dispirited. On June 30, Benedek sent a desperate telegram to the emperor: "I beg Your Majesty urgently to make peace at any price. Catastrophe for the army is unavoidable." Franz Josef was puzzled by this defeatist talk from a general who had yet to fight a major battle. He wired back, "To conclude peace is impossible." And then he (or, rather, one of his ministers) added a barbed question: "Has a battle taken place?"

The demoralized Benedek began to pull back his forces to avoid getting caught in the Prussian pincer. On July 1, his Northern Army occupied a camp site near the fortress of Königgrätz. Benedek's decision to pause here has been the subject of much puzzled discussion by contemporaries and historians alike. As a defensive position, the plain in front of Königgrätz was a curious choice, since the Elbe River was at the army's back, cutting off its lines of retreat. Some suggest that Benedek deliberately chose this position because he felt pressured by the emperor to force a major battle that might never have happened if he had continued retreating over the Elbe. Others argue that Benedek had no such premeditation, that he had no intention of fighting a battle with the river at his rear; he simply hoped to delay the Prussian army to allow his forces to escape. It may be that not even Benedek himself was sure exactly what he was up to. One of his generals later testified that on July 2 the Austrian commander "was physically and morally a broken man." The confusion of the Austrian high command was intensified by Benedek's decision, taken under pressure from the emperor, to sack his chief of staff on the very eve of the battle.

But whatever his motivations—whether to cover his continued retreat or to prepare for a decisive battle—Benedek did order some entrenching work to be done on the left bank of the Elbe. On July 2, artillery batteries were positioned on a series of hills to cover the approaches from the west. As the work continued, Benedek's spirits lifted. He began to think that perhaps he could prevail after all, "if my old luck does not desert me." Not unreasonably. Moltke himself later wrote that the Austrians occupied "an extremely strong position."

BLUNDERING INTO BATTLE

At the time, the Prussian high command had no idea of the enemy's dispositions. For all the vaunted capabilities of the telegraph, the Prussians were operating in an information blackout. Their scouts had entirely lost contact with the main body of the Austrian army. The Austrians likewise had no idea where the Prussians were, even though by July 2 the two armies' outposts were less than five miles apart.

After having issued orders by telegraph from Berlin to his forces in the field, Moltke decided this was no longer sufficient. "War cannot be conducted from the green table," he believed, referring to the desktop at headquarters where troop dispositions were displayed. "Frequent and rapid decisions can be shaped only on the spot according to estimates of local conditions." With a battle brewing, Moltke decided he had to be on the spot himself. On June 30, the king, Bismarck, Moltke, and a large retinue journeyed in six railway cars from Berlin to join the First Army at Gitschin, a Bohemian town that more than two centuries earlier had served as Wallenstein's headquarters.

Upon his arrival, Moltke discovered an exhausted, badly supplied force. For all the General Staff 's voluminous planning, logistics had been badly bollixed up. The staff had placed too much reliance on railways. They expected, for instance, that army bakeries in Cologne would be able to send their bread all the way to the Bohemian border by rail, not realizing that this would take so long that the resulting loaves would be inedible. The staff also had not made adequate arrangements for supplying the army once it moved out of reach of the railroads, which happened once it started crossing the Giant Mountains. There was inadequate provision for horse-drawn vehicles to ferry supplies from railheads to troops who were advancing on foot and horseback.

The staff had expected that the troops could live off the land, but they had not reckoned on the massive increase in army size. Living off the land worked tolerably well for a seventeenth-century army of thirty thousand men; it was much more difficult for the quarter of a million men of the Prussian army in 1866. Moltke tried to ameliorate this problem by spreading his troops out over hundreds of miles, but this offered only a partial solution. Prince Friedrich Karl and his First Army staff were reduced to surviving on potatoes and a few bottles of champagne. On the day of the battle itself, Moltke told his wife he had nothing to eat beyond "two chocolate bonbons and a small piece of bread." Luckily for the Prussians, Austrian logistics were just as bad.

On the evening of July 2, some Prussian cavalrymen galloped into the First Army headquarters with news that they had discovered the main Austrian force at its campground in front of Königgrätz. Without consulting anyone or seeking instructions, Prince Friedrich Karl resolved on an attack the next morning with the First Army and the Elbe Army. He sent a dispatch to the Second Army headquarters, asking his cousin, Crown Prince Friedrich Wilhelm, to dispatch a corps to cover his flank in the coming battle. Only then did he bother to send another rider to the king's headquarters to notify Moltke of his intentions.

When the dispatch arrived at 11 P.M., Moltke was awoken to read it. Groggy though he was, the chief of staff made an instantaneous and accurate evaluation of the situation. He realized that Friedrich Karl was facing the entire Austrian Northern Army, and that it would take not just one corps but all of the Second Army to prevail against such a powerful foe. Moltke immediately sent riders of his own to the Crown Prince's headquarters with fresh orders: "Your Royal Highness will be good enough immediately to make the necessary arrangements to be able to advance with all your forces in support of the 1st Army against the right flank of the enemy's probable advance, and in so doing to come into action as soon as possible."

It is upon such snap judgments, made with incomplete information and without the leisure of reflection, that wars—and empires—are won or lost.

The vanguard of the Second Army did not start marching until 7:00 A.M. It was a hard slog. Because the roads were not big enough to contain so many men, they were forced into fields that were turning muddy in the driving rain. Pulling the artillery through the oatmeal-like goop proved especially difficult. Not until 12:30 P.M. did the Second Army's advance units join the battle, falling on the vulnerable right flank of the Austrian line as Moltke had intended.

Benedek had received intelligence more than an hour before about the imminent arrival of another major Prussian force, but he had not been able to sufficiently strengthen his right wing because so many of the troops from that area had been sucked into the desperate fight in the Swiepwald against Fransecky's 7th Division. Most of the Austrians on the right were not even aware that another Prussian army was about to hit them. When the Second Army arrived, many Austrians were so surprised they surrendered without a fight. "Good God! Where do they come from?" startled Austrian officers wondered.

The Prussian high command knew exactly where these late arrivals had come from; their only wonder was why it had taken them so long to show up. Watching from the hill at Roskosberg, the king and his chief of staff breathed a sigh of relief once they saw through their spyglasses the dark blue coats and black *Pickelhauben* of the Second Army coming into contact with the white-coated Austrians. "The success is complete," Moltke supposedly told the king. "Vienna lies at Your Majesty's feet." Not quite. Much hard fighting remained. But to a strategist of Motlke's caliber, who had played out such scenarios in many war games, it was not hard to project forward several hours and realize the inevitable result of his flanking maneuver now that it was finally under way.

The strong point of the Austrian lines, the key to the whole position, was the hill at Chlum. From here a plethora of artillery batteries had been pounding the Prussian First Army to the west all morning. The Hungarian troops who were manning the small village of Chlum were caught utterly by surprise when they were assaulted from the *north*. Their regimental commander, upon being told that blue-coated troops were advancing up the hill, thought they must be Saxon allies. "You are seeing phantoms!" he roared to one subordinate who insisted they were actually Prussians. Around 2:30 P.M. the phantoms burst from the fog and tall grass that had shielded their advance and routed the panicked Hungarians. Officers tried to stop the retreat, beating soldiers with the flats of their swords, yelling, "You cowards! Stand there, you yellow dogs!" But those who stood and fought were methodically cut down by the needle guns. One Austrian artillery battery put up a particularly brave resistance, but they did not last long. A few minutes of rifle fire killed fifty-two men and sixty-eight horses. Today, visitors to Chlum can see an obelisk monument commemorating the "Battery of the Dead." The Prussians that afternoon did not give much thought to the corpses around them. They wheeled their own cannons to the top of the hill and turned them on the Austrian troops spread out on the plain before them like sheep ready for slaughter.

Benedek was informed of the fall of Chlum at 3:15 P.M. In the next several hours he threw everything he had into an attempt to retake the heights. The Austrian reserve artillery opened up a withering bombardment of Chlum and the neighboring village of Rosberitz. "The air was literally filled with shells, shrapnel and canister," wrote a Prussian officer; "branches of trees, stones, splinters flew around our ears and wounded many . . ." Then came the Austrian infantry, fighting their way house by burning house through Rosberitz. At heavy cost, the Austrians retook Rosberitz and then moved on to Chlum.

Wave after wave they came, Austrian troops charging madly up the hill, bayonets pointed, legs pumping, voices screaming hoarsely. And down they went, wave after wave, scythed by the relentless roar of the Prussians' needle guns and cannons. Within less than an hour one Austrian corps lost more than six thousand men. The relatively small number of Prussian guards who were holding the hill might nevertheless have been overwhelmed by the sheer weight of the Austrian assault were it not for the arrival at 4:30 P.M. of another entire army corps. The Prussians now had overwhelming numbers to hold Chlum. All hopes of a successful Austrian counterattack had ended.

Across the battlefield, nine miles to the south, the other side of the Austrian line was also crumbling. After having been held at bay all morning, the Elbe Army managed to turn the flank of the Saxons who were holding the left side of the Austrian position. Seeing what was happening on the two flanks, at 4 P.M. Moltke gave the long-awaited order for the First Army to advance in the center. The Prussian soldiers, who had been shot at all day, surged forward, a war correspondent noted, "with loud cheers and drums beating."

For the Austrians, the battle was lost. Benedek wired the emperor: "The catastrophe I warned you of two days ago happened today." The only remaining question was whether Moltke would be able to complete the grand envelopment that he had envisioned. Thanks to the suicidal bravery of the Austrian cavalry and artillery, he could not quite pull it off. Benedek, displaying a calmness in defeat that he had not exhibited when victory was still within his grasp, sent his horsemen forward to stop their Prussian counterparts from turning what was already a retreat into a rout. The Austrians lost more than 1,100 men and 1,600 horses to a barrage of Prussian artillery and rifle fire, but in the process they bought a precious half hour that allowed the Austrian reserve artillery to wheel their batteries forward. The gunners kept up a steady fire for hours, sacrificing themselves to prevent the Prussians from pressing the fleeing infantry too closely.

Hordes of Austrian soldiers streamed east, losing all semblance of order along the way. Hundreds drowned in the Elbe. Others made it over makeshift bridges only to find the gates of the Königgrätz fortress closed against them. It took hours to convince the fortress commander, who was terrified of being invaded by the Prussians, to open up. In the meantime, Austrian troops were crushed and suffocated in the rush to get inside the walls. In the next few days the Austrians continued their headlong retreat. The Prussians let them go. They were too tired themselves to set off in pursuit; "many of them had been 19 hours on the march and 10 hours engaged with the enemy," according to the official Prussian General Staff history. The victorious army made camp on the battlefield, eating whatever crumbs they could find before sinking into sleep.

Some commentators have faulted Moltke for letting the bulk of the Austrian army slip from his grasp, but it was no great loss. The Austrians and Saxons left behind all their artillery and most of their supplies. Even a month later, a reporter found the fields "literally covered with knapsacks, scabbards, cartridge pouches, cooking pots, and all the various articles which soldiers carry with them." In addition to their materiel, the Austrians left many men behind. They lost 19,800 taken prisoner and another 24,400 dead, wounded, or missing—44,200 in all, or 21.5 percent of their total strength. The Prussian casualties were much lower: 9,172 men killed, wounded, and missing, just 4 percent of their total force. The losses were highest among those on both sides who had fought in the Swiepwald and on Chlum hill.

On a percentage basis the casualties were lower than in many battles of years past, but because the overall number of men engaged was so large (Königgrätz was one of the biggest battles fought in Europe to that point), the amount of suffering was still very great. One reporter noted, "Every cottage in the neighborhood that had not been burnt was full of wounded." There were not enough doctors to tend to them all, and many did not survive. Anyone motoring today near the Czech city of Hradec Králové (as Königgrätz is now known) can still see gravestones and monuments with their inscriptions in German to gallant, long-dead soldiers of both sides.

The Austrians had been so badly roughed up that they would fight no more. The day after the battle, a Habsburg field marshal was sent to seek a truce from the Prussians. "My Emperor no longer has an army," he said. "It is as good as destroyed."

Moltke wanted to follow up by marching on Vienna, but Bismarck restrained him. The prime minister was fearful that if the war continued, France (then considered the most powerful nation in Europe) would ally itself with Austria and force Prussia to fight a two-front campaign for which it was not prepared. To avoid a lengthy conflict and to produce a lasting peace, Bismarck showed magnanimity toward the defeated Austrians. In the armistice negotiations, he did not ask the Habsburg empire to give up one foot of its soil beyond the province of Venetia, which went as a reward to Berlin's ally, Italy; Prussia gave back the occupied lands of Bohemia and Moravia. In return, Austria had to agree to dissolve the German Confederation and to renounce all hope of a leadership role in Germany. Berlin annexed Schleswig and Holstein along with Hanover, Hesse, Nassau, and Frankfurt, thus uniting the two halves of Prussia and swelling its population to over twenty-three million, roughly the same size as France. The rest of the northern German states and free cities—Saxony, Hamburg, Lübeck, Bremen, Thuringia, Mecklenburg—were pressed into the new North German Confederation, ceding control of their foreign and military policies to Berlin. Within a couple of years, Bismarck forced similar treaties on the south German states. The formal announcement of the German Reich only awaited the defeat of France in 1870–71. However inevitable the process looks in retrospect, the creation of a German state would never have occurred, at least not in the form it took, absent the Prussian victory at Königgrätz.

The Austrian state suffered a near-fatal blow in 1866, as intelligent observers perceived at the time. Within a few months of signing the armistice with Prussia, the Habsburgs had to agree to share power with the Hungarians, turning what had once been simply the Austrian Empire into Austria-Hungary, the Dual Monarchy. Its prestige wounded, the sprawling and decrepit empire would be slowly picked apart in the years ahead by its various national components.

While Austria plummeted to new lows, the London *Spectator* observed, "Prussia has leaped in a moment into the position of the first Power of

Europe." And the military itself had become the first power within the Prussian state. The prestige that accrued from its role in uniting Germany by "blood and iron" gave the General Staff almost complete autonomy from the rest of the government. Before long, it was common to joke that Germany, like Prussia before it, was not a state with an army but an army with a state. The price of such military dominance proved to be high. It was the army, with its precise timetables for mobilization, that helped transform a Balkan crisis in August 1914 into a world war. A few years of total war would cost Germany the last vestiges of civilian government. The chief of staff, Field Marshal Paul von Hindenburg, who as a young officer had earned his baptism of fire at Königgrätz, and his ruthless lieutenant, General Erich Ludendorff, would become military dictators. Their stranglehold was broken only by the ignominy of defeat in 1918.

PLANNING TO WIN

While there is general consensus about the momentous consequences of Königgrätz, there is no agreement about why the battle went the way it did. Historians continue to debate whether it was Prussian competence or Austrian incompetence that accounted for the outcome. The safe answer, which has the added advantage of being true, is that it was a bit of both.

There is no question that Benedek played his hand badly. He failed to decisively attack the Prussians when they were at their weakest, while wending their way through the passes of the Giant Mountains. He then failed to take advantage of his initial success at Königgrätz on the morning of July 3. Until the early afternoon, the Austrians had a large numerical advantage over their Prussian opponents. If Benedek had gambled on an all-out offensive, he might have been able to crush the Elbe Army and the First Army before the Second Army arrived. Failing that, Benedek should have retreated while the going was still good. Instead he stayed in place, leaving his right flank dangerously exposed, and suffered the consequences. Moltke later commented that "[n]o one, of course, dreamed" that the enemy would open themselves up in this fashion.

The needle gun further contributed to the Austrian defeat. Time after time, their superior breechloaders allowed small Prussian units to maul more numerous adversaries. Fransecky's heroic 7th Division never could have held its pivotal position in the Swiepwald were it not equipped with rapid-fire rifles. As Friedrich Engels wrote afterward: "It may be doubted whether without [the needle gun] the junction of the two Prussian armies could have been effected; and it is quite certain that this immense and rapid success could not have been obtained without such superior fire, for the

Austrian army is habitually less subject to panic than most European armies."

Railways also helped determine the outcome. In just twenty-one days, the Prussians transported 197,000 men and 55,000 horses. Their rapid concentration and advance caught the Austrians off guard. The whole conflict lasted just seven weeks. The war against France four years later took slightly longer, because the French people refused to admit they were beaten, but the major combat was also relatively brief, in large part because the Prussians once again were able to transport and concentrate troops more rapidly than their foes.

Of course, railroads and guns do not operate themselves. The human factor must never be lost sight of. Observers generally agreed on the superior motivation and training of Prussian troops. An English journalist found that, in contrast to the peasant conscripts who made up the Austrian ranks in 1866, the Prussian "rank and file are men of education, who know what they are fighting about, not mere machines drilled to mechanical perfection." They were thus "more in earnest, more thoughtful, more willing to risk their lives for a principle, whether false or true, more imbued with a sense of duty."

Prussian commanders were also more thoughtful and more earnest than their Austrian counterparts. Above all, the genius of Helmuth von Moltke towered over the battlefield. As a reward for his service, the chief of staff was awarded a large cash grant by his grateful king, which allowed him to purchase a thousand-acre estate in Silesia. A few years later, his victory over France would earn him promotion to field marshal and the title of count. All those accolades were fully deserved. It is true that Moltke's plans did not work perfectly in 1866; the General Staff did a particularly poor job of meeting the army's logistical requirements. But Moltke and his "demigods" were far ahead of anyone else in harnessing industrial technologies to the demands of warfare. Their only rivals in this respect were Lincoln, Grant, Sherman, and the other Union leaders who had marshaled superior manpower, factories, railroads, and riches—though not superior technology per se (the North and South had roughly comparable weapons)—to crush the Confederacy. But their knowledge was lost after 1865, when the United States disbanded what had been for one brief moment the most powerful army in the world, whereas the German Army continued to develop Industrial Age warfare in accordance with Moltke's high-risk credo: "Great successes in war cannot be gained without great dangers."

The most important military innovation associated with Königgrätz—and the most lasting—was the superb planning that went into the Prussian victory. War had become too complex to be managed by one person, even a great captain like Napoleon, Frederick the Great, or Helmuth von Moltke.

A whole management system was now required, and all across Europe, states copied the Prussian model of the General Staff. All of the continental armies adopted such Prussian innovations as staff rides, war games, and, above all, the writing of complex war plans in peacetime. Many even imitated Prussia's spiked helmets. The Prussians still retained an edge, however, because no other general staff enjoyed the kind of unfettered power that theirs had. This advantage proved much more lasting than any technological edge, which did not—could not—last long in an age of frenetic weapons innovation.

Within days of Königgrätz, every army in Europe was rushing to buy its own breech-loading rifles, many of them superior to the thirty-year-old needle gun. Four years later, when Prussia went to war against France, its infantrymen were at a disadvantage in small arms. The French were armed with a newer breech-loader called the Chassepot that had three times the range of the needle gun. They even had a primitive machine gun called the Mitrailleuse, capable of firing 150 rounds a minute, though they were never able to make very effective use of it. Just as Austrian infantry had been slaughtered charging Prussian rifles in 1866, so Prussian infantry was slaughtered charging French rifles in 1870. Prussia prevailed anyway, ironically enough, because of its artillery. While its cannons had been inferior in 1866, they were superior by 1870. Having seen that muzzle-loaders were outdated, the Prussians scrapped them after 1866 and reequipped their entire force with Krupp's breech-loading rifled cannons made of cheap and durable cast steel. France continued to rely on old bronze muzzle-loaders. Better artillery gave Prussia a crucial edge in 1870 that allowed its gunners to annihilate the French army from long range.

Such technological flip-flops were to become common in the Industrial Age, when plummeting manufacturing costs allowed a state to completely reequip an army of hundreds of thousands within a relatively short period. No army could now afford to wait twenty years or more to field a new weapon, as the Prussians had waited for the needle gun. It was even more unthinkable that any army could rely on the same gun for a century and a half, as the British army had relied on the Brown Bess from the 1690s to the 1840s. In the Industrial Age inventions could transform the battlefield within months.

In those circumstances it proved impossible for any state to develop and maintain a lasting technological edge over equally sophisticated adversaries. Europe was swept by arms races in the decades before World War I, but no power gained an enduring advantage. By 1914 all the major combatants had repeating rifles, machine guns, quick-firing artillery, railroads, telegraphs, and all the other inventions that had transformed warfare.

They also had large conscript armies to operate those weapons. Following

Prussia's victories in 1866 and 1870–71, every major European state except Great Britain copied the Prussian system of putting a large portion of its young men through military training, conscripting them for a few years of active-duty service, and then keeping them in reserve for wartime. Just as Prussia had been subdivided into military districts, so too France, Russia, Italy, and other states were subdivided to facilitate mobilization. And just as Prussian mobilization plans had been drawn up by officers schooled in its War Academy, so other European states set up staff colleges of their own (France's École Supérieure de Guerre was founded in 1880) to bring their officers up to Prussian standards.

Amid this growing parity in personnel and weapons, not even the acknowledged excellence of Germany's General Staff was enough to deliver victory against the multitude of foes their country faced. Their previous triumphs led to a dangerous hubris among senior German officers. Moltke's successors—in particular Alfred von Schlieffen, author of the famous war plan that bears his name, who ran the General Staff from 1891 to 1905—thought their forces could defeat any combination of enemies, forgetting that Bismarck and Moltke had been careful to fight their foes one at a time and to avoid protracted conflicts by pursuing limited war aims. In 1914, the German army, its General Staff commanded by the elder Moltke's nephew and namesake, Helmuth von Moltke the Younger, tried an 1866-style strategic envelopment simultaneously against both Russia and France. The Schlieffen Plan came close to succeeding, but after initial victories in both east and west, the war settled down into a vicious stalemate. On the Western Front, parallel trenches stretching 475 miles from the North Sea to Switzerland produced the kind of attritional fighting that Moltke had been determined to avoid in 1866. On the Eastern Front, the sheer vastness of the Russian steppe made it impossible to finish off the czar's battered army until, in 1917, a revolution brought to power a new government in Moscow committed to exiting the conflict.

Everywhere, machine guns and artillery proved much more potent as weapons of defense than offense. This was mainly because they were bulky and hard to move; when troops advanced rapidly, the only weapons they could carry were rifles, which proved pitifully inadequate against the heavy firepower of dug-in defenders. The attackers might have called in artillery support, but it was impossible to accurately coordinate such fire without portable two-way radios, which did not yet exist. Lacking radio communications with advancing troops and afraid of hitting their own men, gunners had to suspend or shift their fire well before the first waves of infantry reached the enemy lines. Defenders simply burrowed underground like moles and then emerged, once the initial artillery barrage had lifted, to riddle the exposed attackers. The triumph of defensive technology consigned an entire generation of European men to slaughter on a hitherto unimaginable scale. Offensive technologies capable of breaking this deadlock (the tank, submachine gun, radio, and airplane), along with the tactics to take advantage of them, finally emerged in the later stages of World War I, but they would not truly revolutionize warfare until World War II.

While the Industrial Revolution did not give any European power a lasting edge over its rivals, this did not mean that its geopolitical effect was negligible. The skillful use of industrial techniques made possible Prussia's rise from relative weakness into the most powerful state on the Continent. Outside Europe, the new technology enabled the white man to complete his conquest of the world. For while industrialization was leading toward military parity among European states, it was exacerbating the growing disparity between the West and the Rest.

CHAPTER 5

MAXIM GUNS AND DUM DUMS:

Omdurman, September 2, 1898

First came the noise, "a mighty rumbling as of tempestuous rollers and surf bearing down upon a rock-bound shore." Or so it seemed to *Daily Telegraph* correspondent Bennett Burleigh as he listened to the fierce incantations to Allah draw closer. He did not have long to ponder what he was hearing, for a few seconds later the men who were making all that noise came into view from behind the black Kereri hills a few miles north of Sudan's capital, Omdurman.

There were tens of thousands of devout Muslim warriors arrayed ten ranks deep, and they were moving fast and steady across the sandy plain. They were mostly infantrymen wearing patched cotton shirts known as jibbahs and carrying swords and spears. Their commanders, the emirs, rode ahead of them on magnificent Arab stallions. "Gigantic banners fluttered aloft, borne on lofty flagstaffs," wrote another war correspondent, Ernest Bennett of the *Westminster Gazette*. "The rising sun glinted on sword blades and spearheads innumerable, and as the mighty host drew nearer, black heads and arms became visible among the white of the massed jibbahs."

Burleigh, Bennett, and everyone else who was present just outside Omdurman at 6 A.M. on Friday, September 2, 1898, agreed that the spectacle was magnificent. But their awe was tinged with a certain incredulity. Hardly a man among the Anglo-Egyptian forces could believe that the Dervishes, as the Sudanese warriors were called by Westerners, would be bold enough to try a frontal assault over open ground in broad daylight. Did they not know what modern weapons could do?

The Dervishes had more than their share of rifles and artillery, most captured more than a decade earlier when they had chased the Egyptians and their British overlords out of the Sudan. A few of these firearms were now discharged from such long range that their bullets kicked up harmless spouts of sand far in front of the British position. The British gunners waited for the advancing army to get within 2,800 yards (about 1½ miles) and then opened fire with their field artillery, some on gunboats floating in the nearby Nile, the rest arrayed ashore. "About twenty shells struck them in the first minute," observed Lieutenant Winston Churchill of the 21st Lancers. "Some burst high in the air, others exactly in their faces. Others, again, plunged into the sand and, exploding, dashed clouds of red dust, splinters, and bullets amid their ranks. The white banners toppled over in all directions. . . . About five men on the average fell to every shell: and there were many shells."

Yet these followers of the Mahdi, the messiah, who had promised to rouse the entire Muslim world to vanquish the infidels and install his one true faith, could not be stopped by a few artillery rounds. They kept on going, breaking into a run, chanting in Arabic, "There is no God but Allah and Muhammad is the messenger of Allah."

Next to open fire were the Maxim guns. The British had forty-four of them, each capable of spitting out more than five hundred rounds a minute—enough to hit a man multiple times before he keeled over in blood-soaked agony. They did their work with great noise and even greater brutality, strewing perforated and twisted bodies all over the rocky ground. Among those struck early on was Ibrahim al Khalil, one of the commanders of this assault. But still the Dervishes kept coming, to the amazement of their enemies. "Surely there never was wilder courage displayed," marveled Burleigh. "In the face of a fire that mowed down battalions and smashed great gaps in their columns they flinched not nor turned."

The British infantrymen waited for them behind a low breastwork, called a zeriba, made of thick and prickly mimosa scrub laid out in a mile-long crescent around the western bank of the Nile. They were in the classic two-deep formation used by Wellington almost a century before, the front rank kneeling, the second rank standing behind them. But there had been major changes since the Battle of Assaye. No longer did the queen's soldiers wear scarlet. Now the "redcoats" were kitted out in khaki uniforms topped off with sun helmets. More important, they no longer fired muzzle-loading flintlocks. Tommy Atkins, as the common British soldier was known, now hefted the Lee-Metford MK II, a repeating rifle capable of firing ten .303 caliber rounds without reloading and with great accuracy over one thousand yards. The men were trained to fire fifteen aimed shots a minute—one every four seconds.

The stopping power of many of their bullets was augmented by filing off the tips. These hollow-points, known as dum dums because they had first been made at the British arsenal in Dum Dum, India, expanded on impact, carving through living flesh like a serrated knife through a cooked chicken breast. They were such a terrible weapon that they would shortly be outlawed in European warfare, but they were considered fair game against African "savages."

As the Mahdists drew within range, the British riflemen were ordered to fire by section, starting with the Grenadier Guards. They displayed the calm precision of the long-service veterans that they were. An admiring officer later wrote: "The fire discipline of the British throughout the action was a treat to watch; exactly as on parade they changed from volley-firing to independent and back to volley-firing as might be ordered, coolly and without any hurry. Their shooting too was admirable; they began knocking them over at nine hundred yards, and within three hundred nothing could live." The charging Dervishes got closest to the part of the line manned by Egyptian troops armed with the slightly older Martini-Henry rifle, whose range and rate of fire was less than that of the Lee-Metford, but even here they could not get close enough to use their swords and spears. A few of the survivors kept crawling forward until finally their last reservoirs of strength were gone. Then they lay down to die. The Dervishes, who had come so far, came no farther.

By 8:00 A.M., the initial Dervish assault had been broken and the voice of the Anglo-Egyptian commander, Major-General Horatio Herbert Kitchener, sounded over the din. "Cease fire, please! Cease fire! Cease fire! What a dreadful waste of ammunition."

The battle was far from over, and the Dervishes still had a trick or two left up the sleeves of their jibbahs, but already the vast disparity in firepower between the two sides had become painfully apparent. At least 4,000 of the 12,000 attackers had been killed or wounded, to only 160 of the defenders. Lieutenant Angus McNeill of the Seaforth Highlanders wrote: "The whole plain was a fearful sight—simply plastered with dead and dying men and horses—the wounded attempting to crawl away." It was, concluded Lieutenant Robert Napier Smyth of the 21st Lancers, a "Regular inferno."

The outcome at Königgrätz had been very much in doubt. At Omdurman there was no suspense at all. In journalist G. W. Steevens' apt phrase, "It was not a battle, but an execution." This slaughter represented the culmination of centuries of developments in weaponry, transportation, communications, and medicine that brought Europeans to the pinnacle of their worldwide power. With their machine guns, breech-loading rifles, gunboats, railroads, and telegraphs arrayed against an army that still relied on swords and spears, the troops under Kitchener's command enjoyed a preponderance of power seldom equaled before or since.

THE MERCHANTS OF DEATH

A good deal had happened in the world of weapons since the Prussian victory at Königgrätz in 1866. The needle gun had gone from state-of-the-art to obsolete within a few years. It was superseded by better-designed breech-loading rifles such as the British Martini-Henry (introduced in 1869) that leaked less gas, had more durable firing mechanisms, and possessed greater range and velocity. The efficiency of many of these new guns was increased by the introduction of cartridges that contained a percussion cap, powder charge, and bullet in one metal case—no more fumbling with a separate lead ball and paper cartridge. While much easier to use than their paper predecessors, metal rounds still had to be manually loaded after every shot, a time-consuming process.

Repeating rifles were the wave of the future. Successful models were manufactured in the 1860s by two Connecticut companies, Winchester and Spencer. The Turkish army, equipped with lever-action Winchesters, repulsed successive Russian assaults at the Battle of Plevna in Bulgaria in 1877. Within a few years, every army in Europe was buying repeaters. The British version was known as the Lee-Metford. "Lee" referred to James Paris Lee, a Scottish immigrant to America, who, while employed at the Springfield Armory in Massachusetts, had invented a bolt-action magazine system for rifles. "Metford" was William Metford, an English railway engineer who devised a superior method of rifle grooving. The British Small Arms Committee combined their innovations and began reequipping the British army with the resulting rifles starting in 1888.

At first, the .303 Lee-Metford, like all firearms since the dawn of the Gunpowder Age, relied on black powder (a mixture of saltpeter, sulfur, and charcoal) to ignite the explosion that propelled its bullets out the barrel. In 1886, French chemist Paul Vieille invented a superior alternative, a smokeless powder known as Poudre B made from nitrocellulose (also known as guncotton: cotton dipped in nitric and sulfuric acids) mixed with ether and alcohol, then rolled into sheets, and cut into flakes. Getting rid of black powder not only increased the velocity of the shot but also did away with much of the impenetrable smoke that had obscured battlefields for hundreds of years. This made it easier for soldiers to aim and for officers to direct them. The British, along with every other European army, quickly followed the French lead, causing black powder to all but disappear after more than five hundred years.

In 1892 the Lee-Metford was adapted to fire with cordite, a smokeless powder made of nitroglycerin, guncotton, acetone, and mineral jelly that was derived from the work of Alfred Nobel. (Nobel was famous for his invention of dynamite, a nitroglycerin-based explosive that was one of many products of the rapid advances of chemical engineering in the late nineteenth century.) The size of its magazine was also increased to ten shots from five. The resulting weapon was called the Mark II, and it was carried by the British troops at Omdurman in 1898. With a modified barrel and a new name (the Lee-Enfield), it would continue to serve as the primary rifle of British infantry in both World War I and World War II. (An updated version was used by some British soldiers as recently as the Iraq War in 2003.)

The principal advantage of the Lee-Metford and its peers was their speed. A soldier no longer had to insert a new bullet every time he fired. He simply shoved back the bolt-action mechanism to move another bullet from the magazine into the breech. An entire magazine of ten bullets could be expended in seconds. This was blindingly fast compared to older guns. But it was positively slow compared to the new wonder weapon, the machine gun.

Like most other firearms innovations, machine guns had their antecedents in the early years of the Gunpowder Age. As early as the fourteenth century there had been experiments with ribaudequins, or organ guns, made by strapping a number of barrels together. They had little success because each barrel still had to be loaded individually, a laborious, time-consuming process. The first practical machine guns did not appear until the 1860s.

Their most famous designer and most tireless promoter was Richard Jordan Gatling, a dentist from North Carolina who had moved to Indiana in the hope of finding better opportunities to finance and manufacture his brainstorms. Notwithstanding his own sympathy for the Confederate cause, Gatling tried hard to sell his six-barrel, crank-operated weapon to the federal government during the Civil War. But the conservative Union Ordnance Department was even more resistant to machine guns than to repeating rifles. It did not adopt the Gatling gun until 1866, after the South had already surrendered. Gatling had more luck peddling his wares at a tidy profit to Britain, Russia, Spain, Japan, and other countries.

The Gatling gun was an impressive enough weapon, but it still required an operator to turn the handle in order to rotate the barrels. This did not satisfy another American inventor who wanted to create a fully automatic gun: Just point and pull the trigger. Hiram Maxim was not a gunsmith by training. He was a talented, self-taught inventor who at twenty-six received his first patent for a hair-curling iron and then went on to create devices ranging from an electric lightbulb to a better mousetrap. He later recounted what drew him to gun-making: "In 1882 I was in Vienna, where I met an American Jew whom I had known in the States. He said: 'Hang your chemistry and electricity! If you want to make a pile of money, invent something that will enable these Europeans to cut each other's throats with greater facility.' "

At his workshop in London, Hiram Maxim came up with a gun that would use the force of the recoil from the first shot to operate its ejection, loading, and firing mechanisms until an entire ammunition belt (another of his inventions) had been expended. To keep the barrel from overheating, Maxim added a bronze water jacket that had to be kept constantly replenished lest the coolant turn to steam. The whole contraption weighed sixty pounds and required three men to operate. With firepower equivalent to that of a hundred rifles, it was the most efficient instrument of mass slaughter in the world. But developing it was one thing; selling it another.

At first Maxim was stymied by the hardball marketing tactics of a competing Swedish company, Nordenfelt. Their gun was a ten-barrel, hand-cranked affair that was roughly similar to the Gatling and far inferior to the Maxim. But Nordenfelt's chief salesman, a sinister figure of Balkan origin named Basil Zaharoff, stopped at nothing to promote his product at the expense of Maxim's. He was even accused of bribing Maxim's workers to sabotage machine guns sent for trial runs in Austria. Not being able to beat the competition, Maxim joined forces with them. In 1888 the two firms merged to form the Maxim-Nordenfelt Guns and Ammunition Company, whose chief salesman was—who else?—Basil Zaharoff. The new armaments company was big enough to compete against the other giants in the field: Krupp, Winchester, Schneider, Armstrong, and all the rest.

Most of these companies had close relationships with their home governments but also marketed abroad. In the second half of the nineteenth century, a time when Europeans fought few major wars, capitalist competition among these large corporations combined with political competition among various states to drive weapons development at a feverish pace. No one, whether a government or a company, could afford to let a rival get an advantage; all had an incentive to develop and adopt new technology as quickly as possible. This is not to suggest, as pacifist critics did after 1914, that "merchants of death" made a world war inevitable; the decision to use the weapons still rested with statesmen, not salesmen. But the activity of these entrepreneurs did allow Europe to develop its military technology very rapidly through a combination of private enterprise and public spending: what would later become known as a military-industrial complex.

"WHATEVER HAPPENS WE HAVE GOT . . ."

At first, machine gun manufacturers found their wares a tough sell in European defense ministries dominated by conservative aristocrats who continued to regard the bayonet charge as the ne plus ultra of combat. Most generals were intensely suspicious of these industrial death machines that upset traditional military doctrines and seemed to leave little room for individual feats of valor. Sales did not take off until the start of the Great War, when the machine gun was revealed to be a necessity. Both the German and British armies wound up fighting with the Maxim gun, the British version licensed by Vickers, the German one by Krupp. Skeptical as most European generals were of these weapons prior to 1914, they were willing to use them in colonial warfare against "savages" who did not adhere to European notions of chivalry. The results were predictably devastating.

The British general Sir Garnet Wolseley had Gatlings with him when he invaded Ashantiland in 1873 and Egypt in 1882. Another British general, Lord Chelmsford, took four of the guns to fight Zulus in 1879. The British imperialist Frederick Lugard had Maxim guns with him when he occupied Uganda in 1892 and northern Nigeria in 1902–3. Cecil Rhodes's men used five Maxims to crush resistance from the Matabele tribe in 1893. The Matabele expressed bewilderment of "the white man . . . with his guns that spat bullets as the heavens sometimes spit hail," asking plaintively, "Who were the naked Matabele to stand up against these guns?" Some 3,500 Matabele warriors were mowed down, while the invasion force lost only four men. The conquered territory became known as Rhodesia. Such slaughters led to an understandable complacency among European colonists—a mind-set perfectly satirized in Hilaire Belloc's famous verse:

> Whatever happens we have got
> The Maxim Gun, and they have not.

Important as machine guns were, one should not exaggerate their role. They were not hauled along on every imperial campaign, and even when they were deployed, early models had a frustrating tendency to jam in the thick of the action. Rifles, on the other hand, were ubiquitous wherever the white man advanced in the late nineteenth century, and in the hands of steady, well-trained soldiers they were usually enough to defeat just about any opposition. On September 8, 1895, in Mozambique, six hundred Portuguese troops arrayed in a square confronted six thousand Tonga warriors. The two Portuguese machine guns jammed, but they still repelled the attackers with their French-made Kropatschek magazine rifles, at a cost of only five dead. A more famous example, celebrated in the 1964 movie *Zulu*, occurred at Rorke's Drift in southern Africa on January 22–23, 1879, when a British garrison of just 140 men used Martini-Henry rifles to beat back more than four thousand Zulus. No Gatling guns were present at Rorke's Drift.

The Europeans did not always prevail. Just before the Zulu impis reached Rorke's Drift, 40,000 of them had wiped out 1,300 British and

African troops under Lord Chelmsford twelve miles away at Isandhlwana. There was always a possibility of disaster whenever a small European column was sent into hostile territory against a much larger enemy force. The British suffered notable setbacks on numerous occasions, such as the First Afghan War (1842), when a column of 4,500 troops, 700 of them Britons, was wiped out during a retreat from Kabul. Every imperial power had similar experiences. At the Battle of the Little Big Horn in 1876, 256 troopers of the U.S. 7th Cavalry led by Lieutenant Colonel George Armstrong Custer were killed by 2,000 Sioux and Cheyenne warriors, many of them armed with repeating rifles.

Usually these defeats were due to excessive arrogance on the part of Western troops who failed to exploit all their advantages because they held the enemy in such low regard. Chelmsford might have avoided a catastrophe if he had not divided his forces during the invasion of Zululand and if the troops he left behind at Islandhwlana had possessed enough screwdrivers to open their ammunition boxes. Custer might have defeated Crazy Horse and Sitting Bull if he had not left behind four Gatling guns at his camp.

Such defeats were much written about at the time and they remain justly famous today, but we should remember that their notoriety owes much to the fact that they were the exception, not the rule. Even when European armies lost a battle (and this typically happened only when they were badly outnumbered), they usually took a much higher toll among their attackers and then came back to win the war. The British marched back into both Zululand and Afghanistan, in the former case to stay, in the latter case to establish a buffer state. Likewise, the U.S. government was so shocked by Custer's Last Stand that it redoubled its efforts to defeat the Indians. Within a few years, the northern Great Plains would be subdued. Around the world, the number of territories that succeeded in preventing a European takeover, or at least a substantial degree of European penetration, could be counted on one hand.

Europeans in the colonial period liked to pretend that their conquest of the globe was due to their "moral ascendancy" over supposedly inferior races. But they knew that, in the final analysis, it was not superior virtue that brought them victory after victory. It was superior firepower. And they were careful to safeguard their advantage. European armies on colonial campaigns made sure that machine guns were operated only by Europeans; they did not want "natives" learning how to use them. Thus, even when machine guns fell into unfriendly hands they usually did not know what to do with them; the Sudanese Dervishes had a number of captured machine guns they did not employ at the Battle of Omdurman. Seeking to "lock in" their firepower edge, the European powers signed a series of treaties, starting with the Brussels Act of 1890, banning the sale of breech-loaders in equatorial

Africa—an early version of a nonproliferation pact designed to stop the spread of weapons of mass destruction.

In part because of these treaties, but mainly because gun-making was so advanced in Western Europe and North America compared to everywhere else, the Western advantage in firepower steadily increased during the nineteenth century. Many non-Europeans like the Marathas had been able to achieve firepower parity with the Europeans in the age of muzzle-loading flintlocks. Lacking industrial production facilities, few were able to make the transition to machine guns and breech-loaders fast enough to prevent their colonization.

One of the few exceptions were the Ethiopians. Under Emperor Menelik II (r. 1889–1913), they acquired a formidable array of repeating rifles, cannons, and machine guns, mainly from French and Italian suppliers. This was bad news for the Italians when they attempted to conquer Ethiopia. On March 1, 1896, at Adowa, twenty thousand Italian and Eritrean troops, half of them armed with repeating rifles, confronted one hundred thousand Ethiopians, 70 percent of whom had repeating rifles. The latter-day Roman legions were routed, suffering 50 percent casualties. This defeat may be ascribed in large part to the poor logistics, leadership, intelligence, and training of the Italian forces, which were made up inexperienced conscripts, not of professional, long-service regulars as in the other European armies in Africa. But Ethiopia would never have become one of only two African states to keep its independence (Liberia, founded as a colony for freed American slaves, was the other) if it had not accumulated such a powerful arsenal and developed an army capable of making effective use of it.

There have been many examples throughout history of forces that were outgunned and outnumbered but nevertheless persevered for a long time by employing hit-and-run tactics. Several leaders utilizing such guerrilla tactics caused considerable trouble for European colonizers during the nineteenth century. Two of the more skillful were Abd el-Kader (1807–83) and Samori Touré (1830–1900), charismatic Islamic leaders who resisted French invasions of their homelands, Abd el-Kader in Algeria, Touré in a West African empire he had carved out in present-day Guinea and adjoining territories. Both had considerable success in buying Western firearms and ammunition and in setting up workshops to manufacture their own copies. Both managed to hold out for about fifteen years and to inflict heavy casualties on the French forces. But in the end both were captured by French generals skilled in "irregular" warfare (what today would be called counterinsurgency).

Abd el-Kader and Samori Touré's strategies were not widely emulated, because in the nineteenth century few non-Western peoples had sufficient national consciousness to make a decentralized guerrilla war possible. While people have always been attached to their native soil, the political ideology known as *nationalism*—which holds that every national group deserves its own state—is of fairly recent vintage. It was invented by Enlightenment philosophes in the eighteenth century and spread throughout Europe and the Americas during the nineteenth century, but it did not reach Asia and Africa until the twentieth century. Once nationalism and other "isms" (such as communism) had become firmly established, they would inspire numerous guerrilla struggles around the world, from Vietnam to Afghanistan. In the nineteenth century, by contrast, the ideological fuel for waging "people's war" was much scarcer. Most Third World armies chose to resist Western invasion with more or less conventional tactics, like those employed at Omdurman, which proved disastrous against better-armed and better-organized white invaders.

Europeans enjoyed other technological advantages beyond weaponry that facilitated their nineteenth-century conquests. Steamships, railroads, telegraphs, medical science—all were important contributors to the success of colonial campaigns by putting Europeans in a position where they could use their improved guns.

The spread of steamships in the 1830s made it much cheaper and easier to run an empire by cutting the time and cost of transoceanic travel. The voyage from London to Cape Town shrank from forty-two days in the 1850s to nineteen days in the 1890s. Colonial officials were able to move soldiers and supplies much more readily from one domain to another, for instance rushing reinforcements from England to South Africa in 1899 during the Boer War. More important, steam power allowed small gunboats to navigate previously inaccessible waterways, especially in Asia and Africa. The English explorer Macgregor Laird was scarcely exaggerating when he wrote an encomium to "the immortal Watt" in 1837: "By his invention every river is laid open to us, time and distance are shortened."

As early as 1824 the steamboat *Diana* helped British forces move up the Irrawaddy River during the first Anglo-Burmese War (1824–26). Gunboats played an even bigger role a few years later in enabling Britain to win the First Opium War (1839–42) and open up China to Western influence. Employing iron-hulled gunboats built by the East India Company, British forces were able to steam up the Pearl and Yangtze Rivers, two of the most important commercial arteries in China, brushing opposing war junks out of the way. The paddle steamer *Nemesis* even entered Canton through its narrow channels and terrorized the populace into surrender. Sailing ships were used too, but they had to be towed by steamers through inland waters in order to bring their guns to bear. Royal Navy gunboats played an equally important role in winning the Second Opium War (1856–60), when they attacked Canton and the Taku Forts near Peking. As a result of these two conflicts, the imperial court was forced to kowtow before European "barbarians" who demanded not only preferential terms for trade but also outright control over a growing list of "concessions" that became quasi-colonies.

In the years ahead, armored gunboats were used by the British on the Niger River to subdue what later became known as Nigeria; by the French on the Mekong River to police Indochina; by the Portuguese on the Incomati and Zambesi Rivers to pacify Mozambique; and by the Belgians on the Congo River to grab a lucrative slice of central Africa. Such was their fame that ramshackle steamers played a prominent role in works of colonial fiction such as C. S. Forester's *The African Queen* (1935) and Joseph Conrad's *Heart of Darkness* (1902). They even spawned a new term: gunboat diplomacy.

Trains, unlike gunboats, seldom provided direct fire support for troops, although a few armored trains were used by the British during the Boer War (1899–1902). But railways were vital for expanding colonial control inland. India was the best example. Starting in the 1840s, the British provided subsidies to railway promoters who wanted to crisscross the subcontinent. By 1900 India had almost twenty-six thousand miles of track, more than any other country except the United States, Canada, and Russia. This dramatically cut the cost of passenger and freight travel. It also enabled the British to concentrate their troops wherever they were needed, which came in handy during the Indian Mutiny of 1857–58. Ultimately, all of those rail lines proved a mixing blessing from the British perspective because, by helping to knit disparate Indian states closer together, they helped foster a national identity that would lead to demands for independence.

Europeans were less ambitious railway builders in Africa and East Asia. The Belgians did lay a rail line from the Atlantic coast to Leopoldville (now Kinshasa) that proved vital to the Congo Free State. The French built a line from Hanoi to Saigon. The British built a line from Mombasa on the Indian Ocean to Lake Victoria in Uganda that helped to create a new colony called Kenya. But more grandiose imperial visions—the British Cape-to-Cairo Railway, the German Berlin-to-Baghdad Railway, the French Trans-Saharan Railway—were never realized.

A few massive rail projects did become a reality, with important strategic consequences. The Trans-Siberian Railroad completed in 1905 consolidated Russian control over Siberia, just as the American transcontinental railroad helped the United States to consolidate its control over western North America. Indeed the completion of the Union Pacific–Central Pacific Railroad in 1869 probably did more than any other single event to defeat the Plains Indians by facilitating the migration of settlers whose sheer numbers overwhelmed the nomadic natives.

Often running parallel with railroads were electric telegraph lines that allowed orders and information to flow back and forth across empires even faster than a train or steamboat could travel. Using this information, the authorities could send military forces wherever they were needed. After the Indian Mutiny, one British colonial official exclaimed: "The telegraph saved India." This success hastened interest in laying a submarine cable between Britain and India, a project that was completed in 1870. In 1902 Britain completed a worldwide cable network that ran entirely through its own territory. Control of the thick copper strands along the ocean floor was now deemed as important for the security of the empire as control of the waters above them. Luckily for London, British telegraph companies were as dominant in their field as the Royal Navy was in its.

Conquering Africa and other inhospitable climes required more than simply defeating the limitations of time and space. It also involved defeating germs. White troops who were stationed in the "Dark Continent" paid a heavy price—in the 1820s, notes historian Daniel Headrick, "77 percent of the white soldiers sent to west Africa perished, 21 percent became invalids and only 2 percent were ultimately found fit for future services." They were struck down by dysentery, yellow fever, cholera, typhoid, and numerous other ailments, but the most dangerous disease was malaria. It was not until 1897 that scientists figured out that malaria was transmitted by mosquitoes, but even before then, a reliable cure had emerged by a process of trial and error. In 1820, two French chemists succeeded in distilling alkaloid of quinine from cinchona bark, a traditional folk remedy. By 1850 the new drug was routinely being prescribed as prophylaxis for Europeans venturing into Africa. As a result, Headrick continues, "the first-year death rates among Europeans in West Africa dropped from 250–750 per 1,000 to 50–100 per 1,000." Quinine pills led to similar drops in mortality in other tropical climes.

These technological developments help to answer one of the central questions of the nineteenth century: Why did Europeans suddenly acquire so many colonies? Western empires had been spreading steadily across the world for centuries, ever since the dawn of the Gunpowder Age, but the trend accelerated considerably in the late 1800s. In 1876 Europeans ruled less than 10 percent of Africa; by 1914 they had more than 90 percent. Many reasons have been adduced for the "Scramble for Africa," from the search for profits and the competition among European states to social Darwinist and Christian evangelical ideas. But many of these motives had been around for a long time. They were similar to the reasons why Europeans had overrun most of the Americas long ago. Why not Africa and Asia? Because the cost had been too high. "What the breechloader, the machine gun, the steamboat and steamship, and quinine and other innovations did," Headrick writes, "was to lower the cost, in both financial and human terms, of penetrating, conquering and exploiting new territories."

Indeed, the bill to Whitehall for maintaining the largest empire in history was staggeringly low. In 1898, Britain had only 70,000 troops watching over 250 million Indians and another 41,000 soldiers stationed in the rest of the empire, which included 41 million subjects in Africa alone. The total military budget was just 2.5 percent of gross domestic product, considerably less than the United States spends today. "This was world domination on the cheap," writes economic historian Niall Ferguson, and because the marginal cost of acquiring new colonies was so low, Britain and other European nations were ready to grab them even if there was no particular necessity to do so. The Victorian historian J. R. Seeley was driven to joke: "We seem, as it were, to have conquered and peopled half the world in a fit of absence of mind."

The Sudan offers a perfect illustration of how the process worked.

CRUSADE IN THE SUDAN

Sudan was an Egyptian colony before it was a British one. It had been conquered by Muhammad Ali, the reformist ruler of Egypt, in 1823; Sudanese spears proved no match for his gunpowder army. Britain inherited responsibility for this vast region (over one million square miles) in 1882 when it occupied Egypt. At that very moment, a growing number of Sudanese were resisting a foreign occupation that had inflamed local sentiment by, among other steps, trying to stamp out the slave trade. The leader of the Sudanese uprising was Muhammad Ahmad, son of a boat builder, an ascetic holy man who in 1881 began calling himself the Mahdi, or the Divinely Guided One. Inspired by a Sufi strain of Sunni Islam, the Mahdi hoped to purify the faith and defeat its enemies. His followers enjoyed considerable success against the low-quality Egyptian army. By 1883, Mahdists controlled virtually the whole country. The British government sent Major-General Charles "Chinese" Gordon to evacuate the remaining Egyptian garrisons, but instead he got trapped in Khartoum. A relief expedition under Garnet Wolseley arrived too late to prevent the fall of the city and the death of Gordon on January 26, 1885.

The Mahdi went to his own reward six months later. He was succeeded by his leading disciple, Abdallahi ibn Muhammad, who became known as the Khalifa, or Successor. The Khalifa forged a complex, if repressive, state complete with treasury, courts, and armory. A fundamentalist interpretation of Islam was strictly enforced in which slavery was legal but women going out unveiled was not. His capital became Omdurman, a ramshackle city constructed across the Nile from Khartoum, the old center of Egyptian administration. The Khalifa built up a sizable and well-disciplined army equipped with captured Anglo-Egyptian weapons as well as ammunition made by his own craftsmen. His forces defeated the neighboring Ethiopians, but they were repulsed in their attempt to spread jihad into Egypt in 1889. The British, who had more pressing concerns elsewhere, were content to leave the Mahdists largely to their own devices, retaining only the Red Sea port of Suakin.

Many Britons, especially colonial officials in Egypt, were anxious to avenge the death of Gordon, but the British government was not eager to undertake a costly military expedition simply for revenge. A variety of motives eventually led Lord Salisbury, the prime minister, to overcome his initial hesitation and order a slow-motion invasion of the Sudan in 1896. Perhaps paramount among these motives was his fear of the French, who considered the Sudan theirs and had sent an expedition from the west coast of Africa to claim it. Salisbury feared that letting a hostile power control the headwaters of the Nile would imperil the security of Egypt, which in turn would threaten the Suez Canal, the gateway to India, which had opened in 1869. The prime minister also fretted that resurgent Islam, with its "formidable mixture of religious fanaticism and military spirit," could threaten an empire that ruled over so many Muslims, especially in India. The destruction of the Mahdist state, he hoped, would not only safeguard Egypt but also demoralize pan-Islamists hoping to rouse their coreligionists against the West.

Appropriately enough for an expedition with overtones of the Crusades, its commander was a devout Christian who was itching to avenge his murdered friend, "Chinese" Gordon.

"Your Country Needs YOU" proclaims the famous World War I recruiting poster. A giant finger points at the viewer. Steely eyes stare out from above a luxuriant mustache. A granite jaw hovers over a dark tunic. That is how posterity has remembered Horatio Herbert Kitchener. By 1914 he was an earl, field marshal, minister of war, and England's foremost soldier. He was far from famous when, eighteen years earlier, he embarked on the campaign in the Sudan; it was the Battle of Omdurman that would turn him into a semi-deity: "K of K," Kitchener of Khartoum.

Herbert was born in 1850 to a retired army lieutenant colonel who ran his household as if it were a miniature regiment. His sickly mother, to whom he was passionately attached, died when he was fourteen, increasing his tendency toward shyness and loneliness. At seventeen he entered the Royal Military Academy at Woolwich, where cadets were prepared for a career in the artillery or the engineering corps, considered the two most intellectually demanding occupations for an army officer. It was here that Kitchener met, and grew to admire, Charles Gordon, an older engineering officer and a fellow

Christian who had won fame leading a mercenary army to put down the Taiping Rebellion in China in 1862. Kitchener joined the Corps of Royal Engineers upon his graduation, but spent much of his early service surveying Cyprus and the Holy Land. In the process he learned Arabic and Turkish.

He first came to Egypt in 1882 on his own initiative to offer his services to the invading British force. He stayed as one of the twenty-five officers selected to train a new Egyptian army after the old one had been defeated and disbanded. He often operated on his own, gathering intelligence among the Arab tribes while riding a camel through the desert dressed in a turban and flowing robes. After stints as an intelligence officer in the Gordon Relief Expedition and as commander of the Suakin garrison, he was appointed commander of the entire Egyptian army in 1892, already a brigadier at age forty-two. Neither his early success nor his reputation for arrogance, aloofness, and ambition won him many friends.

The Sirdar's very appearance (*sirdar* was an Anglo-Indian word for "leader") was calculated to frighten. He had poor vision; his two eyes did not focus properly and often seemed to look in different directions. After being wounded in the face by a Mahdist bullet in 1888, the resulting scar made him look even more severe. G. W. Steevens of the *Daily Mail* offered this forbidding description of the Sirdar: "He stands several inches over six feet, straight as a lance, and looks out imperiously above most men's heads. . . . Steady passionless eyes shaded by decisive brows, brick-red rather than full cheeks, a long moustache beneath which you divine an immobile mouth; his face is harsh and neither appeals for affection nor stirs dislike."

Kitchener never acquired a wife who might have softened his hard edges. He thought a wife was a "permanent encumbrance" to a hard-working officer and forbade his officers from becoming betrothed. This, along with his closeness to some of his staff officers and his fondness for collecting ceramics, furniture, and flowers, later led to a widespread assumption that he was homosexual. There is, in fact, no record of any sexual liaisons, gay or otherwise. Hard as it is to believe in this sexualized age, he may simply have been celibate. Kitchener, an exponent of a muscular Christian faith, fit the medieval stereotype of the warrior monk. Mere sinners found his presence a bit disconcerting. Officers who kept native mistresses were careful not to mention this fact in front of their disapproving chief.

Although Kitchener was not generally beloved by his subordinates or the correspondents who covered him, he did earn their grudging respect. "His precision is so inhumanly unerring, he is more like a machine than a man," Steevens wrote. (Machine analogies were as popular then as computer metaphors are today.) "You feel that he ought to be patented and shown with pride at the Paris International Exhibition. British Empire: Exhibit No. I, *hors concours*, the Sudan Machine."

True to his reputation, Kitchener, by now a major general, conducted the Sudanese operations with industrial precision. He was not an impetuous cavalier in the old mold; like Helmuth von Moltke, he was very much a modern manager whose talent lay not in personal heroism or charisma but in his organizational abilities. But he lacked Moltke's audacity and brilliance. Also, unlike Moltke during the German Wars of Unification, who feared that any delay would allow neighboring states to gang up on Prussia, he had no particular reason to hurry: Sudan had no potential allies. He could afford to take his time, and he did. Two and a half years elapsed between the time he first entered Sudan—March 15, 1896—and the fall of the Mahdist state. Part of the delay can be explained by political hesitation in London; it took a while for Salisbury's government to agree to the occupation of the entire Sudan. But much of the time was taken up with Kitchener's meticulous preparations.

The key to the entire advance was the Sudan Military Railway, rightly dubbed by G. W. Steevens "the deadliest weapon that Britain ever used against Mahdism." The previous British expedition into the Sudan, under General Garnet Wolseley, had been forced to pull boats down the Nile, with its treacherous cataracts. Kitchener decided to bypass a long, winding section of the river by building a railroad straight across the Nubian Desert into the heart of enemy-held territory. The line began on January 1, 1897, in the Nile River town of Wadi Halfa, just south of the Sudanese border, and eventually ended 385 miles away at Atbara. When everything was complete on July 14, 1898, troops could reach Atbara from Cairo in eleven days, a trip that had previously taken four months.

This was a prodigious feat of engineering that had widely been considered impossible until Kitchener showed that it could be done. The manual labor was performed by Egyptian troops and convicts who were driven mercilessly by Kitchener. "Once he saw a large engine standing idle," his biographer, John Pollock, relates. "The civilian works manager explained that the boiler was cracked and could be dangerous. The Sirdar said sadly that dangerous or not it might haul a heavy load to the railhead. 'After all, we aren't particular to a man or two!' " (Pollock treats this as a joke, but with the Sirdar it was hard to be sure.)

Atbara, the terminus of the line, was nearly two hundred miles from Omdurman. The rest of the journey would have to be completed by foot, camel, and steamboat. Kitchener had specially designed gunboats constructed in London and sent in sections to Egypt. They were transported by rail deep into the Sudan and reassembled on the Nile. By the time he was ready for his final showdown with the Dervishes, Kitchener had at his disposal ten armored gunboats bristling with a total of thirty-six artillery pieces and twenty-four Maxim guns—more firepower than any army had ever deployed in Africa.

The Dervishes did not allow this invasion to go uncontested. Kitchener had to fight numerous skirmishes as he advanced deeper into the Sudan. The biggest of these occurred at Atbara on April 8, 1898. Here the Sirdar faced fifteen thousand Dervishes, most of them frizzy-haired Hadendoa tribesmen. During battles in the 1880s, the British had acquired great respect for these courageous foes whom they dubbed "fuzzy-wuzzy." Kipling had written a tribute to fuzzy-wuzzy: "You're a pore benighted 'eathen but a first-class fightin' man." But not even first-class fighting men could stand for long against Maxim guns, artillery, and Lee-Metfords. Kitchener routed the Dervishes in half an hour. The Sudanese suffered more than 3,000 casualties; the Anglo-Egyptian army only 560.

The road to Omdurman now lay open.

"WE FIGHT IN THE MORNING"

When he had started his advance, the bulk of Kitchener's force was made up of Egyptians and Sudanese, the latter being black Christians and animists from the south who had little affection for the Muslim Arabs who ran the Mahdist state and enslaved their brethren. They had been commanded by British officers, but there had been almost no British troops among Kitchener's forces initially. Before the battle of Atbara, he had asked for and gotten a brigade of British soldiers from London. After the battle, as he prepared his final advance on Omdurman, the cautious Sirdar asked for and got another British brigade. Some of the last troops to arrive, including the 21st Lancers, reached Atbara only on August 15. By the time Kitchener started his final advance on Omdurman on August 24, 1898, he had 25,000 troops altogether, 8,200 of them British—an unusually large force by the standards of colonial campaigns. The British were outnumbered only two to one, a far cry from the ten-to-one odds faced by Wellesley at Assaye.

The Khalifa had an estimated 50,000 men in his army, armed with 15,000 rifles and more than fifty cannons. But their weapons were mostly outdated and poorly maintained; many were encrusted with dirt and dust. Unlike Emperor Menelik of neighboring Ethiopia, the Khalifa had not been updating his armory by dealing with European arms salesmen, even though the French and probably others would have been happy to supply him with weapons to fight their British rivals. He had also rebuffed offers of an alliance from the Christian Menelik. The Khalifa remembered the Mahdists' success against the Egyptian army in the early 1880s and thought he could stage a sequel without any outside aid. He did not realize that the invading force this time was considerably stronger than the one he had defeated more than a decade before.

The smartest strategy for the Mahdists would have been to retreat into the desert, drawing the invaders far from their trains and steamboats. Then they could have raided the overstretched British supply lines. They might not have won, but at least they could have delayed defeat for a number of years. Failing that, they could have somewhat minimized the British firepower advantage by confronting them in the winding alleys of Omdurman. But the Khalifa had no intention of letting the infidels enter his capital. So he adopted the worst possible strategy: a frontal assault on the entrenched ranks of his foes.

He even chose the worst possible time for the attack. The British had first appeared within sight of Omdurman on Thursday, September 1. Kitchener signaled their arrival by having his gunboats blast the Mahdi's tomb, a "yellow brown pointed dome" that was the largest structure in Omdurman, with shells packed with a new explosive called Lyddite that was far more destructive than traditional black powder. The Khalifa massed his army to meet this challenge, but made no move to attack. The British feared that he would wait to pounce until nightfall, when the Dervishes had the best chance of getting close enough to overwhelm them in hand-to-hand combat. To dissuade the Mahdists from this course of action, Kitchener's intelligence chief, Reginald Wingate, cleverly circulated a rumor that the *British* were planning a night attack. As an added precaution, the Sirdar ordered his gunboats to keep their searchlights shining all night.

Some of the Khalifa's commanders were eager for a night assault, but he foolishly rejected their advice, fearing that he would not be able to control his men in the dark. "The best course is what God chooses," he declared. "We fight in the morning after prayers." Thus the dictates of Allah jibed neatly with the preferences of the British, who also wanted a fight in the light.

The Khalifa's opening gambit—sending fourteen thousand men screaming straight at the British lines—seemed suicidal, as it indeed it was for many fighters who eagerly embraced martyrdom, but there was a certain logic to his strategy. Even if the initial assault was repulsed, the Khalifa hoped that it would draw Kitchener out from behind his zeriba (breastwork). Once in the open, the British force could be ambushed by the two larger Dervish divisions, the Green Flag and the Black Flag, which were hidden behind the Kereri hills.

Kitchener seemed to be playing right into the Khalifa's hands when, after the initial assault had been annihilated, he ordered his infantry to leave the zeriba and march at once toward Omdurman. Kitchener was anxious to cut off the remaining Dervishes from their capital, lest he be forced into messy urban combat, which he wanted to avoid at all costs. To make sure that there were no large bodies of the enemy left in his path, the Sirdar sent the 21st

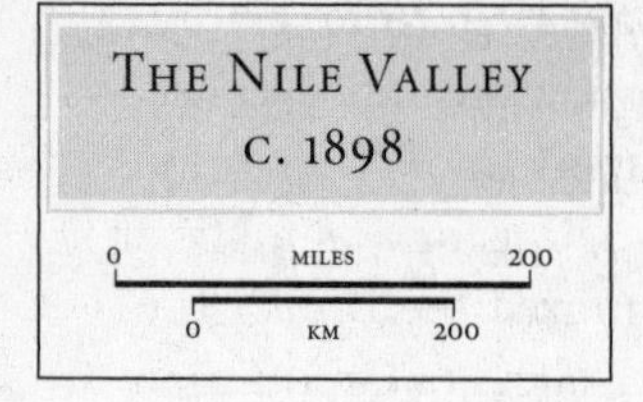
THE NILE VALLEY
c. 1898
0 MILES 200
0 KM 200

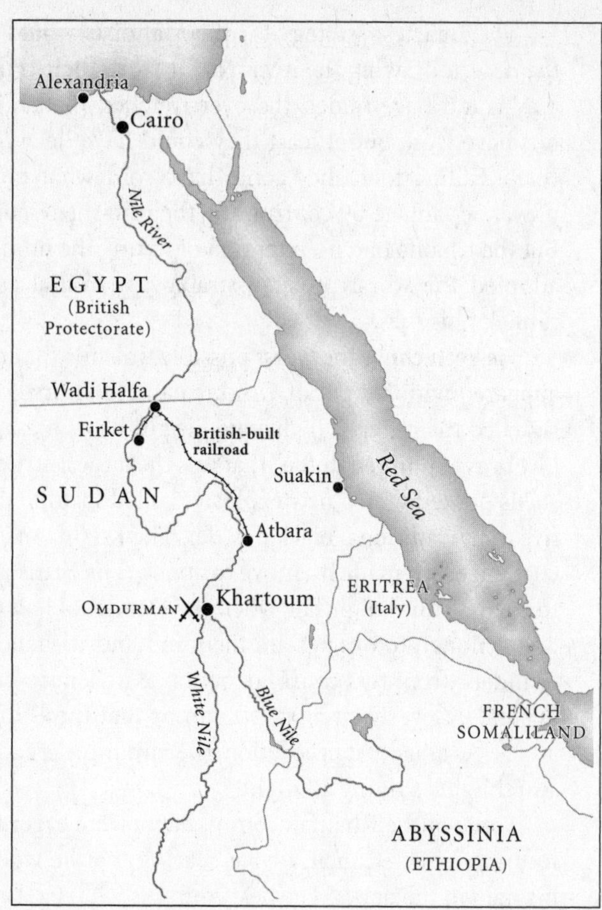
Alexandria
Cairo
Nile River
EGYPT
(British
Protectorate)
Wadi Halfa
Firket
British-built
railroad
Suakin
Red Sea
SUDAN
Atbara
OMDURMAN
Khartoum
ERITREA
(Italy)
White Nile
Blue Nile
FRENCH
SOMALILAND
ABYSSINIA
(ETHIOPIA)

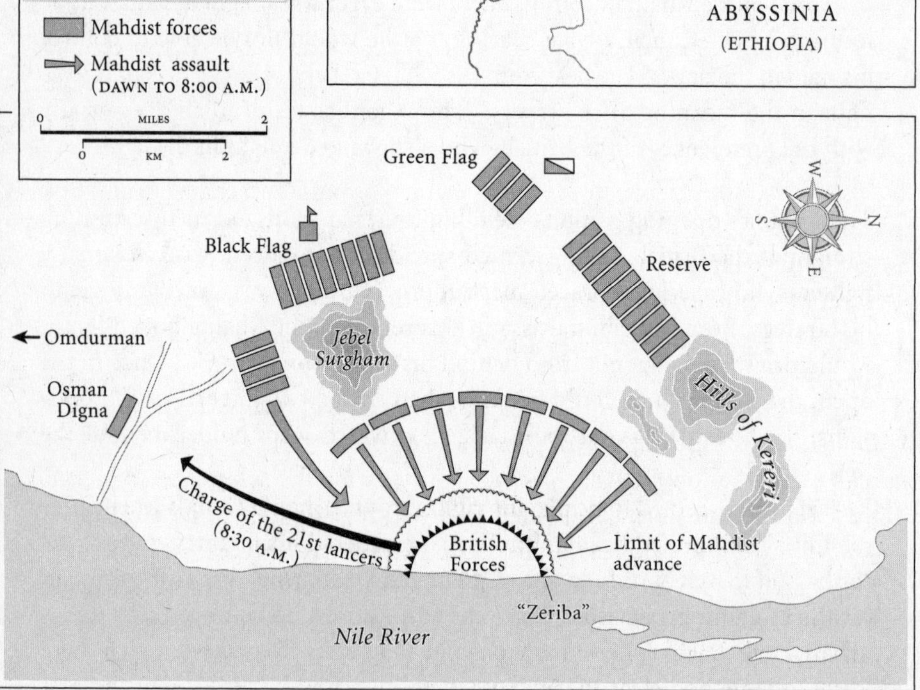
THE BATTLE OF
OMDURMAN, 1898
Mahdist forces
Mahdist assault
(DAWN TO 8:00 A.M.)
0 MILES 2
0 KM 2
Green Flag
Black Flag
Reserve
W
N
S
E
Omdurman
Jebel
Surgham
Osman
Digna
Hills of Kereri
Charge of the 21st lancers
(8:30 A.M.)
British
Forces
Limit of Mahdist
advance
"Zeriba"
Nile River

Lancers on a scouting mission. Among the four hundred cavalrymen who rode out of camp that morning was Lieutenant Winston Churchill, a twenty-three-year-old part-time war correspondent who had just joined the expedition despite the best efforts of the Sirdar (who had no time for the press or for glory-seeking young aristocrats) to keep him away. Partly because of his participation, the subsequent action has become the most famous part of the Battle of Omdurman.

The 21st Lancers had the ignominy of being the only regiment in the British army without any battle honors; wags joked that their motto was "thou shall not kill." Their commander, Colonel R. H. Martin, was determined to erase those slights. So when he saw several hundred Dervishes standing in the open, he did not hesitate. He ordered his trumpeter to sound the charge, and sent his men galloping straight into a trap set by the Khalifa's shrewdest commander, Osman Digna. Hidden behind the small number of Dervishes in the open was a khor, or rocky depression, in which were crouching some two thousand warriors. By the time the Lancers realized what they were charging into—"a dense white mass of men nearly as long as our front and about twelve deep," in Churchill's words—it was too late to stop. They increased their pace and came crashing into the khor. "The collision was prodigious," Churchill wrote. "Nearly thirty Lancers, men and horses, and at least two hundred Arabs were overthrown. The shock was stunning to both sides, and for perhaps ten wonderful seconds no man heeded his enemy." Then a desperate fight broke out in close quarters. "The hand-to-hand fighting on the further side of the khor lasted for perhaps one minute," Churchill continued. "Then the horses got into their stride again, the pace increased, and the Lancers drew out from among their antagonists."

Many of the Lancers were eager to turn around and charge again, but cooler heads prevailed. Instead of attempting another cavalry charge, they dismounted and opened fire with their Lee-Metford carbines (a shorter version of the infantry rifle)—precisely what they should have done in the first place. This withering rifle fire forced the Dervishes to retreat, at no cost to the horsemen.

The charge of the 21st Lancers was glorious and romantic. It was also a foolish anachronism. Martin's men won three Victoria Crosses, Britain's highest decoration, but they paid a high price for their desire to recapture the cavalry's lost glory. This brief clash resulted in seventy men killed or wounded (and 119 horses), nearly half of the total British losses that day. Many of the casualties were disfigured quite hideously. Of one sergeant, Churchill wrote, "His face was cut to pieces, and as he called on his men to rally, the whole of his nose, cheeks and lips flapped amid red bubbles." The regiment as a whole was debilitated for the rest of the battle.

Kitchener did not wait for the 21st Lancers to complete their work before ordering his infantry to start marching for Omdurman at 9 A.M. He did not seem to realize that thirty-five thousand Dervishes—two-thirds of the Khalifa's army—still lurked on the battlefield, not having taken part in the earlier action. The Sirdar was leaving his right flank wide open to attack by a superior enemy force. If the Khalifa had been a bit more adept in marshaling his forces or if his firepower disadvantage had been a little less, he might have made the British pay dearly for their carelessness.

As it was, the British received a nasty shock. Colonel Hector Macdonald's 1st Brigade, made up mostly of Sudanese soldiers, was on the extreme right of the British army as it wheeled left toward Omdurman. They became separated from the other five brigades and ran straight into one of the Khalifa's reserve divisions, the Black Flag, twelve thousand strong. Macdonald coolly deployed his men into line and opened a hot fire on the charging Dervishes. Kitchener, meanwhile, shifted two other brigades around to enfilade the Black Flag's flank. But Macdonald's brigade was almost a mile from the others, and it could easily have been overwhelmed and wiped out, were it not for the heroic work of "Fighting Mac" and his African soldiers.

Macdonald was one of the great tragic figures in the Victorian army. He was one of the few enlisted men who rose to become a general officer. Five years after the Sudan campaign, while serving as military governor of Ceylon, he was accused of having sexual relations with schoolboys and committed suicide rather than face a court-martial. It was a severe blow to the army, for as he proved on this day near Omdurman, as he had on so many previous occasions, the crofter's son from Scotland, whatever his sexual appetites, was "the bravest of the brave."

No sooner had Macdonald finished repulsing a Dervish attack upon his front than he became aware of another large Mahdist force advancing on his rear. If the Green Flag, fourteen thousand strong, had appeared a few minutes earlier, the 1st Brigade would have been caught in a pincer from which it might not have escaped. But the Khalifa, who relied on messengers on horseback, was unable to coordinate a simultaneous two-pronged attack. By 10:10 A.M., when the Green Flag began to fire on 1st Brigade, the attack from the Black Flag had already petered out. Macdonald was able to turn most of his men around in time to deal with this fresh threat. This would have been a difficult maneuver to perform even on the parade ground; that his Sudanese soldiers were able to carry it out while under fire is a testament to their abilities—and Macdonald's leadership. The Dervishes again displayed their trademark relentlessness, charging time after time, but each human wave attack was scythed down by Macdonald's three thousand Lee-Metfords, eighteen field guns, and eight Maxim guns.

The attack on Macdonald's isolated force had been the Mahdists' best bet to inflict substantial casualties on the British forces, but they had failed. Kitchener's reaction was a masterpiece of understated self-satisfaction: "I think we've given them a good dusting, gentlemen!"

The remains of the Mahdist army now faded into the desert. The Khalifa and his bodyguard retreated into Omdurman but made no attempt to defend the city. They were more concerned with escaping. Kitchener marched into the enemy capital virtually unopposed.

In a single morning, Mahdism had been defeated, the Sudan reconquered, Gordon avenged.

WHO'S THE SAVAGE NOW?

It is almost superfluous to examine why the British won the Battle of Omdurman. How could they have lost, given the disparity in firepower between the two sides? At best the Mahdists could have hoped to stave off defeat for a little while or to increase the costs to the British of occupying their country. Outright victory seemed unthinkable, given their shortfall of modern weaponry. When combined with the Khalifa's misguided strategy and the reckless courage displayed by his followers, this technological gap helped produce a catastrophe for the Mahdist cause. At least 10,800 Dervishes were killed and another 16,000 wounded, and many of the latter wound up dying. Five thousand more were taken prisoner. In all, the Mahdists suffered losses of 64 percent or more. The Anglo-Egyptian force lost 48 killed and 382 wounded, or 1.7 percent of their total strength. As a French general would later observe after the Battle of Verdun (1916), "Three men and a machine gun can stop a battalion of heroes."

The British rightly attributed their victory to their advanced technology—Winston Churchill called Omdurman "the most signal triumph ever gained by the arms of science over barbarians"—but, as they were shortly to discover, Britain had no monopoly over the weapons in question. The following year, 1899, British soldiers, including many veterans of the Sudan campaign, were sent to fight Boers who were as adept at using magazine rifles as they were. Even more frustrating was that this enemy would not stand still and let themselves be clobbered by superior forces. The Afrikaners had a maddening habit of eluding British pursuers and striking where least expected. Kitchener, who became the third commander of the British army in the Boer War, was driven to complain, "The Boers are not like the Soudanese who stood up for a fair fight, they are always running away on their little ponies." The British won anyway, but it took far longer and required far more troops than they had originally expected.

The Boer War dented the British aura of infallibility. The Great War would dispel it altogether. This time it would be the British and the other "civilized" nations of Europe who would suffer the unspeakable agonies endured by the "savages" at Omdurman and other imperial killing grounds. It turned out that the best generals in Europe had no better idea than the Khalifa of how to defeat an army equipped with machine guns, magazine rifles, and quick-firing artillery. One of the leading British theorists on the subject, Colonel F. N. Maude, writing in 1905, sounded positively Mahdist as he proclaimed that "the true strength of an Army lies essentially in the power of each, or any of its constituent fractions to stand up to punishment, even to the verge of annihilation if necessary." Annihilation is just what the armies of Europe would suffer as they sent wave after wave of soldiers charging across "no man's land."

The frontal assault tactics that had worked in the age of flintlocks turned out to be a prescription for machine-assisted suicide in the age of Maxim guns. A Napoleonic battalion in 1815 armed with 1,000 flintlock muskets could fire 2,000 rounds a minute to a range of one hundred yards. A century later, a battalion armed with 1,000 magazine rifles and four machine guns could fire 21,000 rounds a minute to a range of 1,000 yards. This meant that, in a bayonet assault, a comparable unit could expect to receive two shots per soldier in 1815 and two hundred shots per soldier in 1915.

The radical implications of that hundredfold increase in firepower for infantry tactics should have been obvious to anyone who watched the Battle of Omdurman or countless other clashes that occurred between the time of the American Civil War (1861–65) and the Balkan Wars (1912–13). But it took a long time for the appropriate lessons to sink in among European armies. The British casualties on July 1, 1916, the first day of the Battle of the Somme, amounted to 57,470 killed, wounded, and missing—enough to make the Dervish losses at Omdurman seem modest by comparison.

All this would have offered scant comfort to the Khalifa even if he had lived long enough to hear about the Great War. He did not. In November 1899 he was hunted down and killed by a mobile column led by Kitchener's intelligence chief, the indefatigable Reginald Wingate. The remnants of the Mahdist movement were suppressed. An Anglo-Egyptian administration was set up that would rule Sudan until 1956. The first governor-general was Kitchener; the second, Wingate.

In a way, the Mahdi did get the last laugh. For in the 1980s, the spiritual descendants of the Mahdists, known as the National Islamic Front, began to exercise a substantial degree of power in the Sudan. For a time the Mahdi's great-great-grandson even served as prime minister. In the 1990s, under the influence of this movement, Sudan gave shelter to Osama bin Laden, another jihadist who would do far more damage to the infidels than the Mahdi ever could have imagined. But that is getting ahead of our story.

In the age of the Mahdi and his immediate successors—the late nineteenth and early twentieth centuries—most of the peoples of Asia and Africa had scant chance of inflicting anything more than temporary setbacks on the steadily encroaching armies of the West. Thanks to the Industrial Revolution, the disparity in military might between the West and the Rest was greater than it has ever been before or since. But the gap was closing fast, as Russia was to discover to its dismay six years after the Battle of Omdurman.

CHAPTER 6

STEEL AND STEAM:

Tsushima, May 27–28, 1905

The buglers sounded reveille at dawn on Saturday, May 27, 1905. It did not take the sailors long to roll out. Knowing that a battle was imminent, many had not bothered to sling their hammocks the night before but had simply slept in a convenient corner. A cloud of expectation hung over the Russian fleet as thick as the gray morning mist. Waves of dread buffeted the ships as strongly as the choppy waters of the Tsushima Strait between Korea and Japan. They had traveled eighteen thousand miles and eight months to reach this point. Upon their actions in the next few hours would rest the fate of two empires.

Russia had been at war with Japan since the previous year to determine which power would dominate Korea and Manchuria. Russia's Pacific Squadron had not fared well against the Japanese navy. The bulk of its ships had been bottled up by a Japanese blockade at their anchorage in Port Arthur, which Russia leased from China. They had been helpless to prevent the Japanese from landing troops on the Korean peninsula and then on Manchuria's Liaotung Peninsula on which Port Arthur (now known as Lushan) sat. When the Russian ships ventured out into the Yellow Sea on August 10, 1904, Admiral Heihachiro Togo had sent them reeling back with heavy losses. When this news reached Czar Nicholas II's glittering court in St. Petersburg, it was decided to send another fleet to the Far East to give the "yellow monkeys" a sound thrashing for daring to intrude into the lair of the Russian bear.

The Second Pacific Squadron, as Russia's Baltic Fleet had been renamed, set off from the Gulf of Finland on October 15, 1904. It was attempting a feat that many observers doubted it could pull off: to sail halfway around the world without having guaranteed access to bases and coal supplies along the way. The voyage was not an easy one. Early on, there had been a mishap in the North Sea, when, one dark night, the Russians opened fire on what they thought were Japanese torpedo boats. In the cold light of dawn, they turned out to be British fishing trawlers. One had been sunk, others damaged. Britain, an ally of Japan, almost declared war over the Dogger Bank incident. Russia agreed to pay an indemnity and the fleet steamed on. And on and on: down the Atlantic coast of Africa, around the Cape of Good Hope, stopping at Madagascar, then straight across the Indian Ocean, through the Strait of Malacca and a brief rest at Indochina's Cam Ranh Bay before heading north through the South China Sea, intending to reach the last remaining Russian port in the Far East, Vladivostok.

France, an ally of Russia, allowed the fleet to make temporary use of its anchorages in Senegal, Madagascar, and Indochina, but for the most part the Russians were on their own. Coal had to be carried in transport ships, some of them leased from the Hamburg-America steamship line, and refueling had to be done at sea. For the twelve thousand officers and crew, this long voyage was marked by suffocating coal dust and enervating tropical diseases, oppressive heat and driving rain, nasty accidents and annoying breakdowns, putrid meat and free-flowing vodka, fraying tempers and volatile relations between officers and sailors.

Perhaps the worst blow of all was the news, delivered after they docked at Madagascar in late December 1904, that Port Arthur had fallen to a Japanese infantry assault and that the fleet that had been based there had ceased to exist, the ships either destroyed by Japanese artillery or scuttled to prevent them from falling into Japanese hands. "It was like a cemetery," seaman A. Novikoff-Priboy wrote of the mood in the fleet when the news arrived. "No laughter, no smiles."

Now there was no hope of meeting Togo with an overwhelming force. On paper, the forty-two ships of the Second Pacific Squadron were, even by themselves, stronger than the entire Japanese navy, but the Russians knew how poorly maintained and how badly run their ships were, and they had heard all too many rumors of the enemy's dauntless courage and accurate marksmanship. "Ah!" grumbled one ship's surgeon. "One doesn't have to be a pessimist to see that only shame and dishonour are in store for us." Discipline among the bluejackets plummeted as fast as their prospects of success. More than one mutiny broke out.

The squadron's commander, Rear Admiral Zinovy Petrovich Rozhdestvensky, was under no more illusion about his prospects of success than the

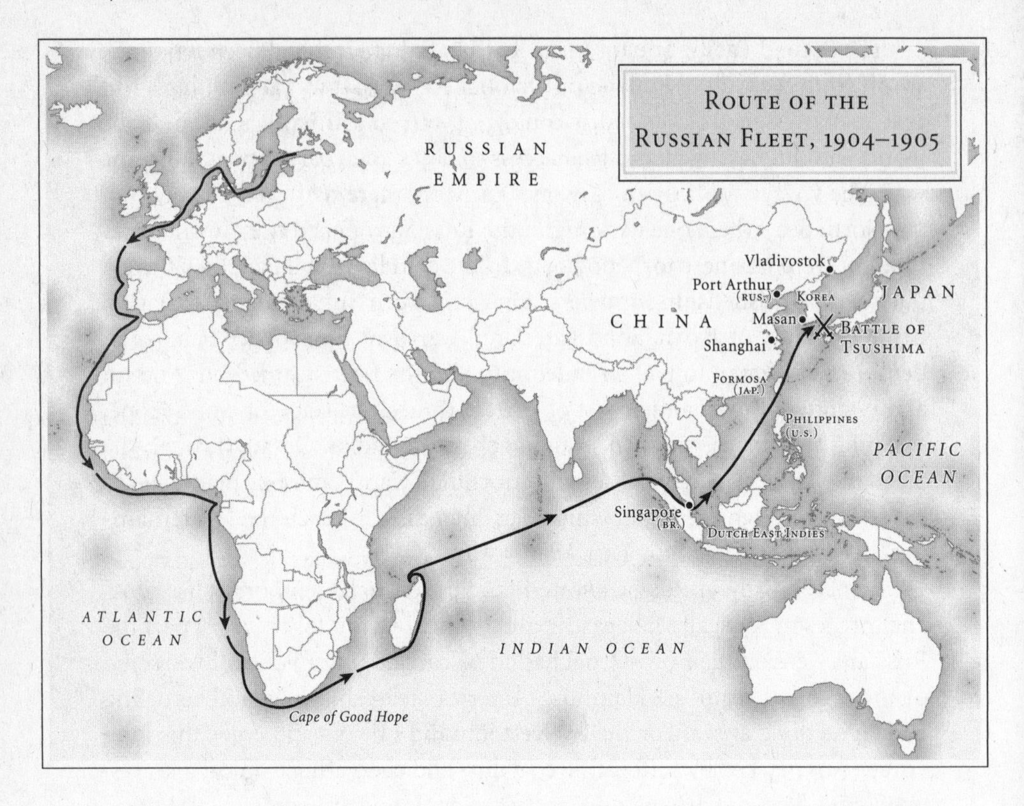

Duke of Medina Sidonia had been on a similar, if much shorter, voyage three centuries earlier. But, like his doomed predecessor, the Russian admiral felt he had no choice but to do his best for his sovereign or die trying. As Rozhdestvensky approached a certain clash with the enemy fleet in the strait between Korea and Japan, his morale and his men's actually lifted. Outwardly, Novikoff-Priboy wrote, the men seemed be "calm and cheerful," often "joking and laughing," though he thought that "despair gnawed at their hearts."

For many days preceding their final clash, neither fleet had an accurate idea of the other's position. That changed at 2:45 A.M. on May 27, 1905, when, through heavy fog, a Japanese light cruiser spotted a Russian hospital ship steaming through the Korea Straits east of the Tsushima islands. A little while later more Russian ships appeared through the mist, their hulls painted black, their funnels a brilliant yellow, their masts flying the blue-and-white St. Andrew's flag. At 4:45 A.M., the Japanese cruiser sent a message with the enemy's coordinates using a new invention, the wireless telegraph (a radio capable of transmitting only Morse code). Admiral Togo received the news at his anchorage on the southern coast of Korea at 5:05 A.M., and within an hour his fleet had steam up and was moving out.

Even as the Japanese fleet crept closer, Rozhdestvensky insisted that the normal routines of the day be performed. May 27 was the anniversary of the czar's coronation, and the monarch had to be properly honored. Crews gathered on their ships' decks, where bearded Orthodox priests with holy-water sprinklers led them in prayers followed by a singing of "God save the czar." Then everyone sat down to lunch, a meal washed down for the men with rum or vodka and for the officers with champagne. (Alcohol consumption before battle, though frowned upon among most modern militaries, was commonplace in centuries past; it was often considered vital in screwing up the men's courage.)

In the wardroom of the battleship *Suvorov*, the officers were just toasting "the health of His Majesty the Emperor and Her Majesty the Empress" when an ominous noise cut through the din of inebriated celebration.

Action Stations! All Hands! Action Stations!

Japanese cruisers were closing in fast. The main Japanese fleet could not be far behind. The climactic clash of the Russo-Japanese War was about to begin.

SAIL AWAY

It was a battle that would be fought with technology, if not tactics, that had changed beyond all recognition since the last great fleet action had occurred off Cape Trafalgar a hundred years before. Vice Admiral Horatio Nelson, like Hawkins and Drake in 1588, had commanded wooden sailing ships armed with muzzle-loading cannons firing solid iron balls. By 1905, not only had sails had given way to steam engines, wooden hulls to iron and steel, and muzzle-loaders to breech-loading artillery firing high-explosive shells, but two entirely new classes of weapons—mines and torpedoes—had become commonplace. Navies had also developed new kinds of vessels: battleships, cruisers, destroyers, torpedo boats, submarines, mine layers, and sweepers. This was the biggest change in naval warfare in four hundred years, and it took many decades for its full implications to be realized.

The first steam warship, the *Demologos*, had been started in 1814 by the American steamboat pioneer Robert Fulton, but the War of 1812 ended before it could be commissioned. In the following decades, paddle wheelers came into common use among navies bent on exploiting or defending rivers. But not until the late 1830s would steam engines become reliable and economical enough to power oceangoing ships. The advent of screw propellers, located inconspicuously beneath a ship's hull, made steam more attractive for navies because, unlike cumbersome paddle wheels, they did not take up a lot of deck space—space that was needed for guns.

In the 1840s and 1850s most of the world's navies adopted steam power, but usually only as a supplement for sails, and they remained committed to wooden hulls. That dedication became increasingly untenable as solid cannonballs were replaced by exploding shells—a design perfected by the French army officer Henri-Joseph Paixhans in 1824. At the Battle of Sinop in 1853 shortly after the outbreak of the Crimean War, a Russian fleet armed with Paixhans shell guns demolished a wooden Turkish squadron while losing no ships of its own. That spelled the end of unarmored ships.

In 1859 France launched the first oceangoing ironclad warship, *La Gloire.* It was still made of wood and had rigging for sails, but it also had a steam engine and a hull plated with four and a half inches of armor that rendered impotent any existing naval gun. As the world's No. 2 naval power throughout much of the nineteenth century, France was often in the forefront of technological innovation. Paris hoped to gain a qualitative edge over the dominant Royal Navy, but its hopes were dashed time after time. The British Admiralty might have been hidebound, sentimental, and anxious not to lose its large investment in old-fashioned ships (all sentiments perfectly captured in a famous 1828 memo in which the leading admirals wrote, "Their Lordships feel it is their bounden duty to discourage to the utmost of their ability the employment of steam vessels, as they consider the introduction of steam is calculated to strike a fatal blow to the naval supremacy of the Empire"), but Britain was such an industrial powerhouse that it was always able to match any worthwhile innovation and build it better and faster than the French themselves could. Thus, one year after *La Gloire* was launched, Britain had its answer: HMS *Warrior*, a bigger, faster ship powered by both sail and steam but made of iron, not wood. Since iron is much stronger (if less buoyant) than wood, its use as a building material allowed ships to become much bigger.

Although Britain and France were the world's naval leaders, the first clash between ironclads did not involve their ships. This celebrated engagement pitted two American ships against one another—the Union ironclad *Monitor* versus the Confederate ironclad *Virginia* (née the *Merrimack*). The *Monitor*, designed by the Swedish-American engineer John Ericsson, was the more revolutionary vessel. It was built entirely of iron and had two main guns housed in a central revolving turret. The *Virginia* merely had iron plates over a wooden hull, and its twelve guns were deployed in fixed positions around its armored casement. Both rode so low in the water that they looked like swimming turtles. Their four-hour clash off Hampton Roads, Virginia, on March 9, 1862, was anticlimactic. Despite pounding each other from point-blank range, neither ship suffered serious damage; their projectiles simply bounced off the armor. Before the year was out, both vessels would be lost: The *Monitor* sunk in high seas, the *Virginia* scuttled as federal troops closed in on its anchorage.

Ironclads were only one of many new weapons tested during the Civil War. Another Robert Fulton invention, the floating mine, was also unveiled. Confusingly, mines were at first called "torpedoes." Hence Admiral David Glasgow Farragut's celebrated cry as he led a Union flotilla into Mobile Bay, Alabama, in 1864: "Damn the torpedoes, full speed ahead."

The forerunners of modern torpedoes were spar torpedoes: explosive charges attached to a long pole protruding beneath the water from a small vessel. Spar torpedoes sank two naval ships during the Civil War, one from each side, but, not surprisingly, the attacking vessels were lost too. One of these was the Confederate vessel *H. L. Hunley*, powered by a hand-cranked propeller, which in 1864 became the first submarine to sink an enemy ship.

Self-propelled torpedoes were developed by the British expatriate engineer Robert Whitehead at a factory in the Habsburg port of Fiume (now part of Croatia) in 1866. By the mid-1870s, Whitehead torpedoes had been bought by all the major naval powers of Europe. The initial model had a top speed of only seven knots, a warhead of eighteen pounds, and a range of 600 yards. Several decades of steady improvement created a much more formidable weapon. By 1905, Whitehead torpedoes could travel 2,190 yards (more than a mile) at a maximum speed of twenty-nine knots while carrying 220 pounds of explosives—enough to blast a gaping hole in any hull. By 1914 their range had grown to 10,000 yards (five and a half miles).

To launch torpedoes, various countries began to build small, fast surface boats as well as crude submarines. This, in turn, led to the development of torpedo-boat destroyers designed to protect large capital ships from these pesky marauders. Destroyers, as they later became known, were equipped with torpedo tubes of their own, light deck guns, and electric searchlights. They came to be used for attacking as well as defending larger vessels.

Capital ships underwent rapid and bewildering changes between the 1860s and the 1890s. New inventions were continually being grafted onto old designs. The result was a series of comic-looking mutant vessels that combined funnels and masts, steam plants and canvas sails. Such unsightly vessels—the military equivalent of a duck-billed platypus—were typical of periods of transition. Just as shipbuilders in the sixteenth century had been loath to give up oars, leading to a temporary fad in galleases that combined oars and sails, so shipbuilders in the nineteenth century were reluctant to discard sails. Britain, the No. 1 naval power, did not stop building warships with sails until the 1880s. This was not entirely the product of mindless conservatism, because it did not make sense to give up the old propulsion system until the new one had been perfected.

The performance of steam plants continually improved as single-cylinder reciprocating engines gave way in the 1850s to high-pressure compound engines and, by the early 1900s, to turbines. Reciprocating steam engines had cylinders with pistons inside pumping up and down; turbines had discs mounted on a single shaft spinning in one direction, which wasted less energy. By the eve of World War I the most modern warships would run on "noiseless" turbines fueled by oil, which offered greater thermal efficiency than coal. Thanks to these souped-up engines, the top speed of even the biggest ships would approach thirty knots (thirty-four miles per hour).

Ship design was in constant flux because of a nonstop contest between defensive and offensive technology. Better weapons led to the development of better armor and vice versa. The introduction of steel—a metal made from an alloy of iron, carbon, and a few other agents that was both more malleable and harder than cast iron—improved both ordnance and armor.

The Bessemer process (which involved blowing air through pig iron), invented in 1856 by the Englishman Sir Henry Bessemer, reduced the cost of making steel so that it became an economical substitute for iron in mass production. In the late 1880s, France's Schneider Company added three to four percent nickel to steel to reduce its tendency to crack. The American inventor Hayward Harvey made the nickel steel tougher still by adding more carbon. Engineers at Germany's Krupp improved on his design by adding a tiny dollop of chromium and manganese, carburizing the plates with gas, and hardening them with a high-pressure water spray. Krupp's cemented armor, which became standard by the fin de siècle, was almost three times sturdier than the wrought iron plates that had been used on early ironclads like the *Warrior* and *Monitor*.

By allowing ship designers to get more protection from thinner armor, Krupp freed up room for heavier artillery. By the 1890s, Armstrong, Krupp, Vickers, Canet, and other large companies were manufacturing breech-loading, rifled naval guns made of nickel steel. These guns fired high-explosive steel shells, and they did not emit much smoke because they used smokeless powders such as cordite. Capital ships began to carry titanic main guns mounted in armored, revolving turrets, turned by hydraulic machinery and served by electric hoists that brought up shells from the ship's depths. Range steadily improved until, by 1914, some ships in the British navy were routinely taking target practice at fourteen thousand yards (eight miles). Poor marksmanship, however, made it difficult to get many hits at such long range—a problem that was to some extent ameliorated by telescopic range finders and incessant gunnery drills but that was never completely solved.

All these trends came together in the 1880s to give birth to the steel battleship, whose name derived from the old line-of-battle sailing ship. Until the launch of the *Dreadnought* in 1906, the standard for battleships was set by the *Majestic* class built by Britain starting in 1895. The best battleships in both the Russian and Japanese navies in 1905 were essentially of this type. To gain a perspective on how much things had changed in the past century, compare the *Majestic* with Nelson's flagship *Victory*, one of the biggest warships of its day.

The *Majestic* displaced 14,900 tons; *Victory*, 2,500 tons. The *Majestic* had 9 inches of Harvey steel armor; *Victory* had no armor at all. The *Majestic* had a top speed of 16.5 knots; *Victory*, 9 knots. *Majestic's* main armament consisted of 4 twelve-inch guns each weighing 46 tons and capable of hurling an 850-pound explosive shell seven miles. In addition, *Majestic* had 12 six-inch, quick-firing guns designed for protection against torpedo boats and various lesser guns, plus 5 torpedo tubes. *Victory* had many more guns (104), but the largest of them weighed only three and a half tons and fired a thirty-two-pound iron ball at a maximum range of one and a half miles.

To sum up: *Majestic* was almost six times bigger and two times faster than *Victory*, while its main guns could discharge three and a half times as much metal almost six times as far with every broadside. Of course, this is a very crude measurement that understates *Majestic*'s considerable qualitative advantages, such as its ability to maintain speed even when there was no wind, to traverse its guns from port (left) to starboard (right), and to discharge shells that exploded on impact. But the fundamental point is clear and indisputable: There had been a revolution in warship design between 1805 and 1895.

While battleships were the most conspicuous and costly symbols of this revolution, they were nearly matched by a new class of ships known as cruisers, which were slightly smaller and less armored but even faster. They were intended for duties such as scouting and commerce raiding—the roles played in the days of sail by frigates—but eventually the most powerful of them took their place in the battle line too. The survivability of these "battle cruisers" was supposed to lie in their superior speed, which was meant to keep them out of range of heavy battleships.

A proliferation of new guns and armaments naturally led to confusion about the best way to employ them. As Britain's senior admiral wrote in 1885, "We are in a period of transition, even as regards guns and shipbuilding. No two naval officers will agree as to what in a few years will be the fighting ships of that time." Only actual clashes between ironclad fleets could provide guidance to puzzled officers, but between the end of the American Civil War in 1865 and the start of the Sino-Japanese War in 1894, there was only one major battle at sea. It occurred on June 20, 1866, off the island of Lissa in the Adriatic. Italy and Austria had been drawn into battle as an offshoot of the Austro-Prussian War, since Bismarck had inveigled Italy into an alliance against the Habsburg empire. The gunnery duel was a draw, but Austria won the battle after one of its ships rammed and sank the Italian flagship.

For several decades thereafter, all major warships were built, in a throwback to premodern galley warfare, with hardened rams on their prows. In one of the least perspicacious tactical pronouncements ever penned, the author of a Royal Navy textbook wrote in 1874 that "the ram is fast supplanting the gun in import." In reality, the increasing effectiveness of artillery and torpedoes was making it nearly impossible for one warship to get close enough to ram another.

A slightly more long-lived naval doctrine was associated with France's Jeune École. The adherents of this school were convinced that cheap torpedo boats and unarmored cruisers were the ultimate offensive weapons; they would make more expensive battleships obsolete. Under their prodding, France and some other nations stopped building capital ships for several years in the 1880s. Unfortunately for the Jeune École, heavy guns soon outranged torpedoes, which meant that large ships could destroy torpedo boats before they could discharge their ordnance. Improvements in torpedo and submarine design would lead to a revival of this school in the early years of the twentieth century. By hiding underwater, submarines had a much better chance than conventional torpedo boats of getting close enough to sink large ships. Submarines would prove especially effective against merchant shipping in both World War I and World War II. But in neither war was the submarine as decisive as the Jeune École might have hoped.

Long before 1914, the Jeune École had been consigned to oblivion by the triumph of a blue-water school of naval strategy associated with the American admiral Alfred Thayer Mahan. His 1890 masterpiece, *The Influence of Sea-Power Upon History, 1660–1783,* with its argument that fleets of capital ships were needed to win command of the seas, became gospel in all the major navies of the world. An important corollary of Mahan's theory was the need to develop a network of coaling bases to support a fleet in wartime. This helped spur the drive by Britain, France, Germany, the United States, and other nations to secure colonies around the turn of the twentieth century. But Mahan's biggest impact was simply in providing a rationale for the construction of battle fleets, which would help foster a naval race in the years before World War I, primarily between Germany and Britain.

Throughout most of the nineteenth century, Britain had adhered to a two-power standard (formally enshrined in law in 1889), meaning that its navy had to be stronger than the next two powers combined. That policy became difficult to maintain after Germany embarked on a major naval buildup in 1898. Britannia could no longer rule the waves by itself. It needed allies. In the Western Hemisphere, London reached a tacit understanding to cede the Royal Navy's policing duties to the U.S. Navy. In the

Far East, it concluded a formal alliance with Tokyo in 1902. That Britain would align itself with the United States, its progeny, was unexceptionable; that it would become to some extent dependent on Japan was an extraordinary testament to the growing power of a country once regarded with contempt in Europe.

Japan's half-century voyage toward the Great Power ranks can be traced through the career of one man: Heihachiro Togo.

THE RISE OF JAPAN—AND TOGO

Togo was born in 1848 in a small village near the port of Kagoshima on the southern Japanese island of Kyushu. He spent his first few years in a country that had barely changed in more than two hundred years. The only powered machinery consisted of a few waterwheels used for tasks like pounding rice and crushing ore. No formal army or navy existed. Military power was wielded by samurai, hereditary warriors who were the only ones allowed to carry swords or commit ritual suicide. The country was ruled by a shogun, or warlord, overseeing a network of feudal chieftains known as daimyo; Togo's family served the daimyo of Satsuma province. Since a previous shogun had totally sealed Japan in 1639, the only officially sanctioned point of contact with Europe had been a small island in the harbor of Nagasaki where a few Dutch traders were permitted to live. Japanese were forbidden on pain of death from leaving the country. Some European science seeped in (it was known as "Dutch learning"), but Japan remained for the most part resolutely traditional.

That all changed when Togo was five years old. On July 8, 1853, Commodore Matthew Calbraith Perry of the United States Navy sailed into Edo Bay (as Tokyo was then called) with four "black ships," two of them steamships that had never before been seen in the Land of the Rising Sun. Upon Perry's return the following year, the cowed shogunate was forced to sign a treaty ending Japan's long isolation and granting the U.S. trade rights. European countries rushed in to demand similar privileges. This naturally led to tensions between the Japanese and the "blue-eyed devils" newly arrived in their midst.

One day in 1862, some samurai hacked to death an English merchant who had failed to dismount from his horse while the regent of Satsuma rode by—a fatal breech of etiquette. After Satsuma refused to execute the culprits and pay an indemnity, a Royal Navy squadron arrived off Kagoshima the following year. Togo was then a fifteen-year-old boy who been educated in Confucian classics, fencing, calligraphy, and other subjects appropriate for a young samurai. Now, along with his father and two brothers, all samurai of Satsuma, he was called upon to man one of the ten forts defending the port. Togo and the other Japanese gunners fought valiantly, but their antiquated smoothbores were no match for the Royal Navy's rifled cannons. Six hours of bombardment leveled not only the forts but much of the town as well; its wood-and-paper houses burned easily. Frustrated, the samurai ran to the water's edge and brandished their swords, daring the British to fight them like men of honor. Naturally the British refused to oblige.

Such humiliations made clear the need to develop a modern military capable of fending off the big-nosed barbarians. The shogunate began the process of modernization but did not progress quickly enough for the more reformist daimyos of the southwestern provinces of Choshu and Satsuma. They were modernizers but not liberals—they wanted to learn foreign ways in order to make Japan better able to resist foreigners. By 1868 they had deposed the shogun. A new government was proclaimed in the name of Emperor Meiji, but real power during the Meiji Restoration was vested in a group of ambitious young samurai.

The first priority of these reformers was to increase Japan's military might. For guidance on the army, they turned first to France, and then, following France's defeat in 1871, to Germany. Britain became the model for Japan's navy. A British naval mission arrived in Japan in 1873. Two years before, the first twelve Japanese cadets had departed for study in England. Among them was Heihachiro Togo, who had initially enrolled in the Satsuma navy at age eighteen and then, after its dissolution, joined the imperial service in 1868.

Togo, who had never left home before, spent the next seven years living among the English. He attended naval schools and trained aboard a sailing ship. Living in the port of Plymouth, he became thoroughly steeped in Britain's naval traditions and came to worship Admiral Nelson. The taunts of his fellow cadets, who called him "Johnny Chinaman," only made Togo more determined to excel at his studies.

In 1877, while still in England, Togo received disturbing news from home. Some samurai of Satsuma had revolted against the new imperial government, which had abolished their traditional caste privileges, including the right to wear swords. The rebels were brutally crushed by the new national army. Among those killed was Togo's brother. Togo had no desire to go home and die in a hopeless cause. He remained in the emperor's service and came home in 1878 aboard one of three British-built warships purchased by the Japanese government.

The thirty-year-old Sublieutenant Togo returned to a land that was being transformed beyond all recognition from the days of his childhood. Japan was vaulting straight from feudalism to the Industrial Age. The initial lead in fostering economic growth was taken by the government, which financed Japan's first railroads, telegraphs, shipyards, and factories using technology imported from abroad. Most of these enterprises proved to be commercial failures, and in the 1870s many were sold off to large domestic corporations. But the government kept key arsenals and shipyards in its own hands.

Japan's leaders were keenly aware of the link between economic and military power. A popular Meiji slogan proclaimed: "Rich Country, Strong Army." The foundation of both lay in an educated populace. Primary schooling was instituted for all, and numerous institutes of higher learning were opened. All young men also became subject to conscription—a radical break with the past, when the samurai elite had enjoyed a monopoly on military power. The army became a "school of the nation," introducing many recruits not only to Western-style clothing and fighting but also to the ideological props of the new regime: Shinto, the official state religion of ancestor worship, and bushido, the samurai code of loyalty, self-sacrifice, and courage. A strong sense of nationalism and xenophobia pervaded the armed forces, which, under Japan's Prussian-style constitution, developed almost completely outside of civilian control.

NIPPON'S NEW NAVY

Important as the army was, for an island nation like Japan the navy was even more vital. It took impressive strides forward from the 1870s to the 1890s while Togo slowly progressed through its ranks. A naval academy was established in 1869, followed by a staff college, a naval engineering school, and other educational institutions. Ordinary sailors were a mixture of conscripts and volunteers. Officers were chosen based on competitive examinations and promoted (in theory at least) on merit. Rigorous technical training combined with the bushido spirit to create a formidable group of naval fighters.

Most of Japan's warships until the early twentieth century were built abroad, but the country was rapidly developing its own manufacturing capacity. In 1885 Togo supervised the construction of one of the first warships made in Japan, and when it was completed he took over its command as a newly minted thirty-nine-year-old captain. His future seemed limitless. Two months later, his career seemed in ruins. Togo was felled by a paralyzing bout of rheumatoid arthritis that kept him more or less bedridden for several years. After a slow recovery, he finally returned to duty in 1890 at Kure naval base, just in time for a visit from two hulking German-built battleships of the Chinese navy. Disguised in plainclothes, Togo made a careful inspection of these ships—considerably bigger than any in Japan's fleet—and he was not impressed by the quality of the Chinese crews, who hung their laundry to dry on the main guns. "Nothing to be afraid of," Togo told a fellow officer, and he was soon proven right.

In 1894 Japan went to war with China in order to wrest Korea from its sphere of influence. Japan reversed the usual order of things by first attacking Chinese forces and then declaring war, a modus operandi it would employ regularly in the future. A small Japanese naval squadron, including the cruiser *Naniwa* commanded by Captain Togo, commenced hostilities on July 25, 1894, by attacking and sinking some Chinese ships near the Korean port of Chemulpo (now called Inchon). Following this initial success, the Japanese confronted the main Chinese fleet in the Yellow Sea near the mouth of the Yalu River on September 17, 1894. The Chinese armada was larger, and the Japanese still had no vessels as big as the two Chinese battleships, but their ships were newer and faster, and their officers and crews were much better trained. In a six-hour battle, the Japanese lost no vessels, while five Chinese ships were sunk and the rest were badly damaged. Japan had won the first major naval victory in its history, and Togo had emerged as a hero even before he led the subsequent successful invasion of the Chinese island of Formosa (Taiwan).

Under the Treaty of Shimonoseki, which ended the first Sino-Japanese War in 1895, China was forced to give up all claims to Korea, pay a large indemnity, and cede Formosa and the Liaotung Peninsula to Japan. But Russia, Germany, and France stepped in and forced Japan to disgorge the Liaotung Peninsula, including Port Arthur. Russia then swooped in and bullied China into giving *it* the Liaotung Peninsula under a twenty-five-year lease. Russia also flexed its muscles by occupying much of the rest of Manchuria and grabbing timber concessions in Korea. It became clear to Tokyo that if it wanted to secure its economic interests in Asia and become an imperial nation in its own right, it would have to push Russia out of the way.

Taking on the czar was a more formidable undertaking than defeating the decrepit Manchus who ruled China. In 1903 Russia's population was three times bigger than Japan's. Its iron and steel production was sixty times greater. It spent three times more than Japan did on defense, and it had the world's largest military, with 1.1 million soldiers and sailors—almost six times more men than Japan. Japan could never hope to close the quantitative gap, but it could build a higher-quality force.

In 1896 Japan launched an ambitious ten-year naval program. By 1904, six battleships had been bought from England. The newest of them, such as the battleship *Mikasa*, were actually slight improvements over the *Majestic* class. In addition, Japan acquired twenty-four cruisers, twenty destroyers, and fifty-eight torpedo boats.

Chosen to lead this formidable armada in 1903 was Vice Admiral Togo.

He was fifty-five years old, a short, slightly stooped, soft-spoken figure, with close-cropped black hair and a beard that was turning white. He was known for being cautious, stubborn, and, above all, laconic; Helmuth von Moltke would have seemed a chatterbox by comparison. Togo had enjoyed a solid career, but he was far from Japan's preeminent admiral. The navy minister, when asked by the emperor why he had been picked for this all-important post, reportedly replied, "Because Togo is lucky."

Not the least of Togo's luck was his choice of enemies. The Russian navy proved to be conveniently and consistently inept.

This might, on the surface, be surprising, since Russia had much more experience with war at sea than Japan did. Peter the Great (r. 1682–1725) had created the Russian fleet in the 1690s, and since then it had scored some impressive victories, principally against the Ottomans. Russia had adopted ironclads early on, and it was considered a world leader in torpedo and submarine warfare. In the decade before the Battle of Tsushima, Russia, like Japan, had undertaken an ambitious naval construction program, producing the world's third-largest fleet—one that had more battleships in the Far East than Japan had in its entire navy.

But this paper strength was deceptive. Russia had not managed to modernize as effectively as Japan had. The czar's armed forces had fallen far from the pinnacle of power they had reached in 1814, when they had marched into Paris as conquerors. Russia's 1856 defeat in the Crimean War showed that it lagged far behind the other Great Powers in taking advantage of the Industrial Revolution. This setback triggered a series of reforms, the most notable being the abolition of serfdom in 1861. But, despite the best efforts of skilled bureaucrats such as Finance Minister Sergei Witte (who served from 1892 to 1903), Czar Nicholas II's government remained a ramshackle affair that was better at repressing dissent than at mobilizing Russia's almost limitless reservoirs of men and materiel. In particular, the czar's navy suffered from three crucial disadvantages in confronting the Imperial Japanese fleet.

The first of these was fleet speed. The Russian navy was made up of a polyglot of ships. The newest were as fast as anything in the Japanese navy, but the older ones were much slower. Because the Russians chose to maximize their total numbers by using old as well as new ships, their fleet was limited by the speed of its slowest members. This gave the Japanese an advantage of at least three to four knots (3.5 mph to 4.6 mph) in any engagement, which proved as important as having the wind gauge in the days of sail: Whoever had superior speed could take the initiative and dictate the pace of the battle.

A second Japanese advantage was their ordnance. The Russians relied on armor-piercing shells. Their delayed fuses were not supposed to explode until after they had penetrated armor plate. In practice many of them turned out to be duds, and when they did explode they did not cause that much damage to the heavily armored Japanese ships. The Japanese, on the other hand, had developed an explosive powder known as Shimose that was placed in thin-skinned shells with fuses designed to explode at the slightest contact. These shells could not penetrate heavy armor, but they could inflict great damage on the unarmored parts of a ship. With four times the explosive power of Russian shells, they would take a heavy toll on Russian crew members not only with their shrapnel but also by releasing clouds of toxic gas. The Japanese advantage in this regard was magnified by the fact that Japanese crews could fire almost twice as fast as their Russian counterparts.

This was a consequence of the third and perhaps most important Japanese advantage: the high quality of their personnel. Many of Japan's sailors were volunteers, while all of Russia's were conscripts. All else being equal, volunteers generally fight better than draftees because they are more motivated and spend more time in the service. The Russian navy was not quite as dominated by the aristocracy as the army—middle-class naval officers could rise based on their performance—but it was less meritocratic than the Japanese service. There was also a much wider gulf between officers and men in the czar's navy. Japanese officers slept on mats, ate the same simple food as their men, and, "[w]hen off duty," a British correspondent wrote, "the officers fraternise with the men almost as if they were equals." Russian officers, by contrast, acted as if their sailors were serfs. Officers ate elaborate meals on fine china washed down with expensive wines, while their men were forced to swallow what one sailor described as "rotten biscuits and stinking decaying meat."

Japanese sailors, indoctrinated in the cult of bushido and the religion of Shinto, were fiercely loyal to their emperor. A growing number of Russian sailors, inspired by socialist calls for revolution, were disenchanted with their czar. One of them, seaman Novikoff-Priboy, later savagely denounced his officers: "These noblemen's sons, well-cared for and fragile, were capable only of decking themselves out in tunics with epaulettes. They would then stick their snouts in the air like a mangy horse being harnessed, and bravely scrape their heels on polished floors or dance gracefully at balls, or get drunk, in these ways demoralizing their subordinates. They didn't even know our names." A Russian engineering officer named Pleshkov fully returned this contempt. "I was dumbfounded by the convict-like appearance of the crew," he wrote of the sailors aboard the cruiser *Aurora*. "Their clothing is almost always terribly dirty. Their faces are pale and puffed and often wear the expression of an idiot."

Whatever the literal truth of these broad generalizations, suffice it to say that any fleet that included both Paymaster's Steward Novikoff-Priboy and Lieutenant Pleshkov was not likely to be a terribly happy one. The sorrows of the Russian fleet would only grow after their first meetings with the Japanese navy.

THE WAR BEGINS

Once again, Japan began a war with a surprise attack. On the night of February 8, 1904, Togo sent his destroyers racing into Port Arthur to sink the Russian Pacific Squadron at its anchorage. Only three out of twenty torpedoes struck home and no ships were sunk. Several other torpedo attacks were no more successful, due mainly to the difficulty of aiming from long range. Mining the harbor entrance proved more successful. When Russian ships came out to challenge the Japanese blockading squadron on April 13, two of the Russian battleships struck mines and sank. Among the dead was the best admiral in the Russian navy. But the mine was an equal-opportunity menace. A month later, two Japanese battleships were sunk by these infernal devices. Within a few hours, Japan had lost one-third of its battleship strength.

While Togo had not yet succeeded in destroying the Russian fleet, he had managed to keep it from interfering with the landings of Japanese troops in Korea. In May 1904, having secured Korea, a Japanese army landed on Manchuria's Liaotung Peninsula to besiege Port Arthur and the smaller port of Dalny from the landward side. Although Port Arthur's fifty thousand Russian defenders held firm at first, the Russian fleet was ordered to break out of the embattled base and head for its other Far Eastern anchorage at Vladivostok, 1,500 miles to the north.

On August 10, 1904, the Russian squadron sortied out, led by six battleships. Togo raced to intercept them with only four battleships of his own. In the resulting gun battle fought at long range, Togo was almost killed by a shell fragment that felled an officer standing next to him. But the Russian admiral had worse luck. Two Japanese shells struck the bridge and conning tower of his flagship, the *Czarevitch,* killing the admiral and jamming the ship's steering mechanism. The *Czarevitch* began turning in circles, forcing the other Russian ships to veer away. Togo might have finished off the Russian squadron then and there, but, with night falling, he was too cautious to move in for the kill. Fearing mines or torpedoes, he allowed the bulk of the Russian fleet to escape back to Port Arthur, thus prolonging the war and leaving it to the Japanese army to finish off the Russian navy.

After four more months of hard, blood-drenched combat, Japanese infantry finally took the fortified hills overlooking Port Arthur. From there, they could shell the ships in the harbor. One by one, the remaining Russian ships were destroyed by field artillery. The last battleship of the once-mighty Pacific Squadron was scuttled on January 2, 1905, the day that Port Arthur finally fell. Russian cruisers based at Vladivostok had enjoyed more success than the Port Arthur squadron—they had managed to sink some Japanese merchantmen—but by this time they, too, had been hunted down and either destroyed or bottled up by the Nipponese navy.

The focus of the war then turned to Manchuria, where hundreds of thousands of troops slugged it out with artillery, rifles, mortars, machine guns, and grenades: the full panoply of industrialized destruction. The Japanese army overran the Russian positions at Mukden on March 10, 1905, but failed to trap and annihilate the opposing army. Thereafter, land operations petered out in an uneasy stalemate.

The war would be decided at sea.

Before the fall of Port Arthur, Russia's Baltic Fleet had been ordered to sail eastward. Its commander, Admiral Rozhdestvensky, had never before led a fleet in action. His only combat experience had come twenty-seven years before as the skipper of a torpedo boat fighting the Turkish navy in the Black Sea. Rozhdestvensky was a brave, handsome, fifty-five-year-old officer with an iron will, "piercing black eyes," and a "well-trimmed pepper-and-salt beard." The son of a military doctor, he had risen based on professionalism, energy and incorruptibility—rare traits in the czar's navy. Even Novikoff-Priboy, hardly a fan, wrote, "Tall, grave, virile, his head bent forward a little, as if heavy with thought, he had so imposing an appearance that it seemed as if he could command success."

Around court circles, Rozhdestvensky was known for his womanizing (he had taken up with another admiral's wife as well as a naval nurse) and his explosive temper. His subordinates quailed before his stream of abuse. He would denounce senior captains as "manure sacks" or "fatheads" or "brainless nihilists." The targets of his opprobrium might have been mortified, but the czar was happy with the results he achieved. At a joint German-Russian naval review in 1902, Rozhdestvensky hosted Czar Nicholas II and Kaiser Wilhelm II aboard his flagship. In the following few hours, his ships proceeded to demolish every target in sight, leading Kaiser "Willy" to exclaim appreciatively to his cousin "Nicky," "I wish I had such splendid admirals as your Rozhdestvensky in my fleet." A Russian sailor later alleged that the targets had secretly been rigged "to fall to pieces at the mere wind from a shell, without a direct hit." No matter. Rozhdestvensky was on his way to the top. Before long the czar appointed him head of the Naval General Staff, and then, when things seemed to be falling apart in the Far East, looked to him as a savior.

Rozhdestvensky accomplished a great deal simply by getting his fleet intact halfway around the world while carrying most of his supplies with him—a feat that presaged the long voyages of self-contained U.S. naval task forces across the vast expanses of the Pacific during World War II. But the forty-odd ships of the Second Pacific Squadron arrived in the Far East in poor shape, their men exhausted, the ships' decks piled high with heaps of coal, their bottoms encrusted with barnacles—all factors that would hamper their performance in battle. Rozhdestvensky made little attempt to improve his men's gunnery or to coordinate tactics. His captains were astounded that he gave them only the vaguest instructions before they headed for their rendezvous with the enemy.

Still, they made for a formidable force, if only in theory. Once Rozhdestvensky had been joined in Indochina by still more reinforcements from home (the Third Pacific Squadron contained the dregs of the Russian navy), he had a total of eight battleships under his command, four of them built since 1900.

Togo had only four battleships of his own, all of them modern, but, with more light ships (including eight cruisers in the main battle line) and light guns, he had an overall advantage of 127 guns to the Russians' 92. Moreover, the Japanese fleet was in much better shape. During the eight months that it took Russia's Second and Third Pacific Squadrons to arrive, Togo was able to retrofit his damaged ships, rest his men, and train them to a high pitch of readiness. He was especially intent on improving their gunnery and carefully coordinating battle plans.

Using wired and wireless telegraphy, the excellent Japanese intelligence service had kept close tabs on the Russians' progress until the final leg of their voyage. There had been ten nerve-racking days while the Russian fleet had dropped out of sight in the South China Sea; Togo fretted that they had given him the slip. He was relieved to learn early on the morning of May 27, 1905, that the Russians had finally been spotted in Tsushima Strait. Within a couple of hours, the bulk of the Japanese fleet set off to intercept them from its base on the southern coast of Korea, opposite the island of Tsushima.

At 1:39 P.M., Togo was standing on the bridge of his flagship, the *Mikasa*, surrounded by his staff officers, all wearing black uniforms buttoned up to their chins and trimmed with silk braid stripes, when he finally spotted the black hulls and yellow funnels of the enemy fleet about eight or nine miles away. (His own ships were painted gray, which made them harder to discern at sea.) He increased speed and ordered a Nelsonian signal hoisted aboard the *Mikasa*: "The fate of the empire rests on today's battle. Let every man do his utmost."

"I HAD . . . NEVER WITNESSED SUCH A FIRE"

Peering south through his binoculars, Togo could see that the Russian battleships were steaming toward him at ten knots. They were in two columns, with the four most modern battleships, led by Rozhdestvensky's flagship, the

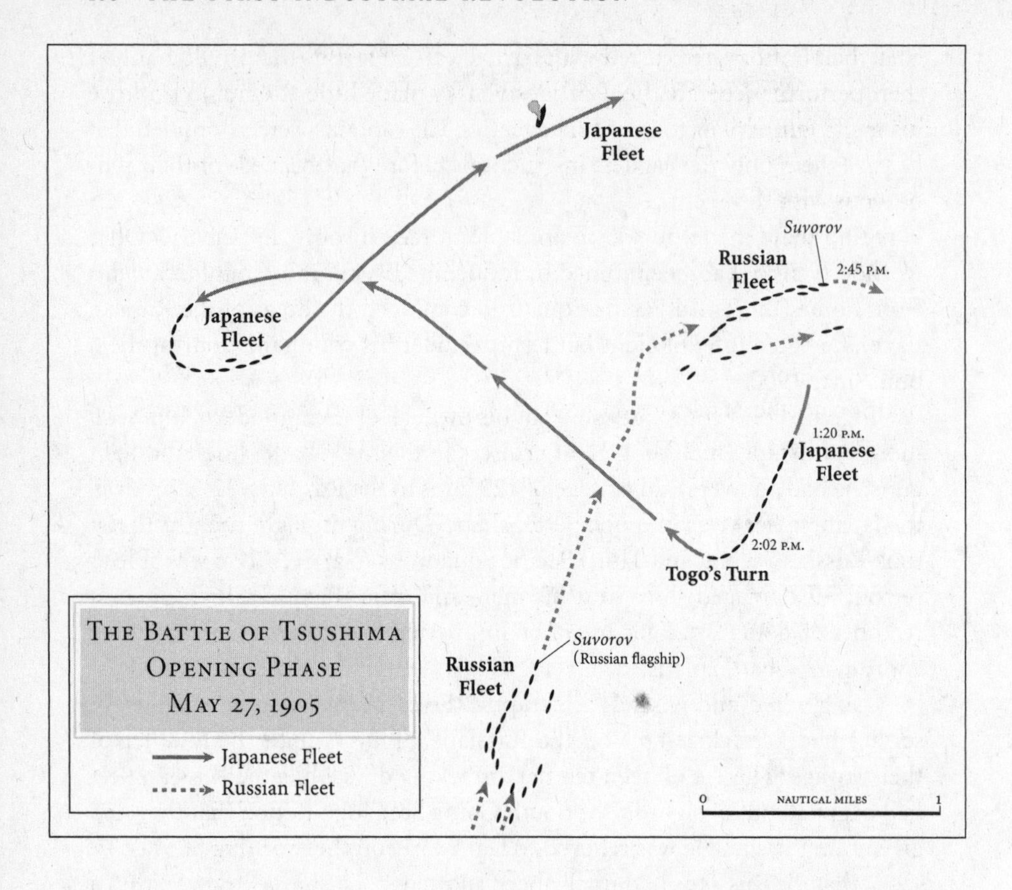

Suvorov, followed closely to the west by four older battleships, led by the *Oslyabya*. If Togo continued heading south, he would slide right by the Russian fleet, which might offer Rozhdestvensky a chance to escape, something that Togo was determined to avoid at all costs. He did not want another inconclusive engagement like the Battle of the Yellow Sea, and he was willing to risk everything to secure a decisive victory. At 2:02 P.M., Togo boldly ordered his ships to undertake a 180-degree turn from south to north. He might have ordered all his ships to turn in place, but then his weakest ships would have been exposed to the strongest ships at the front of the Russian line. Togo wanted to position his battleships against those of the enemy, and he wanted to keep his flagship in the lead where he could exercise control over the battle. So he signaled his ships to turn in succession, circling around to the north with the *Mikasa* still in the lead.

When the Russians saw what Togo was doing, they were astonished. Every Japanese ship would turn at the same point, allowing Russian gunners to zero in their guns. Even at a brisk fifteen knots (seventeen miles per hour), the twelve Japanese capital ships (four battleships, eight cruisers) would take at least fifteen minutes to complete this complicated maneuver.

During this time, they would be extremely vulnerable. "How rash!" exclaimed an officer aboard the *Suvorov*. "Why, in a minute we'll be able to roll up the leading ships."

Hoping to do just that, the *Suvorov* opened fire, followed by the other Russian battleships. But they were firing from such long range—about seven thousand yards, or almost four miles—and their gunnery was so poor that their shells did little damage. Togo completed his U-turn without losing a single ship.

It was his masterstroke. The battle was all but over by the time the Japanese fleet finished coming around in perfect order. Now the Japanese were running northeast, closing in on the Russian fleet, and the initiative was entirely in Togo's hands. The Russian ships, in the meantime, were in considerable confusion. Rozhdestvensky had ordered them to form into a single line—the standard tactical formation employed by most navies for centuries—but their ship-handling was so clumsy that some Russian vessels had to come to a stop to avoid collisions as the two lines merged.

The Russians still had not completed their maneuver when Togo opened fire at 2:10 P.M. At that point the *Mikasa* was about 6,500 yards (3.7 miles) from the nearest Russian ships. Togo kept closing the range to 6,000 yards, 5,000 yards, 4,000 yards (2.2 miles), concentrating his ships' fire on the two leading Russian battleships, the *Suvorov* and *Oslyabya*. The closer the Japanese got, the more their superiority in short- and medium-range guns became apparent. Commander Vladimir Semenoff, a veteran of the Battle of the Yellow Sea, recorded his impressions aboard the Russian flagship:

> I had not only never witnessed such a fire before, but I had never imagined anything like it. Shells seemed to be pouring upon us incessantly, one after another. . . . They burst as soon as they touched anything—the moment they encountered the least impediment in their flight. . . . The steel plates and superstructure on the upper deck were torn to pieces, and the splinters caused many casualties. Iron ladders were crumpled up into rings, and guns were literally hurled from their mountings. . . . In addition to this, there was the unusual high temperature and liquid flame of the explosion, which seemed to spread over everything. I actually watched a steel plate catch fire from a burst. Of course, the steel did not burn, but the paint on it did. . . . At times it was impossible to see anything with glasses, owing to everything being so distorted with the quivering, heated air.

The Japanese ships were suffering too, but, thanks to the faster fire of their crews and their higher-quality shells, the Russians were getting the worst of it. Less than half an hour into the battle, the Russian fire began to slacken. The *Suvorov* fell out of line at 2:40 P.M. Its helm was jammed, and it began circling out of control. Both Admiral Rozhdestvensky and the ship's captain lay in the infirmary after having been wounded by flying splinters that had penetrated the armored conning tower. The *Suvorov* managed to stay afloat, beating off attack after attack, until that evening. The flagship finally sank at 7:30 P.M. after Rozhdestvensky and some staff officers had been transferred to a Russian torpedo boat.

The battleship *Oslyabya* did not last that long. Its guns were put out of action, leaving her defenseless as six Japanese cruisers closed in. Some men were simply pulverized; others had their chests torn open or their arms and legs blown off. Blood mingled with seawater on the decks, and the screams of the wounded filled the dying ship. More shells buckled one of the armor plates on the side, wrote a sailor, "making a hole through which a coach and horses might have been driven." As the ship began sinking, men desperately tried to save themselves. But there was no escape for those trapped in the engine rooms and stokeholds. The hatches had been battened down, and they could be opened only from the outside. Stokers and engineers shouted for help, but not a single sailor, running by in panic, bothered to stop and release them. Those who managed to make it into the water were not much better off; the Japanese continued firing at them. At 3:10 P.M., the *Oslyabya* flipped over entirely—"turned turtle"—and sank to the bottom. Russian destroyers bravely tried to save the survivors, but most of the ship's company did not make it.

Around this time, thick fog began rolling in, which combined with smoke from burning ships and bursting shells to reduce visibility around the battle area and provide a temporary respite for the beleaguered Russians. The battle slackened as ships lost contact in the haze. Somehow, out of this confusion, Rear Admiral Nikolai Nebogatov, who had taken over command for the wounded Rozhdestvensky, managed to regroup the remnants of the Russian fleet into a battle line. Led by the battleship *Borodino,* the Russians fled north, hoping to reach the safety of Vladivostok. Togo took off in pursuit. With his superior speed, he had no trouble catching his prey. A few minutes after 6:00 P.M., the bombardment began in earnest again.

The Russians still had six battleships—two more than Togo—and they were able to put up a stiff fight. But once again the Japanese got the better of the gunnery duel. The battleship *Alexander III*, already badly damaged, sank at 6:55 P.M. with all hands. Half an hour later it was the *Borodino*'s turn. Only one man out of her entire crew was saved. "The loss of the *Borodino*, which happened before our eyes, was so unexpected that we were stupefied," wrote a Russian officer on a nearby ship, "and, uncovering our heads, we gazed on the foaming grave of the heroic ship."

The sun had now set, and Togo decided not to risk his big ships in a night action. He led his battleships and cruisers to the north, leaving twenty-one destroyers and thirty-two torpedo boats behind to finish off the remnants of the Russian fleet. They did not find it easy going. The seas were rough and visibility poor. In the dark, several Japanese destroyers and boats collided with one another, and almost all their torpedoes missed their targets. The Japanese also suffered losses from Russian shell fire. But several already ailing Russian ships did receive a coup de grâce. The battleship *Navarin* was lost to linked mines dropped in its path; the battleship *Sisoi Veliky* succumbed to torpedoes.

When dawn broke over the Sea of Japan on Sunday, May 28, 1905, the once-mighty Russian fleet had been reduced to two badly damaged battleships, two coastal defense ships, and one light cruiser. By midmorning they were surrounded by twenty-seven Japanese warships. As they saw the ring of steel closing around them, "[b]oth officers and men lost hope," wrote one Russian officer. "They fell on their knees and cried that all was lost."

After being fired upon, Admiral Nebogatov saw no choice but to surrender—a humane act for which he was subsequently court-martialed by the czar and sent to prison. A few hours later, the Japanese intercepted a destroyer carrying the wounded Admiral Rozhdestvensky. He was transported to Japan, where he was treated as an honored guest before being repatriated to Russia along with his fellow prisoners of war.

A few Russian ships made it to neutral ports, but only three reached Vladivostok as planned. Out of thirty-eight Russian warships that had entered Tsushima Strait, thirty-one had been captured or sunk, including all of the battleships. Russian losses amounted to 4,830 dead, 5,907 captured, and an unknown number wounded. The Japanese lost only 117 officers and men killed (another 583 were wounded). A number of Japanese ships, including the *Mikasa*, were damaged, and three torpedo boats had been sunk, but not a single battleship or cruiser had been lost.

Togo's triumph was complete.

SMALL DIFFERENCES, BIG CONSEQUENCES

As the news filtered out from the Sea of Japan, the world reacted with astonishment. "ANNIHILATE THE FOE. . . . JAPS WIPE OUT THE RUSSIAN FLEET." So screamed the headline of the New York *Journal* (price one cent, owner W. R. Hearst) on May 27. Two days later, the *Journal*'s more sedate competitor, the New York *Herald*, declared: "TOGO VICTORIOUS IN THE GREATEST OF ALL NAVAL BATTLES." Trying to put recent events into perspective, *Harper's Weekly* opined: "We must recognize that the success attained by Admiral Togo was the most overwhelming recorded in modern naval history. For

Japan it was even more momentous and conclusive than was Lepanto for Venice or Trafalgar for England. It deserves, indeed, to be ranked with Salamis, with Actium and with the series of disasters that proved fatal to the Spanish Armada among the decisive events in naval annals."

Usually, contemporary hyperbole is discounted by scholars of a later age, but this is a verdict that few modern historians would dispute. Who can deny the impact of Tsushima?

The Treaty of Portsmouth, which ended the war on September 5, 1905, granted Japan dominion over Korea, the Liaotung Peninsula, and the southern half of Sakhalin Island. This was quite a colonial haul. Although many Japanese wanted even more, they could only shake their heads and marvel at how far they had come since 1853. A half century earlier they had been humiliated by a handful of Western ships; now they had bested one of the mightiest powers of the West and changed forever the balance of power in Asia. It is hard to think of any nation that has risen faster from obscurity to military glory.

While Japan was rising, Russia was falling. The czar's government barely survived the disasters of the war. As early as January 1905, following the fall of Port Arthur, there had been uprisings in St. Petersburg. The defeat at Tsushima, when Russia's navy all but ceased to exist, added further fuel to the revolutionary embers. Throughout the summer of 1905, Russia was rocked by riots and protests, culminating in a massive general strike in October. Nicholas II kept his throne only by agreeing to turn Russia into a constitutional monarchy. However, the czarist regime—much like the shogunate in Japan—was not able to reform itself fast enough to forestall a successful revolution twelve years later, triggered by another series of even more calamitous military setbacks. "The war with Japan, one of the most terrible blunders made during the reign of Nicholas II, had disastrous consequences and marked the beginning of our misfortunes," the Russian expatriate aristocrat Felix Yusopov would write many years after Tsushima. The battle marked the decline not only of the Russian empire but of all European empires. Jawaharlal Nehru, the Indian independence leader, later traced his epiphany that Asians could defeat Europeans to the Russo-Japanese War.

It is amazing to think what big consequences can flow from such small differences in military capacity. Both Japan and Russia had tried to take advantage of the Industrial Revolution at sea. At first blush, the two sides had produced fleets with roughly equal capacity, the slight edge going to Russia. But what appeared to be an advantage from the safety of a drawing room in St. Petersburg became a major disadvantage when seen from the wardroom of a ship under fire.

The Japanese superiority in speed, firepower, and personnel has already been noted. That Japan developed such a formidable navy within the space of a single generation is a tribute to the leaders of the Meiji Restoration, who had done a better job of westernizing a non-Western country than just about anyone else before or since. No one deserves more credit than Admiral Togo, who, with his daring 180-degree turn at the start of battle, had earned the right to be ranked alongside his idol, Admiral Nelson. This samurai's son exemplified the operational excellence of the Japanese military.

The comparison between Togo and Nelson is apt in more ways than one, because, while the ships at Tsushima were much more powerful than those at Trafalgar, they were employed in roughly similar fashion. Both opposing admirals in 1905 tried to form their ships into a single line and batter the enemy with broadsides—the tactics that had been pioneered by Hawkins and Drake more than three hundred years before. The great historian Garrett Mattingly has described the Spanish Armada as "the beginning of a new era in naval warfare, the long day in which the ship-of-the-line, wooden-walled, sail-driven and armed with smooth-bore cannons, was to be queen of battles; a day for which the armor-plated, steam-powered battleship with rifled cannon merely marked the evening."

Not even the commissioning in 1906 of the super-battleship HMS *Dreadnought* led to a new dawn. The *Dreadnought* had ten twelve-inch guns, six more than the *Mikasa* or any other previous ship, each one capable of discharging an 850-pound shell. With a top speed of twenty-one knots, it was also faster than its predecessors, none of which could go faster than eighteen knots. All existing capital ships were instantly rendered obsolete, and Germany, America, Japan, and all the other naval powers rushed to build Dreadnoughts of their own. By the time World War I rolled around, the original *Dreadnought* itself had become outdated; newer ships were even faster and more powerful.

Yet the commanders of these behemoths, Beatty and Jellicoe and Hipper and Scheer, still fought in a manner that would have been recognizable to their predecessors of the sixteenth century. This was true even though they had to deal with weapons—the mine, torpedo, and submarine—that were utterly different from anything that had come before. (The torpedo perhaps bears some resemblance to the fire ships used by sixteenth-century navies.) While these new underwater menaces gave captains pause, they did not fundamentally invalidate the logic of the battle line. Destroyers, torpedo nets, depth charges, hydrophones (forerunner of sonar), and other inventions—along with simple expedients like convoying—served to blunt the offensive potential of these new weapons so beloved by devotees of commerce-raiding.

Some critics rail against woolly-headed admirals who insisted on building battleships, allegedly for purely sentimental reasons. This criticism has a good deal of validity for the post–World War I era, as we shall see in Chapter

Eight, but the preference for battleships was perfectly logical before and during World War I.

Some historians have speculated that perhaps Germany could have won the Great War at sea if it had not spent so much of its resources building a battleship fleet that made no difference to the war's outcome. What if, the argument goes, the Germans had concentrated on building submarines and had not waited to declare unrestricted submarine warfare until 1917? Might they have starved Britain into submission? Certainly the submarine warfare of 1917–18 took a large toll on Britain's economy. But there was a perfectly good reason why Kaiser Wilhelm II had been loath to unleash his U-boats sooner. He knew they would sink neutral shipping, including some flying the U.S. flag, and this would in all likelihood bring the colossus from across the Atlantic into the conflict. That is exactly what happened in 1917, and Congress's declaration of war sealed Germany's doom. Military analyst Robert O'Connell argues that a submarine blitz in 1915 could have knocked Britain out before American intervention made a difference, but the likelier result would have been to stimulate the earlier development of the techniques that ultimately defeated the U-boats in 1918. In any case, as a prominent naval historian notes, the "submarine was not an effective weapons system before 1914"; the technological and tactical advances that made it so deadly in 1917 (such as the widespread adoption of diesel engines, which allowed greater stealth and endurance than the old kerosene-burning power plants) occurred only after the onset of war.

Germany's decision to build a big surface fleet, therefore, was not a crazy one; nor was Britain's determination to stay ahead of Germany in capital ships. Both sides knew that the Royal Navy was Britain's lifeline; if it lost control of the seas, the home islands could be blockaded or even invaded. And the fact remains that, as of 1914 or even 1918, while the Royal Navy could have been (and was) hurt and harassed by torpedoes, mines, and submarines, the only way it could have been defeated was by a fleet of battleships in a conventional surface engagement.

The great sorrow of Britain's Grand Fleet and Germany's High Seas Fleet was that neither one was able to score a victory to rank with Japan's at Tsushima. Their only all-out clash, at Jutland in 1916, ended inconclusively. The British lost more ships, but the German fleet remained caged up in its ports, where it had little impact on the rest of the war. This does not mean that battleships were outdated, only that the two sides had achieved parity, which is what usually happens when two countries of roughly equal scientific sophistication and economic resources engage in a prolonged arms race. Japan had shown in 1905 what a country could do with the new industrial technology if it were fighting an opponent that was even slightly inferior. Because steel ships sink faster and more easily than wooden ones, warfare could be more decisive in the new age, provided that one side had an edge in technique or technology.

The weapons platform that would break the hegemony of the battleship was in its infancy in World War I. It was not the mine or the submarine or the torpedo. It was the airplane, and when it was combined with the aircraft carrier, the result would be another revolution in naval warfare—a new day that would not take nearly as long to dawn as the three centuries that elapsed between the triumph of sail-and-shot and its replacement by steam-and-steel.

The Consequences of the Industrial Revolution

The First Industrial Revolution transformed warfare between the end of the Crimean War (1856) and the start of World War I (1914). To sum up, the principal developments were:

LAND WARFARE: Conscription increased the size of armies. Industrialization made it possible to equip these growing forces. Railroads and steamships made it possible to move them farther and faster than ever before. Telegraphs made it possible to coordinate their movements. General staffs made it possible to plan their operations. And rifled, breech-loading artillery and small arms—including, eventually, repeating rifles and machine guns—made it possible for armies to mete out unprecedented destruction.

The extent of change, sweeping as it was, should not be exaggerated. Once armies moved past the railhead, they were still, in many respects, back in the age of Gustavus Adolphus, moving no faster than men could march and communicating no faster than horses could gallop. And although cavalry greatly declined in importance, because increases in firepower made it suicidal to charge infantry positions, horsemen continued to be used effectively as late as World War I for maneuvering outside the static Western Front.

NAVAL WARFARE: Machine guns played little role at sea because most naval battles took place outside their effective range, but the transformation of artillery had an even bigger impact at sea than on land. The introduction of high-explosive shells fired by rifled guns rendered wooden ships obsolete, while improvements in steam power created a superior source of energy to replace sails. This led to the emergence of steam-driven ships made of steel. The old categories of sailing warships—which were classified according to the number of guns they carried, as first-rates, second-rates, third-rates, and so forth—were replaced by entirely new classes of vessels: battleships, cruisers, destroyers, gunboats. New underwater weapons—torpedoes and submarines—added another dimension to warfare that previously had been confined to the water's surface.

The electrical telegraph could not be employed to communicate between ships at sea because initially it was dependent on wires. Wireless telegraphs, or radios, introduced at the turn of the twentieth century, were better suited to ships, where their bulkiness and need for large amounts of energy were not nearly as serious a drawback as on land. But early radios were so unreliable, and captains were so afraid of giving away their positions with radio emissions or having their signals jammed, that most resorted to traditional signal flags and semaphores to communicate in the heat of battle.

As noted above, the most immediate and obvious impact of all these developments—along with some others, such as the manufacture of quinine tablets to fight malaria—was to accelerate the European conquest of the world. While a handful of nations and tribes escaped Western occupation during the Industrial Age, the only one that managed to become a Great Power in its own right was Japan. Its rise, crowned at Tsushima, unsettled Western designs for dominance in East Asia and forced the Occidental nations to accept an Oriental state as a peer. The Japanese were under no illusion as to why they were being included in the ranks of the Great Powers. A Japanese diplomat invited to the 1899 Hague Conference commented sardonically, "We showed ourselves at least your equals in scientific butchery, and at once we are admitted to your council tables as civilized men."

Another star was rising on the other side of the Pacific. By the turn of the twentieth century the United States had the biggest and richest economy in the world. For many decades after the Civil War, it had not chosen to translate industrial power into military might. But once it embarked on a naval buildup in the 1890s, the U.S. rapidly constructed a formidable fleet—one that defeated the decrepit Spanish navy in 1898 at the battles of Manila Bay and Santiago Bay as decisively as Japan bested Russia seven years later. The development of the U.S. Army took longer, but it, too, swelled to impressive size and effectiveness in 1917–18 (2.8 million men drafted), before being cut to the bone again in 1919. U.S. influence on world affairs waxed and waned along with the size of its armed forces.

Among European states, Prussia was an early leader in harnessing industrial technology for land warfare and using it to defeat larger rivals—Austria and France. Prussia's successes in the German Wars of Unification led eventually to two world wars whose repercussions are still being felt. The Habsburgs, not the Hohenzollerns, might have dominated the new German state but for their inability to raise effective Industrial Age armed forces of their own. This failure led to Austria-Hungary's final collapse in 1918, following an even more calamitous military defeat. The Ottoman and Russian empires were other examples of states that had been able to generate impressive amounts of military power in the Gunpowder Age but were unable to modernize their armed forces and society in the late nineteenth and early twentieth centuries, and wound up collapsing as a result. Out of their ruins sprouted a panoply of new states from Poland to the Persian Gulf.

TOTAL WAR

The broader impact of the Industrial Revolution was contradictory: It was both boon and bane. Industrial technology allowed countries to produce more food, more medicine, more clothing, more of everything than ever before, and thus increase their populations without triggering a Malthusian crisis. After an initial, painful period of adjustment, industrialized countries became wealthier than any previous societies in history. But at the same time that industrialization made life better for millions, it also resulted in the deaths of millions. The new technologies could be used to extinguish life as effectively as they could be used to support it.

Industrial weaponry did not cause World War I, except indirectly to the extent that it fostered the rise of Germany. But it did make the conflict, paradoxically, both shorter and far more catastrophic than previous "world wars" such as the Seven Years' War (1756–63) or the French Revolutionary and Napoleonic Wars (1792–1815). The figures boggle the mind: From 1914 to 1918, sixty-three million men were mobilized, eight million were killed, and twenty-two million were seriously wounded or disabled. Millions of civilians also died. Those losses pale by comparison with World War II, which would leave some fifty-five million dead, but they were many orders of magnitude greater than those of any previous conflict. Preindustrial states could not possibly have fed, clothed, equipped, moved—or slaughtered—so many individuals. Germany and France had 20 percent of their populations under arms. Britain mobilized only 13 percent, but this was still far higher than the 7 percent that Napoleon had been able to marshal with the *levée en masse*. Not only were there more soldiers than ever, but each one now had at his disposal more firepower than his ancestors could have imagined. A single machine gunner or artilleryman in 1914 could rain down more death than an entire regiment a hundred years before.

The unleashing of total, industrialized warfare wrought profound and lasting changes across the world. The most obvious consequence of World War I was the toppling of old governments and the creation of fresh grudges that led, after a twenty-year armistice, to World War II. A more subtle but even more pervasive consequence was to change the nature of government itself.

Laissez-faire structures had survived largely intact for most of the Industrial Age prior to 1914. States had been strengthened and centralized by the military demands of the Gunpowder and Industrial Revolutions, but they remained, for most of the nineteenth century, fairly minimalist affairs. Social welfare programs, such as insurance for accident victims and pensions for the elderly, were pioneered by Germany's Otto von Bismarck in the 1880s, but they were still rudimentary prior to World War I. Even in the most advanced nations, taxes were low (the state spent 13 percent of British GDP in 1910, compared to 37 percent in 2003), and most governmental departments were staffed by a mere handful of clerks.

The demands of total warfare brutally exposed the shortcomings of this approach. All of the major combatants in World War I had greatly underestimated the amount of ammunition that would be expended by modern armies. By early 1915, generals on both sides were complaining about shell shortages, and the old procurement system, with governments contracting with private companies, was found to be inadequate to meet the burgeoning demand. Already governments had nationalized railroads and shipping lines to facilitate mobilization. Now the armaments industry was taken under the state's wing to keep armies fighting.

Before long, other vital commodities, ranging from sugar to cotton to fuel, became scarce due to the Allied naval blockade of the Central Powers and Germany's U-boat campaign against Allied commerce. Governments responded by nationalizing more industries, controlling prices and currencies, and rationing goods. The private sector hardly ceased to exist; in fact, big industrial corporations working in close cooperation with governments made large profits from the war effort. But all across Europe and North America, governments became much more deeply enmeshed in the economy than they had been prior to 1914.

Total government expenditures in Britain, France, and Germany spiked almost 1,200 percent between 1913 and 1918; military expenditures shot up 2,000 percent. Armed forces that, before the war, had consumed no more than 4 percent of national income were now taking more than 25 percent. The expansion of government was not limited to fiscal affairs. Civil liberties were curtailed in all the combatant states. The process was most drastic in Germany, where by 1916 a military dictatorship was in control. But even in more liberal nations, such as the U.S., antiwar activists like Eugene Debs were subject to arrest and summary confinement.

While the Great War spawned centralized control, it also spurred egalitarianism, if not always democracy. The privileges of aristocrats had traditionally been founded on their domination of the means of war—something that had been especially obvious in the case of Europe's knights and Japan's samurai. Warfare had been somewhat democratized in Europe with the rise of infantry by the 1600s, but the professional armies of the seventeenth and eighteenth centuries remained largely dominated by the nobility. Aristocratic control was broken in France by the French Revolution, but it was reestablished after the fall of Napoleon. While many states adopted peacetime conscription in the nineteenth century, most still kept landed elites in charge—a trend exemplified by Germany, where the world's most modern army was dominated by the Junker nobility.

That hierarchical control was hard to maintain in the face of total war. Old social divisions broke down when millions of men were called to the colors and all classes lived and bled together in the trenches. This ghastly carnage seemed to discredit old institutions such as monarchies and old ways of thinking such as liberalism at the same time that it gave fresh impetus to new ideologies such as communism and fascism. Ancient dynasties—the Romanovs, Habsburgs, Hohenzollerns—fell from power, but even countries that did not see a revolution were transformed. With millions of men called away to the front, the resulting labor shortages gave greater power to the workers who remained. As a result, trade unions saw their membership and power soar during the war. Women also gained more rights as they took jobs vacated by men sent to fight. It was no coincidence that during or shortly after the war the vote was given to women for the first time in most developed countries.

World War I led, then, to the growth of government and the demise of traditional social structures. These trends were to some extent slowed or even reversed after 1918, but no government returned to its pre-1914 level. In the early 1920s, the British and French governments were spending 100 percent more than they had been before 1914. Simply paying off wartime debts meant that tax levels had to stay high. Laissez-faire economic policies made a partial comeback in many countries in the 1920s—even in the Soviet Union, where the New Economic Policy introduced market incentives in agriculture—but everywhere after 1929 the Great Depression led to a return to the wartime socialist model. In Britain, the U.S., France, and other liberal democracies, economic crises resulted in the rise of welfare states; in Germany, Japan, Russia and other less liberal countries, to the rise of totalitarian states.

The connection between wartime exigencies and peacetime norms was particularly strong in the dictatorships. Nazism, fascism, and communism all explicitly sought to expand and make permanent the societal mobilization of the 1914–18 period. As one Nazi put it, "The National-Socialist ethos grew entirely logically from the soldier ethics of the World War." The connection between wartime and peacetime was more tenuous in the liberal democracies but no less real. The National Recovery Administration set up by President Franklin Roosevelt in 1933 was, for instance, modeled on the War Industries Board created by President Woodrow Wilson in 1917.

There was, of course, a world of difference between Stalin's Five-Year Plans and Roosevelt's New Deal, between harsh communism and half-hearted socialism, but both were inspired to some degree by the experience of World War I—a conflict that could never have been waged on such a titanic, transformative scale were it not for the changes in warfare that had occurred in the previous half-century. This was the bittersweet legacy of the Industrial Age.

PART III

THE SECOND INDUSTRIAL REVOLUTION

The whole future of warfare appears to me to lie in the employment of mobile armies, relatively small but of high quality and rendered distinctly more effective by the addition of aircraft, and in simultaneous mobilization of the whole defense force, be it to feed the attack or for home defense.

General Hans von Seeckt (1930)

The Rise of the Second Industrial Age

The First industrial Age was powered by coal and steam, the Second by oil and electricity.

Electricity came first. Its applications had been studied since the late eighteenth century by such scientists as Benjamin Franklin, Alessandro Volta, André-Marie Ampère, and Michael Faraday. But it was not until the 1860s that a practical dynamo—an electricity-generating machine—was developed. This was seized upon by Thomas Alva Edison, who had already gained fame for his invention of the phonograph, the "talking machine," in 1877. Within two years, the Wizard of Menlo Park, New Jersey, became equally renowned as the inventor of the incandescent lightbulb. But his contribution was broader than that. He invented a whole system of lighting centered around an improved generator of his own design whose energy could be transmitted via wires to multiple locations. Edison set up his first electrical system in lower Manhattan in 1882. He was not only a brilliant inventor but also a shrewd businessman. He launched a variety of businesses that merged to form Edison General Electric Company, later simply General Electric.

Within a few years, electricity was not only replacing gas lights and candles for illumination, it was also running factories, elevators, and light rail systems, making possible the rapid growth of urban areas with high-rise buildings. Within a few decades, electricity was running an increasing number of new appliances that were developed to take advantage of its ready availability and that transformed the way people lived. Washing machines, dishwashers, and vacuum cleaners liberated women from a good deal of household drudgery and made domestic servants expendable; air conditioners opened up the southern United States and other torrid climes for faster economic development and population growth; refrigeration dramatically reduced food spoilage and made possible richer, healthier diets.

Important as electricity generation was, it was not a primary source of energy; a dynamo requires an outside energy source—coal, water, wind, natural gas, oil, or, later, nuclear fission—to work. The same is true of the internal-combustion engine, which uses the explosive combustion of fuel to push pistons within a cylinder. The first practical model was created in 1859 by Jean-Joseph Étienne Lenoir, a Belgian working in Paris, and perfected in 1876 by Nicolaus August Otto, a German traveling salesman with no formal technical education. By 1886 engines had been installed on four-wheeled carriages by two competing German engineers, Gottlieb Daimler (a former employee of Otto's) and Karl Benz, giving birth to the automobile. In the next few years, a variety of other important inventions—the carburetor, radiator, steering wheel, crank-starter, pedal-brake control, pneumatic tire—were added by various mechanics.

It was not immediately apparent what type of engine would be best for "horseless carriages." Many early models ran on electricity or steam. Both had their advantages, but by the early 1900s the gasoline motor had triumphed decisively. It may have been noisy and noxious, but it was also small, cheap, and efficient. Best of all, it ran on a new and seemingly inexhaustible source of energy.

In 1901 the first major oilfield was discovered in Texas, and a great boom was on. Previously the oil industry, which had been centered in Pennsylvania, had primarily sold kerosene for lamps. With the spread of electricity, that market was going dark. But a vast new demand was growing as gasoline was used to power automobiles and fuel oil was used to run boilers in ships, factories, and trains. The growing value of petroleum would transform modern life. Men would kill each other in the future for control of this vital resource. In 1933, Interior Secretary Harold Ickes declared, without exaggeration, "There is no doubt about our absolute and complete dependence on oil. We have passed from the stone age, to bronze, to iron, to the industrial age, and now to an age of oil. Without oil, American civilization as we know it could not exist."

Luckily for the United States, it had ample supplies of this precious fuel; it produced two-thirds of the world's petroleum in 1941. It needed all that fuel because its motorization was more advanced than in other countries. European firms, such as Peugeot, Benz, Fiat, and Morris, had taken the early lead in automobile production, but by the early 1900s the Americans were catching up fast, led by a restless engineer named Henry Ford, who tinkered with engines in his spare time while working for the Edison Electrical Company in Detroit. In 1903, after several false starts, he founded the Ford Motor Company. Five years later, he launched what would become the most popular car of its era—the Model T. Simple, inexpensive, and reliable, the hand-cranked "Tin Lizzie" was an instant hit. But Ford's most significant contribution was not what he produced but how he produced it.

Ford took several innovations pioneered by others, notably interchangeable parts (used by the Springfield Armory) and a moving production line (used by Chicago slaughterhouses), and combined them to create the most efficient factory in the world. In 1913 his Highland Park, Michigan, facility introduced a moving, electric-powered assembly line. Previously, gangs of workers had moved from one car to another. Now the cars came to them. Workers stood in one spot, endlessly repeating the same simple motion, whether adding a tire or tightening a bolt. The amount of time needed to assemble a car fell by 88 percent and the price of a finished Model T by more than 70 percent. The only problem was that workers hated this soul-deadening system that turned them, almost literally, into cogs in a machine. To attract and keep laborers, Ford doubled wages, to $5 a day. This not only made it easy for him to staff the assembly line but also, as an added benefit, made his workers well-paid enough to buy the very products they were producing.

Ford's techniques were widely copied in other industries, and, along with the expansion of advertising, they helped to usher in a new era of mass production and mass consumption. The American auto industry led the way. By the mid-1930s, the U.S. and Canada accounted for nearly 90 percent of the world output of trucks, cars, and tractors, and over half of all American families owned a car. This was impressive enough, considering that only a few years before, the horse had been the primary means of personal transportation. Even more amazing was that by the 1930s numerous people were soaring through the clouds.

The age-old dream of flight had first been achieved by the Montgolfier brothers in a hot-air balloon in 1783. Throughout the nineteenth century various inventors tried to create a reliable heavier-than-air flying machine, but none succeeded until 1903, when Orville and Wilbur Wright took their famous flight at Kitty Hawk, North Carolina. The internal combustion engine was as important for the Wrights' rickety Flyer as it was for the Model T; without it, they would not have had a compact power source for their propeller.

Although automobiles were utilized in World War I, starting with the use of Paris taxicabs to ferry French soldiers to the Battle of the Marne in 1914, the war did not play a major role in their development. (The war did lead to the invention of the tank, a close relative of the tractor, which will be examined later.) But, for the airplane, the Great War was crucial. In August 1914, Britain, France, the U.S., Russia, and Germany had all of 774 aircraft among them.

During the next four years, the belligerents produced more than two hundred thousand aircraft. Each new design had the longevity of a monarch butterfly: the average fighter plane went from introduction to combat to obsolescence in less than a year. This made for rapid improvements. In 1914 the typical airplane was a biplane or triplane made of wood, cloth, and wire. By the war's end in 1918 all-metal monoplanes were being produced that were recognizably modern in their design. The top speed of airplanes increased from 126 miles per hour in 1913 to 171 mph in 1920. (By 1939 the record was up to 469 mph.)

Surplus warplanes were put to good use after the Armistice. By 1919 regularly scheduled air service had begun between Paris, Brussels, and London, and the foundations had been laid for Lufthansa, British Airways, Air France, and the other great European state-owned airlines. The U.S., by contrast, developed its aviation industry through a mixture of public and private initiatives. The original impetus came from the U.S. Post Office, which paid private operators to carry airmail. To supplement their income, these new companies ferried passengers as well. By the early 1930s, a number of major airlines—Pan Am, Eastern, American, TWA, and United—had emerged. They were supplied with equipment by a growing manufacturing sector led by Boeing, Curtiss, Douglas, Pratt & Whitney, and Lockheed. Ford even got into the business in the 1920s with its popular Tri-Motor airplane. The 1930s introduced most of the features associated with modern air travel: everything from stewardesses and meals served on trays to cabin pressurization and in-flight movies.

The airline industry became, as it has remained, a ready reserve of pilots and aircraft that could be called upon in wartime. The link between civil and military aviation was especially close in Germany, Italy, Japan, and the Soviet Union, which were all heavily militarized, but it was also notable even in the least militarized major nation, the United States. The ubiquitous Douglas DC-3, a workhorse that carried fourteen passengers, doubled as the C-47, a U.S. military transport. The Boeing 307 Stratoliner, another popular aircraft that cut the time of a journey across the U.S. to a mere fourteen hours, was based on the B-17 bomber.

By the mid-1930s, American airlines were the world's biggest, but aircraft production in the U.S. lagged far behind that of the other Great Powers, whose output was mainly designated for military needs.

The airline industry, along with many others, benefited from great leaps in electronic communications. By the 1930s passengers were routinely inquiring about and booking flights using the telephone, a device which, since its invention by Alexander Graham Bell in 1876, had become a familiar sight in North American and western European offices and households.

The European states brought their telephone networks under government ownership. The U.S. preferred to have a state-regulated but privately owned monopoly.

By 1939, the Bell system, with assets worth an estimated $5 billion, had become the wealthiest corporation in the world. It was at the forefront of a development that had been gathering momentum since the nineteenth century: the split between ownership and management. In 1929, American Telephone & Telegraph, the Bell parent company, had half a million stockholders, none of whom owned more than 1 percent. This left its executives pretty much free to conduct its affairs as they liked. Business schools were springing up to train this new managerial elite, and consultants and gurus—the most famous was Frederick Winslow Taylor, the "Father of Scientific Management"—were coming along to advise them. Bureaucratization and professionalization, which had begun in the armed forces centuries earlier, had now spread to the business world. Many of the management techniques refined in the private sector would then be imported into the military by admirals and generals eager to make their forces as efficient as AT&T, Sears Roebuck, or General Motors—or later IBM, Wal-Mart, or Microsoft.

Coming along just after the telephone, another major technology that transformed communications was the radio. In 1896 Guglielmo Marconi, a wealthy twenty-two-year-old Anglo-Italian inventor, patented a "wireless telegraph" for transmitting dots and dashes through the ether. Before long, his invention gained the capacity to transmit voices and music too. Marconi's Wireless Telegraph Company took an early lead in harnessing this technology, but by the early 1920s it had been overtaken by Germany's Telefunken, France's Compagnie Générale de Télégraphie Sans Fil, and the Radio Corporation of America, all formed to free their nations of dependence on British communications networks. Britain, which had dominated cable telegraphy and pioneered wireless telegraphy, became an also-ran in the radio age. Not even Marconi's discovery in 1924 of a method for bouncing shortwaves off the ionosphere—which allowed radio signals to be sent across the globe simply and cheaply—could resurrect Britain's once-dominant position.

The world's first commercial radio station was opened in 1920 (KDKA in Pittsburgh), and before long radio had transformed the culture by transmitting everything from political addresses and sporting events to comedies like *Amos 'n' Andy*. A medium invented by Edison, motion pictures, completed the transformation of popular entertainment, particularly after the addition of sound in the mid-1920s. Both radio and film would become invaluable propaganda tools in wartime, motivating the masses to keep fighting and keep making sacrifices.

While broadcasting was the most popular radio application for the masses, point-to-point communications proved more important for military and civil transportation. Two-way radios allowed ships and airplanes to update their coordinates and send distress signals if the need arose. Indeed, the radio first gained worldwide fame when it was used to transmit news of the *Titanic*'s sinking in 1912. In the 1920s, radio signals came to be used as aviation navigational aids. By listening to a series of beeps in their earphones, pilots could tell whether they were on course. It became possible to fly by instruments alone at night or in foul weather. Air-traffic control systems created in the 1930s using radios and other navigational aids dramatically reduced the incidence of airplane crashes.

As radio signals proliferated, many people began to notice that passing ships or airplanes would temporarily interfere with transmissions. The Scottish scientist Robert Watson Watt took advantage of this phenomenon to lay out a method of (to quote the title of his groundbreaking 1935 memorandum) "Detection of Aircraft by Radio Methods." The method was simplicity itself: use a directed antenna to transmit a radio pulse into the ether and time how long it took for it to bounce back. Since a radio wave moves at a constant speed (186,281 miles per second), the range of whatever object had caused it to bounce back could easily be computed. The location of the airplane could then be displayed as a blip on a screen using a cathode ray tube, which had been refined for a new medium called television. The entire system came to be known as radar (short for radio detecting and ranging), and it would prove to be of great benefit not only for guiding friendly airplanes but also for shooting down hostile ones. A similar system for using sound waves to detect objects in the water was also developed: sonar (sound navigation ranging).

Electrical generators, internal combustion engines, motor vehicles, airplanes, radios, telephones, radar: All of these technologies that contributed to the growth of the world economy between 1919 and 1939 (most of that growth occurring before the Great Depression started in 1929) would help to devastate a large portion of the world in the six years that followed. While the U.S. was a leader in exploiting these innovations in peacetime, it had not done nearly enough to apply them to the demands of warfare. It spent a mere 1.5 percent of its GDP on defense in 1937. Britain and France spent more—5.7 percent and 9.1 percent, respectively—but their economies were increasingly second-rate. Britain, which had accounted for 22.9 percent of world manufacturing output in 1880, was down to just 10.7 percent in 1938. The U.S. was the undisputed leader in manufacturing in 1938, with 31.4 percent of the global total. The Soviet Union was in second place with 17.6 percent. Ominously, Germany was the No. 3 industrial power, almost as big as Britain and France combined, with 13.2 percent of world output. Japan lagged farther behind, with 3.8 percent of world output, but its technology in key military areas was as advanced as any in the world. The one bright spot for the Western democracies was the parlous economic condition of Italy, the least fearsome of the predatory states, which had a mere 2.9 percent of world output.

But while Mussolini was not a formidable adversary, Hitler and Tojo most definitely were. Their countries had spent a decade building the most advanced armed forces in the world, utilizing the full fruits of the Second Industrial Revolution to realize their mad dreams of conquest. The Western democracies were hard-pressed to match them. While Americans tooled around in Packards and flew on DC-3s, the Germans and Japanese were building tanks and dive bombers. The latent potential of the U.S. and the USSR was almost limitless, and in any prolonged war their greater resources would be likely—though far from certain—to prevail against smaller nations like Germany and Japan. But that was scant comfort to the Allied servicemen who, in the early years of World War II, would have to face the onslaught of well-oiled military machines that had revolutionized the art of war as thoroughly as Henry Ford had revolutionized individual transportation.

CHAPTER 7

TANKS AND TERROR:

France, May 10–June 22, 1940

On the morning of Friday, May 10, 1940—"a most glorious spring day"—the American playwright Clare Boothe was staying as a guest at the U.S. ambassador's ornate residence in Brussels. "I was sleeping so soundly I did not hear the alarm at dawn," she wrote, "but a maid shook me violently by the shoulder and said: 'Wake up! The Germans are coming again!' "

The French and British high commands had gotten a similarly rude awakening just a few hours earlier. There had been plenty of indications that an attack was coming, but the early warnings had been missed. The Allies were caught utterly by surprise when the German invasion of Luxembourg, Belgium, and Holland began at 4:35 A.M. So ill-prepared was the British Admiralty that it did not even have a night watchman available to open its doors for an emergency meeting of the armed services' chiefs of staff at 7 A.M. General Sir Edmund Ironside, the hulking chief of the Imperial General Staff, found the "door double and treble locked," and had to squeeze his six-foot-four-inch frame through an open window. "So much for security," he commented sardonically.

Yet there was no panic. Not at first. The United States ambassador in Brussels, John Cudahy, a veteran of World War I and heir to a Milwaukee meatpacking fortune, proceeded with a scheduled dinner party that night. "We sat down to a very good dinner under an enormous crystal chandelier," wrote Clare Boothe with studied nonchalance, "which only shook very slightly when an anti-aircraft gun went off or a bomb landed in the region of the airport."

Allied generals were equally unperturbed. The French commander in chief, General Maurice Gamelin, strutted up and down his headquarters humming to himself, a staff officer noted, "with a pleased and martial air which I had never seen before." At last there was an end to the Phony War—the period of relative inaction known to wags as the Bore War or the *Sitzkrieg* that had followed Germany's invasion of Poland in September 1939—and the French army would have a chance to show what it could do! "This is the moment that we have waited for," exulted another French general.

At 6:30 A.M., in accordance with the detailed plans worked out over the preceding months, orders were given for the bulk of the British and French armies to advance into formerly neutral Holland and Belgium to block what they thought would be the main German thrust. Everyone expected a replay of 1914, with the *Boches* once again bogging down and a stalemate emerging. The Allies hoped that eventually an economic blockade would squeeze Herr Hitler into making peace.

Things began to go wrong almost immediately. By the end of the first day, the German air force, the Luftwaffe, had destroyed half of the Dutch air force on the ground, while a daring assault by German paratroopers had taken the bridges over the Moerdijk causeway, splitting Holland in two and preventing the Dutch army from linking up with advancing French forces. The Dutch were forced to retreat north toward Amsterdam and Rotterdam. On Tuesday, May 14, following the terror bombing of Rotterdam, the Netherlands surrendered.

The Belgians fared little better. The linchpin of their defense was the massive Eben Emael fortress on the Albert Canal. This powerful redoubt had only one point of vulnerability: its undefended roof. That was precisely where German glider troops landed on May 10. The German capture of Eben Emael the next day, along with two bridges over the Meuse River, severely disrupted Allied plans, which called for the fort to hold out for at least five days while French and British troops set up a defensive line.

Anglo-French forces arriving in Belgium discovered growing chaos. Ridiculous rumors, the first signs of panic, swept through the French ranks—"everywhere swarms of parachutists, saboteurs and spies, sometimes dummies and sometimes purely imaginary," wrote Captain André Beaufre. There was even a tale circulating of German soldiers dressed as nuns. This may have been pure fiction, but the reality was bad enough. The Germans' air-and-armor offensive, popularly known as *blitzkrieg* (lightning war), was hitting with devastating psychological force.

Junkers Ju-87 Stuka dive bombers were particularly terrifying because of the suddenness with which they could appear from a clear sky and swoop down with their sirens screaming. "Nobody who has ever heard the whistling scream made by dive bombers before releasing their load is ever likely to forget the experience," wrote Marc Bloch, a prominent medieval historian and French army reservist who would be executed by the Gestapo in 1944 for his work in the French Resistance. He recalled getting up after one attack in which no one was killed: "[I]t left me profoundly shaken, and when I crept out of the ditch where I had been crouching I was trembling pretty badly." This caused him to reflect on how "this dropping of bombs from the sky has a unique power of spreading terror. . . . It seems to crush the very air with unparalleled violence, and conjures up pictures of torn flesh which are only too horribly borne out in fact by the sights one sees."

Allied morale was further shaken by the dismaying tendency of panzers to show up where they were least expected. German columns would cover thirty, forty, even sixty miles a day—far faster than troops had moved just a generation earlier and far faster than British or French forces could cope with. Bloch described how a French officer ran into a mysterious column of tanks in the main street of a northern French town:

> They were, he thought, painted a very odd colour, but that did not worry him overmuch, because he could not possibly know all the various types in use in the French army. But what did upset him considerably was the very curious route that they seemed to be taking! They were moving [west] in the direction of Cambrai; in other words, away from the front. But that, too, could be explained without much difficulty, since it was only natural that in the winding streets of a little town the guides might go wrong. He was just about to run after the commander of the convoy in order to put him right, when a casual passer-by, better informed than he was, shouted—"Look out! They're Germans!"

As such unexpected and unsettling encounters multiplied, a mass migration of refugees began. "Somewhere in the north of France a boot had scattered an ant-hill, and the ants were on the march," wrote Antoine de Saint-Exupéry, the French aviator and author of *The Little Prince*. "Laboriously. Without panic. Without hope. Without despair. On the march as if duty bound." Entire villages packed up and fled, taking their priest, constable, mayor, and livestock with them. Invalids and babies rode on hay wagons. Men pushed perambulators stuffed with their belongings. Every type of vehicle was harnessed for escape, including hearses and ice cream wagons. British General Alan Brooke wrote of the refugees: "They are the most pathetic sight, with lame women suffering from sore feet, small children worn out with traveling but hugging their dolls, and all the old and maimed struggling along." The refugees did not know where they were going; they just knew they had to get away from the *Boches*. Gustave Folcher, a French army reservist, tried to talk to some but found "[t]he people are half mad, they don't even reply to what we ask them. There is only one word in their mouth: evacuation, evacuation."

Eventually hordes of dispirited French soldiers would join the rabble of the damned. But discipline held for the first few days. British and French troops in Belgium even enjoyed a few limited successes in the early going. On Sunday, May 12, the lead elements of the French army clashed with German troops near the Belgian town of Hannut, not far from Waterloo. In the first major tank-on-tank battle in history, two French light armored divisions fought hard and managed to temporarily block the advance of two panzer divisions before pulling back with heavy losses to prepared defensive positions on Tuesday, May 14.

With a series of such holding actions, the Allies might have been able to slow down and eventually stymie the German invasion, as they had in 1914. If, that is, the Germans were bent on replicating the Schlieffen Plan. But they were not. The center of gravity, the *Schwerpunkt*, of their offensive did not lie in the north—a fact that was not grasped by the slow-witted Allied high command until far too late. From the start there had been reports of German movements in the Ardennes Forest on France's eastern frontier, but their significance had been discounted. How could anyone possibly attack through this hilly and inaccessible area?

By Monday, May 13—three days after the invasion began—it had finally begun to dawn on the French generals that perhaps the Ardennes sector constituted the main German effort. But even then they reacted with what historian Alistair Horne has aptly termed "fatal leisureliness." "Our own rate of progress was too slow and our minds were too inelastic," Marc Bloch ruefully noted, "for us ever to admit the possibility that the enemy might move with the speed which he actually achieved." André Beaufre, then a captain, later a general, agreed: "All our doctrine was founded on faith in the value of defence. . . . [W]e were facing an adversary who waged a kind of war for which we were ill prepared."

As the Allies belatedly realized, the Germans had developed a new kind of warfare made possible by the Second Industrial Revolution emphasizing dazzling speed and daring maneuver, while French and British thinking was still mired in the more static approach of the First Industrial Age. This was more than a bit ironic, because, until just a few years before, Britain and France had actually led the world in the development of armored warfare. The story of how they lost that lead, and with it the early battles of World War II, demonstrates how two sides with equal access to essentially the same technology can put it to very different uses.

THE PROPHETS OF ARMOR

One of the oddities of history is that the tank, the ultimate weapon of land warfare, was originally developed by the British navy. Under the goading of First Lord of the Admiralty Winston Churchill, who was eager to find a way to break the stalemate of the Western Front, the Royal Navy began experimenting in early 1915 with steel-plated "landships" built on a tractor chassis. They were first referred to as "tanks" as part of a cover story which held that they were merely portable water containers.

On November 20, 1917, the Royal Tank Corps launched the first large-scale tank offensive in history. Four hundred and seventy-six British tanks shattered the Germans lines at Cambrai to a depth of six miles. Stupefied German soldiers fled in terror before these movable, smoke-belching forts. But the British were not able to keep going, because too many of their fragile machines broke down. The Germans mustered their courage and drove the slow and ungainly tanks back with an artillery barrage. The French also built many tanks and employed them in offensives that were equally inconclusive, largely because the tank was not yet a very sturdy or reliable contraption.

When the war ended in November 1918, the British and French were planning to field thousands of tanks in a massive blitz in 1919 that would have anticipated many of the innovations employed by the Germans twenty years later. The architect of these plans was (in the words of the Tank Corps' official historian) "a little man, with a bald head, and a sharp face and a nose of Napoleonic cast, his general appearance, stature and feature earning him the title of Boney." Colonel J. F. C. "Boney" Fuller was chief of staff of the Tank Corps, a career infantry officer, graduate of Sandhurst, and veteran of the Boer War who, in the estimation of the corps' official historian, "stood out at once as a totally unconventional soldier, prolific in ideas, fluent in expression, at daggers drawn with received opinion, authority and tradition."

Fuller, who titled his autobiography *Memoirs of an Unconventional Soldier,* positively reveled in his reputation as "a heretic." Before the Great War, he had been a disciple of Aleister Crowley, a cult leader who advocated drug use, sex rituals, Satanic worship, and the occult. Fuller even wrote a 327-page book expounding the principles of "Crowleyanity." In 1911 Fuller finally severed ties with Crowley, whose behavior he found increasingly destructive and megalomaniacal. He did not abandon his interest in the unconventional, but it was not as an enthusiast for yoga or Kabbalah mysticism—both subjects on which he wrote books in the 1920s and 1930s—that he would make his mark on history. It was as the high priest of the cult of armor.

Fuller thought the tank was a wonder weapon that would render obsolete not only the horse but also the infantryman. He envisioned tanks supported by airplanes knifing through front lines to destroy command and communications centers in the rear. Having been "shot through the brain," the enemy's body—the mass of its army—would then collapse into a leaderless, demoralized rabble that could easily be rolled up. Fuller hoped that this strategy could be used to avoid the ghastly stalemate of the First World War: "To attack the nerves of an army, and through its nerves the will of its commander, is more profitable than battering to pieces the bodies of its men."

This was, in essence, the theory that the Wehrmacht put into practice in 1939. But in the 1920s the British army did more than any other to make it a reality. In 1927, at a time when Germany did not have a single tank, Britain set up the Experimental Mechanized Force, the prototype of the armored division, equipped with medium and light tanks, armored cars, self-propelled guns, and motorized infantry in trucks and half-tracks. Through its elaborate exercises on Salisbury Plain, the Experimental Mechanized Force showed that it was possible to effectively maneuver large bodies of tanks utilizing two-way radios (or "radio telephones," as they were known). Its maneuvers may have been sniffed at by traditionalists—after witnessing one exercise, Rudyard Kipling complained, "It smells like a garage and looks like a circus"—but they were closely followed abroad, especially in Germany and the Soviet Union. Also studied were the official British publications on armored warfare: a booklet on *Mechanized and Armored Formations* (1929) and *Lectures on Field Service Regulations, Part III* (1932).

The *Lectures* were written by Fuller, but by the early 1930s his influence was rapidly waning. He had missed his big chance when he had turned down the command of the Experimental Mechanized Force over a petty matter of bureaucratic politics. He was promoted to major general in 1930 and then summarily retired. Boney Fuller continued to be a prolific writer after leaving the army but an increasingly marginal one. He joined Oswald Mosley's British Union of Fascists and took to denouncing democracy ("nothing more than a pluto-mobocracy") and the Jews ("the Cancer of Europe"). His intemperate blasts at the "know nothings" and "mineralized intellects" who ran the army alienated most of his potential supporters within the ranks. One of his favorite targets, Field Marshal Sir Archibald Montgomery-Massingberd, complained, with some justice, that all of Fuller's books "are written with the intention of annoying someone."

Fuller's cause was effectively taken up by a younger, less acerbic, and more genteel champion who looked (to a novelist's eye) "far more like the cartoonist's idea of a learned professor than a military man at all." Basil Liddell Hart had been a British army captain in the Great War and later became an influential historian and defense correspondent for *The Daily Telegraph* and *The Times*. His most famous contribution to strategy was his advocacy of the "indirect approach," which he claimed had been utilized by all "great captains" to strike where their enemies least expected. He believed that tanks and airplanes could make this strategy even more effective in the 1930s than it had been in Scipio Africanus's day or William Tecumseh Sherman's. Although Liddell Hart was a skilled and shameless self-promoter who became a close adviser to War Minister Leslie Hore-Belisha (1937–40), he did not manage to make any more of a dent in the British army of the 1930s than Fuller did.

The army disbanded the Experimental Mechanized Force after a couple of years. It created the 1st Tank Brigade in 1931, but did not form an armored division until 1939. By then, Britain had long lost its lead in mechanized warfare. The island nation continued to build tanks in the 1930s, but they were mostly light models designed for scouting, and instead of massing them together, as Fuller, Liddell Hart, and other strategists advocated, the army chose to parcel them out to infantry divisions where they would be incapable of achieving a decisive breakthrough. Moreover, the British did not continue developing a coherent doctrine of tank warfare—i.e., a set of widely shared and understood principles about how these weapons should be employed. They did not continue exercising large tank formations in the field to gain operational experience with newer and faster models. And they did not practice coordinating air and armor operations.

When war came, the tiny British Expeditionary Force sent to France was entirely motorized (relying in part on requisitioned civilian vehicles), but it had to make do with just a single tank brigade and some armored cavalry regiments. "In September 1939," wrote Bernard Law Montgomery, who as a general commanded a division in France, "the British Army was totally unfit to fight a first-class war on the continent of Europe."

Fuller, Liddell Hart, and other critics were quick to ascribe this failure to the stupidity of the "military mind," but this is only partly correct. There were indeed some purblind Colonel Blimps who continued to insist that tanks would never replace the good old horse. As late as 1926, Field Marshal Sir Douglas Haig would write: "Aeroplanes and tanks . . . are only accessories to the man and the horse, and I feel sure that as time goes on you will find just as much use for the horse—the well-bred horse—as you have ever done in the past."

Such sentiments existed in all armies. Britain primarily resisted the creed of armor in the 1930s, however, not because of the woolly-headed influence of the Haig mind-set but simply because of the overall neglect of its armed forces. The bulk of the paltry defense budget in the interwar years went to the Royal Navy and Royal Air Force, which were, not unreasonably, judged vital to keep Britain safe from invasion. The army, such as it was, was mainly devoted to colonial policing in places like Ireland, India, and Palestine, where there was no need for large tank formations; a few armored cars were quite sufficient to put down "native revolts." Many Britons thought they would never again fight another land war in Europe. Ironically, Liddell Hart and Fuller were among those who argued against a continental commitment. This proved more in tune with the popular mood in the decade of appeasement than their calls for strong armored and air forces.

France, too, had its prophets of armor but they were even more lonely and embattled than their British counterparts. General Jean-Baptiste Estienne, inspector of tanks from 1921 to 1927, was an early and vociferous proponent of deploying heavy tanks independently of infantry formations. Thanks in part to his influence, France became a leader in tank design in the 1920s and 1930s. The Char B1 was the best heavy tank in the world; German shells would simply bounce off its armor. The Somua S35 medium tank was also world-class. It combined speed, protection, and hitting power better than any rival. But France was slow to mass-produce these tanks, and even slower to figure out how to make the best use of them.

A fiery young captain named Charles de Gaulle was impressed by "the idea of the autonomous operation by armoured detachments—an idea," he later wrote, "whose advocates were General Fuller and Captain Liddell Hart." In 1934, while on the faculty of the St. Cyr military academy, he published a book called *Toward a Professional Army* in which he proposed the creation of a volunteer, highly trained "army of shock troops" that would "move entirely on caterpillar wheels." These mechanized units, supported by their "indispensable comrades in arms," the air force, could "rapidly move round far in the rear of the enemy, strike at his sensitive points, throw his whole system into confusion."

The leftist Popular Front government was aghast at the notion of creating a professional military that it feared would become a den of reactionary conspiracy. The military high command was no more impressed by de Gaulle's ideas. Instead of concentrating its armor in the kind of elite outfits that he favored, the army dispersed it among infantry and cavalry divisions composed of indifferently trained draftees. As late as July 1939, General Maurice Gamelin, the French army commander, told the parliament, "One must not exaggerate the importance of mechanized divisions. They can play an auxiliary role in enlarging a breach, but not the major role the Germans seem to expect of them."

In keeping with this philosophy, France in the 1930s concentrated on light armored forces designed to support infantry and artillery, not to fight tanks on their own. On paper, France created the world's first armored division in 1933: the *division légère mécanique* (DLM), or light mechanized division. But it was not fully operational until 1938. The first heavy armored division (*division cuirasée*, or DCR) was not ready until 1939. When the German invasion came in May 1940, France had three DCRs and three DLMs, though without enough modern tanks to fully equip them. A fourth DCR was hastily set up on May 15, in the middle of the campaign, and given to de Gaulle, by then a general, to command. It was too little, too late.

The French had forfeited an opportunity to maintain a lead in armored warfare because their post–World War I mind-set was so defensive. Traumatized by the slaughter in the trenches, they wanted to avoid casualties at all costs. The most famous symbol of this mentality was the Maginot Line, a system of fortifications completed in 1937 along France's eastern border between Switzerland and Belgium. The Maginot Line (named after Minister of War André Maginot) has been much mocked and little understood. In and of itself, it was not a bad idea: The Germans never did penetrate most of its formidable bunkers. But France did not extend it all the way north to cover the Belgian border where the German army had invaded in 1914. Here they had to rely upon mobile forces to fill in the gap. And that is where the trouble occurred. The French mode of operations was primarily passive, designed to minimize casualties, not to achieve decisive results. Their doctrine of "methodical battle" emphasized defense over offense, firepower over maneuver, rigid planning over freewheeling improvisation, centralization over local unit autonomy, and maintaining a continuous front instead of opportunistically trying to punch holes in the enemy lines.

It was, in every respect, the opposite of what the Germans were training for. The French essentially anticipated a replay of 1914–18, expecting that the Germans would charge their defenses and get slaughtered en masse. As a result, they made no serious effort to attack Germany from the west while most of its armed forces were engaged in Poland in September 1939. France's defensive orientation, de Gaulle would complain, "handed the initiative over to the enemy, lock, stock, and barrel."

BUILDING THE BLITZKRIEG

Germany's ability to seize the initiative so completely was rather surprising at first glance, given how badly its military was handicapped until the early 1930s. The 1919 Treaty of Versailles limited Germany to an army of just one hundred thousand men (one-third the size of Poland's), and forbade it from having tanks, armored cars, artillery, or airplanes. Even the fabled General Staff was abolished. These seemingly insuperable difficulties actually proved a blessing in disguise, for they forced the Germans to keep only their very best soldiers, noncommissioned officers, and officers in an all-volunteer, long-service force. It also forced them to think hard about the nature of future war, planning for the day when they would be free of the shackles of Versailles.

Germany began circumventing the Versailles Treaty from the start, first of all by preserving the core of its General Staff under the thin fiction of the *Truppenamt* (Troop Office). The first postwar chief of the General Staff was General Hans von Seeckt, a Pomeranian nobleman from an elite regiment who looked the part, replete with monocle, mustache, stiff carriage, and cold stare. He went on to serve as commander of the army, the Reichsheer, from 1920 to 1926. One of his first moves was to commission a comprehensive staff study of the lessons of World War I—something that the Germans, as the defeated power, undertook much more thoroughly than the victors did.

The French had decided, based on the trauma of the trenches, that the defensive was now the dominant force in warfare. The Germans concluded, to the contrary, that the eternal principles of war had not changed: The offensive still remained the only way to achieve victory. They reasoned that the Schlieffen Plan for the invasion of France had come close to working in 1914, and there was no reason to abandon such audacious schemes in the future. The General Staff studies pointed out that on the southern and eastern fronts of the Great War—where Seeckt and many other prominent officers had served—the fighting had remained mobile, with cavalry dashing hither and yon. Even on the Western Front, the Germans could look with some satisfaction to the infiltration or "storm trooper" tactics they had pioneered in 1917–18: Small infantry units armed with rifles, flamethrowers, and light machine guns used speed and surprise to pierce weak spots in the Allied lines, going around strongholds in an all-out dash for the unfortified rear. In the great offensive of March 1918, storm troopers achieved a major penetration of the Allied front, but the Germans were unable to move up reinforcements fast enough to keep going. If they had, they probably would have won the war.

The advent of armored vehicles and tanks—which the Germans had foolishly neglected in World War I, perhaps their biggest mistake of the war—offered a way to use storm trooper tactics to achieve more decisive results in the future. "The whole future of warfare appears to me," wrote Seeckt, "to lie in the employment of mobile armies, relatively small but of high quality and rendered distinctly more effective by the addition of aircraft." In Britain and France, those were the views of heretical officers of limited influence; in Germany this became received wisdom among the military high command.

Naturally, the German armed forces, like all others in the interwar period, had their share of troglodytes: cavalry officers devoted to the horse and cautious officers devoted to the defensive. One general rejected the idea of using trucks to carry troops into battle, reportedly shouting, "To hell with combat, they're supposed to carry flour!" What set the Germans apart was the intellectual rigor with which they tested competing doctrines utilizing the methods pioneered by Moltke the Elder: historical studies, sand table exercises, war games, and field maneuvers.

Since they could not field actual tanks, they had to make to do with canvas-and-plywood mockups mounted on bicycles and automobiles. They may have looked like something out of a vaudeville skit, but these dummy tanks showed the utility of armor for future operations. The Germans were further able to hone their expertise through a secret arrangement under which a small number of soldiers tested tank and airplane prototypes in the Soviet Union, far from the prying eyes of Allied inspectors. The Germans were also willing to learn from the work of others; Boney Fuller's theories had a particularly large influence on them. Liddell Hart, despite his later attempts to hog the lion's share of the credit, was a lesser influence.

As long as Germany formally pledged to abide by the Versailles Treaty, it could not field a full-fledged tank force or air force of its own. (It did produce small quantities of banned equipment through secret contracts with domestic and foreign suppliers.) But those limits came off after Adolf Hitler became chancellor in 1933. In the following six years the German armed forces grew from 100,000 men to 3.7 million. Tanks and aircraft were at the forefront of the newly expanded force. Hitler had endless enthusiasm for both technologies, which jibed with fascist visions of the future and his own longstanding fear of horses. Around 1934, Colonel Heinz Guderian, chief of staff of the Inspectorate of Motorized Troops, gave the Führer a short tour d'horizon of tank warfare. "Hitler," Guderian wrote, "was much impressed by the speed and precision of movement of our units, and said repeatedly, 'That's what I need! That's what I want to have!' "

What the Führer wanted, the Führer got, though it must not be imagined that his desires conflicted with those of the armed forces. They were just as enthusiastic about mechanization, and just as eager to be free of the constraints of Versailles. In 1935 Germany created its first three armored divisions, with command of one of them going to Guderian, by now a major general.

The son of a Prussian officer, Guderian had started off as an infantryman. He served in the Great War as a signals officer, General Staff officer, and infantry battalion commander. He had not previously been one of the army's leading armor theoreticians; others, such as the now-forgotten Ernst Volckheim, had been more influential. It was only after the war that he gained his first exposure to tanks. As a General Staff officer in the Motorized Transport Department in the early 1920s, he took a crash course in armor theory, studying the works of Fuller, Volckheim, and other "far-sighted soldiers

[who] were even then trying to make of the tank something more than just an infantry support weapon."

With his command of one of the first panzer divisions and his publication in 1937 of a best-selling book on tank warfare, *Achtung—Panzer!*, "Hammering Heinz" vaulted past his mentors. He became the most visible champion of armor within the Wehrmacht (as the German army was renamed); by 1938, he had been promoted to head of a whole armored corps. He was part of that rare breed: a thinker and a doer, someone capable of both coming up with daring ideas and implementing them in the field.

While Guderian and other German strategists were influenced by their English counterparts, they departed from their teaching in important ways. Fuller and Liddell Hart had envisioned tanks operating more or less autonomously. The Germans realized that tanks on their own could not consolidate the gains they made, nor overcome every obstacle in their path. Tanks could be vulnerable to artillery or even to infantrymen throwing grenades on their tracks. And they could be slowed by rivers or other obstacles. All of these limitations required tanks to have supporting forces that could keep up with them. "I became convinced," Guderian later wrote, "that . . . tanks would never be able to produce their full effect until the other weapons on whose support they must inevitably rely were brought up to their standard of speed and of cross-country performance."

Thus the panzer division that Guderian took over was not made up exclusively of tanks (though it had a lot of those: more than 550). It was an all-arms organization of fourteen thousand men complete with infantry in trucks, motorcycles, and armored personnel carriers to mop up resistance and hold ground; reconnaissance units in armored cars to scout ahead; signals technicians to maintain communications links; engineers to dismantle or surmount obstacles; and towed artillery, antiaircraft guns, and antitank guns to deal with foes that the tanks themselves could not handle. While the division did not have its own airplanes, it did work closely with the Luftwaffe, which provided air support.

This was another major difference between the Germans and their enemies in France and Britain. The Allied air forces concentrated their efforts mainly on "strategic" bombing, designed to shatter the enemy's infrastructure and morale. The Royal Air Force and the French Armée de l'Air refused to buy large numbers of dive-bombers, the aircraft best suited for close air support. Nor did they practice coordinating their operations with their own armies. The Luftwaffe was far more adept at "tactical" bombing designed to help the Wehrmacht shatter enemy armies. Strategic bombing could take a fearsome toll on a country, as we will see, but it was a weapon for a long-term war of attrition. Only tactical bombing—which the Royal Air Force disdained as tantamount to "prostitution of the air force"—could help ground forces achieve a rapid victory.

In a series of maneuvers in the late 1930s, the German army showed that it could successfully employ panzers supported by warplanes for deep penetrations. Of vital importance was the use of radios mounted in vehicles—something that Guderian, with his World War I communications background, particularly emphasized, and yet another difference from the French and British armies, which did not have adequate communications with their field units. The Germans were able to maneuver much more adroitly because most of their tanks had at least a radio receiver and every command vehicle had a transmitter. This also made possible the German generals' preferred style of leadership—from the front. Commanders moving in the vanguard of their forces could respond quickly to unforeseen events while staying in touch with their far-flung subordinates. This was quite a change from World War I, when generals had frequently set up shop in luxurious chateaux far behind the lines, a style of leadership that many Allied generals still practiced in 1940.

The German leadership advantage extended to the lower ranks, which, in the best Moltkean tradition, were taught to exercise their own initiative. German commanders continued to issue spare *Auftragstaktik* (mission-type orders) and push authority down to the lowliest NCO, while the French forces relied on a more centralized style of command that made it difficult for them to deviate from elaborate plans prepared before the start of fighting. Contrary to cultural stereotypes, the soldiers of the Nazi regime often displayed more individual initiative than the soldiers of the liberal democracies. (This would later change when Hitler's paranoia and irrational orders limited his subordinates' freedom of action.) The German army's traditional strengths—decentralization, improvisation—proved a perfect fit with the emerging technologies of armored warfare.

Because of his fascist sympathies, J. F. C. Fuller was one of only two Britons invited to Hitler's lavish fiftieth birthday party in Berlin on April 20, 1939. The celebration included a three-hour parade of mechanized forces. After the panzers had finished roaring past, Hitler came up to Fuller and said, "I hope you were pleased with your children?"

Fuller replied, "Your Excellency, they have grown up so quickly that I no longer recognize them."

CASE WHITE AND CASE YELLOW

Hitler was now ready to put his new army to a real-life test. He was able to annex Austria (1938) and Czechoslovakia (1939) peacefully and thereby add their substantial arsenals to his own. Spoiling for a fight, he attacked Poland on September 1, 1939. Case White (*Fall Weiss*) began with air strikes on Polish airfields, followed by two giant panzer spearheads knifing swiftly through Polish defenses. The Poles, who had not yet fully mobilized, were totally outmatched. They had few tanks of their own, and although it is a myth that Polish lancers deliberately charged panzers (they knew better than that), it is a fact that on some occasions Polish cavalry did wind up getting mauled by German tanks. The Wehrmacht needed only eight days to reach the outskirts of Warsaw, with Guderian's corps in the vanguard. Poland capitulated at the end of the month after Soviet forces invaded from the east to join hands with the Nazis.

In its coverage of this campaign, *Time* magazine gave a new word to the world: *Blitzkrieg*. The Germans themselves did not use this term (they preferred *Bewegungskrieg*, or "war of movement") but the name stuck, conjuring up an image of fast-moving mechanized columns. This rather obscured the reality: Due to the severe limitations of German industry, only a small portion of the German force was actually motorized. Almost all of their tanks were either confiscated Czech models or lightweight Panzer I's and II's, not the more modern and powerful Panzer III's and IV's. And, far from being unstoppable, many of their vehicles broke down on the plains of Poland. The bulk of the Wehrmacht moved as their ancestors had, marching on foot and pulling supplies in horse-drawn wagons. But, small as it was, the mechanized component was sufficient to annihilate the woefully equipped Polish forces. Would it be enough to conquer the armies of France and her Western European allies as well?

Most German generals doubted it. They thought Hitler's desire to take on the Great Powers of the West—Britain and France declared war over the invasion of Poland—was madness. It was one thing to conquer Poland or even Denmark and Norway, whose turn came in April 1940. But was not the French army "the strongest in Europe," as the chief of the German General Staff proclaimed in 1938? The German generals were all too aware that they were outnumbered by the combined forces of Britain, France, Belgium, and Holland. By May 1940, Germany had 135 army divisions to 152 for the Allies; 2,439 tanks to 4,204 for the Allies; 3,369 fighters and bombers to 4,981 for the Allies; and 7,378 pieces of heavy artillery to 13,974 for the Allies. Even if all the Allied forces were not on the Western Front, these were hardly the kind of odds that would give an attacker confidence of winning.

The German generals followed orders, however, and began preparing for the next invasion as soon as the conquest of Poland was completed. The initial version of Case Yellow (*Fall Gelb*) called for a virtual replay of the Schlieffen Plan, with the bulk of German forces, as in 1914, once again sweeping through Holland and Belgium into northern France. This thrust would be the responsibility of Army Group B under Colonel General Fedor von Bock. Army Group A under Colonel General Gerd von Rundstedt would mount a minor operation farther south, around the Ardennes Forest, to distract French forces from the main blow.

Rundstedt's chief of staff, General Erich von Manstein, a brilliant strategist, was not satisfied with this conservative plan, which he feared would result in another 1914-style stalemate. He suggested moving more of the panzer force through the dense and hilly Ardennes, where the French would not expect them to strike.

He would not have had his way if Hitler had ordered the invasion to proceed, as originally intended, in the fall of 1939. But due to bad weather and other glitches, the Führer kept postponing the start date until the following spring. In the meantime, there occurred one of those fateful and utterly random events that can alter the course of history. On January 10, 1940, a German transport plane crashed in Belgium carrying an officer who had details of Case Yellow with him. His documents were seized before he could destroy them. The German high command now had to assume that their scheme was known to the Allies. This gave Hitler the impetus to order a change in plans. The Führer was lobbied by Manstein, with support from Rundstedt and Guderian, for the Ardennes plan, and he told his planners to make it happen. War games confirmed the viability of this strategy.

The final version of Case Yellow called for 112 divisions to be hurled against France and the Low Countries. Seven panzer divisions—the bulk of the armored punch—were to be concentrated under Rundstedt's Army Group A in the Ardennes. Bock's Army Group B would invade Holland and Belgium from the north with the remaining three panzer divisions and pin down Allied forces there while Rundstedt swooped around to their south, trapping them in a giant vise that could then be squeezed tight. It was a bold gamble, an inverted version of the Schlieffen Plan that depended on deceiving the Allies even after the start of operations about the plan's true *Schwerpunkt* (center of gravity). General Franz Halder, chief of the General Staff, put the odds of it working at ten to one against but concluded that all the other alternatives were worse. Only Hitler remained supremely confident of victory because he had smelled out the inability of "the systematic French or the ponderous English . . . to operate and to act quickly."

Hitler's reading of the enemy proved all too accurate. General Gamelin, the elderly commander of the French army, had drawn up plans that played right into the Führer's hands. The Germans wanted the Allied armies to advance into Holland and Belgium, where they could be trapped and outflanked, and that is exactly what Gamelin intended. The French plan assumed that the major German thrust would be coming from the north and therefore assigned the best Allied infantry and armored divisions to this sector.

The Ardennes region was regarded as a backwater. It was held by the French Ninth and Second Armies, composed mainly of older, "B-series" reservists who had done their military service many years earlier and had received scant training or new equipment in recent years. These troops had been called up in September 1939, but they did not use the next nine months to make up for their serious deficiencies. (The Germans, by contrast, spent this period correcting mistakes made during the Polish campaign.) Most of them were put to use digging fortifications, not learning to maneuver and fight in the open field. Their equipment remained for the most part antiquated, and they had few antitank guns, much less tanks. With this troop deployment, General Gamelin could not have done more damage to his country if he had been a paid Nazi agent.

THE ROAD TO DUNKIRK

When the offensive unfolded on Friday, May 10, 1940, Bock's Army Group B made faster progress than expected with its assault through Holland and Belgium. While all Allied eyes were fixed on the north, Rundstedt's Army Group A met only scattered resistance as it crossed into Luxembourg and then into southern Belgium and northeastern France. By Saturday, May 11,

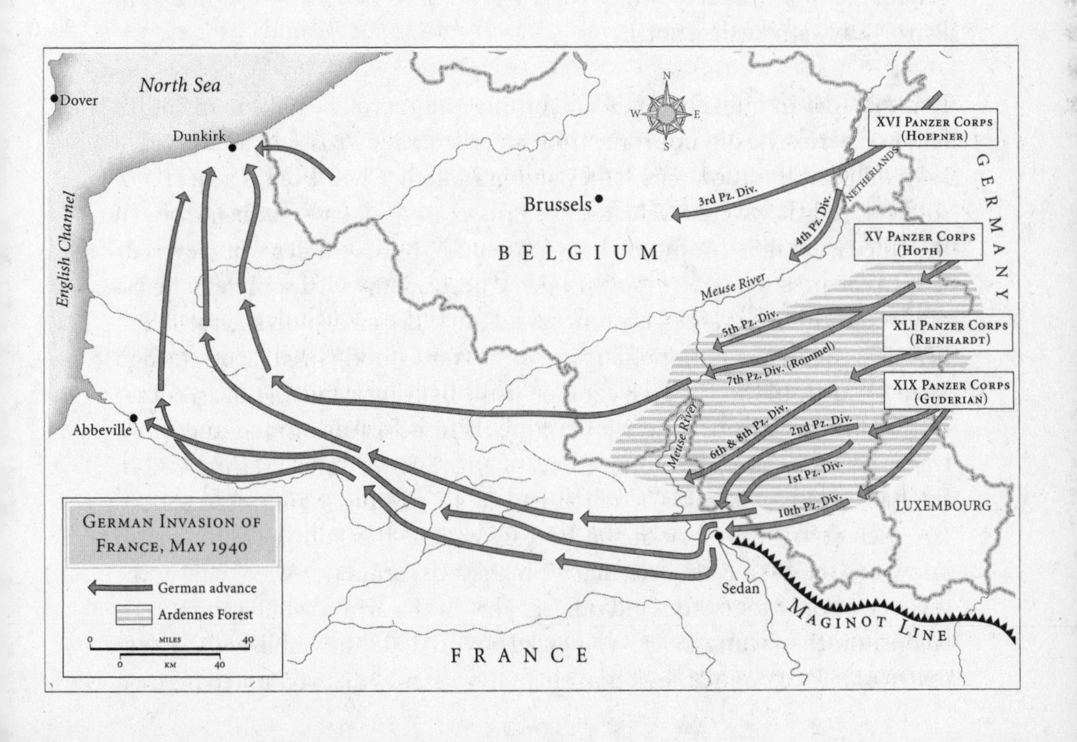

the panzers were deep into the Ardennes, where they found it slow going along the narrow, winding mountain roads. A colossal traffic jam developed with thousands of vehicles backed up for miles. If the Allies had hit the Germans hard from the air, they could have snarled up the advance in a hopeless mess. But most Allied planes were committed to northern Belgium. When the few fighters and bombers in this sector tried to attack they were decimated by heavy fire from German antiaircraft guns and Messerschmitt Bf-109 fighters, both among the finest of their kind in the world.

By late afternoon on Sunday, May 12, just two and a half days after the start of the invasion, the Germans had reached the Meuse River and occupied the town of Sedan virtually without a fight—the very spot where in 1870 Moltke had defeated an earlier French army. The town itself had little strategic value, but the river was of incalculable significance. It was the most prominent geographical obstacle in the path of the German advance. Beyond it lay virtually nothing but plains all the way to the English Channel.

French troops managed to blow every bridge across the Meuse. They dug in on the steep western bank, secure in their bunkers and foxholes, confident that it would be days before the Germans could accumulate sufficient forces to hazard a crossing. That would have been the case if this were still 1914. But it was twenty-six years later, and the enemy the French were facing was far more mobile than the German army of World War I. While the main job of crossing the river had been assigned to Guderian's XIX Corps around Sedan, the first unit across the river turned out to be Erwin Rommel's 7th Panzer Division farther north.

Rommel was an unlikely candidate for this leading role. Unlike many of his fellow generals, he did not come from an aristocratic Prussian family with a long military tradition. His father and grandfather had been teachers, not soldiers, and they were Swabians, not Prussians. And, far from being one of the elite General Staff officers, he was a muddy-boots soldier who never attended the prestigious *Kriegsakademie* (War Academy). To complete the list of incongruities, he was by training an infantry, not an armor, specialist. But his background as a junior officer in World War I stood him in good stead in World War II. In the 1914–18 conflict, while fighting against France, Romania, and Italy, Rommel had shown himself to be a fearless and audacious leader who was skilled in infiltrations of enemy lines. In one daring raid in the Italian mountains, his battalion had captured nine thousand prisoners. For such feats he won both the Iron Cross and the Pour le Mérite (also known as the Blue Max), Germany's highest decorations. After the war, he was one of only four thousand officers kept in the army. While serving as a colonel on the faculty of the War Academy in 1938, he published *The Infantry Attacks*, a popular textbook on infantry tactics that recounted his own war experiences. It was probably this book that first drew him to Hitler's attention. Although Rommel, like Guderian, was never a member of the Nazi Party (and, unlike Guderian, he would be implicated in the 1944 plot against Hitler), he, like many of his fellow generals, admired Hitler for his dynamism and his desire to make Germany great again.

Rommel was considered reliable enough to serve as commander of Hitler's personal protective detail during the occupations of Czechoslovakia and Poland in 1939. As a reward for loyal service, he asked the Führer to give him command of a panzer division. Hitler obliged, and forty-eight-year-old Major General Rommel took over the 7th Panzer Division in February 1940. He had only three months to get ready for the invasion of France, and he had no previous armored experience to draw upon. Yet when Case Yellow unfolded, Rommel proved to be the most successful panzer leader in the entire German army. He maneuvered tanks in a bold and unconventional style reminiscent of the way he had led infantry raiding companies in the Great War.

By the afternoon of Sunday, May 12, Rommel's men were desperately looking for a way across the Meuse. Near the village of Houx, some infantrymen spotted part of an ancient dam that had not been destroyed by the French. A small group of dismounted motorcyclists sneaked across a stone weir and lock gate under the cover of darkness. Not long after midnight, the Germans had their first, tenuous toehold on the western bank of the Meuse. But could they keep it?

Once they discovered what had happened, the French defenders opened up a murderous fire on riflemen who were trying to row across in rubber boats to reinforce the motorcyclists who were already on the other side. "Our boats were being destroyed one after the other by the French flanking fire," Rommel found when he arrived on the scene at 4 A.M. on Monday, May 13, "and the crossing eventually came to a standstill." Other attempted crossings by his division were faring no better, so Rommel decided to take matters in hand personally. After calling up some panzers to provide covering fire across the river, he jumped into a dinghy himself and rowed across with some men. On the other side of the river, he saw French tanks approaching. Since his men had no antitank weapons with them, he ordered them to fire their rifles, machine guns, even a flare pistol to scare away the enemy. Amazingly enough, it worked. Rommel then crossed back to the eastern bank and organized the construction of the first pontoon bridge across the Meuse. When it was ready, he was among the first across in his eight-wheeled command vehicle. He found the situation on the western bank "looking decidedly unhealthy," so he crossed back yet again and oversaw the ferrying of tanks across the river. Fifteen panzers had arrived by the morning of Tuesday, May 14—"an alarmingly small number," Rommel noted—but it was only through his inspirational leadership that even that many had made it across.

A few miles farther south, around Sedan, Guderian was similarly struggling and just barely succeeding in crossing the Meuse. His way was smoothed by a ferocious series of Luftwaffe attacks. All day on Monday, May 13, some 1,500 warplanes pummeled French positions on the west bank. The inaccurate Stukas did not cause much damage, but they took a heavy psychological toll on the terrified French infantrymen who had never been under air attack before. A French general later described how "[t]he gunners stopped firing and went to ground, the infantry cowered in their trenches, dazed by the crash of bombs and the shriek of the dive-bombers. . . . Their only concern was to keep their heads down. Five hours of this nightmare was enough to shatter their nerves, and they became incapable of reacting against the enemy infantry."

At 3 P.M., the air strikes lifted and the moment of truth arrived. German soldiers pushed rubber boats into the river and began rowing the longest sixty yards of their lives. Many were shot out of the water, but some managed to gain the other bank. A handful of infantrymen fired automatic weapons to get close enough to toss grenades into a bunker. The defenders came out to surrender and the Germans draped a swastika flag over the captured bunker, to the sound of loud cheers from the eastern bank. Then the assault troops crawled on to the next bunker.

On it went: hard fighting for a few yards of dirt. By day's end, May 13, the Germans had gained another foothold across the Meuse. A pontoon bridge was prepared, and around midnight a few panzers began to roll across.

The German bridgeheads remained extremely precarious as long as they had few tanks on the other side of the river. If the French had reacted quickly and decisively, they could have wiped out the exposed salients and perhaps altered the whole course of the war. But once again the Allied high command proved too confused and too lackadaisical. It did not help that the French commander in chief, General Gamelin, was in one headquarters just outside Paris, while the commander of the critical northeastern front, General Alphonse Georges, was in another building thirty-five miles away, and his chief of staff was in yet a third location midway between the other two, designated the headquarters of land forces. These multiple command posts were so worried about having their telephone conversations intercepted that they communicated only by motorcycle dispatch riders. Gamelin didn't even have a radio in his headquarters. Similar signals lapses, exacerbated by German cutting of telephone lines, prevented efficient coordination among the principal Allied armies in the field. "In fact," wrote General Bernard Law Montgomery, "it may be said that there was no co-ordination between the operations of the Belgians, the B.E.F. [British Expeditionary Force], and the First French Army; the commanders of these armies had no means of direct communication except by personal visit."

This lack of synchronization showed in the Allies' inability to scramble badly needed reinforcements to the critical sector. Three French tank divisions had been dispatched on the evening of Sunday, May 12, but they were still not ready to counterattack by the morning of Tuesday, May 14. The French 3rd Armored Division should have been thrown at the Germans' exposed flank at once. But nervous commanders refused all day to give the order to attack. In the early evening of May 14, they broke up this powerful division into small packets of tanks that were dispatched to guard various roadways, where they could do nothing to delay, much less stop, the panzer onslaught.

Allied warplanes fared no better. The Royal Air Force (RAF) bravely tried to destroy the pontoon bridges across the Meuse, but their aircraft ran into a wall of flak. Of 71 Fairey Battle light bombers that set out on May 14, 40 did not return, making this the single costliest raid in the RAF's entire history.

As the German foothold on the western bank of the Meuse grew larger and larger, whole French divisions began to crumble, some of them breaking apart at the mere rumor of panzers nearby. On Wednesday, May 15, the Wehrmacht broke through the final remnants of the French Ninth and Second Armies. The panzers were well into the French rear, just as the prophets of armor had predicted, and the battle of France was all but decided just five days after it began. "A perfect road stretches before us and no enemy fliers over us," one German tanker exulted. "The air is completely dominated by our own fliers. A wonderful feeling of unconditional superiority."

It is an exaggeration to say that the Germans simply drove unimpeded to the Channel coast—but not by much. The Allies could muster little in the way of counterattacks. Brigadier General Charles de Gaulle, at the head of the newly formed 4th Armored Division, struck Guderian's panzers on May 17 and May 19. He inflicted some casualties, but, contrary to Gaullist mythology, this did not amount to more than a pinprick.

During an emergency meeting at the French Foreign Ministry on Thursday, May 16, Prime Minister Winston Churchill (who had taken over from Neville Chamberlain only six days before) asked General Gamelin where the strategic reserves were. The French commander in chief shrugged and replied, "*Aucune,*" meaning he had none. When Churchill glanced out the window, he could see clouds of smoke rising from giant bonfires in the garden and "venerable officials pushing wheel-barrows of archives on to them." The evacuation of Paris was already under way.

The only thing that could slow the Wehrmacht was the caution of its own commanders. As the panzers cut a long, narrow swath through France, the higher ranks of the German army became increasingly nervous about vulnerable supply lines and exposed flanks. What if the Allies counterattacked? Couldn't they cut off and annihilate the panzer vanguard before the supporting infantry could catch up?

The commanders on the spot were not worried. Rommel and Guderian could see demoralization spreading in the enemy ranks as they raced past. "Near Senzeille," Rommel wrote, "we met a body of fully armed French motorcyclists coming in from the opposite direction, and picked them up as we passed. Most of them were so shaken at suddenly finding themselves in a German column that they drove their machines into the ditch and were in no position to put up a fight."

But on Friday, May 17, the leading panzer units were ordered to halt. Guderian was so furious that he threatened to quit on the spot. A compromise was reached: He would be allowed to carry out a "reconnaissance in force." Guderian took advantage of this loophole to send his entire armored corps speeding ahead, carefully concealing this movement from army headquarters by not transmitting his orders by radio, which he feared would be intercepted by his own side. On the evening of Monday, May 20—just ten days after the start of the invasion—the first of Guderian's panzers reached the Channel coast, effectively encircling the bulk of the Anglo-French army in Belgium.

The next day, elements of the British Expeditionary Force, which was retreating from Belgium, attacked Rommel's 7th Panzer Division and a Waffen SS infantry regiment near Arras. There was some tough fighting before the British were repulsed and most of their tanks destroyed. This was hardly a critical setback, but it proved a turning point by heightening Hitler's growing timidity. Three days later, on May 24, he ordered the panzers to halt for good. There was no disobeying a directive from the Führer. Guderian's men were within sight of Dunkirk, where the remnants of the northern Anglo-French armies were penned in, but they were not able to finish what they had started. The job of wiping out this pocket was entrusted to the Luftwaffe. The fliers failed miserably. A total of 337,000 Allied soldiers, two-thirds of them British, were successfully evacuated to England between May 27 and June 4.

For the allies, this was the only saving grace from a humiliating defeat. But, as Churchill reminded the House of Commons, "We must be very careful not to assign to this deliverance the attributes of a victory. Wars are not won by evacuations."

This was not quite the end of the fighting. There was more combat when the Germans pivoted south to finish off the remnants of the French army, portions of which continued to fight tenaciously. On June 14 the Germans entered Paris, and three days later the French government, now led by Marshal Henri

Pétain, asked for an armistice. The formal treaty of surrender was signed on June 22 in the very railcar where German delegates had capitulated in November 1918.

The Germans had suffered 27,074 killed, 111,034 wounded, and 18,384 missing during their six-week conquest of France. "The overall total of just over 150,000," notes Alistair Horne, "was equivalent to not much more than a third of the number of men Germany lost at Verdun in 1916, in one single battle of the First World War." Allied casualties—the total number of missing, sick, wounded, dead, and captured from the British, French, Belgian, and Dutch armed forces—amounted to over 2.2 million. Britain still had the bulk of its army, albeit without much of their equipment, but Hitler was left in control of virtually the entire continent from the Vistula River to the English Channel.

WHY HITLER WON A BATTLE BUT LOST THE WAR

Once the immediate shock had passed, the search for explanations began. Why had Napoleon's heirs fallen so quickly? Many Frenchmen chose to blame insidious Fifth Columnists: Nazi and Communist agents who had supposedly undermined the homeland from within. (Nazi Germany and the Soviet Union were allies at this point.) Visiting Paris in May 1940, Clare Boothe heard constant talk of *trahi* (betrayed): "At first it was no more than a whisper. . . . And then the whisper became a great wail that swept through France, a great wail of the damned: 'Trahi . . . trahi. . . .' " Although a number of people were lynched as suspected enemy agents, in reality there were few German spies operating behind enemy lines, and their meager efforts can hardly explain the magnitude of the disaster that befell France.

A more plausible version of the "stab in the back" thesis was that France was undone not by active treason but by passive indifference: Following the carnage of the Great War and two decades of political turmoil pitting right against left, the French had simply lost the stomach to fight another costly war. This theory was widely held by those who participated in the actual events, but it has met with skepticism from modern historians who point to contemporary records showing that French morale had made a considerable recovery by the time Germany invaded Poland in September 1939. Hitler's aggression convinced most French people of the rightness and necessity of war. They were "resigned but resolute," in the words of the British ambassador. And in the war that followed, while some French units crumbled without a fight, many others fought hard even while suffering crushing casualties. In six weeks of combat, France lost an estimated 124,000 men killed and 200,000 wounded, more than the American casualties in the Korean and Vietnam Wars combined over the course of many years.

Even more damning to any attempt to ascribe France's defeat to its loss of will is the fact that morale on the other side was by no means as high as a zeppelin. Many Germans were as reluctant to fight as the French, not least among them generals who were so afraid that Hitler was leading them to disaster they had discussed mounting a coup to topple him. Only with the successful conclusion of the campaign against France did real enthusiasm for the war break out in Germany. French spirits would have been equally ebullient if their soldiers had won any victories to boast of. The low state of morale among the French cannot be entirely dismissed as an explanation for their downfall; there is no denying that most Frenchmen did not fight till the bitter end and that few joined the Resistance or the Free French forces. But the dominant view of recent historians is that the French loss of will was more the consequence, rather than the cause, of battles lost.

Why, then, did the Germans win those battles? The prevalent impression of the time—that the Allies were outnumbered—is false. As we have seen, the Allies enjoyed an advantage in the overall number of divisions, tanks, aircraft, and artillery pieces. The one critical area where they were deficient was in the number of bombers and fighters actually deployed on the Western Front. The Germans had 2,779, the Allies 1,448. But this was not due to some inherent deficiency on the Allied side; it was mainly because the British and French did not commit many of their aircraft to the fight. The British understandably chose to keep the bulk of their air force at home for self-defense. Less understandable, indeed inexplicable, was the French decision to keep many of their planes in southern France and North Africa, where they could do no good. The problem, in sum, was not how many aircraft the Allies had but how they were utilized.

Another popular misconception—that the Germans had superior weapons—does not stand up to scrutiny either. The best tanks belonged to the French, not the Germans; the best Allied aircraft were as good as the best German models. The only technical area where the Germans had a major edge was in their widespread use of radios, which gave them operational flexibility and the ability to concentrate their mechanized forces and warplanes at the decisive point of attack. This made up for the fact that the vast bulk of their troops walked to the front. (Out of more than one hundred German divisions mobilized for the campaign in the West, only ten were tank divisions and another ten were motorized.)

If the Germans did not have material superiority, what accounted for their easy victory? Quite simply, their decisive edge in doctrine, training, planning, coordination, and leadership. Writing in 1942, two years before his death,

Marc Bloch convincingly argued that "the German triumph was, essentially, a triumph of intellect." The Germans had adapted their methods of warfare to the Second Industrial Revolution, which had transformed "the whole idea of distance." The French had not. "The ruling idea of the Germans in the conduct of this war was speed. We, on the other hand, did our thinking in terms of yesterday or the day before. Worse still: faced by the undisputed evidence of Germany's new tactics, we ignored, or wholly failed to understand, the quickened rhythm of our times. So true is this, that it was as though the two opposed forces belonged, each of them, to an entirely different period of human history."

Still, there was nothing inevitable about the outcome. "As I looked at the ground we had come over," Guderian wrote, "the success of our attack struck me as almost a miracle." It is not hard to fathom why even this most self-confident and swashbuckling of generals would be agog at his own success. It could easily have gone the other way, especially if the Germans had stuck to their original version of Case Yellow or if something had gone wrong during their journey through the Ardennes or across the Meuse. That the Germans prevailed so quickly owes something to luck and even more to their meticulous preparation and inspired execution. While the daring use of panzers got most of the attention, the key breakthrough was due to courageous infantrymen rowing across a river under fire—a maneuver that the Germans had practiced meticulously beforehand on the Moselle River. When the time came for the actual crossing, Guderian was able to issue the same orders used in the exercises with only the dates, times, and locations changed. Thus the final victory was a tribute not to panzers alone but to the skillful employment of the combined-arms concept.

From a historical perspective, the German victories in Poland, Norway, Denmark, Belgium, Holland, and, above all, France helped to reestablish the possibility of the decisive campaign. The ability to achieve clear-cut results on the battlefield had been in decline since the mid-nineteenth century, when the firepower and sheer size of armies had increased beyond the ability of transportation and communications networks to cope. Generals could barely find their foes (recall how blind both Moltke and Benedek were on the eve of Battle of Königgrätz), much less maneuver effectively to destroy them. Faced with machine guns or even rifles, cavalry could no longer perform its traditional role of hunting down and destroying the tattered remnants of defeated armies. This meant that the losing side on the battlefield could usually make good its escape, as Lee did after Gettysburg, and return to fight another day. The growing indecisiveness of war reached its apotheosis in World War I, where, on the Western Front at least, combat became a senseless struggle for a few yards of ground. Now, with the rise of mechanized forces, the art of maneuver could once again be practiced as skillfully as it had been by Frederick the Great or Napoleon Bonaparte. With their lightning victories, Hitler's legions had shown that force of arms could win wars, not just battles. Or so it seemed in 1940.

In the warm, heady afterglow of victory, the Germans tended to forget all the doubts that had plagued them before and during the invasion of France. More and more generals joined Hitler in concluding that their war machine was invincible and unstoppable. Not even their failure to knock Britain out of the war could disabuse them of this illusion. Weakened by its losses over France, the Luftwaffe could not establish air superiority over southern England in the summer and fall of 1940, and Hitler had to call off his planned invasion, Operation Sea Lion. Germany returned to the path of conquest in 1941 with the swift occupation of Yugoslavia and Greece. Rommel's Afrika Korps also enjoyed steady success in North Africa against British forces from the time of its arrival in February 1941 until the battle of El Alamein in October 1942. By then Rommel's operations had become a mere sideshow to the much larger war being fought in Russia.

Hitler invaded the Soviet Union on June 22, 1941, with 3.2 million soldiers, 3,600 tanks, and 700,000 horses. No matter that he faced an enemy with millions more men, three times more aircraft, and five times more tanks. His armies enjoyed swift and stunning success against the ill-prepared Russian troops deployed on the frontier. The Soviet tank and air forces were almost completely annihilated. The Nazis advanced deep into Russia, arriving on the doorstep of Leningrad and Moscow by the winter of 1941. Then the offensive stalled out, partly as a result of stout Soviet resistance but mainly due to the inherent limitations of the Wehrmacht.

The blitzkrieg had proved a devastating weapon in the relatively confined spaces of western Europe. Its force was considerably dissipated on the nearly endless steppes of Mother Russia. Warfare in the Second Industrial Age required moving not only tons of food and ammunition but also tons of fuel and lubricants to keep tanks and trucks on the go. German logisticians were simply not able to keep their armies supplied more than a few hundred miles beyond the frontier (Sedan to Dunkirk is 170 miles); in Russia, German armies quickly found themselves more than a thousand miles from their bases.

These difficulties were compounded by the onset of the harsh Russian winter, for which the Nazis had not prepared; they had expected the entire campaign to be over in four months. The Germans found themselves trapped deep inside Russia, freezing, hungry, exhausted, running low on fuel and ammunition, and facing an adversary that was growing stronger by the day. The turning point was the Battle of Stalingrad. By the time it was over in January 1943, the Germans had lost 209,000 men killed and 91,000 captured. Hitler tried one last major offensive at Kursk in July 1943. His forces were repulsed in the largest battle of the war, pitting more than two million men and six thousand tanks against each other.

The Soviet victory at Kursk represented a hard-won armored renaissance. The Russians had been among the leaders in developing mechanized forces in the 1920s and early 1930s. Under Marshal Mikhail Tukhachevsky, they had invented the doctrine of "deep battle," a variant of blitzkrieg, which called for masses of tanks and airplanes to penetrate hundreds of miles behind the enemy's front lines to isolate and encircle opposing forces. In 1937, Stalin had executed Tukhachevsky in his purge of Red Army officers. The "deep battle" doctrine was discredited along with its founder, only to be revived in 1942–43 after the Soviets had suffered severe setbacks at the hands of Nazi tank forces. On the Eastern Front, then, the Germans eventually faced an adversary that fought much as they did—only with far more men and tanks and airplanes to throw into the fray.

The same thing happened in the West. The fall of France alerted Britain and America to the need to develop better mechanized forces. The U.S. had deployed a Tank Corps in World War I, but it was disbanded in 1920 over the anguished objections of two of its leading officers—Colonel George S. Patton and Major Dwight D. Eisenhower. In the interwar years, the U.S. Army spurned the innovative American tank designer J. Walter Christie, who sold his work to Russia, where it formed the basis of the workhorse T-34 tank. In the 1930s the U.S. Army limited its mechanization to one cavalry brigade. The first armored divisions were not formed until after the fall of France, in July 1940. They were tested in war games in Louisiana and Tennessee in 1941, and dispatched the following year to fight in North Africa. U.S. troops did not perform well at first, but by the time of the Normandy invasion in 1944 the U.S. possessed formidable armored forces grouped into all-arms divisions equipped with the serviceable if not spectacular M4 Sherman medium tank and led by generals like Patton whose abilities rivaled those of Rommel and Guderian. And, unlike the Germans, the Americans managed to motorize most of their army, rather than just its spearhead.

The British, likewise, fielded effective armored forces after the fall of France, starting with the 7th Armored Division, which under the tank pioneer General Percy Hobart thrashed Italian troops in North Africa in 1940–41, and continuing on to the much larger forces under Field Marshal Montgomery's command during the drive into Germany in 1944–45.

Despite the considerable achievements and painful sacrifices of the Allied armies in their quest for victory, the Germans ultimately did not lose the war because they faced forces superior in the quality of men or materiel. The Panzer V (Panther) and Panzer VI (Tiger), developed in 1942, were probably the best tanks of the war; Sherman tank rounds would simply bounce off their frontal armor, while they could wreck a Sherman with one shot.

The German soldier, too, was in all likelihood the best of the war. A postwar study by Trevor Dupuy, a retired U.S. Army officer, found that, right up until the end, German units had at least a 20 percent "combat effectiveness superiority per man" over Anglo-American forces, meaning that "[o]n the average, a force of 100 Germans was the combat equivalent of 120 American or 120 British troops." The German advantage over the Russians was even greater. According to Dupuy, one hundred German soldiers were the equivalent of two hundred Russians. While Dupuy's findings have been questioned, there is little doubt that the Germans were at least as effective as their enemies, if not more so.

But few armies, no matter how effective, can prevail when outnumbered as badly as the Germans were by the later stages of World War II. They faced a crippling deficit not only in manpower but also in materiel. As early as 1942, the United States was outproducing all of her enemies combined—in historian Richard Overy's summation, "47,000 aircraft to 27,000, 24,000 tanks to 11,000, six times as many heavy guns." Add in Soviet production, which recovered rapidly after the catastrophes of 1941, and the disparity became almost insuperable.

The Allied weapons may not have been as technologically sophisticated as some German models, but they were cheap, durable, and plentiful. Henry Ford, with his mass production techniques, was more valuable to the Allied cause than any general: "The Ford company alone," Overy notes, "produced more army equipment during the war than Italy." The Allies also had access to vast pools of natural resources that the Axis could not match; most critically, they controlled 90 percent of the world's natural oil production.

If the Allies had fought as incompetently as they had in 1939–41, they might have frittered away these considerable material advantages. Luckily for them, by 1943 their tactical skills had improved enough—if still not perhaps to the German level—to make effective use of the products being churned out by their factories.

This goes to show the limits of a military revolution. If Hitler had possessed the sagacity of a Bismarck and made peace following the victories of 1939–40, as Bismarck made peace following the victories of 1864, 1866, and 1870, he might have consolidated the conquests won by his peerless war machine. By choosing to push the blitzkrieg farther than it could reasonably go—by taking on both the U.S. and USSR—the Führer consigned Germany to a war of attrition that it would have been hard-pressed to win, unless, perhaps, it had developed a true *wunderwaffe* (wonder weapon) like the atomic bomb. (Hitler's ersatz wonder weapons, the V-1 cruise missile and V-2 rocket, were not enough.)

THE FALL OF FRANCE IN RETROSPECT

Its ultimate failure should not cause us to discount the historical importance of the blitzkrieg. Its early success in France had profound and lasting consequences—and not only for the millions of European Jews consigned to the bloody grasp of the Nazis. "For, more than anything else," historian David Reynolds writes, "it was the fall of France which turned a European conflict into a world war and helped reshape international politics in patterns that endured for nearly half a century, until the momentous events of 1989."

Reynolds makes a compelling case. The unexpectedly quick victory of the Germans in 1940 encouraged greater aggression on the part of the Axis. Indeed, it led to the formation of the Axis in the first place. Until the invasion of France, Italy had sat out the war, waiting to see which way the wind would blow. The gale-force success of the blitzkrieg led Mussolini to declare war on France and Britain on June 11, 1940, in the hope of getting a share of the spoils. The outcome in France also tipped the political debate in favor of the pro-German party in Japan. On September 27, 1940, Tokyo joined Berlin and Rome in the Tripartite Pact which divided the world among them.

Encouraged by this alliance, Italy tried to annex Greece and Egypt, with disastrous consequences that were ameliorated only with German help. Mussolini's blundering opened up a new theater of operations in the Mediterranean that would preoccupy British and American forces throughout 1942–43 and delay Operation Overlord, the invasion of France, long enough to ensure that Soviet forces would wind up occupying Eastern Europe.

The defeat of France also meant that its colonies in the Middle East fell into the hands of Vichy regimes closely associated with the Nazis. "It was at that time that the ideological foundations of what later became the Baath Party were laid," writes historian Bernard Lewis, "with the adaptation of Nazi ideas and methods to the Middle Eastern situation." Baathists would come to dominate Syria and Iraq for decades and destabilize the entire region with their virulent nationalism and despotism.

The events of 1940 helped to reshape not only Europe and the Middle East but the Far East as well. The fighting around the Mediterranean tied down the British navy, leaving it unable to reinforce Singapore and other Asian bastions. The RAF was also unable to provide much help because it was fighting off Luftwaffe attacks on the British Isles. British weakness in Asia, along with that of the French and Dutch colonists whose homelands were occupied, spurred Japan to undertake greater acts of aggression. It began by occupying northern Indochina in September 1940. Hitler's early success in the Soviet Union, which blunted a potential adversary of Japan's, led Tokyo to intensify its offensive with the attack on Pearl Harbor in December 1941 followed by a sweep of Western outposts in East Asia. Japan might well have taken these actions anyway regardless of what happened in Europe, but Hitler's victories certainly offered encouragement by showing how weak their mutual enemies were. A French journalist living in Tokyo wrote that the Japanese "were dazzled by the early German victories, fascinated by Hitler, bewitched by the coming partition of the world."

Japan's rapaciousness, in turn, ensnared Hitler in conflict with yet another adversary. Four days after the raid on Pearl Harbor, he honored the spirit, if not the exact language, of the Tripartite Pact by declaring war on the United States. (The treaty required its members to come to each other's defense only if they were attacked, not if they did the attacking.) It is a close call whether this or the invasion of the Soviet Union was his biggest blunder of the war. Either way, Hitler's actions were influenced by the events of 1940. He probably would never have opened another front in the East if Case Yellow had proved indecisive and Germany had gotten bogged down in a stalemate in the West. The operation's quick, glorious conclusion encouraged him to think that he could replicate its success in the Soviet Union and win the war before the "mongrel" Americans—"half Judaized, half Negrified," he sneered—could intervene against him.

Thus, as Reynolds argues, the indirect effect of the fall of France was to drag both the U.S. and USSR into the war and thereby help usher in the bipolar world that lasted until 1989. Admittedly, that is a long chain of causation, some of whose links are formed of nothing more than informed speculation. But there is no denying the basic conclusion that the fall of France—brought about by the success of the blitzkrieg—radically changed the course of World War II and, hence, of the postwar world.

King Charles VIII of France ushered in the modern age with his invasion of Italy in 1494. Italy's ineffectual mercenary armies were not ready for the speed and ferocity of the French advance. (PRINT COLLECTION, THE NEW YORK PUBLIC LIBRARY)

French artillery made short work of old-fashioned castle walls. "Cities," wrote a Florentine politician, "were reduced with great speed, in a matter of days and hours rather than months." (GENERAL RESEARCH DIVISION, THE NEW YORK PUBLIC LIBRARY)

The English navy (left) battling the Spanish Armada in the English Channel. (Print Collection, The New York Public Library)

The Ark Royal, the English flagship in 1588, had four masts, multiple sails, and heavy cannons behind lidded gun ports. This was the basic shape of warship design for over three hundred years. (Ayer Company)

Above, left: Sir John Hawkins, a sometime pirate and slave trader, helped revolutionize naval construction when he became England's treasurer of the navy in 1577. (Library of Congress)

Above, center: Sir Francis Drake, Hawkins' cousin and fellow privateer and vice admiral. He was a master of sail-and-shot tactics. (Library of Congress)

Above, right: King Philip II of Spain was a micromanager and a religious zealot. But faith could not make up for his lack of an effective military bureaucracy. (Library of Congress)

March with the musket shouldered, the rest in your hand

Cock your match

Ram home your charge

Give fire

To get the most out of primitive arquebuses and muskets, the Dutch learned to drill their soldiers. These are panels from Jacob de Gheyn's *The Exercise of Armes* (1607), the first modern drill manual. (Greenhill Books)

King Gustavus Adolphus. His reforms and inspired leadership made Sweden, improbably enough, one of the leading powers of the seventeenth century. (PRINT COLLECTION, THE NEW YORK PUBLIC LIBRARY)

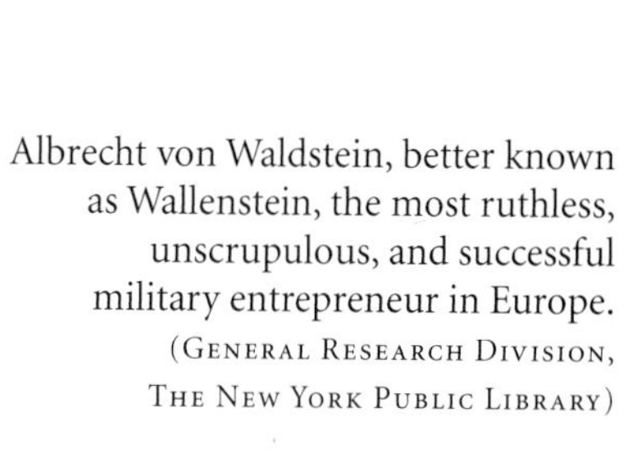

Albrecht von Waldstein, better known as Wallenstein, the most ruthless, unscrupulous, and successful military entrepreneur in Europe. (GENERAL RESEARCH DIVISION, THE NEW YORK PUBLIC LIBRARY)

Wallenstein and Gustavus had their final showdown at Lützen in 1632. Gustavus won the battle but lost his life. His death is depicted in this painting. (MID-MANHATTAN PUBLIC LIBRARY)

Sepoys (Indian soldiers serving the British East India Company) shown in 1790. The British success in forging native troops into European-style regiments made possible their conquest of South Asia. (Royal Collection, London)

Arthur Wellesley, the future Duke of Wellington, won his first notable victories in India. (Library of Congress)

Wellesley on horseback, leading troops at Assaye into "a most terrible cannonade." (National Army Museum, London)

Prussian troops build a railway and string telegraph lines. (Werner Company)

Helmuth von Moltke helped revolutionize military planning during his years on the Prussian General Staff. (Library of Congress)

Ludwig von Benedek, commander of the Austrian Northern Army, thought "the only talents required of a staff chief are a strong stomach and a good digestion." (Fontane, *Der Deutsche Krieg von 1866*)

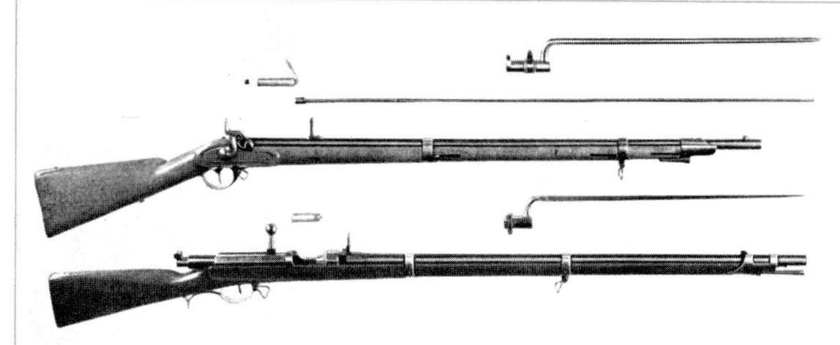

The Prussians' secret weapon, the *Zündnadelgewehr* (needle gun), displayed beneath the Austrian Lorenz musket. The needle gun was loaded through the breech, not the muzzle, allowing troops to fire faster. (Heeresgeschichtliches Museum, Vienna)

King Wilhelm I and his retinue survey the battlefield at Königgrätz. The Prussian victory made possible the rise of a militarized Germany run from Berlin. (Picture Collection, The New York Public Library)

Hiram Maxim poses with his most lethal invention. Along with other products of the Industrial Revolution, machine guns allowed Europeans to control 84 percent of the world by 1914. (BETTMANN/CORBIS)

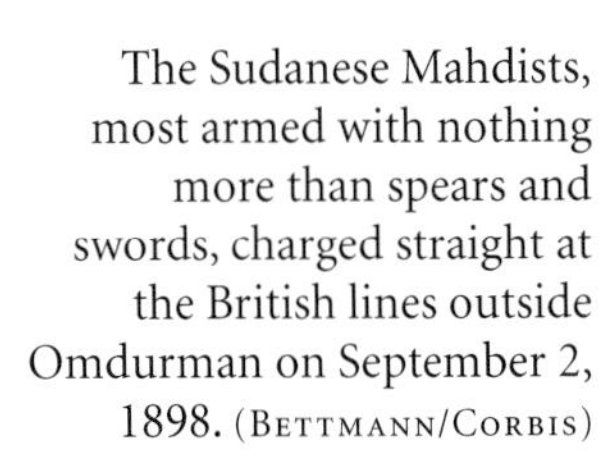

General Horatio Herbert Kitchener. "His precision is so inhumanly unerring, he is more like a machine than a man," wrote one journalist. (LIBRARY OF CONGRESS)

The Sudanese Mahdists, most armed with nothing more than spears and swords, charged straight at the British lines outside Omdurman on September 2, 1898. (BETTMANN/CORBIS)

British infantrymen calmly firing Lee-Metford MK II repeating rifles from behind their zeriba (a breastwork of mimosa scrub) at the advancing Sudanese. (HULTON ARCHIVE/GETTY IMAGES)

A Japanese naval gun firing at the Battle of Tsushima, May 27, 1905. Japan, which had no modern guns a half-century earlier, learned better than any other non-Western nation how to wage industrial warfare. (TOPHAM/THE IMAGE WORKS)

Left: Heihachiro Togo commanded the Japanese fleet at Tsushima. His audacious turn in front of the Russian warships sealed their doom. (LIBRARY OF CONGRESS)

Right: Zinovy Petrovich Rozhdestvensky, Togo's opposite number, was known for his womanizing and his explosive temper. (*THE ILLUSTRATED LONDON NEWS*)

The sinking of the Russian battleship *Borodino.* Only one sailor survived. (HULTON ARCHIVE/GETTY IMAGES)

J. F. C. Fuller, enthusiast for yoga, fascism, and armored warfare. His advocacy of tanks in the 1920s was prescient but he alienated the British army establishment with his abrasive manner. (CASSELL)

General Heinz Guderian in France, 1940. An influential armor strategist, he also proved to be a successful field commander. (NATIONAL ARCHIVES)

A panzer rolls into France. German tanks were no more numerous and no better than those of the French and British, but the Germans were much more adept in employing them to punch holes in the enemy lines. (NATIONAL ARCHIVES)

German signalmen in France. The Germans made more extensive use than the Allies did of two-way radios. This allowed them to move faster and to better coordinate air and armored forces. (National Archives)

The key breakthrough in the battle of France was accomplished by German infantrymen rowing across the Meuse River under fire. (National Archives)

The British built the world's first aircraft carriers in World War I. Here a Sopwith Pup takes off from HMS *Furious.* By 1942, the British lead in naval aviation was long gone—and so was most of their Asian empire. (IMPERIAL WAR MUSEUM, LONDON)

Rear Admiral William Moffett, first head of the Navy's Bureau of Aeronautics, brought the U.S. Navy into the air age. (LIBRARY OF CONGRESS)

Admiral Isoroku Yamamoto commanded the strike force that hit Pearl Harbor. He told a colleague, "I like games of chance. You have told me that the operation is a gamble, so I shall carry it out." (NATIONAL ARCHIVES)

Artist's rendition of the aircraft carrier *Akagi* (Red Castle) steaming toward Hawaii. By 1941, Japan had the world's best aircraft carriers and naval aircraft. (TOM FREEMAN/ARIZONA MEMORIAL MUSEUM ASSOCIATION)

B-29 Superfortresses (seen over Japan in 1945) were the biggest and most sophisticated aircraft of World War II. (National Archives)

B-29 cockpit on a bombing run. The pilot (on the right) switches on the auto pilot while the bombardier (center) adjusts his sights. (National Archives)

Curtis LeMay, commander of the B-29s which pulverized Japan. His men thought he was a "hardheaded bastard" but also "a great damn field commander." (Library of Congress)

This photo shows the impact of B-29 raids on Tokyo. At least 84,000 people died on March 9–10, 1945. (National Archives)

Saddam Hussein looked like a formidable foe when he occupied Kuwait in 1990. His army was the fourth-largest in the world and it was equipped with the best weapons his oil wealth could buy. (ANONYMOUS/SYGMA/CORBIS)

On the first night of Desert Storm, the USS *Missouri* launches a Tomahawk cruise missile. (DEPARTMENT OF DEFENSE)

An F-117 stealth fighter drops a 5,000-pound "smart bomb." An airpower strategist estimated that one aircraft flown by one pilot could now do the work that in World War II had required 1,000 bombers flown by 10,000 men. (DEPARTMENT OF DEFENSE)

An important American advantage was the quality of the all-volunteer armed forces. General Norman Schwarzkopf is shown addressing the troops. (DEPARTMENT OF DEFENSE)

General Tommy Franks faced a formidable challenge in responding to 9/11. He adopted an unconventional approach in Afghanistan because nothing else was possible on short notice. (Department of Defense)

The Predator, an unmanned aerial vehicle, made its combat debut in Afghanistan carrying Hellfire missiles. (General Atomics Aeronautical Systems)

Special Forces troops in Afghanistan rode horses and carried satellite radios, laptops, and laser designators. Their improvisation revealed the potential of Information Age warfare. (Department of Defense)

Many of the U.S. units that invaded Iraq in 2003 had portable computer terminals known as Blue-Force Trackers.
(U.S. Army)

Army and Marine armored units took Baghdad in a three-week blitzkrieg—exactly the kind of campaign they had trained for. Then the real war began.
(Department of Defense)

American troops were not prepared for the bombings and ambushes they faced in Iraq starting in the summer of 2003. "Insurgents don't show up in satellite imagery very well," a general lamented.
(Mohammed Khodor/Reuters/Corbis)

A diagram shows the Airborne Laser mounted in a 747 to shoot down ballistic missiles. It may herald a new era of high-energy weapons. (Department of Defense)

The X-47B, one of the first drones built especially for combat, is designed to operate off aircraft carriers. (Northrop Grumman)

Possible uniform of the future. Using nanotechnology, it may be possible to create "battlesuits" that change colors to blend into the environment and turn impermeable when danger is detected. (Natick Soldier Center)

CHAPTER 8

FLATTOPS AND TORPEDOES:

Pearl Harbor, December 7, 1941

The most powerful squadron of aircraft carriers yet assembled turned east into the wind at 5:50 A.M. and increased its speed to twenty-four knots to make sure that there was enough wind velocity for takeoff. The six Japanese flattops, along with their escort vessels, were just 230 miles northeast of Oahu. The seas were rough; high waves pitched the giant vessels from side to side as if they were corks bobbing in the water. If this had been a training mission, they would never have tried to launch in this kind of weather. But there was no turning back now.

The pilots ate a special celebratory breakfast of *sekihan* (red beans and rice), attended their final preflight briefing, and then climbed into their cockpits. Within 15 minutes, 182 aircraft—43 Zero fighters, 51 Val dive bombers, 48 Kate high-level bombers, 40 Kate torpedo bombers—were safely airborne. As mechanics frantically worked to prepare a second wave for launch, the first wave gathered into formation and headed south by the light of the rising sun.

Commander Mitsuo Fuchida, senior flight leader of the entire Hawaii Strike Force, led the attack personally aboard a three-man Kate bomber. The thirty-nine-year-old aviator was among the best that Japan, or any other nation, had. A graduate of the Eta Jima Naval Academy (Japan's Annapolis), the Kasumigaura flight school (Japan's Pensacola), and the Japanese Naval War College, as well as a veteran of the war in China, Fuchida was a combat-hardened professional with more than three thousand hours of flight time under his belt—ten times more than the average American pilot. He had put on red underwear and a red shirt for this mission; in case he was wounded, he did not want the blood to show and alarm any fainthearted comrades. Wrapped around his helmet was a white samurai *hachimaki* headband signed by his maintenance crew. On it was a single word: *Hissho* (Certain Victory).

Though he was a kind, warmhearted man who bore no personal animus toward America, Fuchida was also an ardent nationalist who worshipped the emperor and believed in the need to establish Japanese dominance across East Asia, which would not be possible as long as the United States barred the way. Along with many other Japanese, Fuchida was an admirer of Adolf Hitler; he even grew a toothbrush mustache and let a lock of his dark hair fall over his forehead in imitation of his hero. He particularly admired the German success at *Dengeki Sakusen* (thunder and lightning operations), as the blitzkrieg was known in Japan. As he sat in his cockpit, the canopy open and the wind rushing against his face, he was proud to now have the chance to match the exploits of Hitler's samurai.

It was not easy to see through the dense cloud cover, but Fuchida had no problem finding his way; he simply homed in on the signal of a Honolulu radio station. Suddenly the clouds cleared to reveal the sparkling blue surf and white sands of Kahuku Point on the northern tip of Oahu. Farther south, Pearl Harbor loomed before them, "glistening in the sun," as Fuchida later recalled. Through his binoculars he could see the awe-inspiring might of the U.S. Pacific Fleet at rest. He was disappointed that the two aircraft carriers were missing, as intelligence reports had indicated they would be. But he was thrilled to see seven battleships anchored in neat rows alongside Ford Island in the middle of the harbor. An eighth battlewagon was in dry dock. "What a majestic sight!" he muttered. "Almost unbelievable."

The Japanese pilots feared that they would be detected, forcing them to fight their way in. But there was no sign of alarm as they swooped closer. Not a single American patrol plane, nor a single puff of antiaircraft fire rose to challenge them. At 7:53 A.M. Fuchida had his radio operator send the prearranged signal—*Tora, Tora, Tora* (Tiger, Tiger, Tiger)—to alert their superiors that complete surprise had been achieved. Two minutes later, the bombing and strafing runs began.

As the first explosions went off, most people on the ground could not understand what was going on. Sunday was a lazy day in the peacetime armed forces, the one time of the week when the men could sleep in. A number were nursing hangovers from a wild Saturday night among the bars and bordellos of Honolulu's Hotel Street. The first reaction of many was to groan in exasperation and demand to know why some so-and-so had scheduled a training exercise on a day of rest. Even seeing airplanes with scarlet disks on their wings streak across the sky so low that, as one sailor put it, "you could chuck potatoes at them," did not convince some groggy servicemen that they were under attack. One sailor was heard to exclaim appreciatively, "This is the best goddamn drill the Army Air Force has ever put on!"

The possibility of air attack had always seemed remote on this idyllic tropical isle. American commanders had been more worried about sabotage by Hawaiians of Japanese ancestry. To make them easier to guard, aircraft were neatly lined up, wingtip-to-wingtip, on the tarmac; as it turned out, this also made them easier to destroy from above. Although some studies had warned of the danger of air attack, the prevailing opinion was that the Japanese were simply incapable of such a feat. Weren't they those funny little men with big eyeglasses and buck teeth? Surely they could not travel across an entire ocean to hit "the strongest fortress in the world," as General George C. Marshall had described Oahu less than eight months before. Vice Admiral Wilson Brown spoke for many when he asserted, with the certitude of the myopic, that "Japanese fliers were not capable of executing such a mission successfully." Even after Japanese fliers executed just such a mission, suspicion was rife in the United States that they could not have pulled it off alone; many Americans falsely assumed they must have had German help.

The Japanese success in catching Pearl Harbor unawares would have seemed even more inexplicable if the public had known about the top secret intelligence that their government possessed. Army and navy cryptanalysts had penetrated Japan's diplomatic codes and knew that an offensive was coming, but the general assumption was that it would hit elsewhere—probably the Philippines or Malaya. Certainly not Pearl Harbor. A last-minute alarm might have been raised by two army privates experimenting with a mobile radar set near Kahuku Point, one of five recently deployed around the island. They detected a large number of blips headed straight for them, only to be told by a bored duty officer not to worry about it; it must be a flight of B-17 bombers that was expected from the mainland. (The B-17s did arrive, right in the middle of the air raid.)

Generations of cynics have assumed that there must have been a cover-up somewhere, that U.S. government decision-makers must have known what was coming and chosen to conceal it in order to lure the U.S. into war. The reality is more prosaic. America fell victim to complacency, not conspiracy. Although the winds of war had been gathering for some time, most Americans preferred to think of more pleasant things. Nineteen-forty-one was the year of top-selling musical hits from the bandleaders Tommy Dorsey and Glenn Miller; of film comedies like Abbott and Costello's *Buck Privates* and Bob Hope and Bing Crosby's *Road to Zanzibar*; of Joe Louis's fight against Billy Conn for the heavyweight boxing title; and of the two great

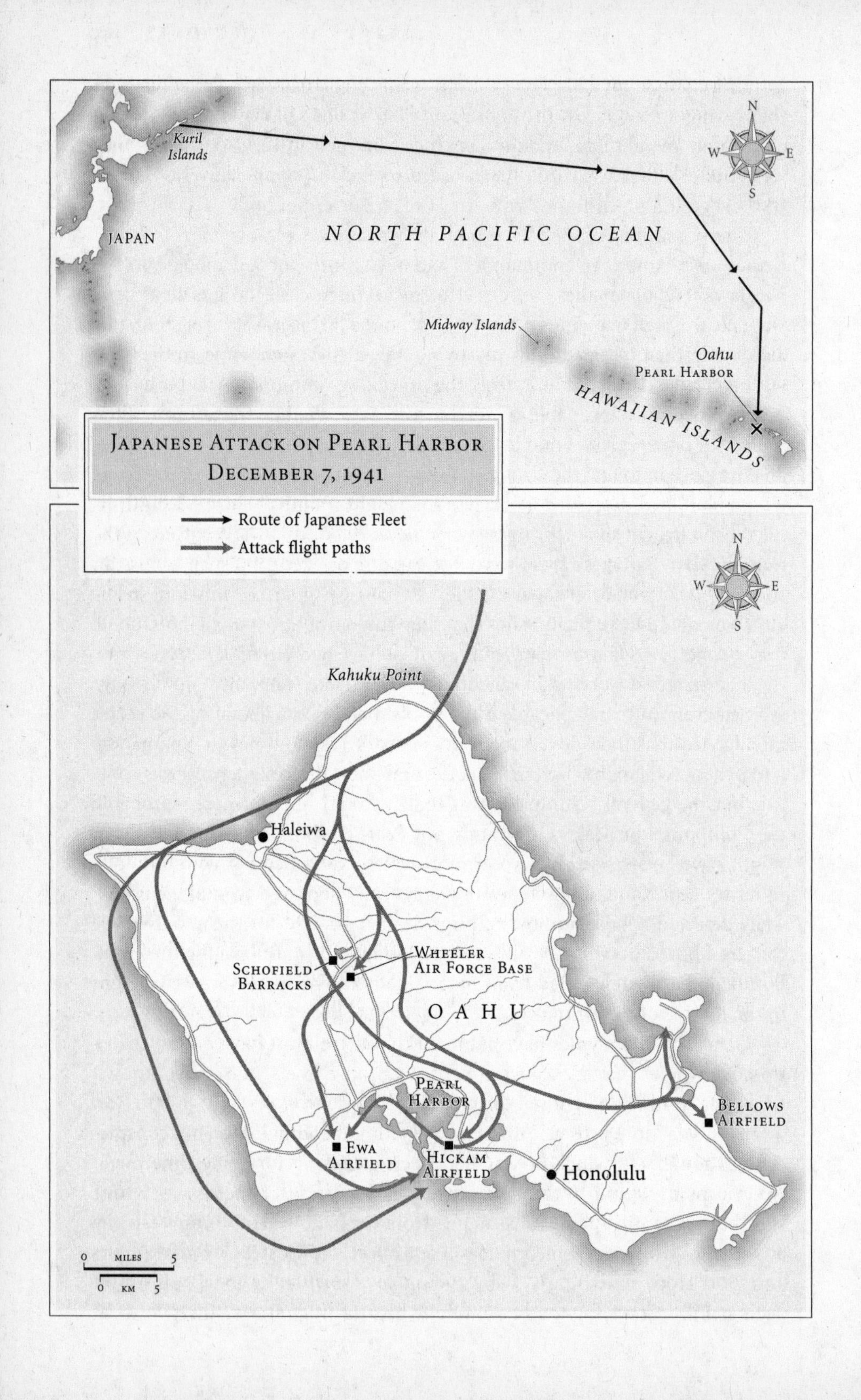

Kuril Islands
JAPAN
NORTH PACIFIC OCEAN
N
W
E
S
Midway Islands
Oahu
Pearl Harbor
Hawaiian Islands
Japanese Attack on Pearl Harbor
December 7, 1941
Route of Japanese Fleet
Attack flight paths
Kahuku Point
Haleiwa
Schofield Barracks
Wheeler Air Force Base
OAHU
Pearl Harbor
Bellows Airfield
Ewa Airfield
Hickam Airfield
Honolulu
0 miles 5
0 km 5

baseball records, Joe DiMaggio's 56-game hitting streak and Ted Williams's .406 batting average. The draft, first introduced in 1940, had been extended in October 1941 by just one vote. So unready was the army for war that some conscripts drilled with broomsticks and eggs in lieu of rifles and hand grenades.

A somnolent country was finally jarred out of its reverie by a startling message sent at 7:58 A.M., December 7, from Ford Island Naval Air Station: "AIR RAID, PEARL HARBOR—THIS IS NO DRILL." Amid the hoarse wail of Klaxons, the men aboard the battleship *California* received an earthier version of the same alert over their public address system: "Everyone get to your battle stations! This is no shit!"

Across the island, soldiers, sailors, and marines, their "adrenaline . . . pumping about a thousand miles an hour," as one later recalled, broke into locked ammunition lockers, grabbed whatever weapons happened to be available—ranging from .45 caliber pistols to .50 caliber machine guns—and began to blaze away at the sky. They managed to down some Japanese aircraft, especially in the second wave that began to appear around 9:00 A.M., but they were hard put to mount an effective defense because so many of their fighter aircraft were caught on the ground. The only notable success was enjoyed by two brand-new P-40 Warhawks flown by army lieutenants George Welch and Kenneth Taylor, who had been awake when the attack occurred because they had stayed up all night playing poker. Between them, they downed seven Japanese aircraft, mostly slow dive bombers. But most of the raiders reached their targets unimpeded.

Decades later, men who were there could vividly recall the sickening sound made by bombs and torpedoes as they tore open the steel decks of anchored ships. "It wasn't a Bang. It wasn't a Boom. It was Whoosh!" said Marine Private First Class James Cory of the USS *Arizona*. "You could hear the bombs whistling down and feel them hit and penetrate, and then this Whoosh." Fireman Dan Wentrcek on the USS *Nevada* said there was "a rending of metal and a concussion that's huge—just like something was picking the ship out of the water and bobbing it around. . . . It's a sensation you can't really describe."

Worst hit was the battleship *Arizona*, which sank in a five-hundred-foot fireball after a bomb penetrated its deck and detonated the forward magazine. Almost three-quarters of the 1,511 officers and crew were killed. One young sailor was horrified to see "steel fragments in the air, fire, oil—God knows what all—pieces of timber, pieces of the boat deck, canvas, and even pieces of bodies. . . . I saw a thigh and a leg; I saw fingers; I saw hands; I saw elbows and arms." Private Cory had equally gruesome memories of men whose "hair was burned off; their eyebrows were burned off; the pitiful remnants of their uniforms in their crotch was a charred remnant; and the insoles of their shoes was about the only thing that was left on those bodies." Those who escaped had to swim to shore through a burning oil slick—a "sea of fire"—while being strafed by Zeros.

In two hours of marauding, 350 Japanese airplanes managed to sink or cripple eighteen ships. In addition to Pearl Harbor, all of the major air bases on the island—Kaneohe and Ford Island Naval Air Stations, the Marines' Ewa Field, the Army Air Corps' Wheeler, Hickam, and Bellows Fields—were hit. One hundred eighty-eight airplanes were destroyed, another 159 badly damaged—86 percent of the total. Even worse, 2,413 Americans were killed, 1,178 wounded. All this at a cost to Japan of just 29 lost airplanes and 55 dead airmen. One full-size Japanese submarine and five midget submarines, which contributed nothing to the attack, were also sunk.

The destruction was not as horrific as it first appeared. All but two of the battleships were salvaged and repaired. Only the *Arizona* and *Oklahoma* never saw action again. The raiders largely left alone the infrastructure of Pearl Harbor, the oil tanks, workshops, and dry docks upon which the Pacific Fleet depended. And the U.S. Navy's two Hawaii-based aircraft carriers, the *Lexington* and *Enterprise*, were unscathed; both had been at sea during the attack, ferrying airplanes to Midway Island and Wake Island, respectively. So it could have been much worse. But it was bad enough.

When he reported to Vice Admiral Chuichi Nagumo aboard the aircraft carrier *Akagi* (Red Castle) shortly after noon, Commander Fuchida could proudly report that his mission had been accomplished: "The main force of the U.S. Pacific Fleet will not be able to come out within six months." Fuchida wanted to take another crack at Pearl, but as far as the more cautious Nagumo was concerned, this was all he had been asked to accomplish.

Six months was judged long enough for Japan to grab for itself all of the prime European colonies in the Pacific and Southeast Asia, from the Philippines, Singapore, and Hong Kong to the Dutch East Indies (Indonesia). The U.S., its fleet crippled, would be helpless to interfere. Britain and the Netherlands would pose even less of an obstacle. It was to make possible this audacious campaign of conquest that the Imperial Japanese Navy struck Pearl Harbor. As soon as this assignment had been carried out, Nagumo turned around and headed for home, leaving smoldering ruins in his wake.

The attack on Pearl Harbor is one of the most famous, or perhaps simply infamous, events of the twentieth century. It has been endlessly relived in countless books, articles, and movies, ranging from the sublime (*From Here to Eternity*) to the workmanlike (*Tora! Tora! Tora!*) to the simply execrable (the 2001 film *Pearl Harbor*). What most popular accounts miss is the element of novelty in the whole enterprise. Today it seems natural for war to be fought by aircraft carriers and airplanes, torpedoes and high-explosive bombs, with radar sets and code breakers in a vital supporting role. But this is only natural from the vantage point of the early twenty-first century. Seen from the perspective of 1941, this was an entirely new way of war.

Twenty years earlier, no one would have worried about aircraft being launched from a floating airstrip thousands of miles from home to attack a heavily fortified naval base. Indeed, many admirals the world over still thought the battleship was the sine qua non of naval power, while the aircraft carrier was dismissed as little more than an adjunct. That mind-set was damaged on December 7 as severely as the battlewagons of the Pacific Fleet. It would finally be consigned to Davy Jones's locker by the Battle of Midway six months later in which two fleets pummeled each other from far outside gun range.

A brave new world was dawning in which the four-hundred-year-old preeminence of naval artillery was being eclipsed by an invention that was less than forty years old—the airplane. It was not a revolution that the Japanese had pioneered. But it was one in which by the late 1930s they had become the world leaders, just as the Germans had become leaders in tank warfare. In the six months that followed the raid on Pearl Harbor, America, and even more Britain, would learn just how far behind they had fallen in certain key technologies and techniques. The result was some of the largest and most humiliating surrenders and setbacks ever suffered by the armed forces of the United Kingdom and the United States. Only by the narrowest of margins was the U.S. able to recover sufficiently to stave off an even more catastrophic defeat.

BRITAIN'S ABORTED TAKEOFF

Naval aviation was born in the U.S. but grew to maturity across the Atlantic. In 1910, just seven years after the Wright brothers' first flight, a biplane designed by the entrepreneur Glenn Curtiss and flown by civilian test pilot Eugene Ely took off from a wooden platform affixed to the deck of a U.S. Navy cruiser anchored in Hampton Roads, Virginia. The following year the stunt was performed in reverse: Ely landed on the deck of a battleship anchored in San Francisco Bay. Also in 1911 a Curtiss seaplane landed on the water near San Diego and was hoisted aboard a navy ship by crane. Thus were born in crude prototype the aircraft carrier and the seaplane tender, two entirely new types of ship. It was not clear at first which of them would predominate or even whether they would be necessary at all; perhaps land-based aircraft or lighter-than-air dirigibles could provide sufficient support at sea.

Starting in 1914, the center of naval aviation shifted to Britain, whose navy had to improvise under the pressure of war. Prodded by First Lord of the Admiralty Winston Churchill (whose enthusiasm for new weaponry also led, as previously noted, to the development of the tank), the navy commandeered some small passenger liners and converted them into seaplane tenders. On Christmas Day 1914, three of these vessels transported floatplanes near the North Sea port of Cuxhaven to bomb a German zeppelin base. The first air strike in history launched from the sea was completely ineffectual, owing to the pilots' inability to locate their targets, but an important precedent had been set. Two British seaplanes armed with torpedoes did better the following year; they managed to sink three Turkish merchant ships in the Dardanelles.

As the war went along, the Royal Navy pioneered all sorts of uses for flying machines, starting with reconnaissance and gunfire spotting and moving on to attacking enemy ships and shore installations and defending their own vessels and bases. It quickly emerged that seaplanes were too unwieldy and cumbersome to be the primary instruments of aviation at sea; dirigibles were too vulnerable to attack; and land-based airplanes lacked sufficient range for many missions. Aircraft that could take off and land on floating runways were found to be the most useful.

A number of British warships had wooden platforms affixed to permit the takeoff and landing of a few airplanes. There was not much runway room because steamships were split down the middle by funnels, masts, and a bridge. One of these jerry-rigged ships, the converted cruiser HMS *Furious*, is considered the world's first aircraft carrier. The first flush-deck carrier, HMS *Argus*, was completed in 1918, too late to be used in the war. From photographs, the *Argus* looks strikingly similar to modern aircraft carriers save for the notable omission of an island superstructure on the side of its flight deck; that was introduced by HMS *Eagle* in 1920.

By war's end, the Royal Navy had the world's only aircraft carriers along with 3,500 aircraft and 55,000 men to operate them. The British completely dominated naval aviation. Yet, just as in tank warfare, another field they had pioneered, their lead would evaporate long before the next world war broke out. The reasons again come down to lack of funding, foresight, and organization.

Perhaps the heaviest blow suffered by naval aviators was the decision in 1918 to subordinate them to the newly formed Royal Air Force, the world's first independent military air service. The RAF's institutional culture favored strategic bombing, which it alone could perform, and disdained support of troops or ships, which, it feared, would subordinate it to the older services. As an unwanted stepchild, the Fleet Air Arm suffered from chronic neglect during the interwar wars. It did not attract first-class officers, nor did it purchase first-class aircraft. The navy finally regained control of its aviation in 1937, but by then it was too late to make radical adjustments before the onset of World War II. Unlike in the U.S. or Japan, there were no "air admirals" to push the cause of aviation; the "gun club" of battleship admirals had an especially strong grip in the Royal Navy.

The Royal Navy was also hampered by the parsimony that afflicted all of Britain's military services between the wars. So loath was the government to spend money on defense that it voluntarily gave up the Royal Navy's centuries-old predominance at sea. At the Washington Naval Conference in 1922, Britain, Japan, the United States, France, and Italy agreed to limit naval armaments. Britain settled for parity with the U.S.; each was given a limit of 525,000 tons of capital ships and 135,000 tons of aircraft carriers. Japan settled for 315,000 tons of capital ships and 81,000 tons of aircraft carriers. (Japan was grudgingly willing to accept overall inferiority because Britain and the U.S. had many other commitments around the world, leaving it dominant in the Pacific.) These treaty limits would remain in effect until 1936 and hamper the modernization of all the major navies. Britain was especially hamstrung because it had more existing ships than its competitors; it simply kept aging aircraft carriers in service while Japan and the U.S. built more modern models.

The result was that by 1939 the Royal Navy looked decidedly antiquated. It had six aircraft carriers to go along with thirty battleships and cruisers, but most were small and obsolete; only one was less than ten years old. The aircraft that flew off their decks were even more outmoded. The standard torpedo bomber, the Fairey Swordfish, was designed in the mid-1930s but, with its double wings and open cockpit, looked as if it had flown through a time warp from 1916. The standard fighter was another antiquated biplane, the Gloster Sea Gladiator, which was so underpowered that, against modern German or Japanese fighters, it was a flying coffin. What makes this backwardness all the more astounding is that at they very same time the RAF was fielding two of the best fighters in the world, the Spitfire and Hurricane—both, needless to say, monoplanes. The RAF hoarded these precious planes, however, eventually forcing the Royal Navy to buy aircraft from the U.S. This was yet another consequence of budgetary constraints: Whitehall decided to conserve its resources for defense of the homeland, which did not require modern naval aircraft, while neglecting the defense of its Far Eastern possessions, which did.

For all its woes, the Fleet Air Arm showed its worth during the first two years of the war against the Germans and Italians, who had no aircraft carriers of their own. (Aside from the U.S. and Japan, the only other nation with a carrier was France, which had one.) In 1940 land-based British bombers sank a German cruiser during the invasion of Norway. The following year, carrier-based Swordfish torpedo bombers crippled the German battleship *Bismarck*, allowing it to be finished off by pursuing battleships. The most spectacular feat of Britain's naval aviators was the nighttime attack on the Italian fleet in Taranto harbor in southeastern Italy on November 11, 1940, a raid closely studied by the Japanese as they planned their own attack on Pearl Harbor. At Taranto, twelve Swordfish sank one battleship and heavily damaged two others. But Britain paid a stiff price for such successes. By the time the Pacific war started on December 7, 1941, three British fleet carriers had been sunk and one heavily damaged in the Mediterranean and North Sea. Four new flattops ordered before the war offset these losses, but Britain nevertheless found itself without enough carriers to meet all of its needs. As a direct consequence of this shortfall, the Royal Navy was chased out of the Pacific and Indian Oceans, which it had dominated for centuries.

THE REBEL VS. THE INSIDER

By this time, the U.S. Navy had overtaken the British, at least in aviation. Its success was largely due to one man: Rear Admiral William A. Moffett, head of the navy's Bureau of Aeronautics during its first, formative decade. Moffett was an old battleship man, a graduate of the Naval Academy, a veteran of the Spanish-American War, and recipient of the Medal of Honor for his role in helping to provide gunfire support for the U.S. occupation of Veracruz, Mexico, in 1914. He had had almost no contact with aviation until he was forty-eight years old. As commander of the Great Lakes Naval Training Station near Chicago in 1917–18, he established a flying school to complement the one in Pensacola, Florida, which was overwhelmed by wartime demand. In 1920 he was put in charge of naval aviation by the Chief of Naval Operations and immediately began lobbying to create a full-fledged aviation bureau on a par with such longstanding naval fiefdoms as the Bureau of Steam Engineering and the Bureau of Navigation. His success in 1921 meant that aviators had an institutional base of support in the U.S. Navy that they lacked within the Royal Navy. It also meant, not incidentally, that Moffett got a promotion to rear admiral in order to head BuAer, as the Bureau of Aeronautics was dubbed in the U.S. military's abbreviation-mad culture.

With his craggy face, bushy eyebrows, and meticulously tailored uniforms, Moffett was a consummate bureaucrat and political infighter. He needed all of his considerable cunning to defeat two sets of enemies: on the one hand, battleship admirals skeptical of the utility of airplanes; on the other, air enthusiasts skeptical of the utility of ships. The most colorful advocate of the latter view was Brigadier General William "Billy" Mitchell, who gained fame as head of the American Expeditionary Force's Air Service during World War I. Mitchell came back from France with a chestful of medals and an evangelical zeal for air power, convinced that all the other military services should be relegated to the sidelines, if not put out of business altogether. Pilots, and especially bomber pilots, would provide all the defense the nation needed.

This was not, needless to say, the navy's view. While some admirals were sympathetic to the claims of the air advocates, most hooted in derision. "I don't want to hear any more about sinking battleships with air bombs," snorted Navy Secretary Josephus Daniels. "That idea is so damned nonsensical and impossible that I'm willing to stand bareheaded on the bridge of a battleship while that nitwit [Mitchell] tries to hit it from the air!" Luckily for Daniels, no one took him up on the offer.

A series of tests was arranged in 1921: Target ships would be anchored off the Virginia Capes and army and navy aircraft would try to sink them. The highlight was an attack on the captured German battleship *Ostfriesland*. With a bevy of Washington notables in attendance, Billy Mitchell's fliers sent the old battlewagon to the bottom of the ocean with a series of 2,000-pound bombs. Senior navy officers were said by a reporter to have openly sobbed as they watched the big ship roll over "like some immense, round, helpless sea animal." An exultant Mitchell bragged to his fliers afterward: "Well, lads, I guess we showed old Admiral Tubaguts today!" The admirals, after they had gotten over their initial shock, did not think the fliers had shown them much at all. They noted that the *Ostfriesland* had been stationary, unmanned, and unable to defend itself. It was also obsolete. But the public had no doubts that the bombing had been, in the words of the *New York Times*, "an epoch-making performance." Emboldened, Mitchell redoubled his advocacy for the creation of an RAF-style air force that would control both army and navy air assets.

Moffett would have none of it. He, too, believed in air power, but he thought it had to be combined with fleet action and kept within the navy's hands. Mitchell had nothing but scorn for this view and its advocate, who was not even a pilot. Although the head of BuAer was obligated to be a certified aviator, Moffett met this requirement by taking a five-week course to qualify as a navigator. Mitchell, by contrast, was described by one newspaper, with only slight hyperbole, as "the most competent and intrepid pilot in America."

Mitchell was also one of the greatest showmen in America. Decked out in his nonregulation pink breeches, flared tunic, and gleaming cordovan boots, he was described by one reporter as "a plumed fellow with the aura of banner, spear and shield." Outspoken, handsome, and rich, Billy Mitchell was a crowd pleaser who would be portrayed posthumously by Gary Cooper. During his lifetime, he cultivated the public with such stunts as sending a squadron of army aircraft to barnstorm across the country, staging mock bombing runs of major East Coast cities, and taking the humorist Will Rogers up for a spin.

In 1925 Moffett tried to counter these propaganda coups with some of his own. The admiral ordered three navy flying boats to go from California to Hawaii and a navy dirigible to tour the Midwest. Alas, one of the flying boats disappeared at sea and the airship *Shenandoah* crashed with the loss of thirteen crew members, including its captain, who was a friend of Mitchell's. A distraught Mitchell unleashed a public tirade. "These accidents," he thundered in a statement splashed across newspaper front pages, "are the direct result of the incompetency, the criminal negligence, and the almost treasonable administration of our national defense by the Navy and War Departments." Admiral Moffett was a Southern gentleman who was seldom prone to public outbursts. But he felt compelled to return fire. "The most charitable way to regard these charges," he declared, "is that their author is of unsound mind and is suffering from delusions of grandeur."

Many of Mitchell's superiors agreed with that harsh assessment. He was charged with insubordination and a variety of other offenses. After a celebrated 1925 court-martial that rivaled the contemporaneous Scopes "monkey trial" for public interest, Mitchell was convicted and suspended from active duty for five years. He chose to retire instead, making him in the popular imagination a martyr for air power. In reality, most of the generals and admirals he condemned with such harsh invective were not Luddites; they merely thought aircraft had to be used in a balanced way with other elements of military power.

William Moffett, in his less flamboyant way, actually did more to advance air power within the military than the brash Billy Mitchell. He became one of the navy's most successful peacetime innovators during the twentieth century, rivaled only by Admiral Hyman Rickover, father of the post–World War II "nuclear navy." The Royal Navy suffered greatly in the interwar period for not having a comparable figure. While Moffett's goal was to turn aircraft carriers into an offensive striking force, he was clever enough to disguise his intentions by telling the "gun club" that aircraft would merely serve as scouts for their beloved battlewagons. As an old battleship man himself, Moffett did not seem as threatening to naval traditionalists as an outsider like Billy Mitchell did—yet the implications of Moffett's innovations were just as radical.

Moffett's initial success was the completion in 1922 of the *Langley*, an 11,050-ton converted collier that became America's first aircraft carrier. Moffett then successfully lobbied for the conversion of two cruisers into the

36,000-ton aircraft carriers *Saratoga* and *Lexington.* Among the biggest and fastest in the world when they were commissioned in 1927, they had eight-hundred-foot flight decks and could carry eighty-three planes each.

Moffett inherited a navy aircraft factory in Philadelphia. He turned it into a flourishing research and development center and got out of the business of aircraft production, which he contracted out to private firms. Under his guidance, the Philadelphia center worked out innovations such as turntable catapults to launch airplanes and tail hooks and restraining wires to stop them from sliding off the flight deck. His researchers also developed an air-cooled radial engine that greatly enhanced aircraft performance.

In 1926 Moffett helped push legislation through Congress, based on the recommendations of a presidential commission (the Morrow Board), that authorized the creation of an assistant navy secretary for aeronautics and the procurement of one thousand naval aircraft over the next five years. Perhaps his most lasting legacy was convincing Congress that all air commands, including aircraft carriers, had to be restricted to fliers—a requirement that lured a number of ambitious captains, such as Ernest J. King and William F. Halsey, Jr., into taking courses to qualify as aviators or observers. This corps of air-minded admirals would prove invaluable in World War II.

Moffett had few blind spots, but one of them was a killer. He was passionately attached to airships, those silvery, egg-shaped, helium-filled behemoths that were all the rage in the 1930s. He even envisioned using them as airborne aircraft carriers. One rainy day in 1933, he took off aboard one of the navy's dirigibles, the *Akron*, and never came back. The airship crashed off the New Jersey coast, killing seventy-three people in the nation's worst aviation disaster to that point. A final dirigible accident two years later closed the books on this ill-fated program.

Though Moffett was no longer around to guide it, the Aeronautics Bureau continued to develop aviation as best it could during the hard-luck years of the 1930s. Its biggest innovation, in conjunction with the Marine Corps, was the invention of dive bombing—a technique that allowed greater accuracy against moving targets than traditional level bombing.

Naval rearmament began in earnest in 1938 and accelerated considerably after the fall of France in 1940. By the time war came in December 1941, the navy had seven major aircraft carriers, second only to Japan, which had ten. The aircraft on their decks were better than Britain's, not as good as Japan's. The primary U.S. fighter was the Grumman F4F Wildcat; the primary torpedo bomber, the Douglas TBD Devastator; the primary dive bomber, the Douglas SBD Dauntless. Only the Dauntless would prove good enough to last most of the war. The others had to be replaced because they were bested by Japanese competitors. Nor was the Curtiss P-40, the latest army fighter, any better. The Catalina PBY flying boat, on the other hand, a long-range, sea-based bomber and reconnaissance airplane, was world-class.

There was still no agreement over the proper role of aircraft carriers. Were they designed to support the traditional battle line or supplant it? It might not seem much of a contest because even the most powerful battleship guns had a maximum range of twenty miles, while carrier-based aircraft could easily fly ten times as far. But admirals were not sure that aircraft could sink fast-moving battleships encased in heavy armor and protected by batteries of antiaircraft guns. They also fretted that aircraft carriers would not be able to defend themselves, especially at night and in foul weather when air operations would be almost impossible. This fear was not entirely misplaced. With their large quantities of aviation fuel and munitions stored beneath unarmored, wooden flight decks, American aircraft carriers were floating Molotov cocktails. (Armor was thought to add too much weight and thereby reduce airplane capacity. Britain did heavily armor its carriers, reducing their effectiveness.) Four of the seven prewar U.S. carriers were sunk in the first year of fighting. But exposed as carriers were, battleships turned out to be in even greater jeopardy—and they did not have a comparable long-range offensive punch to compensate for their vulnerability.

This was not immediately apparent in May 1940, however, when President Franklin Roosevelt ordered the battleship-heavy U.S. Pacific Fleet to stay in Hawaii rather than return to its home port in San Pedro, California, following an exercise. Roosevelt, a former assistant secretary of the navy (1913–21), hoped that this force of eight battleships and two aircraft carriers, along with various support vessels, would deter Japan from further aggression. All it did was to bring an alluring target into the enemy's crosshairs.

SAMURAI OF THE SKIES

The Imperial Japanese Navy began experimenting with aviation as early as the British and Americans. But because Japan did not see much combat in World War I, it had fallen behind the other powers by 1918. To catch up, it turned to its traditional mentors: for the army, the French; for the navy, the British. A British naval mission arrived in 1920 complete with over one hundred demonstration aircraft in a bid to boost the British aviation industry. British pilots formed the first faculty of the newly established Japanese naval aviation school at Lake Kasumigaura. British naval architects helped Japan complete its first aircraft carrier, the *Hosho*, in 1922. British aircraft designers helped Mitsubishi design its initial carrier aircraft. Winston Churchill, Secretary of

State for War and Air, was confident Britain and Japan would never go to war—"I do not believe there is the slightest chance of it in our lifetime," he exclaimed in 1924—so what was the harm?

While the Japanese were always happy to learn from *gaijin*, they sought to achieve self-sufficiency as soon as possible. By 1941, they had succeeded—spectacularly so. At the time of the Pearl Harbor attack, Japan had the finest naval aircraft, pilots, and aircraft carriers in the world, all overseen by its Naval Aviation Department, created in 1927.

Japan not only had more aircraft carriers than any other navy—ten—but the most modern of them, the *Shokaku* (Soaring Crane) and *Zuikaku* (Happy Crane), built after the lifting of treaty limits in 1936, were superior to anything the U.S. Navy would deploy until 1943. These 29,800-ton monsters could carry seventy-two aircraft and steam over eleven thousand miles without refueling—easily enough to get to Hawaii and back—with a top speed of over 34 knots (39 mph). Their completion by the end of September 1941 made the raid on Pearl Harbor possible, and their subsequent absence at Midway may have tipped the outcome of that critical battle against Japan.

The Japanese navy had at first tried building aircraft itself, but by the early 1930s it had settled on a better division of labor: Navy engineers would come up with specifications for airplanes and private firms would compete to build them. Japan did not have a large civil aviation industry, but three major firms—Mitsubishi, Nakajima, and Aichi—developed a high degree of sophistication as they became the primary suppliers for the navy. (The army, which rarely spoke to the navy, acquired its aircraft separately, mainly from these same firms.) Japanese industry boosted its airplane production from 1,181 in 1936 to more than 5,000 in 1941. This was still only a fifth of the U.S. total that year, but the Japanese navy deployed more aircraft on the eve of Pearl Harbor—over three thousand—than either the British or Americans, and their aircraft enjoyed, on the whole, a substantial qualitative edge.

The planes that would devastate Pearl Harbor were designed in the mid-1930s. The Aichi D3A1 Type 99 dive bomber, dubbed "Val" by the Allies, was similar to the Stuka on which it was modeled. The Nakajima B5N2 Type 97 (Kate) was a versatile three-man bomber that could drop either one torpedo or several bombs. Its maximum speed was 100 mph faster than its British counterpart, the Swordfish, and 30 mph faster than its U.S. counterpart, the Douglas Devastator. To go along with these carrier-based attack aircraft, the Japanese navy developed two potent land-based bombers, each with twin engines, a crew of seven, and the capacity to carry either bombs or torpedoes. The Mitsubishi G3M2 Type 96 (Nell) was adopted in 1936; five years later came the Mitsubishi G4M1 Type 1 (Betty), with a phenomenal range of 3,700 miles—greater than the B-17, though it lacked the Flying Fortress's bomb capacity. They were not used at Pearl Harbor, but they would be employed with deadly efficiency in the western Pacific. All of these attack aircraft struck fear into the hearts of Allied seamen in the war's early days as they sank one ship after another.

The most feared of all Japanese aircraft was the Mitsubishi A6M2 Type O (Zero) fighter, which entered service in the summer of 1940. The Zero's brilliant designer, Jiro Horikoshi, created a sleek airplane that was faster, more nimble, and had greater range than any contemporary fighter, land- or sea-based. Its armaments—two 7.7 mm machine guns in the nose, two 20 mm cannons in the wings—were also more formidable than those of any comparable aircraft. This lethal combination of firepower and high performance was made possible by the use of a newly developed zinc-aluminum alloy that was stronger and lighter than the materials used to build other airplanes.

Upon its introduction, the Zero allowed the Japanese to wipe the Chinese air force from the sky. In the early years of the war in the Pacific, it also ran rings around British and American warplanes. Not until 1943 did the U.S. produce a superior aircraft. By that time the Zero's weaknesses, which it shared with other Japanese planes, had become apparent: Built to maximize offensive power, it lacked basic defensive elements such as armor and antiexplosive, self-sealing gas tanks. This was in accordance with the *bushido* ethic which placed a low priority on individual warriors' self-preservation. (For the same reason, many Japanese pilots disdained wearing parachutes in combat because they did not want to risk the disgrace of being captured.) It meant that, once hit, Japanese airplanes did not have much ability to survive; the Betty bomber was later nicknamed "Zippo" by U.S. fighter pilots for its tendency to go up in flames. But in the war's early days this was not much of a concern, because Allied defenders generally lacked airplanes capable of keeping up with, much less hitting, their attackers.

Japan's edge in the quality of its personnel was even greater than its edge in the quality of its airplanes. Naval aviators, known as the Sea Eagles, formed a small, elite corps of volunteers. Unlike in the U.S. or British navies, most were not commissioned officers. They were generally either NCOs drawn from the surface fleet or teenage boys recruited straight out of civilian schools. Competition for flight training was ferocious, and cadets were disqualified for the slightest failing. In the 1930s the navy graduated only one hundred pilots a year. The crème de la crème were selected for aircraft carriers; landing on a bobbing strip of steel in the middle of the ocean was rightly considered the most demanding task a pilot could perform.

The pilots, and the rest of Japan's navy, conducted tough drills in harsh conditions, including stormy weather and darkness, leading many to comment afterward, "War is so easy, compared with peacetime exercises!" Through relentless practice, Japan's naval pilots attained unparalleled accuracy in dive bombing, high-level bombing, and aerial torpedoing, as well as learning how to coordinate these different modes of attack into a coherent tactical framework. The performance of many pilots was further enhanced by their participation in Japan's war in China, which began in 1937. This taught the Japanese, for instance, about the need to have fighters escort long-range bombers to their targets—a seemingly obvious point, but one that the British and Americans would not grasp until they had suffered horrific bomber losses during the first few years of the war.

The fliers who attacked Pearl Harbor had an average of eight hundred hours of flying time, almost three times as much as the average U.S. Navy pilot, and most had combat experience that the Americans lacked. There was no question that Japanese aviators were vastly superior; the problem was that there were not enough of them. On the eve of war, the U.S. Navy and Marine Corps had 8,000 active-duty pilots; the Japanese navy had only 3,500, and just 900 of them were carrier-qualified. This was not because of America's larger population size (which did not prevent Japan from having almost twice as many men in uniform overall in 1941); it was mainly because the U.S. Navy emphasized quantity over quality. Japan made the opposite decision, which meant that if its samurai of the skies could not win a quick victory, they would be bled dry in a war of attrition.

This was one of many dilemmas confronting the commander in chief of the Imperial Japanese Navy's Combined Fleet as he contemplated the prospect of conflict with the United States. Since August 1939 that job had been held by Isoroku Yamamoto, an unlikely candidate to be one of the leading Axis commanders. Yamamoto had become familiar with America as a student at Harvard, 1919–21, and as naval attaché in Washington, 1925–28. He admired the American people—Lincoln ranked high in his personal pantheon—and disliked Japan's new allies, the Nazis. Moreover, he was well aware of the vast advantages the U.S., with its larger population, richer economy, and greater industrial capacity, possessed in any confrontation with Japan. He counseled Tokyo to avoid awakening this sleeping giant. "If I am told to fight regardless of the consequences," he warned Japan's premier, prophetically, in 1940, "I shall run wild for the first six months or a year, but I have utterly no confidence for the second or third year." On another occasion he wrote, "A war between Japan and the United States would be a major calamity for the world."

Such views, though widely held within the upper ranks of the more cosmopolitan navy, were heresy to the narrow-minded, nationalistic army officers who dominated the government. While serving as vice minister of the navy from 1935 to 1939, Yamamoto's life was in constant jeopardy from right-wing assassins; there was a price of 100,000 yen on his head. The navy appointed him commander of the Combined Fleet, rather than navy minister, in large part simply to get him out of Tokyo and out to sea, where he would be safe from attack by his own countrymen.

By 1941, Yamamoto's views were in a decided minority in the government. After President Roosevelt embargoed all oil and scrap metal sales to Japan in July in retaliation for the occupation of southern Indochina, Tokyo decided it had no choice but to go to war in order to, as the Foreign Ministry put it, "secure the raw materials of the South Seas." Because all the decision makers assumed (perhaps wrongly) that the U.S. would not stand by as Japan gobbled up Dutch and British colonies, it was decided that war against the U.S. was inevitable. And since the Imperial Navy had only enough fuel for eighteen months of operations, the sooner the better.

The fifty-seven-year-old Yamamoto would be at the forefront of the war effort. Like his hero, Admiral Togo, he was not very big, even by Japanese standards—only five feet three inches, 125 pounds—but his broad shoulders, shaved head, and thick chest conveyed an impression of strength. As a young ensign at the Battle of Tsushima in 1905, he had been severely wounded by an exploding gun. For the rest of his life he walked around with two fingers missing on his left hand and the lower half of his body badly scarred. "Whenever I go into a public bath, people think I'm a gangster," he good-humoredly complained. Among the geishas of Tokyo, whose establishments he liked to frequent, the admiral was jocularly known as "Eighty Sen," "since," a biographer writes, "the regular charge for a geisha's manicure—all ten fingers—was one yen." That he would gladly take this kind of ribbing suggests that Yamamoto was notably lacking in the pomposity that often comes with high rank. He had a good sense of humor as well as a tendency to speak his mind.

Yamamoto gave up alcohol as a young man, making him a rarity in the hard-drinking world of the Imperial Navy. His only weakness, other than the geishas (one of whom became his mistress), was an obsessive love of games of chance. He would bet on anything, from bowling to blackjack. He was skilled at *shogi* (Japanese chess), bridge, and especially poker, which he would gladly play for thirty or forty hours at a stretch. He often told his subordinates that if he retired from the navy he would move to Monaco to become a professional card player. He would apply this gambler's mentality—always carefully calculating the odds and not being afraid to risk everything on one roll of the dice—throughout his naval career.

Although not a pilot himself, Yamamoto had spent much of his career around naval aviation. After a brief stint as second-in-command of the Kasumigaura flight school, he went on to command the aircraft carrier *Akagi* and then two carriers arrayed in a carrier division. These sea commands were interspersed with stints as technical director of the navy's Aviation

Department and head of the entire department. In these assignments, he came to the conclusion that in the next war, carriers would be the most important elements of sea power.

This view did not win the assent of many other admirals. In the 1930s the navy continued building battleships, including two of the biggest ever made, the *Yamato* and *Musashi*. Their advocates boasted that these 72,000-ton behemoths, with their eighteen-inch guns, were virtually unsinkable and unstoppable. Yamamoto, who noted that each one cost the same as one thousand airplanes, was not impressed. He echoed fliers who jeered that the "three great follies of the world were the Great Wall of China, the Pyramids, and the battleship *Yamato*." Indeed, both the *Yamato* and *Musahi* would be sunk during the war without ever getting a chance to inflict a single blow on the enemy.

HITTING HAWAII—AND BEYOND

The traditional Japanese plan for fighting America had been conceived by battleship enthusiasts in thrall to Alfred Thayer Mahan. It called for luring the U.S. Navy across the Pacific, wearing it down along the way with submarine and aircraft attacks, and then annihilating it in a decisive surface engagement—the Great All-Out Battle—near the home islands. In other words, essentially the same strategy that had worked against the Russians in 1905. (U.S. war plan Orange was a mirror image: It called for the U.S. fleet to advance across the Pacific, rescue the Philippines, and destroy the Japanese fleet.) Yamamoto rejected this passive approach in favor of trying to inflict a ferocious blow against the U.S. on the very first day of the war, thereby giving Japan enough time to grab the rich resources of southeast Asia and, he hoped, force the U.S. to negotiate a peace settlement. It is not clear when he first came up with the idea of striking Pearl Harbor; the evidence suggests that the plan formed in his mind around the middle of 1940, shortly after the U.S. fleet moved to Hawaii. His letters show that he sought to emulate, and improve upon, Togo's 1904 attack on Port Arthur that opened the war with Russia.

There were considerable obstacles to pulling this off. Simply steaming all the way to Pearl Harbor and back—a round-trip of 6,788 miles from Yokohama—was considered a difficult feat. Most of the ships lacked the fuel capacity to make it that far. They would have to refuel at sea, something the navy had little experience with. And once they arrived, how much damage could they do? The waters of Pearl Harbor were only forty feet deep, too shallow for standard torpedoes to work properly, and U.S. battleships were too heavily armored to be penetrated by standard bombs. In any case, aerial ordnance might miss altogether; in initial trial runs, only 10 percent of bombs hit their targets.

The problems of refueling and ordnance accuracy were largely solved the old-fashioned way: practice, practice, practice. During their voyage to and from Hawaii, Japanese ships would manage to refuel eighteen or nineteen times at sea from accompanying tankers. And after months of hard work to improve their aim, Japanese bombardiers would achieve astonishing hit rates on December 7: 90 percent for torpedoes, 59 percent for dive bombing, 37 percent for high-level bombing. All of those munitions were able to cause considerable damage because of two technical breakthroughs that were completed just days before the Hawaii Strike Force assembled on November 16, 1941.

Japan already had the best aerial torpedo in the world, the Type 91, with a longer range (1.6 miles), faster speed (42 knots), and bigger warhead (330 pounds) than its American counterparts. Best of all, it was reliable, whereas the latest U.S. torpedoes, the Mark 13 and Mark 14, had a dud detonator that made them almost completely ineffective. The trick was to utilize the Type 91 in shallow waters. Normally it sank one hundred to three hundred feet after being dropped in the water. The solution, discovered in the fall of 1941, was to add wooden fins to the tail, which kept the torpedo closer to the water's surface. While torpedoes would be the primary weapons employed against the battleships that had outside berths near Ford Island, they could not touch the battlewagons anchored on the interior line. To penetrate their decks, the heaviest naval artillery shells, weighing more than 1,700 pounds each, were converted into aerial bombs. The undercarriages of attack aircraft had to be modified to carry these ship-busters; this work was still being performed by technicians on board while the task force headed for Hawaii. Their success can be judged from the fact that it was one of these bombs that destroyed the *Arizona*.

The biggest innovation of all wasn't technical. In April 1941, under the prodding of the brilliant young airman Menoru Genda, a Japanese Billy Mitchell, Yamamoto created the First Air Fleet with five carriers (others would be added later) grouped together in one potent strike force. Organizing a fleet around carriers, with battleships and cruisers in an auxiliary role, "marked," in Genda's words, "an epoch-making progress in the field of naval strategy and tactics"—a "revolutionary change" that immediately vaulted the Imperial Navy ahead of any competitor. Genda had been advocating such a move since 1936 but had continually met stout resistance from navy traditionalists who openly questioned his sanity. If Yamamoto had not used his clout to overcome their opposition, the attack on Pearl Harbor would not have been possible.

Even with this formidable striking force, the attack was considered a great risk. The Navy General Staff was flatly opposed, as were many of Yamamoto's subordinates, including Admiral Chuichi Nagumo, a torpedo expert who commanded the First Air Fleet despite his lack of familiarity with air power. They pointed out all sorts of reasons why the attack might fail—bad weather, bad luck, bad intelligence—and suggested that the consequences—the loss of a sizable portion of the navy's aviation strength—would be unacceptable. A tabletop simulation at the Naval War College in early September concluded with the loss of two carriers and a third of the strike aircraft. Rear Admiral Ryunosuke Kusake, chief of staff of the First Air Fleet, bluntly told Yamamoto: "You are an amateur naval strategist and your ideas are not good for Japan. This operation is a gamble." Yamamoto wasn't discouraged. "I like games of chance," he replied insouciantly. "You have told me that the operation is a gamble, so I shall carry it out."

To get the cautious Naval Staff to go along, Yamamoto had to threaten to resign. But even he did not fully trust airplanes to do the job all by themselves. Over the objections of some of his airmen, he insisted on adding thirty submarines to the strike force, including five newly developed miniature subs that were supposed to sneak into Pearl Harbor prior to the air attack.

One of those subs was discovered before the main attack. That the main task force remained unseen was a gift from the revered ancestors that Yamamoto and other Japanese warriors worshipped. It was no small thing to sneak more than thirty ships across the Pacific Ocean. A northern path starting in the Kurile Islands was chosen to avoid busy sea lanes, and radio silence was rigorously maintained, with communications officers going so far as to remove the transmit keys from shipboard radio sets. This was a wise precaution because the United States, like Japan, had an efficient network of radio interception stations that were used to triangulate ship locations based on their transmissions. This was known to the Japanese. What they did not realize was that the Americans were reading many of their encrypted messages.

U.S. cryptanalysts had been breaking some Japanese naval codes since 1926, and most of the Purple diplomatic code since August 1940. Unfortunately neither one provided sufficient warning. The Purple decrypts contained some hints about what was to come, but they merely suggested that Japan was about to initiate hostilities; they did not pinpoint the exact location. And for good reason: The Japanese Foreign Ministry itself did not know what was coming. The army and navy were not in the habit of sharing their plans with diplomats, or any other civilians, for that matter. The naval codes might have provided more of a clue, but they were changed on December 1, 1941. American code breakers were not able to penetrate them again until the early spring of 1942.

In the weeks leading up to December 7, the intelligence advantage rested with the Japanese, who had good information about the disposition of the U.S. fleet thanks to observations transmitted by their consular personnel in

Honolulu. American intelligence analysts, by contrast, had no idea where the bulk of Japan's carriers were. The result was that much of Oahu was literally asleep when Commander Mitsuo Fuchida led the first wave in at 7:55 A.M.

Although December 7 was to become, in President Franklin Roosevelt's words, "a date which shall live in infamy," it was immediately overshadowed by other blows in a wide-ranging offensive, which, for sheer geographic scope, has never been equaled. Japan's armed forces struck simultaneously at targets from Hawaii to Singapore, a distance of almost seven thousand miles. On December 8, Betty and Nell bombers, escorted by Zeros with specially modified engines, flew all the way from Taiwan to destroy the bulk of the U.S. air force in the Philippines Islands—the strongest outside the U.S.—while it was still on the ground. Why General Douglas MacArthur, the American commander in the Philippines, was not better prepared remains a mystery; but surely one of the reasons is that he, along with most other Westerners, underestimated the range and capabilities of Japanese aircraft. He did not expect that they could make a round trip of more than one thousand miles. That miscalculation allowed Japanese aircraft to strike undetected and Japanese troops to subsequently land virtually unopposed.

Two days later, the Imperial Japanese Navy dealt its former mentors a crippling blow. Faced with the likelihood of Japanese aggression, Prime Minister Winston Churchill had insisted on dispatching a naval task force to Singapore as a deterrent in December 1941. The Admiralty wisely insisted that an aircraft carrier accompany the surface squadron, but the only one available, the newly commissioned *Indomitable*, was out of service due to a training accident. In a fateful decision, Churchill sent HMS *Prince of Wales* and HMS *Repulse* to the Far East on their own. *Prince of Wales*, nicknamed HMS Unsinkable, was one of Britain's newest and most powerful battleships; its ten fourteen-inch guns could hit targets more than twenty miles away. *Repulse* was older, a battle cruiser built in 1916 with six fifteen-inch guns and a top speed of twenty-nine knots. Both ships were copiously equipped with antiaircraft guns that were supposed to keep them safe. Or so Churchill reckoned. As it turned out, the usually farsighted prime minister was trapped in an earlier age of warfare.

When Japanese troops began landing on the Malay peninsula, Task Force Z—composed of *Prince of Wales* and *Repulse*, along with a handful of destroyers—was moored in Singapore harbor. The task force commander, Vice Admiral Sir Tom Phillips, sortied out on December 8 to try to sink the Japanese invasion force. Phillips had in the past deprecated the threat posed by airplanes, and, although he had modified his views somewhat based on recent experience (e.g., the sinking of the *Bismarck*), he still did not grasp the full capabilities of Japanese aircraft. He thought that as long as he stayed two hundred miles away from any Japanese base he would be safe, and so he did not bother to request land-based air cover. Phillips was in for a nasty shock on December 10 when his task force was spotted by ninety-six Betty and Nell bombers operating more than four hundred miles from their bases near Saigon. Within a few minutes, both British ships had been sent to the bottom of the Gulf of Siam by a flurry of bombs and torpedoes. More than eight hundred crewmen died, including Admiral Phillips. Six RAF fighters arrived too late to help, but even if they had been on the scene, there was little these antiquated Brewster F2A Buffaloes could have done to stop the Japanese onslaught. "Against the Zero fighters," the Zero's designer proudly noted, "the Buffalo pilots literally flew suicide missions."

More than two capital ships died that day. So did the whole idea of the battleship as the queen of the seas. For the first time in history, a battleship in motion had been sunk by aircraft alone. Almost four hundred years of naval tradition—a tradition dominated by the Royal Navy—was lost at sea. As the London *Evening Standard* wrote two days after the sinking of Task Force Z: "It is a tremendous and terrible vindication of those who argued that supremacy at sea could only be retained if the fullest possible profit were extracted from the new naval weapon in the air. The lesson is driven home with the impact of a torpedo." Neither the battleship nor the Royal Navy would ever recover from this blow.

When combined with the attacks on the U.S. fleet in Hawaii and the U.S. air force in the Philippines, the sinking of *Prince of Wales* and *Repulse* opened the door to a Japanese campaign of conquest across the width and breadth of the Pacific. General Sir Alan Brooke lamented in his diary that from "Africa eastwards to America through the Indian Ocean and Pacific, we have lost command of the sea." Western possessions fell one after another like overripe plums: Wake Island, Guam Island, Hong Kong, Borneo, New Britain Island, the Solomon Islands, the Gilbert Islands, northern New Guinea—the list was long, and, from the European and American perspective, thoroughly depressing, if not entirely unexpected. That independent Thailand shared their fate was no surprise either. More shocking was the surrender of Singapore, the supposedly impregnable "Gibraltar of the East," whose ninety thousand defenders were overcome by just sixty thousand Japanese attackers. Next came the downfall of the Dutch East Indies. Meanwhile, Admiral Nagumo's carrier armada bombarded Port Darwin, Australia, and then proceeded into the Indian Ocean, where the Royal Navy had held sway for more than two centuries. British ports in Ceylon were bombed in April 1942, and two British heavy cruisers and an aircraft carrier were sunk. The remainder of the British Eastern Fleet had to beat an ignominious retreat to East Africa. Much of Burma was lost to the Japanese army by May 1942. At the same time, the last-ditch defenders of the Philippines were forced to surrender in their island redoubt on Corregidor.

With all of these new acquisitions added to Japan's previous conquests—Korea, Taiwan, Manchuria, Indochina, northern and eastern China, the Marshall, Caroline, and Mariana Islands—the Greater East Asia Co-Prosperity Sphere had become a reality. The Rising Sun flag flew across East Asia. Heady talk circulated in Tokyo of invading India and Australia. "The Japanese," reported a French journalist living in Tokyo, "were giddy." But Japan, like Germany before it, had been made overly cocky by its early success. Nemesis waited in the wings.

THE TURN OF THE TIDE

Japan's immediate comeuppance would be delivered by the five American aircraft carriers that were now operating in the Pacific. (Following the attack on Pearl Harbor, two carriers had been rushed from the Atlantic and one from the West Coast to reinforce the two already in the central Pacific.) The damage suffered by most of the U.S. battleships at Pearl Harbor liberated the naval brass from the battleship-centric gospel of Mahan. With few battleships left, the newly appointed commander of the Pacific Fleet, Admiral Chester Nimitz, had no choice but to organize a series of carrier task forces (aircraft carriers surrounded by a screen of cruisers, submarines, and destroyers), thus establishing the dominant form of naval organization lasting into the twenty-first century. The remaining battlewagons were relegated to low-priority patrol and escort duty in the waters between California and Hawaii. They would continue to play a role in the war but a marginal one, primarily by providing fire support for amphibious landings.

Existing American carriers and their aircraft were, as previously noted, inferior to their Japanese counterparts, but the gap was not wide. With a bit of luck, skill, and good planning, the Americans could hold their own. It helped that they enjoyed an advantage in electronic warfare—an entirely new realm of competition created by the widespread use of radios and related devices. The U.S. Navy, with help from the British, was more advanced than the Japanese in the deployment of radar and sonar. It also had access to decrypts of Japanese codes; by April 1942, American cryptanalysts were once again reading Japanese naval codes. Thus U.S. commanders knew a great deal more about what their adversaries were planning than the Japanese knew about them.

Admiral Nimitz first put this knowledge to good use by sending the carriers *Yorktown* and *Lexington* to attack a Japanese invasion fleet off New Guinea. In the ensuing Battle of the Coral Sea (May 7–8, 1942), U.S. airplanes sank the outdated Japanese carrier *Shoho* and badly damaged the ultramodern *Shokaku*. The Japanese, in turn, sank the *Lexington* and badly damaged the *Yorktown*. This battle—the first in history where the opposing fleets never caught sight of each other—was essentially a draw, but one that set the stage for the first U.S. victory of the war by destroying so many aircraft belonging to the *Shokaku* and *Zuikaku* that they were temporarily taken out of commission.

Following this engagement, Yamamoto sought to surprise the Americans near the U.S.-held island of Midway, but, thanks to his superior intelligence, Nimitz was able to ambush the ambushers. At the Battle of Midway (June 4–7, 1942), four Japanese carriers confronted three American carriers, including the hastily repaired *Yorktown*. Dauntless dive-bombers managed to surprise the Japanese ships, which lacked radar. Many Japanese airplanes were caught on deck in the process of refueling and rearming, as helpless as U.S. aircraft had been on Oahu six months earlier. All four Japanese flattops were lost, to only one for the U.S. As significant, the Imperial Navy lost one-third of its elite corps of Sea Eagle fliers.

This victory has been dubbed a "miracle," because U.S. forces were outnumbered and outclassed. The subsequent turn of the tide need not be ascribed to divine intervention. After Midway, the U.S. Navy canceled plans to acquire more battleships and put all of its shipyards' energies into building carriers. By 1943 the United States was deploying *Essex*-class fast carriers that were superior to those in the Japanese navy. New U.S. fighters—the Grumman F6F Hellcat, the Lockheed P-38 Lightning, the Vought F4U Corsair—were also more than a match for the dreaded Zero.

Aside from its increasingly high quality, there was the sheer quantity of U.S. weaponry. In 1944, the United States produced 96,318 aircraft versus 28,180 for Japan, and 415 warships versus 287 for Japan. By war's end the U.S. Navy had commissioned more than 1,200 ships, including more than 100 aircraft carriers—27 fleet carriers and 77 small escort carriers. By 1945, the U.S. had 17,976 frontline aircraft in the Pacific versus 4,600 for Japan. A historian puts U.S. output into context: "[A]t peak production in March 1944, when her factories were producing one aircraft every 294 seconds, the United States would have been able to replace all that were lost on Oahu on 7 December 1941 in something like sixteen hours."

The Imperial Japanese Navy was inexorably crushed by this overwhelming agglomeration of military might. Even more than airplanes, the Japanese lacked skilled pilots. By the end of 1942 the best Japanese aviators—the ones who had staged the Pearl Harbor attack and its daring aftermath—had mostly been lost in battle. Their hastily trained replacements found themselves no match for veteran U.S. pilots flying first-class aircraft. The climactic battle of the war in the Pacific occurred in Leyte Gulf off the Philippines. Here, on October 23–26, 1944, in the largest naval battle in history, the Japanese lost four major carriers, three battleships, and ten cruisers. Thereafter the Imperial Navy all but ceased to exist. The only way the Japanese could continue to attack the U.S. Navy was to send unskilled young kamikaze pilots to ram their airplanes into American ships (a crude but effective forerunner of "smart" bombs), a tactic that only reinforced the Americans' icy determination to do whatever was necessary to destroy continued resistance.

Admiral Yamamoto did not live to see the defeat he had anticipated. He was killed in 1943, a victim of American code-breaking that uncovered his schedule and allowed P-38 fighters to shoot down a Betty bomber in which he was traveling near the Solomon Islands.

As with Germany, so with Japan: Its eventual defeat did not mean that its early victories were thereby rendered irrelevant. On the contrary, Japan's initial successes dealt the European empires in Asia a blow from which they never recovered. The physical loss of the colonies may have been only temporary, but the loss of face was permanent. For the second time—the first was the 1905 Russo-Japanese War—a "yellow" race had beaten a "white" one. That was not something that anyone in Asia or Africa would soon forget. Indeed, the Pearl Harbor raid is still referred to in Japan as "the greatest single victory that disproved white supremacy." Such perceptions can change history.

"The British Empire in the Far East depended on prestige," Australia's minister to China wrote in May 1942. "This prestige has been completely shattered." This was even more true of the Dutch and French empires. All of them attempted to reclaim their colonies after the war but found it nearly impossible to do so. In China, Indonesia, Malaya, Indochina, Burma, the Philippines, and elsewhere, the war had energized anticolonial movements. Some of them cooperated with the Japanese conquerors and were armed by them: Sukarno, the first leader of independent Indonesia, was a leading example. Others, such as the future Philippine president Ramon Magsaysay, fought against the Japanese and received arms from the Allies. In either case, these fighters did not relinquish their weapons when the war ended. The days of docile natives who could be bossed around by a handful of European plantation owners and district commissioners were at an end. Even in India, which the Japanese did not invade, the war hastened independence because Britain had to make political concessions in order to win the support of native politicians for the war effort.

Some of the most successful anti-Japanese guerrilla fighters were communists such as Ho Chi Minh and Mao Zedong. The disruption caused by the war allowed "Reds" to come to power in China, North Korea, and North Vietnam, and to foment serious rebellions from the Philippines to Malaya. Historian Arthur Herman may or may not be right to assert that "without the sinking of the *Prince of Wales* and *Repulse* there might not have been an India or Pakistan—or a Vietnam War." But there is no denying that the course of history would have been vastly different if Japan had not been so successful in 1941–42 in taking advantage of the aviation revolution that had transformed warfare at sea.

America's ability to match Japan's innovations in short order meant that Tokyo could not win the war. But could America bring a proud and determined nation to its knees without suffering crippling casualties in the process? Destroying the Imperial Japanese Navy was a necessary precondition for success. It was not by itself enough to guarantee victory. That ultimately required bringing overwhelming force to bear on Japan's home islands. And that, in turn, required the application of another military revolution: the revolution in long-range bombardment.

CHAPTER 9

SUPERFORTRESSES AND FIREBOMBS:

Tokyo, March 9–10, 1945

The sirens wailed around midnight on Friday, March 9, 1945. *Enemy bombers on the way*, the radio reported. This did not cause undue alarm. The people of Tokyo had gotten used to being bombed since the killer "Bees"—the B-29 Superfortresses—had first appeared over their city the previous November. Mostly, they had dropped their bombs in the daytime and from high altitude, aiming for factories and other industrial targets. The raids were not terribly destructive, and they had quickly become part of the normal routine in a country at war. No one had any inkling of what was about to unfold as they looked up at the sky that night, bedazzled by what a reporter described as the "long, glinting wings, sharp as blades" that could be seen in the dark sky, "glittering blue, like meteors, in the searchlight beams spraying . . . from horizon to horizon."

The men flying the B-29s knew this raid was different from all the previous ones, and they were, to put it bluntly, scared. When the briefing officers had announced that morning what they were planning to do on the night of March 9–10, the crewmen uniformly concluded that the "Big Cigar"—their nickname for Major General Curtis E. LeMay, commander of the 21st Bomber Command, who always had a fat stogie stuffed in his mouth—had "just gone stark crazy." LeMay had not been satisfied with the results of the previous raids, and he insisted on a radical change of tactics in order to eradicate the heart of the enemy capital.

Instead of dropping standard high-explosive bombs, he decided to switch to a new jellied gasoline mixture known as napalm, designed to burn, not blast, the paper and wood houses of Japan. And instead of coming in at 25,000 to 30,000 feet—a high altitude that made U.S. bombers virtually impervious to air defenses—pilots were told to fly at just 5,000 to 8,000 feet. This was supposed to make their bombing more accurate and more deadly, but it would also leave the giant bombers, each one nearly as wide as a football field, hopelessly exposed to fighters and flak.

To make the mission even more unnerving, LeMay decided to increase the ordnance load beyond the recommended level—each airplane would carry 12,000 to 16,000 pounds of bombs—by removing almost all defensive armaments and the gunners who operated them. A B-29 normally had ten .50 caliber machine guns and a 20 mm cannon; on this raid, only a single tail gun would remain. There would be no fighter cover either, because U.S. fighters lacked the navigational equipment to operate at night. The disarmed B-29s would be on their own over the enemy's capital with no friendly forces within a thousand miles if they got into trouble. The crews knew that if they were shot down and captured they would likely be tortured and killed by the Japanese, who regarded them as war criminals. Their only protection would come from the element of surprise and the cloak of night.

That did not satisfy the airmen, one of whom later recalled how "the guys cursed General LeMay" and sat in "stunned . . . disbelief" as they contemplated "such a 'suicide' raid." "We'll get the holy hell shot out of us," one officer complained. Lieutenant Colonel Robert Morgan, who had flown a B-17 over Europe (the famous *Memphis Belle*) before transferring to a B-29 called *Dauntless Dotty*, "remembered the low-level missions that had been tried in Europe, and the results—whole squadrons of B-17s blown out of the sky." Another officer marveled at "the unprecedented, daring, almost unbelievable decision to go in at low level instead of at twenty-five to thirty thousand feet."

LeMay himself had no idea whether this gamble—which he decided to take without consulting his superiors and against the advice of almost all his subordinates—would pay off. He feared that he might lose three-quarters of his aircraft, as his flak experts were warning him. Visions of writing letters to families of dead airmen filled his head.

By 9:00 P.M. on March 9, nearly three hours after the first heavy airplane had lumbered into the sultry sky, all 334 Superfortresses had taken off from their airfields in the Mariana Islands (Guam, Tinian, Saipan). For their commander left behind at his Guam headquarters, there was nothing to do but wait until they had completed their fifteen-hour, 3,200-mile round-trip. When his public relations officer wandered into the operations room at 2 A.M., he found LeMay alone with a handful of clerks, sitting on a bench in the Quonset hut, nervously smoking a cigar. "I can't sleep," the jowly thirty-eight-year-old general explained. "I usually can, but not tonight. . . . I'm sweating this one out myself. A lot could go wrong."

A lot already had gone wrong with the B-29, or, as it was known within the U.S. Army Air Forces, "the $3 billion gamble." It was undoubtedly the most powerful instrument of long-range destruction ever devised up to that point. Boeing's engineers had worked wonders with its design, but it had shown a disconcerting tendency to catastrophically malfunction in flight; engine fires were a common hazard. Figuring out how to make effective use of the super-bomber had stumped one commander after another. Many bombs had missed their targets and many bombers had been lost due to foul weather, mechanical glitches, and the sheer, unprecedented challenge of flying thousands of miles over open water. "We'd flown more than 2,000 missions over Japan, with no decisive damage to any important target," Lieutenant Colonel Morgan recalled ruefully. "Clearly the [21st] Bomber Command was at a crisis-point."

"The Big Cigar" hoped that a new approach would finally achieve his goal of reducing Japan's cities to ashes. He knew that if the Tokyo fire raid worked, the war might be shortened; if it did not work, the likelihood would grow of an invasion of Japan that would almost certainly result in hundreds of thousands, possibly millions, of American casualties. The stakes could not be higher, not only for the United States but also for bomber advocates like LeMay, who had spent the interwar years forging what they thought would be a war-winning instrument.

The bombers had not lived up to the hype so far. This was their best, and perhaps last, chance to show what they could do.

THE RISE OF STRATEGIC BOMBING

Strategic bombing, like tank warfare and naval aviation, was born in World War I and grew to maturity in World War II. The earlier conflict was only a few weeks old when, on August 30, 1914, a German pilot in a flimsy monoplane dropped a few bombs and a derisive note on Paris. By 1915 German zeppelins and airplanes were targeting London. The British capital suffered fifty-two air raids during the war that killed or injured nearly 3,000 people. The British gave as good as they got: The Royal Flying Corps bombed Cologne, Stuttgart, and other German cities, killing or wounding over 1,900 people. Although alarming enough at the time, the damage caused by these raids was, in retrospect, minimal, and it had no impact on the outcome of the war. It did, however, provide a taste of things to come. In the interwar years, there was a widespread expectation that "strategic" bombing—i.e., targeting urban and industrial infrastructure, not armies in the field or ships at sea (the purview of "tactical" aviation)—would be much more significant in the future.

The strategic bombing choir was led by three high-strung divas: Air Marshal Hugh "Boom" Trenchard, commander of the Royal Flying Corps in World War I and then of the Royal Air Force until 1929; Brigadier General Billy Mitchell of the United States, who had been profoundly influenced by Trenchard during their work together in France in 1917–18; and General Guilio Douhet, commander of Italy's first aviation units and author of the influential 1921 book *The Command of the Air*, in which he argued that air forces had rendered the rest of the military obsolete. These visionaries believed that bombing would be so catastrophic that any future war would be over in a matter of days, and thus targeting civilians, far from being inhumane, would actually *lessen* the horrors of war. It was not so much physical damage that the bomber visionaries foresaw. It was mental damage. They thought that any population under bombardment would crack under the strain. J. F. C. Fuller (who was almost as enthusiastic about airplanes as he was about tanks) predicted, with spectacular lack of foresight, that if Britain were attacked by airplanes the government "will be swept away by an avalanche of terror. Then will the enemy dictate his terms, which will be grasped at like a straw by a drowning man. Thus may a war be won in forty-eight hours and the losses of the winning side may be actually nil!"

The bomber had a particular attraction for interwar Britons who did not want to fight again on the Continent and saw it as a low-cost alternative to large and expensive standing armies. The RAF even took over one of the army's traditional functions—imperial policing—by bombing and strafing rebels in Somaliland, Iraq, and Afghanistan. The British were aware, of course, that just as they planned to bomb their enemies, so their enemies would bomb them. But, based on the superior performance of bombers over fighters in the late 1920s and early 1930s, they concluded that there was little that could be done to protect their isles. "The bomber will always get through," former (and future) Prime Minister Stanley Baldwin proclaimed in 1932. "The only defence is in offence, which means you have got to kill more women and children quicker than the enemy if you want to save yourselves."

Thus was born the theory of mutual assured destruction that would blossom fully during the Cold War. Fear of the bomber was to cast a long shadow over diplomacy in the 1930s: Britain and France did not want to trigger a war that, they feared, would lead to the incineration of their cities. Indeed, Harold Macmillan would later say that "we thought of air warfare in 1938 rather as people think of nuclear warfare today."

The United States was also drawn to strategic bombing in the 1920s and 1930s, largely for the same economic reasons that attracted the British. But the U.S. Army Air Corps was less enthusiastic than some in the RAF about indiscriminate "area" bombing designed to shatter enemy morale. American airmen feared this would be not only immoral but ineffective, because it would simply spur a greater desire to resist among the targeted population. Rather than slaying civilians, they preferred to focus on immobilizing the industrial infrastructure that supported any modern war effort.

Instructors at the Air Corps Tactical School (located, after 1931, at Maxwell Field near Montgomery, Alabama) were impressed by a mishap that occurred in 1936: A flood in Pittsburgh incapacitated a plant that made special springs needed for variable-pitch propellers. Because no other factory manufactured the same component, this accident temporarily disrupted aircraft production across the country. To the air strategists, this was a powerful demonstration of how fragile any modern nation's "industrial web" was, and how easily it could be disrupted if a few vital "chokepoints" were targeted. Based on an analysis of the U.S. economy, the Tactical School officers—many of whom would go on to become senior air force generals in World War II—concluded that the top priorities of any bombing campaign had to be electric power systems, transportation (chiefly railroads), fuel refining and distribution, food distribution and preservation, and steel manufacturing. "Loss of any of these systems would be a crippling blow," a Tactical School study concluded. "Loss of several or all of them would bring national paralysis."

Yankee ingenuity was applied to the challenge of inducing paralysis in enemy nations. In 1934 the Air Corps asked manufacturers to design a heavy, long-range bomber. The Boeing Aircraft Company, building on its work for commercial airlines, responded with what would become the B-17: a four-engine aircraft that could carry 2,400 pounds of bombs about six hundred miles from its base at a top speed of 268 mph. Bristling with machine guns mounted in Plexiglas turrets, the B-17 was nicknamed the Flying Fortress. It was the most advanced bomber of its day when it went into production in 1938, but it was soon slightly surpassed in range, speed, and bomb capacity by Consolidated Aircraft Company's B-24 Liberator, which first flew in 1939. Though less glamorous than the B-17, the B-24, with its distinctive twin tail, became the most widely produced American aircraft of the entire war.

Both bombers were equipped with a top-secret bombsight originally commissioned by the U.S. Navy in 1932 from an eccentric inventor named Carl Norden. The gyroscopically stabilized Norden bombsight incorporated a crude mechanical computer that used data entered by a bombardier for ground speed, air speed, and crosswinds to determine a bomb's optimal release point. It could even be linked with an autopilot so that the bombsight could briefly take control of the plane on its bombing run to make sure that it was lined up correctly on the target. No other nation had a bombsight as advanced as the Norden, which cost $1.5 billion to develop and produce—almost as much as the atomic bomb. So vital was this technology that Norden was constantly accompanied by bodyguards during the war lest Axis kidnappers try to acquire his secrets. Combined with the B-17 and B-24, the Norden bombsight gave the U.S. the best strategic bombing technology in the world, making this the only field of military endeavor where America was the clear leader prior to World War II, though even here the U.S. lagged far behind other combatants in the sheer number of airmen and airplanes. (Only thirteen B-17s were operational in 1939.)

Britain was not much better prepared. In 1939, it had no heavy bombers and only 536 light bombers that lacked enough range or bomb capacity to do much damage to Germany. The initiative in the air, as in most other areas, passed by default in the early stages of the war to the Nazis.

THE AIR WAR IN EUROPE

"When the war finally broke out at the beginning of September 1939," notes historian Richard Overy, "the expected knock-out blow from the air did not materialize." The Luftwaffe attacked Warsaw, Rotterdam, Belgrade, and other cities during the blitzkrieg across Europe from 1939 to 1941, but the airmen's main contribution to victory was in the tactical, not the strategic, realm—that is, in support of ground troops, not on their own. When they tried to wipe out the British Expeditionary Force at Dunkirk in May 1940 and then to prepare the way for an invasion of Britain, Hermann Göring's fliers utterly failed to achieve their goals. Germany was not well prepared to undertake a long-range bombing campaign, and Britain was well prepared to defend against it.

Whitehall had belatedly made up for years of neglect in the mid-1930s by ordering two high-performance fighter planes: the Supermarine Spitfire and Hawker Hurricane. With their all-metal, streamlined construction, retractable landing gear, powerful Rolls-Royce Merlin engines, and eight wing-mounted machine guns, they were as good, if not better, than the Luftwaffe's top-of-the-line fighter, the single-engine Messerschmitt Bf 109, and far superior to the Nazis' heavier, twin-engine Bf 110 escort fighter. Still, the British were badly outnumbered at the start of the Battle of Britain—some 750 Spitfires and Hurricanes pitted against more than 1,600 German bombers and 1,000 fighters based within striking distance of England. Air Marshal Sir Hugh Dowding, the head of RAF's Fighter Command, compensated for this numerical inferiority with skillful use of electronic technology. Starting shortly after Robert Watson Watt's invention of radar in 1936, the British had deployed a series of early warning stations around their coastline that were linked to RAF's Fighter Command, which in turn was linked to fighter bases. When enemy bombers were detected, fighters could be scrambled and vectored via radio to intercept them. Together, fast fighters and radar shifted the advantage in the air from offense to defense.

The Germans played into British hands by not targeting air defenses. They initially set out to attack the RAF, but they did not stick with it. The Luftwaffe's failure to take down British air defenses, which rested on extremely vulnerable 350-foot steel radar towers, cost it the possibility of victory in the Battle of Britain. Instead, in September 1940, the Luftwaffe switched to bombing British population centers, starting with London. The Blitz killed more than forty thousand civilians and seriously injured more than fifty thousand but, contrary to the expectations of prewar theorists like J. F. C. Fuller, it did not dent British determination to continue fighting. The offensive finally petered out when, in another example of the growing importance of electronic warfare, the British realized that German bombers were being guided to their targets by radio waves and began to jam the signals. By the spring of 1941, Hitler needed to redirect his aircraft in any case toward his coming war with the Soviet Union.

The lack of success enjoyed by the Luftwaffe did not shake the faith of British or American airmen in the potential of strategic bombing. It merely convinced them that the Luftwaffe, which lacked a four-engine bomber, did not have the right tools for the job. The terror attacks on London, Coventry, and other cities also fired up British determination to mete out similar punishment to German cities. When Winston Churchill toured the East End of London in September 1940 to inspect bomb damage, the crowds told him, "We can take it, but give it 'em back."

That's just what RAF's Bomber Command, soon to be joined by the U.S. Army Air Forces (as the Air Corps had been renamed in 1941), aimed to do. Unfortunately, the Allies had not learned the prime lesson of the Battle of Britain: the need to begin any air campaign by suppressing enemy air defenses. Their failure to heed this lesson would cost them dearly.

The RAF first bombed a German city—Berlin—in August 1940. Until 1942, however, the RAF kept its focus mainly on industrial targets in western Germany (the Ruhr) and occupied territories. The results were execrable. The bomber forces suffered heavy losses while, according to one internal study, fewer than 25 percent of bombs fell within five miles of the designated target.

Under the leadership of Air Marshal Arthur "Bert" Harris, who took over Bomber Command in February 1942, the British offensive achieved better results. "Bomber" Harris has often been credited (or blamed) for switching from "precision" to "area" bombing; in fact, the targeting of enemy population centers predated his assumption of command. What Harris did was to make area bombing more effective and more ruthless by utilizing newly delivered four-engine, long-range bombers, the Handley Page Halifax and

Avro Lancaster (Britain's answer to the B-17 and B-24), which would be guided to their targets by radio navigation devices with cute codenames ("Gee," "Oboe") and, later, H2S radar sets.

Harris was intent on massing enough bombers to deliver a crushing load of explosives and incendiaries against urban centers. The first of his thousand-bomber missions hit Cologne on May 30–31, 1942, inflicting more damage in a single night than London was to receive during the entire war. Hamburg suffered worse. A raid on July 27, 1943, generated a 1,000-degree-Fahrenheit firestorm that ravaged three-quarters of the city and killed at least sixty thousand people. By the time the war in Europe ended on May 8, 1945, Harris was well on the way to his stated goal of "the elimination of German industrial cities": Half of Germany's urban area was laid waste by Allied bombers.

The RAF's deliberate targeting of civilians—something that was denounced as "barbarism" when it was done by the Germans, Italians, or Japanese—was hardly cloaked by the lame explanation offered by the British government: The raids, Winston Churchill claimed, were intended to "dehouse" German industrial workers. It was never explained how a bomber could be counted upon to deprive a worker of his house without also depriving him of his life. The reality was that British bombers had proved incapable of hitting factories. If they were to be utilized at all, they had to go after entire cities, and never mind any ethical qualms.

The U.S. Army Air Forces would move in the same direction, if more slowly and with even less candor. When U.S. B-17 and B-24 bombers first began operating out of England in the summer of 1942, they were used strictly for daylight raids against military and industrial targets, first in occupied Europe, then, starting in 1943, in Germany itself. Bert Harris insisted this was a complete waste; he snorted derisively at the search for "panacea" targets. But the USAAF commander, General Henry "Hap" Arnold, was wedded to his prewar doctrine of "precision."

A division of labor was later agreed upon at the Casablanca summit between Winston Churchill and Franklin Roosevelt in January 1943. The result was Operation Pointblank, more popularly known as the Combined Bomber Offensive: The British would bomb cities by night, the Americans would try to hit specific targets by day. "If the R.A.F. continues night bombing and we bomb by day," wrote General Ira Eaker, commander of the U.S. 8th Air Force, "we shall bomb them round the clock and the devil shall get no rest."

That was the theory, anyway. In practice, the RAF and USAAF found penetrating German airspace to be no easy feat. After a period of early neglect, the Nazis assembled a formidable air-defense system built, like the British one, around a radar early-warning net. The Germans fielded high-tech interceptors, including night fighters equipped with portable radar sets, fighters armed with air-to-air rockets, and eventually the first jet fighter, the Messerschmitt Me-262, which became operational in mid-1944. None of these aircraft was ever available in sufficient numbers to stem the bomber tide, but the Germans did have plentiful and extremely effective flak (an abbreviation for *Flieger-Abwehr Kanone*, or airplane defense cannon). The high-velocity 88 mm gun could propel a twenty-pound shell as high as 31,000 feet, and it could fire fifteen to twenty times a minute. At their peak, the Germans deployed 55,000 antiaircraft guns.

Flying into the teeth of such defenses, Allied heavy bombers initially suffered catastrophic casualties. On August 1, 1943, 178 B-24s from the 15th Air Force based in North Africa were dispatched to hit the oil refineries in Ploesti, Rumania, which kept the Nazi war machine humming. The Germans managed to get advance warning of the raid by intercepting and decoding Allied communications. One-third of the bombers were lost. To make this raid even more frustrating, the Ploesti refineries had not been running at full capacity when they were hit. They were able to make up the losses suffered from the raid so quickly that, within weeks, more fuel was being refined than before the attack.

This was typical of Allied bombing attacks in 1943 and early 1944. American fliers sent after targets like the Messerschmitt works at Regensburg and the ball-bearing plants at Schweinfurt (both in Bavaria) took heavy losses without achieving the decisive results their commanders had hoped for. The German ability to repair damage and restart production proved greater than the Allied ability to keep the pressure on. The Allies had not realized that Germany had not been fully mobilized at the start of the war. There was so much slack in the system that Armaments Minister Albert Speer was able to triple production between 1942 and 1944 despite incessant bomber attacks.

While the Allies were uncertain about how much damage their raids were causing the enemy, they were well aware of how much their own fliers were suffering. A study of 2,051 airmen in the U.S. 8th Air Force found that by the end of the war 63 percent had been killed, captured, or missing, and 10 percent wounded—higher casualties than the marines suffered in their desperate fighting in the Pacific. Knowing the odds, many crewmen succumbed to a fatalism later portrayed, in exaggerated form, in Joseph Heller's comic novel *Catch-22*. They avidly counted down the days until they had completed their assigned twenty-five missions only to be dismayed, in many cases, to learn that the requirement had been increased to thirty or thirty-five. "Nobody expected to live through it," said one 8th Air Force veteran.

Casualties were so high because bombers were attempting to enter enemy airspace on their own, without benefit of fighter escort. Prior to the war, the RAF and USAAF had assumed that it would be impossible to build a long-range escort fighter: "[T]he fighter experts asserted," recalled General Haywood Hansell, one of the USAAF's pioneer tacticians, "that a fighter with the range to accompany bombers would be so large and heavy that short-range interceptors could easily outfly and outfight them." Instead of trying to solve this engineering problem, the British and Americans preferred to load up their bombers with defensive armaments. This solution failed the test of battle, so in late 1943 the USAAF borrowed an ingenious expedient from the Japanese: adding drop tanks to fighter planes. These extra fuel canisters, which could be jettisoned in flight, dramatically extended the range of escort aircraft. North American Aircraft's new P-51 Mustang, which arrived in Europe in late 1943, had its combat range increased from 400 miles to 1,800 miles—easily enough to cover most of Germany from bases in England. The P-51 was complemented by the older Lockheed P-38 Lightning and Republic P-47 Thunderbolt. Together, these long-range fighters would turn the tide of the air war. An 8th Air Force study found that escorted bomber missions suffered seven times fewer losses than unescorted ones.

Not only did the fighters allow the bombers to reach their targets safely, but they also took a heavy toll on Luftwaffe interceptors. Destroying German aircraft in the skies proved easier than destroying the fortified factories in which they were built. In the first three months of 1944, the Luftwaffe lost 3,450 fighters. Even more important was the loss of experienced pilots. In the first five months of 1944, the Luftwaffe had almost all of its fighter pilots killed, seriously wounded, or captured (2,262 of 2,395). Their replacements did not have nearly as much training. The Germans began sending pilots with just 80 hours of flight time to take on British and American fliers with an average of 225 hours in the cockpit.

The Allies continued to suffer heavy bomber losses during the first half of 1944; on some raids, 10 percent of the bombers failed to return home. But they were easily able to make good their losses at a time when one factory alone—Ford's famous Willow Run, Michigan, plant—was churning out a new B-24 every hour. The Germans had no such manufacturing capacity to call upon. By sheer dint of attrition, the Luftwaffe was ground down into nothingness.

By D-Day, June 6, 1944, Allied domination of the skies was complete; the invasion of Europe would not have been possible otherwise. On the first day of Operation Overlord, the USAAF would fly 8,722 sorties, compared to 250 for the Luftwaffe. Important as it was to secure air control over the Normandy beachheads, it was a job disdained by the bomber barons of the RAF and USAAF, who wanted to focus all their efforts on the strategic air campaign. General Dwight D. Eisenhower was able to get the air cover he needed only by threatening to resign if the RAF and USAAF were not temporarily subordinated to his command. Under the direction of Eisenhower and his deputy, Air Marshal Arthur Tedder, the RAF and USAAF were so successful at toppling bridges and blowing up railroad marshaling yards that the

Wehrmacht was unable to rush massive reinforcements to Normandy. Much of the credit belonged to fighter-bombers from Major General Elwood R. "Pete" Quesada's U.S. 9th Fighter Command. His nimble airplanes could not carry nearly as much ordnance as the heavy bombers but, because they were smaller and faster, they could fly lower without as much worry about being hit. Hence their bombing was generally more accurate.

On those few occasions when heavy bombers were employed in direct support of troops, the results were usually a fiasco. American bombs hit their own side so often that some G.I.'s grimly joked that the 8th and 9th U.S. Air Forces should be renamed the "8th and 9th Luftwaffe." One misguided raid killed more than one hundred U.S. soldiers, including Lieutenant General Leslie McNair, the highest-ranking American officer killed in the war.

Bomber accuracy improved during the course of the war, but it never attained the levels claimed by prewar theorists who boasted that a B-17 equipped with a Norden bombsight could "put a bomb in a pickle barrel from twenty thousand feet." Even under ideal conditions, a B-17 would be lucky to drop its load within one thousand feet of that mythical barrel. But conditions were seldom ideal. Many bombing missions were flown at night or in overcast conditions, forcing bombardiers to rely on radar or radio directional finders to locate their targets. "Blind bombing" was about as accurate as it sounded: Bombs could be counted upon to fall within two miles of their target; anything closer was sheer luck. Overall, only 20 percent of bombs fell within one thousand feet of the aim point. On one awful occasion in 1945, thirteen U.S. bombers sent out to hit a German town hit a Swiss town by mistake. (A Luftwaffe squadron had committed an even more embarrassing error on May 10, 1940, when it set out to bomb the French town of Dijon and instead hit the German town of Freiburg-im-Breisgau, killing fifty-seven civilians.)

The only way bombers could be sure of hitting a target was by obliterating the entire neighborhood. This led inevitably to the USAAF adopting the same kind of area bombing as the RAF. On February 13–14, 1945, the USAAF joined the RAF in the destruction of Dresden, which killed between 25,000 and 40,000 people. (The initial estimate of 250,000 dead, which gave the raid much of its infamy, was later found to be considerably exaggerated.) Yet the Americans never admitted what they were doing. They stuck to their story that they were concerned solely with "military objectives."

In a way, it was true, because, as Churchill said in 1942, "Morale is a military target." But the German will to fight was never broken by bombing, despite the death of as many as 300,000 civilians in air raids. While Germans were disheartened by the attacks, there was precious little they could do to affect the policies of their fanatical Führer, who was determined to resist to the end.

Much more effective than attacks on civilians were attacks on oil refineries and other fuel facilities. These raids, which began in May 1944, starved the Wehrmacht and Luftwaffe of fuel that they desperately needed and seriously hindered their defensive efforts on both the Western and Eastern Fronts. Also effective were attacks, beginning in September 1944, on Germany's railroads and other transportation links. Coal that was desperately needed to run factories could not be moved from mines in the Ruhr. The sheer weight of Allied air power was crushing: 83 percent of the total bomb tonnage dropped on Germany (1.4 million tons) fell during the last year of the war. This contributed to a general collapse of the German economy during the harsh winter of 1944–45. The Germans were willing to fight on, but by early 1945 they lacked the means to do so. When the war ended, the Germans still had plenty of tanks and aircraft but the fuel needed to keep them operating had run out.

Postwar skeptics of the efficacy of strategic bombing could point to the fact that German war production continued to rise until 1944, but they ignore the question of how much higher it would have gone were it not for the bombing. (Albert Speer estimated that in 1944 the output of tanks, trucks, and airplanes had been more than 30 percent below what was planned.) They also overlook the fact that the bombing campaign diverted a substantial portion of the German war effort into self-defense. Roughly a million Germans were not available to fight at the front because they were involved in air defense; another million were kept busy doing cleanup and rescue work after air raids. Ten thousand 88 millimeter guns that could have been used to knock out Allied tanks (the 88 was the best antitank gun as well as the best antiantiaircraft gun of the war) were instead deployed against Allied aircraft. Half of the electronics and a third of the optical instruments produced by German industry were denied the army and navy because they were reserved for antiaircraft work. So was 20 percent of all ammunition production. Hitler wasted further resources trying to develop revenge weapons—the V-1 cruise missile and V-2 ballistic missile—to retaliate for the destruction of his cities. Neither one ever justified the investment poured into them. The U.S. Strategic Bombing Survey later estimated that the resources put into these rockets could have been used to produce twenty-four thousand fighter planes.

By the time Allied armies entered Germany, the country lay devastated. Its cities resembled a moonscape, its people scrounging in the rubble for cigarette butts or scraps of food or a place to keep warm. More than 300,000 Germans had been killed by bombs. Another 780,000 had been wounded and 7.5 million made homeless. The Combined Bomber Offensive did not, contrary to the hopes of the bomber theorists, win the war all by itself, but, as a recent study concludes, it was "essential to the defeat of Nazi Germany." Where the bomber theorists had been most wrong was in thinking that bombers could inflict massive damage at scant cost to themselves. Allied air forces paid a fearful price for their success. More than 16,000 Allied bombers were lost over Europe, and more than 29,000 American and 47,000 British airmen killed. American air crews suffered a 40 percent higher fatality rate than did American ground forces in Europe.

IRON ASS

One of the outstanding combat commanders to emerge from the European bomber campaign was Curtis L. LeMay. Much later on, in the 1960s, LeMay was to gain notoriety as a supposed militarist and extremist. He was widely suspected of wanting to start another world war during the Cuban Missile Crisis in 1962 when he was Air Force chief of staff. His statement in 1965, following his retirement, that the U.S. should bomb North Vietnam "back into the Stone Age," would become emblematic in certain circles of Neanderthal thinking. He would cement this reputation by running as vice presidential nominee on the third-party ticket headed by Southern segregationist George Wallace in the 1968 presidential election. It is hard, if not impossible, to recover his pre-1960s reputation, but it is worth remembering that before he was regarded as a character out of *Dr. Strangelove* he was widely seen as a "boy wonder," a daring and innovative aviator who had distinguished himself in World War II, led the Berlin Airlift in 1948, and over the next nine years transformed the Strategic Air Command from a dispirited rabble into a fearsome, highly disciplined instrument of deterrence.

About the boy-wonder part there can be no doubt: LeMay rose stratospherically. A mere lieutenant in 1939, within five years he had become, at thirty-seven, the youngest major general in the entire U.S. Army. Quite an accomplishment for an officer who never went to West Point (the Air Force Academy would not be established until 1955) and came from a hardscrabble background. LeMay was born in Columbus, Ohio, the oldest of seven children, but spent much of his childhood moving around the country while his ne'er-do-well father, Erving, pursued various odd jobs. Curt was a rebellious youngster, often running away from home, but he matured into a serious young man. In his autobiography, he lamented how as a high school and college student in the 1920s he never had time for the parties or extracurricular activities that his peers enjoyed: "I had to spend my extracurricular hours earning money to maintain myself at the time." Upon entering Ohio State University, he signed up for the Reserve Officers Training Corps once he discovered that it offered a way to underwrite part of his education. Even as an ROTC cadet, however, he still needed the extra income he derived from a job working at a steel foundry. So after finishing with his schoolwork and ROTC commitments at 5 P.M., he would rush off to the foundry and work until 2 or 3 A.M. Not surprisingly, he often slept through his early-morning classes and never quite finished his civil engineering degree.

Like many young men of his generation, LeMay was entranced by the romance of the air and the aces of World War I—the Red Baron (Manfred von Richthofen), Billy Bishop, Eddie Rickenbacker, and all the rest. Charles Lindbergh, Amelia Earhart, and other daring fliers kept aviation in the headlines in the 1920s. After LeMay paid five bucks to spend five minutes in the air with a barnstorming pilot at a carnival in the early 1920s, he was hooked, determined to spend his military service in a cockpit. Through a complicated series of bureaucratic maneuvers, he got his wish and reported in 1928 for flight training at March Field, near Riverside, California, to be followed by advanced training at Kelly Field near San Antonio, Texas. The washout rate for novice pilots was around 75 percent; the death rate was also substantial. LeMay had some close calls, but he passed his tests in a cloth-covered biplane and earned his wings in 1929. Starting out as a fighter pilot, he transferred to bombers in 1937 and instantly fell in love with the new Flying Fortress. As a B-17 pilot and navigator, he was responsible for some notable feats in the late 1930s: navigating a group of B-17s on a long-range flight to South America and leading another squadron in a mock intercept of an Italian ocean liner six hundred miles off the East Coast of the United States.

With the coming of war, the Air Corps expanded rapidly from a small, elite group of professionals into a mass organization composed of civilians in uniform. Career officers like LeMay won rapid promotion but experienced plenty of headaches trying to break in new recruits. Newly promoted to lieutenant colonel and given command of the 305th Bombardment Group in 1942, he had only a few months to train his men in the California desert before they were dispatched to Europe. In spite of a hard training schedule that earned him the nickname "Iron Ass," LeMay observed that "we were still a sorry outfit when we left . . . to go overseas. Most of our navigators had never navigated over water. Most of our gunners had never fired at a flying target from an airplane." Yet within a month of their arrival in Scotland in October 1942, they were flying their first missions over occupied France with their thirty-five Flying Fortresses.

LeMay had no more idea than his men of what to do in combat, but he learned fast. He was responsible for several tactical innovations that were widely copied in the 8th Air Force. He borrowed from the British the idea of using pathfinders—specially selected and trained crews who would lead the rest of the bombers to their target. He also decided that there was no point in zigzagging over the target to avoid getting hit by flak; the only way to get bombs on target was to fly straight and level, and damn the consequences. To reduce the danger from marauding fighters, LeMay came up with the idea of tightly arraying eighteen to twenty-one bombers so they could provide interlocking fields of fire. The Combat Box became the standard formation in the USAAF.

LeMay was no rear-echelon commander. He loved nothing better than to go out on missions with his "boys." (Later, in the Pacific, he would chafe when USAAF headquarters would not let him go out on raids because he had knowledge of the atomic secret.) On August 17, 1943, by now in charge of an entire air division, he personally led the perilous attack on the aircraft plants at Regensburg, Bavaria. In the process, he lost 24 of his 146 bombers (16 percent) and most of the rest were damaged. But LeMay kept the pressure on and achieved more success as the war progressed. His accomplishments attracted the eye of General Hap Arnold, who had a big project starting in the Pacific and needed someone like Curt LeMay—a seasoned combat leader, a big-time "operator"—to take charge of it.

THE SUPERFORTRESS TAKES OFF

"The B-29," writes historian Eric Larrabee, "was the greatest U.S. gamble of the war—greater even than the atom bomb ($3 billion invested, as opposed to $2 billion for the bomb, on a similar absence of hard evidence)." Its genesis lay in a decision in 1940 to commission a Very Long Range Bomber that would exceed the range of the B-17 and B-24, neither of which had yet flown in combat. Boeing had already been working on such a project internally, so it was ready with detailed blueprints within weeks of getting the USAAF notice. Consolidated Aircraft later came up with a competing plane designated the B-32, but it never saw much use. General Arnold was so impressed by Boeing's B-29 that he ordered 1,664 before the first prototype had even flown.

The B-29, the biggest airplane used in World War II, was a dramatic improvement over its predecessors. It had a tactical radius of 1,600 miles and could carry 15,000 pounds of bombs—seven and a half tons—at a top speed of 357 mph, with a ceiling of 38,000 feet. Four Wright R-3350 Cyclone engines delivered almost twice the horsepower and twice the range of the B-17. It had not only the most powerful aircraft engines in the world but also the longest wings and the biggest propellers; its tail fin was as tall as a three-story building. With a cigar-shaped, metallic silver fuselage, the Superfortress looked sleek, shiny, futuristic. The cockpit was enclosed in Plexiglas that offered stunning views in three directions. Crewmen painted a name and logo on the airplane nose (or, to be exact, they usually paid an artist in uniform to do the painting). In remembrance of girls back home, the name was often female; typical examples were *Geisha Gertie, Texas Doll*, and *Coral Queen*. The accompanying nose art often depicted a woman either scantily clad or altogether naked—a practice that boosted crew morale but was dropped in 1945 after complaints from prudish church groups back home.

The standard crew of eleven consisted of a pilot, copilot, flight engineer, bombardier, navigator, radio operator, radar observer, and four gunners. They had a luxury unimaginable to crewmen of older aircraft: They did not have to wear bulky sheepskin-lined flight suits to keep warm and oxygen masks to breathe at high altitudes, because the B-29 was the first combat plane with pressurized cabins. It also had a revolutionary centralized fire-control system. Its twelve .50 caliber machine guns (or, in some cases, ten machine guns and a 20 mm cannon) were mounted in power-driven turrets that could be controlled remotely from a single location using an analog computer known as the Black Box. It also featured the most advanced radar system installed on any U.S. airplane during the war—the APQ-13, developed by Bell Labs. All of these systems required vast amounts of electricity. The B-29 had 150 electric motors powered by seven generators and connected by eleven miles of wiring. The airplane's systems were so complicated that it took a year to train a flight engineer to run them.

The Superfortress project, begun two months after the German invasion of Poland, was rushed through development and production in record time. The first prototype rolled out of Boeing's Seattle plant on September 11, 1942, to be taken aloft by test pilot Eddie Allen, who offered a laconic assessment: "She flies." Within five months, Allen was dead along with ten other B-29 experts, killed when the prototype they were testing caught on fire and crashed in Seattle—or, in USAAF parlance, "pranged." This did not augur well for the airplane's future, and indeed the B-29 was plagued by problems throughout its service life. LeMay was to comment later in a ghostwritten quip: "B-29s had as many bugs as the entomological department of the Smithsonian Institution. Fast as they got the bugs licked, new ones crawled out from under the cowling." Many of these bugs related to its unreliable air-cooled radial engines, which had a disconcerting tendency to overheat and catch fire. (The B-29 caught fire at four times the rate of the B-17 or B-24.) But Hap Arnold was not deterred; he needed an ultra-long-range bomber to strike Japan, and he was determined to get it.

It was no easy feat to mass-produce a machine as complicated as the B-29. The typical car had 15,000 parts; the typical bomber over 1.5 million. Each Superfortress required more than 27,000 pounds of sheet aluminum and over 1,000 pounds of copper. New factories had to be built specially for the job, and Boeing did not have the resources to manufacture all the needed aircraft on its own. Other major defense contractors, including North American Aviation, Bell Aircraft, and the Glenn L. Martin Company, were called in to help. Assembly plants were built or converted for B-29 production in Wichita, Kansas;

Marietta, Georgia; Omaha, Nebraska; and Renton, Washington. With so many men in uniform, more than 40 percent of the workers were women. Rosie the Riveter's efficiency was one of the secrets behind the U.S. war effort: By 1944 American workers were 82 percent more productive than Japanese workers and 55 percent more productive than German workers.

Since the B-29 was in effect an experimental aircraft, it underwent major upgrades after it left the factory. This work stretched out so long that it became doubtful whether the Superforts could become operational by the spring of 1944 as President Roosevelt had demanded. In early March 1944 General Arnold demanded an all-out push to get the planes ready. The work was performed at "modification centers" spread across the Great Plains of Kansas in the middle of a harsh winter. Because there were no hangars big enough to house the giant bombers, mechanics had to work out in the open, wearing high-altitude flight suits to keep warm amid blizzards. The "Battle of Kansas" was finally won when the first B-29s were ready to go at the end of March 1944. By the time the war was over, 3,432 of them had rolled off the production lines.

No serious consideration was given to sending the B-29s to Europe, where the B-17s and B-24s already had the air war well in hand. All along the determination was to use them in the Pacific. The question was how to get them close enough to hit Japan—i.e., within 1,600 miles. Because the Soviet Union would not give permission to base the bombers in Siberia, and because, as of early 1944, the U.S. had not yet acquired any island bases within range, the decision was made to stage the bombers out of China. Generalissimo Chiang Kai-shek undertook prodigious efforts to build airfields around Chengtu in Szechwan province. More than a quarter of a million Chinese peasants moved dirt and rocks by hand to carve out landing strips. But because Japanese conquests had cut off Chiang's Nationalists from the coast, the only way to get the B-29s and all their logistics into China was to fly them in from India. This required a harrowing 1,200-mile journey over the Himalayas, the tallest mountain range on earth, known to fliers as the Hump. It took an average of six B-29 supply flights to accumulate enough ordnance and fuel to fly one combat mission.

Once the bombers started going out on raids in June 1944, other problems quickly emerged. Bomb accuracy was woeful; flying at high altitudes in little visibility, the B-29s had trouble hitting targets such as railroad yards in Bangkok and steel plants in Manchuria. While Japanese fighter and flak opposition was not heavy, mechanical mishaps and bad weather resulted in a number of Superfortresses going down and a number of others returning to base without completing their mission.

The B-29s that were part of the 20th Bomber Command in India were not assigned to any of the U.S. commanders presiding over different parts of the Asian theater: Admiral Chester Nimitz (Pacific Ocean Area), General Douglas MacArthur (Southwest Pacific), and General Joseph Stilwell (China-Burma-India). As a weapon that transcended traditional theater boundaries, the B-29s were under the direct control of General Hap Arnold back in Washington, thus presaging the era of global air power that would be inaugurated by the Strategic Air Command in 1946.

Although his nickname was short for "Happy," and he was in fact usually in good spirits—he was the one senior military commander whose company President Roosevelt genuinely seemed to enjoy—Arnold was hardly indolent or laid-back. He was a West Point graduate and a pioneer airman, taught to fly in 1911 by the Wright brothers themselves. He had survived the early days of army aviation when there was no more dangerous job in the entire military. (Of the twenty-three army officers who qualified as pilots between 1909 and 1913, eighteen died in crashes.) He had presided over the USAAF's expansion from 20,000 men and a few hundred airplanes in 1938 to 2.3 million men and 78,000 airplanes by 1944. The job was taking a severe toll on his health; during the war he would suffer four heart attacks. (He had a fifth heart attack in 1948 and died two years later.) But, despite General George Marshall's insistence that he take a vacation, he would not relax or retire; he continued working twelve-hour days. He had already won elevation to the Joint Chiefs of Staff and status for the air force as practically a coequal of the older services. He dreamed of making his service entirely independent, and wanted to show what it could do on its own to "destroy the capacity and the will of the enemy for waging war." The B-29s were far from achieving that goal in 1944 and he was determined to see improvements—or else. He sacked the first commander of the 20th Bomber Command and dispatched Curt LeMay from Europe to get better results.

But not even LeMay could radically improve the Superforts' performance. No matter what he did, the China bases would be too far from their targets and from their logistics. The best they could manage was four combat missions a month. Finally seeing that it was hopeless, General Arnold decided to shift his focus of operations from China to a trio of newly captured islands in the central Pacific.

MARSHALING IN THE MARIANAS

In early 1944 the Joint Chiefs of Staff had ordered Admiral Nimitz to invade the Mariana Islands—Guam, Saipan, Tinian—to create bases that would bring B-29s within range of Japan. Nimitz was not enthusiastic about this operation, because he had just been through a bloodletting on Tarawa and he knew the Japanese defenses were formidable, comprising fifty-nine thousand troops spread among the three islands, buttressed by a large naval armada offshore—the First Mobile Fleet, with nine aircraft carriers, six battleships, and five hundred aircraft.

In the Battle of the Philippine Sea (June 19–20, 1944), the U.S. Navy's Task Force 58, composed of fifteen fast carriers and seven battleships, all equipped with advanced radar equipment, managed to drive off the Japanese fleet, sinking three carriers and badly damaging three more, while suffering only light losses. The marines and soldiers who hit the beaches were not so lucky. Slogging their way inch by agonizing inch through the Marianas' thick jungles, limestone caves, and mountain peaks, they suffered a

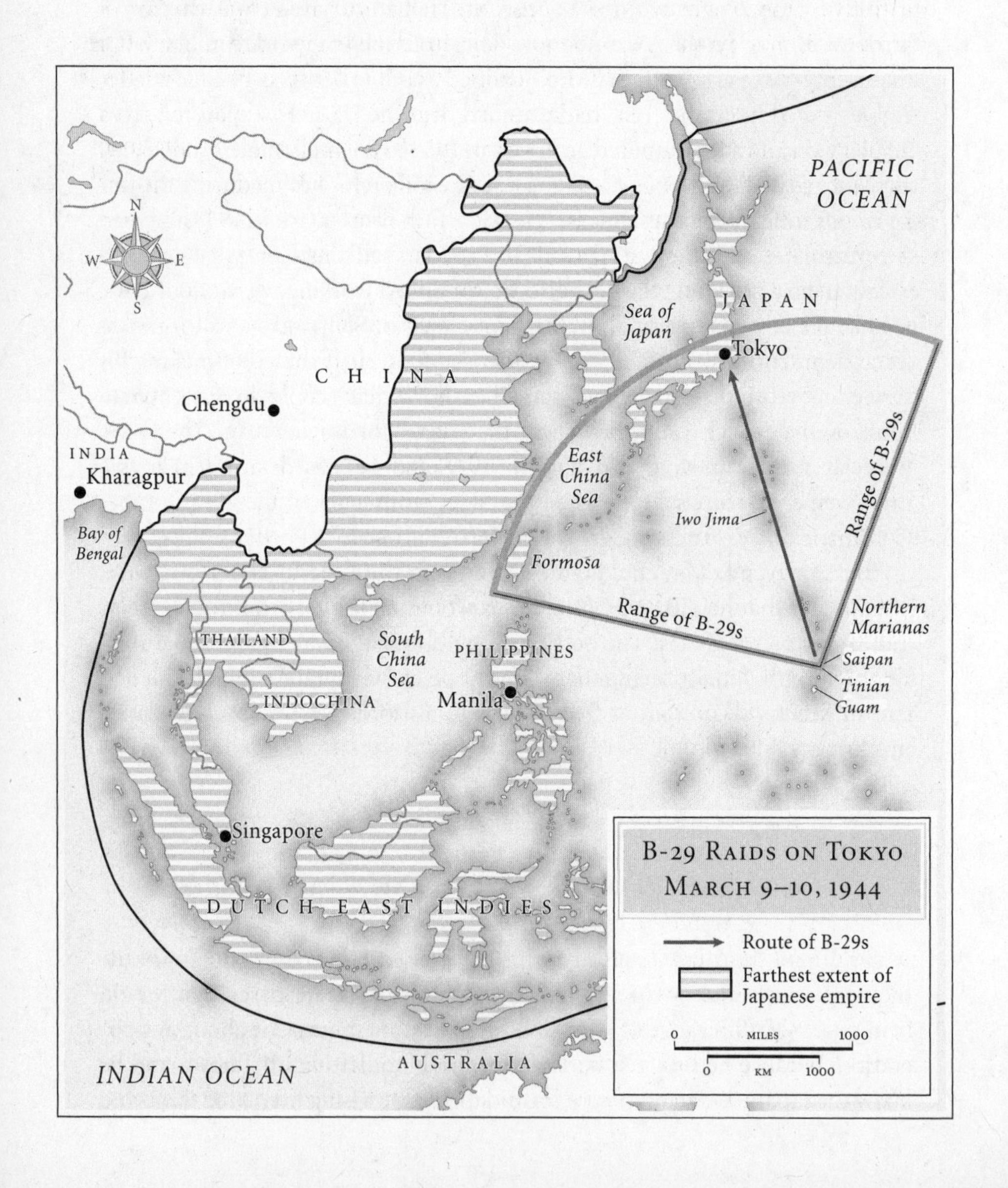

total of twenty-three thousand casualties in clearing out the determined defenders. Japanese soldiers preferred suicide to surrender. So did a thousand or so Japanese civilians on Saipan, including women and children, who threw themselves off the cliffs and blew themselves up with grenades rather than fall into the hands of the white barbarians. By August 1, 1944, almost three months after the first invasion, all three islands had finally fallen, and the work of building air bases could begin.

On October 12, 1944, the first B-29 landed on Saipan, piloted by Brigadier General Haywood "Possum" Hansell, Jr., newly designated commander of the 21st Bomber Command. He found conditions shockingly primitive. Only one runway had been constructed on Saipan and only two-thirds of it was paved. There were "no other facilities whatever," he later wrote, "except for a bomb dump and a vehicle park with gasoline truck-trailers." Ground crews were living in tents on the edge of the jungle. The problem was that the navy had given the construction of army air bases low priority. Hansell's successor, Curt LeMay, would fume: "They had built tennis courts for the Island Commander; they had built fleet recreation centers, Marine rehabilitation centers, dockage facilities for inter-island surface craft, and every other damn thing in the world except subscribing to the original purpose in the occupation of those islands." The airmen initially had to hack away the jungle with their own knives and put up their own tents while enduring tropical downpours and subsisting on chow—generally canned rations—that, according to Hansell, "was simply awful." Early arrivals also had to face occasional *banzai* charges from a handful of surviving Japanese soldiers who had hidden in the jungle as well as air raids from Betty bombers based on the island of Iwo Jima, which was still in Japanese hands.

Eventually, Navy Seabees (the name comes from "C.B.", short for "construction battalion") brought in hundreds of bulldozers, power shovels, cranes, trucks, and other equipment to build better facilities for the B-29s. All three islands became dotted with tin Quonset huts and long airstrips. Tens of thousands of soldiers, sailors, marines, and airmen would turn them into boomtowns complete with churches, theaters, baseball diamonds, and traffic jams.

The first mission from the Marianas took off on November 24, 1944, bound for a Nakajima aircraft plant in Tokyo. Simply getting airborne in a sixty-five-ton B-29 was a challenge. The Saipan runway ended at the edge of a cliff with a five-hundred-foot drop to the water below. The heavy bombers would dip down after takeoff and then try to regain altitude; not all of them made it. Long-distance navigation over the Pacific Ocean presented its own difficulties: If something went wrong, there was no emergency landing field available and no certainty that the crew would be rescued by naval vessels. "My greatest worry was thinking of the possibility of ditching in that dreary ocean," one crew member wrote. When they climbed up to their strike altitude of thirty-two thousand feet, the pilots discovered a new problem: the jet stream. Caught in winds of over 140 mph, the B-29s were whipped past their targets so fast they had almost no time to aim. This difficulty was compounded by heavy clouds covering their targets. There were only about seven days a month when the skies were clear enough to do high-level visual bombing of Japan, but without any weather stations in the area it was impossible for the USAAF to know *which* seven days. Out of 111 B-29s that took off on this first raid on November 24, only twenty-four managed to bomb the primary target, and the damage they caused was minimal. The bombers did not suffer much from flak or fighters—their high altitude and high speeds generally kept them out of danger—but they did not achieve much, either.

This was to be the pattern of the next few months, turning the Pacific bombing campaign into what Lieutenant Colonel Bob Morgan, the *Memphis Belle* veteran, described as "an exercise in futility." Although "Possum" Hansell managed to improve his crews' effectiveness with practice, he remained wedded to the prewar doctrine of high-altitude precision bombing against industrial chokepoints and never managed to get the decisive results that were widely expected. In early January 1945, General Arnold fired Hansell and installed LeMay, telling him that if he did not get results fast, he too would soon be gone.

When LeMay arrived on Guam to take over on January 19, he was not impressed. "This outfit has been getting a lot of publicity without having accomplished a hell of a lot in bombing results," he proclaimed, and he set out to get better results. The men quickly found that their new commander was, as one of them put it, "a hardheaded bastard" and a "great damn field commander." His new public relations officer described him as a "big, husky, healthy, rather stocky, full-faced, black-haired man." He was not a yeller or a table-pounder; one of his pilots later referred to his "chillingly soft voice." In fact he was so soft-spoken, his PR officer wrote, that he "couldn't make himself heard even in a small room except when you bent all your ears in his direction." He had an especial tendency to mumble when he had a pipe or cigar stuck in his mouth, which was most of the time. Before coming to the Marianas he had been a pipe smoker but, finding that pipe tobacco mildewed fast in the tropical climate, he switched to cheap cigars from the navy store.

LeMay's most notable feature—the one that helped establish his "aura as a borderline sociopath" (to quote Lieutenant Colonel Morgan)—was the scowl invariably plastered across his face. In 1942 he had contracted Bell's palsy, a virus that disabled the nerves on the right side of his face. Although he made an almost complete recovery, the right side of his mouth remained immobile for the rest of his life. Aides had to learn that, as one of them put it, "the grimace LeMay frequently made when talking to his people was not intended to express disgust. The grimace is a smile."

That wasn't always the case, of course. Sometimes the grimace really was a grimace. It certainly was in January 1945, when LeMay first contemplated how to change the "stinko" results the 21st Bomber Command was getting. He responded as he had in Europe by enhancing crew training, designating pathfinder planes, instituting the Combat Box formation, and demanding straight-and-level bombing runs. He began to get better results but still not enough to satisfy himself or his demanding bosses in Washington. This led to LeMay's momentous decision to switch to targeting cities from low altitudes at night. This new tactic was designed to save fuel and improve accuracy by getting under the jet stream. To enhance bomb loads, LeMay ordered defensive armaments jettisoned. And to maximize the impact of the bombing he decided to switch from high explosives to a new jellied gasoline mixture developed under the leadership of Harvard chemist Louis Fieser. It was made with extracts of aluminum naphthanate and aluminum palmitate; hence it became known as napalm. Stateside experiments on a mock Japanese village had shown that these bombs would burn down wood and paper houses with ease. But while napalm had been used in infantry flamethrowers, it had never been tried in a major way in bombing until LeMay ordered his daring and dangerous nighttime raid over Tokyo on March 9–10, 1945.

"WE WERE IN HELL"

LeMay's audacious scheme worked to perfection. The sky over Tokyo was clear that night, and the pathfinders had no trouble scattering their hundred-pound M-47A2 incendiaries to carve out a giant flaming X in the heart of the city's eastern, working-class district. The bulk of the bomber fleet then swooped in to drop their M-69 clusters around this densely populated aim point. Each cluster was composed of 38 six-pound napalm sticks, and they were dropped every fifty feet in order to inundate the target area with jellied fire. More than 1,600 tons of bombs—3.2 million pounds—were dropped altogether.

Robert Guillain, a French reporter who was one of the few foreigners remaining in Tokyo, wrote, "Bursts of light flashed everywhere in the darkness like Christmas trees lighting their decorations of flame high into the night, then fell back to earth in whistling bouquets of jagged flame." The beauty of this fireworks display was quickly eclipsed by its terrible consequences. A strong wind was blowing that night, the *Akakaze* (Red Wind), and, Guillain wrote, "Barely a quarter of an hour after the raid started, the fire, whipped by the wind, began to scythe its way through the density of that wooden city."

As the walls of flames raced ahead, consuming flimsy houses and sucking in more oxygen, they grew in intensity, eventually reaching over 1,800 degrees Fahrenheit. Despite having seen Tokyo ravaged in the great earthquake and fire of 1923, the authorities had done little to prepare for a calamity of such magnitude. They had ordered some houses torn down for fire breaks, and they had organized the populace into neighborhood civil defense associations, but there was not nearly enough firefighting equipment, firefighters, or concrete shelters to go around: all oversights for which the populace now paid a dreadful price. "We were in Hell," a woman who was then a sixth-grade schoolgirl recalled many years later. "All the houses were burning, debris raining down on us. It was horrible."

Some people tried to take shelter in a few concrete or stone buildings, usually temples or theaters. These turned out to be death traps. "Inside," writes historian Hoito Edoin, "the people were being baked as if in a casserole." Many more hid in shallow dirt trenches near their homes that provided no protection whatsoever. They were incinerated or asphyxiated. Others fled across bridges, only to meet the same gruesome fate as the flames and heat enveloped them and melted the steel girders. Some of those who jumped into the rivers were boiled alive. Within a few hours, charcoal-black corpses were piled everywhere like cordwood. Many were so badly burned that it was no longer possible to tell if they had been male or female, young or old. Horribly disfigured survivors, some with their skin peeling, tried, in vain, to find aid stations. Many hospitals had been destroyed, and those that were left did not have much by way of medicines, which had become scarce because of the U.S. naval blockade. As a result, many of those who survived the initial conflagration would perish in the days that followed from lack of lifesaving drugs.

The firestorm became so intense that the B-29s were buffeted like ping-pong balls by turbulence from rising thermal updrafts, and their fuselages were blackened by soot. The crewmen were sickened by the smell of roasting human flesh that reached their cabins thousands of feet above the burning city. Some threw up. Yet intermingled with the horror was a pervasive sense of satisfaction among the American crews that Pearl Harbor was finally being avenged many times over. Most of the crewmen did not wallow in introspection. They viewed what they were doing as a job, and they were simply glad to survive another night at the office.

As LeMay had suspected, the Japanese antiaircraft guns were not designed to hit low-altitude targets, and they had precious few fighter planes capable of nighttime operations. "We were bombing with damn near impunity," one U.S. pilot later wrote. The raiding force suffered much lighter losses than many had feared: Out of 334 Superforts, only 14 were lost (4 percent), mainly from mechanical malfunctions and the superheated air currents created by their own incendiary bombs. A small price to pay for the level of damage inflicted: A total of 16.8 square miles of Tokyo, more than 10,000 acres—equivalent to two-thirds the land area of Manhattan—was burned out that night. More than 260,000 buildings were destroyed, more than 1 million people left homeless.

The total toll will never be known. The official Japanese estimate was 83,793 people killed, 40,918 injured. Other estimates put the death toll at over 100,000. The firebombing of Tokyo on March 9–10 set a horrific new standard of destruction that would be rivaled in the annals of war only by the atomic bombing of Hiroshima five months later. Brigadier General Thomas Power, who personally led the raid, was scarcely exaggerating when he called it "the single greatest disaster incurred by any enemy in military history. . . . There were more casualties than in any other military action in the history of the world."

LeMay did not take time to celebrate, and it would never have occurred to him to mourn. He kept the pressure on. Within the next week, three more major cities—Osaka, Nagoya, Kobe—received similar treatment. All the raids were devastating, though none caused as many casualties as the attack on Tokyo because the conditions that had been present on March 9–10 (high winds, little firefighting apparatus, heavy population density, a critical mass of bombs) were not replicated elsewhere. The blitz ended on March 18 when LeMay temporarily ran out of incendiaries. Five raids had left four of Japan's biggest cities in ruins (Nagoya was hit twice). "LeMay's brutal act of faith in low-level bombing marked the turning-point in our aerial campaign over the home islands," Lieutenant Colonel Morgan wrote. "It amounted to nothing less than the beginning of the end of the war."

In late March 1945, the 21st Bomber Command was diverted, much against LeMay's wishes, to support the invasion of Okinawa, which was to begin on April 1. The B-29s were put to work disabling Japanese landing strips from which kamikaze pilots might take off, as well as the factories that could produce more airplanes for them to fly. Notwithstanding their efforts, plenty of suicide planes managed to plow into U.S. ships.

A more fruitful use of the bombers, but one that LeMay also resisted, was dropping mines in Japanese waters. American submarines enjoyed great success in torpedoing Japanese merchant shipping, thereby cutting off the home islands from the raw materials they needed to sustain the war effort. Submarines and surface ships also laid mines for the same purpose. Starting in late March 1945, B-29s joined the mine-laying effort; by war's end, they had dropped 12,054 of these deadly devices. Their work made Japan's busiest waterways and harbors impassable for weeks at a time, sank or crippled three hundred vessels, and completed the task of starving Japan of such basic necessities as food, oil, coal, and iron ore.

Although mining was effective, LeMay was anxious to get back to his city-busting campaign. "I feel that the destruction of Japan's ability to wage war lies within the capability of this command," he wrote to General Lauris Norstad in April 1945, and he set out to make good on his boast. By this time, LeMay had over one thousand bombers, allowing him to send out four hundred to five hundred planes at a time.

The Joint Chiefs sought to assist him by ordering the Marine Corps to capture Iwo Jima, a volcanic island midway between the Marianas and Japan that could serve as a base for P-51 fighter escorts. Although the Marines waged their most bloody and storied battle of the entire war to secure Iwo Jima, losing 6,821 killed and almost 20,000 wounded, their heroism was mostly for naught. It turned out that the island was too far away to serve as an effective fighter base. And, unlike in Europe, fighters were not needed to protect the bombers over Japan because the defenses were so weak. To justify the cost of this operation, senior officers later pointed to Iwo Jima's role as an emergency landing field for B-29s, but their claims relied on exaggerated statistics.

More effective than land-based fighters were those that flew off U.S. and British aircraft carriers that ventured near the coasts of Japan in the summer of 1945. They took to bombing and strafing to complement the work of the big bombers. For the most part, however, they were simply making the rubble bounce. By June 1945, the Superforts had razed Japan's major cities so completely that they were able to move on to medium-sized cities, those with populations under 300,000, and destroy them too. They were so sure of their domination of the skies that, like a pool player calling his shot, they dropped leaflets warning Japanese civilians which cities would be next on their hit list, confident that there was nothing the Japanese could do to stop them.

Nor could the Japanese retaliate against the American homeland, since they lacked a long-range bomber of their own. The best they could do was to load thousands of hand-made rice-paper balloons with small incendiary devices and send them wafting across the Pacific. Some actually reached North America but the damage they caused was negligible—a few minor forest fires in the Pacific Northwest and a total death toll of six people in rural Oregon. Even as these comically ineffectual devices were drifting eastward, a weapon of unimaginable magnitude was coming from the west.

In June 1945, a new squadron of specially modified B-29s arrived on Tinian. The 509th Composite Group, led by twenty-nine-year-old Lieutenant Colonel Paul Tibbets, was secretive about its mission. On August 6, the world learned its secret when a B-29 known as the *Enola Gay*, piloted by Tibbets, dropped an atomic bomb over Hiroshima, leaving a mushroom cloud and utter annihilation in its wake. Three days later, a B-29 named *Bock's Car* dropped another A-bomb on Nagasaki. The former raid killed at least 130,000 people, the latter more than 60,000. (Radiation poisoning would increase the long-term toll.) The unleashing of this unprecedented firepower finally forced Japan into surrendering.

"AWFUL YET NECESSARY"

President Truman's decision to drop the A-bombs sparked a debate that continues to the present day. Yet the atomic bombs did not mark a radical break with the past. They merely represented the natural culmination of the strategic bombing campaign begun in 1944. It would not have mattered much to the average Japanese whether she was killed by atomic bombs or napalm bombs. Indeed, the March 9–10 firebombing of Tokyo caused more death and destruction than the atomic raid on Nagasaki. The difference, of course, was that it had taken 334 B-29s to destroy central Tokyo. Only three were needed over Hiroshima and Nagasaki: the bomber, a chase plane, and a weather plane.

By war's end, the Superforts had leveled sixty-six of Japan's biggest cities. The only major city spared the wrath of fire was Kyoto, which Secretary of War Henry Stimson declared off-limits because of its cultural value. Half of Tokyo, by contrast, was burned to the ground, including the Imperial Palace, which was hit by accident. The bombers devastated 178 square miles of Japan—three times the size of Washington, D.C. Just 3.5 percent of the total area damage was inflicted by the two atomic bombs. The U.S. Strategic Bombing Survey would later put the total death toll from the bombing campaign at 330,000 and the number of injured at 476,000; the actual figures were probably greater. In addition, 8.5 million people became refugees. All of this was achieved at a fraction of the casualties suffered by the USAAF in Europe. The entire bombing campaign against Japan resulted in the loss of 359 B-29s and 3,415 airmen killed and missing.

Even before the dropping of the atomic bombs, major elements of the Japanese government had been looking for a way to end the onslaught. Prince Fumimaro Konoye was later to say, "Fundamentally the thing that brought about the determination to make peace was the prolonged bombing by the B-29s." Admiral Kantaro Suzuki, the moderate premier who took over in April 1945, was of a like mind: "It seemed to me unavoidable that in the long run Japan would be almost destroyed by air attack so that merely on the basis of the B-29's alone I was convinced that Japan should sue for peace." But such sentiments may not have prevailed over the determination of hard-line army officers to resist to the bitter end without the shock administered by the appearance of mushroom clouds over Hiroshima and Nagasaki.

The deliberate destruction of Japanese and German cities has become more controversial the farther that World War II recedes into the mists of memory. At the time, there were precious few qualms among either the politicians or publics of the Allied nations about killing German or Japanese civilians. The USAAF was always careful to justify its bombing by claiming that it was going after dispersed industrial targets, such as the cottage industries that helped supply Japan's war effort, but even LeMay had to admit this was a "pretty thin veneer" for what amounted to the wholesale killing of men, women, and children. LeMay spoke for many when he wrote that "we just weren't bothered about the morality of the question. If we could shorten the war we wanted to shorten it." He went on to argue: "Actually I think it's more immoral to use *less* force than necessary, than it is to use *more*. If you use less force you kill off more of humanity in the long run, because you are merely protracting the struggle." LeMay suggested that he wanted to avoid the misguided altruism of "The guy who cut off the dog's tail an inch at a time so that it wouldn't hurt so much."

Bob Morgan, one of the pilots who flew the Tokyo fire raid of March 9–10, later admitted that "the smell of burning human flesh" was "an odor that I will never be able to get completely out of my nostrils," but he, too, defended the raid as "something awful . . . and yet necessary."

Whatever the morality of strategic bombing, its efficacy was clear. It did not win the war by itself, but it played a prominent role in the Allied victory in both the European and Pacific theaters. Naval and ground forces contributed significantly to the defeat of the Japanese Empire, but mainly by making possible the ultimate success of the air campaign. Even now, more than sixty years later, Japan represents perhaps the only example of a major belligerent forced to surrender through the application of air power without ever having its home territory (with the exception of Okinawa) invaded. The closest parallel, on a much smaller scale, was Serbia's capitulation in the Kosovo campaign of 1999, but Slobodan Milosevic was forced to give up only one province, not his entire country. In September 1945, by contrast, the Japanese allowed their whole country to be occupied without firing a shot. Strategic bombardment came close to living up to the hopes of its most fervent champions, but only because it had been applied with a ruthless disregard for civilian casualties never equaled before or since.

The Consequences of the Second Industrial Revolution

The three preceding chapters have focused on three of the most important innovations of the Second Industrial Revolution (1917–45)—tanks, aircraft carriers, and heavy bombers—but there were many others. Not everything changed, to be sure, but the application of new technologies, especially the internal combustion engine and the radio, along with new ways of utilizing them, redefined military operations for decades to come.

LAND WARFARE: The most obvious change was the introduction of tanks and other motor vehicles, which helped to restore mobility to a battlefield that had turned increasingly static in the period between the Crimean War and World War I. Armored vehicles were most effective when closely coordinated via radio with ground-attack aircraft, a combination that was at the heart of the Germans' widely copied blitzkrieg. Air-and-armor operations would eventually be used by most of the major belligerents of World War II and also by a number of postwar states. The Israeli Defense Forces proved particularly adept at blitzkrieg-type offensives in 1956 and 1967. The Egyptian and Syrian armies, in turn, had some success with this same approach in 1973 before finally being defeated in the Yom Kippur War. The Indian army staged a lightning offensive of its own in 1971. Its conquest of East Pakistan in just fourteen days gave birth to the new state of Bangladesh.

Airborne operations were another new element of warfare, allowing troops to be moved far behind enemy lines using transport aircraft, gliders, and parachutes. Large-scale airborne operations ranged from the Germans successfully dropping paratroopers on Crete in 1941 to the British and Americans unsuccessfully dropping paratroopers into Holland in 1944 (Operation Market Garden). Victorious or not, airborne troops invariably suffered heavy casualties because they lacked much firepower or mobility once they arrived. For this reason, airdrops generally fell out of favor for large-scale offensives after 1945; with a few exceptions, such as the U.S. invasion of Panama in 1989, they would be used primarily to infiltrate small groups of commandos or spies behind enemy lines.

Amphibious operations were less novel—Julius Caesar had invaded England from the sea in 55–54 B.C.—but they were put on a sounder basis after the failure of the Gallipoli landings in 1915. In the interwar period, the U.S. Marine Corps studied how to seize advanced island bases. This was mostly a matter of coordinating assaults with supporting fire from sea and air, but it also involved developing new equipment such as shallow-draft landing craft with quick-release ramps. The Marines' 1934 *Tentative Manual for Landing Operations* evolved into the basis of amphibious doctrine not only for the Marine Corps but also for the U.S. Army and Navy. Amphibious warfare proved absolutely vital in the Pacific, where the Marines engaged in a campaign of "island hopping," as well as in the European theater, where Allied troops stormed ashore in North Africa, Sicily, Italy, and finally in Normandy and the south of France. All of the major Allied amphibious attacks succeeded but at heavy cost. (The first thirty minutes of Steven Spielberg's film *Saving Private Ryan* is a harrowing dramatization of the dangers involved in sending unprotected infantrymen against entrenched shore positions.) The ability to insert troops by helicopter or fixed-wing aircraft led to a decline in amphibious operations after World War II. Although U.S. Marines bluffed Saddam Hussein into thinking that they were going to land on the beaches of Kuwait in 1991, their commanders concluded that such an assault would be too costly. The last great landing occurred forty-one years earlier, in September 1950, when General Douglas MacArthur's men disembarked at Inchon to cut off the North Korean advance.

While the means of getting infantrymen into battle had improved during the Second Industrial Age, once engaged they had to make do with weapons that were not terribly novel. The typical grunt carried a rifle roughly similar to the one his father had used in World War I; some were actually issued exactly the same rifles. The Japanese armory was particularly antiquated; their main rifle dated from 1905. The best rifle of World War II was probably the U.S. M-1 Garand, a semiautomatic weapon that could fire eight rounds from a single clip with eight pulls of the trigger. General George S. Patton reportedly called it "the greatest battle implement ever devised." Infantrymen were also armed with handheld submachine guns (such as the American Tommy gun, the British Sten gun, and the German and Russian "burp guns") and crew-served heavy machine guns (such as the German MG34 and MG42). The most innovative weapons issued to infantrymen were shoulder-fired antitank rockets such as the American bazooka and the German *Panzerfaust*.

This was part of a general resurgence of a weapon that had last played a major role in warfare in 1812 when British Congreve rockets had inspired

Francis Scott Key's mention of "the rockets' red glare" in "The Star-Spangled Banner." In addition to the German V-1 and V-2, the Red Army deployed the Katyusha, a short-range, unguided rocket that could be fired from mobile launchers in bursts of more than forty at a time. Artillery also improved with the American introduction of proximity fuses, employing a tiny radar set in the nose cone to set off the shell whenever it got close to a large object. (Previously a shell would go off only if it hit the target—a contact fuse—or at some predetermined point after firing—a time fuse.) Some field guns also became self-propelled, moving forward on armored chassis in an echo of tank warfare.

NAVAL WARFARE: For surface vessels and even submarines there was much continuity between the First and Second World Wars. The battleships, cruisers, destroyers, and submarines of the 1939–45 period were generally bigger, faster, and better armed than their 1914–18 predecessors but not fundamentally different. Indeed, they had not changed much since the Russo-Japanese War of 1905. Yet naval warfare was nevertheless transformed by the introduction of aviation. Fleets that were once built around battleships came to be built around aircraft carriers instead.

Aircraft proved superior not just to conventional surface ships but also, in the Battle of the Atlantic, to submarines as well. German U-boats preying on Allied shipping were foiled through a variety of means including convoying of merchants ships and the use of radar and sonar. But the weapon that proved most effective was an aircraft dropping depth charges. The dispatch of long-range B-24s equipped with the latest radar to patrol the North Atlantic in 1943 helped turn the tide against the U-boats. The proliferation of small escort carriers also allowed air cover for convoys even in the middle of the ocean. Submarines proved more effective in the Pacific, where the vast distances precluded effective patrolling by aircraft and where the Japanese did not develop the types of advanced antisubmarine techniques employed by the Allies in the Atlantic. U.S. submarines took a heavy toll on Japanese merchantmen and warships alike once they managed to fix the problems that bedeviled their Mark 14 torpedo early in the war. "A force comprising less than 2 percent of U.S. Navy personnel," naval historian Ronald Spector would write of U.S. submariners, "had accounted for 55 percent of Japan's losses at sea." The torpedo, whether launched by submarines, surface ships, or airplanes, proved the biggest ship-killer of the war.

AERIAL WARFARE: Aircraft as a component of ground and sea warfare proved indispensable, and victory often went to whichever side was more adept at integrating them into its operations.

Heavy bombers suitable for strategic bombing were utilized mainly by the U.S. Army Air Forces and the Royal Air Force. After briefly flirting with strategic bombing in the Battle of Britain, the Germans largely abandoned it. The Soviets, Japanese, French, and Italians never embraced it to begin with—a decision that some of them would come to regret. A Japanese naval aviator later wrote, "Had Japan developed such bombers as the B-17, I believe the war would have taken a different course. We did not have a single warplane comparable to these aircraft, and Japan paid a heavy price for this lack." Germany did develop a four-engine bomber, the Messerschmitt Me-264, dubbed the Amerika because it was intended to bomb New York, but it never got past the prototype stage.

Most of the bombs used in the war were iron canisters filled with high explosives that were guided to their targets by nothing more than wind, gravity, and sheer luck. The standard armaments of all aircraft were machine guns and cannons, supplemented later in the war by rockets. By 1945, jet aircraft began to appear that were much faster than their propeller-driven predecessors, but none was deployed in sufficient numbers to make much of a difference. Nor did the long-range missiles developed by Germany, the V-1 and V-2, affect the war's outcome. These would be the weapons of the future. Into this category also falls the helicopter, which was still in the prototype stage during World War II but would be employed on a large scale by the French in Algeria in the 1950s and by the Americans in Vietnam in the 1960s.

ELECTRONIC WARFARE: In addition to aerial warfare, the Second Industrial Age added another element to the traditional clash of armies and navies: electronic warfare. Since the dawn of organized warfare, armies had always tried to intercept the plans of the other side, usually by employing spies or capturing messengers. What had historically been a hit-or-miss business was placed on a much more scientific basis with the birth of signals intelligence. The widespread use of radios afforded virtually limitless opportunities for learning the enemy's secrets, provided their codes could be broken. The British were pioneers at this art. The Admiralty's Room 40 succeeded during World War I in breaking many of the Imperial German Navy's communications, but the Royal Navy did not take full advantage of this windfall because its code breakers were not well integrated with operational commands. British code breakers did, however, score a historic success in 1917 when they intercepted and decoded a telegram from German Foreign Minister Arthur Zimmermann offering Mexico the return of the American Southwest in return for making war on the United States. The publication of the Zimmermann telegram helped convince the U.S. to enter World War I.

In World War II, the big British coup was replicating the Enigma machine used to encode most German messages. British scientists also pioneered radar and sonar. All of these inventions were shared with the United States. Meanwhile, U.S. Navy and Army cryptanalysts on their own had already broken the Japanese naval and diplomatic codes. Ultra and Magic—the code names given to the cracking of German and Japanese ciphers, respectively—gave the Allies an invaluable edge over the Axis, who never developed the same level of success in penetrating Allied communications.

Electronic warfare turned into a game of cat and mouse, pitting teams of scientists and engineers against one another. Many lives depended on the outcome of what became known as the "wizard war." In 1942 U-boats gained a major edge in their attacks on Allied commerce when the German naval intelligence office, B-Dienst, was able to break the ciphers used by the Americans, British, and Canadians to route transatlantic convoys. This information helped the submarines sink more than 1,600 ships that year. The Allies were operating blind for most of the year because the German navy had changed its Triton cipher. It was no coincidence that the advantage in the Atlantic shifted to the Allies at the end of 1942 when they managed once again to crack the German codes and changed their own so that the U-boats would no longer be privy to information about convoy movements.

Similar back-and-forth battles raged around radar, sonar, and radio-directional navigation equipment. For instance, the RAF was able to bomb Hamburg so effectively in 1943 because it came up with a clever system of dispersing aluminum chaff (known as Window) to disrupt German radar. When the Germans developed alternative methods of detecting incoming bombers (for instance, by homing in on their "Identification, Friend or Foe" radio beacons), they in turn were able to devastate the attacking formations. Ultimately, the Allies won the "battle of the beacons"—and with it the war.

THE BUTCHER'S BILL

Because weaponry reached new heights of destructiveness during the Second Industrial Revolution, it is not surprising that the number of fatalities in World War II—some fifty-five million dead—was greater than in all previous wars combined. That horrific figure is somewhat deceptive, however, for two reasons: first, more civilians were killed than in any other conflict, whether by Allied bombs or Nazi death squads; second, more men fought in World War II than in any other conflict. When looking at *military casualties as a percentage of the entire armed forces,* World War II was considerably less destructive than the average war of the eighteenth or nineteenth century. The United States military, for instance, suffered 6.7 percent dead, wounded, and sickened in World War II, down from 29 percent for Union forces in the Civil War. (These figures include both battle and nonbattle deaths—i.e., those caused by disease, accident, and suicide.)

The percentage of soldiers killed and injured declined in part because of the increasing tendency to disperse troops to blunt the devastating effects of modern firepower. The average number of soldiers deployed per square kilometer of front declined from 100,000 in ancient times to 3,883 during the American Civil War, 404 in World War I, and just 36 in World War II. (It declined further to 25 in the 1973 Yom Kippur War and has kept on falling since, down to 2 in the 1991 Gulf War.) The high casualties in the wars of the First Industrial Revolution (e.g., the Civil War, the German Wars of Unification, the Russo-Japanese War, and World War I) can be ascribed largely to the fact that weapons' lethality had increased faster than the ability of commanders to adjust their tactics. The increasing use of motor vehicles, airplanes, and radios after 1917 allowed troops to spread out over much greater distances, decreasing their vulnerability to modern firepower.

From a long-term perspective, the lethality of weapons increased roughly two-thousand-fold between the Peloponnesian Wars and World War II, between the days of spears and those of tanks, but the dispersion of forces increased by a factor of over four thousand. The fact that "dispersion has actually increased more rapidly than lethality"—in the words of Trevor Dupuy, who generated these statistics—meant that the impact of modern technology was somewhat blunted. Men remained in control, no matter how destructive machines became.

Advances in medicine also kept fatalities lower than might have been expected, given the rise in weapons' lethality. Battlefield medical care was vastly improved in the mid-nineteenth century by Florence Nightingale and her team of nurses in the Crimean War. Armies thereafter did a much better job of organizing litter-bearers and field hospitals. Following Louis Pasteur's announcement of the germ theory of disease in 1878, surgeons began to sterilize medical instruments and wear gowns, masks, and rubber gloves. Such simple steps helped to stop the transmission of diseases that had once been rampant in operating rooms. The invention of safe and effective anesthesia, in the form of ether or chloroform (first used on a wide scale in the American Civil War), revolutionized surgery by making it possible for surgeons to work more slowly and methodically while lessening the chances that the patient would die of shock. The X-ray machine, invented in 1895, allowed doctors to better diagnose the nature of injuries. The discovery of different blood groups in 1901–2 made possible safer blood transfusions in World War I, which reduced deaths from blood loss. In World War II blood transfusions were employed on a much greater scale by mobile surgical teams. So, too, were antibiotics like penicillin, discovered in 1928 by Alexander Fleming but not mass-produced until 1943.

The result of all these advances was a rapid falloff in deaths from wounds and especially disease. The U.S. Army in World War II suffered 99 percent fewer deaths from disease than in the Civil War. And the number of soldiers who died of their wounds after receiving medical care declined from an average of 15 percent in nineteenth-century wars to 4.5 percent in World War II. (The figure would fall to 2.5 percent in the Korean War, thanks partly to the introduction of helicopters as aerial ambulances, which allowed casualties to be treated faster than ever before.) In military medicine, as in so many other fields, the German army took the lead, but their work was matched and possibly exceeded by the British and American medical corps.

WHAT PRODUCED VICTORY?

There is a tendency to see the outcome of World War II as foreordained: How could the Axis possibly have prevailed over the much greater industrial resources of the Allies? Richard Overy rightly cautions against this "hindsight bias" in his invaluable book, *Why the Allies Won*. As he points out, "Economic size as such does not explain the outcome of wars." Indeed this book offers many examples of significant battles—e.g., the Spanish Armada, Assaye, Königgrätz, and Tsushima—won by the side with the smaller economy. World War II further proves the point.

The German armed forces were far more starved of resources in the 1920s than their rivals in Britain or France, yet they were able to develop the blitzkrieg while they could not legally buy a single tank. The Germans out*thought* their enemies in the interwar period, which is why in 1939–41 the Third Reich was able to out*fight* the countries of Western and Eastern Europe even though they were in aggregate much richer than Germany itself. By 1942 Hitler was in control of most of the economic resources of Europe from the English Channel to the Urals. On paper, at least, this gave the Third Reich the potential to compete against the U.S. and USSR; today the European Union, which encompasses roughly the same area, has a GDP greater than that of the United States. Japan, too, grabbed a vast empire for itself in Asia that should have given it greater ability to hold its own. Yet by 1942 the U.S. was outproducing all of the Axis states combined. The USSR, too, staged a remarkable recovery from its devastating losses in 1941 and was soon outproducing Germany, even though the Germans had overrun two-thirds of its coal and steel industry.

The outcome of the battle of production lines was no more predetermined than the outcome of Midway, El Alamein, or Stalingrad. The Allies produced more tanks, aircraft, and ships because they were more skilled than their enemies at mobilizing their industrial base. Hitler did not even start to put his economy on a full war footing until 1942, and by the time his armaments minister, Albert Speer, was starting to implement his major reforms in 1943, Germany was already losing the war. For most of the conflict the German economy was a mess, characterized by wasteful production of too many models of everything (at one point 425 different kinds of aircraft were being made) and not enough standardization. The Japanese were even more backward. The Soviets were less advanced technologically than the Germans, but they were able to churn out greater quantities of simple and reliable equipment under the aegis of their economic planning agency, Gosplan. The United States was able to combine quality and quantity by using the mass-production techniques pioneered by Henry Ford. American factories under the direction of the War Production Board and other bureaucracies achieved stunning leaps in productivity during the war years. And many of the machines they were producing, ranging from *Essex*-class fast carriers to B-29 heavy bombers, were, by the end of the war, the best in the world.

The growing complexity of warfare in the Second Industrial Age made it imperative to integrate complex systems—not only military but also industrial and scientific—into a smooth-running whole. No country had a perfect track record: The Germans did well with tank warfare and submarines but not with strategic bombing or aircraft carriers; the Americans did well with long-range bombers and aircraft carriers but not tanks. On the whole, however, the Allies pulled off the difficult feat of war management far better than the Axis. Nazi Germany was plagued by the erratic and often irrational decision-making of Adolf Hitler, who fostered an atmosphere of bureaucratic chaos and infighting. While Japan had no single leader of comparable power, it was handicapped by the lack of coordination between its army and navy. The British and Americans, by contrast, set up a Combined Chiefs of Staff Committee that, despite some inevitable friction, capably coordinated their joint war effort. Even Stalin, who often fell prey to the same megalomania and delusions as Hitler, learned as the war went along to defer to his increasingly competent general staff and to such gifted commanders as Marshal Georgi Zhukov.

This underscores a theme running throughout this volume: Having an efficient bureaucracy is the key determinant of whether a country manages to take advantage of a military revolution. Just as England was better organized than Spain in 1588, Sweden better than the Holy Roman Empire in 1631–32, Britain than the Marathas in 1803, Prussia than Austria in 1866, Britain than the Sudan in 1898, and Japan than Russia in 1905—so too the Allies were better organized than the Axis by the end of World War II. The reason German armies were able to reach the gates of Moscow and Japanese armies the borders of India before being defeated was that the Axis had done a better job of organizing *before* the war. This gave them an important initial advantage that they allowed to slip away through catastrophic miscalculations, which once again goes to show that the early movers in a military revolution are not necessarily the long-term winners.

The German and Japanese failure to make better use of their resources and to set reasonable war aims consigned them to defeat. So total and traumatic was their collapse that after 1945 they were forced to renounce the very idea of power politics and pledge to use force only for self-defense. Britain and France were also losers, even though they were on the winning side. They emerged in 1945 much weakened and unable to hold on to their colonies—a fate they might have avoided or delayed had they been more prepared for war. Britain, Holland, Belgium, and France would have had a good chance of defeating the 1940 Nazi offensive if they had done a better job of building armor and air forces. The British might also have been able to stave off the Japanese onslaught in the Pacific if they had invested more in modern aircraft carriers and naval airplanes. Britain was able to avoid total defeat in 1940–41 partly because of its geography but mainly because it had made some wise investments in the late 1930s—particularly in fighters, radar, and code breaking—that helped to offset some of its deficiencies. Poland, Czechoslovakia, and the other nations of central and eastern Europe were not so lucky. Lacking either a favorable geographical position or modern armed forces, they were enslaved in turn by the Germans and the Russians and did not taste freedom until the 1990s.

The Soviet Union and the United States were the biggest beneficiaries of the Second Industrial Age. Their rise to global power ended the period of Western European dominance that began around 1500. Victory in the Great Patriotic War gave communism enhanced legitimacy inside the Soviet Union and made it a more attractive model for other countries from China to Cuba. Likewise the outcome of the war enhanced the power and prestige of the heavily bureaucratized American government. Federal civilian employment increased nearly fourfold between 1939 and 1945, and even after five years of demobilization the total number of civilians employed by Washington in 1950 was 100 percent higher than it had been in 1939. Total federal expenditures were almost three times higher in 1948 on an inflation-adjusted basis than they had been in 1938. Giant corporations such as Boeing, General Motors, and General Electric, which had grown even bigger to manufacture military materiel, continued to dominate the U.S. economy (indeed, the global economy) after the war.

In other words, World War II reinforced the trends toward statism and corporatism that had been given such a big boost by World War I. The enhanced power of the U.S. government could be used for ends both good (desegregation) and bad (McCarthyism). In the case of the Soviet government, the ills were much greater (the Gulag) and the benefits much less apparent. What united the two Cold War rivals was that in both the U.S. and USSR, the state reached the pinnacle of its power and popularity in the late 1940s and 1950s, largely based on its success at waging war in the Second Industrial Age. It would take two losing guerrilla struggles—in Vietnam and Afghanistan—to crack the aura of state invincibility.

PART IV

THE INFORMATION REVOLUTION

It's a constant struggle of one-upsmanship. We adapt, they adapt. It's a constant competition to gain the upper hand.

Major John Nagl (2004)

The Rise of the Information Age

In the aftermath of World War II, it was widely believed that the atomic bomb had wrought a revolution in warfare. The United States, followed by the Soviet Union and a few other countries, rushed to add these fearsome new weapons to its arsenal.

By the mid-1950s, nuclear bombs had gotten smaller, with advances in warhead miniaturization, and more lethal, with the development of the hydrogen bomb. As part of the Eisenhower administration's "New Look" defense policy, the U.S. attached nuclear explosives not only to missiles and rockets but also to bombs, depth charges, torpedoes, artillery shells, and mines. The U.S. Navy deployed nuclear-armed submarines and aircraft carriers capable of sinking the largest enemy warship with one blast. The "Pentomic" army deployed a range of atomic weapons, including the Nike antiaircraft missile and the Davy Crockett short-range rocket launcher. The U.S. Air Force—an independent service as of 1947—toyed with building atomic-powered aircraft. While that didn't work out, it did put nuclear ordnance on its fighter aircraft. But the bulk of America's atomic arsenal, which by 1959 comprised 12,305 warheads, was in the Strategic Air Command, led by Curtis LeMay. By 1959, SAC had almost two thousand jet bombers, mainly B-47s but also newer B-52s, both essentially longer-range, higher-speed, bigger-payload versions of the B-29.

If the Soviet Union were to initiate hostilities, the U.S. planned "massive retaliation," laying waste to Russian cities in a nuclear version of the firebombing of Tokyo. But how credible was this threat? With the help of information stolen from the Manhattan Project, the Soviet Union had tested its first atomic bomb in 1949. And it also had the means of delivering its A-bombs—first the Tu-4 bomber, a copy of the B-29 (several of which had fallen into Russian hands when they crash-landed in Siberia in 1944), then various jet bombers. Unlike in World War II, the U.S. could not rain death on its enemies secure in the knowledge that its own cities would remain safe. World War III would likely lead to the destruction of Washington as much as of Moscow.

This balance of terror hardened in the age of intercontinental ballistic missiles, inaugurated by the Russian launch of the Sputnik I satellite in 1957. Both sides deployed nuclear-tipped missiles that were not very accurate but that were capable of devastating any city on the other side of the world within half an hour of launch. Some of these missiles were deployed aboard nuclear-powered submarines—the first was the USS *George Washington* in 1960—that were virtually impervious to attack because, unlike diesel submarines, they did not have to surface to recharge their batteries.

By the time of the Cuban Missile Crisis in October 1962, it was apparent to both sides that their nuclear arsenals were too destructive to be used as casually as "normal" weapons. Nuclear weapons added an element of stability to the Cold War by dissuading either side from directly fighting the other. But they turned out to have almost no impact on the wars that actually broke out. Struggles such as the Korean and Vietnam wars, to say nothing of the Yom Kippur War (1973), the Indo-Pakistan War (1971), the Afghanistan War (1979–89), the Falkland Islands War (1982), the Iran-Iraq War (1980–88), or the more numerous civil wars that would rage around the world, were conducted without the appearance of mushroom clouds. Unfortunately for the United States, its focus on nuclear war led to a neglect of the skills and equipment needed to fight and win "limited" wars like those in Korea and Vietnam.

The history of atomic weapons serves as a cautionary example of the danger of too avidly embracing a new technology. Far from being determined to fight the last war, the generals after World War II were intent on fighting a war that never arrived. While the whole world was focused on the implications of the Manhattan Project, it is possible to argue with the benefit of hindsight that an event with even greater implications for the future was occurring at an obscure research laboratory at the University of Pennsylvania.

ELECTRONIC BRAINS

The first electronic digital computers were built in the late 1930s and early 1940s. The most famous and influential was the Electronic Numerical Integrator and Computer (ENIAC) constructed at the University of Pennsylvania between 1943 and 1945. With its blinking lights, whirring tape drives, and clanking teletypes, ENIAC occupied a fifty-by-thirty-foot room and weighed thirty tons. At the heart of this "electronic brain" were more than seventeen thousand vacuum tubes, a close cousin of the lightbulb, that were used to amplify and switch electrical currents. Already widely employed in radio sets and telephone relays, vacuum tubes now allowed ENIAC to perform five thousand calculations per second by representing numbers through binary, on-off signals. In the process, the tubes emitted so much heat that the temperature in the room sometimes reached 120 degrees Fahrenheit.

ENIAC's designers were physicist John Mauchly and engineer John Presper Eckert, Jr. The funding for their work came from the U.S. Army's Ordnance Department, which needed to perform complex calculations involving wind speed, temperature, and many other factors to get bombs and artillery shells on target. Massive data-crunching was also required for other wartime tasks such as code breaking and atomic bomb design. Both the British and American governments funded research into electronic computers that could produce faster and more accurate results than the clerks—the human computers—who had been utilized in the past. The British version, designed by mathematician Alan Turing and known as Colossus, was used to help the cryptanalysts at Bletchley Park decipher German communications. But Colossus remained a secret after the war and all the existing machines were destroyed in the name of security, thus costing Britain a chance to replicate its Industrial Age supremacy in the Information Age. German engineer Konrad Zuse also built a digital computer, the Z3, but it was obliterated in an Allied air raid.

ENIAC survived the war and helped spawn a computer industry in the United States because the University of Pennsylvania's Moore School of Electrical Engineering, where it was built, held a series of seminars in 1946 to share its secrets with a large roster of scientists and engineers. ENIAC, and even more its successor, EDVAC (Electronic Discrete Variable Automatic Computer), designed with the aid of the mathematical genius John von Neumann, became the models for all future computers with their central processors, stored memory, and input and output units.

Until the 1970s, most computers remained in the ENIAC mold—large, expensive, and hard to use. Small, mass-market computers would be made possible by three ingenious devices—the transistor, integrated circuit, and microprocessor—each invented about a decade apart.

The transfer resistor, or transistor, was developed in 1947 by three scientists at Bell Laboratories: William Shockley, John Bardeen, and Walter Brattain. It could perform the same amplifying and switching functions as a vacuum tube but without the need for a bulky glass dome filled with winding coils and filaments. The transistor was composed of nothing more than a small metal cylinder with three tiny gold wires connected to a crystal of germanium, a grayish white element that, when specially treated, was found to be good at both insulating and conducting electricity; hence it was known as a semiconductor. The whole contraption was just half an inch high, gigantic by today's standards but fifty times smaller than a vacuum tube. The transistor was also less prone to burn out and did not emit as much heat as a vacuum tube. Its invention made possible much smaller electronic devices such as portable radios. Shockley was eager to exploit the commercial possibilities by leaving Bell Labs in 1954 to form his own semiconductor company in his hometown, Palo Alto, California, thus spawning Silicon Valley.

But while a big improvement over vacuum tubes, transistors were still costly and delicate. They had to be wired together by hand along with other parts, such as resistors, capacitors, and diodes, to make electronic instruments work. In 1959 Jack Kilby, a scientist at Texas Instruments in Dallas, patented a new technique for reducing an entire electronic circuit onto a semiconductor wafer the size of a fingernail. He could fit five transistors on a single crystal (or chip) of germanium. Working independently, Robert Noyce, one of eight defectors from Shockley's firm who started Fairchild Semiconductor, came up with an even better design that utilized silicon instead of germanium. Kilby and Noyce are credited as coinventors of the integrated circuit, better known as the microchip.

Noyce would go on to start Intel (short for Integrated Electronics) along with another Shockley alumnus, Gordon Moore. Between 1969 and 1971, a young Intel engineer named Ted Hoff came up with a powerful new type of microchip that could be programmed to perform a variety of functions, unlike earlier chips which were made for specialized purposes and could not be altered once they left the factory. Hoff 's invention was an entire "computer on a chip"—a microprocessor that could serve as the central processing unit of electronic devices. Intel's 4004 chip and its more powerful successors made possible a variety of gadgets such as the digital watches, video games, and calculators that became popular in the 1970s.

The spread of microprocessors also led to a radical reduction in computer size. The first successful desktop computers came out in 1977 from Tandy, Commodore, and Apple, utilizing widely available peripherals (video monitors, keyboards, audiotape memory storage). Not far behind were feisty software upstarts such as Microsoft, with its soon-to-be-ubiquitous operating systems (first MS-DOS, then Windows).

This marked a major inflection point in the history of the computer industry. Until then, electronic innovation had been driven largely by government, and especially military, spending—a rarity in the history of technology. Most previous advances, whether steamships, machine guns, radios, or airplanes, were the work of independent inventors. The period when government-supported research and development (R&D) was primarily responsible for expanding the frontiers of electronic knowledge started in the 1940s and ended in the 1970s. The leaders of the new microcomputer industry—long-haired hippies and dropouts like the founders of Apple, Stephen Wozniak and Steven Jobs—were the last people on earth one would associate with the "military-industrial complex." They brought a counterculture sensibility to the computer business that has not totally dissipated even as it has come to be dominated by multibillion-dollar behemoths.

COMMUNICATIONS UNBOUND

Like the electronic computer, the Internet would be launched by Uncle Sam but fully developed by the private sector. The original impetus came from the Defense Department's Advanced Research Projects Agency (forerunner of DARPA), which was created in the wake of Sputnik. In the 1960s it set out to devise a communications network that would allow scientists across the country to share precious computer resources. (Contrary to popular legend, the primary goal was not to create a network capable of surviving a nuclear war.) The ARPANET was based on the principle of "packet switching"—breaking each computer message into smaller components and routing them via the fastest available connection to another computer, where they would be reassembled. Starting in 1970 with "nodes" at four West Coast research universities, the ARPANET grew so rapidly during the next decade that by 1983 the Pentagon decided to spin off a separate network—Milnet—and leave civilians with their own system, which came to be known as the Internet. Early users created bulletin boards and added e-mail (including the now-ubiquitous @ sign) almost as an afterthought.

The Internet really took off with two further inventions. First, between 1989 and 1991, Tim Berners-Lee, a researcher at a physics lab in Switzerland, invented hypertext markup language (HTML) and hypertext transfer protocols (HTTP) that would allow anyone with an Internet connection to access information stored on a computer server simply by typing in the correct address, or "URL" (uniform resource locator). He called his creation the World Wide Web. In a related development, in 1993 Marc Andreessen and Eric Bina, two young programmers at the University of Illinois, Urbana-Champaign, created the first easy-to-use browser for displaying text and graphics. They called it Mosaic; its commercial version became known as Netscape Navigator.

The ingredients had now been assembled for the Internet to become a mass consumer medium. The number of Internet users worldwide soared from an estimated 4.4 million in 1991 to 665 million in 2003, and it continues to grow exponentially. The number of Web sites has increased during that same period from zero to some forty million.

Another powerful communications medium, wireless telephony, was on the ascent at the same time as the Internet. Mobile telephones—essentially two-way radios—had existed since the 1940s, but they had been so bulky that they had to be carted around in cars, and their capacity had been so limited that, as late as 1981, only twenty-four people in New York City could use their mobile phones at once. The problem was that each phone call took up an entire electromagnetic frequency, as if it were a radio station, and there was a limited amount of spectrum available. Bell scientists had figured out a way around this problem as early as 1947: divide a large area into numerous "cells," each one served by low-frequency transmitters. As a user moved, calls would get automatically switched from one cell to another. Thus multiple conversations could occur on a single channel. But this idea was not implemented until the 1980s, partly because of regulatory roadblocks but mainly because the electronics needed to run the phones and switch the calls automatically did not become practical until the development of low-cost microprocessors.

The first commercial handheld portable phone, Motorola's DynaTAC, came out in 1983. It weighed two and a half pounds, cost $3,500, and was nicknamed "the Brick." Thirteen years later, in 1996, Motorola introduced the StarTAC, which weighed just 3.1 ounces and cost under $500. As the size and price of cell phones plummeted, the number of users around the world soared from 16 million in 1991 to 1.3 billion in 2003.

All of this traffic put an immense strain on existing telephone networks: The number of calls carried by AT&T on an average business day skyrocketed from 37.5 million in 1984 to 270 million in 1999. Microwave radio relays (invented for radar systems in World War II) had been used to transmit long-distance telephone calls and television signals since 1947; satellites had been added in 1962. In the 1970s an even better technology came along—hair-thin strands of glass filament that could transmit pulses of light emitted by tiny lasers the size of a grain of sand. Because light is a much more efficient transmission medium than either electric signals or radio waves, a single strand of fiber-optic cable could carry 65,000 times more information than a copper wire. The first practical optical fibers were produced by researchers at Corning Glass in 1970 using lasers that had been invented a decade earlier by Theodore Maiman of Hughes Research Laboratories. A fiber-optic cable was laid underneath the Atlantic in 1988. Since then every continent except Antarctica has been hooked up and more than 80 percent of long-distance voice and data traffic is now carried over fiber-optic lines.

The story of the Information Revolution is far from complete. It continues to change shape as computer power and telecom bandwidth continue to increase. State-of-the-art microchips went from having 64 transistors in 1965 to a billion transistors in 2002, a rate of sustained improvement far faster than the railroad or the airplane or any other previous invention. This is an illustration of what has become known as Moore's Law: The number of transistors on a microchip will double every eighteen to twenty-four months. Such exponential growth presumably cannot continue forever, but it has not stopped yet.

Bandwidth capacity is increasing at a similar clip; by 2001, more than 280 million miles of fiber-optic cable had been laid, enough to circle the earth 11,320 times. Costs are falling commensurately, with the price of silicon dropping 50 percent every eighteen months. In 1965 a single transistor cost $5. In 2002 you could buy five million transistors for $5.

THE WIRED WORLD

The most obvious impact of the Information Revolution has been felt on the economy, where it has been reshaping one industry after another, much as the First and Second Industrial Revolutions did. Boeing offers a good example. In 1962, it unveiled the Boeing 727 after almost seven years of work by five thousand engineers. Thirty-two years later, in 1994, it rolled out the 777, a larger, more sophisticated airplane but one that took fewer than four years to produce thanks to computer-assisted design software. Computers reduced the development cycle by 36 percent while improving quality.

Simply buying computers isn't enough. The most successful firms, like the most successful armies, innovate organizationally as well as technologically. Consider how Wal-Mart in 2002 became the biggest company in the world measured by sales. Part of the answer may be found in its early decision to bypass wholesalers and deal directly with manufacturers, which allowed it to get better deals than its rivals, and its more recent decision to open up "big box" stores known as Sam's Clubs, which have proven popular with shoppers. But Wal-Mart has also been an early adopter of barcode scanners, radio frequency identification tags, satellites, and computers to tie its stores and suppliers closer together and enable them to respond to shifting consumer demand. Former CEO David Glass calls "the early commitment to technology . . . a major driver in the success of this business." Such stories can be multiplied a thousandfold—as can tales of once-successful companies such as Sears, Woolworth, and Montgomery Ward that owed their rise to railroads or other industrial technologies but failed to "reengineer" themselves for the Information Age and fell by the wayside.

The Information Revolution has had a similarly mixed impact on individual workers. Millions of new jobs have been created in software engineering, semiconductors, and other industries that are of relatively recent vintage.

At the same time, millions of clerical and blue-collar jobs that date from the First or Second Industrial Revolutions—jobs such as bank cashier, secretary, and steelworker—have been lost. Cheap computers and communications links allow companies to eliminate or "outsource" marginal jobs, sometimes to the other side of the world where they can be performed more cheaply. It has become common for companies to have headquarters in the United States, call centers in India, factories in China, and sales outlets all over the world. Information technology has helped spawn a new wave of globalization, just as industrial technology did in the nineteenth century.

In the emerging global division of labor, the richest economies have prospered based not on the possession of natural resources or vast populations, as in the past, but by creating a favorable climate for innovation and investment—for the exploitation of human, not physical, capital. This explains why resource-poor city-states like Hong Kong and Singapore have grown so fast while countries with much greater material resources like Russia and Nigeria have stagnated. Or why Israel, with no oil, has become wealthier per capita than Saudi Arabia, which has the world's largest oil reserves. A corollary is that the desire for territorial expansion—once a prime cause of wars—has all but disappeared among developed states. "The competition for the best information has replaced the competition for the best farmland or coal fields," noted the late Citibank chairman Walter Wriston, though competition for at least one natural resource—oil—remains keen.

By the twentieth century, economies in the developed world had moved away from agriculture; by the twenty-first century they had moved away from manufacturing, too. Roughly 70 to 75 percent of workers in developed countries are employed in "service industries"—a category that includes everyone from wealthy doctors, lawyers, and executives to the working poor: cashiers, janitors, security guards, waiters. For those left behind, the Information Revolution has been as painful as the Industrial Revolution was for dispossessed agricultural workers. Overall, however, the Information Revolution has been a great economic boon. High-tech firms have been growing much faster than those in other industries, driving world economic growth since 1980.

The performance of the United States during that period has been particularly impressive. Since the early 1980s, the U.S. has consistently experienced high growth without high unemployment or inflation, a combination economists once considered virtually impossible. Even the bursting of the high-tech stock market bubble in 2000 proved to be only a temporary disruption in this long boom. What explains this prolonged expansion? Primarily the growth in worker productivity. "From 1995 through 2003," says former Federal Reserve vice chairman Roger Ferguson, "average annual productivity growth was 3 percent, double the 1½ percent rate of growth that prevailed between 1973 and 1995"—a development that he attributes in large part to "substantial . . . investments in high-tech equipment in the late 1990s."

One should not exaggerate the impact of the Information Revolution. It has almost certainly not made as much of a difference in the lives of ordinary people as the coming of electricity or even indoor plumbing. (Would you rather give up your toilet or your Internet connection? Your refrigerator or your cell phone?) Nor has it lived up to all of its initial hype, particularly when it comes to the impact of computers on society.

Optimists once dreamed that, as Ronald Reagan put it in 1989, "The Goliath of totalitarianism will be brought down by the David of the microchip." And, indeed, cell phones with text messaging and computers with Internet access played an important role in helping democracy activists organize bloodless revolutions in Georgia (2003), Ukraine (2004), and Lebanon (2005). But other Goliaths, from China to Cuba, have proven adept at censoring the Internet and manipulating it for their own purposes. Even if computers and attendant technologies are not going to usher in a new golden age, however, they have clearly transferred a large measure of power toward individuals and away from big institutions, whether governments or corporations.

No one is forced any longer to get news exclusively from a handful of major media outlets; you can learn all you want, whenever you want, from the Internet and satellite television. (You just have to be careful not to believe it all.) "We've moved from a media oligarchy to a media democracy," says David Westin, president of ABC News. Also a financial democracy. Investors no longer have to rely only on what they learn from their stockbroker and from day-old stock quotations in *The Wall Street Journal* or the *Financial Times*. They can now trade online and gain real-time access to information that was once the exclusive preserve of a handful of giant financial institutions.

Those financial firms, in turn, are no longer as much at the mercy of governments as they once were. Their improved ability to gather information and move money (roughly $1.5 trillion is traded every day in world currency markets) has given financiers the ability to savagely punish governments that take actions they do not like, as the leading nations of East Asia discovered to their dismay during the 1997–98 financial meltdown. Walter Wriston suggested that the gold standard of the nineteenth century has been replaced by the "information standard": "The electronic global market has produced what amounts to a giant vote-counting machine that conducts a running tally on what the world thinks of a government's diplomatic, fiscal, and monetary policies. That opinion is immediately reflected in the value the market places on a country's currency." Politicians, he added, are powerless to "opt out of the information standard"—at least not without paying a penalty of extreme penury, as have Sudan, Myanmar, North Korea, and other nations. Thus, all over the world, states have been acceding to the markets' inexorable push for economic liberalization.

Financiers are not the only ones utilizing technology to enhance their own power at the expense of the state. So do pornographers, bookies, money launderers, and all sorts of nongovernmental organizations, ranging from antiglobalization activists and human-rights campaigners to terrorist groups. Just as industrial technology proved a powerful instrument of centralization, so information technology has proven to be an equally powerful instrument of decentralization. The instantaneous movement of information across the globe blurs the traditional boundaries of corporations and countries alike. It may be premature to suggest, as Walter Wriston did, that we are seeing the "twilight of sovereignty," or to argue, as law professor Phillip Bobbitt does, that we are seeing the emergence of a new type of political entity, "the market state," but there is no question that the proliferation of information is having a profound political as well as an economic effect.

While much is still murky, one impact of the Information Age so far is reasonably clear: Even while decreasing the importance of traditional nation-states, it has given a substantial boost to the American position in relation to that of other states. In the late 1980s many commentators were predicting the decline and fall of the United States. Instead it was the Soviet Union that collapsed, in large part because it was unable to keep up with the accelerating economic and military power accrued by the U.S. through the deployment of information technology. By the end of the 1990s the United States also had vaulted far ahead of Germany and Japan, which had been widely seen as being on the ascent in the late 1980s. Their declining fortunes, while the United States continues to experience robust economic growth, cannot be ascribed entirely to the effects of information technology, as both Germany and Japan are quite sophisticated technologically; indeed, in some fields of consumer electronics, Japan may even be ahead of the U.S. A large part of the explanation lies in the fact that their economies have been hobbled by heavy-handed regulations, high tax rates, and stifling cultural restrictions. But there is no doubt that America has greatly benefited from its lead in information technology, just as Britain benefited in the nineteenth century from its lead in industrial technology.

Since 1980, the United States has been the world's leading producer of high-tech products, accounting for about 32 percent of the global total in 2001. The U.S. is also the biggest market for high-tech goods, with Americans owning roughly a third of the 650 million computers in the world. Most key information technologies, as we have seen, were Made in the USA, from transistors and microchips to the Internet and fiber-optic lines. Despite growing competition from other nations, American-based (if not born) scientists are still awarded more patents than those of any other country—42 percent of the global total. And two-thirds of the world's most influential scientists still work in the U.S. Moreover, the United States, with just 4 percent of the world's population, accounts for almost 50 percent of global research and development spending—more than $250 billion annually, split about evenly between the government and the private sector. English remains the lingua franca of the Internet, and the U.S. is home to more Web sites than any other country.

And, as we shall see, American weaponry remains at the cutting edge of military developments.

CHAPTER 10

PRECISION AND PROFESSIONALISM:

Kuwait and Iraq, January 17–February 28, 1991

The helicopters lifted off into the moonless Arabian night at 12:56 A.M. The small squadron—two U.S. Air Force MH-53J Pave Lows specially equipped for unconventional operations and four heavily armed Army AH-64 Apaches—took an hour and twenty minutes to reach the Iraqi border from their desolate base in the Saudi desert. A few minutes behind them flew an identical squadron of two more Pave Lows and four more Apaches. The helicopters continued pushing north, penetrating Iraqi air space at a rate of 138 miles per hour. In order to avoid tipping off the enemy, they flew just seventy-five feet above the ground and used no running lights, a feat made possible by satellite navigation systems, night vision goggles, and terrain-following radar.

Their mission in the early morning hours of January 17, 1991, was to fling the first lightning bolts of Desert Storm, the U.S.-led campaign to expel the Iraqi army from Kuwait. General H. Norman Schwarzkopf wanted to penetrate Iraqi airspace and wipe out key targets without alerting Baghdad. But Iraq deployed a sophisticated air defense system designed with the help of French and Russian advisers and known as Kari (*Iraq* spelled backward in French). The country's perimeter was ringed by radar stations that would give the alarm long before conventional aircraft could reach their targets. Once they were alerted, more than 700 fighter aircraft, 7,000 antiaircraft guns, and 16,000 surface-to-air missiles stood ready to defend Saddam Hussein's domain. Hence this bold commando raid, designated Task Force

Normandy, to wipe out two Iraqi air defense stations twenty miles apart. Because the two were linked, they would have to be destroyed simultaneously, before either could give the game away.

The Pave Lows, which had the world's most sophisticated electronic navigation systems, were meant to be pathfinders, leading the lower-tech Apaches over the unmapped terrain to their prey. Nine miles from their targets, the Pave Lows swooped down close to the desert floor and dropped green phosphorescent sticks to mark the spot, then began their long flight back to base. The Apaches went on alone, pulling up about four miles south of the Iraqi radar stations. Having flown this far in a straight line, with just five rotor discs' separation between each aircraft, they now spread out side by side, like an eighteenth-century formation of musketeers deploying for battle.

At one of the sites, code-named Objective Oklahoma, men began running and lights began to flick off. The Americans feared they had been detected. There was not a moment to lose. At ten seconds before 2:38 A.M., one of the pilots broke radio silence. "Party in ten," he intoned.

Ten seconds later, AGM-114 Hellfire missiles roared out of their pods and followed a laser beam toward a generator, which vanished in smoke and flames. "This one's for you, Saddam," muttered an Apache pilot.

The attackers kept firing, Hellfire after Hellfire. Closing in, they switched to unguided Hydra rockets and 30 mm cannon fire. The attack lasted just four minutes, long enough to turn every structure in the Iraqi ground station into a smoking wreck. Similar havoc was being wreaked at Objective Nebraska twenty miles away. The commander of Task Force Normandy, Lieutenant Colonel Dick Cody of the 101st Airborne Division, told his base that the outcome was "Alpha Alpha"—the prearranged signal to indicate that maximum destruction had been inflicted with no American casualties. A twenty-five-mile gap had been blasted in the early-warning line, though the Iraqis had managed to get off a phone call warning Baghdad that an attack was under way.

With the path now clear, a strike package of two dozen U.S. F-15E Eagle and British GR-1 Tornado fighters streaked into western Iraq, accompanied by two EF-111A Ravens stuffed full of electronic jamming equipment intended to disrupt enemy radar. Their targets were Scud missile launchers that could be used to attack Saudi Arabia and Israel.

The first bomb was dropped on Iraq thirteen minutes later and thirty miles to the north. The attacker was an F-117A Nighthawk, which unleashed a pair of two-thousand-pound laser-guided bunker-busting bombs, called GBU-27s (GBU stands for "guided bomb unit"), on the air defense operations center at Nukhayb, a vital link between the radar posts on the border and the air defense headquarters in Baghdad. The sky immediately lit up with the multicolored flashes of antiaircraft fire, but the F-117 escaped unscathed; its special stealth design made it virtually invisible to radar.

Eight other F-117s streaked farther north, toward Baghdad, the most heavily defended city ever attacked from the air. The F-117 was designed to evade such defenses, but it had never had a real test in combat before. (It had flown unopposed in Panama in 1989.) No one knew how well it would perform. Even the most optimistic scenarios estimated that a few would be chewed up in the teeth of Saddam's fierce defenses. The gloomiest prognosticators feared a repeat of the disastrous 1943 raid over Schweinfurt, Germany, in which sixty B-17s had been lost.

Back in Riyadh, Lieutenant General Chuck Horner, the coalition air commander, knew that one of the first targets scheduled to be struck by his F-117s was the Baghdad International Telephone Exchange (known within the American military as the "AT&T building"), the hub of much of Iraq's civil and military telephone traffic. He also knew that CNN's live reports from Baghdad were relayed through the building. A few minutes before 3 A.M., when the AT&T building was supposed to be obliterated, Horner sent a staff officer upstairs from his underground command bunker to monitor CNN. At 3 A.M. on the dot, the network went off the air, and "backslapping and boisterous talk" spread through the Allied command bunker. The stealth fighters had done their job. "Shit hot!" an enthusiastic officer exclaimed.

Next, like a carnival performer trying to top the previous act, another new weapons system made its debut. The American cruise missile was a direct descendent of the German V-1. It was powered, like an airplane, by an air-breathing jet engine. (By contrast, ballistic missiles are powered by a rocket that does not need air to operate.) While slower than a rocket, a cruise missile's low, level flight path, skimming close to the surface, made it difficult to detect, and it was able to achieve pinpoint accuracy with its sophisticated navigational computers.

The first cruise missiles fired in the Gulf War were unleashed from an armada of naval ships in the Persian Gulf and the Red Sea. Their BGM-109 Tomahawk cruise missiles guided themselves to Baghdad, hundreds of miles away, using terrain-contour matching radar. Arriving at the outskirts of the capital, they switched to a different navigation system, using optical sensors to compare the passing scenery with a digital map stored in onboard computers. The Tomahawks found their targets unerringly, plowing into the presidential palace, Baath Party headquarters, a missile complex, and Iraq's largest power plant, putting all of them out of action with their thousand-pound warheads. Other "high-value" targets were struck by air-launched AGM-86C cruise missiles fired by seven lumbering old B-52G Stratofortress bombers that took off and landed in Louisiana, traveling a total of 14,000 miles with multiple in-flight refuelings to complete the longest bombing mission in history.

Many of these initial strikes were designed to disrupt the elaborate electronic network that controlled the Kari air-defense system. The coup de grâce was a carefully planned deception. Small aerial drones, normally used for target practice, were equipped with radar reflectors and beacons to replicate the radar "signature" of much larger warplanes. As soon as they appeared over Baghdad, eager Iraqi defenders "painted" them with their radars and unleashed salvos of radar-guided missiles and artillery shells to knock them down. Unbeknownst to the Iraqis, flying thirty miles south of Baghdad, just outside of their radar range, were coalition aircraft equipped with AGM-88 High-Speed Antiradiation Missiles (HARM) designed to home in on radar emissions. A large number of Iraqi radar dishes were obliterated in the ensuing ambush. For the rest of the war, Iraqi defenders became understandably reluctant to turn on their radars, which they now knew spelled certain death. In a few hours the coalition air forces had achieved what the Luftwaffe had been unable to do during the many months of the Battle of Britain: neutralize enemy air defenses.

What was remarkable was not only how effectively Allied airplanes were able to hit their targets but also at what scant cost. Allied commanders had expected to lose as many as fifty aircraft during the first night of the war. Only one—a Navy F/A-18C Hornet—actually went down. (During the entire war, thirty-eight fixed-wing coalition aircraft would be lost to enemy action.) The biggest challenge turned out to be avoiding midair collisions among nearly seven hundred friendly airplanes swarming over the skies above Iraq. But this challenge, too, was surmounted, thanks to the guidance provided by Air Force E-3 AWACS (Airborne Warning and Control System) aircraft and Navy E-2C Hawkeyes. General Michael Dugan, a former Air Force chief of staff, was right to call this "the most awesome and well-coordinated mass raid in the history of air power."

Viewers all over the world were suitably impressed in the days that followed when the U.S. military released carefully edited gun-camera video showing missiles and bombs hitting bridges and buildings dead-on. Such displays of wizardry were slightly misleading—even the most sophisticated "smart bombs" did not hit every target, finding the right targets for them to hit was no easy task, and even when they landed on target some of them turned out to be duds. Attempts to decapitate the Iraqi leadership on the first night by killing Saddam Hussein and his top lieutenants failed miserably (a failure repeated twelve years later). Nevertheless, something extraordinary happened on the night of January 17–18, 1991. It was the opening night not only of Operation Desert Storm but, arguably, of a whole new era of warfare.

Most of the public's attention was riveted on precision-guided munitions that made possible a quantum leap in bombing accuracy over the unguided projectiles of World War II. Just as impressive, though less noted, were the high-tech command, control, and reconnaissance systems that undergirded the entire coalition war effort. Behind these technological feats was an important breakthrough in the composition of the American armed forces. Iraq's military was still organized on the mass, draftee model of the Industrial Age; the United States had made the transition to an all-volunteer, professional force better able to utilize the sophisticated equipment of the Information Age. It was not just new technology, then, but also new forms of recruitment, training, and organization that made possible the one-sided outcome on the first night of the Gulf War—and in the days and nights that followed.

ALL THAT YOU CAN BE

Just as the Prussian victory at Königgrätz in 1866 and the German victory in France in 1940 grew out of previous defeats, so the American victory in the Gulf War was the result of the regenerative process launched after the end of the Vietnam War. As General Barry McCaffrey put it: "This war didn't take 100 hours to win, it took 15 years."

Two decades before they arrived in the sands of Arabia, the U.S. armed forces had stumbled out of the jungles of Indochina, dazed, defeated, and demoralized. In the aftermath of America's most disastrous war, the armed forces were wracked by racial tensions, rampant drug use and alcoholism, and a pervasive culture of sloppiness and insubordination. Some units came to be dominated by criminal or racial gangs. Many officers would not venture into the enlisted barracks without a sidearm; between 1969 and 1971 there had been eight hundred "fraggings," or incidents in which soldiers attacked their own officers or NCOs.

Things only got worse when the draft was abolished in 1973. For the first time since 1940 the armed forces had to compete in the labor market, and they had no idea how to do it. Given the poor reputation of the post-Vietnam military—only 32 percent of Americans in 1973 had a "great deal of confidence" in those running the armed forces—few of the best and brightest were willing to sign up. Defense spending plummeted and recruiting quotas could not be met. Half of the Marine Corps and Army came to be composed of high-school dropouts. Many officers, who remembered the poor quality of the pre–World War II volunteer army depicted in James Jones's classic novel, *From Here to Eternity,* doubted that an all-volunteer force could ever be viable.

America's weakness was exposed by the humiliating seizure of U.S.

embassy personnel in Tehran in 1979, which led to a bungled rescue mission. The army chief of staff, General Edward "Shy" Meyer, did not mince words. "Basically what we have," he said in 1980, "is a hollow army."

In retrospect, 1980 was the nadir of the post-Vietnam military. Already significant changes were under way that would, within a decade, bring the armed forces to new heights of power and prestige. The most visible sign of change was President Jimmy Carter's decision to significantly increase defense spending—a trend that was accelerated by his successor, Ronald Reagan, who also boosted soldiers' spirits with his unabashedly patriotic and pro-military rhetoric. Just as significant as these political changes, if not more so, were major internal transformations within the armed services that had been set in motion in the late 1970s, and in some cases even earlier.

One of the first priorities for post-Vietnam military reformers was increasing the quality of those in uniform. This task was tackled with gusto by Major General Maxwell Thurman when he took over the Army's Recruiting Command in 1979. A Vietnam veteran, a devout Catholic, and a lifelong bachelor who, in the words of one journalist, "approached each assignment in the army with the fervor and devotion of a Trappist monk," Thurman was determined to save the institution he loved. He pushed Congress to approve a major military pay increase as well as a new version of the G.I. Bill that would offer college scholarships to soldiers after they left the service. He dispatched the best officers and NCOs on recruitment duty. And he began to market the Army as an ideal place to learn valuable skills, an approach crystallized in a new advertising slogan he developed with a New York advertising agency: "Be All You Can Be." Accompanied by a snappy theme song (an advertising trade journal later ranked it as the second-best jingle of the twentieth century, after McDonald's' "You deserve a break today"), the army's new slogan helped spark a recruiting rebirth.

The other services followed the Thurman approach, while also working closely with Hollywood to help prepare pro-military movies such as *Top Gun* (1986) and *The Hunt for Red October* (1990). As the armed forces began to rack up small victories, such as the 1983 invasion of Grenada (depicted in the 1986 Clint Eastwood film *Heartbreak Ridge*), their popularity continued to rise, and recruiters no longer had any trouble filling their quotas. They actually began to turn away low-quality applicants. By 1990, 97 percent of army recruits were high school graduates. The glut of recruits allowed the military to raise standards and crack down on troublemakers. The navy led the way in 1981 by instituting a "zero tolerance" policy for drug use backed up by random urinalysis tests, a policy soon emulated by the other services. The number of people in uniform using illicit drugs fell from 27.6 percent in 1980 to 3.4 percent in 1992.

At the same time, the military made a conscientious attempt to improve the integration of African-Americans and women. This was not always a smooth process, and it gave rise to its share of scandals (mainly involving sexual harassment), but through a combination of outreach, mentoring, and crackdowns on discrimination, the military proved largely successful in creating a racially and sexually diverse force. A symbol of this achievement was the elevation of General Colin Powell to become the first black chairman of the Joint Chiefs of Staff in 1989. "The military had given African-Americans more equal opportunity than any other institution in American society," Powell proudly wrote in his autobiography.

Overall, the American military did a remarkable job between the Vietnam War and the Gulf War of creating an all-volunteer force—something that, while not unprecedented, was certainly unusual by historical standards. Most military organizations have been composed of some volunteers, usually in the higher ranks, leavened by larger numbers of men forced into service, whether as citizen militia (Greek hoplites), tribesmen (the Mongol horde, Zulu impis), slave soldiers (Ottoman janissaries), feudal levies (Japanese samurai, European knights), or modern conscripts. Once conscription became entrenched in the mid-nineteenth century, most countries were reluctant to move away from it, as Charles de Gaulle discovered in the 1930s, when his advocacy of a professional French army was met with withering scorn from his own government.

Britain and America remained the only major Western states that continued to rely on volunteers until the twentieth century. Britain finally broke down and instituted a draft in 1916 during the First World War. The U.S., which had already tried a draft during the Civil War, returned to it in 1917. Both countries abandoned conscription in the interwar period and then instituted it again when another war loomed—Britain in 1939, the U.S. in 1940. Britain was the first to return to its volunteer tradition. It abolished the draft in 1962, a decade before the U.S. But the British army, while of high quality, remained tiny—its total strength in 1992, 153,000 men, made it smaller than the U.S. Marine Corps. The pre-WWI volunteer U.S. Army had been even smaller, comprising as few as 25,000 men during the Indian wars of the late nineteenth century. By contrast the post-Vietnam military had two million active-duty personnel (and as many again in the Reserves and National Guard), making it by far the largest all-volunteer force ever fielded. This was nothing less than a revolution in personnel practices.

Yet, despite the growing quality of its soldiers, the U.S. armed forces remained badly outnumbered by their communist adversaries. The USSR, which had not abolished the draft, fielded a force of more than 3.5 million men in the 1970s, increasing to over 5 million by the early 1980s. The Soviet advantage was equally great in numbers of tanks, artillery, and aircraft. Back in the 1950s or 1960s, the U.S. could confidently rely on its nuclear advantage to deter Soviet aggression, but that edge had disappeared by the end of the 1970s amid a Soviet ICBM buildup.

It became increasingly clear to strategists in the Ford, Carter, and Reagan administrations that they would have to develop a new generation of conventional weapons to offset the Soviets' numerical advantage. It was at just about this time that microprocessors were revolutionizing the computer industry. The Soviet Union, needless to say, had no Silicon Valley of its own. Here was one advantage that the U.S. still had, and the Pentagon was intent on exploiting it.

GET SMART

Since the dawn of the Gunpowder Age, projectiles had been on their own once they were expelled from a gun barrel or, later, from an airplane's bomb bay. No matter how carefully a gunner or bombardier aimed prior to firing, once the trigger had been pulled he no longer had any control over where the munitions went. They were at the mercy of the laws of ballistics and gravity, and hence not very accurate.

That first began to change in World War II. The Germans took the lead; their Fritz X, a radio-controlled bomb, was used against the Allied landing fleet at Salerno, Italy, in 1943. But most of their efforts were not terribly successful. More than half of the V-2 rockets aimed at London missed the metropolitan area altogether because of the inaccuracy of their clumsy gyroscopic steering mechanisms. U.S. scientists did not fare much better. Most of their guided-bomb projects—one program was designed to convert old B-17s and B-24s into unmanned missiles, another to copy the V-1 and V-2—did not get off the ground. Their only success was the VB-1 Azon, a thousand-pound bomb with a radio receiver that could be steered toward its target by a bombardier. More than 450 Azons were dropped in the war; they proved particularly effective in knocking out bridges in Japanese-occupied Burma.

Although a few radio-controlled bombs were also used in the Korean War, the whole field languished in the 1950s and early 1960s. The U.S. Air Force was the natural outlet for smart bombs, but until the mid-1960s its bomb development was delegated to the army and navy ordnance departments, making it a bureaucratic orphan. Who needed accurate munitions anyway, if (as the working assumption had it) the bombs of the next war would be atomic? It took the Vietnam War to revive interest in precision-guidance technology and to spark a general renaissance in air warfare.

The U.S. Navy and U.S. Air Force, which had put all of their energies into getting ready for nuclear conflict against the Soviet Union, were woefully ill-prepared for the type of conventional combat they encountered in the skies over North Vietnam. Heavy jet fighters such as the F-105 Thunderchief were not agile enough for dogfighting against Soviet-built MiG-17s and MiG-21s. U.S. guided air-to-air missiles, like the AIM-4 Falcon and AIM-7 Sparrow, designed to hit big and slow Soviet bombers, proved ineffective against small and fast Soviet-built North Vietnamese fighters. American fighter pilots, who had enjoyed a lopsided advantage in the latter days of World War II and throughout the Korean War, found themselves barely able to hold their own in air-to-air combat over Vietnam.

They had even worse luck in dealing with ground fire, which had been revolutionized by the development of surface-to-air missiles after World War II. This new weapon first proved its effectiveness in 1960, when a Soviet SA-2 radar-guided battery shot down Gary Francis Powers's U-2 spy plane. The Soviets amply supplied their North Vietnamese allies with SA-2s and radar-controlled flak guns, later supplemented by SA-7 shoulder-fired missiles. The U.S. Air Force and U.S. Navy, both of which operated aircraft over North Vietnam, initially had neither the equipment nor tactics to deal with this menace. As the war went along, American pilots learned to avoid enemy batteries with evasive maneuvers and to disrupt them with radiation-seeking missiles and electronic jamming equipment, giving birth to the techniques that would be utilized with such success against Iraq decades later. The United States paid a heavy price for these lessons: More than 1,500 of its aircraft were downed in Indochina, 95 percent by ground fire.

Besides leading to the death or capture of many pilots, heavy ground fire disrupted bombing patterns and made it hard for U.S. aircraft to achieve their objectives during the Rolling Thunder campaign against North Vietnam from 1966 to 1968. Pilots were further handicapped by the fact that, unlike in World War II or Korea, they could not simply undertake indiscriminate area bombing. The Johnson administration was sensitive to the political ramifications of "collateral damage" and enforced strict limitations on where and when U.S. aircraft could strike. Unfortunately, with bomb accuracy only slightly improved since World War II, U.S. aircraft lacked the capacity to execute pinpoint raids.

The solution did not come from an intensive Manhattan Project–style crash program of the kind that had produced the B-29 and the atomic bomb. Rather it was a stroke of serendipity that inspired Air Force Colonel Joe Davis, Jr., to set in motion the first laser-guided bomb in 1964. As deputy commander of an Air Force armaments laboratory at Eglin Air Force Base in Florida, Davis was dazzled by a demonstration of a newfangled laser. The scientists who showed off the device had no intention of using it to guide bombs, but that was the first thing Davis, an old World War II and Korean

War fighter ace, thought of. He even went aloft with a handheld movie camera to prove that a beam from a cockpit could be consistently directed at a fixed point on the ground. Using discretionary funds that did not need approval from the cumbersome Pentagon procurement bureaucracy, Davis awarded a $99,000 contract to Texas Instruments to develop a laser bomb–aiming system. The result was the Paveway, which initially required two aircraft to deliver—one to drop a bomb with small, movable wings, the other to aim a laser beam at its target. Eventually a single aircraft was equipped with both the laser-guidance pod and the bomb. As soon as they had proven their effectiveness, Paveways were rushed to Vietnam.

It was later determined that 48 percent of Paveways dropped in 1972–73 around Hanoi and Haiphong achieved direct hits on their targets vs. only 5.5 percent of unguided bombs dropped on the same area a few years earlier. The average Paveway landed within 23 feet of its target, compared to 447 feet for a "dumb" bomb. The leap in accuracy brought about primarily by laser guidance made it possible to take out tough targets that had withstood earlier air raids. The most dramatic example was the Thanh Hoa (Dragon's Jaw) bridge seventy miles south of Hanoi, a crucial supply artery for the North. Starting in 1965 U.S. pilots had flown 871 sorties against it, losing eleven aircraft without managing to put it out of commission. In 1972 the bridge was attacked for the first time with Paveway bombs. Just 14 jets managed to do what the previous 871 had not—send the span crashing into the Red River.

The U.S. wound up employing 28,000 Paveways in Southeast Asia, more smart bombs than have been used in any conflict before or since. They did not save the U.S. from defeat, partly because they were introduced late in the war (only 0.2 percent of all munitions dropped were precision-guided), but mainly because a guerrilla foe hiding in the jungles was not very vulnerable to air attack. Still, the Vietnam experience set the U.S. military on the path to future smart bomb developments. Better microelectronics led to the invention of improved bombs and missiles with aiming systems utilizing radar, lasers, thermal sensors, satellite navigation, inertial guidance, and electro-optical sensors, often in combination to make up for each individual mechanism's shortcomings.

By the time of the Gulf War, the most common ground-attack precision munitions in the U.S. arsenal were laser-guided Paveway III bombs, guided missiles like the Maverick and Hellfire, and ship-launched cruise missiles guided by internal computers programmed with precise target coordinates. Although laser-guided bombs and cruise missiles were relatively few in number, they would have a disproportionate impact in the war's early days in hitting Iraq's best-protected targets with unprecedented accuracy. As one airpower strategist explained:

> [D]uring World War II, an average B-17 bomb during a bombing run missed its target by some 2,300 feet. Therefore, if you wanted a 90 percent probability of having hit a particular target, you had to drop some nine thousand bombs. That required a bombing run of one thousand bombers and placed ten thousand men at risk. By contrast, with the new weaponry one plane flown by one man with one bomb could have the same level of probability. That was an improvement in effectiveness of approximately ten-thousand-fold.

NEW WEAPONS, NEW CONTROVERSIES

The most revolutionary weapons system of all in 1991 was a stealth attack aircraft equipped with two thousand-pound laser-guided bombs. Its genesis lay in Lockheed's famed Skunk Works, a top-secret research lab in Burbank, California, that had produced such cutting-edge Cold War aircraft as the U-2 and SR-71 high-altitude spy planes. In the mid-1970s, Skunk Work engineers figured out how an airplane could be made virtually invisible at night by using special composite materials and flat panels that absorbed rather than reflected radar emissions. President Carter's Defense Secretary, Harold Brown, a physicist by training, and his Undersecretary for Research and Engineering, William J. Perry, another Ph.D. scientist (and later Defense Secretary, 1994–97), grasped the possibilities immediately and gave the project their enthusiastic support as part of their strategy to develop a qualitative advantage that would offset the Russians' quantitative edge.

Because it was so highly classified, the stealth work (known initially as Project Harvey, after the invisible rabbit in the 1950 Jimmy Stewart movie) cut through normal Pentagon red tape. The first prototype of the F-117A stealth fighter was ready to fly in 1977 and the first production-line model was delivered in 1981—a remarkably fast procurement cycle. That same year, the Air Force gave the go-ahead for Northrop Grumman to develop a stealth bomber which could carry more munitions. The first B-2 Spirit flew in 1989, but it was not employed in the Gulf War. By contrast the F-117A played a pivotal role, particularly, as we have seen, in the first hours of Desert Storm. Both aircraft were very much the progeny of the Information Age: They required tremendous computational power to design and they were so ungainly that they require computer assistance ("fly by wire") to operate.

The stealth aircraft were only the most advanced of many new weapons systems that were developed in the 1960s and 1970s and joined the U.S. arsenal in the 1970s and 1980s. The Air Force procured two agile new fighter-bombers, the F-16 Fighting Falcon and F-15 Eagle, the B-1 Lancer bomber, and an aircraft for close support of ground forces, the A-10 "Warthog." The navy had its own super-fighters, the F-14 Tomcat and F/A-18 Hornet, as well as Aegis guided-missile cruisers (the first was the *Ticonderoga*, commissioned in 1983), *Los Angeles*–class nuclear submarines, and *Nimitz*-class nuclear-powered aircraft carriers. The army bought a main battle tank, the M1 Abrams; an armored personnel carrier, the M2/M3 Bradley Infantry Fighting Vehicle; a utility vehicle called the Humvee (High Mobility Multi-Purpose Wheeled Vehicle); the AH-64 Apache attack helicopter and the UH-60 Black Hawk transport helicopter; an air defense system called the Patriot; and a mobile surface-to-surface missile launcher, the M270 Multiple Launch Rocket System.

With the exception of the stealth aircraft, which remained a tightly guarded secret until 1988, all of these systems were extremely controversial when they were in development. Virtually all were plagued by delays and cost overruns that led to embarrassing stories in the press. A few were even canceled outright. Critics hammered the military for an unwieldy procurement bureaucracy that, they claimed, was producing costly weapons that would not stand up to the rigors of battle. Journalist James Fallows, in his influential 1981 book *National Defense*, derided the Pentagon's "pursuit of the magic weapon" encumbered with "more and more complex computer systems, whether or not there is reason to think that computers will help on the battlefield, and often when there is reason to think they will hurt." In a similar vein, *Time* magazine ridiculed the Pentagon's "slavish devotion to the latest high technology." Such criticisms were echoed by Congress's Military Reform Caucus, a bipartisan group of more than one hundred lawmakers led by Senator Gary Hart, who pushed for simpler, cheaper weapons in greater numbers.

Luckily, the Pentagon did not follow their advice. If it had, the U.S. would have fought Iraq in 1991 with equipment roughly equivalent to the enemy's, instead of having weapons at least a full generation ahead.

What the self-styled reformers did not realize was that adding sophisticated electronics did not have to make weapons systems less reliable and harder to operate. Thanks to advances in solid-state electronics, new aircraft such as the F-15 and F-16 were not only far more lethal than their predecessors but also easier to fly and less prone to malfunction. Far from being an encumbrance, advanced electronics gave such weapons a vital edge over less sophisticated adversaries.

Consider the M1 tank built by General Dynamics starting in 1980. It had a gas turbine engine that allowed it to go nearly 45 mph and Chobham ceramic armor that could survive frontal hits from even the most advanced Soviet tanks, the T-72 and T-80. Its 120 mm main gun fired 45-pound Sabot rounds tipped with depleted uranium (a substance that is more than twice as dense as steel) that could penetrate a T-72 at two and a half miles, well outside the T-72's own range. But its true advantage lay in a fire-control system that employed laser range-finders, thermal and optical sights, and digital ballistics computers to allow its 120 mm main gun to hit targets while on the move and notwithstanding fog, night, or other conditions that would have rendered earlier tanks useless. In World War II the average tank needed seventeen shots to kill an enemy tank; in the Gulf War, the Abrams would come close to achieving the ideal of one shot, one kill.

The M1's ability to operate at night was a key advantage shared by most U.S. weapons systems in 1991. Night-vision equipment had been developed by the U.S. Army starting in the 1950s. It came in two versions: image-intensifying devices that amplify small amounts of ambient light and thermal forward-looking infrared detectors that sense differences in temperature between an object and its environment. The former are generally carried by soldiers as goggles; the latter usually come in more cumbersome systems attached to vehicles and aircraft. Since Iraqis had few if any comparable devices, the U.S. military "owned the night."

In addition to their night-vision devices, the U.S. armed forces benefited from their unrivaled electronic warfare and reconnaissance capabilities. The U.S. Air Force and U.S. Navy operated a variety of aircraft designed to keep an eye on the "battle space," the most famous of which was the AWACS, a Boeing 707–320B equipped with a huge rotating radar dome that could identify low-flying objects from more than 250 miles away. Onboard sat thirteen to nineteen mission specialists who could analyze information and coordinate air operations in real time, allowing hostile aircraft to be intercepted as soon as they were airborne and friendly aircraft to avoid hitting each other or shooting at each other.

What the AWACS did for air operations, the E-8A JSTARS (Joint Surveillance Target Attack Radar System) did for the ground war. Also housed in a 707 airframe, the JSTARS's synthetic-aperture radar, located in a canoe-shaped appendage under the fuselage, could locate and track moving vehicles over more than two hundred miles. It was still in the experimental stages when Iraq invaded Kuwait in August 1990, but two prototypes were rushed to Saudi Arabia and they proved invaluable in locating Iraqi ground forces.

The AWACS and JSTARS were complemented by numerous other aircraft designed to listen in on enemy communications (RC-135 Rivet Joint), jam enemy radars (EA-6B Prowler, EC-130H Compass Call), or photograph enemy positions (TR-1/U-2). High above all these aircraft a constellation of U.S. satellites monitored the battlefield from space.

Although the Soviet Union beat the U.S. into space with the launch of Sputnik in 1957, the U.S. quickly surpassed its rival to gain military predominance on this high ground—a lead it has never relinquished. The Gulf War has sometimes been called the "first space war" because it utilized the full panoply of U.S. satellites to aid allied military forces. The exact details remain shrouded in secrecy, but satellites are known to have performed myriad functions, including providing meteorological data, creating detailed maps, offering early warning of Scud missile launches, relaying communications, and spying on enemy forces.

The most novel and important use of satellites was to provide navigational help to coalition forces. The Global Positioning System (GPS) was based on a simple premise: A user could determine his exact location by timing how long it took a radio beam to travel from his position to several satellites in fixed orbit. Navstar GPS, begun in 1973 by the Pentagon, was designed to orbit at least twenty-four satellites which would give anyone anywhere line-of-sight to at least four satellites at one time—the minimum needed to get an accurate fix. Only sixteen of the satellites had been deployed by the time the Gulf War began, so they did not provide continuous coverage. Another major limitation was the lack of GPS receivers. By the time Desert Storm began, following a last-minute shopping spree, the coalition had 840 military GPS receivers and 6,500 commercial models. Even with its limited availability, GPS made possible much more accurate maneuvering and striking than had ever been feasible before. Allied tank forces would not have been able to move through the vast deserts of Iraq without the aid of GPS. Nor would the opening night raid by Task Force Normandy have been possible without it. The reason why Pave Low helicopters had to lead Apaches to the Iraqi radar stations was that they had GPS and the Apaches didn't.

Space systems helped provide the American-led coalition with a lopsided advantage in what military professionals called C^3I (pronounced "see three eye")—command, control, communications, and intelligence. (With the addition of computers, surveillance, and reconnaissance, this formula has now become C^4ISR—"see four eye-ess-are".) Against this vast array of air and space sensors, the Iraqis had no satellites of their own and no way to fly air reconnaissance because of the Allies' domination of the skies. Nor could they buy satellite time from private firms; the U.S. had bought up all the available capacity. It was almost as if, in the Second Industrial Age, a country with tanks and aircraft was fighting an enemy without any. The disparity was that profound. In the tools needed to fight in the Information Age, Iraq was critically deficient.

"EVERYONE HAD BEEN THERE BEFORE"

Developing all this high-tech gadgetry was one thing. Learning to utilize it properly was another. The U.S. would not have done so well in the Gulf War had not its armed forces transformed their training and doctrine since the Vietnam War. The training revolution began in 1969 when the navy, concerned about the poor showing of its aircraft over North Vietnam, established the Fighter Weapons School at Miramar Naval Air Station in San Diego. Better known as "Top Gun," it offered pilots realistic training in dogfighting that significantly improved their combat performance.

The air force took note and in 1975 opened its own version of Top Gun. Red Flag exercises at Nellis Air Force Base in Nevada allowed pilots to compete against an Aggressor squadron emulating the tactics and equipment of Soviet adversaries. Here a new generation of aviators learned how to put together elaborate "strike packages" designed to penetrate enemy air defenses. Experience showed that a pilot was most likely to be shot down during the first ten combat sorties; Red Flag was designed to ensure that those missions occurred on a training range, not over a real battlefield.

The army set up a realistic training center of its own at Fort Irwin, California, amid the barren scrubland of the Mojave Desert. Starting in 1981, mechanized battalions would travel to the National Training Center to fight a simulated engagement against a highly skilled "Opfor" (Opposing Force) modeled on a Soviet motorized rifle regiment. Lasers simulated the effects of actual gunfire and computers kept track of the action for later analysis. Umpires delivered brutal after-action reports on what went right and wrong. The visitors usually got whipped by the first-rate Opfor, but they learned a good deal from the experience.

At the start of previous wars, American soldiers had been thrown into battle without much combat experience or realistic training to draw upon, and they usually paid a steep price for their inexperience. For instance, the 1st Armored Division was mauled by veteran German units at Kasserine Pass, Tunisia, in February 1943, losing more than six thousand men. That didn't happen this time. "Desert Shield and Desert Storm went so easily," wrote General Chuck Horner, "because everyone had been there before."

It also went well because the armed forces had worked out a doctrine ideally suited for operations against a foe like Iraq. One of the U.S. Army's most important innovations after Vietnam was the creation in 1973 of the Training and Doctrine Command to fashion an intellectual renaissance. Its first commander was General William DePuy, a veteran of World War II and Vietnam who proceeded to obliterate the traditional American approach toward war.

In his first operations manual, which came out in 1976, DePuy noted that traditionally the U.S. was "accustomed to victory wrought with the weight of materiel and population brought to bear after the onset of hostilities." This had worked in the Industrial Age but was no longer suitable for the dawning Information Age. Given the lethality of modern weapons—demonstrated in the 1973 Yom Kippur War, the first conflict in which guided munitions played a major role—he did not think it was possible to lose the first battles and still stagger to victory. "Today the U.S. Army must, above all else, prepare to win the first battle of the next war."

This was an important innovation that was eagerly greeted by the army. So was DePuy's emphasis on realistic training, which led to the creation of the National Training Center. The actual strategy DePuy crafted, known as Active Defense, was less popular: As its name implied, it was an essentially reactive approach that called for falling back in the face of a Soviet onslaught in Europe. Other ideas bubbled up at various military institutions, including advanced schools devoted to the operational art opened by the army, marines, and air force in the 1970s.

The eventual result was a new doctrine prepared by DePuy's successor, General Donn Starry, and adopted in 1982. His approach, known as AirLand Battle, was anything but static. It was essentially a variant on the German blitzkrieg or Russian "deep battle" and a far cry from the attritional strategy utilized by U.S. forces in all of the country's major conflicts going back to the Civil War. AirLand Battle called for attacking Red Army rear echelons, seizing the initiative, outmaneuvering the enemy, and using a variety of weapons simultaneously to produce a counteroffensive that would be "rapid, unpredictable, violent, and disorienting to the enemy." It was predicated upon the assumption that the U.S. had superior weapons and superior personnel that could compensate for its inferiority in total numbers. The air force bought into this doctrine and the Marine Corps came up with its own version, known as Maneuver Warfare.

This was in essence the strategy that was put to use in Desert Storm. Originally developed to counter Soviet tank armies on the plains of Europe, AirLand Battle proved ideally suited to counter Soviet-style tank armies in the deserts of the Middle East.

The final element needed for an American victory in the Gulf War was having the right organizational structure in place. Chaotic operations such as the 1980 Iran hostage rescue and the 1983 invasion of Grenada had revealed the pitfalls of interservice rivalry. This gave a boost to military reformers on Capitol Hill who wanted to create a more unified command structure. After several years of debate, Senator Barry Goldwater, an Arizona Republican, and Representative Bill Nichols, an Alabama Democrat, managed in 1986 to push through the most significant shake-up of the Pentagon since the creation of the Department of Defense in 1947.

The Goldwater-Nichols Defense Reorganization Act established a clear chain of command running from the president to the defense secretary to a unified field commander. The entire world was broken up into vast regions that were placed under the command of a single four-star general or admiral who had complete authority over all U.S. forces within his jurisdiction. There were separate commands for Europe and Africa (European Command), the Atlantic (later eliminated), the Pacific, Latin America (Southern Command), and the Middle East, the Horn of Africa, and Central Asia (Central Command), joined in 2002 by Northern Command for homeland defense. Other commands were established for responsibilities that transcended geographical boundaries. By 2002, after some reshuffling, these were the Joint Forces Command (responsible for preparing forces for war), Special Operations Command, Strategic Command (responsible for ballistic missiles, cyberwar, and space programs), and Transportation Command.

An emphasis was put on "jointness": The chairman of the Joint Chiefs of Staff was made principal military adviser to the president and defense secretary, sidelining the individual service chiefs with their more parochial concerns, who were left with the responsibility to equip and train the forces that would be assigned to a unified commander. To assist him, the chairman was given a vice chairman (another four-star) and an expanded Joint Staff of more than one thousand officers. Service on a joint staff became mandatory for any officer seeking promotion to flag rank.

The original geographical commands all preceded the Goldwater-Nichols Act. The European Command could trace its origins back to World War II; the Pacific Command had even earlier antecedents, in the Philippine War of 1899–1902. The most recent, Central Command, was created in 1983. What the 1986 legislation did was to give the combatant commanders unprecedented authority within their domains. It was power that General H. Norman Schwarzkopf was ready to seize when he took over as Centcom's third chief in 1988 with responsibility for operations in twenty-five countries stretching from Kazakhstan to Kenya.

STORMIN' NORMAN

Schwarzkopf would emerge as the biggest media star of the Gulf War, and no wonder: He was big, colorful, outspoken, blustering, profane, and he had an unexpected flair for deadpan wit. At one press conference, he mordantly announced, "Now you're going to see the luckiest man alive," just before playing a video clip of an Iraqi vehicle that had barely cleared a bridge before it was brought down by a laser-guided bomb.

What television viewers did not see was his volcanic temper, which scalded his staff with frequent eruptions. When displeased, which was often, the general would turn a deep shade of purple and holler, "This is dumb! This is stupid!" Many who worked with him came to see him as, in Chuck Horner's words, a "screaming, tantrum-throwing prima donna." "He thrived on confrontation," Horner explained. "His temper was famously quick and violent, and he was notorious for verbally hanging, drawing, and quartering those who didn't reach his standards." Schwarzkopf himself acknowledged that he was "not an easy commander" to work for. Luckily, General Colin Powell, who served as an intermediary between the Centcom chief and his boss, Defense Secretary Dick Cheney, had sufficient tact and diplomatic skill to handle "Stormin' Norman."

For all his faults, the imperious Schwarzkopf also brought many virtues to the table. He was smart, hardworking (he routinely worked 18 to 20 hours a day during Desert Storm), well-versed in the ways of the Middle East, and absolutely devoted to the welfare of his soldiers—all qualities developed during a lifetime in uniform.

Schwarzkopf dated his military service all the way back to 1946, when he was just twelve years old. His father, also named H. Norman Schwarzkopf, was a former superintendent of the New Jersey State Police and a reserve army officer who was called to active duty in World War II and sent to train a gendarmerie for America's ally, the shah of Iran. Young Norm's mother, left stranded at home, had turned into a terrifying alcoholic, so he jumped at the chance to join his father in Tehran after the war. Iran in the late 1940s provided an early introduction not only to military life but also to the ways of the Middle East. Norm spent the rest of his childhood bouncing around various schools in Europe and the United States before ending up at West Point. The years after his graduation in 1956 were relatively uneventful until in 1964, when, as an engineering professor at his alma mater, he saw a chance to go to Vietnam and eagerly grabbed it. Few ambitious and dedicated officers would pass up an opportunity to gain combat experience, and this general's son was no exception.

Schwarzkopf spent the next year in Vietnam serving as an adviser to a South Vietnamese airborne brigade. He participated in seven major combat operations, was hospitalized twice for illness, and wounded once—but never lost his enthusiasm. The war in those days had a sense of high purpose that Schwarzkopf shared. After coming home he married a TWA stewardess, and then promptly volunteered for another tour of duty in Vietnam.

By now it was 1969 and morale in the draftee army was sinking faster than an armored vehicle in quicksand. Colonel Schwarzkopf was given command of what he discovered was "probably . . . the worst battalion in the United States Army." Men walked around firebases without weapons or helmets, their uniforms in disarray, their faces unshaven. In whipping them into shape, Schwarzkopf developed his trademark style: "I had to be a complete son of a bitch to get any results, which often entailed losing my temper five or six times a day. Being calm and reasonable just didn't work."

Near the end of Schwarzkopf 's second tour, he was hit by shrapnel while trying to save a wounded enlisted man trapped in a minefield. This feat earned him a Silver Star for gallantry (one of many decorations he would win) and established his reputation as an officer who would do anything for his soldiers.

Back home, he found an army that was in "bad shape," but his own career took off like a turbocharged jet. He progressed rapidly from staff jobs, which he hated, to field commands, which he loved. By 1983 he had his dream job as a two-star general commanding the 24th Mechanized Infantry Division at Fort Stewart, Georgia. The 24th Division was designated to be the first heavy U.S. force sent to the Middle East in the event of a conflict. Schwarzkopf prepared the division for the war it would one day wage, and in the process prepared himself for his future command. Just as important was his role as the senior army officer in Operation Urgent Fury, the bungled, if ultimately successful, 1983 invasion of Grenada. He saw the results of interservice rivalry firsthand and found that it cost lives.

Schwarzkopf developed a reputation as an inspirational troop leader but also as someone who lacked diplomatic finesse. He probably would never have made it to four-star rank were it not for the support of his mentor, the army chief of staff, General Carl Vuono, who in 1988 lobbied for him to get the Centcom job.

With the Soviet threat receding, Centcom needed a new raison d'être to replace its previous mission of defending Iranian oil fields from the Red Army. Schwarkzopf developed a war plan designed to counter the hegemonic ambitions of Saddam Hussein, the aggressive Iraqi dictator who had already attacked Iran and made no secret of his desire to get his hands on his neighbors' oil.

Schwarzkopf tested his new strategy in a staff simulation called Internal Look from July 23 to July 28, 1990. As the command exercise proceeded, Saddam Hussein turned up the volume of his threats against Kuwait. Reality began to run in parallel with Centcom's fictional scenario when on August 2 the Iraqi army rolled into Kuwait. Schwarzkopf now had a real war on his hands.

THE GATHERING STORM

The most immediate concern for President George H. W. Bush was that Saddam Hussein would keep going into Saudi Arabia. There was precious little to stop him. The Iraqi army was the fourth-largest in the world, considerably bigger than America's, much less Saudi Arabia's. Saddam Hussein had more than 900,000 soldiers, and they were equipped with the best equipment his oil riches could buy—Soviet T-72 tanks and MiG-29 fighters, South African 155 mm artillery, French Exocet cruise missiles, and F1 Mirage fighters. He was also known to have stockpiles of chemical weapons and more than six hundred Scud-B missiles capable of delivering them. Moreover, Saddam's army was battle-tested: It had just concluded an eight-year war in which it had fought Iran to a standstill and it had overrun Kuwait in less than two days. The hundreds of thousands of troops that Saddam massed in Kuwait and southern Iraq could easily have gone all the way to Riyadh if he had given the command.

After a high-level U.S. delegation won permission from Saudi King Fahd to send U.S. troops to defend his kingdom, Centcom began flowing units into Saudi Arabia, starting on August 7–8 with a squadron of F-15Cs and a lightly armed brigade from the 82nd Airborne Division. By November, a much more formidable force was in place: 60 navy ships, 1,000 aircraft, 250,000 troops, 800 tanks. Schwarzkopf now was able to guarantee the defense of Saudi Arabia, but the White House pressed him to go farther and plan for the expulsion of Iraqi forces from Kuwait.

This Schwarzkopf was reluctant to do. He felt he had adequate capacity for an air offensive but not for a ground attack. The plan he drew up and sent to Washington called for an unimaginative plunge straight into the teeth of Iraqi defenses in southern Kuwait. Wags at the National Security Council called it the "hey diddle diddle, straight up the middle" option. Schwarzkopf made clear that he had little faith in the plan himself. He demanded that at least two more armored divisions be dispatched to supplement the one he already had. Only then would he guarantee the success of an attack.

Schwarzkopf may not have expected Bush to grant his request, but in fact the president gave Schwarzkopf more than he requested. The entire VII Corps, a heavy tank force, arrived from Germany to supplement the XVIII Airborne Corps already in Saudi Arabia. Eventually the total U.S. force would consist of 550,000 personnel, along with 2,000 tanks, 1,990 aircraft, and 100 warships. Additional forces from France, Britain, Egypt, Syria, and various Gulf states brought the overall coalition total to around 800,000 personnel. They would face an entrenched Iraqi force of 550,000 soldiers, 4,280 tanks, 2,880 armored personnel carriers, and 3,100 pieces of artillery deployed in Kuwait and southern Iraq.

While the Iraqis had traveled, at most, only a few hundred miles from their bases, the bulk of coalition troops came from thousands of miles away. Moving all of them and keeping them supplied was a monumental feat of logistics. Lieutenant General William "Gus" Pagonis, Schwarzkopf 's logistics chief, used computers and cell phones along with old-fashioned 3x5 cards to coordinate a vast infrastructure network stretching from the shores of America to the ports and landing fields of Saudi Arabia. Between August 1990 and August 1991, his logisticians moved more than 12,000 tracked vehicles and 117,000 wheeled vehicles. In addition, they served more than 122 million meals, pumped 1.3 billion gallons of fuel, and delivered 70 million pounds of mail. It was as if they had taken a city the size of Atlanta, along with everything needed to keep it running, and moved it eight thousand miles into the middle of a desert.

The capabilities of Pagonis's logisticians turned out to be vital to the final plan for the ground offensive. With some prodding from Pentagon civilians, Schwarzkopf and his planners decided to put the main axis of the offensive deep into the Iraqi desert where the enemy had not prepared any defenses. While a force of U.S. Marines and Arab soldiers would attack straight into Kuwait to fix Iraqi forces in place, a heavy left hook, two full army corps, would swing around from the west and catch the Iraqi army in a pincer movement. It was a recipe for a double envelopment leading to the enemy's utter annihilation—the ideal of all generals since Hannibal's crushing victory over the Roman legions at Cannae in 216 B.C. (Cannae had provided inspiration for the German offensives against France in both 1914 and 1940.)

To execute this encirclement properly over such vast distances would stretch supply lines to their limits. Making the logisticians' job even harder, Schwarzkopf did not want to deploy forces toward the west until the air campaign had disrupted the Iraqis' ability to track U.S. movements. That meant moving the entire VII Corps and the XVIII Airborne Corps—more than 250,000 troops and 64,000 vehicles—hundreds of miles westward in less than fourteen days. And that was just to get them to the starting line. Once the ground offensive began, all of those units would have to be supplied on the move: no easy feat because a single armored division needed 400,000 gallons of fuel, 213,000 gallons of water, and 2,400 tons of ammunition daily.

Schwarzkopf was so doubtful that the supply services could keep up that he made Pagonis swear in writing that he could do it. Pagonis, who had a healthy ego, did not hesitate. "Logisticians will not let you or our soldiers down," he wrote Schwarzkopf on December 29, 1990, and he proved as good as his word.

INSTANT THUNDER

While Schwarzkopf took personal charge of the ground war, he knew little about air warfare and was content to delegate that task to his air component commander, Lieutenant General Chuck Horner. A thickset Iowan with a bulldog face, Horner liked to play the country rube to the casual observer, but he had been an outstanding fighter pilot in Vietnam and had acquired plenty of command savvy in the years since. He was part of a wave of fighter pilots who had broken the bomber barons' hegemony over the air force that had stretched back to the days of Curtis LeMay. By the early 1990s not a single bomber pilot could be found among the ranks of air force four-stars. This was mostly a matter of numbers—there were many more fighters than bombers and hence more fighter pilots than bomber pilots. Once the World War II generation retired, the fighter pilots naturally rose to the fore—a development with important implications. Many of the new breed were contemptuous of bomber purists in the Billy Mitchell mold who made exaggerated claims for strategic air power and disdained attacks against enemy troops. Horner called such people "airheaded airmen."

Much to his chagrin, the initial air plan for Iraq was drawn up by someone who, in his opinion, fit that pejorative description. Colonel John A. Warden III was a noted air power theorist and head of a Pentagon planning cell known as Checkmate. Warden believed that the proper use of air power was to strike five centers of gravity (or "rings"), starting with the enemy leadership—the most important—and proceeding to economic production, infrastructure, population, and, last and least, the enemy's military forces. Just as the air power theorists of the 1930s had exalted the long-range bomber as a super-weapon capable of paralyzing a modern state, so Warden now did the same thing with precision-guided munitions and stealth aircraft.

At Schwarzkopf's request, Warden drew up a preliminary air plan that called for six days of bombing designed to hit eighty-four targets, mainly around Baghdad. He dubbed it Instant Thunder to distinguish it from the ineffectual Rolling Thunder campaign during Vietnam. Warden believed that Instant Thunder would paralyze the Iraqis, obviating the need to attack Iraqi forces in Kuwait. When Horner was briefed on this plan in Riyadh on August 20, 1990, he made no effort to conceal his exasperation. He threw Warden out of his office and put him on the next plane back to Washington, although he kept some of Warden's assistants to help design a new air plan.

What finally emerged was a campaign that would last six weeks, not six days, and that would direct the bulk of air strikes against Iraqi ground forces, not leadership targets in Baghdad. But some of Warden's initial emphasis on the "inner ring" survived (his list of eighty-four strategic targets was expanded to more than four hundred), leading to the mistaken perception in some quarters that he was the author of the air plan that won the Gulf War.

The initial success of the Allies in suppressing enemy air defenses in the first few days of Desert Storm—using tactics modeled on those that the Israeli Air Force had employed to cripple Syrian air defenses in 1982—guaranteed that the rest of the campaign would flow smoothly. Few of Iraq's eight hundred fighter aircraft rose to challenge Allied warplanes, and when they did, they were quickly shot down. Remaining Iraqi aircraft either stayed in hardened hangars, where they were picked off one by one, or fled to Iran.

The only response Iraq was able to muster was to fire some inaccurate Scud-B missiles, not much more advanced than the German V-2, at Saudi Arabia and Israel. Except for one missile that hit a U.S. barracks, killing twenty-eight soldiers and wounding ninety-seven, the damage was negligible, but it did cause some political anxiety for the coalition by raising the possibility that Israel would enter the war and the Arab allies would leave it. Patriot batteries were deployed to defend Saudi Arabia and Israel, which had a calming psychological effect even if later evidence showed that they probably had not shot down any Scuds. More and more coalition forces, ranging from British and American special operations teams to F-15Es, were also dispatched to take out the mobile Scud launchers in western Iraq. Because the Scud launchers could "shoot and scoot" in just six minutes, there were no confirmed kills, but the coalition pressure made launches increasingly infrequent and erratic.

Two weeks into the air war, on the night of January 29–30, 1991, Saddam launched a surprise ground attack from the "heel" of Kuwait into northeastern Saudi Arabia, probably hoping to draw the coalition into a premature ground war that would inflict unacceptable casualties on Western forces. Elements from three Iraqi heavy divisions, with four hundred armored vehicles, managed to mass unnoticed. Part of their attack was blunted by a thin screen of U.S. Marines along the border, but another column advanced eighty miles into Saudi Arabia and occupied the deserted oil town of Ra's al-Khafji. The situation might have been dire were it not for the ability of JSTARS to track enemy movements and direct against them a withering array of air attacks. One captured Iraqi officer later testified that his brigade suffered more in thirty minutes than during the eights years of the Iran-Iraq War. The air barrage allowed Saudi and Qatari forces to stage a counterattack and retake Khafji on February 1.

There was no respite for Iraqi forces once they escaped back to Kuwait. Allied aircraft aimed, first of all, to isolate Iraqi troops in the Kuwait Theater of Operations from resupply and reinforcement. Interdiction strikes on highways, bridges, and railroads proved highly effective. Before long, front-line Iraqi units were experiencing shortages of everything from food to ammunition.

Hitting entrenched forces in the desert was harder. The Iraqis dug their armored vehicles into the sand and camouflaged them effectively. But they had not reckoned with U.S. technology. Metal tanks cooled less rapidly at night than the surrounding desert, making them visible on infrared scopes. Aircraft equipped with infrared systems and laser-guided bombs, mainly Vietnam-era F-111F Aardvarks, were able to destroy more than 1,300 armored vehicles. Iraqi tank crews became so terrified of these air raids that many were afraid to even enter their vehicles.

Their demoralization was heightened by the dropping of more than twenty-eight million leaflets urging Iraqis to surrender, a job undertaken by psychological warfare specialists from the U.S. Special Operations Command. Many of these "bullshit bombs" were skillfully coordinated with real B-52 bombs to terrify Iraqi soldiers. (Saddam tried to counter this propaganda with some of his own: His inept radio broadcasts warned American troops that while they were at the front their wives at home were being seduced by Robert Redford, Sean Penn—and Bart Simpson.)

Although Allied commanders knew they were inflicting severe punishment on the Iraqis, it was hard to know exactly how much. Even with all their vaunted technology, bomb damage assessment remained more an art than a science. It was hard to tell from the air whether a tank had been put out of operation. Schwarzkopf had said that he wanted air power to destroy 50 percent of Iraqi tanks, artillery, and armored personnel carriers before launching his ground offensive. There was no consensus among intelligence analysts about whether that goal was being reached, but by early February the Centcom commander, with a little prodding from Washington, decided to proceed anyway.

No one knew what to expect. Iraqi engineers had constructed an elaborate defensive line of barbed wire, bunkers, trenches, mines, artillery. Even many American officers thought it would take months and cost tens of thousands of casualties to liberate Kuwait. More gloomy commentators warned of a Somme-style quagmire.

THE GATES THAT DID NOT CLOSE

As usual, the main ground offensive was preceded by Special Forces infiltrations and conventional reconnaissance probes. The ground war officially

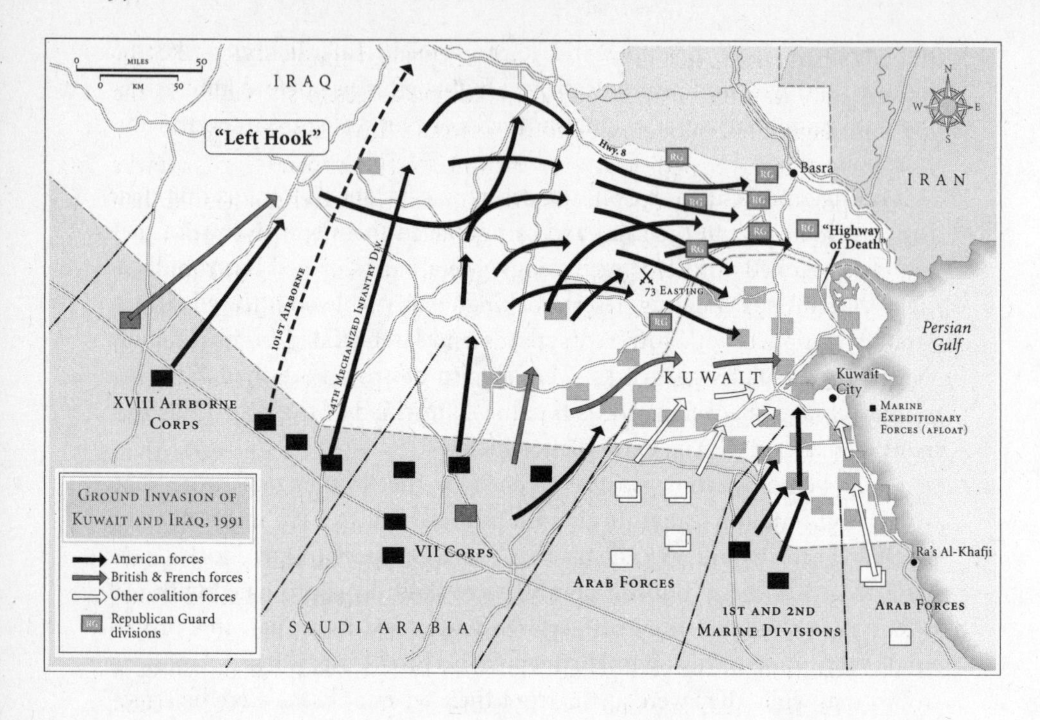

began at 4 A.M. on February 24, 1991, a cloudy, rainy, miserable Sunday morning, when the 1st and 2nd Marine divisions and an army tank brigade—eighty-five thousand troops in all—knifed straight into the heart of Iraqi defenses along the Kuwait–Saudi Arabia border. Iraqi defenders did not set oil-filled trenches on fire. Nor did they unleash chemical weapons. They did fire some artillery, but this was quickly extinguished by return fire directed by counter-battery radars that traced enemy shells back to their point of origin. The frontline Iraqi infantry divisions that were supposed to slow the Allied advance and give their mechanized forces time to counterattack were collapsing.

Three hundred miles to the west, the XVIII Airborne Corps found even fewer obstacles. The 82nd Airborne Division, the 101st Airborne Division, and the French 6th Light Armored Division were supposed to seize bases in western Iraq that could be used as jumping-off points to grab control of the Euphrates River Valley and seal off Iraqi troops in Kuwait from retreat or reinforcement. All the first-day objectives fell faster and easier than expected. Even the 101st's potentially risky air assault ninety-three miles into Iraq—the largest helicopter operation ever attempted—went off virtually without a hitch after being delayed by early morning fog.

Casualties were practically nonexistent among coalition forces on the first day. The biggest problem was dealing with hordes of Iraqi prisoners. Thirty-eight days of air attacks had taken more of a toll than all but the giddiest strategists had anticipated. Many Iraqi conscripts had simply deserted. Others were only waiting for the arrival of coalition forces to surrender. In the next four days, more than eighty thousand would throw up their hands.

The 1st Marine Expeditionary Force and XVIII Airborne Corps attacks were supposed to be first-day diversions. The main Allied thrust would then come the next day from the VII Corps, with more than 1,400 tanks and 146,000 soldiers, which would slam into the Iraqi flank between the two initial thrusts. With the unexpected success of the preliminary attacks, Schwarzkopf decided to move up VII Corps' advance to that very afternoon.

At 2:30 P.M. on Sunday, February 24, VII Corps unleashed one of the most intense artillery barrages in history, utilizing 155 mm howitzers and Multiple Launch Rocket Systems to pulverize Iraqi lines with more than 11,000 shells and rockets in half an hour. At 3 P.M., the rain of steel stopped and the ground advance began, led by the 1st Infantry Division (Mechanized), the "Big Red One" of World War I and II fame. As tanks with plows and rollers cleared paths through the minefields, any Iraqis who did not surrender were buried alive in their trenches by M9 armored bulldozers.

The coalition encountered its first serious resistance on the morning of Monday, February 25. The marines, who did not have enough night-vision goggles, had halted for the night at the edge of the vast Al Burqan oil field, roughly halfway to Kuwait City. The oil wells had been set on fire by Iraqi saboteurs. Masked by heavy smoke, two Iraqi armored brigades massed for a counterattack on the marines' right flank. General Mike Myatt, commander of the 1st Marine Division, was busy directing his riposte when artillery shells began to explode near his own command post. "What the hell is that?" Myatt demanded. It was an attack by Iraqi armored forces, and they were in danger of overrunning the division's mobile headquarters.

The command post's protective detail consisted of one platoon armed with TOW (Tube-launched, Optically-tracked, Wire-guided) antitank missiles and one company of light-armored-vehicles (LAVs) armed with 25 mm chain guns that could fire two hundred rounds a minute. In charge of the defense was a hulking former college football player named Eddie Ray. A product of the South Central Los Angeles ghetto, Ray had enlisted in the Marine Corps once his playing career had been cut short by injuries because, like many African-Americans, he expected a fair deal in the military. He now amply repaid the Corps' confidence in him. Dashing hither and yon in his LAV, Captain Ray directed a desperate defense against a force many times larger than his own. Backed up by Cobra gunships, Ray managed not only to save the command post but also to destroy most of the attackers. Fifty enemy armored vehicles were left in smoking ruins and more than 250 Iraqis were taken prisoner, while Ray's company lost not a man—a feat for which Ray would earn his service's second-highest decoration, the Navy Cross.

Not far away, the rest of the Iraqi attack was stopped by equally determined marines. The Iraqis had managed to delay the coalition advance by a day through the sacrifice of two armored brigades, but now the road to Kuwait City lay open. By the afternoon of Tuesday, February 26, the Kuwaiti capital was completely encircled. The next morning, the first Arab units entered the capital, followed by the marines. Schwarzkopf, who proved adept at coalition management, wanted to give the Arabs the honor of liberating an Arab capital, just as Eisenhower had allowed the Free French forces to liberate Paris in 1944.

Meanwhile, the 24th Mechanized Infantry Division, part of XVIII Airborne Corps, was charging ahead as fast as any tank unit in history. Under the leadership of Major General Barry McCaffrey, one of the army's most gung-ho commanders, the 24th was supposed to reach the Euphrates and then wheel to the east to engage Iraqi forces. By pushing day and night through sandstorms and downpours, McCaffrey managed to reach Highway 8, almost 190 miles from his starting point, by the evening of February 26. The Screaming Eagles of the 101st Division, light infantrymen who moved by helicopter, were already astride this vital artery from Basra to Baghdad. The Iraqis could no longer hope to retreat to the west. The only possible escape route was to the north, along the Tigris River Valley.

The one part of the coalition force that appeared to be plodding behind was VII Corps. Schwarzkopf had told its mild-mannered commander, General Fred Franks, he did not "want a slow, ponderous pachyderm mentality. . . . I want VII Corps to slam into the Republican Guard. The enemy is not worth shit. Go after them with audacity, shock action, and surprise." But Franks, who had lost part of his left leg in Vietnam, knew the risks of precipitate action. He preferred a carefully calibrated advance that would allow all five of his divisions to hit Saddam Hussein's best units, the Republican Guard, at the same time.

Eight Republican Guard divisions had been kept back from the front lines. They were better supplied and better protected than the low-level conscript forces the Americans had run into so far. Despite the pounding they had taken from the air, they still had 75 percent of their armor and they were able to maneuver as cohesive units. They assumed blocking positions in front of VII Corps' advance to allow the bulk of the Iraqi army to escape from Kuwait. This resulted in the heaviest ground fighting of the war.

On Tuesday, February 26, the 2nd Armored Cavalry Regiment, a reconnaissance force at the head of VII Corps' advance, stumbled onto the Tawakalna Republican Guard Division in the desert about fifty miles west of the Kuwait-Iraq border. A blinding sandstorm hindered helicopter operations, but the armored forces kept moving with the help of GPS devices. At

4:15 P.M., Captain H. R. McMaster, a West Point graduate who was as brainy as he was brawny, was in the lead tank of Eagle Troop when he crested a rise and his gunner shouted out, "Tanks direct front!" Eight T-72s were hidden behind a sand berm, making them almost invisible to the naked eye, but McMaster's gunner had spotted them with his thermal sights. McMaster, who had been standing up, peering out of the open hatch, dropped down into the tank and yelled, "Fire, fire Sabot!" As the tank continued to bounce along at 20 mph, the gunner hit the button on his laser range finder. The digital display showed that enemy tanks were 1,420 meters away. In the next instant the M1A1's gyro-stabilized main gun boomed out and the turret of a T-72 flew off its hull in a hail of sparks. It took a crew member only two seconds to slam another depleted-uranium Sabot round into the breech. The ballistics computer instantly calculated crosswinds, temperature, velocity. Seconds later, another Iraqi tank exploded in flames. McMaster's tank nailed a third T-72 before the rest of his troop, with eight more Abrams tanks, caught up and joined the battle. In thirty seconds, eight Iraqi tanks were obliterated at no cost to the Americans.

The other troops of the 2nd Armored Cavalry Regiment were enjoying similar success in their own battles against the Tawakalna Division. By the time the battle of 73 Easting (named after a map grid coordinate) was over at 11 P.M., more than two hundred Iraqi vehicles had been destroyed and an entire Iraqi brigade, along with most of an armored division, had been rendered useless. This was a microcosm of what happened when VII Corps slammed into the Republican Guard that day and the next. It was, in a word, a rout.

The Iraqis had started evacuating Kuwait in the early morning hours of February 26. For the next two days a long line of traffic streamed north out of Kuwait City as Iraqi soldiers took whatever vehicles they could find, loaded them up with whatever they could steal, and tried to get home. They didn't get very far. JSTARS tracked their movements and vectored warplanes to hammer them. It was like going hunting at the zoo. More than 1,900 vehicles were destroyed along the two paved roads out of Kuwait City.

Journalist Michael Kelly visited Highway 6 a few days later and found a catalog of horrors. "One poor soul had tried to make his escape in a Kawasaki front-loader, swiped I guess from some construction site in Kuwait City. He had been sitting up there in the open cab, chugging along at a few miles an hour, when the cluster bomb, or whatever it was, ripped him in half. The left upper side of him was hanging upside down from the yellow frame of the cab, but his whole right side and bottom half had been torn off." Not far away, in the middle of the road, "was a medium-size flatbed truck that had been bombed and roasted. There were ten bodies in and about it, and all of them had been cooked to the point of carbonization, leaving shriveled, naked mummies; black, charcoaled husks with bared rictus grins and hands that had become claws. Their skin was stretched taut and shiny, heatshrunk over their skulls."

Sympathy for the fleeing soldiers must be tempered by knowledge of the atrocities they had committed in Kuwait—raping women, killing babies, releasing oil slicks into the Persian Gulf. In any case, for all the destruction evident on the "Highway of Death," most of the Iraqis who had been in the bombed vehicles escaped alive; coalition pilots did not fire on individuals fleeing on foot.

But senior policy makers in Washington, including General Powell, worried that the violence was turning gratuitous and could sully the U.S. victory. On Wednesday, February 27, Powell called Schwarzkopf and told him that President Bush wanted to announce a cease-fire effective at midnight eastern standard time (8 A.M. Thursday in the Persian Gulf). Schwarzkopf's "gut reaction was that a quick cease-fire would save lives," he recalled. He told Powell, "I don't have any problem with it."

Yet problems would quickly emerge. President Bush made his decision to end the ground war after just one hundred hours—a figure that had a nice ring to it—because he had been assured that no significant Iraqi forces could escape with their equipment. As Schwarzkopf boasted at a press conference on the evening of February 27, "The gates are closed . . . on their military machine."

Not quite. The Iraqis managed to get about a quarter of their tanks and half of their armored personnel carriers out of Kuwait, largely because the marines had advanced so rapidly that they had pushed many Iraqi units out of Kuwait before VII Corps could arrive to cut them off. Field commanders knew that the Cannae-style envelopment was not yet complete, but no one with sufficient authority was willing to challenge the autocratic Schwarzkopf, and he in turn was not willing to challenge the politically powerful Powell. The cease-fire became a fait accompli even though U.S. forces had not achieved Schwarzkopf's stated goal of "destroy[ing] the Republican Guard."

Schwarzkopf compounded this mistake at the ceasefire talks on March 3. Although the coalition prohibited Saddam Hussein from operating warplanes, Schwarzkopf acceded to an Iraqi request that they be permitted to fly armed helicopters, ostensibly to ferry senior officials around the country. Within days, Iraqi gunships and tanks were being used to suppress rebellions among Kurds and Shiites who, at American instigation, had risen against Saddam's tyranny. The subsequent slaughter would bolster the power of the Baathist regime and tarnish America's reputation.

THE DRUBBING IN THE DESERT

Among military officers and experts, debate still rages about whether the Gulf War represented a revolutionary advance in the art of war. Skeptics argue that the clash of tank armies in the desert did not look all that different, at least outwardly, from similar battles in World War II or the Arab-Israeli wars. A good deal of the equipment employed by Allied forces was far from revolutionary: The U.S. Army depended for short-range communications on forty-year-old FM radios that didn't work very well; commanders still marked troop movements on laminated maps with grease pencils; few troops had access to high-tech gadgets like GPS transponders; 92 percent of munitions were "dumb" bombs only a bit more accurate than those used in World War II; and even smart bombs did not find their targets every time.

Moreover, for all their vaunted intelligence capabilities, the coalition forces made several serious mistakes. The Allies were not able to track down Saddam's Scuds, much less Saddam himself. And, near the end of the conflict, senior American policy makers "did not have," in the words of National Security Adviser Brent Scowcroft, "a clear picture of exactly what was happening on the ground." Even more embarrassing was the number of American deaths caused by American weapons. A quarter of the 147 U.S. combat fatalities were the result of "friendly fire." The Information Revolution, suffice it to say, had not dispelled the fog of war.

Perhaps the biggest cause for skepticism about whether the Gulf War was truly "transformative" is the simple fact that there was no surprise about the outcome. Who could doubt that a superpower, backed up by medium-sized powers like Egypt, Saudi Arabia, France, and Britain, would defeat a relatively impoverished nation of twenty-five million people?

Yet, if the outcome was never in doubt, the manner in which it was achieved was. Before the war, eminent experts had confidently predicted that the coalition would suffer tens of thousands of casualties. Based on computer models of past conflicts, the Pentagon stockpiled 16,000 body bags. In the event, the coalition lost only 240 soldiers killed in action and 776 wounded out of a total force of 795,000. (The U.S., with 550,000 troops in theater, lost 147 killed in action and 467 wounded.) That is a staggeringly low fatality rate of 0.03 percent—a tenth of the per-capita losses suffered by Israel in the Six-Day War of 1967 and a twentieth of the losses suffered by Germany in the invasion of France in 1940, the two previous operations that were considered the gold standard of twentieth-century military excellence. U.S. aircraft losses, at one-twentieth of 1 percent of all sorties, were 42 percent lower than in Vietnam and 95 percent lower than in World War II. Statistics show that American men between the ages of twenty and thirty were actually safer on average fighting in Desert Storm than staying at home in the U.S. The coalition victory, according to the U.S. Army's official history, came "at the lowest cost in human life ever recorded for a conflict of such magnitude."

What accounted for this unprecedented outcome? Not simply numerical superiority. At the start of Desert Storm, the Allies had some 620,000 ground troops vs. an estimated 550,000 Iraqis. By the time the ground offensive began, the Allies enjoyed a substantial advantage because the Iraqi force had been depleted through death and desertion to no more than 350,000. Yet there are many examples in history of similarly outnumbered defenders inflicting much heavier casualties on their attackers. The coalition lacked the three-to-one advantage traditionally considered necessary for attacking forces to be certain of victory.

The Iraqis' failure to fight better had much to do with the low quality of their personnel. Most Iraqi officers showed themselves to be plodding, unimaginative, and overly reliant on direction from the top. This was hardly surprising, given that in any totalitarian regime individual initiative is discouraged as a potential threat to the dictator's absolute authority. While this centralized system allowed Saddam Hussein to keep control, it did not make for an effective fighting force. The Iraqis, like other victims of blitzkrieg (the French in 1940, the Egyptians and Syrians in 1967, the Pakistanis in 1971), were locked into rigid plans that called for set-piece engagements and could not be altered fast enough to adjust to a rapidly changing battlefield.

The Iraqis planned on a war of attrition. What they got was a war of maneuver executed by a force that, in the best Moltkean tradition, tried to push decision-making down to the lowest levels, so that, for instance, Captains H. R. McMaster and Eddie Ray could use their personal initiative during the battles of 73 Easting and Al Burqan. (On the other hand, the war's unsatisfying conclusion was due to the failure of military and civilian leaders in Washington and Riyadh to adjust quickly enough to changing realities on the ground.)

The disparity in quality was even greater between enlisted personnel. The U.S. troops, volunteers all, were professionals with esprit de corps to match their high level of competence. The average Iraqi soldier was a badly trained and unmotivated conscript with little love for the brutal regime in Baghdad. Many were quite happy to surrender after a few weeks of bombardment.

Yet the Iraqis did not simply refuse to resist. Most regular armored divisions and all of the Republican Guard fought tenaciously at battles like 73 Easting. Even after the mauling they received, many units were able to withdraw in good order and then go into action to put down Shiite and Kurdish uprisings. Nor were the Iraqis completely incompetent. Senior Iraqi commanders made some shrewd decisions, such as maneuvering the Republican Guard to block VII Corps' advance, thereby allowing the remnants of the Iraqi forces to escape. The Iraqis likely would have been able to put up a much more credible defense were it not for the fact that they faced an enemy of unsurpassed sophistication.

The average U.S. weapon used in Desert Storm was introduced in 1973, while for the Iraqis the comparable year was 1961. This twelve-year gap is much wider than in most wars fought by modern, industrialized states. The qualitative disparity was even wider than this figure would suggest because a handful of the most advanced U.S. weapons systems had an impact out of all proportion to their numbers. There were only two JSTARS, but they made Allied attacks on Iraqi ground forces much more effective, especially at Khafji and along the "Highway of Death." F-117A stealth fighters flew only 2 percent of all attack sorties, but they hit 40 percent of Iraq's best-defended targets. And just a few thousand GPS receivers allowed Schwarzkopf's "left hook" to take place.

The Iraqis were aware of some of these Allied capabilities and therefore took care to avoid radio or telephone communications that could be intercepted. But the impact of other American advances was magnified because they came as a surprise. Not realizing that coalition forces could navigate their western desert, for instance, the Iraqis left it virtually undefended. Not realizing that coalition sensors could locate targets even through walls of sand, the Iraqis buried their tanks in the desert where they could be picked off by laser-guided bombs. And not realizing that coalition air strikes could target forces on the move, the Iraqis failed to erect air defenses to protect their logistics lines.

The worst Iraqi mistake of all was fighting the coalition in the desert, exactly the kind of conflict the U.S. armed forces had spent decades training for. In retrospect, Saddam Hussein would have been well advised to remove his forces from Kuwait before the onset of hostilities. That he did not do so was due to monumental hubris: He thought he could inflict upon the Americans the kind of losses he had meted out to the Iranians and thereby break the American will to fight. He did not seem to realize that the Iranians had an Industrial Age military and a relatively low-quality one at that, whereas the Americans were developing an Information Age military built around smart weapons and smart people. Although the transformation was far from complete in 1991, the combination of U.S. prowess and Iraqi miscalculation produced a historic drubbing in the desert.

THE MAKING OF A HYPERPOWER

The political consequences of the Gulf War were profound. The Soviet Union collapsed that year in part because its leaders found themselves unable to keep up with American advances in commercial and military technology, leaving America standing alone in the first rank of states. The natural predominance Washington would have enjoyed in any case was greatly magnified by the combat prowess exhibited by the U.S. armed forces in the Gulf War. After the conflict, wrote George H. W. Bush and Brent Scowcroft, "American political credibility and influence had skyrocketed."

This seemingly easy victory washed the bitter taste of Vietnam from the mouths of U.S. policy makers and made the use of force much more palatable once again. During the next decade, U.S. troops would be dispatched to northern Iraq (to protect the Kurds), Somalia (to deliver relief supplies), Haiti (to restore democracy), and Bosnia and Kosovo (to stop ethnic cleansing). The U.S. would also launch punitive air strikes against Iraq, Sudan, and Afghanistan. However well justified, these were wars of choice, not necessity—conflicts undertaken in large measure because the costs of action appeared so low. They call to mind the "liberal imperialism" practiced by Britain and other European states a century earlier. Just as the Europeans in the nineteenth century were willing to act even when they didn't have to, simply because of the huge advantage they had derived from the Industrial Revolution over all potential adversaries in the developing world, so now the U.S. was willing to act because of the huge advantage it had derived from the Information Revolution. The U.S. found it could smite its foes even while its defense budget and the overall size of its armed forces declined after the end of the Cold War.

These small wars proved particularly seductive because few if any U.S. personnel were killed—and when the U.S. did lose soldiers, as in Somalia in 1993, it simply pulled out. The corollary to this aversion to casualties was that the U.S. shied away from messy conflicts in which American ground forces would have been in danger of suffering serious losses. Thus, despite repeated plots against American targets by the al Qaeda network, the U.S. government did not dispatch a military expedition to Afghanistan to defeat the terrorists. Like nineteenth-century Britain, the U.S. had achieved "world domination on the cheap," and it was not willing to pay too high a price in American blood to maintain its hegemony.

In any case, the threats to American power did not loom large during the go-go years of the 1990s. Following its victories in the Cold War and the Gulf War, the U.S. could concentrate on the pursuit of prosperity. U.S. military predominance helped lay the foundation for a global economic expansion, just as British naval predominance in the nineteenth century had underwritten a previous era of "globalization."

But beneath the surface of a placid decade, tension was building. Even many of America's allies came to be wary of this new "hyperpower," as the U.S. was labeled by a French foreign minister in the 1990s. Resentment grew in Europe because the members of the European Union, for all their wealth (their collective GDP was nearly equal to America's), did not generate much in the way of military power. This left them completely dependent on American protection to deal with crises even on their own continent, putting fresh strains on the transatlantic alliance.

Even more dangerous resentment was growing in the Middle East where, following the Gulf War, the U.S. became the most important geopolitical player. This made Uncle Sam the inevitable focus of frustrations felt by a generation of young men trapped in stagnant dictatorships that offered little in the way of social, political, intellectual, or economic opportunity. Many zealots intoxicated by religious extremism came to believe that the way to topple their "near enemy"—hated regimes in Riyadh, Cairo, and Amman, to say nothing of Jerusalem—was to attack the "far enemy" that supported these governments: the United States of America. Yet they could not hope to compete against the U.S. on a conventional battlefield. The more astute among them came to understand that the Information Age offered new opportunities for what strategists call "asymmetric warfare"—the ability to inflict great damage on a powerful adversary by using unconventional weapons.

Like commercial airliners.

CHAPTER 11

SPECIAL FORCES AND HORSES:

Afghanistan, October 7–December 6, 2001

The MH-47E Chinook touched down at Landing Zone Albatross at 2:00 A.M. on Friday, October 19, 2001, halfway around the world from New York City, where the remains of the World Trade Center were still smoldering. It had been a harrowing "infil," or infiltration. The long flight from Karshi Kanabad air base in southern Uzbekistan had forced the lumbering transport helicopter to navigate the towering peaks of the Hindu Kush at night and in a blinding sandstorm. With zero visibility, two MH-60L Black Hawks that had been providing armed escort had been forced to turn back. But knowing the importance of this mission, the pilots from the 160th Special Operations Aviation Regiment (the "Night Stalkers") kept pushing their bird deeper into enemy-held territory, relying on synthetic aperture radar (which uses microwaves to generate a picture of the landscape) and infrared sensors to avoid the numerous obstacles in their paths. The pilots waited with dread for the high-pitched crack of enemy fire. It did not come. The helicopter, bigger than a Greyhound bus, safely reached its destination in the Daria-Suf Valley of northern Afghanistan where the Special Forces team on board was due to rendezvous with the anti-Taliban fighters of the Northern Alliance.

Hundreds of miles to the south, in the Panjshir Valley, another Special Forces team was landing at about the same time to join another faction of the Northern Alliance.

U.S. aircraft had been bombing Afghanistan for twelve days, but, apart from some preliminary reconnaissance work, these were the first U.S. military units on the ground. Rather than attempt a conventional invasion of Afghanistan that risked bogging down like the Soviets and the British who had come before, President George W. Bush had decided on a bold strategy that would rely on massive amounts of U.S. air power and small numbers of U.S. commandos to strengthen the Northern Alliance, which had been fighting the fundamentalist fanatics of the Taliban for years with scant results. Large numbers of conventional American troops would be sent in only if the Northern Alliance failed.

At the core of this strategy were the Army Special Forces, popularly known as the Green Berets (they prefer to be called "the quiet professionals" or simply "operators"), organized into twelve-man A-teams. Each Operational Detachment-Alpha (ODA) is led by a young captain and composed of senior noncommissioned officers who are experts in weapons, combat engineering, intelligence, medicine, and communications. Each ODA specializes in a particular region of the world; the ones who were now entering Afghanistan came from the 5th Special Forces Group, based at Fort Campbell, Kentucky. They routinely operated in the Middle East and Central Asia and spoke the languages of those areas, though they were unfamiliar with Afghanistan itself.

As well trained as these men were, they had only the sketchiest intelligence available in the days after September 11 when they were rushed into secure isolation facilities and told to prepare for a mission against the perpetrators of the deadliest assault ever on American soil. One team asked for a file on the Afghan warlord they were supposed to meet, Attah Mohammed, and was instead given a briefing on the dead hijacker Mohammed Attah. They did not know what to expect once they had penetrated the land that was known as the "graveyard of empires," where the British had been stymied in the nineteenth century and the Soviets in the twentieth.

As soon as the Chinook's rear ramp descended, a dozen operators from ODA 595, each weighed down by more than one hundred pounds of gear, sprinted through the dust cloud kicked up by the helicopter's twin rotors and assumed prone fighting positions. Approaching them out of the dark was a group of Afghans wearing turbans and armed with AK-47s and rocket-propelled grenades. Fingers tensed on the triggers of M-4 carbines. "It was like the sand people from 'Star Wars' coming at you," the team's warrant officer said. Luckily, the strangers turned out to be friendly. They were representatives of the Northern Alliance and their job was to guide the new arrivals to a small compound of mud brick buildings where they bedded down in a cattle stable.

At 9 A.M. the next morning a contingent of twenty horsemen bristling with weapons galloped into camp. A few minutes behind them came thirty more cavalrymen escorting General Abdul Rashid Dostum, one of the leading lights of the anti-Taliban movement. Described by one journalist as "a burly figure with short, spiky salt-and-pepper hair that comes down low above his brow, giving him the appearance of an irritable bear," Dostum was an ethnic-Uzbek warlord with a seventh-grade education who had been fighting for various sides during the long years of war that had convulsed Afghanistan. In the 1980s he had served with the Soviet-backed army against the mujahideen rebels. After the fall of the communist regime in 1992, he switched his allegiance to the mujahideen leader Ahmad Shah Massoud and helped him capture Kabul. The warlords' rule was notorious for its cruelty and criminality and led to the rise of the Taliban, a group of madrassah students who set out to impose a harsh brand of Islamic law. Dostum and his allies were chased out of Kabul by the Taliban and relegated to the northern corner of the country, where they had been fighting the regime ever since.

So far the anti-Taliban Northern Alliance had made little progress toward overthrowing Afghanistan's new rulers. They controlled less than 15 percent of the country's territory, and their fifteen thousand or so fighters were unable to make any headway against the far more numerous and better armed Taliban forces, which stood at least forty thousand strong, counting their legions of foreign (mainly Arab) volunteers. The rebel alliance was in disarray after the assassination of Massoud two days before 9/11 by al Qaeda suicide bombers posing as journalists. Dostum was eager to see if the Americans could finally break the deadlock.

The Uzbek warlord requested that some of the newly arrived commandos proceed with him to his headquarters, where they could plot an immediate offensive. While part of ODA 595 stayed behind to coordinate air supply drops for Dostum's impoverished forces, six operators led by Captain Mark Nutsch, the team leader, rode back with Dostum on tough little ponies. Raised on a cattle farm in Kansas and having competed in collegiate rodeos, the sandy-haired Nutsch was a crack rider. But most of his men had never been on a horse before. It had been more than eighty years since U.S. soldiers had galloped into battle. Now they found themselves learning horse-handling skills while traversing treacherous mountain paths three feet wide, knowing that one slip would result in annihilation. "It was pretty painful," a master sergeant recalled of the six-hour trek. "They use simple wooden saddles covered with a piece of carpet, and short stirrups that put our knees up by our heads. The first words I wanted to learn in Dari were, 'How do you make him stop?' "

Two days after landing, on Sunday, October 21, ODA 595 launched its first attacks in coordination with Dostum using weapons beyond the imagination of any previous generation of cavalrymen. Employing their GPS receivers, the team identified the positions of faraway Taliban bunkers and radioed the coordinates to a B-52 bomber twenty thousand feet overhead. Only a white contrail was visible in the blue sky when satellite-directed bombs began raining down. Those initial munitions missed their targets, because Dostum, fearing for his allies' safety, would not let the Americans get close enough to the front lines to make accurate observations. But Dostum was happy. The presence of American bombs was a big boost to his men's morale and a big blow to the Taliban. He picked up a walkie-talkie tuned to the enemy frequency and told the opposing Taliban commander, "This is General Dostum speaking. I am here, and I have brought the Americans with me."

It was not yet clear, however, what the new arrivals could accomplish. Could a force of a few dozen commandos supported by a few hundred aircraft really turn the tide in a war that had been stalemated for years? Not even U.S. commanders were overly optimistic. They expected that it would take many months of preparation before the Northern Alliance could mount a major offensive in the spring and that Kabul might not fall for at least a year. "They thought they'd let the Special Forces go in and play around for a few months," recalled one officer assigned to Central Command, "and then the real fight would occur when the 101st and the 82nd Airborne arrived."

The lack of results so far was already leading some in the press to invoke the dread specter of Vietnam. Creating "another Vietnam" was the fondest wish of Osama bin Laden and the other leaders of al Qaeda. By attacking New York and Washington, they hoped to draw the world's sole superpower into a quagmire that would result in a victory for the Afghan holy warriors to match their triumph over the Soviet Union in the 1980s. They had not reckoned on the vast differences between the Soviet armed forces, circa 1980, and the U.S. armed forces, circa 2001. The former fought to subjugate Afghanistan, the latter to liberate it. As important, the former was a low-tech relic of the Industrial Age whereas the latter was attempting to remake itself for the dawning Information Age. That transformation would be showcased in Afghanistan.

WAGING NETWAR

It was not only the Americans who were taking advantage of the most advanced computer technology. So were their foes. Al Qaeda sought to promote its brand of religious extremism by employing the most sophisticated tools produced by the secular world it reviled. Its leaders were university-educated professionals: Osama bin Laden was an engineer; his deputy, Ayman al-Zawahiri, a physician. They communicated with one another and with the outside world using cellular and satellite phones, fax machines, e-mail, electronic instant messaging, Web sites, satellite television, CD-ROM disks, and audio- and videotapes. The terrorists even employed an office manager in Kabul who, according to one journalist, "maintained the computer's files in a meticulous network of folders and subfolders that neatly laid out the group's organizational structure and strategic concerns."

Al Qaeda was an organization optimized for what Rand Corporation analysts John Arquilla and David Ronfeldt have dubbed *netwar*. This does not refer only to what happens online but to any kind of conflict waged by a networked, decentralized organization. Al Qaeda's diffuse structure, loosely linking together terrorist cells from Abu Sayyaf in the Philippines to Islamic Jihad in Egypt, made it far different from Nazi Germany, the Soviet Union, and other hierarchical foes previously faced by the United States. Al Qaeda was unable to amass anything close to the power of those nation states, but neither did it present as obvious a target for retaliation. Bin Laden's organization effectively operated within the seams and cracks of an increasingly globalized world to exploit the vulnerabilities of its adversaries. U.S. Central Command came to refer to its adversary as the "Virtual Caliphate."

Advanced technology is not necessary for waging netwar; sometimes a clever guerrilla fighter can employ simple means like runners and drum codes to control his forces, as the Somali warlord Mohamed Farah Aidid showed in 1993–94 while eluding capture by the U.S. military. But the fruits of the Information Age make such a fighter much more lethal and allow him to extend his operations across the entire world, as Osama bin Laden did so successfully. Political scientist Rajan Menon has compared al Qaeda to General Motors. Both, he suggests, are multinational organizations with worldwide reach that "rely on high-speed transportation, computerized global banking networks and information systems that enable instantaneous communication and the marshaling of financial and human resources." (A cynic might add that al Qaeda has been more successful in recent years than GM.)

The United States government, of course, had access to far more powerful computers than the IBM or Compaq machines purchased by al Qaeda. Al Qaeda spent an estimated $30 million a year on its operations prior to 9/11. The United States spent more than $30 *billion* annually on intelligence and another $350 billion on its armed forces. Yet the U.S. government was not organized to make optimal use of its resources. As we have seen, taking advantage of a technological revolution requires major changes in organization, training, and procedures. Those changes had not yet spread throughout the U.S. government. In 2001, most federal agencies were still organized along the hierarchical lines of the Industrial Age, not on the networked lines of the Information Age, and they were still focused on old-fashioned threats from enemy states, not from global terrorist organizations. There was little communication between the FBI and CIA or even within different parts of the same agencies. At the FBI, for instance, intelligence analysts and criminal investigators were forbidden from sharing information. Both sides of the FBI relied on outdated, 1980s-vintage computers that used dial-up modems and required, according to one journalist's account, "eleven keystrokes to complete a search."

Thirty miles north of the FBI's headquarters in downtown Washington, the National Security Agency (NSA) downloaded vast quantities of information from its worldwide network of listening stations, but there were not enough Arabic-speaking analysts to sift through all the material and figure out what it meant. The explosion of communications in the Information Age was overwhelming the eavesdroppers' ability to keep up. And more and more information was traveling over fiber-optic lines, the Internet, and digital cellular phones, all of which proved harder to intercept than traditional analog calls and faxes sent over microwave relays and satellites. The problem was compounded by the NSA's byzantine bureaucracy, with senior managers becoming virtual "warlords" who would not share information with one another.

A game of electronic cat and mouse developed akin to that played by the Germans and the Allies during the Battle of the Atlantic in World War II. U.S. investigators figured out how to track al Qaeda operatives based on their cell phones and even how to target missiles on their transmissions. Al Qaeda operatives became aware of this vulnerability due to a press leak and stopped using cell phones, turning instead to couriers or to coded e-mails and Web postings. Al Qaeda, writes an American intelligence analyst, "had some of the best operational security anywhere." Penetrating such a secretive organization requires spies, not just electronic gadgets, but the CIA was notoriously weak in "humint" (human intelligence).

The U.S. and its allies did have occasional successes against al Qaeda, for instance foiling a plan to bomb multiple airliners simultaneously over the Pacific in 1995 and the millennium plot to bomb Jordanian hotels and the Los Angeles airport in 1999. But, by and large, the West was fighting a losing battle against a ruthless terrorist network that had progressed from success to success.

AL QAEDA'S RISE

Osama bin Laden had been active in channeling foreign fighters and funding to help Afghan mujahideen fighting Soviet occupation in the 1980s.

When the Russians announced in 1988 that they were pulling out, bin Laden and a Palestinian mullah, Abdullah Azzam, decided to establish al Qaeda (The Base) as a headquarters for future jihad. After a falling-out between the founders, Azzam was murdered the following year, leaving bin Laden as the sole leader. His ire turned against the United States when American troops came to Saudi Arabia during the Gulf War. The seventeenth son of a Saudi construction magnate, bin Laden was subsequently exiled from his homeland for plotting against the Saudi monarchy, which he believed was in cahoots with the "Great Satan." Thereafter he established his headquarters in Sudan, which had a sympathetic Islamist government. After Sudan expelled him under international pressure in 1996, he made his way back to Afghanistan, where he became a leading bankroller of the Taliban, who in return offered him the run of their country. The U.S. government later estimated that over ten thousand terrorists were trained in the Afghan camps set up by bin Laden.

Bin Laden and his associates were linked to a wide variety of terrorist attacks, including the 1993 bombing of the World Trade Center (six dead), the 1993 attacks on U.S. forces in Somalia (eighteen dead), and bombings of U.S. military compounds in Saudi Arabia in 1995 and 1996, which together killed twenty-four Americans. In 1996 bin Laden issued a declaration of "War Against the Americans Occupying the Land of the Two Holy Places," a reference to Mecca and Medina in Saudi Arabia. In 1998 he issued another *fatwa* claiming that it was the "individual duty for every Muslim" to kill "Americans and their allies—civilians and military."

Few in the United States took this threat seriously. But later that year the Saudi exile showed that he meant what he said. On August 7, 1998, explosives-laden trucks struck the U.S. embassies in Kenya and Tanzania, killing 224 people, including twelve Americans. In October 2000 came the bombing of the USS *Cole* in the harbor of Aden, Yemen, which killed seventeen sailors. One of the most sophisticated warships in the world, a $900 million *Arleigh Burke*–class guided-missile destroyer, was almost sunk by a cheap speedboat loaded with explosives.

The U.S. response to these blows was weak, dilatory, and ineffectual, epitomized by the cruise missiles fired against Sudan and Afghanistan in 1998. The Clinton administration, and the Bush administration in its first months in office, was not prepared to order a ground offensive that risked American casualties to confront al Qaeda's growing threat. Yet, in the absence of "actionable," real-time intelligence, precision weaponry could do little against such an elusive foe. As a result bin Laden was able to carry his operations to new heights of destructiveness.

The original idea for the 9/11 plot—hijacking airliners and crashing them into prominent American buildings—came from Khalid Sheikh

Mohammed, a native of Pakistan who grew up in Kuwait, received a mechanical engineering degree from North Carolina A&T University, and fought against the Russians in Afghanistan. His scheme was an ingenious recipe for a poor man's smart missile: The guidance would be provided by men, not machines. By early 1999 he had received a green light from bin Laden to mount the "planes operation."

He proceeded methodically, buying flight simulator software, renting movies on hijackings, surfing the Internet to research flight schools in the U.S. Mohammed decided to hijack planes on long flights because they would be full of fuel that would add to the force of the explosion, and to seize Boeing aircraft because they would be easier to fly than those made by Airbus. He recruited his key co-conspirators from an Islamist cell in Hamburg whose members were familiar with Western ways and conversant in English. The remaining thirteen "muscle" hijackers, almost all Saudis, had lesser levels of education and sophistication. The plane plotters reached the U.S. in the summer of 2001 and, like any professional military unit, began to undertake careful surveillance of their objectives and launch trial runs of their operations.

The cell members left plenty of clues that, if put together, would have alerted the U.S. government to their existence. In July 2001, for instance, an FBI agent in Phoenix told his superiors that Middle Eastern men suspected of terrorist ties were enrolling in flight schools. The following month the FBI's Minneapolis field office arrested Zacarias Moussaoui, a French citizen of Moroccan descent with a jihadist background who wanted to learn how to pilot a jumbo jet. But no one at the FBI put two and two together. The CIA was not even alerted to the clues gathered by the FBI. The 9/11 Commission later concluded: "The September 11 attacks fell into the void between the foreign and domestic threats. The foreign intelligence agencies were watching overseas, alert to foreign threats to U.S. interests there. The domestic agencies were waiting for evidence of a domestic threat from sleeper cells within the United States. No one was looking for a foreign threat to domestic targets."

That threat materialized with unexpected and devastating force on the morning of Tuesday, September 11, 2001. Four teams of hijackers boarded four commercial airliners—two at Boston's Logan Airport, one at Washington Dulles, one at Newark. A few minutes after takeoff, the hijackers grabbed control of each of the jets using box cutters they had been allowed to bring on board. At 8:46 A.M. American Airlines flight 11 crashed into the North Tower of the World Trade Center in downtown New York. At 9:03 United Airlines flight 175 hit the South Tower. At 9:37 American Airlines flight 77 plowed into the Pentagon.

By this time, the North American Aerospace Defense Command had scrambled four Air National Guard fighter aircraft (two F-15s from Falmouth, Massachusetts, two F-16s from Langley, Virginia). It was too little, too late. An air defense system designed to deal with enemy bombers simply could not handle this new threat. If it had been left to the military, the 9/11 Commission later concluded, the fourth hijacked airplane might well have hit its target—either the Capitol or the White House. That fate was averted by the brave passengers of United Airlines flight 93, whose uprising led their aircraft to crash in an empty field near Shanksville, Pennsylvania, at 10:02 A.M.

Heroic rescue measures allowed most of those inside the World Trade Center to escape before the collapse of the Twin Towers, but the 9/11 attacks nevertheless resulted in the death of 2,973 people—23 percent greater than the fatalities on December 7, 1941. The economic toll on New York City alone was estimated at over $80 billion. At a cost of less than $500,000, and equipped with nothing more than box cutters and one-way airline tickets, nineteen men had inflicted more damage on the United States than the Imperial Japanese Navy had managed with dozens of warships, hundreds of aircraft, and thousands of sailors.

There was nothing new about religious hatred or terrorism or even about religious terrorists willing to sacrifice their own lives to eradicate the objects of their wrath. All were as old as recorded history. What was new was the ability of a few fanatics to inflict damage on such a massive scale so far from home. When the British were fighting the Mahdists in the Sudan, they never feared that their enemies would seek vengeance in London. In the late 1890s the Sudan was simply too cut off from the rest of the world to make such long-range operations conceivable. A century later, the world had been so tightly knit together with modern communications and transportation technology that al Qaeda and its ilk could wage their insurgency from one end of the planet to another. They could even send their operatives from one of the world's poorest and most primitive countries to wreak havoc in the richest and most powerful nation in history. To respond to this brutal attack, the United States would have to prove itself capable of waging its own brand of netwar.

PREDATORS ON THE PROWL

Increasing U.S. capabilities in this regard was a top priority for Secretary of Defense Donald H. Rumsfeld. He took over the Pentagon for a second time (his first stint had been in 1975–77 under President Gerald R. Ford) determined to transform the military to better respond to the threats, and take advantage of the opportunities of, the Information Age. On September 10,

2001, he had delivered a prescient speech in which he declared, "We must change for a simple reason—the world has—and we have not yet changed sufficiently. The clearest and most important transformation is from a bipolar Cold War world where threats were visible and predictable, to one in which they arise from multiple sources, most of which are difficult to anticipate, and many of which are impossible even to know today."

Rumsfeld was distressed to find that, despite the changing nature of these threats, much of the military still resembled the force that had faced off against the Soviet Union, with large tank divisions sitting in Germany waiting for an enemy that no longer existed. But the armed forces had also changed in the past decade. The biggest and most visible change was a decrease in size; manpower was cut by a third, from two million active-duty personnel in 1991 to 1.4 million in 2001. Defense budgets also shrank in the 1990s as politicians rushed to spend the "peace dividend" from the Cold War's end. While smaller, the military was becoming more lethal largely through the utilization of information technology.

One of the most important advances of the past decade was the development of the Joint Direct Attack Munition (JDAM). Less than 10 percent of the ordnance used in the Gulf War had been precision-guided, partly because of the high cost of these systems: Tomahawk cruise missiles initially cost $1.4 million apiece. Paveway III laser-guided bombs were cheaper, at only $55,000 each, but they could be dropped only by certain specially equipped aircraft and only under the right weather conditions. In 1995, the U.S. Navy and Air Force began manufacturing cheap tail kits that for only $20,000 could turn any iron bomb into a smart weapon—a JDAM that would use GPS signals to steer itself within thirty-three feet of a programmed aim point from up to fifteen miles away. While not quite as accurate as laser-guided bombs, JDAMs were not impaired by clouds or dust storms that would deflect laser beams. And they could be employed by virtually any airplane. Previously smart bombs had been delivered only by low-capacity aircraft like the F-16 or F-111. Now they could be spewed out by bomb trucks like the B-52 or B-1 from high altitudes far beyond the range of enemy air defenses. For the first time, heavy bombers could be employed to provide close support to ground troops without undue risk of friendly casualties.

JDAMs were used initially in 1999 in NATO's campaign to stop ethnic cleansing in the Serbian province of Kosovo. The percentage of precision munitions dropped by U.S. forces during that war jumped to 29 percent of the total. Although ultimately successful, the Kosovo War revealed some of the shortcomings of American technology: The Serbs were able to camouflage much of their military hardware to prevent it from being destroyed. Overhead U.S. sensors were capable of directing pinpoint strikes against buildings in Belgrade, but they had trouble finding military targets in the field, allowing the Serbian army to move much of its equipment out of Kosovo intact at the end of seventy-eight days of bombing. Making the most of precision munitions still required the presence of human spotters on the ground.

But while people remained essential on the ground, they were becoming less important in the air. The use of unmanned aerial vehicles (UAVs) had expanded considerably since the Gulf War. The 1995 NATO military operation over Bosnia saw the first use of the RQ-1A Predator, a slender UAV with an inverted-V tail that made it look like a flying meat fork. With a top speed of just 120 knots (140 mph), the Predator was slower than some sports cars, but it was capable of flying twenty-four-hour missions at up to 26,000 feet with a payload of over 450 pounds. Initially it carried only surveillance equipment—synthetic aperture radar that could see through clouds, infrared sensors that could see at night, and electro-optical sensors that could shoot video. The Predator was piloted remotely by an operator manipulating a joystick in a ground station, and it beamed its pictures back via satellite uplinks. Using a Predator, commanders in Washington or Tampa could watch real-time events unfolding on battlefields half a world away, albeit without a soundtrack. Predator downlinks also became available to AC-130 gunships, JSTARs, and some other aircraft.

After the Kosovo War, the Predator was given an added payload—a laser pod that could be used to designate targets for destruction. Just before 9/11, the Predator became the first UAV with its own offensive capability. Two Hellfire laser-guided antitank missiles were added to some Predators as part of a secret project between the Pentagon and CIA designed to assassinate al Qaeda leaders. The armed Predators, which were controlled by the CIA, were rushed into service in Afghanistan.

Another experimental UAV would also be employed—the RQ-4 Global Hawk. Though not capable of firing its own missiles, the jet-powered Global Hawk had higher resolution cameras than the turboprop-driven Predator and could fly higher (up to 65,000 feet), farther (14,000 nautical miles), and longer (thirty-six hours at a time) while carrying a bigger sensor payload (up to 2,000 pounds). While the Predator could provide a close-up view of one part of the battlefield, Global Hawk offered a more panoramic perspective (it could monitor an area the size of Illinois) that supplemented the work of satellites and U-2 spy planes. "It's like a low-earth orbit satellite that's present all the time," one air force general marveled.

JDAMs, Predators, Global Hawks, and various other UAVs were important, but perhaps the most momentous change since 1991 was less visible—the digital integration of these and many other systems into a war-fighting network that spanned the globe. At its heart was a series of high-tech command posts that resembled the bridge of the *Starship Enterprise* on *Star Trek*.

The overall operation in Afghanistan would be run by Central Command from its headquarters at McDill Air Base in Tampa, Florida. Here, behind cipher-locked steel doors protected by gun-toting sentries, stood the Joint Operations Center. This windowless facility was manned at all times by thirty-two soldiers sitting elbow-to-elbow at computer terminals. Bundles of fiber-optic cables snaked across the floor. On the wall were plasma screens on which commanders could watch live imagery from Afghanistan, plot the location of all warships in the Arabian sea, or simply keep track of what Fox News Channel was reporting. The Combined Air Operations Center at Prince Sultan Air Base in Saudi Arabia, which would run the air war in Afghanistan, had a similar setup, relying on as many as 100 T-1 fiber-optic lines, each one capable of transmitting 192,000 bytes of information per second. All these facilities, as well as many field units, were connected via the Pentagon's secure version of the Internet, the Siprnet (Secret Internet Protocol Router Network). Senior commanders could also talk to one another by encrypted STU III (Secure Telephone Unit) or by secure video-conferencing.

What this meant in practical terms was that the U.S. military had a much greater degree of flexibility and nimbleness than it had a decade earlier. During the Gulf War the process of determining bombing targets generally took seventy-two hours and involved flying an Air Tasking Order from Lieutenant General Chuck Horner's Riyadh headquarters to the navy's aircraft carriers. Now directives could be sent with the push of a button. In fact, because aircraft were connected directly to the information grid, targets no longer had to be determined in advance. For the first time in aviation history, it became commonplace for pilots to take off with only a vague idea of where they were headed and then wait for instructions about where to drop their munitions.

Most ordinary soldiers were not yet plugged into this broadband network, but the Special Forces were, well, special. In addition to getting training not given to regular army grunts, they were able to buy Iridium satellite telephones and Panasonic Toughbook laptop computers with satellite modems. By 2001 they could remain connected to the Siprnet even while crawling around caves in Afghanistan.

"As the Information Age matured in the civilian world, the concept of digital, three-dimensional 'battlespace' became a reality," wrote Centcom's commander, General Tommy Franks. "Plasma screens replaced paper maps in computerized command posts and higher-echelon headquarters. Bandwidth to receive and deliver data was becoming an asset as vital as ammunition and fuel." The result, he continued, with understandable hyperbole, was that "Commanders were acquiring the ability to pierce the 'fog and friction of war,' that evocative metaphor frequently attributed to Clausewitz. . . . The revolution in sensor technology, coupled with flying observation platforms (many mounted on new Unmanned Aerial Vehicles), promised today's commanders the kind of Olympian perspective that Homer had given his gods."

Yet Rumsfeld was not satisfied. The impatient defense secretary felt, with some justification, that the Defense Department was an entrenched bureaucracy resistant to change—"one of the world's last bastions of central planning," he called the Pentagon. When he had left government in 1977 after fourteen years in the legislative and executive branches, he had gone into the corporate world. As chief executive officer of G. D. Searle and Co. he had turned around the pharmaceutical firm by marketing products like Metamucil fiber supplement and NutraSweet artificial sweetener. He then sold the company and went on to become CEO of General Instrument Corp. He saw the changes in business that had occurred over the past two decades, and he did not feel that the Pentagon was keeping pace with "the technology revolution [that] has transformed organizations across the private sector."

Brimming with self-confidence, which a growing number of detractors saw as arrogance, Rumsfeld was not shy about confronting senior generals and admirals and senior lawmakers on Capitol Hill. From his first months in office, he referred to the "Pentagon bureaucracy" as an "adversary that poses a threat, a serious threat, to the security of the United States of America." The bureaucracy fought back so effectively that the consensus in Washington was that Rumsfeld would be the first member of President Bush's cabinet to leave office. On September 7, 2001, the *Washington Post* reported, "The sweepstakes have already begun on who might succeed Secretary of Defense Donald Rumsfeld if and when he steps down."

Four days later everything changed. The 9/11 attacks, which killed so many people, gave Rumsfeld a fresh lease on life—and not only because his wartime press briefings established him as a tough-talking television star adept at sparring with reporters. As Rumsfeld quickly discovered, the military had no off-the-shelf plans for invading Afghanistan, only for firing cruise missiles and dropping bombs, which could not guarantee decisive results. The difficult nature of the problem—rooting out a global terrorist network headquartered in a mountainous, landlocked country of twenty-five million people located far from any U.S. base—almost dictated an unconventional, or "transformational," solution.

TOMMY'S WAR

It was apparently CIA Director George Tenet who first came up with the idea of focusing the Afghan campaign on CIA and Special Forces operators working hand-in-glove with the Northern Alliance. He presented the idea to Bush and his war cabinet at Camp David on the weekend of September 15–16. The president liked what he heard. It was left to General Tommy Franks and his staff to figure out how to put these suggestions into practice. There was little in the Centcom commander's background to suggest he would be amenable to such an unorthodox approach—and yet he was, if only because he had no alternative.

Tall, craggy, and plainspoken, with big ears and a hangdog expression, Tommy Franks had a reputation as a good ol' boy. "Franks," a friend of his said, "is not a whiner. He is not a pontificator. He is not seeking self-glory." He was also not, in the words of another officer who dealt with him, "a deep thinker." Nor, unlike his Centcom predecessor, Norman Schwarzkopf, was Franks a West Pointer who had been groomed for a military career since childhood.

While Schwarzkopf had grown up the son of a general, the man who adopted Franks had been an air force mechanic in World War II who after the war left the military to take a variety of poor-paying civilian jobs. Tommy grew up dirt-poor in small-town Oklahoma and Texas. After graduating from the same high school in Midland, Texas, as future first lady Laura Bush, he entered the University of Texas at Austin intending to be a chemical engineer. Instead he spent most of his time partying with his fraternity brothers at Delta Upsilon. He flunked out after two years and joined the army as a private. After a year as an enlisted man, he passed the exams to enter Officer Candidate School, graduating in 1967 as a second lieutenant of field artillery. He then spent a year in Vietnam performing a variety of jobs ranging from artillery observer to scout. It was in the jungles and rice paddies that Franks, who later described himself as a "spoiled, unfocused and immature" young man, really grew up. He repeatedly volunteered for dangerous assignments and paid the price, earning three Purple Hearts to go along with his Bronze Star.

Upon his return to the United States he completed a bachelor's degree at the University of Texas, Arlington, and embarked on the numerous stops necessary for a rising young army officer. Like other military men of his generation, he was seared not only by the experience of combat in Vietnam but also by the post-combat trauma inflicted on an army where, as he later recalled, "near-mutinous ill-discipline was rife in some outfits." But, unlike many of his peers, he stuck with the army. By the time of the Gulf War he was a brigadier general and assistant commander of the 1st Cavalry Division when it was dispatched to Saudi Arabia.

After the Gulf War, the army chief of staff, General Gordon Sullivan, assigned him to run a task force to figure out how to digitize the army, making it stronger even as it was getting smaller. Franks pushed for, among other things, the purchase of digital simulators that would allow troops to train on systems like the Abrams tank and Apache helicopter in a "virtual" environment without the expense of live-fire exercises.

Following a stint as commander of 3rd Army, a headquarters that oversaw the land forces component of Centcom, Franks was recommended for the top job by the outgoing Centcom commander, Marine General Anthony Zinni. Franks was appointed in 2000 by President Clinton and Secretary of Defense William Cohen. He soon found himself working for a very different boss. He found Rumsfeld to be a "harsh taskmaster" and a "genetically impatient" one. He and the secretary had their share of blowups. Although he lacked Schwarzkopf's flamboyance, Franks was as prone to profane temper tantrums as was his Centcom predecessor. (In his memoir, Franks recounts one incident where he cussed out some members of the Joint Chiefs of Staff for offering him "gratuitous advice" that he thought was inspired by interservice rivalries. "I had no tolerance for this parochial bullshit," he fumed.)

Eventually, Franks and Rumsfeld found a modus vivendi. Franks discovered that Rumsfeld respected someone who stood up to him and did not cower before his intimidating interrogations. Their alliance became battle-tested in the weeks after 9/11 when Franks had to develop and execute a battle plan for what became known as Operation Enduring Freedom. Rumsfeld liked what Franks came up with, though he constantly challenged various assumptions and always pushed for speedier action. Of the defense secretary, Franks wrote, "He not only thought outside the box—he didn't recognize that the box existed." But not even an iconoclastic defense secretary could escape certain inexorable requirements.

Before a single sortie could be flown over Afghanistan, air access rights had to be negotiated with states ranging from Saudi Arabia to Pakistan. Emergency search-and-rescue personnel had to be pre-positioned in Pakistan and Uzbekistan to rescue any downed pilots. Predator drones had to be moved into bases in those neighboring countries to conduct surveillance. Four aircraft carriers had to be shifted into position in the region. (One of them, the USS *Kitty Hawk*, was stripped of most of its fixed-wing contingent and employed as a floating Special Operations base.) Air Force squadrons had to be flown into position at air bases in Kuwait and Oman. And—although this was not a top priority for this administration—allies had to be consulted and coordinated with; the British and Australians, among others, had crack commando units that would prove useful.

The bombardment of Afghanistan finally began on Sunday, October 7, 2001, focusing on the Taliban's handful of command centers and antiaircraft defenses. Along with the bombs came drops of humanitarian relief supplies to Afghan civilians that would eventually total more than two million daily rations. U.S. aircraft had the run of the sky from the start, but they had trouble finding worthwhile targets to hit in a country without any major industry or even many paved roads. "Coming up with a target list was having us all scratching our heads," confessed Franks's deputy, Marine Lieutenant General Michael "Rifle" DeLong. "We had to put our hopes in 'targets of opportunity,' hitting the enemy when they moved, especially when they fled the early air campaign."

In order to keep collateral damage to a minimum, Franks kept a tight leash on the bombing attacks. Targets that involved a high risk of harm to civilians or to a mosque had to be personally cleared by the four-star general, who checked with Centcom's lawyers and often wound up asking Rumsfeld for permission. Rumsfeld, in turn, sometimes turned to the president for a ruling. This led to scenes of Franks watching Predator video and agonizing over whether to take a shot at a convoy suspected of containing Taliban or al Qaeda leaders. It was subsequently claimed that on one occasion a Predator missed a chance to kill Mullah Muhammad Omar because of Franks's caution; whether the Taliban leader was actually in one the vehicles spotted by the Predator may never be known. Al Qaeda's senior military commander, Mohammed Atef, was killed, however, along with a number of other al Qaeda fighters, in a strike that involved F-18 fighters dropping bombs and an MQ-1 Predator firing Hellfires. (These drones would wind up firing 115 missiles in Afghanistan and laser-designating 525 targets.)

This punctiliousness about targeting was an outgrowth of the spread of precision-guided munitions: Now that bombs could be aimed so accurately, any civilian casualties became a major scandal. In the Gulf War, a hit on a Baghdad bunker full of civilians had caused international outrage; in the Kosovo War, it had been a hit on the Chinese embassy in Belgrade. There were several such mishaps in Afghanistan as bombs fell on residential areas of Kabul, a Red Cross warehouse, and a wedding party. Incidents that would have been regarded as routine in 1945 were now front-page news worldwide. Global indignation was one of the few things capable of stopping American air power; therefore, America's enemies, from Slobodan Milosevic to Mullah Omar, took care to trumpet any civilian casualties, whether real or not. "The technology itself invited the sort of political and legal scrutiny that airmen had so railed against in Vietnam," notes journalist Stephen Budiansky.

The early bombing of Afghanistan had little impact in part because of the commanders' caution: They were intent on preventing any unnecessary casualties among American airmen as well as among Afghan civilians. Aircraft were ordered to remain at high altitudes out of range of man-portable antiaircraft missiles such as SA-7s and Stingers. It was, however, hard to identify moving targets from twenty thousand feet. The solution was to have spotters on the ground calling in air strikes, but initially confusion reigned at Central Command and Special Operations Command about which units would be assigned, where they would go, and precisely what they would do.

While the military planned, the CIA acted. The first Americans on the ground belonged to a ten-man CIA team, codenamed Jawbreaker, which landed in the Panjshir Valley north of Kabul on September 26 to rendezvous with Northern Alliance leader Mohammed Fahim. In a scenario that sounded as if it had come out of Hollywood, Jawbreaker's leader was an old clandestine operations hand named Gary Schroen, who was summoned for one last hurrah just days away from retirement. He was given melodramatic instructions by counterterrorism chief Cofer Black: "I don't want bin Laden and his thugs captured, I want them dead. . . . I want bin Laden's head shipped back in a box filled with dry ice. I want to be able to show bin Laden's head to the president."

Jawbreaker brought a vital asset for this task: three cardboard boxes packed with $3 million in hundred-dollar bills that would be useful for buying allies. (Schroen noted that "that amount . . . is bulky and heavy and does not fit into one of those small metal briefcases carried by the bad guys in the movies.") But Jawbreaker had neither the equipment nor the expertise to direct the military strikes that would be necessary to eliminate al Qaeda. That was the responsibility of the Special Operations Command. Colonel John Mulholland, head of 5th Special Forces Group, who as commander of Task Force Dagger had responsibility for Special Forces in northern Afghanistan, was working frantically to get his own people into the country, but it was not easy.

Simply preparing a jumping-off point at the old Soviet air base at Karshi Kanabad (K2), one hundred miles north of the Afghan border, was a major undertaking. The logisticians who first arrived found decrepit buildings with peeling paint and shattered windows; broken-down old Soviet equipment littered the polluted landscape. Virtually everything that U.S. forces would need had to be flown in from Europe. By October 6, a C-17 Globemaster III transport aircraft was landing every two hours at K2, increasing the base's population in a single week from one hundred to two thousand personnel. Soldiers from the 112th Special Operations Signal Battalion installed a super-high-frequency satellite hub, telephone switches, secure videoconferencing equipment, and over two hundred computers. The 528th Special Operations Support Battalion took care of everything else, from digging latrines and distributing mail to setting up a mess hall where hot food could be served.

While all this was going on, Rumsfeld kept calling Franks, demanding, "When is something going to happen, General?"

On the night of October 19–20 something dramatic did happen. One hundred ninety-nine Rangers staged a low-level parachute assault from MC-130 Combat Talon aircraft based in Pakistan to seize a desert landing strip near Kandahar in southern Afghanistan. At the same time, a string of MH-47 Chinooks based on the aircraft carrier *Kitty Hawk* ferried a covert direct-action unit popularly known as Delta Force (technically, 1st Special Forces Operational Detachment–Delta) to seize Mullah Omar's compound in Kandahar. Both strike teams left within a few hours of their arrival and neither one reaped much of an intelligence windfall. But Franks was happy. The next day, the Pentagon released video of the Rangers' mission. This not only satisfied Americans' thirst for tangible ground action, but also diverted the Taliban's attention from more important operations occurring hundreds of miles to the north.

For October 19–20 was also the night when, far from the glare of the world's cameras, the first two Special Forces A-teams landed to rendezvous with the Northern Alliance.

MAZAR-E SHARIF: "LASER, LASER, LASER"

Special Forces operations at first centered around the capture of Mazar-e Sharif, Afghanistan's second-largest city and the center of the northern part of the country. The Northern Alliance controlled territory south of the city in the towering Hindu Kush mountains. Its principal leaders in this region were Abdul Rashid Dostum, the Uzbek militia leader, and Attah Mohammed, who led a force composed mainly of Tajiks. On Saturday, October 20, as we have seen, ODA 595 had linked up with Dostum about sixty-eight miles south of Mazar-e Sharif in the Daria Suf Valley. Thirteen days later and twenty-five miles to the west, ODA 534 landed in the early morning hours of Friday, November 2, in the snowy Daria Balkh Valley to join Attah Mohammed. Both warlords also received small contingents of CIA officers who had better language and cultural skills than their military counterparts and focused on gathering intelligence. In addition, Dostum was assigned a Special Forces lieutenant colonel who acted as his personal adviser.

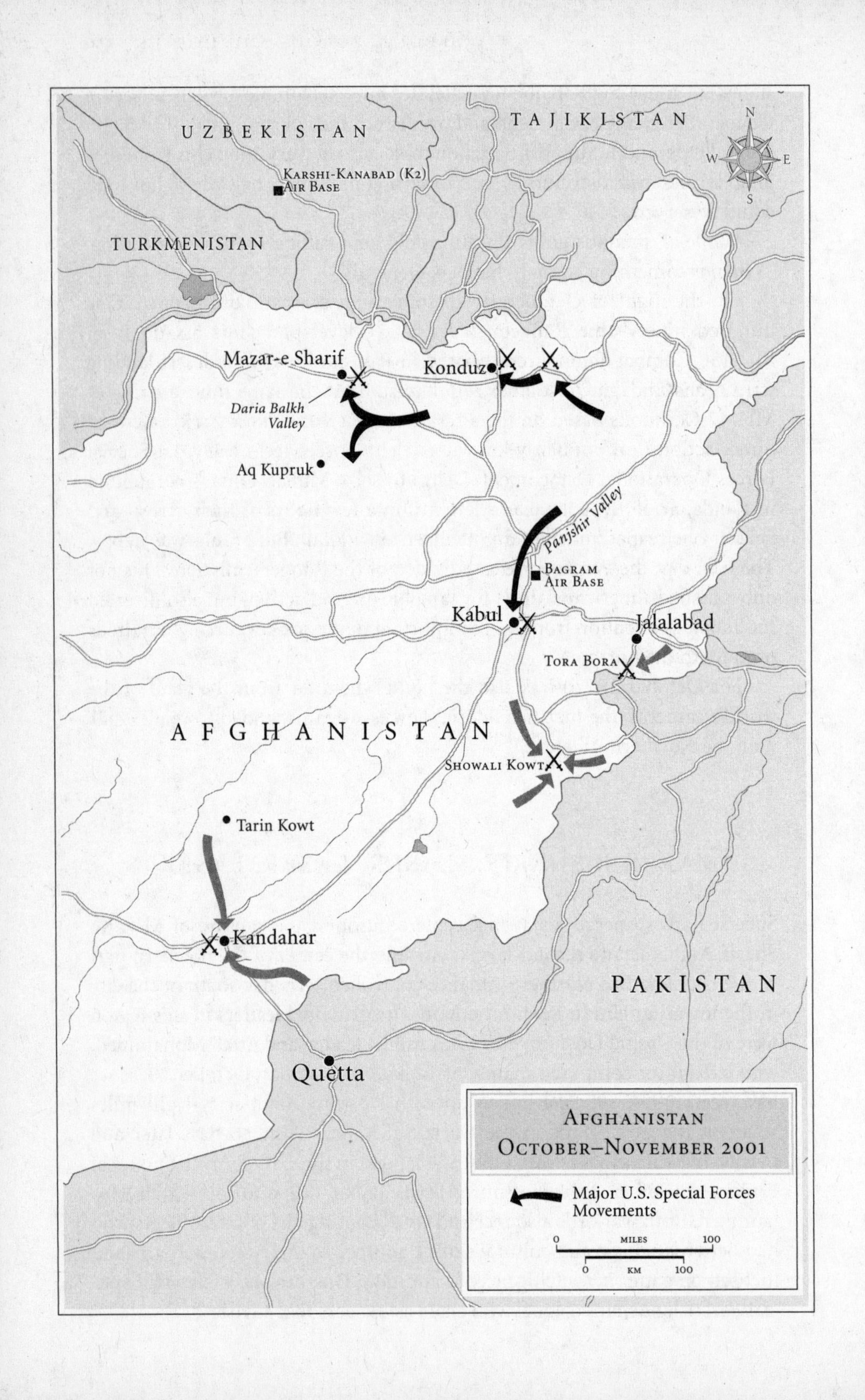
UZBEKISTAN
TAJIKISTAN
N
W
E
S
KARSHI-KANABAD (K2)
AIR BASE
TURKMENISTAN
Mazar-e Sharif
Konduz
Daria Balkh
Valley
Aq Kupruk
Panjshir Valley
BAGRAM
AIR BASE
Kabul
Jalalabad
TORA BORA
AFGHANISTAN
SHOWALI KOWT
Tarin Kowt
Kandahar
PAKISTAN
Quetta
AFGHANISTAN
OCTOBER–NOVEMBER 2001
Major U.S. Special Forces
Movements
0
MILES
100
0
KM
100

What the Americans learned from their Afghan hosts was dismaying. Taliban forces not only had far more men, but they also had artillery, armored cars, and tanks, while the Northern Alliance had no weapons heavier than light machine guns. The Taliban had cars and trucks; the Northern Alliance only horses. The Northern Alliance fighters were dedicated but woefully under equipped. The Green Berets were shocked to see militiamen walking through the snow without socks, their feet clad only in plastic flip-flops, while the Americans were shivering in their heavy mountain gear. Even worse, two-thirds of the Northern Alliance fighters did not have guns.

The Special Forces operators immediately radioed for supplies, ranging from guns and ammunition to boots and blankets. MC-130 Combat Talon cargo aircraft flew fourteen-hour round-trip missions from Incirlik Air Force Base in Turkey to deliver what was required. Airdrops of rations, bundles of cash, and, on one occasion, even a Sony PlayStation requested by a tribal chieftain, were also used to buy entrée with the Northern Alliance. The A-teams won further goodwill by offering free medical care to wounded fighters. But the most important thing the Americans could offer was air power to aid the Northern Alliance's offensive.

The plan was for Dostum and Attah to advance in parallel through the river valleys which met at the Tangi Gap twenty-five miles south of Mazar-e Sharif. For Dostum and ODA 595, the key obstacle would be the Taliban-held village of Bai Beche. For Attah and ODA 534, it would be the Taliban-held village of Aq Kupruk. To maximize their effectiveness, both A-teams split into smaller elements that spread out on horseback to direct air strikes against Taliban forces. To blend in, they grew beards and sported eclectic outfits combining elements of U.S. Army uniforms with articles of Afghan clothing. On several occasions, the small Green Beret detachments were on the verge of getting wiped out by Taliban attacks; only air power saved them. "Frankly," Captain Mark Nutsch of ODA 595 wrote in a dispatch to headquarters, "I am surprised that we have not been slaughtered."

The Americans' stature grew in the eyes of their Afghan allies when they demonstrated the efficacy of their air power. While attacking Aq Kupruk, one of the Green Berets let General Attah look at a target through a tripod-mounted laser designator. "He actually watched the round go in. Attah saw the target go away—a direct impact in real time," recalled Captain Dean, the commander of ODA 534, who does not want his last name used. "And after that all he wanted was 'laser, laser, laser.'"

Such pinpoint raids began to strip away the Taliban's advantage by systematically destroying their heavy weapons, blowing up their command centers, and demoralizing their troops.

Impressive as smart bombs were, they were not enough to dislodge a determined enemy from entrenched positions. The Taliban defenders at Aq

Kupruk and Bai Beche repulsed initial Northern Alliance assaults even after a day or two of bombing. Victory still required the close coordination of air and ground action, just as it did during the German blitzkrieg through France in 1940. Only here the ground element would come from ponies, not panzers.

At Bai Beche the crucial breakthrough occurred by accident. A Green Beret told one of Dostum's lieutenants to get his horses ready for action while they got aircraft into position. This was misinterpreted as a signal to charge. The men of ODA 595 watched in disbelief as 250 horsemen galloped straight at a Taliban position a mile away that was about to be bombed. They were convinced that a "friendly fire" catastrophe was about to occur. No one would ever have intentionally ordered a cavalry charge in such close proximity with an air strike. But it worked out better than anyone could have expected. One of the Green Berets recalled: "Three or four bombs hit right in the middle of the enemy position. Almost immediately after the bombs exploded, the horses swept across the objective—the enemy was so shell-shocked. I could see the horses blasting out the other side. It was the finest sight I ever saw. The men were thrilled; they were so happy. It wasn't done perfectly, but it will never be forgotten."

This key breakthrough in the Daria Suf Valley occurred on Monday, November 5. The next day Attah's forces took Aq Kupruk and broke through in the Daria Balkh Valley. Taliban and al Qaeda forces were now in headlong retreat toward the Tangi Gap, with the Northern Alliance in hot pursuit. Attah and Dostum, ODAs 534 and 595, combined forces to push through this treacherous terrain. The Taliban had mined the area around this mountain pass and positioned rocket launchers and artillery to defend their positions. This chokepoint was shattered by B-52 strikes, allowing the Northern Alliance to pass through on Friday, November 9. Now nothing stood between them and Mazar-e Sharif.

The Northern Alliance forces, accompanied by their American advisers, entered the city the next day. "Entering Mazar-e Sharif was like a scene out of a World War II movie," recalled one member of ODA 534. "The streets, the roadsides . . . were just lined with people cheering and clapping their hands and just celebrations everywhere. It was unlike anything we'd ever seen, other than maybe on a movie screen."

Most of the remaining defenders fled, heading east to Konduz or south to Kabul. But some six hundred al Qaeda and Taliban "dead-enders" would not surrender, instead barricading themselves in a girls' school located next to the city's famed Blue Mosque. ODA 595 called in JDAMs that demolished the school and ended the resistance without damaging the mosque, albeit at the cost of killing some hostages and eleven of Attah's men who, unbeknownst to the Green Berets, were inside negotiating for the hostages' release.

The fighting around Mazar-e Sharif was not quite over. On Sunday, November 25, some six hundred Taliban and al Qaeda prisoners penned up in the ancient mud-brick fortress of Qala-i-Jangi on the outskirts of the city rose up and overpowered their Northern Alliance captors. Two CIA officers were caught inside; one of them, Johnny "Mike" Spann, was killed, while the other escaped to summon help. A motley collection of coalition forces responded to the uprising, including eight members of the British Special Boat Service, the staff of a nearby U.S. Army Special Forces battalion headquarters, and a platoon from the 10th Mountain Division that was rushed in from K2. They immediately called in air strikes. Unfortunately a Navy F-18 pilot entered the wrong coordinates for a JDAM. The two-thousand-pound bomb fell two hundred yards short, wounding nine coalition soldiers who had to be evacuated.

The prison was finally retaken with the aid of two AC-130H Spectre gunships. Although slow and ungainly, these four-engine turboprop airplanes came equipped with a blistering array of ground attack weapons, including five cannons and a 25 mm GAU-12 Gatling gun capable of spitting out 1,800 rounds a minute. Its precise electronic-targeting gear allowed an AC-130 to keep pounding one spot on the ground as it circled overhead. On the night of Monday, November 26, two AC-130s emptied their entire magazines into the prison. One of their last rounds hit a munitions depot, setting off a fireworks show that lasted for hours. Hundreds of rebels were killed. The remainder took refuge in a Soviet-built underground bunker that proved impervious to air strikes. Dostum's men finally flushed them out on Thursday, November 29, by flooding their hiding hole with frigid water. Fewer than one hundred Taliban survived the four-day battle, including an American named John Walker Lindh.

Major Kurt Sonntag, a Special Forces officer, found a scene of utter "carnage" when he entered the prison. "It was pretty bad," he said. "The whole fort was just peppered. Every tree was stripped of all its bark and leaves. All the walls were hammered. All the vehicles were riddled. Every bit of livestock was dead. And there were bodies and pieces of bodies everywhere."

THE ANGELS OF DEATH

The uprising at Qala-i-Jangi, which the Special Forces later concluded had not been a spontaneous occurrence but had been planned to coincide with a Taliban counterattack on Mazar-e Sharif, was one of the last spasms of resistance in the north. The other major cities—Bamiyan, Taluqan, Konduz, Herat—fell in short order to other Northern Alliance forces assisted by Special

Forces teams. While al Qaeda Arabs had nowhere else to go and usually fought to the death, Afghan Taliban fighters, many of them impressed into service, began surrendering en masse once they were convinced of the superiority of American firepower.

The unerring accuracy and awesome destructiveness of American bombs had a powerful psychological effect on the Afghans. They had never seen anything like it. Rumors began to spread that the Americans had a "death ray." While Dostum was negotiating the surrender of Taliban forces in Konduz, he pulled out a radio so that the enemy commander, Mohammad Fazal, could hear a female weapons officer aboard an AC-130 that was turning lazy loops overhead. "Dostum explained that we have the angel of death overhead and that we possess the death ray. If they don't surrender now all their troops will burn in hell," a Green Beret recalled. "Fazal jumped on the radio and his men were surrendering within minutes."

The war in Afghanistan was very different for the men and women pouring all this firepower out of the sky from the one experienced by the operators on the ground. Pilots, whether at bases in the rear or on aircraft carriers, had always enjoyed easier living than infantrymen slogging through the mud. But in past wars, from World War I to Vietnam, they had paid the price by experiencing a casualty rate that rivaled those of the "ground pounders." The primacy of American air power in the post-Vietnam era meant that aviators could now inflict devastating punishment at virtually no risk to themselves. Air-to-air combat was all but extinct and most ground-to-air defenses could be easily suppressed or evaded. Not a single pilot of a fixed-wing aircraft was killed or even wounded during the initial campaign over Afghanistan.

The biggest challenge for most aviators was simply staying alert on such long flights, which often required the ingestion of amphetamine "go" pills. Navy F-14s and F-18s flew five- to eight-hour missions from their carriers seven hundred miles away in the northern Arabian Sea. Air Force B-52s and B-1s flew fifteen-hour missions from the British island of Diego Garcia in the Indian Ocean. B-2 Stealth bombers flew forty-four-hour missions from Whiteman Air Force Base in Missouri. And, most amazing of all, F-15E Strike Eagles flew the longest combat fighter missions in history: ten- to fifteen-hour sorties from Al Jaber air base in Kuwait, requiring multiple in-flight refuelings from KC-10 or KC-135 tankers to cover the three-thousand-mile round-trip.

In each of these two-person strike aircraft the pilot and weapons officer sat atop hard ejection seats without a cushion beneath a Plexiglas canopy so constricting that they could not raise their arms far above their heads. Deep-vein thrombosis—the bane of long-haul airplane travelers—was a real danger. To compensate for these discomforts, crews generally flew only once every three or four days. In between missions, they lived in air-conditioned mobile trailers with cable TV, dined on steak and lobster, surfed the Internet, and drove into Kuwait City to shop or dine. It was a far cry from the austere conditions experienced by the Green Berets, who ate packaged rations (MREs, or Meals Ready to Eat) and bedded down with a blanket in the freezing Hindu Kush.

Yet these two sets of warriors from different services needed each other to accomplish their missions. The unsung heroes of the Afghan campaign were air force combat controllers and terminal attack controllers who were assigned to the A-teams on the ground to serve as liaisons with the aircraft overhead. (Green Berets also called in some air strikes themselves.) Using laser designators, GPS locators, binoculars, and satellite radios, the air force enlisted men directed pilots onto their targets to earn the ultimate accolade—"Shack," slang for a direct hit.

KABUL'S COLLAPSE

While their positions were collapsing in the north, the Taliban remained entrenched around Kabul. The CIA's Jawbreaker team and ODA 555 (the "Triple Nickel") had been operating for weeks out of Bagram airfield, a Northern Alliance outpost in the Panjshir Valley only thirty-five miles north of the capital. The Green Berets and an air force combat controller mounted the airfield's decrepit two-story control tower, its glass long gone, its walls pockmarked with bullet holes, to call in air strikes on Taliban front lines located just a few hundred yards away.

They developed a cunning method of recalibrating their bombing: Following an air strike, a Northern Alliance fighter would get on his walkie-talkie and, pretending to be one of the Taliban, ask the enemy soldiers across the way, "Hey, are you OK? What happened?" One of the Taliban might reply, "Yes, we're fine, but that bomb blew up a car next to our house." Then, recalls Sergeant First Class Scott Zastrow, with a laugh, "We'd call the aircraft and say, OK, take out the house next to the burning car."

Eventually the Taliban caught on and targeted the exposed control tower, the tallest building in the entire area, with their mortars, rockets, and tanks. Zastrow, a wiry, wisecracking medic, recalls that his team had some "hairy moments." Once a tank round slammed into the base of the structure and a huge fireball roared through the control tower. The flames and heat barely passed over the heads of the Americans, who hit the floor just in time. Although some team members wanted to abandon the tower, they decided it was too good a vantage point to give up. For three solid weeks the Triple

Nickel kept on calling in round-the-clock air strikes in an attempt to level the odds between their fifteen hundred Northern Alliance allies and the fifteen thousand Taliban on the other side.

The Special Forces operators and their CIA counterparts were frustrated because they could never get as many bombers as they wanted. The Bush administration did not want to risk shattering Taliban positions in front of Kabul before a provisional government was in place to take over the capital. Many officials in the State Department and CIA feared a repeat of the atrocities that had occurred the last time the Northern Alliance had occupied Kabul. It was not until Sunday, November 11, forty-four days after Jawbreaker's arrival and after repeated pleading from the Americans on the ground, that the Panjshir Valley finally received heavy air strikes, including B-52s whose two-thousand-pound bombs shook the earth.

Taliban defenses quickly crumbled. On Tuesday, November 13, General Mohammed Fahim's forces, accompanied by the Green Berets sporting long beards and full native regalia, entered Kabul to a tumultuous welcome. The next day, the Americans drove to the abandoned U.S. embassy building, which they found surprisingly well preserved, for an emotional flag-raising. "To us," Zastrow said, "that was more significant than taking Kabul."

The remaining Taliban forces either surrendered or fled south to their last remaining stronghold, Kandahar.

WITH KARZAI TO KANDAHAR

The U.S. did not have many allies in southern Afghanistan, a heavily Pashtun region that formed the core of the Taliban's support. The best bet turned out to be Hamid Karzai, a former deputy foreign minister who was a committed liberal and a fluent English speaker (an American officer quipped that Karzai spoke English "better than 50 percent of [my] guys") but no warlord. Unlike Dostum or Fahim, Karzai did not command a sizable militia. He first returned to Afghanistan from exile in Pakistan on October 8 riding a motorcycle accompanied by just two aides. Before long he met heavy Taliban resistance and had to be rescued by a U.S. helicopter, which took him back to Pakistan. Abdul Haq, another Pashtun rebel leader who tried to infiltrate southern Afghanistan, was caught and executed on October 26. When Karzai returned via helicopter on Wednesday, November 14, the day after Kabul's fall, he brought more help in the form of ODA 574 accompanied by a couple of Delta Force operatives and seven CIA officers. The A-Team was led by Captain Jason Amerine, a tall, lean, intense thirty-year-old West Point graduate who spoke Arabic but none of the local languages.

Shortly after their arrival, Karzai got word by satellite phone that Tarin Kowt, the capital of Uruzgan province, had risen up against the Taliban. Karzai rushed there with the Americans, accompanied by about thirty local men who had joined them. They arrived on the night of Friday, November 16, to learn that a Taliban convoy containing hundreds of soldiers was on its way from Kandahar to retake the village. Amerine had only a "weak force" to defend Tarin Kowt. He set up his men on a ridge outside of town that overlooked the mountain pass through which he figured the Taliban would come.

Just after dawn on Saturday, November 17, the pilot of a Navy F-18 fighter reported spotting eight to ten vehicles heading toward Tarin Kowt and asked what he should do. Amerine's laconic reply: "Well, smoke 'em."

The bombs began falling and Taliban pickup trucks began blowing up. "We were just pummeling them," Amerine said, when all of a sudden Karzai's men panicked and started running away. The Green Berets had no choice but to go along with their Afghan allies as they drove back to Tarin Kowt in a frenzy—the Afghans had the keys to the cars. Amerine had to set up a fresh defensive position just outside of town, knowing that if the Taliban penetrated his lines everyone in the village, including his men, would be slaughtered.

"It was a pretty ugly situation," Amerine recalled, but the ferocity of the air strikes managed to stop most of the Taliban. Fewer than twenty Taliban soldiers ultimately made it into town and they were driven off by armed villagers. By the time the battle was over at 10:30 A.M., more than thirty wrecked trucks and three hundred dead Taliban littered the valley floor. "We'd crushed them," Amerine exulted.

Karzai would later say the battle of Tarin Kowt was "the turning point" in the south because it convinced the local mullahs that the Taliban could be beaten and that Karzai was a figure to be reckoned with. Many more men now rushed to join his ranks. While Karzai was building up his militia, Amerine spent the next two weeks coordinating air attacks on Taliban forces in a seventy-mile "kill zone" between Tarin Kowt and Kandahar. Using a telephone and laptop linked to a portable satellite dish, and relying on reports gathered by Karzai's contacts as well as on overhead imagery, the A-team leader fed targeting data to U.S. aircraft so that they could engage Taliban vehicles while leaving civilians alone.

On Sunday, November 25, twelve hundred marines began arriving at Objective Rhino, a dirt airstrip fifty miles southwest of Kandahar that had been briefly occupied by the Rangers on October 20. But the marines limited their activities to patrolling in the vicinity of their new base. This was still the Special Forces' show.

On Friday, November 30, Amerine and Karzai led their men toward Kandahar in an odd procession of beaten-up vehicles that looked like leftover props from a "Mad Max" movie. The men of ODA 574, bearded and scruffy, fit right in.

Their next battle was fought around the village of Showali Kowt, starting on Monday, December 3. The Taliban put up a fierce resistance to prevent their enemies from seizing the only bridge across the Arghandab River just ten miles north of Kandahar. The battle raged back and forth for two days.

By this time, Karzai had been joined by Lieutenant Colonel David Fox, a Special Forces battalion commander who was acting as his personal military adviser; most of the Afghan rebel leaders had one. Early on the morning of Wednesday, December 5, eight more men from Fox's headquarters staff arrived, bringing with them the first mail ODA 574 had received in weeks. The men sprawled out to open their letters and packages. At 9:30 A.M., just as a sergeant was opening some Rice Krispie treats from his wife, the happy scene was obliterated by a deafening roar.

Amerine quickly realized that the Taliban had nothing capable of causing so much havoc. They had been hit by one of their own bombs. It later turned out that a newly arrived air force controller had mistakenly signaled a B-52 to drop a two-thousand-pound JDAM on his own GPS coordinates. No amount of sophisticated computer gear could prevent a stupid mistake on the part of its operators.

Two of Amerine's men were killed instantly. Another died later that day. The rest, including the captain himself, were wounded but survived. At least twenty-five of Karzai's guerrillas were also killed and fifty wounded; Karzai was only slightly nicked. More would have died were it not for the prompt response of Air Force PJ's (Pararescue Jumpers) in Pakistan, who hopped aboard two MH-53J Pave Low III heavy-lift helicopters to evacuate the casualties.

For ODA 574, the war was over. But not for Karzai.

While Karzai had been advancing toward Kandahar from the north, another rebel leader accompanied by another A-team had been moving from the south. Gul Agha Sherzai, described by a journalist as "a squat, bushy-haired man with a big belly and a rubbery face," had been advancing with some eight hundred Pashtun followers and ODA 583 up Highway 4, the main road from the Pakistan border to Kandahar. They ran into heavy Taliban resistance at the village of Tahk-te-pol on the night of Friday, November 23, but blasted their way through with the help of an AC-130 gunship. On Sunday, November 25, Sherzai reached the outskirts of Kandahar airport, a Taliban stronghold. ODA 583 spent the next seven days calling in air strikes to soften up the enemy positions.

While this was going on, Karzai kept busy on his satellite phone, negotiating the surrender of the Taliban garrison in Kandahar. His work bore fruit on Friday, December 7, when Sherzai's men advanced into the city and found that the Taliban had deserted. ODA 583 accompanied them on an impromptu victory parade as the citizenry cheered and tossed marigolds at them. That afternoon Sherzai occupied the governor's palace.

Karzai had higher ambitions. He was chosen by the Bonn Conference convened by the United Nations to become the interim leader of the entire country—an office he assumed on December 22 at a ceremony in Kabul attended by Tommy Franks.

TORA BORA AND BEYOND

Routing the Taliban had not taken a year or more, as originally anticipated. The regime had fallen less than two months after the insertion of the first A-teams.

Surviving al Qaeda members sought refuge in the Tora Bora redoubt in the Spin Ghar mountains near the frontier with Pakistan thirty miles southwest of Jalalabad. U.S. radio intercepts indicated that Osama bin Laden was in this area in early December 2001. During the first two weeks of December, at least three A-teams worked with Delta Force, the CIA, the British Special Air Service, and thousands of Afghan gunmen to root out the enemy fighters. The Green Berets called in massive air strikes, including the biggest bomb in the U.S. arsenal—a fifteen-thousand-pound BLU-82 Daisy Cutter so large that it had to be dropped with a parachute out of a C-130 transport aircraft. But the munitions had limited effect against an enemy hiding in mountain caves and bunkers. U.S. intelligence later estimated that, while scores of al Qaeda members were killed around Tora Bora, at least one thousand escaped to Pakistan, including bin Laden.

America's Afghan and Pakistan allies, it emerged, had not been terribly zealous or skillful in prosecuting their offensive; some were even said to have taken bribes to let al Qaeda members escape. "Failing to capture these senior [Taliban and al Qaeda] leaders was a blow to US strategic goals," a U.S. Army study later acknowledged.

Critics asked why General Franks had not sent U.S. troops to act as a "blocking force" to prevent the exodus of enemy fighters. In his defense, Franks could point to the difficult terrain, featuring sheer mountain peaks fifteen thousand feet high that had frustrated previous invaders. In the 1980s the Red Army had staged several attacks on Tora Bora with thousands of soldiers. All failed. There was no guarantee that dispatching a larger number of Americans (such as the one thousand marines who were then around Kandahar) would have made a difference. But they might have. The Afghanistan campaign as a whole had shown the effectiveness of relying on proxy forces and smart bombs; Tora Bora showed the limitations of that strategy.

In the continuing war against Taliban and al Qaeda remnants, U.S. forces mounted their first major conventional offensive on March 2, 2002, in the Shah-i-Kot Valley, which had become another guerrilla sanctuary. Operation Anaconda involved fifteen hundred troops from the 101st Airborne and 10th Mountain Divisions in addition to Afghan militiamen and U.S. and allied commandos. The offensive suffered some serious setbacks early on when it became apparent that intensive U.S. reconnaissance had seriously underestimated the strength of enemy forces and had failed to locate at least half of their hiding holes. A SEAL (Sea-Air-Land) team was inadvertently dropped right on top of a Taliban position, leading to the loss of two helicopters and the death of seven men. Other U.S. troops were stymied by heavy mortar fire from surrounding mountaintops. Smart bombs and sensors were not enough to defeat these determined defenders; they had to be rooted out the old-fashioned way, cave by cave, as the marines had once rooted out the Japanese on Iwo Jima and Saipan. By the time Anaconda was concluded on March 16, hundreds of the enemy had been killed while hundreds more had escaped.

The Taliban would continue to wage a low-level guerrilla war against the new government in Kabul, a government bolstered not only by a growing number of U.S. troops but also by peacekeepers supplied by other (primarily European) nations. American-led nation-building efforts got off to a slow start but picked up momentum in 2003 when President Bush committed more soldiers and more money to support the central government. The total number of U.S. troops soared to twenty thousand in 2004 as Afghanistan prepared for a presidential election. Establishing the authority of a new government, it turned out, required a lot more troops than did overthrowing the previous regime.

Despite Taliban efforts to disrupt the voting, millions of men and women lined up on October 9, 2004, to cast their ballots for president. Hamid Karzai won a large majority. Afghanistan continued to be plagued by problems, ranging from the drug trade to grinding poverty, but there was no doubt that it had made considerable progress since the days when the Taliban required the wearing of burkas and banned the flying of kites.

Beyond Afghanistan, the U.S. and its allies kept up the hunt for Islamist terrorists around the world. Abu Ali al-Harithi, the mastermind behind the USS *Cole* bombing, was blown up in 2002 along with six associates traveling in a car across the Yemeni desert. They were killed by a Hellfire missile fired from a CIA-controlled Predator, apparently after a secret U.S. military intelligence unit called Gray Fox had figured out how to home in on Harthi's cell phone transmissions. The same year, Ramzi bin al-Shibh and Abu Zubaydah, two senior al Qaeda leaders, were captured in Pakistan. In 2003, Khalid Sheikh Mohammed, the 9/11 planner, was snared in Pakistan by U.S. and Pakistani forces, while Hambali, a leader of the Indonesian al Qaeda affiliate Jemaah Islamiyah, was taken in Thailand by the CIA working alongside Thai security forces.

By the summer of 2004, President Bush was able to report that two-thirds of al Qaeda's senior leaders had been caught or killed. But Osama bin Laden was not among them, and his organization had not been rendered harmless. Islamist terrorists continued their reign of terror after 9/11, killing thousands of people in a string of gruesome attacks in Baghdad, Bali, Istanbul, Karachi, Riyadh, Casablanca, Moscow, Jerusalem, Madrid, London, Sharm el-Sheikh, Bombay, and elsewhere.

The largely autonomous cells that carried out these atrocities proved difficult to stamp out. As Secretary of Defense Rumsfeld admitted in an October 2003 memo leaked to the press, it was hard to tell whether the U.S. was winning the war: "Are we capturing, killing, or deterring and dissuading more terrorists every day than the madrassas and the radical clerics are recruiting, training and deploying against us?"

"DECENTRALIZED BUT COORDINATED"

In order to better confront this amorphous threat, the U.S. government undertook an extensive intelligence and defense reorganization after 9/11—the most sweeping since the 1947 legislation that set up the CIA, Defense Department, and National Security Council. The creation of a Department of Homeland Security, integrating agencies from the Coast Guard to the Border Patrol; the passage of the USA Patriot Act, which, among other steps, tore down the internal barriers at the FBI between the Criminal and Intelligence divisions; the establishment of a National Counterterrorism Center blending information from the FBI, CIA, and other agencies; the appointment of a director of national intelligence to oversee the entire intelligence community; the designation of U.S. Special Operations Command as the lead military agency pursuing terrorists overseas; and the creation of the U.S. Northern Command to coordinate the defense of North America—all were efforts to better enable the government to fight a netwar against al Qaeda and its ilk. There was also a big increase in spending on homeland defense.

While increases in resources for antiterrorism were important, even more vital was the removal of bureaucratic barriers that had prevented making effective use of all the tools already at the government's disposal. The difficulty of doing this was demonstrated by the fact that almost four years after 9/11 more than thirty federal "watch lists" still had not been completely consolidated into a single, instantly accessible database of suspected terrorists.

Afghanistan showed how successful a netwar could be. "We had accomplished in eight weeks what the Russians couldn't accomplish in ten years," Lieutenant General "Rifle" DeLong noted with pride. It had taken approximately three hundred Special Operations troops and one hundred CIA officers—along with hundreds of aircraft and thousands of native allies—only forty-nine days to bring down the Taliban. Casualties on the U.S. side were minimal. Twelve Americans were killed in Operation Enduring Freedom in the fall of 2001, only one (CIA officer Mike Spann) at the hands of the enemy. As of December 31, 2005—four years after the fall of the Taliban—total U.S. fatalities in the Afghanistan theater stood at 259; roughly half of those not a result of enemy action. By contrast, the Soviets had not been able to defeat approximately the same number of Afghan enemies with 100,000 Red Army troops fighting for almost a decade and suffering 15,000 fatalities.

Although few in number and lacking in armor or artillery, the Special Forces in Afghanistan were, in a sense, the most powerful infantrymen in history because they fought not with shoulder-fired weapons, whose range and power is severely limited, but with GPS locators, satellite radios, laptops, and laser-designators that could summon pinpoint air strikes with the push of a button. Roughly 60 percent of all U.S. munitions employed in Afghanistan were precision guided—six times greater than in the Gulf War.

It must be stressed that air power had not rendered ground troops obsolete. In fact, bombing without ground troops, which took place in Afghanistan between October 7 and October 19, 2001, achieved scant results. Enemy forces could readily go to ground and ride out even the harshest U.S. bombing raids. A ground element was still required to root them out. What Afghanistan showed was not that air power could win a war by itself, but that it could tremendously boost the effectiveness of even a small number of poorly equipped and badly trained ground troops. The Northern Alliance units that took Mazar-e Sharif and Kabul were not much bigger or better armed than the Northern Alliance units that had been stymied for years. They were still tremendously outnumbered and outgunned by the Taliban, but now they could defeat their enemies because of the addition of U.S. air power. It also helped that the A-Teams were able to coordinate feuding rebel commanders who, in ordinary circumstances, would not have spoken to one another.

The Americans, in turn, greatly benefited from their Northern Alliance allies, not only because they offered a fighting force willing to suffer casualties so that U.S. ground forces wouldn't be at risk, but also because they were an invaluable source of intelligence. American soldiers could not tell at a glance if a man was a farmer or a fighter; Afghans could. "If I was a conventional army guy going through there," noted a sergeant from ODA 595, and "that guy stands over there like a meek farmer, [I'd] have gone right by him. An Afghan knew instantly, 'Hey, that guy is not from here. He's a bad guy.'"

Al Qaeda had the information advantage prior to 9/11; its leaders knew about the vulnerabilities of U.S. infrastructure, while the U.S. government did not know about the "planes operation." In Afghanistan the information advantage shifted. Mullah Omar and Osama bin Laden could not figure out what the Americans were up to or where they would strike next, while the Americans generally had a good idea of the location and strength of Taliban units. While al Qaeda had shown itself to be nimble and skillful in staging terrorist attacks around the world, it was clumsy and slow-moving in defending its own base. The Americans had the initiative and they used it bring down the Taliban.

The one aspect of the U.S. campaign that attracted the most attention was the unlikely juxtaposition of horses and laser designators, primitive tribesmen and precision-guided munitions, an anomalous situation that Special Forces Captain Mark Nutsch aptly likened to "the Flintstones meet the Jetsons."

This was indeed unusual and important, but it was only one sign of the ad hoc, improvisational nature of the entire campaign. The lack of preparation time and the remoteness of the battlefield meant that there was little of the bureaucratic infrastructure that too often stifles individual initiative. The key decisions were made by captains and sergeants, not generals and colonels—something that was unusual in any hierarchical military structure. The U.S. armed forces were more decentralized than those of, say, China, Russia, or Iraq, but an internal study in 2002 found that the U.S. Army too often encouraged "compliance instead of creativity, and adherence instead of audacity" among its junior officers, who were "seldom given opportunities to be innovative; to make decisions; or to fail, learn and try again."

In Afghanistan, that mold was broken. Opportunities for audacity were plentiful, and they were eagerly seized by the A-teams.

Dean, the team captain of ODA 534, recalls that when he was sent to Afghanistan his entire mandate consisted of a handful of PowerPoint slides that told him to conduct unconventional warfare, render Afghanistan no longer a safe haven for terrorists, defeat al Qaeda, and "coup" the Taliban. How he accomplished those goals was up to him. "We were given an extraordinarily wonderful amount of authority to make decisions," says Dean. It was not until after the fall of the Taliban that the bureaucracy of the regular army arrived in Afghanistan with its PXs and its paperwork.

Dean credits the "network approach" of the U.S. Special Forces and Air Force for the success of Operation Enduring Freedom. They could be so "decentralized but coordinated," he adds, because advanced communications gear kept them better connected to each other than soldiers had been in any previous conflict.

Almost by accident, U.S. troops had managed to realize some of the potential of the Information Revolution while fighting in one of the least "wired" places on the planet.

CHAPTER 12

HUMVEES AND IEDs:

Iraq, March 20, 2003–May 1, 2005

Lieutenant Colonel Eric "Rick" Schwartz, a trim, compact tank commander with gray hair and stained green Nomex overalls, couldn't believe what he was hearing. He and his men in Task Force 1–64 (1st Battalion, 64th Armor Regiment) had been fighting virtually without rest for sixteen days, ever since they had left Kuwait more than three hundred miles to the south. They had just finished an engagement against the Republican Guard's Medina Division, a one-sided battle but one in which Schwartz had been wounded by a piece of shrapnel in his shoulder. Now that they had reached the outskirts of Baghdad, Schwartz figured that his job was nearly done. He had been told that tank units like his would be assigned to loosely cordon off the capital to allow airborne troops and Special Forces to seize the heart of the Baathist regime. But things had been going so well that the generals decided to throw caution to the winds and send an armored detachment charging straight into Baghdad. This maneuver became known as a "thunder run," a term borrowed from the Vietnam War when armored columns had moved at high speed to keep open supply lines to embattled fire bases known as Thunder I, Thunder II, etc.

On the afternoon of Friday, April 4, 2003, Schwartz was informed that at first light the following day he would lead the first thunder run straight up Highway 8 to central Baghdad and then loop southwest into Saddam Hussein International Airport, which had just been seized by another American unit. No one was sure what kind of resistance the first American troops in the capital would meet, but there was no doubt that it would be heavy. Saddam Hussein had planned to turn Baghdad into another Stalingrad, or at least another Mogadishu.

When he first heard the plan, while standing in a tent pitched in a dusty field eleven miles south of Baghdad, Schwartz blurted out, "Are you fucking crazy . . . sir?"

A moment of silence followed. "No," replied Colonel David Perkins, commander of the 3rd Infantry Division's 2nd Brigade Combat Team, of which Task Force 1–64 was a part. "And I'm coming with you. We have to do this."

As soon as the Rogues (as the 1st Battalion was known) crossed the line of departure at 6:30 A.M. on Saturday, April 5, they ran into heavy fire from bunkers and trenches, rooftops and alleyways. The Iraqis hit them with everything they had—RPG-7 rocket-propelled grenades, AK-47 assault rifles, heavy machine guns, air defense artillery. Cars and pickup trucks even tried to ram the column's heavy armored vehicles. To no avail. Gunners aboard the twenty-nine Abrams tanks and fourteen Bradley Fighting Vehicles cut down the Iraqi attackers with ruthless efficiency. But more and more kept appearing. All manner of vehicles—taxis, sedans, buses, garbage trucks, ambulances—were dropping off gunmen along the route, many of them black-clad Saddam Fedayeen (Sacrificers of Saddam) and foreign jihadists eager to attain martyrdom. They would keep coming until they were cut in half. One gunman on foot got close enough that Colonel Perkins had to whip out his own 9 mm Beretta pistol to kill him.

For all their ferocity, Saddam's men attacked with little coordination; the Americans called their tactics (or lack thereof) "spray and pray." Every vehicle in the convoy was hit multiple times but only two—a Bradley and an Abrams—were knocked out of action, and the convoy's progress was not impeded. Finally the Iraqis tried to bar access to the airport by dragging heavy concrete slabs, three feet tall and two feet thick, into the highway. Seeing this, the commander of the lead tank sped up to 24 mph and rammed the barrier with his seven-ton plow, shearing the top off the concrete. Succeeding tanks ground the barrier into dust under their treads. By 9 A.M., Task Force 1–64 had reached the safety of the airport.

The Rogues, 760 strong, had killed at least one thousand of the enemy while losing only one man, a staff sergeant who was hit in the head while firing from an open tank hatch. Five more Americans had been seriously wounded, but this seemed a remarkably light toll, considering the magnitude of the resistance they had encountered. The crewmen, their faces smeared with grime, their coveralls dark with sweat, were so exhausted and terrified by this ordeal that, when it was over, many cried in relief. "I don't think anyone of us had a dry eye," Lieutenant Colonel Schwartz said.

In their own minds, the Americans had demonstrated that no part of Baghdad was off limits to them. But Saddam Hussein's inventive information minister, Mohammed Saeed al-Sahaf, drew a different lesson. He told reporters assembled at the Palestine Hotel in downtown Baghdad that an American attack into the city had been repulsed. "Today we butchered the force present at the airport," he crowed. "They fled. The American louts fled."

To demonstrate the falsity of "Baghdad Bob's" propaganda, U.S. commanders decided to stage another thunder run. Colonel Perkins, who would also lead this expedition, was privately determined that this time he would spend the night in the city to prove "in no uncertain terms" that the Baathist regime was finished once and for all. Given the success of the first thunder run, he had no doubt of his tanks' ability to penetrate into Baghdad. The question was whether their more vulnerable supply column would be able to follow them in. Unless its gas-guzzling tanks could be refueled, the 2nd Brigade would have to pull out, and Saddam would gain a propaganda victory.

On the morning of Monday, April 7, two armored battalions of the 2nd Brigade had no trouble fighting their way into the governmental district of Baghdad. Just two hours after setting out, sixty Abrams tanks and twenty-eight Bradleys burst through the gates of Saddam Hussein's grandiose palace complex. It was as if an enemy had invaded Washington, D.C., and seized everything from the Capitol to the White House. Only one tank was disabled en route and no Americans were killed. The biggest setback occurred when an Iraqi surface-to-surface missile struck the brigade's operations center on the outskirts of town, killing five and wounding seventeen. But an improvised command post was functioning again within an hour. Standing in front of Saddam's Sujud Palace, an American officer proudly told Fox News, "Saddam Hussein says he owns Baghdad. Wrong! We own Baghdad."

Task Force 3–15 (3rd Battalion, 15th Infantry Regiment), assigned to hold open the 2nd Brigade's supply routes, had a rougher day. Its commander, Lieutenant Colonel Steven Twitty, had only a few hundred soldiers, fourteen tanks, and thirty Bradleys to secure three vital cloverleaf intersections along Highway 8, whimsically dubbed Objectives Larry, Curley, and Moe. All three were attacked by wave after wave of gunmen on foot and suicide bombers in automobiles.

Objective Curley was particularly hard-pressed. Its defense was entrusted to a pickup team of eighty soldiers cobbled together at the last minute under Captain Harry "Zan" Hornbuckle, a twenty-nine-year-old staff officer who had never been in combat before. His soldiers, who had no tanks and only four Bradleys to assist them, were under incessant assault from all directions. By midmorning, twenty Americans had been wounded and ammunition was running low. The situation looked so dire—"There was not a soldier on Curley who did not think he was going to die that day," a corporal commented—that even medics and the chaplain were picking up weapons. One wounded nineteen-year-old private who was being taken away on a stretcher sat up and fired his shotgun when he saw a nearby Iraqi reaching for an AK-47. Despite such determination, Team Zan was saved only by the timely arrival of reinforcements, first a platoon, then an entire battalion.

Team Zan's bravery was fully matched by the soldiers of the 2nd Brigade supply column, who had to drive their soft-skinned vehicles full of ammunition and gasoline past this gauntlet. A single hit could have turned any of these trucks into a fireball. Five did explode, but the rest kept going and managed to reach the city center. Their arrival at dusk meant that Colonel Perkins could spend the night in Baghdad.

On Tuesday, April 8, the resupplied 2nd Brigade fought off numerous counterattacks on its perimeter in the city center. As the Americans consolidated their control in Baghdad, the Iraqi regime began falling apart. Baathist stalwarts stripped off their uniforms and fled the city. By this time the 1st Marine Division was advancing through the southern part of the capital. On April 9, the marines helped a crowd of Iraqis pull down a statue of Saddam in Firdos Square.

The war had been won in just three weeks. Or had it?

In the days and months and years that followed, the U.S. armed forces would come to realize that there was more to victory than seizing an enemy capital and more to warfare than pouring firepower from various "weapons platforms." Securing a favorable political outcome required changing a dysfunctional political culture, revitalizing a battered economy, and bringing order out of chaos. All of these tasks, challenging enough under any circumstances, would prove especially difficult to carry out in Iraq because of incessant assaults from a shadowy army of insurgents. Although the guerrillas had few weapons to match those in the U.S. arsenal—they fought for the most part with RPG-7s, AK-47s, and crude bombs known as IEDs (improvised explosive devices)—their assaults would take a growing toll on the American armed forces. Just as the thunder runs into Baghdad had shown the U.S. military's strengths, so the protracted aftermath would reveal its weaknesses—and the limitations of information technology.

"We have [made] huge leaps in technology," noted Air Force Lieutenant General Lance L. Smith in 2004, "but we're still getting guys killed by idiotic technology—a 155 mm shell with a wire strung out." That painful paradox was not something that the upper echelons of the Bush administration had anticipated when they had first drawn up plans for what was supposed to be a short war to overthrow a vicious dictator.

"A REVOLUTIONARY CONCEPT"

President Bush had demanded serious planning for the invasion of Iraq to begin shortly after the September 11, 2001, attacks. Even if Saddam Hussein had nothing to do with the strikes on Washington and New York (and no convincing evidence ever emerged to establish a link between Iraq and the al Qaeda plotters), the attack had shifted the calculus of risk for the administration. Given widespread concern that Saddam Hussein was building weapons of mass destruction (chemical, biological, nuclear) in violation of United Nations sanctions, Bush wanted to eliminate any possibility of Saddam mounting a catastrophic terrorist strike. The president and his top advisers also believed that replacing the repressive Baathist regime with a democratic government could begin the transformation of an entire region that had turned into a breeding ground of fanaticism and terrorism. In a speech to the American Enterprise Institute on February 26, 2003, Bush summed up his case for war: "America's interests in security, and America's belief in liberty, both lead in the same direction: to a free and peaceful Iraq."

When Tommy Franks was first asked to present a plan in November 2001 to achieve that ambitious goal, he had nothing to offer but Desert Storm Redux. Operations Plan (Oplan) 1003–98, last updated in 1998, called for weeks of bombing followed by an invasion by four hundred thousand soldiers. Even Franks had to concede that this concept was "out of date": "It didn't account for our current troop dispositions, advances in Precision-Guided Munitions, or breakthroughs in command-and-control technology—not to mention the lessons we were learning in Afghanistan."

Rumsfeld and his senior aides were so impressed by those lessons that they pressed for the lightest possible force in Iraq. But Iraq was not Afghanistan. The Baathists were more powerful than the Taliban and there were was no indigenous opposition comparable to the Northern Alliance. Saddam's military was much degraded since the Gulf War, but on paper, at least, it still looked formidable. As former Pentagon official Anthony Cordesman told Congress in August 2002:

> Iraq is still the most effective military power in the Persian Gulf. It still has active forces of over 400,000 men. It still has an inventory of over 2,200 main battle tanks, 3,700 other armored vehicles, 2,400 major artillery weapons. It

> still has over 300 combat aircraft in its inventory, though perhaps less than half of these are truly operational. And it certainly still has some chemical and biological weapons. This is not a force that can be dismissed.

Moreover, even if Saddam's military crumbled as quickly as it had in 1991, there remained the challenge of occupying the country afterward. The Army chief of staff, General Eric Shinseki, a veteran of peacekeeping in Bosnia, warned that "something on the order of several hundred thousand soldiers" would be necessary to establish order. But General Franks, who had never served in Bosnia or Kosovo, did not insist on a force of that size, perhaps because he knew that Rumsfeld would never grant it, and Deputy Secretary of Defense Paul Wolfowitz dismissed Shinseki's estimate—in words that would come back to haunt him—as "wildly off the mark."

The Centcom commander and his civilian masters were acutely, indeed excessively, conscious of the risks involved in sending too many troops: An overly large force would take too long to deploy, limiting the president's diplomatic options and eliminating the possibility of a surprise attack. While they were gathering in Kuwait, large U.S. troop deployments would be vulnerable to attack. And once they deposed Saddam, the presence in Iraq of too many U.S. soldiers would, it was feared, spark a nationalist backlash.

Eventually a compromise plan was worked out that was somewhere between the Gulf War and the Afghanistan War. Oplan 1003V, completed in December 2002, called for a "running start": beginning the invasion with a relatively small ground force of about 150,000 troops, and then flowing more forces into Iraq if necessary. Instead of preceding the invasion with a long bombing campaign, as in Desert Storm, air operations were supposed to start only fifteen hours ahead of ground operations. Utilizing the lessons of Afghanistan, this war plan would rely heavily on Special Forces and precision air power in lieu of heavy ground units. General Franks called this "a revolutionary concept, way outside the box of conventional thinking."

The initial idea was for a classic pincer movement, with the U.S. Army's 3rd Infantry Division and the 1st Marine Division charging up to Baghdad from the south while the U.S. 4th Infantry Division and the British 1st Armoured Division came down from the north. This plan was upset at the last moment when the Turkish parliament refused to give U.S. and British forces permission to move through their country. Franks was forced to turn over the entire northern sector to a small group of commandos and paratroopers who would be flown in to work with the Kurdish peshmerga militia. The 4th Infantry's equipment was kept floating off Turkey in a successful ruse to convince Saddam Hussein that the main thrust of the invasion would come from the north; the division would not enter Iraq until Baghdad had already fallen.

The British 1st Armoured Division, meanwhile, was hastily redirected to the south. Its job would be to isolate Basra, Iraq's second-largest city. The bulk of the armored punch for the blitzkrieg to Baghdad would come from just one division—the 3rd Infantry, with about two hundred M1A1 Abrams tanks and 250 M2/M3 Bradley Fighting Vehicles. The 3rd Infantry's main effort would be supported by the 1st Marine Expeditionary Force, the 101st Airborne Division, and a brigade from the 82nd Airborne Division.

The coalition had a total of just 350,000 personnel in the theater of operations when Operation Iraqi Freedom began. The initial ground force of 150,000 troops was two-thirds smaller than the army that had won the Gulf War, and its mission was much more ambitious: not just to expel the Iraqis from Kuwait but to occupy an entire country the size of California populated by twenty-five million people.

"Speed and momentum are the keys," Franks explained. Like the German invaders of France in 1940, the coalition forces would try to move so fast that their adversaries would be too flummoxed to respond effectively.

A RUNNING START

The ground war was supposed to start on Saturday, March 22, 2003, but on the night of Wednesday, March 19, the CIA received information that Saddam Hussein and his sons were meeting in a bunker underneath the Dora Farms complex south of Baghdad. Bush decided this was too good an opportunity to pass up and ordered an air strike to decapitate the Baathist regime. Before dawn on Thursday, March 20, two F-117 stealth fighters dropped four two-thousand-pound satellite-guided bombs on the complex, followed by a flurry of thirty-nine Tomahawk cruise missiles fired from naval ships. The Dora Farms complex was wiped out, but Saddam was not killed. After the war, it was discovered that there was not even a bunker there. This was the first of at least fifty air strikes on senior Iraqi leaders, none of which succeeded. American forces were again learning that even the best weaponry in the world was not much use without adequate intelligence.

The bulk of the planned air campaign would not start until the night of Friday, March 21, but in the meantime imagery from a Predator drone suggested that saboteurs were setting fire to oil wells in the giant Rumailah oil field in southern Iraq. To forestall the danger of mass destruction, which would deny postwar Iraq crucial oil revenues, General Franks ordered the invasion to start right away. U.S., British, and Australian commandos had already infiltrated the western deserts of Iraq from Jordan and Saudi Arabia. On the morning of Friday, March 21, the rest of the coalition force began to pour across the Kuwait border.

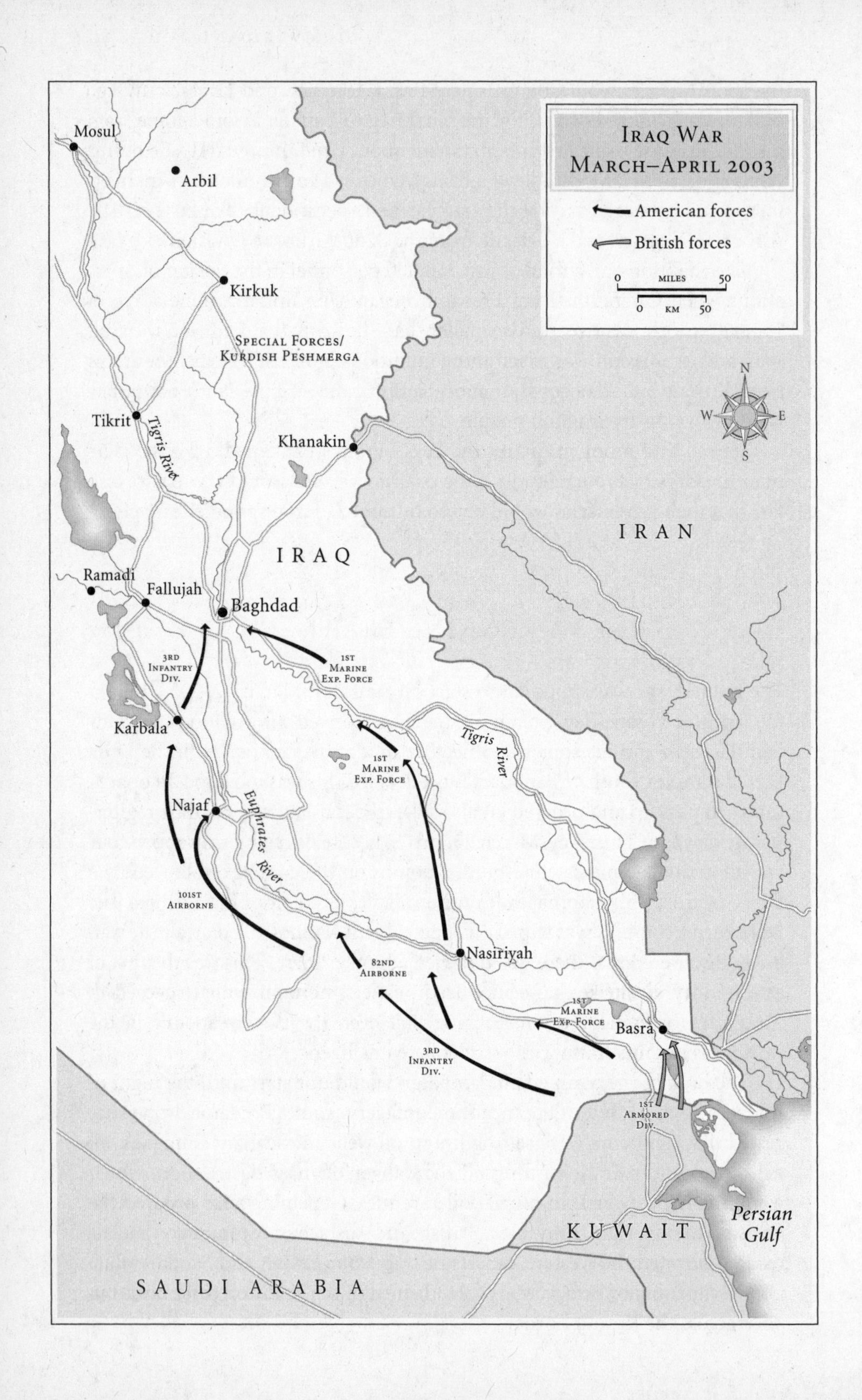
Iraq War
March–April 2003
American forces
British forces
0 miles 50
0 km 50
N
W
E
S
Mosul
Arbil
Kirkuk
Special Forces/
Kurdish Peshmerga
Tikrit
Tigris River
Khanakin
IRAN
IRAQ
Ramadi
Fallujah
Baghdad
3rd
Infantry
Div.
1st
Marine
Exp. Force
Karbala'
1st
Marine
Exp. Force
Tigris
River
Najaf
Euphrates
River
101st
Airborne
Nasiriyah
82nd
Airborne
1st
Marine
Exp. Force
Basra
3rd
Infantry
Div.
1st
Armored
Div.
Persian
Gulf
KUWAIT
SAUDI ARABIA

Sending in ground units before the start of a major bombing campaign was an audacious move—the Iraq War's equivalent of Desert Storm's "left hook." And just like that earlier stratagem, this one caught Iraqi defenders flatfooted. Their ineffectual response was limited to firing five short-range surface-to-surface rockets into Kuwait, all of which either missed their targets or were destroyed by PAC-3 batteries. (These Patriot Advanced Capability missiles were considerably more lethal than the original Patriots deployed in the Gulf War, though the aircraft they shot down sometimes belonged to friendly forces.)

Having broken through the border berm, the 1st Marine Expeditionary Force (which included British forces) headed east to secure the Rumailah oil field. Only nine wells had been destroyed before they arrived. The British 1st Armoured Division then peeled off to besiege Basra, a process that would take almost three weeks, while the 1st Marine Division and Task Force Tarawa, a brigade-size marine unit, headed north through the heavily populated Fertile Crescent between the Tigris and Euphrates Rivers. The 3rd Infantry also achieved its initial objectives easily, seizing Tallil Air Base near An Nasiriyah for use as a forward logistics hub. Making no attempt to subdue the city of Nasiriyah, the division headed straight for Baghdad through the largely empty desert west of the Euphrates.

The invaders first began encountering serious resistance on Sunday, March 23. The day began when an army supply convoy from the 507th Maintenance Company took a wrong turn and drove into Nasiriyah. Iraqis ambushed the lightly armed convoy, killing eleven soldiers and capturing six, including Private Jessica Lynch. Task Force Tarawa, about 6,000-strong, was on the edge of town preparing to seize another bridge over the Euphrates. They did not want to pacify the entire city of some four hundred thousand people, but they soon realized they had no choice because of the vicious resistance they were encountering. This did not come primarily from regular Iraqi army units or even from the elite Republican Guard but rather from fanatical irregulars, some of them foreign holy warriors, organized into various militias.

The most determined fighters came from the Saddam Fedayeen, which had at least twenty thousand men. Modeled on the Palestinian guerrillas fighting Israel, they were armed with Kalashnikovs, RPG-7 rocket-propelled grenade launchers (a Russian-made successor to the Panzerfaust and bazooka used in World War II), and some heavy machine guns mounted on pickup trucks. What they lacked in military hardware or training they made up for with suicidal courage. While most of Saddam's soldiers preferred to slink away, the Fedayeen did not hesitate to charge straight at American tanks. Nor did they have any compunctions about using Iraqi women and children as human shields or about employing hospitals, schools, and mosques as fighting positions. The Fedayeen were mown down by the bushel-load, but their fierce resistance temporarily slowed down the U.S. advance and imperiled strung-out supply lines.

The marines first got a taste of these guerrilla tactics in Nasiriyah, which was controlled by at least three thousand Fedayeen. By the afternoon of Sunday, March 23, eighteen marines had been killed and more than seventy wounded. (Some of the losses occurred when an Air Force A-10 mistakenly bombed a group of marine amphibious vehicles.) It took Task Force Tarawa a week of heavy block-by-block fighting to clear out the town's defenders.

March 23 was a black day for the army as well. That night, two squadrons of AH-64D Apache Longbows were supposed to pulverize the Medina Republican Guard Division, which sat in the path of the 3rd Infantry. These were the most sophisticated attack helicopters in the world, but they were foiled by a rudimentary defense. When the choppers passed overhead, city lights flicked on and off, signaling the defenders to fire small arms and antiaircraft guns into the sky. All thirty helicopters ran into a wall of lead. One was forced down and its two pilots captured. (They would later be freed by advancing U.S. forces.) The others limped back to base with varying degrees of damage. The failure of this raid was a major embarrassment for the U.S. military, coming as it did only a few hours after the ambush of Jessica Lynch's supply convoy and the death of numerous marines in Nasiriyah.

Things seemed to go from bad to worse the next day when a sandstorm of biblical proportions enveloped southern Iraq. For the next three days winds gusted up to 50 mph and rain fell in torrents. Even with goggles on, it was impossible to see more than thirty feet ahead. Helicopters were grounded, convoys slowed to a crawl. Many poetic descriptions have been offered of this trying time when, in the words of one journalist, the sky was "the color of bile—brown tinged with yellow." Army Lieutenant General William S. Wallace was more blunt. "The weather," he said, "really sucked."

The advance of U.S. ground forces, which had been as fast as a Maserati until this point, now became as slow as a jalopy. Supply slowdowns meant that some army units found themselves running low on food and ammunition. All the Americans, after days of nonstop advance, were getting worn out from stress and sleeplessness. To make matters worse, Iraqi irregulars were swarming out of numerous towns along the invasion route to attack U.S. troops. The 3rd Infantry found itself getting sucked into nasty urban fights in Najaf and Samawah similar to the marine dustup in Nasiriyah. This was not something that U.S. commanders had expected or prepared for. "We'd had no warning that Saddam was dispatching these paramilitary forces from Baghdad," General Franks said.

The original idea of bypassing all the obstacles on the road to the Iraqi capital wasn't working. So the senior U.S. ground commanders—in addition to Franks there was Lieutenant General David D. McKiernan, commander of the entire coalition land force, and his principal subordinates, Lieutenant General James Conway of the 1st Marine Expeditionary Force and Lieutenant General William S. Wallace of the Army's V Corps—decided to call an audible. The decision was made on Wednesday–Thursday, March 26–27, to pause for a few days to allow frontline units to rest up and resupply for the final push to Baghdad. The marines were anxious to keep moving but acquiesced in a temporary delay.

In the meantime, the 101st Division and a brigade of the 82nd Airborne Division, which had been held in reserve for the final fight in Baghdad, were committed to defending lines of communication that now stretched 250 miles from the border of Kuwait. The 82nd relieved the 3rd Infantry in Samawah, while the 101st took over in Najaf. Here they would fight for the next couple of weeks while the 1st Marine Division and the 3rd Infantry Division completed their dash to Baghdad. McKiernan thought the decision to pause and regroup was the most important one he made during the entire war.

When news of the "operational pause" leaked out, television commentators, many of them retired generals, began to talk of the offensive "bogging down" and "running out of steam." The Iraqis, who monitored the BBC, CNN, and al Jazeera, believed this story line—much to their subsequent regret. The Fedayeen tried to use the cover of the sandstorm to attack U.S. forces while the Republican Guard began to reinforce their positions south of Baghdad. But while American soldiers may have been blinded, their sensors were not. JSTARs, U-2s, and a lone Global Hawk—as well as Special Forces teams that had infiltrated enemy lines—were able to spot the Iraqi movements and direct various Air Force, Navy, and Marine Corps aircraft to destroy them.

Laser-guided bombs were useless in the sandstorm but satellite-guided JDAMs worked with the usual precision. Radar-guided Paladin howitzers also took a heavy toll on the enemy. And when they did manage to sneak close to U.S. formations, the Iraqis were ripped apart by tanks and armored vehicles employing thermal sights. Franks later called the aerial bombardment during the sandstorm "one of the fiercest, and most effective, in the history of warfare," but it took place out of sight of the international news media, so the devastating effectiveness of these strikes did not become clear until later.

The initial air campaign had been promoted by the news media as a massive onslaught designed to inflict "shock and awe" that would lead to a rapid regime collapse (a concept developed at the National Defense University). In fact, while U.S. commanders hoped that the Baathist government might fall apart fast, their initial targeting was restrained. Coalition commanders were so anxious to preserve a viable postwar Iraq that, in contrast to the Gulf War, they did not even turn out the lights in Baghdad or drop key bridges into the water. Notwithstanding some symbolic strikes on Saddam's palaces, 79 percent of air strikes would be in direct support of coalition ground forces.

The air campaign over Iraq was distinguished by the record percentage of precision munitions employed (68 percent), the almost complete lack of Iraqi resistance (Iraqi air defenses had never recovered from the Gulf War), and the coalition's unprecedented speed and flexibility in making targeting decisions.

THE DIGITAL DIVIDE

The coalition's use of information technology had improved in the year and a half since the beginning of the war in Afghanistan, to say nothing of its advancement since earlier conflicts. Colonel Christopher Haave, an A-10 pilot who was director of combat operations in the Coalition Air Operations Center at Prince Sultan Air Base in Saudi Arabia, recalled how during combat over Kosovo in 1999 he would have to visually identify targets and then locate them on a paper map in his cockpit. While the aircraft was on autopilot, he would take a grease pencil, write down the target coordinates on his canopy, and read them over the radio to an airborne controller, who would then relay them, also by radio, to the operations center, where some officer would examine a giant wall map, try to find out about civilian structures in the target vicinity, and then walk around to various senior officers to seek approval for an attack.

In 2003, by contrast, a Predator or another drone could provide a video feed of the target directly to the air operations center along with its GPS coordinates. A detailed grid map of Iraq had already been entered into the computer network, so an air operations officer could immediately click on the target location to see a picture and get other information, such as whether it was considered to be a "high collateral damage" location or whether any friendly troops were nearby. With a few more clicks, an officer could determine what assets were available to attack the target, which electronic jammers and tankers were available to provide support, and other relevant data. Then, without leaving his desk, the officer could enter an Internet chat room or use Internet telephony to get approval to hit the target, and a weapons controller could instantly relay that approval to an aircraft already on its way. "With networked information and real-time collaboration, we were able to make decisions quite quickly," Haave said.

So quickly that on some occasions less than ten minutes elapsed between the time a target was identified and when it was struck—a process that had taken an average of three days as recently as the 1991 Gulf War. This speed was all the more impressive when one realizes that on many occasions the pilots controlling the Predators and the analysts making sense of their findings were located not in the Middle East but thousands of miles away, at bases in the United States. Secure satellite data links obliterated the limitations of time and space.

Higher-echelon ground commanders were just as plugged in to the information grid as officers like Colonel Haave at the air command. At the Joint Operations Center at Camp As Salilyah in Qatar, Centcom staff could watch Predator video on plasma screens, conduct teleconferences with Tampa or Washington, listen in on intercepted enemy telephone conversations, and learn the latest war news from CNN or Fox News. Best of all, they could watch the progress of coalition forces through Blue Force Tracker, part of the army's Force XXI Battle Command Brigade and Below command-and-control suite. Many coalition vehicles and aircraft carried a transponder that would transmit their GPS location, which would then be displayed as a small blue icon on tracking screens. Enemy units, once their location became known, could be added as red icons.

Before the start of hostilities there was a frantic effort to equip as many units as possible with their own Blue Force Trackers, but the only division to have these devices installed on most vehicles was the 4th Infantry Division, which did not take part in the invasion. The 3rd Infantry and the 1st Marine Division were lucky if their battalion or company commanders had a few terminals; there were none available for platoon leaders.

One of the most useful features of this system was its ability to transmit and receive short, text-only e-mails. Army and marine units were supposed to have broadband satellite access that would allow them to download reconnaissance photos and operations plans even while on the move, but this never proved practical. The broadband system depended on stationary, line-of-sight antennae, but the fast pace of the advance and its vast geographic scope meant that units were seldom in effective range to receive Internet access. Frontline units generally conducted long-distance communications via satellite phone or Blue Force Tracker e-mail while employing FM radios—the same technology used by their forefathers in World War II—for shorter distances. Thus, although the U.S. armed forces utilized many times more bandwidth in 2003 than in 1991, little of it reached the tip of the spear.

The kinds of communications glitches that had been common in previous conflicts were reduced but hardly eliminated. During the early fighting in Nasiriyah, for instance, a regiment from the 1st Marine Division on the outskirts of town found itself unable to talk to Task Force Tarawa inside the city because the West Coast–based 1st Marine Division employed different radio frequencies from the East Coast–based marines of Tarawa.

Combat units often found out where the enemy was just as soldiers have been doing for millennia: by "movement to contact," the military term for bumping into your foes. The difficulty of locating the enemy was exacerbated by the fact that the Fedayeen and other Iraqi irregulars did not employ the kinds of heavy equipment and robust telecommunications networks that U.S. surveillance systems were designed to monitor. They generally drove civilian vehicles, not T-72 tanks, and spread orders by word of mouth, not radio. As a result, the official U.S. Army history concluded, "Every echelon found it nearly impossible to track militia and Fedayeen movements."

But the Fedayeen were so weak and disorganized that lack of information about their movements was not a major impediment to the success of the invasion.

"WE HAD SADDAM BY THE BALLS"

The sandstorm blew itself out on Thursday, March 27. U.S. forces spent the next three days securing their supply lines and pounding the Republican Guard with air power and artillery. By Tuesday, April 1, both the 3rd Infantry and the 1st Marine Division were on the move again, attacking in multiple locations to confuse the Iraqis about where their main effort would come.

This was expected to be the most dangerous part of the entire campaign. Intelligence reports indicated that U.S. troops were entering a "red zone" inside which Saddam Hussein would use chemical weapons to defend his capital. In addition, both American divisions faced formidable geographic obstacles. The 1st Marine Division had to cross over to the east bank of the Tigris River in order to avoid the most heavily defended approaches to the capital. The 3rd Infantry had to shoot through the infamous Karbala Gap—an easily defended bottleneck only a mile wide between Lake Bahr al-Milh and the city of Karbala—and then cross over to the east bank of the Euphrates River. There they would confront three Republican Guard divisions positioned south of Baghdad. Any competent defenders would have blown up all the bridges in the path of the U.S. advance, destroyed dams that would have inundated the area with millions of gallons of water, and zeroed in their rockets and artillery to annihilate the invaders. Luckily for the Americans, the Iraqi defense was far from competent.

The 3rd Infantry roared through the Karbala Gap with minimal resistance. Iraqi defenses stiffened only when the lead U.S. tanks reached the Euphrates River. The key fight was at Objective Peach, the al-Kaed (Leader) bridge where the division planned to cross the Euphrates twenty miles south of Baghdad. The task of securing the bridge was assigned to a reinforced armor battalion known as Task Force 3–69 (3rd Battalion, 69th Armor Regiment) led by Lieutenant Colonel Ernest "Rock" Marcone, a veteran of Desert Storm. Marcone's men arrived on the afternoon of Wednesday, April 2, and made short work of the Fedayeen who were defending the west side of the bridge. But Republican Guard forces on the eastern bank continued to rain down mortar and artillery shells. They had already tried to blow up the bridge, but three lanes remained standing.

Just before 4 P.M., American infantrymen and engineers jumped into six inflatable rubber boats and began rowing across the Euphrates under enemy fire, just as German troops had crossed the River Meuse in May 1940. The engineers reached the far bank and managed to cut the wires on all the remaining explosives. Marcone then sent his men racing across the bridge under the cover of a heavy smoke screen. The Americans had their first major bridgehead over the Euphrates—if they could keep it.

Beginning at 3 A.M. on Thursday, April 3, the Republican Guard mounted a fierce counterattack with more than five thousand soldiers, at least twenty-five tanks, seventy armored personnel carriers, and considerable artillery. To stop them, Marcone had only one thousand men, thirty M1 tanks, and fourteen Bradleys, along with air support from Apache helicopters and various fixed-wing aircraft. It would prove more than enough. In three hours of heavy fighting on a moonless night, the Iraqi forces were annihilated. Thus ended the best-organized Iraqi counterattack of the entire war. Marcone recalled that "for about a mile and a half, you couldn't walk down that road without stepping on a body part and watching tanks and PCs [personnel carriers] burn."

At roughly the same time that the 3rd Infantry was securing a Euphrates crossing, the 1st Marine Division was grabbing a bridge across the Tigris near Numaniyah. The way to Baghdad was clear. "At that point," said General Wallace, the tough-talking V Corps commander, "I was pretty confident that we had Saddam by the balls."

He was right. Tough fighting remained to be done—including an assault on Saddam Hussein International Airport led by "Rock" Marcone and his armored task force, and the thunder runs into Baghdad by other 3rd Infantry units—but the Baathist regime had only days to live. The Stars and Stripes would soon fly over Baghdad.

The U.S. achievement appeared impressive at first blush. No army had ever traveled faster with fewer losses. A force one-third the size of the one employed in Desert Storm had accomplished far more, albeit against a less formidable opponent. Saddam's military machine had never really recovered from the drubbing it received in 1991. In the intervening decade the march of computer technology had largely passed Iraq by. While his military capabilities were atrophying, Saddam's megalomania, paranoia, and delusions were reaching new heights.

As late as April 2, Saddam, who got much of his information from the Internet, was still convinced that the invasion from the south was only a feint; the main coalition attack, he felt, would come from the west and north. Thus he kept many divisions tied up where they were not needed. In another sign of the fantasy world in which he lived, Saddam boasted of his formidable urban fortifications and hinted that he still retained extensive stockpiles of chemical and biological weapons. Both his own generals and the outside world took him at his word, but he did little to turn either threat into reality, because he never believed that an American invasion posed a serious threat to his regime. No one in his inner circle dared to contradict him. In the end Saddam was the victim of the "culture of lies" that he himself had created. As one Iraqi general put it: "People were lying to Saddam, and Saddam was believing them or deceiving himself."

The incompetence of the Iraqi defenders was magnified by the competence of the invaders, whose prowess had grown since the days of Desert Storm. Not only had technology advanced since 1991 but so had the doctrine and organization for utilizing it. General Schwarzkopf had been happy to have the various services fight, in essence, their own individual wars. The lack of army-marine coordination in the Gulf War was a principal reason why the back door was never shut on the Republican Guard. Since then, "jointness" had become a mantra within the armed forces; an entirely new organization had been created in 1999, known as the U.S. Joint Forces Command, to heighten interservice coordination. (By contrast, Iraq's entire command structure was designed to hinder coordination between different units; Saddam did not want any possibility of commanders plotting against him.)

There were still some snafus in Iraq, but on the whole, writes Stephen Biddle of the U.S. Army War College, "Service interactions were broad, deep and profound, ranging from tight integration of close air support (CAS) with ground maneuver to the use of Army logistical assets to support Marine combat units inland, the use of Naval aircraft to support Army Special

Forces, the use of Air Force tankers to refuel Navy aircraft, [and] the use of joint-Service teams to plan and conduct operations. . . ."

PHASE IV

The aura of omnipotence that surrounded American troops after their daring dash to Baghdad would not last long. The fall of Saddam's police state was followed by a disintegration of law and order. Pervasive looting broke out all over the country. Government buildings were stripped of everything down to their wiring and bathroom faucets. Street crime became rampant. For the most part, U.S. soldiers and marines simply stood by and watched. They had no orders to stop this epidemic of criminality. "We paid a big price for not stopping it because it established an atmosphere of lawlessness," American envoy L. Paul Bremer III later said.

U.S. troops also paid a big price for failing to stop the escape of fleeing Baathists. Because of Turkey's refusal to allow transit rights for the 4th Infantry Division (a consequence, in part, of inept American diplomacy), there were few if any regular U.S. troops in northern and western Iraq. Franks had to rely on a handful of Special Operations forces, supplemented in the north by the 173rd Parachute Brigade and the Kurdish peshmerga. This small force did a superb job of pinning down much larger Iraqi army contingents. But once the conventional combat phase ended there were simply not enough Americans to seal off the northern and western escape routes from Baghdad. "By the time Baghdad was taken," Rumsfeld later said, "the large fraction of the Iraqi military and intelligence services just dissipated into the communities."

Most towns in the Sunni Triangle—the area, bounded by Baghdad to the east, Ramadi to the west, and Tikrit to the north, where most of Saddam's support lay—were not occupied until weeks after the fall of Baghdad. And they were never bombed at all. In a perverse way, the coalition paid a price for the precise application of firepower made possible by the advent of "smart" munitions. Not having seen their cities turned into rubble, Iraqis were much less ready than Germans or Japanese to accept foreign occupation. "The people north of Baghdad never understood at a visceral level that they had been defeated," said Lieutenant Colonel Todd A. Megill, chief intelligence officer of the 4th Infantry Division, which eventually occupied most of the Sunni Triangle.

Many critics, especially in the army, argued that the problem was compounded by the lack of coalition troops. The ratio of foreign soldiers to civilians was much lower in Iraq than in the occupations of Bosnia, Kosovo, Germany, or Japan. The Bush administration had hoped to bring in large numbers of allied troops, but there would never be more than 25,000 non-American troops in Iraq. Contractors made up some of the shortfall by importing an army of at least 20,000 to 25,000 private gunslingers to guard buildings and convoys, but the occupiers still found themselves shorthanded for such tasks as securing highways and borders. "While the U.S. could take Iraq with three divisions, it couldn't hold it with three divisions," argued retired diplomat James Dobbins, a veteran of postwar reconstruction efforts in Haiti, Somalia, Bosnia, Kosovo, and Afghanistan. He believed 450,000 troops would have been necessary to secure the country.

The political and military architects of the invasion had decided to use a much smaller force that could reach Baghdad more quickly and avoid many of the problems (such as attacks on Israel or destruction of oil fields) associated with a protracted slugfest. It was a gamble that paid off handsomely in the conventional phase but one for which the coalition would pay a heavy price after the fall of Baghdad.

A multitude of critics would later charge that "no planning" had been conducted for what Central Command called Phase IV—stability operations after Saddam's fall. This was an exaggeration, but it was true that, as in 1991, planners had focused most of their attention on the conventional combat phase without giving enough attention to what came next. "Franks told us he had a Phase IV plan," one administration policy maker later said, "but it was never really fleshed out."

What planning there was for Phase IV had concentrated on a list of what military and civilian officials imagined to be the most awful contingencies, such as massive refugee flows, food shortages, and a currency collapse—none of which came to pass, thanks at least in part to their efforts. Less preparation had been made for keeping order in postwar Iraq. Central Command, acting on estimates provided by the U.S. intelligence community, had assumed there would be an interim government in place with most of the Iraqi army and police at its disposal along with functioning infrastructure. That would not turn out to be the case. The state fell apart much faster and more completely than decision makers at the Defense Department and in the armed forces had anticipated, and there was no provisional government to take its place because Defense Department suggestions to set one up had been blocked by State Department and CIA officials who did not want to place in charge exiles who might lack popular legitimacy.

That was a defensible position, but the upshot was that the Americans were forced to run Iraq themselves—a job they were not prepared for. Prime responsibility for this monumental undertaking fell initially to the tiny Office of Reconstruction and Humanitarian Assistance created under Pentagon auspices just two months before the invasion began. Its head was Jay

Garner, a retired army three-star general who had led efforts to aid Kurds in the aftermath of the Gulf War. He and a handful of aides did not arrive in Baghdad until April 21, 2003, two weeks after the collapse of the regime. They did not find working telephones, much less working ministries, and were not able to achieve much.

Three weeks later, on May 12, 2003, Garner was replaced by L. Paul "Jerry" Bremer III, a tough-talking former Foreign Service officer who favored tan desert boots and blue Brooks Brothers suits. Bremer was an expert on counterterrorism, but he had never run a large organization or spent much time in the Middle East. Yet, as head of the newly formed Coalition Provisional Authority (CPA), he became a veritable viceroy who, within days of his arrival, had to make major decisions with momentous consequences.

Bremer's most controversial steps were to purge most Baathists from the government and to dissolve the Iraqi army. Both decisions—CPA Orders No. 1 and No. 2—involved difficult tradeoffs. Booting the Baathists risked alienating the Sunni minority and depriving the government of critical talent, but leaving them in place would have alienated Shiites and Kurds, who formed 80 percent of the population. Dissolving Saddam's army and security forces made the reestablishment of dictatorship less likely, but it also set loose a large number of unemployed malcontents with military training. Bremer argued that he had no choice, because the mostly Shiite conscripts had simply left their barracks and gone home; there was not much army left. But by not making more of an effort to reconstitute the military (as many U.S. officers were trying to do) or, more important, to build a new internal security force in its place (initial efforts focused on creating a small conventional military designed to fight other armies, not guerrillas), Bremer's actions contributed to the creation of a dangerous security vacuum.

So did lack of reconstruction assistance. Only $2.5 billion had been budgeted initially to rebuild Iraq, an amount that would prove grossly inadequate, given the dilapidated condition in which Saddam Hussein's misrule and a decade of sanctions had left the country and especially its oil industry. (Administration officials would later claim that they had no idea in advance of how run-down everything was, but private experts had foreseen the need for at least $25 billion to $100 billion in reconstruction aid.) In November 2003 Congress voted $18.4 billion in further aid for Afghanistan and Iraq, but the money had to flow through so many bureaucratic brooks and eddies that only a trickle reached its ultimate destination. As of December 2004, just $2 billion had been spent in Iraq—and much of that went for security and overhead costs incurred by American contractors.

In the meantime, coalition military commanders tasked with rebuilding Iraq had to make do with cash seized from Saddam's vaults. The slow pace of reconstruction added further fuel to the fires of discontent.

Even as the Americans were struggling to bring order to Iraq, unbeknownst to them, Baathists were organizing resistance to the occupation. There was no shortage of recruits. In addition to large numbers of former soldiers, there were thousands of criminals released from prison by Saddam Hussein prior to the invasion and thousands of foreign Muslims who had been invited to wage jihad against the "infidels," the most notorious being Abu Musab al-Zarqawi, the Jordanian-born leader of what became known as al Qaeda in Iraq. Some of the foreign jihadis had been killed fighting U.S. forces during the invasion but more were infiltrating every day over borders that had been left largely unguarded.

These disaffected elements—former regime loyalists, criminals, and jihadis—were almost all Sunni Arabs, part of a minority that had ruled Iraq since its inception as an independent nation in 1932. They found sympathizers among the broader Sunni community, which had benefited from Saddam's rule and was now fearful of suffering reprisals from Shiites and Kurds, who had been oppressed for many years.

The money needed to support an insurgency was readily at hand. Saddam and his top lieutenants had siphoned billions of dollars from the state treasury. Some former Baathists took refuge in next-door Syria, which provided an invaluable operations base for them. Arms were also plentiful. Organized gangs looted as many as ten thousand weapons depots. "The terrorists took all the explosives they would ever need," an Iraqi police commander said ruefully.

Later, evidence emerged that these events were not entirely haphazard. An office of Saddam's intelligence service, M-14, the Directorate of Special Operations, had made plans in advance for waging guerrilla warfare against the occupiers. Failure to see this coming was yet another of the many problems that set back the U.S. occupation.

Confusion reigned in the early days after Baghdad's fall, which occurred faster than anyone, including U.S. planners, had anticipated. It took a while for the Baathists to get organized. Despite the prevalence of looting and disorder, coalition forces had a relatively peaceful respite lasting more than a month. But by mid-May 2003 they found themselves coming under attack with growing frequency in Baghdad and the Sunni Triangle.

"There was a time when the insurgency could have been headed off or greatly reduced," said Major General Buford Blount, who commanded the 3rd Infantry Division during the invasion. But by the summer of 2003 that opportunity was slipping away.

At first the guerrillas employed mostly RPG-7s and AK-47s and suffered heavily from return fire. Soon they began using mortars and bombs that were harder to defend against and that took a greater toll. On August 7, 2003, a car bomb blew up the Jordanian embassy in Baghdad, killing seventeen people. On August 19, a cement truck loaded with explosives plowed into the United Nations compound in Baghdad, killing twenty-four people, including special envoy Sergio Vieira de Mello, and all but driving the U.N. out of Iraq. On August 29, another car bomb killed a prominent Shiite ayatollah and at least eighty-five worshippers at the Imam Ali mosque in Najaf.

General John Abizaid, a cerebral Lebanese-American who took over Centcom when Tommy Franks retired on July 7, 2003, was forced to conclude that U.S. forces were facing a "classic guerrilla-type campaign."

HAMMERS AND "HEARTS AND MINDS"

Guerrillas have been giving regular armies fits since at least 329 B.C., when one of Alexander the Great's columns was ambushed and wiped out by Massagetae nomads not far from the modern city of Samarkand, Uzbekistan. While typically not as well armed as their conventional opponents, guerrillas enjoy the advantages of stealth and surprise. They can strike at a vulnerable spot and then disappear to hide among the civilian population or in the wilderness. If an army does not respond swiftly, it can be bled dry. But if it undertakes savage reprisals, it risks driving more supporters into the insurgency's camp.

Not only Alexander but most great captains, from Julius Caesar to Napoleon Bonaparte, have found that putting down an insurrection can be more difficult and time-consuming than defeating a regular army. The prevalence of guerrilla warfare increased in the second half of the twentieth century with the spread of nationalist, communist, and, most recently, jihadist ideology. Both superpowers were humbled by Third World guerrillas—the Americans in Vietnam, Beirut, and Somalia; the Russians in Afghanistan and Chechnya.

In recent years, the Information Revolution had vastly improved the conventional capabilities of the U.S. military, but an insurgency negated many of those technological advantages. As Brigadier General John DeFreitas III, the chief American intelligence officer in Iraq, put it in 2004: "Insurgents don't show up in satellite imagery very well."

Before the war the U.S. intelligence community could not penetrate the Baathist government to determine whether it had weapons of mass destruction. Now it had the much more difficult task of penetrating a far-flung and diffuse insurgency.

The burden of quelling the insurgency and rebuilding Iraq fell squarely on the shoulders of U.S. soldiers spread thinly across the country. The Coalition Provisional Authority was so short on resources, especially in the provinces, that soldiers joked that CPA stood for "Can't Provide Anything." Iraqi security forces were virtually nonexistent at first; the initial units set up by U.S. contractors had such poor training, equipment, and morale that many of them deserted at the first whiff of the enemy's guns. Coalition contingents made a more important contribution, especially the British with their long experience in Northern Ireland; but their troops were stationed only in the Shiite south, which was for the most part peaceful anyway. By default, U.S. soldiers had to become "little potentates"—Lieutenant General Jim Conway's phrase—to run most of Iraq.

Little in their backgrounds had prepared them for this task. With the exception of a few Army civil-affairs specialists and even fewer Arabic speakers (including General Abizaid), American soldiers were not trained in civil governance or schooled in the ways of the Middle East. Such subjects did not much interest a military focused almost exclusively on conventional operations. The year before the Iraq invasion, the army had announced plans to close the Peacekeeping Institute at the Army War College (a decision subsequently reversed in July 2003). The army did not update its decades-old counterinsurgency doctrine until 2004, and West Point did not devote an entire class in the subject until 2005. Most soldiers were exposed only to the kind of tank-on-tank battles practiced at the National Training Center at Fort Irwin.

This outlook—focused on conventional, not irregular, conflict—reflected the views not only of most senior military officers but also of their civilian superiors. President Bush and Secretary Rumsfeld had often expressed their own disdain for "nation-building" and "peacekeeping," which they associated with the foreign policy of the hated Clinton administration. Rumsfeld's technocentric vision of military "transformation," in which the overall number of soldiers would be reduced and resources redirected to long-range, precision-strike systems, left the American armed forces ill-prepared for the challenges they would face in Iraq.

"We're doing a lot of things we didn't know we were getting into," said the commander of a field artillery battalion whose men had to abandon their howitzers and walk the streets like infantrymen. Tankers, logisticians, engineers, and other specialists were also drafted for infantry duty. Infantrymen, in turn, had to learn how to be beat cops. And officers had to turn into politicians, spending time chatting up sheiks and mullahs. The bulk of the burden fell on junior officers and NCOs who had to deal with fast-changing political, economic, and security conditions with little guidance. In the words of an army study: "OIF [Operation Iraqi Freedom] requires junior leaders to be warriors, peacekeepers, and nation-builders—simultaneously."

Many soldiers and marines took to these tasks with great energy and ingenuity, but, not surprisingly, there were some missteps along the way.

Starting in early June 2003, U.S. Army units launched a series of major sweeps through the Sunni Triangle designed to corral insurgents and their sympathizers. With code names such as Peninsula Strike, Desert Scorpion, and Iron Hammer, many of these operations involved thousands of troops backed up by tanks, artillery, and war planes. Often entire villages or neighborhoods were cordoned off and searched house by house. Sometimes suspected enemy safehouses were pounded into rubble. While these operations reaped some valuable intelligence and apprehended some guerrillas, they also alienated many Iraqis who did not appreciate having foreign troops barging into their houses in the middle of the night, pointing guns at them. And these operations were ultimately not effective because the U.S. military lacked the manpower to garrison areas cleared of insurgents. As soon as the Americans pulled out, the guerrillas moved back in.

Many U.S. and British officers thought a softer, less "kinetic" approach might have been more productive. "We made things worse for ourselves by pulling out the hammer as the first tool of choice," said Lieutenant Colonel Bjarne M. "Mike" Iverson, an Arabic-speaking combat engineer who served in 2003–4 as political-military advisor to Lieutenant General Ricardo Sanchez, the senior military commander in Iraq. "We wounded their Arab pride and their tribal pride." Another U.S. Army colonel put it more bluntly: "If I were treated like this, I'd be a terrorist!"

At the same time, coalition military units launched programs designed to win "hearts and minds": handing out soccer balls and treats to kids, setting up city councils and police forces, rehabilitating hospitals and schools, improving sewage treatment, trash collection, electricity generation, and other basic services. Their attempts to rebuild were hampered, however, by lack of resources, the U.S. government's Byzantine procurement bureaucracy, the insurgents' skillful sabotage, and the corruption among many Iraqi (and some American) officials. As a result, three years after the fall of the Baathist regime, Baghdad still did not have reliable round-the-clock power.

INSIDE AND OUTSIDE THE WIRE

From June 2003 on, every week seemed to bring worse news for the coalition: more bombings, more deaths, more crime, more chaos. By September 2003 more U.S. soldiers had died in the "post-combat" phase than in the invasion of Iraq. And still the casualties kept coming—110 coalition soldiers died in November 2003 alone.

The safest areas for U.S. troops were their heavily fortified bases, but even here they suffered a steady stream of casualties from mortar and rocket attacks. More than one hundred of these "little oases" sprang up across the country, many located on the grounds of Saddam's numerous former palaces. They were built by KBR (a subsidiary of Halliburton formerly known as Kellogg Brown & Root) and staffed mainly by laborers imported from such non-Muslim nations as India and the Philippines. The largest bases, located around Baghdad and the northern town of Balad, even featured their own fast-food restaurants, coffee bars, swimming pools, movie theaters, and giant PXs the size of a Wal-Mart. In between raids, soldiers could watch DVDs or work out in the weight room, send e-mail, play video games, and talk with people back home via satellite phone or videoconference.

What they lacked, for the most part, were the warrior's age-old consolations—wine and women. Both alcohol and local women were strictly off-limits in deference to local sensibilities and security concerns. There was almost no "fraternization" of the kind which had been common in past occupations. This made it all the more difficult to gather the intelligence needed to stop the insurgency.

Senior U.S. civilian and military leaders were even more cut off in their Baghdad safe haven known as the Green Zone, which had once been occupied by Saddam and his cronies. Here, surrounded by eight miles of blast walls, concertina wire, watchtowers, and security checkpoints, lived and labored some two thousand officials along with thousands of support personnel. Described by one magazine writer as "a little America embedded in the heart of Baghdad," the Green Zone was replete with mess halls serving American food, televisions showing American programs, and telephones with American area codes. American officials were supposed to venture out only in armored convoys full of bodyguards.

The insurgents sought to further increase the Americans' isolation by intimidating or killing Iraqis who worked with them, whether as translators or janitors. Thousands of "collaborators," ranging from police cadets to senior government ministers, were slain and their families threatened, kidnapped, or attacked.

Whenever Americans went "outside the wire," leaving behind their well-guarded compounds, they were vulnerable to attack. The guerrillas' most effective weapon was the improvised explosive device, accounting for more than half of all U.S. combat casualties. Insurgents became clever at hiding them in roadside trash, rusted gasoline cans, broken-down cars, even dead animals. Sometimes a string of such bombs would be linked together to form a daisy chain that could devastate multiple vehicles. Insurgents also packed explosives into cars driven by suicide bombers, which the U.S. military called VBIEDs (vehicle-borne improvised explosive devices).

Soldiers would have been relatively safe if they patrolled only in M1 tanks. While, by early 2005, eighty Abramses had been disabled by insurgents (designed to deflect frontal hits, the tanks proved vulnerable in their rear and underside), most crew members managed to escape with only minor injuries. But these sixty-eight-ton behemoths were so heavy that they chewed up the pavement, so loud that they made it impossible to surprise anyone, and so forbidding that they prevented their occupants from interacting with locals to build trust and gather intelligence. Armored personnel carriers (such as the new twenty-one-ton Stryker, which ran on tires, not tracks) offered a better balance between protection and agility. But there were not nearly enough to go around.

For the most part, U.S. troops had to rely on aluminum-and-fiberglass Humvees not intended for frontline use. They were vulnerable to IEDs, rocket-propelled grenades, even AK-47s. In 2003, only one in eight Humvees in Iraq was armored. "We're kind of sitting ducks in the vehicles we have," complained a military police commander. In 2003, the Pentagon ramped up production of armored Humvees and trucks, but in the meantime some troops had to affix jury-rigged steel panels, known as "Hillbilly armor," to their own vehicles. Of course even the most heavily armored vehicles could be blown up by a sufficiently large mine or bomb, and as the strength of American armor increased, so did the size of insurgent munitions.

The lack of "up-armored" Humvees was just one of many shortages facing U.S. troops, especially those from the less well equipped National Guard and Reserves who were increasingly called up for occupation duty. There was nothing unusual about this; armies have often found themselves strapped for supplies in wartime, and U.S. troops were infinitely better off in Iraq than their forefathers had been in the Korean War, World War II, the Civil War, or the Revolutionary War. What was striking and unusual in 2003 was that U.S. forces were superbly equipped for high-end combat, but some of them lacked cheaper and more mundane pieces of equipment needed to fight a guerrilla war—equipment that, unlike aircraft carriers, fighter planes, or other expensive weapons system, was not of much interest to the lawmakers, contractors, and senior service officials who set the nation's military spending priorities.

In tank, artillery, and other units where soldiers were not used to operating as infantry, there were sometimes not enough assault rifles, radios, GPS devices, or night-vision goggles to go around. By mid-2004, army units were even running low on small-arms ammunition. Worst of all was a shortage of the most advanced bulletproof vests, a piece of gear that had grown in importance over the past few decades.

Body armor had gone out of fashion in the seventeenth century once guns could reliably penetrate medieval steel mesh or plates. It was only in World War I that armies revived the practice of wearing steel helmets (heads were exposed in the trenches) and only in World War II that some personnel, beginning with air crews, began to be issued nylon flak jackets that could stop shrapnel if not bullets.

In 1965 DuPont chemist Stephanie Kwolek invented a synthetic fiber known as Kevlar to replace the steel belts inside tires. This material was five times stronger than steel and capable of stopping a 9 mm handgun bullet. In the 1970s, after finding that goats draped in Kevlar survived being shot, the Justice Department and the U.S. Army adapted Kevlar for use as helmets and body armor by police officers and soldiers.

The next major advance occurred in the 1990s when the Pentagon's Defense Advanced Research Projects Agency (DARPA) developed carbide ceramic plates for insertion into the front and back of Kevlar vests. These Small Arms Protective Inserts were harder than Kevlar and capable of stopping a 7.62 mm rifle round. A Kevlar vest with ceramic plates, known as the Interceptor Multi-Threat Body Armor System, weighed just 16.5 pounds, as opposed to 25 pounds for its predecessor, making it not only safer but also easier for infantrymen to lug around.

The Interceptor system, first used in Afghanistan, helped to reduce deaths among U.S. forces. So did advanced medical care. In Iraq, Forward Surgical Teams equipped with the latest in trauma care were spread throughout the country so that wounded soldiers could receive surgical attention within thirty minutes of an attack. In World War II, one in three casualties died. From the Korean War to the Gulf War, the figure was one in four. In Iraq it was just one in ten.

An unfortunate side effect of this success was that more soldiers came home permanently injured. Since body armor left the extremities unprotected, a growing number of soldiers had to have arms or legs amputated. But better to lose a limb than a life—especially since high-tech prosthetics made possible a far wider range of activities than had been possible for previous generations of disabled veterans.

Even more would have survived if Interceptor armor were available to all troops in Iraq. But, as with up-armored Humvees, there were critical shortages early on. The Pentagon rushed to increase production when the magnitude of the threat became clear, but the procurement process was so sluggish that some soldiers felt compelled to ask their families to buy vests for them in civilian stores and ship them to Iraq.

A HIERARCHY VS. A NETWORK

Getting better equipment was only one of many adjustments that U.S. forces had to make against an increasingly powerful insurgency whose 10,000 to 30,000 fighters (estimates varied widely) stymied their best efforts to pacify Iraq.

One lieutenant colonel, figuring out that IEDs were being set off with radio transmitters from toy cars, mounted a toy-car transmitter on the dashboard of his Humvee and taped down the levers so that as he drove around it would detonate any bomb on the same frequency within one hundred yards. The Defense Department soon came up with a more high-tech version of this same concept, fielding a variety of jamming devices to disrupt IED transmissions. Bomb-inspection and bomb-disposal robots were also rushed to Iraq. But a member of the Pentagon's Joint I.E.D. Defeat Task Force had to admit, "There is no technology silver bullet." When U.S. forces jammed some bomb-detonation frequencies, insurgents responded by switching to different frequencies or by detonating bombs with wires or infrared beams. The Pentagon had trouble keeping up. Troops found that some of the most useful responses were tactical, not technological: intensively patrolling roads, for instance, or planting snipers at highway overpasses to discourage insurgents from planting bombs. U.S. forces managed to find and defuse 30 to 40 percent of all IEDs, but the remainder ripped into soft-skinned vehicles and their even softer-skinned occupants.

"It's a constant struggle of one-upsmanship," said Major John Nagl, a counterinsurgency expert who was stationed in the Sunni Triangle in 2003–4. "We adapt, they adapt. It's a constant competition to gain the upper hand."

The insurgents had important advantages in this competition because of their vastly greater knowledge of the local area. U.S. troops were hobbled by their rotation policies: Most soldiers spent only a year in Iraq, most marines six to seven months. They found that it could take many months just to figure out what was going on. By the time they were fairly knowledgeable, it was time to go home. Many CIA and State Department officers spent even less time in Iraq, rotating through on sixty- to ninety-day stints.

The military was forced to adjust its training regimen, designed for conventional combat operations, to better prepare incoming troops for the rigors of counterinsurgency. A top priority was to provide more combat training for logisticians whose trucks had become magnets for insurgent attacks. The army hired two information technology companies to develop a simulator known as the Convoy Skills Trainer so that soldiers could hone their skills in fighting off ambushes. Before embarking for Iraq, soldiers could climb into four plywood boxes rigged up like Humvees and stare at virtual-reality screens that gave them the illusion that they were driving along while fighting off insurgent attacks using weapons that fired electronic beams, not bullets. The results of their actions would be displayed on giant video screens for analysis.

Many of the troops who arrived after the insurgency had gotten under way also received more cultural and civil affairs training than their predecessors. Before taking over Baghdad in April 2004, the 1st Cavalry Division based at Fort Hood, Texas, sent its officers to a training session where Austin city officials taught them how to run a modern metropolis. Officers and NCOs rode along with electrical, sewage, and trash workers to see how they did their jobs. Some units could further hone their skills at complexes such as the Joint Readiness Training Center at Fort Polk, Louisiana, which was transformed into a simulacrum of Iraq complete with Arabic street signs and mosques. To heighten the realism, more than 1,200 staff, including hundreds of Iraqi-Americans, would interact with soldiers playing roles ranging from insurgents to reporters.

Some of the most successful learning aids bubbled up from the lower ranks. In 2000, two army captains on their own initiative set up a web site called *companycommand.com,* where their peers could exchange ideas and debate various topics. A year later they set up a similar web site for lieutenants called *platoonleader.org*. Eventually the army began underwriting both sites. By 2004, *companycommand.com* had ten thousand members, or more than a third of all captains in the army. Individual divisions set up their own versions, such as the 1st Cavalry Division's *cavnet.* Even at the most remote base in Iraq, an officer could go online for tips on how to stop IEDs or what to do when a soldier got killed.

Not all U.S. soldiers could adjust to the unconventional nature of this conflict. But many could, and they managed to achieve impressive results.

Acting more as detectives than as soldiers, the 101st Airborne Division managed to locate Saddam's sons, Uday and Qusay. On July 22, 2003, they were killed in a firefight at their Mosul hideout. Less than six months later, the 4th Infantry Division, working with Special Operations forces and intelligence agencies, bagged Saddam himself. On December 12, 2003, the disheveled Iraqi dictator was hauled out of a subterranean "spider hole" near his hometown of Tikrit. This coup was the product of careful intelligence work that involved interrogating Saddam's supporters and drawing an elaborate diagram of their connections.

As important as Saddam's capture was (or the death on June 7, 2006, of Abu Musab al-Zarqawi), it did not spell an end to the uprising. By this time, the guerrillas were so decentralized and so well-established that there was barely a pause in their attacks. In Iraq it was the insurgents who were waging "netwar"—thanks, ironically, to the invaders' success in smashing Saddam's overly centralized command structure. The Americans' bureaucracy remained intact and often proved a hindrance. Simply sorting through all the intelligence reports generated by various U.S. government agencies in Iraq was a major challenge because they were scattered among dozens of different databases. "We are a hierarchy trying to fight a network," conceded General Peter Schoomaker, the U.S. army chief of staff.

"AL JAZEERA KICKED OUR BUTTS"

In their desire to stamp out the uprising, some U.S. troops and private security contractors committed excesses that are, unfortunately, all too common in this kind of dirty war. Fearful of suicide bombers, soldiers were known to open fire on vehicles that got too close to their convoys or did not slow down fast enough in front of their checkpoints. Often these cars were driven not by terrorists but by hapless civilians who paid for their confusion with their lives. There were also reports of interrogations getting out of hand, with soldiers beating and occasionally killing suspects. Given the frustrations of battling a vicious enemy that did not abide by the rules of war, it was not surprising that young soldiers making split-second decisions while in mortal danger had great difficulty in calibrating the proper response. "We're really good at teaching privates to be gung-ho killers," observed Lieutenant Colonel Kenneth E. Tovo, a Special Forces officer who served in Iraq. "It's much harder to teach restraint and judgment—to teach when not to shoot."

The most serious abuses discovered among U.S. forces in the first two years of the war occurred between October and December 2003 at Abu Ghraib prison near Baghdad. A small group of sadistic guards subjected inmates to grotesque humiliations such as stripping them naked and photographing them in a human pyramid. An independent review panel headed by former Defense Secretary James Schlesinger found "no evidence of a policy of abuse promulgated by senior officials or military authorities," but criticized a failure of oversight and leadership. Among other problems, there were simply not enough soldiers, especially not enough trained military police and interrogators, to handle the thousands of detainees who were rounded up. "In Iraq there was not only a failure to plan for a major insurgency," the Schlesinger panel found, "but also to quickly and adequately adapt to the insurgency that followed after major combat operations."

News of the Abu Ghraib scandal broke on April 28, 2004, just three months after weapons inspector David Kay had announced that intelligence estimates about Saddam Hussein's weapons of mass destruction that had been used to justify the war were "almost all wrong." Together, these events helped to undermine American public support for the war effort.

A further shock came on March 31, 2004, when four American civilians employed by the security firm Blackwater drove into Fallujah. They were ambushed and killed, and their corpses strung up and disfigured. This ghastly spectacle, broadcast worldwide, led Rumsfeld and Abizaid to order the local marine garrison to invade Fallujah to try to regain control of the city, rather than continuing their strategy of slowly trying to win over local support.

On April 4, 2004, two marine battalions launched an assault on Fallujah. Five days later, while marines were still battling their way into the city, the offensive was suspended because of inflammatory media coverage, primarily on the Arabic satellite news channel al Jazeera, which claimed that marines were deliberately targeting mosques and civilians. Bremer, Abizaid, and other senior officials feared that, if the operation continued, support for the U.S. would crumble throughout the country.

The U.S. military had done a good job of conducting "information operations" during the major combat phase of Operation Iraqi Freedom. Embedding hundreds of reporters with coalition forces had provided a sympathetic view of the invasion. But now the propaganda initiative had passed to the guerrillas, who cannily produced videos of bombings and beheadings for dissemination on the Internet and satellite television. This was an element of modern warfare created by the Information Revolution that no democratic government could ignore. While the Russian army could carry out a scorched-earth policy in Chechnya safe from much public scrutiny, the U.S. armed forces had to conform to the highest standards of conduct or risk public obloquy that would undermine their mission.

In Fallujah, negative news coverage succeeded in doing what Saddam Hussein's military had failed to do: It stopped the mighty U.S. military in its tracks. As Lieutenant General James Conway, commander of the 1st Marine Expeditionary Force, put it, "Al Jazeera kicked our butts."

The Marines had to leave Fallujah, handing the insurgents their most notable victory.

Even as fighting raged in Fallujah, followers of a fiery young Shiite mullah named Moqtada al Sadr rose up to battle the coalition from the holy city of Najaf to the Baghdad slums known as Sadr City. Other insurgents targeted bridges and roads used by coalition logistics convoys; supplies began to run low at forward operating bases. Many ill-trained Iraqi police officers and troops walked off the job rather than fight.

Gloomy commentators worried that all of Iraq was rising up against the coalition. But most Shiites and Kurds sat on the sidelines. Ayatollah Ali Sistani, the senior Shiite cleric, counseled cooperation, not confrontation, with the occupiers.

After seven weeks of fighting, Sadr's al Mahdi militia, having failed to receive widespread support and having suffered numerous military setbacks, agreed to a cease-fire. The Mahdists rose up again in August 2004, but they were no more successful than their namesakes had been in the Sudan in 1898. The Shiite fighters were decisively defeated by a joint army-marine force in a three-week battle for control of Najaf. Of vital importance was the Americans' ability to avoid hitting the sacred Imam Ali mosque where insurgents were hiding—a tribute to their precision weaponry and disciplined soldiery. After approximately one thousand guerrillas had been killed (to eleven Americans), Ayatollah Sistani negotiated a truce and Sadr agreed to lay down his arms. Sadr City went from being one of the most dangerous areas for U.S. troops to one of the safest.

Having reestablished control over southern Iraq, U.S. commanders were determined that Fallujah not continue as a terrorist stronghold in the west. On November 8, 2004, four marine battalions and two army battalions—about ten thousand men altogether—backed up by two thousand Iraqi soldiers invaded the city in a carefully planned assault. Utilizing tanks and close air support, the Americans retook Fallujah in savage street fighting that demolished much of the city. At a cost of 71 dead Americans and hundreds more wounded, U.S. forces claimed to have killed at least 1,200 guerrillas and captured 1,000 more. They also uncovered 26 bomb factories, 350 arms caches, and 8 safe houses where hostages had been tortured and killed. Many guerrillas escaped the U.S. dragnet, but Fallujah had been cleared out—and wiped out.

Like the siege of Najaf, the battle of Fallujah was a conventional fight in which the U.S. advantages in armor, air power, and artillery could be brought to bear even in an urban environment. But such opportunities were rare. U.S. soldiers generally had the frustration of battling an unseen foe—"ghost fighters," one marine corporal called them. And, unlike in Najaf or Fallujah, progress could seldom be measured by looking at a map. In a counterinsurgency, the goal was to win not territory but—much more difficult—the allegiance of the people who lived there.

This task was complicated by the fact that the new rulers of Iraq were foreigners—Jerry Bremer and his Coalition Provisional Authority staff. The CPA had some notable achievements, such as helping Iraqis to draft an interim constitution, but its lack of legitimacy made it hard to mobilize opposition to the insurgency. The Defense Department had never been happy with this arrangement, and in October 2003 President Bush decided to speed up the transition to self-rule. On June 28, 2004, the CPA went out of business. It was replaced by an interim government led by former exile Ayad Allawi, who could just as easily have been appointed a year earlier. At the same time, the U.S. military put a renewed emphasis on training and equipping Iraqi security forces. The job was assigned to Lieutenant General David Petraeus, one of the army's most respected officers, who had just completed a stint commanding the 101st Airborne Division in northern Iraq.

Seven months later, on January 30, 2005, contested parliamentary elections were held for the first time in many decades. More than eight million Iraqis (60 percent of eligible voters) braved insurgent bombs and bullets to cast ballots. Although at least forty people were killed on election day, the process went more smoothly than many had expected, thanks to the work of Iraqi security forces and 170,000 coalition troops. Most Sunnis boycotted the balloting, but they turned out in large numbers on December 15, 2005, when a total of 10.9 million voters (70 percent of those eligible) cast ballots for parliamentary candidates.

Insurgent attacks did not decrease after these elections. But the formation of an elected government and the increasing effectiveness of Iraqi troops raised hopes that, eventually, the bulk of foreign forces would be free to leave. Most Americans were eager for that day to come, because, as of May 2005, the war had cost them $148 billion and more than 1,600 lives—a toll that continued to grow steadily in the months ahead.

Recruiting shortfalls led many officers to wonder whether the professional military created after Vietnam could stand the strain of such a protracted and deadly conflict. The Army Reserve, its commander warned on December 20, 2004, was "rapidly degenerating into a 'broken' force."

UNCONVENTIONAL CHALLENGES

To see what comes easily to an Information Age military and what does not, compare casualty figures for the initial invasion with those for the subsequent occupation. Between March 20 and April 9, 2003, the U.S. had suffered 139 dead and 542 wounded. (The British had an additional 33 fatalities.) Between May 1, 2003, and May 1, 2005, 1,510 Americans died and 11,346 were wounded. In other words, the first two years of the guerrilla phase were eighteen times more costly than the conventional combat operations which had ended on May 1, 2003.

By historical standards—by comparison with the Algerian War of Independence or the Vietnam War, to say nothing of the carnage of World War I or World War II—the insurgency in Iraq was not particularly costly. Allied losses on D-Day (an estimated 5,400 servicemen died on June 6, 1944) far exceeded losses in two years of fighting in Iraq. Given that some 900,000 U.S. military personnel served in Iraq in 2003 and 2004, the fatality rate was a minuscule 0.15 percent and the total casualty rate was 1.3 percent. But such statistics offered no solace to the families of the fallen or to an American public shocked by the number of body bags coming home and dismayed by the collapse of the chief rationale for the invasion—the alleged presence of massive chemical and biological stockpiles in Iraq.

Americans had gotten used to thinking of war as a bloodless, push-button exercise in which they could zap their enemies at scant cost to themselves. Many people had come to assume that Desert Storm (total U.S. casualty rate: 0.1 percent) was the new norm. The occupation of Iraq showed how savage real war could be—and how little protection information technology offered against a cunning and ruthless foe.

High-tech hardware was hardly useless. Precision munitions came in handy when the U.S. wanted to target terrorist safe houses in Fallujah or Najaf without destroying neighboring structures. A three-dimensional computer modeling tool known as Urban Tactical Planner was useful for preparing troops for raids in cities. JSTARS aircraft circling in the night sky above Iraq coordinated troop movements in response to enemy ambushes or IEDs. And so many unmanned aerial vehicles were employed to search for insurgents—more than seven hundred—that there was a serious risk of midair collisions.

But, even in the conventional phase, brute force was sometimes called for rather than finesse—for instance, during the thunder runs. High-tech gadgets became even less important in the guerrilla phase. The primary task of any army engaged in counterinsurgency is simply to gather intelligence. And, as Major General James Mattis, the Pattoneseque commander of the 1st Marine Division, remarked while garrisoning southern Iraq in August 2003, "We get 95% of our intelligence from humint [human intelligence] and only 5% from sigint [signals intelligence] or elint [electronic intelligence]."

The lesson the United States was learning in Iraq was similar to the lesson that other armed forces had learned on other battlefields: a military machine built for one purpose, no matter how superb, could not easily be redirected for another kind of fight. In Spain's case, an armada that might have been well suited for the placid waters of the Mediterranean did not fare as well in the tempestuous seas of northern Europe. In Germany's case, an army built for a blitzkrieg across relatively confined spaces did not fare as well in the vast steppes of Russia. In America's case, armed forces built for conventional combat in the Arabian desert or European plain did not fare as well in the streets and alleys of Iraq.

In fairness, it is doubtful that any other army could have subdued Iraq at lower cost; there simply is no way to fight this kind of war on the cheap. But that is precisely the point: Defeating an insurgency still requires the kind of messy, block-by-block fighting that many thought had been rendered obsolete by the dawn of the Information Age. Operation Iraqi Freedom showed that there were still some things that not even the most advanced machines could do.

The Consequences of the Information Revolution

It is tempting to argue, based on the difficulties that U.S. forces have encountered in Iraq (and the similar problems of the Israeli Defense Forces in Lebanon, the West Bank, and Gaza Strip), that the effect of the Information Revolution on warfare has been overstated by some enthusiasts. There is some truth to this—there *has* been an awful lot of hype—but it would be a mistake to go too far in dismissing the results of recent advances. That would be akin to denying the effect of the Industrial Revolution in the early 1900s simply because the British had a lot of trouble subduing the Boers. Improvements in technology have always had a more immediate impact on conventional than on irregular warfare. But that doesn't mean that the art of war has remained static. Indeed the tactics of Islamist guerrillas, who rely so heavily on the Internet, cell phones, and satellite television—all of which barely existed in 1980—show just how much things have changed.

Not all of the changes wrought by the Information Age are obvious at first glance, because the basic military systems of the early twenty-first century look roughly similar to their predecessors of the Second Industrial Age. Military analyst Michael O'Hanlon notes that "basic propulsion systems and designs for aircraft, ships, and internal-combustion vehicles are changing much more gradually than in the early twentieth-century, when two of those three technologies had only recently been invented." The average speed of a U.S. Navy destroyer, for instance, has not increased in the past one hundred years. The U.S. Air Force continues to rely on B-52H bombers last built in 1962. And the Marine Corps still uses helicopters that flew in the Vietnam War. What has been changing with great rapidity since the mid-1970s is the communications, targeting, surveillance, and ordnance technology that can make such "legacy" systems considerably more potent.

LAND WARFARE: Advanced armies are still structured, as they have been since the 1940s, around armored forces complemented by infantry troops who move by armored vehicle, truck, and aircraft. The best tank in the world is probably the American Abrams (of which the U.S. has nine thousand), but the British Challenger II, the German Leopard II, the Israeli Merkava Mk. 4, and the Russian T-80 and T-90 come within striking distance. All modern tanks have stabilized turrets, night-vision capabilities, laser range-finders, and targeting computers that allow them to fight in conditions—on the move or in the dark—that would have stymied earlier models. In addition, composite or reactive armor offers far more protection than in years past, and main guns firing depleted uranium rounds have far more penetrating power. Armored personnel carriers and infantry fighting vehicles, such as the American Bradley Fighting Vehicle and Stryker and the Russian BMP and BTR, are essentially light tanks, some running on wheels, others on tracks, that sacrifice armor and armaments for extra room to carry infantry, command-and-control suites, or other cargo. Self-propelled artillery and rocket systems are also mounted on armored chassis.

Armored vehicles have improved over the years. But so have antiarmor weapons. These range from heavy missiles such as the U.S. Hellfire and Russian Ataka-V fired from vehicles or aircraft to handheld versions such as the U.S. Javelin, the Franco-German Milan, and the Russian Kornet. In addition, even the most advanced tanks can be disabled by other tanks, massive mines, aerial bombs, or artillery shells. The full impact of advances in antiarmor technology has not yet become apparent, because the forces that have fought modern tanks in recent years—Iraqis, Palestinians, Chechens—have not possessed the latest defensive weapons. But the U.S. success in wiping out Iraqi tanks from standoff ranges suggests that, in the constant struggle between offense and defense, the advantage may have shifted against heavy armor.

The U.S. Army is responding to these changes by budgeting at least $124 billion to develop a Future Combat System that is supposed to replace much of its current armored force with a family of lighter vehicles, manned and unmanned, that will have stealth designs that will make them harder to detect and hybrid-electric engines that will lessen their fuel requirements—one of the chief disadvantages of the gas-guzzling Abrams, because it increases demands on vulnerable supply lines. Future vehicles will feature advanced composite armor designed to deliver more protection than current models for the same amount of weight, but they will rely for protection less on armor and more on locating and destroying the enemy before they are attacked. Critics believe this is placing too much faith in "perfect situational awareness," and that these vehicles will not be of much use against guerrillas who can strike with no warning.

As usual, the infantryman's tools have changed least of all. A modern soldier has better protection than his forefathers if he wears Kevlar body armor, but his firepower—which comes primarily from a handheld assault rifle like the M-16 or AK-47 and from a variety of crew-served mortars and machine guns—does not vary significantly from that of a G.I. in World War II. A replacement for the M-16 known as the XM29 is under development but it is hardly revolutionary. In addition to shooting the same 5.56 mm rounds as the M-16 out of one barrel, it will have another barrel that can fire 20 mm high-explosive airburst projectiles to a range of half a mile. These minigrenades will come with embedded microchips that will control when they explode, allowing them to kill enemy fighters who might be lying flat on the ground or hiding behind a berm. Alternatively, nonlethal projectiles like rubber balls could be substituted for crowd-control situations. This is not terribly different from the capability afforded by grenade launchers attached to today's M-16s. Electronic guns that are capable of spitting out a million rounds a minute have also been developed. They might permit a soldier to stop an incoming rocket-propelled grenade with a solid wall of lead. But such weapons are years away from being fielded.

Unfortunately for Western infantrymen, the proliferation of small arms can put low-tech foes on an almost equal footing with the representatives of the most advanced militaries. There are 250 million military and police small arms knocking around the world, and more are being manufactured all the time by at least 1,249 suppliers in ninety countries.

The salvation of Information Age infantry, at least when they are conducting conventional operations, is their ability to use a wireless communications device to call in supporting fire on exact coordinates. It is doubtful that any military force will again enjoy the preponderance of power of a Kitchener at Khartoum, but Americans dropping JDAMs on Afghan tribesmen armed with Kalashnikovs—or even on Iraqi soldiers with outdated T-72 tanks—came close. The American edge decreases considerably, however, when its troops have to deploy for peacekeeping or counterinsurgency operations that leave them exposed to low-tech ambushes.

NAVAL WARFARE: Navies remain divided, as they have been since the dawn of the Second Industrial Age, into aircraft carriers, submarines, and surface ships. The major difference is that blue-water naval competition has disappeared after more than five hundred years. No one even tries to challenge the U.S. Navy anymore on the high seas. Virtually every other navy in the world is little more than a coastal patrol force.

The U.S. has twelve aircraft carriers, nine of them *Nimitz*-class, nuclear-powered supercarriers that can carry more than seventy high-performance aircraft such as the F/A-18E/F Super Hornet. A tenth supercarrier is in the works. No one else has a single one. France has the world's only other nuclear-powered aircraft carrier, the *Charles de Gaulle*, but it's half the size of the *Nimitz*. Russia has one aircraft carrier, the *Admiral Kuznetsov*, that rarely leaves port, and it has sold another one, the *Admiral Gorshkov*, to India. Britain has three small *Invincible*-class aircraft carriers that are used only for helicopters and vertical-takeoff Harrier jets. France, Italy, Spain, Japan, and South Korea have similar helicopter carriers in the works. These ships are comparable to the U.S. Navy's twelve amphibious assault ships, which transport helicopters, jump jets, and marines.

Whenever they leave port, U.S. capital ships are surrounded by surface and submarine escorts. Twenty-four *Ticonderoga*-class cruisers and forty-five (and counting) *Arleigh Burke*–class destroyers come equipped with Aegis phased-array radar that can track up to nine hundred targets in a three-hundred-mile radius. These surface combatants can also operate on their own or in conjunction with smaller vessels such as frigates and minesweepers.

In World War II, ships that didn't carry aircraft were limited to firing torpedoes or heavy guns with a range of less than thirty miles. Starting in the 1960s some submarines were equipped with intercontinental range ballistic missiles, but their targeting was so imprecise that it made no sense to equip them with conventional warheads. Ballistic-missile subs became a mainstay of nuclear deterrence. The development of accurate cruise missiles starting in the 1970s allowed submarines and surface combatants to hit land targets hundreds of miles away with conventional ordnance. Improvements in torpedo design, including the development of rocket-propelled supercavitating torpedoes, also allow submarines to do more damage in their traditional ship-fighting role.

The U.S. has the world's largest fleet of nuclear-powered attack submarines (54) and nuclear-powered ballistic-missile subs (16). Russia is next, with 37 attack submarines and 14 ballistic missile subs. Britain has 15 nuclear-powered submarines, followed by France with 10, and China with six. Not only are U.S. submarines more numerous but they are also more advanced, the most sophisticated being three 1990s-vintage Seawolfs described by one defense analyst as "the fastest, quietest, and most heavily armed undersea vessels ever built."

Because of the growing power of each of its vessels and the lack of competitors, the U.S. Navy has consolidated its high seas hegemony even while its fleet has shrunk from almost five hundred ships in the 1980s to fewer than three hundred in the early years of the twenty-first century. The potency of U.S. naval vessels is increased by linking together sensors and weapons systems with a tactical Internet known as FORCEnet.

While the U.S. Navy probably will remain unchallenged in blue waters, it faces greater threats as it gets closer to shore. Here water currents, thermal layers, and various obstacles can interfere with even the most advanced sensors, and a variety of defensive weapons systems lurk in wait.

More than seventy-five thousand antiship missiles are owned by seventy countries. A few are ballistic, but most are of the cruise-missile variety. Their potency was proved in 1987 when French-made Exocets fired by an Iraqi aircraft crippled the USS *Stark*, a Perry-class frigate, killing thirty-seven sailors. Earlier, Argentinian-owned Exocets sank two British ships during the Falkland Islands War (1982). Newer antiship cruise missiles such as the Russian-made Yakhont, Sunburn, and Uran are even deadlier because they have faster speeds, greater stealth capabilities, and more accurate, GPS-enhanced targeting. Russia is selling these missiles to customers abroad and some nations, such as China, are developing their own versions.

U.S. warships have sophisticated defensive systems to guard against air attack: Incoming missiles can be deflected by electronic countermeasures, flares, or chaff, or destroyed by naval aircraft, sea-to-air Standard missiles, or, as a last resort, by rapid-fire, radar-guided Phalanx guns. But, like the *Stark*, a warship could be caught by surprise or overwhelmed by a flurry of missiles coming from different directions.

Even more worrisome from an American viewpoint is the fact that transport ships and fuel tankers which have to replenish a fleet at sea have no protection when they are outside the defensive range of a battle group. They are as vulnerable as supply convoys on the roads of Iraq. Because a supercarrier has only about a three-day stockpile of JP-5 jet fuel (6,500 barrels a day are needed during combat operations), the most powerful warship in history could be rendered useless if its fuel tankers were sunk. The cargo container ships upon which global commerce depends also have no antimissile protection. A country or even a subnational group armed with cruise missiles could wreak havoc on the world's shipping lanes at chokepoints like the Straits of Hormuz or Malacca.

The threat to shipping, civil and military, is increased by diesel submarines. The latest diesel submarines have ultraquiet electric engines that make them hard to detect with sonar, and they are much cheaper to buy or produce than a nuclear-powered submarine. Russia has exported Kilo-class diesel-electric subs to China, India, Iran, and Algeria, among others. China is producing its own *Song*-class diesel submarines in a bid to challenge U.S. naval hegemony by using the same strategy that Germany, with its U-boats, once used to challenge British dominion of the waves. U.S. antisubmarine defenses are quite sophisticated, especially in open waters, but even American sensors can have trouble tracking quiet diesel subs in noisy coastal waters.

Mines, which can be scattered by submarines or other vessels, represent another major threat to shipping. More than three hundred different varieties are available on the world market. They can be triggered by changes in magnetic fields, acoustic levels, seismic pressure, or other factors. Some come equipped with microelectronics that allow them to distinguish between different types of ships, while others have small motors that allow them to move around. This makes it difficult to certify that a shipping channel is free of mines—it may have been safe an hour ago, but not anymore. Demining technology has lagged behind; the U.S. Navy, for one, has never placed much emphasis on lowly minesweepers. It has paid a price for this neglect. In 1987, during operations to prevent Iran from closing the Persian Gulf, an Iranian mine of World War I design nearly sank the frigate *Samuel Roberts*. Four years later, in the Gulf War, the cruiser USS *Princeton* and the amphibious landing ship *Tripoli* were nearly blasted apart by Iraqi mines. And even a cheap motorboat packed with explosives can pose a significant threat to a modern warship. The USS *Cole*, an *Arleigh Burke*–class destroyer, was badly damaged in such a terrorist attack in the port of Aden, Yemen, in 2000.

All of these threats could be largely negated if U.S. fleets were to stay far out at sea, but they have to approach fairly close to land to launch aircraft or missiles with operational ranges of only a few hundred miles. Moreover, the places where the U.S. Navy is likely to fight in the future are dangerously narrow. The Persian Gulf is only thirty miles wide at its narrowest point, the Taiwan Strait only one hundred miles wide.

To maintain its dominance, the U.S. Navy regularly updates the electronics and weapons aboard its warships even as the hulls and propulsion systems remain unchanged. It also plans to build a variety of unmanned vessels along with a CVN-21 aircraft carrier to replace the *Nimitz*-class, a *Zumwalt*-class DD(X) destroyer to replace *Oliver Hazard Perry*–class frigates and *Spruance*-class destroyers, a CG(X) cruiser to replace the *Ticonderoga*-class cruisers, and a smaller and speedier Littoral Combat Ship with no direct parallel in today's fleet that would focus on clearing mines, hunting submarines, and fighting terrorists in coastal waters. All of these new vessels will have improved defenses and information-processing tools as well as "plug and play" capacity that will allow them to be quickly reconfigured for different missions. They will also incorporate composite materials, stealthier designs, and electric propulsion to make them harder to detect, though an aircraft carrier with a 4.5-acre flight deck can never exactly hide.

Whether all of these warships are truly needed, given the U.S. Navy's already substantial lead over all competitors, remains an open question. A program to develop giant sea bases—perhaps akin to offshore oil-platforms—that would allow American ground and air forces to operate overseas might be of greater use, given the growing difficulty the U.S. has had in gaining basing and overflight rights from other countries.

AERIAL WARFARE: Fighters such as the American F-15 and the Russian MiG-29 were designed in the 1970s for air-to-air combat, but this has become almost as rare as ship-to-ship actions. Since the Israelis destroyed much of the Syrian air force in 1982, and the U.S. and its allies made similarly quick work of the Iraqi air force in 1991, few if any aircraft have been willing to challenge top-of-the-line Western militaries. (The U.S. Air Force has not produced an ace—an airman with at least five aerial kills—since 1972.) That may change with the sale to China of the Russian-built Sukhoi Su-30, whose performance characteristics are said to exceed those of the F-15C, but the F/A-22 Raptor, the F-35 Joint Strike Fighter, and the Eurofighter should restore the Western edge. The odds of future aerial dogfights are, however, slim.

Modern surface-to-air missiles pose a more immediate danger, because they are cheaper and easier to operate. The U.S. and its allies have developed effective methods of neutralizing most existing air defenses. In addition to jammers, radar-seeking missiles, and decoys, the U.S. employs stealth technology, first used on the F-117 Nighthawk, then on the B-2 Spirit, and now on the F/A-22 and F-35. Future aircraft may be designed with "visual stealth" technology to render them almost invisible even in daylight.

No other nation has deployed any stealth aircraft. But advanced sensor networks may be able to detect first-generation stealth planes. The Serbs actually managed to shoot down an F-117 in 1999.

None of the most sophisticated surface-to-air missiles, such as Russia's double-digit SAMs (SA-10, SA-15, SA-20), was available to Iraq, Serbia, Afghanistan, or other states that the U.S. has fought in recent years, but they are being sold to other customers. So are shoulder-fired antiaircraft missiles such as the American FIM-92 Stinger, British Starstreak, French Mistral, Chinese Qianwei-2, and Russian SA-7 Grail, SA-14 Gremlin, SA-16 Gimlet, and SA-18 Grouse. There are at least one hundred thousand such systems in the arsenals of over one hundred states and at least thirteen nonstate groups such as Hezbollah, the Revolutionary Armed Forces of Colombia (FARC), and the Tamil Tigers. The best models have a range of 23,000 feet.

The potential of hand-carried missiles was demonstrated in the 1980s when Stingers took a significant toll on Soviet aircraft in Afghanistan. The threat is serious enough for the U.S. to rely increasingly on unmanned drones for high-risk missions and to mandate that manned aircraft in war zones stay above 15,000 or 20,000 feet. SAMs pose an especially great threat to helicopters, which don't have the option of flying that high, and for airplanes taking off or landing. Three cargo aircraft leaving Baghdad International Airport have been seriously damaged by missiles, and, while all of them survived, several U.S. helicopters hit with SAMs in Iraq and Afghanistan did not. An Israeli jetliner was almost shot down in Mombasa, Kenya, in 2002 by al Qaeda operatives firing an SA-7. Only the terrorists' targeting error prevented the deaths of 271 passengers and crew. Other civilian airliners are sure to be less lucky.

Assuming that warplanes can reach their destination, the growing precision of bombs and missiles has made it possible to take out targets with fewer and smaller munitions than ever before. (The U.S. Air Force's latest bomb carries only fifty pounds of explosives.) Weapons are getting smarter all the time. The U.S. Sensor-Fuzed Weapon, first employed in the Iraq War, disperses forty "skeet" antiarmor warheads that use infrared and laser sensors to find and destroy armored vehicles within a thirty-acre area. The Tactical Tomahawk, which entered production in 2004, can loiter up to three hours while searching for targets and receive in-flight retargeting instructions.

The U.S. preponderance in smart bombs and missiles helps to compensate for the relatively small size of its manned bomber force. As of 2005, the U.S. Air Force had only 157 long-range bombers (B-52s, B-1s, B-2s), a considerable reduction not only from World War II (when the U.S. had 34,780) but also from the end of the Cold War (360). While few in number, each B-2 can perform the work of thousands of B-29s by "servicing" eighty "aim points" per sortie.

Tankers such as the KC-10 and KC-135 vastly extend the range and effectiveness of combat aircraft. Cargo-lifters like the U.S. C-5, C-17, and C-130 and the Russian An-70 and An-225 also perform a valuable, if unglamorous, role in projecting military power around the world. The U.S. owns 740 tanker aircraft and 1,200 cargo aircraft, far more than any other country. A lack of such support aircraft makes it difficult for even the relatively sophisticated European militaries to move their forces very far.

A host of other aircraft, ranging from JSTARS and AWACs to Rivet Joint and Global Hawk, perform surveillance and electronic-warfare missions in support of combat forces. Their numbers have been growing: While there were only two JSTARS in the Gulf War, in the Iraq War there were fifteen. But commanders have become so dependent on these systems that there never seem to be enough to go around. These, too, are vital U.S. assets that few other nations possess.

SPACE WARFARE: A growing amount of surveillance, communications, and intelligence work is being performed by unmanned aircraft and satellites. In 2001 the U.S. had an estimated 100 military satellites and 150 commercial satellites in orbit, as much as the rest of the world combined. The U.S. spends more than $15 billion a year on space, some 90 percent of the global total. The most advanced U.S. surveillance satellites are said to be able to pick out a six-inch object from 150 miles above. A new generation of satellites utilizes stealth technology to make it more difficult for other countries to track their progress and thus to hide equipment from American eyes.

The advantage the U.S. military derives from mastery of space is slowly eroding, however. The Global Positioning System developed by the Defense Department is now widely available for countless commercial applications that have spawned a $30-billion-per-year industry. A potential enemy could use GPS signals to locate targets in the U.S. the same way the U.S. military uses it to locate targets in Iraq or Afghanistan. The U.S. could jam or degrade GPS signals in wartime, but it would have to do so very selectively for fear of imposing a severe toll on the economy, because GPS devices are now essential for civil aviation, shipping, and other functions. In any case, the European Union in cooperation with China is launching its own GPS constellation, known as Galileo, that would be outside of U.S. control.

This is part of a trend of more and more countries—at least forty to date—lofting their own satellites. But getting access to space no longer requires having your own satellite. A growing number of private firms with such names as Google Earth, Keyhole, DigitalGlobe, and Space Imaging sell or give away high-resolution satellite photos via the Internet. The best of these offer imagery of sufficient quality to identify objects eighteen inches wide. The Israeli-owned ImageSat International offers customers the opportunity to redirect its EROS-A imaging satellite (launched in 2000 aboard a Russian rocket) and download its data in total secrecy with few if any restrictions. Its CEO boasts, "Our customers, in effect, acquire their own reconnaissance satellite . . . at a fraction of the cost that it would take to build their own." The private satellite industry is becoming so pervasive that the U.S. military has turned to it to provide some of its own imaging (typically low-resolution pictures used for mapping) and much of its communications needs.

Targets identified from space could be attacked either with terrorist (or commando) strikes or with the growing number of missiles proliferating around the world. More than two dozen nations have ballistic missiles and by 2015 at least a dozen will have land-attack cruise missiles. Either type of projectile could be topped with chemical, biological, or nuclear warheads. At least eight countries already have nuclear weapons and more are trying to get them, in part to offset the tremendous U.S. advantage in conventional weaponry.

In response, the U.S. is working on a variety of missile defenses. The most advanced are the ground-based Patriot Advanced Capability 3 and the sea-based Standard Missile 3, which already have been deployed to protect U.S. troops overseas. The deployment of a long-heralded system designed to protect the U.S. homeland against long-range missiles began in 2004 with the installation of interceptors in Alaska. Eventually, the U.S. plans to field a multilayered defense using a variety of sensors and weapons on land, sea, air, and space. Also in the works are systems designed to defeat low-flying cruise missiles, which are hard to distinguish from ground clutter.

FOG FORECAST

The changes wrought by the Information Revolution are still in their early stages and have serious limitations. Even the best surveillance systems can be thwarted by simple countermeasures like camouflage, smoke, and decoys, by bad weather, or by difficult terrain like the deep sea, mountains, or jungles. Sensors have limited ability to penetrate solid objects, so they cannot tell what is happening in underground bunkers such as those that North Korea and Iran use to hide their nuclear weapons programs. Urban areas present a particularly difficult challenge: There are far more things to track (individuals) and far more obstructions (buildings, vehicles, trees, signs) than at sea or in the sky—and figuring out whether a person is a civilian or an insurgent is a lot harder than figuring out whether an unidentified aircraft is a civilian airliner or an enemy fighter. It is harder still to figure out how hard enemy soldiers will resist or what stratagems they will employ. No machine has yet been invented that can penetrate human thought processes. Even with the best equipment in the world, U.S. forces frequently have been surprised by their adversaries.

Some strategists expect that advances in information technology will greatly diminish if not altogether obliterate some of these difficulties. The Pentagon is creating a Global Information Grid that will pool data from all U.S. assets, whether an infantryman on the ground or a satellite in space. The ultimate goal: to provide a perfect operational picture—a "God's-eye view" of the "battlespace."

This ambitious objective could be furthered by the development of better microwave radars that could see through walls, foliage, or soil; cheaper, more pervasive sensors that could provide 24/7 coverage of the battlefield; better data compression and transmission techniques that could allow more bytes to be sent much faster; and more powerful computers that might make it possible to create, for example, a real-time, three-dimensional model of a city showing all the people who reside in it. Not only are electronic computers getting more powerful in accordance with Moore's Law, but even greater leaps in computing power may be achieved using biological computers that rely not on electronic components but on DNA, individual molecules, or proteins; optical computers that use light rather than electricity; and quantum computers that exploit the atomic properties of matter. Some scientists speculate that by using such methods a billionfold increase in computing power is possible. "That means you could hold the power of all earth's present computers in the palm of your hand," says Stan Williams of Hewlett-Packard.

Yet no matter how far information technology advances, it is doubtful that the Pentagon will ever succeed, as some utopians dream, in "lifting the fog of war." The fallibility of American soldiers and the cunning of their enemies will surely continue to frustrate their best-laid plans. Indeed, America's growing reliance on high-tech systems creates vulnerabilities of its own: Future enemies have strong incentives to attack U.S. computer and communication nodes. Strikes on military information networks could blind or paralyze the armed forces, while strikes on civilian infrastructure, such as banking or air-control systems, could cause chaos on the home front.

Safe to say, adversaries will figure out ways to blunt the U.S. informational advantage. From Operation Anaconda in Afghanistan to numerous misadventures in Iraq, they already have.

AMERICA'S UNPARALLELED POWER

While various setbacks in the war on terror underscore the limits of American power, it is important not to lose sight of the bigger picture: The dawn of the Information Age has coincided with a period of American global hegemony. Part of the explanation lies in U.S. economic strength and its "soft power." But Europe and Japan are rich, too, yet their global sway lags far behind. What they lack is America's military strength.

In the early years of the twenty-first century the United States enjoys a preponderance of military power greater than that of any other nation in history. Rome was rivaled in land power by Carthage, Persia, and other states, and its sphere of control never extended far beyond western Europe and the Mediterranean region. Britain was rivaled in naval power by the Netherlands, France, and later Germany, and it was never dominant on land. Today America is rivaled in land, sea, and air power by . . . no one. Although the dominance of U.S. forces can still be challenged when they come into close contact with the enemy on his home turf, they are undisputed masters of the "commons" (sea, air, space), which allows them to project power anywhere in the world at short notice. In the words of journalist Gregg Easterbrook, "The American military is now the strongest the world has ever known, both in absolute terms and relative to other nations; stronger than the Wehrmacht in 1940, stronger than the legions at the height of Roman power."

What accounts for this huge military edge? Not manpower. The U.S. armed forces in 2006, with about 1.4 million active-duty personnel and 1.2 million reservists, were hardly the biggest in the world. NATO's European members and China each had roughly 2.2 million men and women under arms. Other nations—North Korea with 1 million active-duty soldiers and 4.7 million reservists, Russia with 960,000 on active duty and 2.4 million in the reserves—were not far behind. In any case, because the U.S. armed forces typically do their fighting overseas, they seldom enjoy a numerical advantage in their theater of operations. Whether in Iraq or Afghanistan, they have usually been outnumbered.

In the Information Age the U.S. edge lies in quality, not (as it did during the Second Industrial Age) in quantity. Both its soldiers and their equipment are among the best on earth. The U.S. armed forces are not the only Information Age military—Britain, France, Japan, Singapore, Australia, Sweden, and other nations have followed in its wake—but they have gone the furthest fastest.

Among other nations, Israel's success in harnessing the Information Revolution has been particularly consequential. The balance of power in the Middle East shifted sharply in Israel's favor following the Yom Kippur War, because it modernized its military to a far greater extent than its neighbors did. This helps to explain why no Arab nation has mounted a conventional attack on Israel since 1973.

The United States' success in harnessing the Information Revolution produced its profoundest effects without a shot being fired. Aware that the Soviet Union could not keep up with U.S. military advances in the 1980s, Mikhail Gorbachev responded with perestroika and glasnost, reform efforts that spun out of control and ultimately consigned the Soviet Union to the ash heap of history. The demise of the USSR set the stage for the U.S. military dominance that has underpinned the current era of globalization, which has been remarkably devoid of conflict between states.

Wars between states have declined in size and frequency for a variety of reasons, not the least being the integration of old adversaries like France and Germany into the European Union, the international norms propagated by the United Nations Charter, the spread of capitalism and democracy across the world, and the deterrence provided by nuclear weapons. But surely another part of the explanation is that few states feel free to challenge the Pax Americana. Several of those that have tried, including Panama, Serbia, Afghanistan, and Iraq, have seen the downfall of their regimes. "War, let alone 'decisive war' between major states, currently is enjoying an off season for one main reason," writes British strategist Colin Gray. "So extreme is the imbalance of military power in favor of the United States that potential rivals rule out policies that might lead to hostilities with the superpower."

ACCESS DENIED?

It is doubtful that any country will mount a full-spectrum challenge to U.S. military capabilities in the foreseeable future. The entry barriers are simply too high, especially for air, sea, and space systems. *Virginia*-class nuclear submarines cost $2.4 billion, *Nimitz*-class aircraft carriers go for $6 billion, and the F-35 Joint Strike Fighter program will cost at least $245 billion. The U.S. spends around $500 billion a year on its military, almost as much as the rest of the world combined. In fact, the U.S. spends more simply on the research, development, testing, and evaluation of new weapons—$71 billion in 2006—than any other country spends on its entire armed forces. (By way of comparison, the top three spenders after the U.S. are Russia, whose defense budget in 2003 was estimated at $65 billion; China, at $56 billion; France, at $45 billion; and Japan and the United Kingdom, at $42 billion each.)

It is not only U.S. hardware that's hard to replicate; so is the all-volunteer force that makes it work. Operating high-tech military equipment requires long-service professionals, not short-term conscripts. Countries as diverse as Vietnam, China, Germany, and Russia are emulating the Anglo-American model by downsizing their forces and relying less on draftees. Many other nations have abolished the draft altogether. The U.S. military's edge lies not simply in recruiting high-quality personnel but in its methods for training and organizing them. Initiatives undertaken in earlier decades, such as setting up realistic training centers to simulate combat conditions and forcing the services to work more closely together (the Goldwater-Nichols Act), continue to bear fruit. Few other armed forces have made comparable reforms. Writes one scholar: "The superb U.S. all-volunteer military force . . . is a unique human and institutional asset that less capable foreign rivals can neither copy nor steal."

But a potential adversary does not need to duplicate U.S. force structure in order to challenge it. The United States faces a growing "asymmetric" threat both from other states and from substate groups.

As we have seen, a variety of off-the-shelf missiles can threaten U.S. tanks, surface ships, and aircraft, especially when they get close to hostile territory. The power of smart munitions is outstripping the protection afforded by speed or armor. After 2010, write defense analysts Michael Vickers and Robert Martinage, "the survivability of aircraft carriers, high-structure surface combatants [e.g., tanks], and non-stealthy aircraft of all types could increasingly be called into question as maritime, over-the-horizon 'area denial' capabilities and extended-range air defense systems continue to mature."

Also vulnerable are the ports, airfields, and bases that the U.S. uses to project its power overseas. Imagine how much damage Saddam Hussein could have done in 2003 if he had been able to annihilate the one port in Kuwait that was being used to disembark coalition troops or the large desert bases in Kuwait where over one hundred thousand British and American troops gathered prior to the invasion of Iraq. The Pentagon's 2001 Quadrennial Defense Review warned that "[f]uture adversaries could have the means to render ineffective much of our current ability to project military power overseas."

If the U.S. armed forces could not count on safe, assured access to overseas bases they would have to radically change the way they do business. It would no longer be practical to rely on large land armies or lots of short-range combat aircraft operating out of vulnerable forward bases supplied by equally vulnerable cargo ships, trucks, and aircraft. The U.S. Army might be forced to rely on a small numbers of commandos supported by long-range aircraft and missiles—the Afghanistan model. The navy might have to depend more on submarines and the air force on stealth aircraft. All the services might have to make greater use of unmanned vehicles. The battlefield, which has been growing less crowded for centuries, might empty out even further as small units try to conceal themselves from ubiquitous sensor networks, emerging only briefly to launch lightning strikes before they go back into hiding.

This has become known as the "swarming" scenario, and it has attracted support from the likes of military historian Alexander Bevin, who writes, "Large concentrations of troops and weapons are targets for destruction, not marks of power, and [in the future] they no longer will exist. . . . Military units, to survive, must not only be small, but highly mobile, self-contained, and autonomous."

Even if these predictions are accurate, however, no timeline is attached. Timing matters. After all, everyone knows that a bull market will be followed by a bear market, but the key to making money in stocks is knowing when that transition will occur. Likewise, the key to winning future wars is knowing when to move from one form of military to another. A premature decision to change (such as the U.S. Army's Pentomic design in the 1950s) can leave one unprepared to fight and win wars that actually occur, Vietnam being the classic example.

In any case, it is doubtful that a complete shift to "swarming" will ever take place. Winning wars, as distinct from winning battles, will continue to require controlling territory, which in turn will require a substantial presence of ground troops, as the U.S. has learned in Afghanistan and Iraq. No wonder weapon will alter this fundamental reality.

"AMERICAN HIROSHIMA"

Even as strategists look to the future, armed forces must not lose sight of the threats of the moment, and they do not come for the most part from traditional militaries. They come largely from terrorist groups—some with state sponsorship, others without—that utilize the fruits of modern technology to their advantage.

"Irregular" attacks carried out by tribes, clans, or other nonstate actors are as old as warfare itself; they long predate the development of modern armed forces and the nation-state. The religious fanaticism that animates so many of today's terrorists and guerrillas is equally ancient. But technological advances have made such attacks far more potent than in the past. The progeny of the Second Industrial Revolution—assault rifles, machine guns, mortars, rocket launchers, land mines, explosives—long ago spread to the remotest corners of the globe. Fighters who a century ago might have made do with swords and muskets now have access to cheap and reliable weapons, such as the AK-47, capable of spewing out 100 bullets a minute. More advanced technologies, from handheld missiles to chemical, biological, and nuclear weapons, give even a small group of insurgents the ability to mete out far more destruction than entire armies could a century ago. And thanks to modern transportation and communications infrastructure—such as jumbo jets, the Internet, and cell phones—insurgents have the capability to carry out their attacks virtually anywhere in the world.

September 11, 2001, showed the terrifying possibilities of such unconventional warfare. It is easy to imagine that in the future super-terrorists will be able to kill hundreds of thousands, even millions, with effective weapons of mass destruction. All of the materials, as well as the know-how, needed to craft such devices are all too readily available.

Chemical and biological threats will be discussed in Chapter Thirteen, but nuclear weapons, more than any other kind, have the ability to trump U.S. military hegemony. The atomic bomb is more than sixty years old. It belongs to an age of rotary-dial telephones and fin-winged cars. It is a miracle that it has not been used in anger since 1945, but that streak won't last forever. And while Information Age technology offers a reasonable chance of stopping a nuclear-tipped missile, there is much less probability of stopping a terrorist with a nuclear suitcase. (Yes, nuclear bombs have gotten smaller, just like computers.) There is little in theory to prevent al Qaeda from carrying out its oft-expressed desire to create an "American Hiroshima." In the words of a retired four-star general who once ran antinuclear terror programs for the Department of Energy: "It is not a matter of if; it's a matter of *when*."

The most important challenge for the armed forces of the U.S. and its allies in the post-9/11 world is to "leverage" their advantage in conventional weaponry to deal with an unconventional threat. Information technology can be an important part of this task. Embedded microchips can track the 18 million cargo containers moving around the world and prevent terrorists from using them to smuggle weapons. Computerized cameras scanning a crowd may be able to pick out a terrorist based on facial recognition patterns.

Doglike sniffing machines may be able to recognize suspects by their body odor. Powerful computers utilizing artificial intelligence programs can sift vast reams of digital data to pick out information about terrorist plots—if concerns about violating the privacy of innocent people do not get in the way. A variety of unobtrusive sensors can warn of the presence of explosives or chemical, biological, or nuclear weapons. Handheld computer translating devices such as the Phraselator already in use by U.S. troops can bridge some of the language gap between Western operatives and the regions where they operate.

But in the final analysis, having the best technology is not enough to defeat the most ruthless terrorists. In fact, most of the expensive weapons systems being purchased by the U.S. and its allies are almost completely irrelevant to the war against terrorism. Smart bombs can be useful for killing the enemy once he's been located. But figuring out who the enemy is, where he is, and what he is up to—that requires smart *people*.

Various remedies have been proposed to fix the obvious American weakness in this area: Requiring more foreign-language training for the armed services. Recruiting more spies. Setting up a military school devoted to civil governance. Creating an agency tasked with "nation-building." Appointing "cultural scouts"—an enhancement of the existing Foreign Area Officer program, which is widely seen as a career dead end—who would spend long periods of time in foreign lands. These are all good ideas and some are beginning to be implemented, but they are harder to enact than a decision to buy a new weapons system. They require changing the culture of the Defense Department and other government agencies, which have not stressed skills in counterinsurgency, human intelligence, or nation-building.

TOOTH AND TAIL

While the U.S. armed forces have made some major strides in recent decades, especially in getting different branches to work more closely together, they still have a long way to go before they have an organizational structure that makes the most effective use of their high-tech equipment. The army, although putting greater emphasis on more nimble, four-thousand-person brigades, still maintains a divisional structure that dates back to the First Industrial Age, while the basic units of the navy (carrier groups) and the air force (squadrons and wings) date back to the Second Industrial Age. All of the services employ a personnel system developed more than one hundred years ago that treats individuals as interchangeable cogs in a giant machine. Army officers have traditionally spent an average of only eighteen months at each duty station during the course of a twenty-five-year career, making it difficult to build true expertise in such complex areas as foreign cultures.

Another problem is a cumbersome procurement process that costs too much and takes too long to deliver new equipment. The U.S. paid a heavy human cost for those shortcomings in Iraq, where body armor, armored Humvees, and IED-jamming devices were slow to arrive. The time and expense involved in producing new weapons systems has increased severalfold since the 1970s, even as the private sector has been able to produce products like cell phones and microchips faster and more cheaply. "In order to stay ahead of adversaries with access to technologies available in the global marketplace, the DOD [Department of Defense] needs to shorten the timeframe from concept to testing," writes Lieutenant Colonel Augustus Way Fountain III, a chemistry professor at West Point. That's exactly what the best-run companies—the Microsofts and the Wal-Marts—have done, but the U.S. armed forces (and most others, for that matter) have lagged behind.

In general, the U.S. military is hindered by a sluggish, bloated bureaucracy that has resisted countless reform efforts. "We're No. 1 in the world in military capabilities," says David M. Walker, head of Congress's Government Accountability Office. "But on the business side, the Defense Department gets a D—giving them the benefit of the doubt. If they were a business, they wouldn't be in business."

Some statistics buttress the point. The "tooth to tail" ratio in the U.S. military—the number of support personnel as opposed to actual trigger-pullers—varies from 12:1 for the Army and Marine Corps, to 15:1 for the Navy, and 32:1 for the Air Force. "An astounding 70 percent of the defense budget is spent on overhead and infrastructure (the bureaucratic 'tail'). Only 30 percent directly reaches our combat forces in the field (the 'tooth')," writes retired Admiral Bill Owens. "No community would tolerate 7 out of every 10 police officers sitting at their desks pushing paper. The nation should not tolerate such a ratio in the military."

What makes this unwieldy setup so intolerable is that it can stymie the potential of even the most advanced information technologies. Retired Marine Colonel T. X. Hammes notes that

> most commanders must submit their intelligence requirements up the chain of command. Each level validates, consolidates, and prioritizes the requests, which are then fed through the centralized staff system to task the assets that will actually collect against the requests. The information is collected, passed to another section for analysis, then put in the form of a usable product, and finally disseminated through the same cumbersome system. Thus, the premier benefit of the Information Age—immediate access to current intelligence—is nullified by the way we route it through our vertical bureaucracy.

The good news, from the American perspective, is that many potential enemies (e.g., China or North Korea) are encumbered by even more sclerotic bureaucracies. The bad news is that this is not true of nimble, networked groups like al Qaeda. To fight them effectively, the U.S. military will have to display more of the decentralized decision-making that it showed in Afghanistan in the fall of 2001. This will not be easy to do because modern command and control technology is a two-edged sword: It can be used to centralize or decentralize. It will always be difficult for a senior military or political leader to resist the temptation to micromanage operations from afar—a style of leadership that modern communications technology has made easier, but no more effective, than in the past.

Veteran journalist Robert Kaplan rightly calls "the dinosauric, vertical bureaucracy of the Industrial Age . . . the greatest single impediment to America's ability to wage a successful worldwide counter-insurgency." Unless the U.S. government can streamline its Industrial Age bureaucracy and become a networked organization, it may find that even purchasing the latest and best technology will not offer sufficient protection against the country's foes.

PART V

REVOLUTIONS PAST, PRESENT, FUTURE

Historically, victors don't learn nearly as well as losers.

Vice Admiral Arthur K. Cebrowski (2002)

CHAPTER 13

REVOLUTIONS TO COME

Even as the Information Revolution continues to move ahead as fast as a speeding e-mail, other areas of scientific and technological progress—along with attendant changes in tactics, culture, and organization—are also starting to reshape warfare. Most are related, directly or indirectly, to the great leaps in computing power that have occurred in recent decades. Many of these potential revolutions (especially robotics and cyberwar) could be classified as subsets of the Information Revolution. But they are likely to take warfare in strange and unexpected directions, many of which (especially computer, biological, and chemical warfare) will empower small states and substate groups at the expense of large nation-states. What follows is a brief survey of some of the most promising (or menacing) developments on the horizon, whose potential may or may not be realized. The focus on cutting-edge technology is in no way meant to suggest that political or organizational developments will not be important in the future; the nature of war will always be determined by the interaction between warriors and their tools, not by the tools alone.

RISE OF THE ROBOTS

The falling size and cost of electronics has made it possible to decrease the number of people needed to operate major weapons systems or, in some instances, eliminated the need for human operators altogether. Maintaining the engines aboard a ship used to require dozens of sailors to work for extended periods in noisy, grimy, cramped quarters. The new DD(X) destroyer will have an engine room controlled entirely by remote sensors and cameras. Or, to take another example, consider the evolution of the long-range bomber from the B-29, which had a crew of eleven, to the B-2, which can hit many more targets but has a crew of just two—and they spend much of their time supervising the autopilot functions.

The greatest advances in robotics have been made in unmanned aerial vehicles (UAVs), with the U.S. in the lead, Israel following close behind, and at least forty other countries trying to catch up. By the time the Iraq War began in 2003, the U.S. fielded six major UAVs: the Air Force's Predator and Global Hawk, the Army's Hunter and Shadow, and the Marines' Pioneer and Dragon Eye. These ranged in size from the 27,000-pound Global Hawk (comparable to a Learjet) to the five-pound Dragon Eye (more like a model airplane). What they had in common was that they were all designed as surveillance systems. But in a pattern that echoes the history of manned flight, UAVs such as the Predator were soon put to work attacking enemy positions.

Soon to be deployed are drones built especially for combat—Boeing's X-45 and Northrop Grumman's X-47. In one writer's fanciful description, the former is "[f]lat as a pancake, with jagged 34-foot batwings, no tail and a triangular, bulbous nose" that give it the appearance of "a set piece from the television program *Battlestar Galactica*," while the latter is a "a sleek kite-shaped craft with internal weapons bays for stealth and curved air intakes like the gills of a stingray." Both are designed to be almost invisible to radar and to perform especially dangerous missions like suppressing enemy air defenses. The major difference is that the X-45 is supposed to take off from land like the F-15, while the X-47 is to operate off aircraft carriers like the F-18. Also in development is the Unmanned Combat Armed Rotorcraft, which is designed to perform the functions of an attack helicopter like the Apache. An unmanned helicopter, known as Fire Scout, is already being bought by the U.S. Navy and Marine Corps. Unlike the Predator, most of these new UAVs do not require constant control by a human operator; newer UAVs can be programmed to fly themselves and even drop munitions without direct human intervention.

Further into the future may be projects such as a nuclear-powered UAV that could fly at seventy thousand feet and stay on station for months or even years at a time; a UAV tender that could serve as a mother ship for launching and recovering smaller UAVs; unmanned aerial tankers that could refuel other UAVs in flight; and vertical-takeoff UAV cargo-carriers that could supply troops in a combat zone. Many of these UAVs could utilize smart munitions with their own target-recognition systems, thus introducing another layer of robotics into the process. An existing example is the Low-Cost Autonomous Attack System, a one-hundred-pound bomb with fins and a small turbojet engine that allow it to hover over an area for up to thirty minutes, using a laser-radar sensor to search for high-priority targets based on programmed algorithms. Once it picks out a target, it can configure its multimode warhead into the most appropriate form—fragmentation explosives for unprotected soldiers, an armor-piercing projectile for tanks—prior to impact.

The most revolutionary UAVs are the smallest. The Defense Advanced Research Projects Agency (DARPA) is working on aerial vehicles the size of an insect or a hummingbird that could hover undetected and perch on a telephone pole or a window ledge. Some models have no wings at all; others use flapping, bird-style wings. They are designed to be cheap enough that they could saturate a battlefield with sensors.

Unmanned ground vehicles aren't as advanced as UAVs, but they are playing a growing role as well. In Iraq and Afghanistan, the U.S. Army and Marine Corps have used robots with names like PackBot, Matilda, Andros, and Swords to search tunnels, caves, and buildings for enemy fighters and explosives. "Some are as big as a backhoe. Others can be attached to a backpack frame and carried by a soldier," writes the trade industry publication *Defense News*. "They move on treads or wheels, climb over obstacles with the aid of flippers, mount stairs, peep through windows and peer into caves with cameras and infrared sensors, sniff for chemical agents, and even operate a small ground-penetrating radar."

As this description indicates, ground-based robots, like their aerial counterparts, are still used mainly for reconnaissance. But weapons are beginning to be mounted on them, too. The Talon, a two-foot-six-inch robot that looks like a miniature tank and was designed for bomb disposal, was sent to Iraq equipped with grenade and rocket launchers as well as a .50 caliber machine gun. It is controlled remotely by a soldier using a video screen and joystick.

Developing more sophisticated unmanned ground vehicles will be tougher than developing better UAVs because there are so many more obstacles that can impede movement on the ground. But progress is rapidly being made. In 2004, DARPA sponsored a race in the Mojave Desert to see if an autonomous robotic vehicle could complete a 132-mile course. The farthest any competitor got was seven and a half miles, but the following year four vehicles finished the entire course, with the winner (a souped-up Volkswagen

Touareg) claiming the $2 million prize. Buoyed by these results, the Pentagon is pushing ahead with plans for new ground robots such as the MULE (Multifunction Logistics and Equipment Vehicle), a two-and-a-half-ton truck that could carry supplies into battle or wounded soldiers out of it; the Armed Robotic Vehicle, a five-ton minitank that could be equipped with missiles or a 30 mm chain gun; and the Soldier Unmanned Ground Vehicle, a thirty-pound, man-portable scout that comes equipped with weapons and sensors. These are all integral elements of the Army's Future Combat System.

Scientists are also trying to create a self-powered robotic suit—an exoskeleton—that could enable soldiers to carry far heavier loads, move much faster, and conceivably even leap short buildings in a single bound. A prototype developed at the University of California, Berkeley, allows a soldier to carry 180 pounds as if it were less than five pounds.

The U.S. Navy is exploring robotic technology for a variety of its own missions. In addition to carrier-based UAVs (both fixed-wing and rotary), the navy is developing Unmanned Surface Vehicles and Unmanned Undersea Vehicles. Most of these drones would swim, but some might crawl along the ocean floor like crabs. They could perform such difficult missions as antisubmarine warfare, mine clearance, undersea mapping, and surveillance in coastal waters.

All drones, whether operating on soil, sea, or sky, offer major advantages over traditional manned vehicles. They can be deployed for longer periods because robots don't need to eat or sleep; they can undertake maneuvers that might put too much stress on the human frame; they can be made much smaller and cheaper because they don't need all sorts of expensive redundancies and life-support systems (no oxygen tanks, no ejection seats); and they can be much more readily sent on high-risk missions because, should anything go wrong, nobody has to worry about notifying the next of kin. These advantages have persuaded Congress to ratchet up spending on unmanned programs. Lawmakers have mandated that one-third of all U.S. deep strike aircraft be unmanned by 2010 and one-third of all ground combat vehicles by 2015.

There are two chief limitations on the use of robots at the moment. First, computers and sensors are not yet smart enough to deliver anything close to the "situational awareness" of a human being. Second, a shortage of bandwidth limits the number of drones that can be remotely controlled at any one time. Both problems will become less acute with improvements in computer and communications technology but, notwithstanding the *Terminator* movies, there is little cause to think that androids will take over the battlefield of the future. It is doubtful that computers will ever be smart enough to do all of the fighting. But machines increasingly will be called upon to perform work that is dull, dirty, or dangerous.

STAR WARS

In a sense, satellites have been the most successful unmanned vehicles of all. The U.S. military has been reliant on them much longer than on Predators or Global Hawks. Strategists argue that this makes space the new "high ground" and its control a vital object for future military campaigns. A commission chaired by Donald Rumsfeld before he became defense secretary warned in 2001 of a "Space Pearl Harbor" should an adversary disable critical U.S. satellites.

A number of antisatellite weapons—principally missiles, some nuclear-tipped—were developed during the Cold War by both the U.S. and the USSR. It is not hard to hit satellites, especially those in low-earth orbit, which are only 400 to 1,600 miles above sea level, making them vulnerable to medium-range ballistic missiles owned by such countries as China, North Korea, and Iran. (They are also vulnerable to ballistic missile defenses of the kind the U.S. is developing.) A nuclear missile could have a particularly devastating effect, because it would release a pulse of electromagnetic energy that would fry electronic circuits for hundreds, even thousands, of miles. While strong international norms exist against using nuclear weapons on earth, a country might be more willing to detonate one in space. But disabling a satellite doesn't require an atomic blast. Marbles could do it. Small pellets released by a missile in the orbital path of a satellite could destroy it because of the high speed (over 14,000 miles per hour) at which satellites travel.

You don't have to destroy satellites to disable them. You could hit their ground stations or jam their transmissions. A Russian company sells a $4,000 device to jam GPS signals; some were bought by Iraq but destroyed by the U.S. military in 2003. Cuba in 2002–3 periodically jammed the Telsat-12 commercial satellite on behalf of Iran to disrupt Farsi language programming from Iranian émigrés in California. And in 2004 the U.S. Space Command deployed its first mobile teams with gear designed to jam enemy satellite communications.

The most innovative and perhaps most effective antisatellite weapons might turn out to be other satellites. Mini-satellites that weigh as little as a pound or two could shadow bigger satellites and, in the event of war, destroy or disable them. The crudest method would be to simply explode, making them the suicide bombers of outer space. But that would scatter debris that could impede other satellites, including those of the nation that launched the killer satellite. It might be more effective to jam transmissions from the target satellite, blind its sensors with a laser or spray paint, fry its electronics with a burst of microwave energy akin to a lightning bolt, cut its power cable, or nudge it out of its assigned orbit.

Small satellites are being developed by many countries, including the U.S., Russia, Israel, Sweden, and China. Their small size means that they are quite affordable even for a nonsuperpower. In 2000 a British university launched an eleven-pound satellite atop a Russian booster for less than $1 million.

The growing threat to satellites no doubt will lead to the adoption of defensive measures. In the event of attack, a satellite might maneuver out of the way using small jets, deploy shields akin to a medieval suit of armor to protect itself, or fight back with its own missiles or directed-energy weapons. High-priority satellites might even be escorted by defensive satellites, just as aircraft carriers are escorted by smaller vessels.

So far we have been looking at an early version of space-to-space warfare—the extraterrestrial equivalent of biplanes jousting in the skies above France during World War I. Space also could be used to launch attacks against the earth—the equivalent of the first aerial bombing.

The U.S. Defense Department has explored the possibility of orbiting a laser in space to shoot down ballistic missiles. Another idea on the drawing board is to hurl solid tungsten or titanium rods from orbiting satellites at terrestrial targets. The "Rods from God" would travel so fast (about 7,200 miles per hour) that, even without any explosives, they could take out hardened bunkers as effectively as nuclear weapons but without any of the radioactive fallout. These rods could be dropped on any target in the world with just fifteen minutes' notice and without having to get any other country's permission for basing or overflight rights. Conventional air defenses would be useless against such small, fast-moving objects. However, the cost of this system might be prohibitive.

Another space weapon in development is known as the Common Aero Vehicle. A cone-shaped, unmanned capsule that could carry one thousand pounds of sensors or munitions, the Common Aero Vehicle would be launched into space by booster rockets. It would then detach itself and, like the Rods from God, glide back through the atmosphere toward its target. The Pentagon hopes that eventually this weapon will be carried aloft by the Hypersonic Cruise Vehicle, a reusable space plane that could fly at fifteen times the speed of sound. It is slated for completion by 2025 and could also function as a "starfighter" capable of attacking enemy satellites.

In 2004 the U.S. Air Force released a formal doctrine for "Counterspace Operations" designed to ensure "space superiority"—"the freedom to operate in the space medium while denying the same to an adversary." This publication is notable for including chapters on both defensive *and* offensive space operations, concepts that would be instantly familiar to Luke Skywalker or Captain James T. Kirk. Nor would they be surprised to hear that there is talk about cleaving off a separate space force from the air force, as the air force itself was spun off from the army in 1947.

RAY GUNS

No science fiction film would be complete without laser guns. In the real world, the use of lasers on the battlefield has so far been limited mainly to guiding explosive projectiles. The most advanced laser designators are for military use only, but anyone can buy a laser rifle scope that projects a red dot onto a target.

The first laser weapon was used in combat in 2003 when the U.S. Army deployed the Zeus system to Afghanistan. Mounted on a Humvee, the Zeus laser emits two kilowatts of energy. It can neutralize unexploded ordnance or mines from about 1,000 feet away without causing a full detonation. The Zeus destroyed more than two hundred pieces of ordnance in six months' use at Bagram airfield near Kabul. In 2005 the Zeus was sent to Iraq for similar work.

Solid-state lasers powered by electricity, like the Zeus, cannot generate more than twenty-five kilowatts of power at the moment. That is more than enough for commercial applications like DVD players, fiber-optic communications lines, supermarket bar code readers, laser printers, and laser eye surgery. But at least fifty to one hundred kilowatts are needed for most combat applications.

Lasers that derive their energy from chemical reactions are more cumbersome—they require bulky vats of chemicals—but they are more advanced and more powerful. Two have already been tested by the U.S. Defense Department: the Airborne Laser and the Tactical High-Energy Laser.

The Airborne Laser is a 14,000-pound pod mounted on a modified 747 to shoot down ballistic missiles in their boost phase. By mixing oxygen, hydrogen peroxide, chlorine, and iodine, it can generate a beam of one thousand to two thousand kilowatts. (A smaller version that is still under development, known as the Advanced Tactical Laser, could be mounted on an AC-130 gunship and used against cruise missiles, artillery shells, or ground targets.) The Tactical High-Energy Laser was developed jointly by the Israeli Defense Forces and the U.S. Army to shoot down short-range missiles, mortars, and artillery shells. It derives its energy from mixing fluorine, helium, and deuterium. In tests since 2000, it has successfully intercepted more than twenty-five projectiles in flight, including Katyusha rockets of the kind that have been used to attack northern Israel. But it was canceled in 2006 because it was too bulky.

Laser weapons (whose beams, unlike in the movies, are generally not visible to the naked eye) offer several benefits over ordnance propelled by chemical or kinetic energy. First, they can strike their target much faster because laser beams move at the speed of light—186,282 miles per second, or more than 160,000 times faster than a typical bullet. A missile fired from 186 miles away at a jet airplane flying at 621 mph would take six minutes to reach its target, giving the pilot time to escape. A laser beam would arrive in 1/1,000th of a second. Second, lasers can be aimed much more precisely, reducing the risk of collateral damage. Third, while they are expensive to develop, lasers are cheap to fire. The Tactical High Energy Laser costs about $8,000 per shot, compared to $3.8 million for the Patriot 3 missile. Fourth, solid-state lasers have an essentially unlimited magazine—as long as there is fuel to generate electrical power, they can keep firing.

But lasers also have some serious limitations: Their beams can be deflected by water vapor or clouds, and they cannot hit targets over the horizon because of the curvature of the earth. To solve these problems, the Pentagon is studying how to concentrate laser beams and how to relay them across greater distances using mirrors mounted on satellites or high-altitude dirigibles.

Lasers are only one type of directed-energy weapon. Other electromagnetic devices generate microwaves and electromagnetic pulses, which are particularly effective when employed against electronic equipment because semiconductors are extremely sensitive to any spikes in energy. A high-powered microwave gun could fry the circuits aboard an airplane or missile, causing it to crash. When employed against people, the advantage of directed-energy weapons is that their effects can be modulated to avoid inflicting fatal injuries. This is particularly useful for peacekeeping missions where a premium is placed on avoiding civilian casualties.

The U.S. Air Force and Marine Corps have spent $40 million to develop the Active Denial System, which emits millimeter waves that penetrate about 1/64th of an inch into the skin, causing an intense burning sensation. The pain goes away as soon as the device is shut off or an individual steps out of the way of its beam. That makes it useful for dispersing hostile crowds. There are, of course, cheaper nonlethal weapons such as shotguns firing beanbags, canisters emitting pepper spray or tear gas, and Taser pistols firing electrical darts. The advantage of the Active Denial System is that it can cover a much wider area with the flick of a switch.

The Marine Corps and Army have fielded another weapon that tries to achieve the same effect using sonic waves rather than microwaves. The Long-Range Acoustic Device is a forty-five-pound, dish-shaped antenna that can emit earsplitting noise of up to 150 decibels. This weapon is designed to be nonlethal, but it could easily kill at higher intensity levels: Sonic waves above 150 decibels can inflict internal injuries. The day is not far off when, as on *Star Trek*, troops will be able to set their phasers to "kill" or "stun."

While energy can cause damage on its own, it can also be used to hurl projectiles much more efficiently than traditional firearms utilizing chemical reactions. Within a decade or two an electromagnetic gun might be able to fire a thirty-three-pound tungsten rod that would strike a target three hundred miles away in just six minutes. The Navy's DD(X) destroyer, which is being equipped with generators that could power an electromagnetic gun, could carry hundreds of these inexpensive projectiles, as opposed to existing *Arleigh Burke*–class destroyers which carry a maximum of ninety costly Tomahawks.

WHEN COMPUTERS ATTACK

The more that advanced economies and militaries come to rely on information technology, the more their enemies will seek to disrupt those vital networks.

The exploits of computer hackers are well known. Some, such as Kevin Mitnick, have even made the FBI's Ten Most Wanted Fugitives list. While their intrusions have been costly, a coordinated attack by terrorists or a hostile military service could be far more painful. In the early 1980s the CIA allegedly planted a Trojan horse virus inside seemingly innocuous Canadian software imported by the Soviet Union to run its new trans-Siberian natural gas pipeline. According to former Air Force Secretary Thomas Reed, the pipeline software was sabotaged "to reset pump speeds and valve settings to produce pressures far beyond those acceptable to the pipeline joints and welds." The result, Reed writes, was a 1982 explosion and fire that caused serious damage to the pipeline and to the fragile Soviet economy.

More nightmarish possibilities have been imagined in works of fiction. In the 1983 Matthew Broderick movie *Wargames*, a teenage hacker breaks into the computers of the North American Aerospace Defense Command and almost initiates World War III by creating false images of inbound Soviet missiles. In the 1990 Bruce Willis movie *Die Hard 2*, villains take over the air-traffic control computer at Dulles International Airport and send a passenger airliner on a suicide course into the ground. In the 1994 Tom Clancy novel *Debt of Honor*, a malevolent Japanese industrialist hacks into Wall Street computers to cause a multibillion-dollar market crash. And in a

2004 episode of the television series *24*, terrorists use the Internet to penetrate a nuclear power plant's control system, causing a meltdown.

The militaries of many nations take such possibilities seriously. Two Chinese People's Liberation Army colonels write that it is now "possible to start a war in a computer room or a stock exchange that will send an enemy country to its doom." China is one of at least thirty countries, including the U.S. and Russia, developing cyberattack capabilities. No doubt terrorist groups are following suit.

All major companies and governments are, of course, aware of the dangers of cyber-intruders and have built elaborate protections such as the "firewalls" developed by DARPA in the early 1990s. But it's hard to keep out all intrusions. U.S. military computers get probed by hackers thousands of times a year, some of whom succeed in disrupting networks for days at a time. Computers controlling power grids, dams, and other vital infrastructure are also under constant attack. Information warfare specialists try to repel such attacks, but they can never tell if they are totally successful. A "sleeper" virus could infiltrate a network and lie dormant for months or years until given the signal to strike. Conventional antivirus programs would be helpless to spot a Trojan horse virus customized for one particular mission; conventional defenses can only block threats that have already been identified.

Hacker attacks get most of the public attention, but electronic networks are also vulnerable to physical attack. The U.S. has already demonstrated its prowess at such operations. On the opening night of the Gulf War, coalition forces successfully took down Iraqi air defenses by focusing not on individual guns and missiles but on the radars and command centers that made them work. It is rumored that some of the cruise missiles employed by the U.S. had electromagnetic pulse warheads designed to fry electronic circuits with a blast of energy akin to that released by a nuclear explosion. Although details remain classified, the U.S. is known to have developed more sophisticated "e-bombs" since then—and it is not alone. A Russian company has marketed a mobile radio-frequency cannon, the Ranets E, that is supposed to disrupt electronics in a six-mile radius. Terrorists could assemble a radio-frequency weapon of their own by using off-the-shelf components like a TV dish antenna, automotive ignition coils, and copper tape. In one Defense Department exercise, building such a weapon took only a couple of weeks and cost just $500, yet its signal was strong enough to damage military communications equipment.

The U.S. and other advanced countries are particularly vulnerable to such an attack. In the words of Bill Baker, the U.S. Air Force's chief directed-energy researcher: "The smarter the weapon, the dumber HPM [high-powered microwaves] can make it."

NANOTECH:
THE WORLD'S SMALLEST WEAPONS

In 2002 the U.S. Army provided a $50 million grant to set up the Institute for Soldier Nanotechnologies at the Massachusetts Institute of Technology. Researchers hope to produce a new generation of uniform that will have tiny wires embedded in the fabric to keep a soldier connected, 24/7, with others in his unit and with higher headquarters. His helmet would send and receive video, audio, and text messages. It would also contain night-vision and infrared sensors, a GPS system, and laser finders. The suit's sensors could adjust the color and pattern of the camouflage to make the wearer virtually invisible as his surroundings change. This is known as "active camouflage," in contrast to existing "passive camouflage," which stays one color. This "battlesuit" would also have built-in heating and air-conditioning and possibly electronic augmentation to boost the power of human muscles.

In normal wear the uniform would feel as if it were made of light cotton fiber, but if the interwoven sensors detected the presence of chemical, biological, or radiation weapons, it could instantly close its "pores" to become airtight—as would the suits of other soldiers in the vicinity. If incoming bullets or shell fragments were detected, the uniform could become many times harder than Kevlar. If the soldier moved near a bomb, the suit could sound an alert. Should the soldier nevertheless become wounded, the suit could transmit information about his condition to medics while beginning emergency treatment by administering antibiotics, antiseptics, or anesthetics. The suit itself could serve as a temporary bandage to stanch bleeding or as a cast to keep a broken limb in traction.

All in all, the battlesuit would give ordinary mortals many of the attributes of comic book superheroes, from deflecting bullets to turning invisible. This could be possible due to advances in nanotechnology, the science of manipulating objects one to one hundred nanometers in length. (A nanometer is a billionth of a meter.) To put that figure into perspective, the human hair is fifty thousand nanometers wide, while an individual atom is one-fifth of a nanometer in size. Most molecules, including DNA, proteins, and enzymes, are on the nanoscale.

There has been a lot of hype, pro and con, about nanotechnology, which has been billed as the Next Big Thing. Alarmists fear that it could destroy the planet. (Michael Crichton's novel *Prey* features a gray goo made up of self-replicating microscopic machines that run amok, even though no real-world technology allows the creation of robots that reproduce themselves.) Conversely some enthusiasts suggest that nanotech might be as important as electricity—or, in the field of warfare, as important as gunpowder or the atomic bomb. This may go too far, but there is no question that nanotech has important military implications. And battlesuits may be only the beginning.

Nano-sniffers that can detect explosives or biological, chemical, or nuclear weapons could be embedded in the paint of cars and buildings to provide instantaneous warning of danger. Nanomaterials could make lighter yet stronger armor (carbon nanotubes are already used in some armored Humvee turrets) or buildings that would stand up much better to terrorist attacks. Nanoscale electronics could do much to bring about the goal of network-centric warfare in which sensors are ubiquitous on the battlefield.

Many other nanotech applications are currently under development by governments and private investors who are spending about $4 billion annually in this field. (About half of that comes from the United States, evenly divided between the government and the private sector.) Nanotech is such a new area that its implications are barely understood at present. Fantastic possibilities have been proposed, such as implanting nano-devices into the brain and retina in order to enable a person to access the Internet simply by thinking about it. Obviously all such inventions would have military as well as civil applications.

Imagine fighter pilots who could control aircraft with their brainwaves running circles around opposing fliers still reliant on moving sticks and switches with their muscles. That was the scenario in the 1982 Clint Eastwood film *Firefox*. Sound far-fetched? At Duke University researchers have already hooked up tiny probes to the individual neurons of a monkey that allow it to control robotic arms with its thoughts.

In other words, nanotech has already made possible telekinesis—moving objects with the mind. Whatever could be next?

SUPERBUGS AND SUPERSOLDIERS

Germ warfare has an ancient lineage. As summed up in a book written by three *New York Times* reporters: "More than two millennia ago, Scythian archers dipped arrowheads in manure and rotting corpses to increase the deadliness of their weapons. Tatars in the fourteenth century hurled dead bodies foul with plague over the walls of enemy cities. British soldiers during the French and Indian War gave unfriendly tribes blankets sown with smallpox." But the potency of such weapons was dramatically increased by twentieth-century advances in medical science.

In the 1920s and 1930s, all of the major powers stockpiled bombs full of anthrax and other germs, but while British secret agents may have used a botulism-spiked grenade to assassinate the Nazi secret policeman Reinhard Heydrich, only Japan wound up using biological warfare on a large scale. There is considerable evidence of the Japanese armed forces employing anthrax, typhoid, and plague to kill thousands of Chinese civilians.

Both the Soviet Union and the United States had extensive biological weapons programs during the Cold War that were built on wartime Japanese research (just as postwar space programs expanded on the work of Nazi scientists). Although the 1972 Biological Weapons Convention outlawed the use of such weapons, the Soviets continued to produce hundreds of tons of weaponized germs and to use them on occasion. In 1978 a Bulgarian agent employed an umbrella tipped with ricin (a poison derived from castor beans that is twice as toxic as cobra venom) to assassinate Bulgarian defector Georgi Markov in London. Around the same time, allegations surfaced that the Laotian government, a Soviet client state, had sprayed "yellow rain" on Hmong tribesmen. The Reagan administration charged that samples taken from Laos showed evidence of Soviet-produced mycotoxins, though some Western experts believed that it was simply dried bee feces. While the "yellow rain" case remains unsolved, there is no doubt that biological warfare occurred in 2001 when someone mailed powdered anthrax to several U.S. senators and media organizations, infecting twenty-two people and killing five.

Chemicals have been employed as weapons much more extensively than bugs. They were developed in the late nineteenth century as part of the advances in chemistry that also produced dynamite and smokeless gunpowder. In World War I, the Central Powers and the Triple Entente together used an estimated 124,000 tons of chemicals, primarily chlorine, phosgene, and mustard gas, killing 65,000 to 90,000 soldiers and sickening a million more (including a twenty-nine-year-old corporal named Hitler). This was only a small fraction of the war's body count, however, and after the initial shock wore off, the efficacy of chemical weapons rapidly declined because of the deployment of gas masks. Gas was also found to be hard to control; it could easily waft back to one's own lines.

After the Great War, chemicals were employed by the Italians in Abyssinia, by the French and Spanish in Morocco, by the Japanese in China, and possibly by the British in Afghanistan. But, despite the German discovery in the mid-1930s of nerve gas, a compound far more potent than anything employed in 1914–18, chemicals were not used extensively in World War II. This was largely because of the threat of retaliation in kind and the limited utility of chemical weapons against well-prepared troops. Chemical weapons work best against helpless targets such as the concentration camp inmates murdered by the Nazis employing Zyklon B gas.

The use of chemicals in warfare has been frowned upon by international law ever since the 1899 Hague Convention pledged signatories "to abstain from . . . the diffusion of asphyxiating or deleterious gases." The U.S. skirted such dicta in Vietnam when it employed a variety of defoliants—the most famous was Agent Orange—to cut back jungle canopy and CS tear gas to flush Viet Cong fighters out of their bunkers. Although these agents were not designed to kill, the Nixon administration in 1969–70 unilaterally renounced the use of all chemicals or germs.

Saddam Hussein had no such compunctions. In the 1980s, he employed mustard gas and various strains of nerve gas (sarin, tabun, VX) on a massive scale not only against Iranian troops but also against rebellious Kurdish villagers in northern Iraq. On a much smaller scale, the religious cult Aum Shinrikyo released sarin gas into a Japanese apartment building in 1994 and into the Tokyo subway the following year, killing a total of nineteen people and injuring some 5,300.

While never used as weapons as extensively as chemicals, germs have the potential to cause far more damage. Just seventy nanograms of botulinum toxin is enough to kill an adult male; a pound could theoretically kill a billion people. Ricin is marginally less toxic but as little as ten-millionths of a gram can be fatal. It is estimated that 220 pounds of aerosolized anthrax spores released in a major city could produce three million casualties, a toll rivaling that of a thermonuclear bomb. And biological weapons are much easier to produce than nukes, which require centrifuges and other expensive equipment along with rare metals (uranium or plutonium). In 1999–2000 the Pentagon ran an experiment in which officials set up their own germ factory in the Nevada desert using equipment purchased from hardware stores and other commercial suppliers. It cost only $1.6 million to manufacture small quantities of weaponized anthrax.

Once produced, germs could be disseminated by bombs, by missiles, by spraying devices attached to vehicles or aircraft, or by hand. They can be very difficult to trace because of the lag time between infection and the first signs of symptoms. (To this day, no one knows who was behind the 2001 anthrax attacks in the U.S.) Al Qaeda has actively pursued the acquisition of botulinum, salmonella, ricin, and other biological weapons. In addition, at least seventeen countries currently have active chemical and biological weapons programs.

The danger from this quarter is growing because of advances in biological science. Ever since the advent of gene splicing in the 1970s, it has been possible to manufacture germs that are more virulent and more resistant to treatment or to environmental factors (such as heat and cold) than anything found in nature. Soviet scientists working in secret laboratories came up with many such "superbugs," including novel strains of plague and anthrax. Countless possibilities exist for nightmarish hybrid viruses. In Richard

Preston's well-researched novel *The Cobra Event* (1997), a mad scientist combines smallpox with a virus akin to the common cold and an insect virus to come up with a "brainpox" that spreads quickly and melts the brain.

People are not the only possible targets. Crops and livestock could be targeted with pathogens such as swine fever, bird flu, mad cow disease, or foot and mouth disease that could easily cause billions of dollars in damage. Microorganisms could also be deployed to destroy explosives, metals, fuel, adhesives—just about anything. Think of what happens to a sunken ship like the *Titanic:* Its hull is slowly consumed by microorganisms. Such bacteria, if enhanced with gene-splicing techniques, could be deployed to eat enemy vehicles on land. Chemicals could be used in the same way: Superacids billions of times more aggressive than hydrofluoric acid could stop an enemy offensive by melting tires and disabling engines.

Of course, while modern science makes biochemical warfare more potent, it also provides defenses in the form of vaccines and antibiotics. DARPA, for instance, is funding work on gene vaccines that could be far more effective against a wider range of pathogens than traditional vaccines. Someday nanorobots might be injected into the bloodstream to identify and fight infections. This is why, all things considered, nuclear weapons probably still pose a greater danger: There is no drug that can prevent you from being incinerated by an atomic explosion or irradiated by its fallout.

DARPA is using advances in biological science to better protect soldiers from bullets as well as from germs. One group of DARPA-backed researchers is working to create a pain vaccine: A soldier would still feel the initial shock of getting shot but after that the pain would disappear, and inflammation and swelling would be substantially reduced. Other DARPA-funded scientists are trying to utilize the body's natural chemicals to enable a soldier to stop hemorrhaging and seal a wound through, believe it or not, willpower. Still another set of DARPA researchers is trying to emulate the regenerative qualities of tadpoles by figuring out if humans can grow back missing arms or legs.

This is part of a broader DARPA initiative to enhance human performance. Researchers are working on cellular tuning that, if successful, may allow every G.I. to display the strength and endurance of a world-class athlete like Lance Armstrong. Actually better than that: These super-soldiers may be able to go a week without sleeping if scientists can figure out how to copy the genetic wiring of whales and dolphins, which rest only part of their brain at a time. Another bit of cellular rewiring may allow a soldier to avoid eating for days at a time by living off stored body fat. Joe Bielitzki, head of DARPA's Metabolically Dominant Soldier program, aims to create nothing less than an "Energizer Bunny in fatigues."

Future soldiers might have super-intelligence to go along with their super-endurance, because researchers are working to implant microchips into the brain to enhance cognitive performance. "Soldiers having no physical, physiological or cognitive limitations," proclaims Mark Goldblatt, head of DARPA's Defense Science Office, "will be the key to survival and operational dominance in the future."

If such work succeeds, Bielitzki jokes, the army will have to change its slogan to "Be all that you can be and a lot more."

Of course, such research is highly speculative, and even if it produces results, it could easily benefit not its American originators but their enemies. Imagine the damage that could be done by super-resilient, super-strong, super-smart terrorists.

EPILOGUE:

Five Hundred Years and Counting: What the Past Teaches About the Future

Any one of the innovations described in the preceding chapter has the potential to radically change the nature of warfare and with it the nature of the international system. While the U.S. has been dominant so far in the Information Age, there is no guarantee that its streak will continue. A challenger, whether a rival state like China or even a nonstate group like al Qaeda, could utilize new ways of war (or, in the case of nuclear weapons, not so new) to alter the balance of power. Cheap to produce and easy to disseminate, germs, chemicals, and cyber-viruses are particularly well-suited for the weak to use against the strong. If any of them become common and effective tools of warfare, especially terrorist warfare, the U.S. and its allies could be in deep trouble.

History is full of examples of superpowers failing to take advantage of important Revolutions in Military Affairs: The Mongols missed the Gunpowder Revolution; the Chinese, Turks, and Indians missed the Industrial Revolution; the French and British missed major parts of the Second Industrial Revolution; the Soviets missed the Information Revolution. The warning that appears at the bottom of mutual fund advertisements applies to geopolitics: *Past performance is no guarantee of future returns.* The end can come with shocking suddenness even after a long streak of good fortune.

Perhaps *especially* after a long streak of good fortune. The longer you are on top, the more natural it seems, and the less thinkable it is that anyone will displace you. Complacency can seep in, especially if, like the United States, you enjoy power without peer or precedent.

Israel discovered the dangers of primacy in 1973 when it almost lost the Yom Kippur War to Egyptian and Syrian forces that it had handily defeated just six years before. The Israelis were caught off guard by new antitank and antiaircraft missiles supplied by the Soviet Union—a foretaste of what the Information Age had in store.

In hindsight, the ability of the Egyptians and Syrians to bounce back from their humiliation in the Six-Day War (1967) should not have been so surprising. Defeat has often been a spur to innovation, from the Prussians' humiliation in the Napoleonic Wars, to the Germans' humiliation in World War I, to the Americans' humiliation in the Vietnam War. In the case of Japan in 1853, it did not take actual defeat but the mere threat of it, made explicit by the arrival of Commodore Perry's "black ships," to catalyze wide-ranging reforms. Out of all these setbacks were born new ways of fighting that led once-vanquished forces to victory on future battlefields.

It is much less common to see dominant powers innovating. More typical is the case of the Ottoman Empire, which mastered only one major military revolution—gunpowder—and then only in its early years. In their heyday in the fifteenth and sixteenth centuries, the Turks' gun-wielding armies and fleets carved out and defended a vast empire encompassing Asia Minor, North Africa, and the Balkans. By the eighteenth century, however, their glorious record of martial triumphs had become a major obstacle to making the innovations necessary to keep up with European competitors. The Sublime Porte's modernization was so belated and half-hearted that by the nineteenth century the onetime scourge of Christendom had become known as the "Sick Man of Europe." Early success set up the Turks, like so many others, for later defeat.

One of the few partial exceptions to this rule is Britain, whose Royal Navy stayed No. 1 from the age of sail to the age of steel. But not even the Royal Navy could successfully navigate the next major shift, from battleships to aircraft carriers—a failure that hastened the fall of the British Empire.

Business history is replete with the same story. Not a single maker of sailing ships made a successful transition to steam power. And not a single maker of minicomputers—not Digital Equipment Corporation, not Data General, not Prime, not Wang, all seen as invincible giants as recently as the 1980s—made a successful transition to personal computers. More recently, Sony, which was supremely successful in the era of transistor radios and portable cassette players, has struggled to adjust to the digital age, its Walkman having been superseded by Apple's iPod.

After a company fails, its employees, customers, and investors can always move on to a different firm. It is much harder for citizens to move on to another country after their homeland fails to keep up in the struggle for security. The stakes are obviously much greater in the geostrategic realm, making it all the more important to understand why some states fail and others succeed.

UNCONTROLLABLE CREATIVITY

History, alas, does not offer a blueprint of how the process of military innovation occurs. There is no single model that covers all cases, and this book has made no attempt to develop one. As James Q. Wilson noted in his magisterial study of bureaucracies:

> Not only do innovations differ so greatly in character that trying to find one theory to explain them all is like trying to find one medical theory to explain all diseases, but innovations are so heavily dependent on executive interests and beliefs as to make the chance appearance of a change-oriented personality enormously important in explaining change. It is not easy to build a useful social science theory out of "chance appearances."

To the limited extent that we can generalize about five hundred years of history, it seems fair to say that the most radical innovations come from outside of formal military structures. There are some recent exceptions, such as the atomic bomb, the satellite, and the stealth airplane, but most of the key inventions that changed the face of battle since the Middle Ages—the cannon, handgun, three-masted sailing ship, steam engine, machine gun, rifled breech-loader, telegraph, internal combustion engine, automobile, airplane, radio, microchip, laser, wireless telephone—were the products of individual inventors operating more or less on their own: geniuses such as Robert Fulton, Hiram Maxim, Johann Nikolaus von Dreyse, and Guglielmo Marconi. Some had military applications in mind; most did not. For instance, the casting techniques that made cannons more effective in the fifteenth century were originally developed to make church bells.

Even where government has played a big role in the development process, as with the Internet and the electronic computer, the key advances were usually made by people not on its payroll: William Shockley, John Bardeen, and Walter Brattain (the transistor); Jack Kilby and Robert Noyce (the microchip); Ted Hoff (the microprocessor); Paul Allen and Bill Gates (MS-DOS and Windows); Tim Berners-Lee (the World Wide Web); Marc Andreessen and Eric Bina (the Mosaic browser); and many others. The laser is another case in point: It was invented in 1960 by Theodore Maiman, an unknown young physicist at Hughes Aircraft Company who had no Pentagon funding and who worked on the project despite, not because of, the wishes of his superiors.

While government and corporate R&D programs have grown exponentially since World War II, fundamental technological innovation (as opposed to small-scale, incremental improvement) is simply too erratic and mysterious a process to be at the beck and call of any institution. "We can no more 'explain' the breakthroughs inside the minds of a Montgolfier or a Westinghouse," notes economist Joel Mokyr, "than we can explain what went on inside the head of a Beethoven when he wrote the *Eroica*."

Because creativity is so unpredictable, no country can count on making all, or even most, major scientific and technological breakthroughs.

DISSEMINATION AND NULLIFICATION

Moreover, few if any technologies, much less scientific concepts, will remain the property of one country for long. France matched the Prussian needle gun less than four years after the battle of Königgrätz. Germany matched the British *Dreadnought* three years after its unveiling. The USSR matched the U.S. atomic bomb four years after Hiroshima and Nagasaki. It is a truism that new technology, if it proves effective, tends to disseminate quickly.

One exception, and it is a notable one, is that technology was slow to move from the West to the rest of the world in the late stages of the Gunpowder Age and during the First Industrial Age. This created a yawning imbalance of power that allowed Europeans to conquer much of the world on the cheap. But by the mid-twentieth century the balance had righted itself, and Asians and Africans in possession of modern weaponry were able to win their independence from European states weakened by two world wars and the collapse of assumptions of racial superiority. Some analysts may discount the importance of technology in determining the outcomes of battles, but there is no denying the central importance of advanced weaponry in the rise of the West.

Military revolutions usually favor the offense at first, but, with the major exception noted above, that initial edge soon dissipates. The spread of *trace italienne* fortresses in the sixteenth century negated the advantage that heavy artillery had conferred on besieging forces. The spread of trenches starting in the American Civil War offset the impact of rifles and then machine guns. Today, sophisticated radar systems may be negating the edge enjoyed by first-generation stealth aircraft. More significantly, enemies are increasingly resorting to guerrilla tactics to avoid the full fury of American firepower.

The process of technological dissemination and nullification has speeded up since the rise in the mid-nineteenth century of such major arms manufacturers as Krupp, Winchester, and Armstrong, which were happy to sell to just about anyone. Thus, German troops were killed during the

Boxer Rebellion in 1900 by Chinese soldiers firing Mauser rifles and Krupp artillery pieces.

Contemporary arms manufacturers such as Lockheed Martin, Northrop Grumman, and the European Aeronautic Defense & Space Company operate under greater export restrictions but still seek to market the latest technology around the world. Even more pervasive today are firms that sell dual-use devices such as computers, night-vision goggles, and GPS trackers which can have both military and civil applications. Thanks to their success, many of America's key Information Age advantages are rapidly passing into the hands of friends and foes alike. This is part of a longer-term trend—the Westernization of the world increasingly puts the peoples of Asia, the Americas, and Africa on a par, economically as well as militarily, with those of Europe and North America.

As important as technological nullification is psychological nullification. The first time an army faces a major new weapon—the needle gun at Königgrätz, the machine gun at Omdurman, the tank in Poland and France, the smart bomb in the Gulf War—it is likely to be caught off guard. The resulting panic can be as damaging as the physical effects of the weapon itself. The next time, if there is a next time, the other side is likely to be less impressed. Thus the coalition bombing campaign of Iraq in 2003 did not induce the same "shock and awe" as its predecessor in 1991. Having been bombed more or less continuously for a decade, Iraqis had become inured to the effects of precision munitions. The speed and ferocity of the U.S. armored advance, by contrast, came as a surprise.

The way to gain a military advantage, therefore, is not necessarily to be the first to produce a new tool or weapon. Often it is to figure out better than anyone else how to utilize a widely available tool or weapon.

WEAPONS OF MASS DISRUPTION

Faster aircraft, heavier tanks, more accurate rifles, and longer-range artillery are all examples of what management guru Clayton Christensen calls "sustaining" innovations—i.e., products that don't require major organizational adjustments in order to assimilate. They are part of a process of continual improvement undertaken by most modern armed forces. But other innovations, the ones that are the focus of this book, are "disruptive" breakthroughs that occur less frequently and profoundly unsettle the status quo.

Taking advantage of these major advances usually requires what James Q. Wilson calls a "change-oriented personality"—someone like John Hawkins, Gustavus Adolphus, or Curtis LeMay who is not afraid to shake up conventional ways of doing things. Fundamental changes can be preached from the outside but seldom imposed by civilians on a professional military. Consider the lack of success that J. F. C. Fuller and Basil Liddell Hart had in the 1930s preaching the gospel of armored warfare to the British army. John F. Kennedy had equal lack of success when he tried to get the U.S. Army interested in guerrilla warfare in the early 1960s—and he was the president!

The most successful innovators have tended to be people like Field Marshal Helmuth von Moltke, Admiral William Moffett, General Hap Arnold, and Generals Hans von Seeckt and Heinz Guderian: insiders, not outsiders. At best, civilians can play a supporting role in aiding military mavericks against their bureaucratic foes, though popular accounts tend to overstate the influence of flamboyant rebels such as Billy Mitchell. As political scientist Eliot Cohen writes: "To the extent that civilians can control the Department of Defense, they do so less by issuing edicts (which can be evaded, watered down, or delayed by the military and civil service bureaucracy) than by grooming and selecting internal leadership."

Change usually begins in peacetime, but it often takes a major battle—a Spanish Armada, Breitenfeld, Tsushima, Pearl Harbor, or Gulf War—to cause militaries to decisively shift how they do business. Sometimes battles take place during a time of great ferment but neither side can effectively implement a revolutionary operational concept, or both sides come up with the same concept, thereby forfeiting a chance to seize a decisive advantage. This was true of the U.S. Civil War and World War I, which is why both conflicts were such fearful, protracted slaughters and why neither one is considered at length in this book.

STRATEGY AND INNOVATION

Culture, geography, politics, and other factors greatly affect how receptive a military is to proposed changes. Especially important is a country's strategic situation, a combination of its location and its fears and ambitions. Geography is not destiny, or else it would be impossible to explain why Britain was a naval power for centuries while Japan—another island nation off the coast of a major continent—was not. Or why Prussia, rather than another nearby state such as Saxony or Bavaria, became a Great Power starting in the eighteenth century. Or why Sweden rose from obscurity to prominence in the seventeenth century and then fell back into obscurity in the eighteenth century—all without changing its geographic position. But, even though it is only one factor among many, there is no doubt that geography has influenced which nations are more receptive to which military revolutions.

Early in the modern period, Italian states like Venice were content to rely on galley fleets that served them well in the placid waters of the Mediterranean, the chief source of their foreign trade, but that proved utterly unsuited for a new way of naval warfare pioneered by Atlantic powers like Holland, Portugal, and England. Spain and France were caught in the middle, with one foot in the Mediterranean and another in the Atlantic. They never managed to become as effective at sail-and-shot tactics as the more single-minded English and Dutch. Centuries later, Germany became a leader in utilizing panzers because it planned to fight a fast-moving land war against numerous enemies on its frontiers. The Nazis did little to develop aircraft carriers or four-engine bombers because they did not think they needed them against their continental rivals. The United States was the mirror image: It led the way in the development of long-range bombers and aircraft carriers because it expected to fight a naval and air war against enemies far removed from its borders, but it did little to develop tank units because it did not expect to fight a major land war. Such expectations may turn out to be ill-founded (Germany could have used B-17s; the U.S. could have used Tiger tanks), but they powerfully affect decisions about the allocation of scarce resources.

It helps to have relatively few scenarios to prepare for. Germany in the interwar years had the luxury of preparing only for a land war in Europe, whereas Britain had to prepare not only for that contingency but also for naval wars in the Atlantic and Pacific as well as for imperial policing in its colonies. The U.S. had the advantage of focusing on a single foe after the Vietnam War. The concepts and technologies created to fight the Red Army just happened to be perfectly suited to battling the Iraqi army.

Today the U.S. faces a much bigger challenge because it has many potential foes, ranging from nonstate actors (al Qaeda and its ilk) to medium-sized powers (North Korea, Syria, Iran) and a rising great power (China). Because the U.S. has chosen to be strong in every sphere of combat (sea, land, air, space, cyberspace), in every type of warfare (from peacekeeping to high-intensity conflict), and in every corner of the globe, it faces pressure to invest and innovate in many different fields at once, or else to rein in its ambitions.

IS THERE A DEMOCRATIC ADVANTAGE?

Western states have been the most successful military innovators over the past five hundred years. There was something about western Europe (and its overseas offspring) that made it much more dynamic and open to change than other civilizations. Having a relatively liberal political and intellectual climate, of the kind that the West developed toward the end of the Middle

Ages, helps to create an atmosphere in which innovation can flourish. The Soviet Union's lack of freedom ultimately sabotaged its attempts to keep pace in the Information Age, just as the lack of freedom in Spain and France made it difficult for them to keep pace in a naval arms race with first the Netherlands and then Britain.

But we should be wary of simple-minded democratic triumphalism. History has offered many examples of autocratic states that proved more adept than their democratic rivals at exploiting military revolutions. The success of the Prussian and German armed forces between 1864 and 1942 and of the Japanese between 1895 and 1942 shows how well even relatively undemocratic systems can innovate. All that is required is some degree of openness to change, a commitment to meritocracy, and an ability to critically examine one's own mistakes—all disciplines in which the German General Staff, however illiberal, excelled. In fact, most democracies, which tend to be less militaristic than autocracies, face a disadvantage in taking advantage of military innovations because they are less inclined to be generous to their armed forces in peacetime—a problem that beset liberal Western nations during the 1930s.

Nor is there much evidence to suggest that soldiers fight better for a democracy than for a dictatorship. Man for man, the Wehrmacht was probably the most formidable fighting force in the world until at least 1943, if not later. German soldiers were even known for showing more initiative than the soldiers of democratic France, Britain, and the United States. Meanwhile, Soviet soldiers stoically endured privations and casualties far beyond anything suffered by their Western allies. North Vietnam is another modern state that fielded superb armies despite a notable democracy deficit.

In any case, the differences between the armies of dictatorships and democracies are less significant than they may appear at first blush. Even the most liberal states have to employ command-and-control methods in their armed forces. And even the most autocratic states have to pay attention to troop morale and to allow room for individual initiative.

But if democracies do not have an advantage in creating formidable war machines, they do seem to have an intrinsic edge in figuring out how to use them. Autocracies tend to run amok because of the lack of internal checks and balances. Philip II, Gustavus Adolphus, Louis XIV, Frederick the Great, Napoleon, Wilhelm II, Hitler, the Japanese leaders of the early twentieth century—they all built superb militaries but ultimately led their nations into ruinous wars. (So did Saddam Hussein with his less impressive but nevertheless formidable army.) They had no sense of limits, and no politician was strong enough to stop them. Their tactics may have been superb, but their grand strategy was lousy, the best examples being Napoleon's and Hitler's foolhardy invasions of Russia. Democracies sometimes overreach too (witness the Boer, Algerian, and Vietnam wars), but they tend to avoid the worst traps because they have a more consensual style of decision-making. As two political scientists note, "being vulnerable to the will of the people restrains democratic leaders and helps prevent them from initiating foolhardy or risky wars."

BUILDING BETTER BUREAUCRACIES

The key to successful innovation, whether for a dictatorship or a democracy, is having an effective bureaucracy. This was the chief advantage enjoyed by Elizabeth I over Philip II, George III over Daulat Rao Sindia and Raghuji Bhonsle II, Emperor Meiji over Czar Nicholas II, Adolf Hitler over Édouard Daladier and Paul Reynaud, and the two George Bushes over Saddam Hussein. Prussia's secret weapon in the nineteenth century was not the needle gun or the railroad or the telegraph. It was the General Staff, which figured out how to utilize these innovations.

America's secret weapon today is not the stealth airplane or the Predator but the agency that was responsible for their development (and much else besides)—the Defense Advanced Research Projects Agency. Ever since its forerunner was set up in 1958 during the Sputnik crisis, DARPA has shown how a government agency can push the frontiers of innovation by allocating grants to universities, think tanks, and private companies for high-risk ventures. To the limited extent that innovation can be systematized, DARPA has done it. Other nations trying to compete with the United States are hobbled by not spending as much as the U.S. does on research and development, but even spending more money would be no guarantee of success. If it were, then the European Union, whose collective defense budget is two-thirds the size of America's and which has even more soldiers under arms, would be much closer to the U.S. in military capabilities than it actually is. The problem is that most European spending is unfocused, duplicative, and inefficient, whereas DARPA has been smart about allocating its $2 billion annual budget.

Bureaucracies are so important because, as this book has stressed, the realization of a Revolution in Military Affairs requires far more than simply revolutionary technology. It also requires revolutions in organization, doctrine, training, and personnel. That was achieved by the Swedes when they crafted mixed-arms formations made up of pikemen and musketeers; by the Prussians when they figured out how to rapidly mobilize and move large numbers of riflemen by railroad; by the Japanese when they decided to group aircraft carriers together in strike groups; and by the Americans when they integrated smart bombs, sensors, stealth, and professional soldiers in the AirLand Battle doctrine.

Bureaucratic innovation can seldom be limited to the military alone because armed forces are always a reflection, however refracted, of the broader society. Each military epoch comes with its own distinctive system of governance. The rise of the Gunpowder Age fostered the growth of absolute monarchies. The First and Second Industrial Ages fostered giant welfare and warfare states. The Information Age is leading to a more decentralized, flatter form of government and the rise of more powerful nongovernmental groups. States that fail to keep up with these transformations risk getting run over by those that do.

For powers that lack effective bureaucratic structures, the possession of modern weaponry is of dubious utility, as states from eighteenth-century India to the twentieth-century Middle East have found. The Arab nations are particularly egregious in this regard: Their long record of military futility since 1945 comes despite having access to copious stocks of modern arms from such outside suppliers as France and the Soviet Union. No matter how great the Arab preponderance in men and materiel—and against Israel in 1948, 1956, 1967, and 1973 their advantage appeared, on paper at least, to be insuperable—they have continually contrived to snatch defeat from the jaws of victory. In one of the lesser-known episodes of this long record of ignominy, the well-armed Libyan military was routed by ill-armed Chadians in 1986–87 after Moammar Qadafi tried to annex northern Chad. The only military strategies (if such they are) that Arabs have been able to employ with any success are terrorism and repression. "The incompetence of Arab tactical leadership, their severe problems managing information, and the inability of their personnel to properly employ and maintain their military hardware," writes former CIA analyst Kenneth M. Pollack, "left the Arab states highly vulnerable to most potential adversaries."

The armies of Russia and the United States were far more competent, but in Afghanistan and Chechnya, Vietnam and Iraq, they, too, found themselves stymied by smaller, poorer adversaries, largely because their armed forces were not properly configured for counterguerrilla warfare. This does not mean that modern military hardware is useless—only that by itself it is not enough to guarantee victory against a clever, determined adversary. When combined with the right organization, doctrine, training, and leadership, however, sophisticated weaponry can confer a decided advantage even in battling irregular foes.

DREAD OF INNOVATION

It is no surprise that the authoritarian Arab states have not, for the most part, managed to make the changes necessary to harness modern military power. No Arab dictator can afford to have a military that is too strong for fear that it will be employed against him. But even for more liberal polities, which generally need not fear a military coup d'état (though France faced such a prospect as recently as the early 1960s), transitions from one military system to another can be wrenching, because they require uprooting existing career patterns and deeply held beliefs. Officers trained in cavalry charges were not happy about the advent of tanks, any more than sailors trained in battleships were happy about the arrival of aircraft carriers, or knights trained in sword fighting were happy about the spread of firearms. Militaries are inherently conservative organizations. As a British army colonel noted in 1839: "In no profession is the dread of innovation so great as in the army."

Successful adaptation to major technological shifts requires overcoming that dread and changing the kinds of people who are rewarded within a military structure. The rise of railroads and steamships in the nineteenth century elevated the importance of logisticians and engineers, technocrats who were initially despised by traditional army and navy officers. In the Imperial German Navy, engineering officers were sent to a separate school, they wore less gaudy uniforms, without the sashes and imperial crowns sported by regular officers, and they were not allowed to dine in the officers' mess. They even had to endure the indignity of having their wives addressed as "women," not "ladies." It was not until 1899, half a century after the adoption of steam power, that the U.S. Navy merged "line" and "engineering" officers. In the more aristocratic British and German navies the process took even longer.

Today, the U.S. Air Force faces difficult dilemmas as it figures out how to integrate unmanned aerial vehicles: Should someone controlling a Predator from a trailer thousands of miles away be a certified pilot? Should control time count as "flying" hours? This may seem a picayune matter, but it looms large in a service where the fastest advancement has always gone to fliers. If the air force doesn't give greater promotion opportunities to UAV operators, it cannot attract and keep the best people for these jobs, but if it advances them its fighter-jock culture will inevitably change.

This is part of a broader challenge confronting all Information Age militaries: how to make room for those who fight with a computer mouse, not an M-16. Will traditional warriors—men with shaved heads and hard bodies—continue to run things, or will nerds with bad posture and long hair, possibly even women, assume greater prominence? Two Chinese strategists write that "it is likely that a pasty-faced scholar wearing thick eyeglasses is better suited to be a modern soldier than is a strong young lowbrow with bulging biceps," but, even if that is true, reordering any military along those lines presents a far more profound and problematic challenge than questions about which tank or helicopter to buy. As Eliot Cohen writes: "The cultural challenge for military organizations will be to maintain a warrior spirit and the intuitive understanding of war that goes with it, even when their leaders are not, in large part, warriors themselves."

THE DANGER OF TOO MUCH CHANGE—AND TOO LITTLE

This book surveys many instances where militaries had to change or die. Those armed forces that did not successfully integrate the gun, the long-range bomber, precision-guided munitions, or other important innovations experienced the agony of their members dying in great numbers. But we have also looked at some instances of militaries too eager to change in the wrong way. In the 1930s, the U.S. Army Air Corps and the Royal Air Force placed too much faith in the ability of unescorted bombers to win a future war—a doctrinal mistake that cost the lives of tens of thousands of air crews over Europe. In the 1940s Hitler poured vast resources into the development of the V-1 and V-2, resources that might better have been employed on his conventional forces. And in the 1950s the U.S. Army, Navy, and Air Force did so much to rearrange themselves around the demands of the nuclear battlefield that they were not ready for the actual threat they would soon confront in the jungles of Vietnam.

Arguably a similar phenomenon has occurred in Iraq, where the Information Age armed forces of the United States have found themselves frustrated by less sophisticated adversaries. Many now ask: Why did the Defense Department not invest in more linguists, more MPs, more civil affairs specialists, more soldiers in general, rather than in more JDAMs and JSTARS? The answer is that senior leaders, such as Donald Rumsfeld, believed that the future of warfare lay in high-tech information systems, not in lowly infantrymen. This does appear to be a mistake in light of events in Iraq—but it may not turn out to be so mistaken if the U.S. finds itself in a clash with China or North Korea.

There is no rule of thumb to suggest how much or how little a military should change in response to technological developments. Each revolution raises painful questions of prioritization such as those that the United States and other countries confront today: Should they pay for more traditional infantrymen, or push resources into "transformational" programs like surveillance satellites, wireless broadband networks, and directed-energy weapons? Should they continue to build traditional tanks, aircraft, and ships or switch to unmanned platforms? Each path has major risks and trade-offs: Paying for larger standing forces can make it easier to respond to today's threats; cutting force strength and using the savings to pay for high-tech hardware can make it easier to respond to tomorrow's threats. (The only strategy that definitely won't work is the one that many European countries are pursuing: cutting existing forces but not investing the savings in any other defense program.) It would be nice to be able to do everything at once. But no one, not even the Pentagon, has enough money for that.

History indicates that the wisest course is to feel one's way along with careful study, radical experimentation, and freewheeling war games. Paradoxically, revolutionary transformation often can be achieved in evolutionary increments. The Germans did not shift over their entire army to panzer divisions in the interwar years. In 1939–40 only about 10 percent of German forces were composed of armored units, and the Wehrmacht had more ponies than panzers, but this was enough to produce breakthroughs from Poland to France. Likewise, General H. H. Kitchener did not have many machine guns when he confronted the Mahdists at Omdurman, but the few he had produced utterly devastating results. Nor did the U.S. in the 1980s convert its entire air force to stealth aircraft, but having even a few F-117s had an outsize impact on the outcome of the Gulf War. A little cutting-edge technology can go a long way against a less advanced foe.

A corollary is that a military revolution does not necessarily sweep aside all old weapons and old ways of doing things. Battleships may have been dethroned as queens of the sea in 1941, but they continued to perform a valuable auxiliary role as a shore bombardment platform into the 1980s. Gustavus Adolphus did not simply toss out pikes in order to make way for muskets and cannons; he used a combination of weapons, old and new, to achieve the maximum shock effect. Indeed, bayonets continue to be fixed onto rifles (though rarely used) hundreds of years after edged weapons lost their primacy on the battlefield. And nuclear weapons have hardly rendered conventional weapons obsolete.

This offers a counterpoint to skeptics who deny the existence of an Information Revolution simply because not everything has changed: It never does. On the other hand, this also offers a cautionary lesson that some modern-day J. F. C. Fullers or Billy Mitchells anxious to scrap the tank, the aircraft carrier, or the manned airplane should keep in mind: Introducing "transformational" systems does not necessarily mean getting rid of all "legacy" platforms. Rather, it means readjusting the balance between the two. "You need to think about how to make a transition," counsels Andrew Marshall of the Pentagon's Office of Net Assessment, "not about how to eliminate current weapons."

While no one would wish for more combat, the U.S. armed forces are helped in this process by having so many wars to fight that can serve as field laboratories for the testing of new technologies. The first Predator was rushed into service for the Kosovo War. Having performed well there, an armed version made its debut two years later in Afghanistan. This, in turn, spurred the development of purpose-built unmanned combat aerial vehicles that will no doubt be tried out in another conflict before long.

The U.S. armed forces would be even more adept at innovation if they were willing to stage more realistic war games in which adversaries could use unconventional tactics instead of fighting in the way that American admirals and generals would prefer. They would also be helped if defense spending could be allocated according to a rational judgment of strategic priorities, not based on the political muscle of major defense contractors and their allies on Capitol Hill. That, however, seems unlikely to change as long as America remains a democracy.

SILENT SPUTNIK?

The quickening pace of scientific and technological progress, which has been speeding up ever since the start of the Industrial Revolution more than two hundred years ago, puts a greater premium on having scientists and engineers who can stay at the frontiers of change. It is neither necessary nor possible to invent everything yourself. But it is vital, at a minimum, to be able to keep up with breakthroughs made elsewhere and take advantage of their military potential. The Royal Navy did not have to be the technological leader in the nineteenth century, but it did need the capability to quickly match and surpass any innovation by the No. 2 naval power, France.

In this regard, many experts note that U.S. hegemony might be endangered by its failure to produce more math, science, and engineering graduates. The U.S. has made up for this shortfall in the past by importing smart people from abroad (38 percent of science and engineering doctorate-holders in the U.S. are foreign-born). But that has become harder to do in the wake of post-9/11 visa restrictions and booming economies in eastern Europe, India, China, South Korea, and Taiwan—the major sources of American scientific and engineering talent—that discourage immigration to America. If China can keep more of its geniuses at home, it will be easier for Beijing to challenge U.S. power. Some scientists warn that the U.S. is facing a "Silent Sputnik" crisis which could imperil U.S. leadership.

Remedying this looming shortfall will probably require more funding for math, science, and engineering education, and that won't be cheap. It will be even more expensive to translate the resulting ideas into actual military programs. It does not necessarily take a lot of money to innovate; breakthroughs like the blitzkrieg and carrier warfare emerged out of paltry military budgets in the interwar years. But it does take a lot of money to bring inventions to fruition, especially today, when each new weapons system costs several times more than its predecessor. It also costs a lot to field high-quality soldiers able to cope with the complexity of modern war. The annual cost to the United States for each member of its armed forces more than doubled in constant terms over the past thirty years—from $125,000 per person in 1970 to $264,000 per person in 2003—and it continues to increase.

There is no free lunch: Despite the fervent hopes of some transformation advocates, there is no way to significantly increase military power while cutting costs. Today, even more than in the fifteenth century, more military capability requires "money, more money, and again more money." With the U.S. facing budget deficits and looming bills for social welfare programs, questions inevitably arise about whether it can afford to keep spending so much on defense. Other countries confronting tight budgets face similar debates: Is it necessary to stay at the forefront of military change? Is it vital to take advantage of the Information Revolution and its successors?

The answer will vary from country to country. As one security analyst notes, "try as they might, countries like Burkina Faso or Paraguay . . . will never be candidates for the exploitation of the RMA." But for the United States and other countries that aspire to be first- or second-tier military powers, there is no alternative but to stay abreast of the changes—tactical and technological, conventional and "asymmetric"—transforming the modern battlefield.

WHY RMA'S MATTER

The major theme that runs throughout this book is the importance of not missing out on the next big change in warfare. History is driven by many factors, but while academia focuses on economics, race, class, sexuality, geography, germs, culture, and other influences on the course of human events, it would be foolish and short-sighted to overlook the impact of military prowess and especially aptitude in taking advantage of major shifts in war-fighting. Of course, a country's success or failure in harnessing change cannot be divorced from such underlying factors as its economic health, scientific sophistication, educational system, political stability, and so forth. But, contrary to Napoleon, God is not necessarily "on the side of the big battalions." Even big and wealthy countries often lose wars and head into long-term decline through a lack of military skill.

The considerable gains of the Axis during the early years of World War II came, after all, against a coalition of Allied states that in aggregate had 40 percent greater GNP and 170 percent larger population. That the Axis ultimately lost shows that military skill can sometimes be trumped by greater resources if a war drags on long enough *and* if the side with greater resources shows sufficient wisdom in their employment. But even in a long coalition war, the side with the greater resources does not always prevail. The alliance of Britain, Hanover, and Prussia was dwarfed in economic and demographic resources by its adversaries in the Seven Years' War (1756–63)—France, Austria, Russia, Sweden, Saxony, and (near the end) Spain—but still managed to win largely because of the superior skill of the Prussian army and the British navy. This book chronicles many other examples of the poorer side emerging victorious: Britain beat the Spanish Armada, Sweden beat the Holy Roman Empire at Breitenfeld and Lützen, Prussia beat the Habsburg Empire at Königgrätz, and Japan beat the Russian Empire in 1904–1905. More recent instances might be cited, such as North Vietnam's defeat of the United States or the Afghans' defeat of the Soviet Union.

These were not anomalies. In a statistical analysis of twentieth-century wars, the side with the larger GNP, population, armed forces, and defense expenditures won only a little more than half the time, making these factors about as useful in predicting military outcomes as flipping a coin. Political scientist Stephen Biddle, who analyzed these statistics, writes, "Superior numbers can be decisive or almost irrelevant depending on the two sides' force employment. This in turn means that states' relative economic, demographic, or industrial strength are poor indicators of real military power: gross resource advantages matter only if they can be exploited via modern-system force employment, and many states cannot do so. . . . How forces are used is critical."

Indeed, while some states translate riches into military power, as the U.S. did in the early years of the twentieth century, just as many states translate military power into riches. That is what England did when it sent its navy to conquer colonies and carve out trade routes, and what Prussia did when it sent its army to overrun the neighboring German principalities. Some states are drained by war, but many others attain Gustavus Adolphus's ideal of making war "pay for itself"—a feat achieved most recently by the United States when it succeeded in making its allies foot most of the bill for liberating Kuwait in 1991.

The ongoing proliferation of destructive technology means that the link between economic and military power is more tenuous than ever. Al Qaeda, whose entire budget would be insufficient to buy a single F-22, can inflict devastating damage on the world's richest country. Advances in biological and cyberwar promise to put even more destructive potential into the hands of ever smaller groups, as does the continuing proliferation of nuclear weapons.

Imagine the devastating consequences of a mega-attack by terrorists. Not only could millions die, but international travel and commerce—the lifeblood of the global economy—could be severely disrupted. Such a scenario reveals the falsity of economic determinist arguments which counsel that military strength is unimportant and that it is feasible to stint on military preparedness in order to strengthen the economy. On the contrary, there can be no long-term prosperity without security. The entire world today depends, no matter how begrudgingly or unwittingly, on the protection provided by the United States, whose armed forces keep open air and sea lanes, safeguard energy supplies, and deter most cross-border aggression.

Dreamers can convince themselves that military power no longer matters, that economic interdependence has consigned war to the dustbin of history, that a country need wield only "soft power," but history is likely to deliver a stark rebuke to such wishful thinking. As a matter of fact, it already has. The attacks of September 11, 2001, put an end to a decade of talk about the "end of history," a "strategic pause," the inexorable flow of "globalization," and the "peace dividend." The incidence of war may have declined for the moment, but great dangers still loom ahead. Santayana had it right: "Only the dead have seen the end of war."

FIGHTING WILDCATS AND RODENTS

Technological advances will not change the essential nature of war. Fighting will never be an antiseptic engineering exercise. It will always be a bloody business subject to chance and uncertainty in which the will of one nation (or subnational group) will be pitted against another, and the winner will be the one that can inflict more punishment and absorb more punishment than the other side. But the way punishment gets inflicted has been changing for centuries, and it will continue to change in strange and unpredictable ways.

In assessing the future conduct of conflict, most analysts tend to fall into one of two camps. One group stresses the dangers of terrorists and guerrillas who use cheap, simple weapons like AK-47s, machetes, or explosives. Another group stresses the danger of high-tech weapons such as cruise missiles and killer satellites proliferating around the world and into the hands of states such as China and North Korea. The former school (associated with ground-combat arms) stresses the need for better warriors; the latter school (associated with air and naval forces), the need for better machines. The reality is that both high-intensity and low-intensity threats are real and that both top-notch people and first-rate equipment are needed to counter them.

Michael Evans of the Royal Military College of Australia offers sage advice when he writes: "In a dangerous and unpredictable world, military professionals and their political masters must . . . be ready to tame the big wildcats and not simply the vicious rodents, to be able to fight troops like Iraq's former Republican Guard as well as Taliban, al-Qa'ida militia, and terrorists."

Today, the U.S. is much farther along in figuring out how to tame the Republican Guard than al Qaeda, and it needs to place more emphasis on making up for its deficiencies in irregular warfare rather than simply enhancing its already substantial lead in conventional warfare. While the Information Revolution has decreased the number of weapons and soldiers needed to defeat a conventional adversary, occupation duty and nation-building—the prerequisites for turning a battlefield triumph into a long-term political victory—continue to demand lots of old-fashioned infantry. Therefore, the U.S. and its allies would be making a mistake if they were to seriously stint on force size in order to procure more high-tech systems.

But that doesn't mean that the U.S. can ignore the dangers of major warfighting or the dictates of technological change. That was the mistake Britain made before 1914 and again before 1939. The British had the world's best "small war" force—an army well-trained and equipped for fighting bandits and guerrillas—but it was ludicrously insufficient to deter German aggression or to defeat Germany once a world war broke out. That mistake, symbolized by deficiencies in tanks and aircraft carriers, hastened the end of the Pax Britannica.

Today, the possibility of conventional interstate war is lower than at any time in the past five hundred years, but it has not disappeared altogether. Because Americans and other citizens of Western democracies no longer seem willing to suffer the same level of casualties experienced by their ancestors, their armed forces must be able to defeat adversaries at scant cost in lives. That argues for keeping the qualitative edge that the U.S. gained in the Information Age—an edge that cannot be preserved by standing still. It will be necessary to keep innovating because, as previously indicated, some of the technologies and techniques employed by the U.S. are starting to be negated by their dissemination around the world. Innovation must be organizational as much as technological, and it needs to focus on potential threats across the entire spectrum, from low-intensity guerrilla wars to high-intensity conventional conflicts.

In any case, the boundaries between "conventional" and "unconventional," "regular" and "irregular" warfare are blurring. Even nonstate groups are increasingly gaining access to the kinds of weapons—from missiles and land mines to chemicals and perhaps even atomic bombs—that were once the exclusive preserve of states. And even states will increasingly turn to unconventional strategies to blunt the impact of American power.

Two colonels of the Chinese People's Liberation Army envision "unrestricted warfare" encompassing not only traditional force-on-force encounters but also financial warfare (subverting banking systems and stock markets), drug warfare (attacking the fabric of society by flooding it with illicit drugs), international law warfare (blocking enemy action by using multinational organizations), resource warfare (seizing control of vital natural resources), even ecological warfare (creating man-made earthquakes, tsunamis, or other disasters). In a clever bit of jujitsu, many of these strategies turn the strengths of Information Age countries against them. Al Qaeda is pursuing similar strategies.

Countering such threats will require much more than simply buying more advanced aircraft, tanks, or submarines. Such traditional weapons systems may be almost entirely useless against adversaries clever enough to avoid presenting obvious targets for precision-guided munitions. To fight and win the wars of the future—wars that may more closely resemble a series of terrorist attacks or hit-and-run raids than traditional force-on-force armored, aerial, or naval engagements—will require reorganizing conventional militaries to emphasize such skills as cultural awareness, knowledge of foreign languages, information operations, civil affairs, and human intelligence. It will also require cutting away the bureaucratic fat to turn bloated Industrial Age hierarchies into lean Information Age networks capable of utilizing the full potential of high-tech weapons and highly trained soldiers.

The United States' readiness for such challenges will determine whether it can keep its position as the lone superpower or whether the world will see another power shift of the kind that accompanied the Gunpowder Revolution, the First Industrial Revolution, the Second Industrial Revolution, and the early stages of the Information Revolution. The course of future history will turn on the outcome.

ACKNOWLEDGMENTS

Like my previous book, this one grew out of my desire to shed historical light on a current military controversy. I wrote *The Savage Wars of Peace: Small Wars and the Rise of American Power* (2002), after becoming interested in the debate over American military interventions in the 1990s in such places as Somalia, Bosnia, Haiti, and Kosovo. I wrote *War Made New* after becoming interested in the debate over "defense transformation" and Revolutions in Military Affairs that began after the 1991 Gulf War and that has been going on in one form or another ever since. The major difference is that while I had to work on my last book at night, in the morning, and on weekends around the margins of my full-time job at *The Wall Street Journal*, I was able to devote most of my energies to this project over the past four years.

This freedom was made possible by a serendipitous change of jobs. After *Savage Wars* came out in 2002, Leslie Gelb convinced me to leave the *Journal* and become a senior fellow at the Council on Foreign Relations, of which he was then president. Although I loved working at the *Journal*, it is a choice that I have never regretted, in no small part because Les and his director of studies, Michael Peters, were such congenial bosses. I have been just as delighted to work with their successors—Council president Richard Haass and director of studies James Lindsay. I owe an immense debt not only to Les, Mike, Richard, and Jim, but also to many past and present colleagues at the Council—especially Walter Mead, Rachel Bronson, Steven Cook, Gideon Rose, Michael Doran, Bill Nash, Raj Menon, Steve Sestanovich, and Isobel Coleman—who have made it such a stimulating and supportive environment.

Among the many benefits of working at the Council is having the support of a first-rate research associate. This book has benefited from the work of a succession of intrepid researchers—Reihan Salam, Victoria Thompson, and Ian Cornwall—whose labors were out of all proportion to their meager compensation. Ian worked for me the longest, and to him I owe particular thanks, not only for supporting the writing of this book but also for enabling its author to deal with various other responsibilities. Other CFR staff members—Christine Quinn, Avery Alpha, Leigh Gusts, Marcia Sprules, Michelle Baute, and Connie Stagnaro come to mind—also provided crucial support.

Another benefit of the Council is that every year some of the most distinguished colonels and naval captains in the American armed forces take up positions as visiting fellows. I have gained more than I can say from the insights, advice, and (pardon the pun) fellowship of the CFR's military fellows—Ron Bailey, John Boggs, Mark Bucknam, Sal Cambria, Jeff Fowler, Dan Gerstein, Waldo Givhan, Chris Haave, Pete Henry, Pete Mansoor, David Marquet, and Marty Peatross. Chris, Pete Mansoor, and Ron shared wonderful stories about their experiences in Operation Iraqi Freedom. Dan, Ron, and Chris led a memorable tour of domestic military installations in 2005. Pete Mansoor and Mark, who happened to be in residence when I finished my manuscript, were kind enough to read a draft and offer extremely useful feedback. Some other military officers affiliated temporarily with the Council, especially Jim Gavrilis, Laurence Spinetta, and Ike Wilson, were also important sources of insight.

In addition, the Council arranges meetings of members known as study groups to critique drafts of book chapters produced by its staff. Professor Paul Kennedy of Yale came down from New Haven to chair several such sessions. A number of Council members were interested enough to read multiple chapters and to make valuable suggestions. Mallory Factor was particularly important in helping me to shape the first part of the book.

Bernard "Mick" Trainor, a retired marine general and former Council fellow, took time off from completing his own book on the Iraq War to chair a couple of study group sessions in Washington.

Irina Faskianos, the indefatigable director of the Council's National Program, kindly arranged for additional study groups to meet in Chicago, Los Angeles, and San Francisco. I am grateful to everyone who turned out for those meetings, and especially to the CFR members who chaired them: Henry Beinem, president of Northwestern University; Michael Nacht, dean of U.C. Berkeley's School of Public Policy (who chaired two sessions); Kal Raustiala, professor of law at UCLA; and Jim Thomson, president of RAND.

Yet another meeting was convened by another distinguished institution—the Carnegie Council on Ethics and International Affairs in New York—to offer me guidance on the final part of the book. I am immensely appreciative of Nick Rizopoulos, who has chaired these sessions for years, for providing such a stimulating forum for debate and discussion.

Finally, the Council commissioned reviews of the manuscript from five anonymous scholars; I thank them for their close readings and numerous suggestions.

The freedom and support made possible by the Council does not come cheap, and I would be remiss if I did not thank those who have funded my work: the Smith Richardson Foundation, the Bradley Foundation, the Olin Foundation, the Sloan Foundation, and the Randolph Foundation. At these institutions, I am especially grateful to Marin Strmecki, Nadia Schadlow, Doron Weber, Jim Piereson, Heather Higgins, Dusty Rhodes, and Diane Sehler. Roger Hertog has been a source of all sorts of crucial support, not least moral and intellectual.

During the four years that I spent on *War Made New*, some of the book's themes were previewed in the pages of various newspapers and magazines. I am grateful to a number of editors who ran my work, especially Nick Goldberg and Susan Brenneman at the *Los Angeles Times*; Bill Kristol, Claudia Winkler, Richard Starr, and Phil Terzian at *The Weekly Standard*; Jim Hoge and Gideon Rose at *Foreign Affairs*; David Shipley, Barry Gewen, and Katy Roberts at *The New York Times*; Gwen Robinson and John Gapper at the *Financial Times*; Fred Hiatt and Warren Bass at *The Washington Post*; Tunku Varadarajan, Eric Gibson, and Erich Eichman at *The Wall Street Journal*; Adam Garfinkle at *The American Interest*; and Richard Snow at *American Heritage*.

I benefited from conversations with a number of experts during the course of my work, including Andy Marshall, Steve Rosen, Robert Kaplan, Bob Scales, Newt Gingrich, Eliot Cohen, Mike Vickers, Andy Krepinevich, Barry Watts, Wick Murray, Harvey Sapolsky, Victor Hanson, the late Art Cebrowski, Gary Anderson, and Walter Mead. Jonathan Tucker of the Monterey Institute of International Studies provided me with guidance on chemical and biological warfare. Andy Kessler was good enough to send me his manuscript on the history of the Industrial and Information Revolutions, and Fleming Saunders sent me his draft paper on the history of smart weapons. Thanks to all of them.

Naturally, a book this large in scope must rely on the work of numerous scholars. A full list is available in the Bibliography, but I must mention the particularly valuable work of William McNeill, Paul Kennedy, Stephen Budiansky, the late Trevor Dupuy, John Lynn, Michael O'Hanlon, Dennis Showalter, Daniel Headrick, Richard Overy, Williamson Murray, Michael Vickers, and especially Geoffrey Parker, the dean of early modern European studies. My conclusions may not always agree with theirs, but this book would have been much harder to construct were it not for their intellectual spadework.

As part of my research into the Information Revolution, I interviewed military personnel who took part in operations in Iraq and Afghanistan. A number of people went to considerable trouble to arrange those interviews. I thank Colonel Glenn Vavra, Kathy Devine, Lieutenant Colonel Roger Carstens, Major Jim Gavrilis, Theresa Brinkerhoff, Captain Kathleen Turner, and Colonel Kevin Weddle. Dean, a Special Forces captain who served in Afghanistan, not only provided much valuable information but also read Chapter Eleven and corrected a few mistakes.

I conducted various field trips for further research. Major Generals Jim Mattis and David Petraeus (they have since earned additional stars) were instrumental in arranging my first visit to Iraq in 2003. I thank them for their hospitality and openness. Marine Sergeant First Class Brad Lee was a delightful guide and Bing West a boon travel companion. My second trip to Iraq, in 2006, was the work of many individuals. I thank Major Joseph Breasseale, Captain Lyle Gilbert, First Lieutenant Michelle Lunato, Lieutenant Colonel Michael Negard, Major Robert Palmer, Major Jeffrey Allen, Major Michael Humphreys, and Lieutenant Colonel Hugh "Trey" Cate. I was able to visit Afghanistan in 2006 thanks to General Jim Jones, the Supreme Allied Commander Europe, and his subordinates, James Snyder and Commander Martyn Reid.

Two Davids gave me hands-on experience with historical firearms. Dave Miller of the Smithsonian Institution's Museum of American History gave me a tour of a locked storeroom packed with hundreds of years of firearms—including an actual needle gun! And David J. DeLucia of the Westchester County Department of Parks, Recreation, and Conservation allowed me to test-fire some of the muskets and rifles—ranging from a Brown Bess to an M1—from his magnificent private collection.

Thanks to Boeing for a 2004 briefing on the Future Combat System, and to the MIT Soldier Institute of Nanotechnology for a 2003 briefing on its battlesuit.

My education has been considerably enhanced and my spirits buoyed by the time I have spent lecturing at military schools. I thank all those who have invited me to spend time at such outstanding institutions as the Army Command and General Staff College, the Naval War College, the Army War College, West Point, the Naval Academy, the National Defense University, and the John F. Kennedy Special Warfare School. I can't vouch for what the students learned from me, but I learned an awful lot from them. So, too, I have learned a great deal from serving on the U.S. Joint Forces Command's Transformation Advisory Group. I thank Admiral Ed Giambastiani for extending an invitation to join this distinguished panel and his successor, General Lance Smith, for keeping this stimulating group going.

A number of people contributed to the production of this book. Glen Hartley and Lynn Chu are first-rate literary agents who know the world of ideas as much as they know the book business. They are also valued friends and advisers without whose support this book would never have been written. Brendan Cahill was an inspired and enthusiastic editor whose suggestions improved the final product in all sorts of ways. Bill Shinker was a terrific publisher. Patrick Mulligan did much to shepherd this book through its final production stages at Gotham. David Lindroth drew the handsome maps. In addition to all his other duties, Ian Cornwall tracked down and acquired the illustrations.

Last but far from least there is my family to thank. My mother, Olga Kagan, and my stepfather, Yan Kagan, read the manuscript and provided much encouragement and some important suggestions. To them the book is dedicated. My kids—Victoria, Abigail, and William—were thoughtful enough to leave Daddy alone while he was working. Well, sometimes. My mother-in-law, Kathleen McCarty, helped in all sorts of ways. And then there is my wife, Jeannette K. Boot, Esq. Words cannot express my gratitude to her, not only for her unstinting love and support, but also for more specific contributions, such as arranging to turn our attic into a home office, where much of this book was written. She even accompanied me on several tours of battlefields, feigning interest with remarkable grace. And all the while holding down jobs far more demanding than mine. I am grateful to one and all.

BIBLIOGRAPHY

ORAL HISTORY INTERVIEWS[1]

Amerine, Jason. Interview with author. West Point, New York, November 9, 2004. Captain Amerine was team leader of ODA 574 during Operation Enduring Freedom in the fall of 2001.

Anderson, Joseph. Interview with author. Mosul, Iraq, August 28, 2003. Colonel Anderson was commander of the 2nd Brigade, 101st Airborne Division, in Iraq in 2003.

Bartholomees, James. Interview with author. Fort Leavenworth, Kansas, January 23, 2006. Captain Bartholemees commanded a Ranger company in Afghanistan in 2004–5.

Dean (Last Name Withheld). Interview with author. West Point, New York, November 22, 2004. Captain Dean was the team leader of ODA 534 during Operation Enduring Freedom in the fall of 2001.

Dillow, Shane. Interview with author. Fort Leavenworth, Kansas, January 23, 2006. Captain Dillow served with Army Special Forces in Afghanistan.

Feith, Douglas. Interview with author. Washington, D.C., July 25, 2005. Feith was undersecretary of defense for policy during Operation Enduring Freedom and Operation Iraqi Freedom.

Frontline, "Campaign Against Terror: Interviews." www.pbs.org/wgbh/pages/frontline/shows/campaign/interviews. [Documentary aired September 8, 2002.]

———."The Invasion of Iraq: Interviews." http://www.pbs.org/wgbh/pages/frontline/shows/invasion/interviews. [Documentary aired March 4, 2005.]

———."Bioterror: Interviews." www.pbs.org/wgbh/nova/bioterror/biow_popov.html [Documentary aired in 2001.]

Gavrilis, James. Interview with author. Washington, D.C., February 3, 2005. Major Gavrilis was a Special Forces company commander in the western Iraqi desert in March and April 2003.

Haave, Christopher. Interview with author. New York City, January 13, 2005. Colonel Haave was director of combat operations at the Combined Air Operations Center in Saudi Arabia from February to May 2003.

Hagenbeck, Franklin. Oral History Interview 130 EF I 133, conducted March 15, 2002. U.S. Army Center for Military History.

Hopkins, Michael. Interview with author. Washington, D.C., December 1, 2004. Major Hopkins was a Special Forces company commander in Afghanistan in early 2002.

Iverson, Bjarne (Mike) M. Interview with author. Carlisle, Pennsylvania, March 15, 2005. Lieutenant Colonel Iverson was political-military adviser to Lieutenant General

[1]All ranks given for the time period under discussion. Interview subjects who requested anonymity are not named here.

Ricardo Sanchez, Commander of Combined Joint Task Force 7 in Iraq from July 2003 to June 2004.

Johnson, James. Interview with author. Fort Leavenworth, Kansas, January 23, 2006. Air Force Major Johnson was a combat controller in Afghanistan.

Kirkpatrick, Michael. Interview with author. Fort Leavenworth, Kansas, January 23, 2006. Captain Kirkpatrick was a company commander with the 2nd Armored Cavalry Regiment when it garrisoned a zone north of Baghdad's Sadr City in 2003.

Linnington, Michael. Interview with author. Along the Iraq/Syria border, August 27, 2003. Colonel Linnington was commander of the 3rd Brigade, 101st Airborne Division, in Iraq in 2003.

Lopez, Matthew. Interview with author. Karbala, Iraq, August 22, 2004. Lieutenant Colonel Lopez was commander of the 3rd Battalion, 7th Marine Regiment.

Marshall, Andrew. Interview with author. Washington, D.C., December 20, 2002. Marshall is head of the Pentagon's Office of Net Assessment.

Mattis, James. Interview with author. Camp Babylon, Iraq, August 21, 2003. Major General Mattis was commander of the 1st Marine Division during two tours in Iraq, in 2003 and 2004.

Megill, Todd A. Interview with author. Carlisle, Pennsylvania, March 15, 2005. Lieutenant Colonel Megill was the intelligence officer (G-2) of the 4th Infantry Division when it garrisoned the Sunni Triangle from July 2003 to June 2004.

Morris, Michael F. Interview with author. Carlisle, Pennsylvania, March 16, 2005. Lieutenant Colonel Morris served on the staff of the 1st Marine Expeditionary Force during Operation Iraqi Freedom, with responsibility for Phase IV planning.

Peterson, Cory. Interview with author. Fort Leavenworth, Kansas, January 23, 2006. Major Peterson was a navigator on an Air Force Special Operations MC-130H in Operations Iraqi Freedom and Enduring Freedom.

Petraeus, David. Interview with author. Mosul, Iraq, August 27, 2003. Major General Petraeus was commander of the 101st Airborne Division in Iraq and then commander of Multi-National Security Transition Command—Iraq.

Sonntag, Kurt. Interview with author. Fort Bragg, North Carolina, January 19, 2005. Major Sonntag took part in Operation Enduring Freedom in the fall of 2001 as executive officer of the 3rd Battalion, 5th Special Forces Group.

Tovo, Kenneth E. Interview with author. Carlisle, Pennsylvania, March 15, 2005. Lieutenant Colonel Tovo was a Special Forces battalion commander in northern Iraq from March to May 2003.

Walker, Keith C. Interview with author. Fort Hood, Texas, April 13, 2005. Colonel Walker was chief of staff to the 1st Cavalry Division when it garrisoned Baghdad from March 2004 to March 2005.

Zastrow, Scott. Interview with author. Fort Bragg, North Carolina, January 19, 2005. Sergeant 1st Class Zastrow was the senior team medic of ODA 555 during Operation Enduring Freedom in the fall of 2001.

BOOKS, ARTICLES, REPORTS

Abbate, Janet. *Inventing the Internet*. Cambridge, Mass.: The MIT Press, 1999.

Abulafia, David, ed. *The French Descent into Renaissance Italy, 1494–95*. Aldershot, U.K.: Variorum, 1995.

Ackerman, Robert K. "Special Operations Forces Become Network-Centric." *Signal* magazine (March 2003).

Adams, Eric. "Is This What War Will Come To?" *Popular Science* (June 2004).

Addington, Larry H. *The Patterns of War Since the Eighteenth Century*, 2nd ed. Bloomington: Indiana University Press, 1994.

Agawa, Hiroyuki. *The Reluctant Admiral: Yamamoto and the Imperial Navy*. Trans. John Bester. Tokyo: Kodansha International, 1979.

Ahnlund, Nils. *Gustav Adolf the Great*. Trans. Michael Roberts. Westport, Conn.: Greenwood Press, 1983 [1940].
Alanbrooke, Field Marshal Lord. *War Diaries, 1939–1945*. Alex Danchev and Daniel Todman, editors. Berkeley: University of California Press, 2001.
Alexander, Bevin. *How Wars Are Won: The 13 Rules of War From Ancient Greece to the War on Terror*. New York: Crown Publishers, 2002.
Alexander, John B. *Future War: Non-Lethal Weapons in Twenty-First Century Warfare*. New York: St. Martin's Press, 1999.
Alibek, Ken, with Stephen Handelman. *Biohazard: The Chilling True Story of the Largest Covert Biological Weapons Program in the World—Told from the Inside by the Man Who Ran It*. London: Arrow, 2000 [1999].
Allison, Graham. *Nuclear Terrorism: The Ultimate Preventable Catastrophe*. New York: Times Books, 2004.
Anderegg, C. R. *Sierra Hotel: Flying Air Force Fighters in the Decade After Vietnam*. Washington, D.C.: Air Force History and Museums Program, 2001.
Anderson, Gary. "Baghdad's Fall May Not Be the End." *Washington Post* (April 4, 2003).
Anderson, Jon Lee. *The Fall of Baghdad*. New York: Penguin Press, 2004.
———. "Letter from Afghanistan: After the Revolution." *New Yorker* (January 28, 2002).
Apple, R. W., Jr. "A Military Quagmire Remembered: Afghanistan as Vietnam." *New York Times* (October 31, 2001).
Arkin, William M. "The Pentagon's Secret Scream." *Los Angeles Times* (March 7, 2004).
Arnaldi, Girolamo. *Italy and Its Invaders*. Cambridge, Mass.: Harvard University Press, 2005.
Arnold, H. H. *Global Mission*. New York: Harper & Brothers, 1949.
Arnold, Thomas. *The Renaissance at War*. London: Cassell, 2002.
Arquilla, John, and David Ronfeldt. *Networks and Netwar: The Future of Terror, Crime, and Militancy*. Santa Monica, Calif.: National Defense Research Institute, RAND, 2001.
———. *Swarming & the Future of Conflict*. Santa Monica, Calif.: National Defense Research Institute, RAND, 2000.
Ashton, T. S. *The Industrial Revolution*. Oxford: Oxford University Press, 1997 [1948].
Asprey, Robert B. *War in the Shadows: The Guerrilla in History*. Revised and updated. New York: William Morrow, 1994.
Atkinson, Rick. *Crusade: The Untold Story of the Persian Gulf War*. Boston: Houghton Mifflin, 1993.
———. *In the Company of Soldiers: A Chronicle of Combat*. New York: Henry Holt and Company, 2004.
"The Austrian Defeat." *The Times* (London) (July 11, 1866).
Aylwin-Foster, Nigel. "Changing the Army for Counterinsurgency Operations." *Military Review* (November–December 2005).
Bacevich, Andrew J. *The Pentomic Era: The U.S. Army Between Korea and Vietnam*. Washington, D.C.: National Defense University Press, 1986.
———. "Preserving the Well-Bred Horse." *National Interest* (Fall 1994).
Bacon, Lance M. "To hell and back." *Air Force Times* (March 24, 2003).
Bailey, J. B. *Field Artillery and Firepower*. Annapolis: Naval Institute Press, 2004.
Baker, A. D. III. "World Navies in Review." *United States Naval Institute Proceedings* (March 2004).
Baldwin, Neil. *Edison: Inventing the Century*. New York: Hyperion, 1995.
Bamford, James. *Body of Secrets: Anatomy of the Ultra-Secret National Security Agency*. New York: Anchor Books, 2002 [2001].
Bank, David, and Riva Richmond. "Where the Dangers Are." *Wall Street Journal* (July 18, 2005).
Barlone, D. *A French Officer's Diary (23 August 1939–1 October 1940)*. Trans. L. V. Cass. New York: Macmillan, 1943.
Barnes, Julian E. "Road with a Bad Rep: The Army Got the Bad Guys off Baghdad's Airport Route. Now, About the Good Guys . . ." *U.S. News & World Report* (February 21, 2005).

Barzun, Jacques. *From Dawn to Decadence: 500 Years of Western Cultural Life, 1500 to the Present*. New York: HarperPerennial, 2001 [2000].

Baum, Dan. "Battle Lessons: What the Generals Don't Know." *New Yorker* (January 17, 2005).

Beason, Doug. *The E-Bomb: How America's New Directed Energy Weapons Will Change the Way Future Wars Will Be Fought*. New York: Da Capo Press, 2005.

Beaufre, André. *1940: The Fall of France*. Trans. Desmond Flower. New York: Alfred A. Knopf, 1968.

Beaver, Patrick. *The Crystal Palace, 1851–1936: A Portrait of Victorian Enterprise*. London: Hugh Evelyn, 1970.

Beebe, Gilbert W., and Michael E. De Bakey. *Battle Casualties: Incidence, Mortality, and Logistic Considerations*. Springfield, Ill.: Charles C. Thomas, 1952.

Belasco, Amy. "The Cost of Operations in Iraq, Afghanistan, and Enhanced Security." Washington, D.C.: Congressional Research Service, March 2005.

Belloc, Hilaire. *The Modern Traveler*. London: Edward Arnold, 1898.

Belotes, James H., and William M. Belote. *Titans of the Seas: The Development and Operations of Japanese and American Carrier Task Forces During World War II*. New York: Harper & Row, 1975.

Benjamin, Daniel, and Steven Simon. *The Age of Sacred Terror: Radical Islam's War Against America*. New York: Random House, 2003 [2002].

Bennell, Anthony S., ed. *The Maratha War Papers of Arthur Wellesley: January to December 1803*. Phoenix Mill, England: Sutton, 1998.

Bennett, Ernest N. *The Downfall of the Dervishes, or the Avenging of Gordon: Being a Personal Narrative of the Final Soudan Campaign of 1898*. New York: Negro Universities Press, 1969 [1899].

Bennett, Geoffrey. *Naval Battles of the First World War*. New York: Penguin Books, 1968.

Berenson, Alex. "Fighting the Old-Fashioned Way in Najaf." *The New York Times* (August 29, 2004).

———. "After the Siege, a City of Ruins, Its Dead Rotting." *The New York Times* (August 28, 2004).

Bergen, Peter. "The Long Hunt for Osama." *The Atlantic Monthly* (October 2004).

Bergner, Daniel. "The Other Army." *The New York Times Magazine* (August 14, 2005).

Berkowitz, Bruce. *The New Face of War: How War Will be Fought in the 21st Century*. New York: Free Press, 2003.

Berntsen, Gary, with Ralph Pezzullo. *Jawbreaker: The Attack on Bin Laden and Al-Qaeda: A Personal Account by the CIA's Key Field Commander*. New York: Crown Publishers, 2005.

Berry, Henry. "*This Is No Drill!*" *Living Memories of the Attack on Pearl Harbor*. New York: Berkley Books, 1992.

Biddle, Stephen. *Afghanistan and the Future of Warfare: Implications for Army and Defense Policy*. Carlisle, Penn.: Strategic Studies Institute, U.S. Army War College, November 2002.

———. *Military Power: Explaining Victory and Defeat in Modern Battle*. Princeton: Princeton University Press, 2004.

———. *Toppling Saddam: Iraq and American Military Transformation*. Carlisle, Penn.: Strategic Studies Institute, U.S. Army War College, April 2004.

———, and Stephen Long. "Democracy and Military Effectiveness: A Deeper Look." *Journal of Conflict Resolution* (August 2004).

Biddle, Tami Davis. *Rhetoric and Reality in Air Warfare: The Evolution of British and American Ideas About Strategic Bombing, 1914–1945*. Princeton: Princeton University Press, 2002.

Black, Jeremy. *European Warfare, 1494–1660*. London: Routledge, 2002.

———. *European Warfare, 1453–1815*. New York: St. Martin's Press, 1999.

———. "On Diversity and Military History," *Historically Speaking* (April 2003): 7–9.

———. *War: Past, Present & Future*. New York: St. Martin's Press, 2000.

Blackiston, John. *Twelve Years' Military Adventures in Three Quarters of the Globe*, 2 vols. London: Henry Colburn, 1829.

Blake, Mariah. "Tin Soldier: An American Vigilante in Afghanistan, Using the Press for Profit and Glory." *Columbia Journalism Review* (January/February 2005).

Bloch, Marc. *Strange Defeat: A Statement of Evidence Written in 1940*. Trans. Gerard Hopkins. New York: W.W. Norton, 1999 [1946].

Block, Leo. *To Harness the Wind: A Short History of the Development of Sails*. Annapolis: Naval Institute Press, 2003.

Blond, Georges. *Admiral Togo*. Trans. Edward Hyams. New York: Macmillan, 1960.

Bodley, R. V. C. *Admiral Togo: The Authorized Life of Admiral of the Fleet, Marquis Heihachiro Togo, O.M.* London: Jarrolds, 1935.

Bolia, Robert S. "Overreliance on Technology in Warfare: The Yom Kippur War as Case Study." *Parameters* (Summer 2004).

Bone, Elizabeth, and Christopher Bolkcom. *Unmanned Aerial Vehicles: Background and Issues for Congress*. Washington, D.C.: Congressional Research Service, April 25, 2003.

Bonin, John A. *U.S. Army Forces Central Command in Afghanistan and the Arabian Gulf During Operation Enduring Freedom: 11 September 2001–11 March 2003*. Carlisle, Pa.: Army Heritage Center Foundation, 2003.

Boot, Max. "History Can Offer Bush Hope." *Los Angeles Times* (September 23, 2004).

———. "Hurry Up and Wait: Among the Rear Echleons in Iraq." *The Weekly Standard* (March 20, 2006).

———. "The Iraq War's Outsourcing Snafu." *Los Angeles Times* (March 31, 2005).

———. "Reality Check—This Is War." *Los Angeles Times* (May 27, 2004).

———. "Reconstructing Iraq; With the Marines in the south and the 101st Airborne in the north." *The Weekly Standard* (September 15, 2003).

———. *The Savage Wars of Peace: Small Wars and the Rise of American Power*. New York: Basic Books, 2002.

———. "The Struggle to Transform the Military." *Foreign Affairs* (March/April 2005).

———. "What We Won in Fallouja." *Los Angeles Times* (December 2, 2004).

Boothe, Clare. *Europe in the Spring*. New York: Alfred A. Knopf, 1941.

Bowden, Mark. "The Kabul-Ki Dance." *Atlantic Monthly* (November 2002), 66–87.

Boyer, Peter J. "The New War Machine: How General Tommy Franks Joined Donald Rumsfeld in the Fight to Transform the Military." *New Yorker* (June 30, 2003).

Bradford, Sarah. *Cesar Borgia: His Life and Times*. New York: Macmillan, 1976.

Brady, Thomas A., Jr., Oberman, Heiko A., and Tracy, James D., eds. *Handbook of European History 1400–1600: Late Middle Ages, Renaissance and Reformation, Vol I: Structures and Assertions*. Grand Rapids, Mich.: William B. Eerdsman, 1994.

Braiker, Brian. "Master Blaster: A New Noisemaker." *Newsweek* (July 12, 2004).

Branigan, William. "A Brief, Bitter War for Iraq's Military Officers." *Washington Post* (April 27, 2003).

Brasher, Nathan. "Unmanned Aerial Vehicles and the Future of Air Combat." *Naval Institute Proceedings* (July 2005).

Bremer, L. Paul III, with Malcolm McConnell. *My Year in Iraq: The Struggle to Build a Future of Hope*. New York: Simon & Schuster, 2006.

Brett-James, Antony. *Life in Wellington's Army*. London: George Allen & Unwin Ltd., 1972.

———. *Wellington at War, 1794–1815: A Selection of His Wartime Letters*. London: Macmillan, 1961.

Brewer, John. *The Sinews of Power: War, Money and the English State, 1688–1783*. New York: Alfred A. Knopf, 1989.

Brinkley, Douglas. *Wheels for the World: Henry Ford, His Company, and a Century of Progress, 1903–2003*. New York: Viking, 2003.

Brodie, Bernard, and Fawn M. Brodie. *From Crossbow to H-Bomb: The Evolution of the Weapons and Tactics of Warfare*. Rev. ed. Bloomington: Indiana University Press, 1973.

Brookes, Peter. "The Art of (Cyber) War." *New York Post* (August 29, 2005).

Brooks, John. *Telephone: The First Hundred Years*. New York: Harper & Row, 1976.

Brooks, Risa. "Making Military Might: Why Do States Fail and Succeed?" *International Security* (Fall 2003).

Brown, Drew. "U.S. Lost Its Best Chance to Decimate Al-Qaida in Tora Bora." *Knight Ridder/Tribune News Service* (October 14–15, 2002).

Bruce, Robert V. *Lincoln and the Tools of War*. Indianapolis: Bobbs-Merrill, 1956.

Brzezinski, Matthew. "The Unmanned Army." *New York Times Magazine* (April 20, 2003).

Brzezinski, Richard, and Richard Hook. *The Army of Gustavus Adolphus, I: Infantry*. Oxford: Osprey, 1991.

———. *The Army of Gustavus Adolphus, II: Cavalry*. Oxford: Osprey, 1993.

Bucholz, Arden. *Moltke and the German Wars, 1864–1871*. Hampshire, England: Palgrave, 2001.

Buderi, Robert. *The Invention That Changed the World: How a Small Group of Radar Pioneers Won the Second World War and Launched a Technical Revolution*. New York: Simon & Schuster, 1996.

Budiansky, Stephen. *Air Power: The Men, Machines and Ideas That Revolutionized War, from Kitty Hawk to Gulf War II*. New York: Viking, 2004.

———. *Battle of Wits: The Complete Story of Codebreaking in World War II*. New York: Free Press, 2000.

Bullitt, Orville H., ed. *For the President: Personal and Secret. Correspondence Between Franklin D. Roosevelt and William C. Bullitt*. Boston: Houghton Mifflin, 1972.

Burdick, Charles, and Hans-Adolf Jacobsen, eds. *The Halder War Diary, 1939–1942*. Novato, Calif.: Presidio Press, 1988.

Burleigh, Bennet. *Khartoum Campaign 1898, or the Re-Conquest of the Soudan*. London: Chapman & Hall, 1899.

Burns, John F. "U.S. Checkpoints Raise Ire in Iraq." *New York Times* (March 7, 2005).

Burrell, Robert S. "Breaking the Cycle of Iwo Jima Mythology: A Strategic Study of Operation Detachment." *The Journal of Military History* (October 2004), 1143–86.

Buruma, Ian. *Inventing Japan, 1853–1964*. New York: Modern Library, 2003.

Bush, George, and Brent Scowcroft. *A World Transformed*. New York: Vintage Books, 1999 [1998].

Cahlink, George. "Better 'Blue Force' Tracking." *Air Force Magazine* (June 2004).

Caidin, Martin. *A Torch to the Enemy: The Fire Raid on Tokyo*. New York: Bantam Books, 1992.

Callander, Bruce D. "Controllers: Modern airpower owes much to the elite USAF commandos who hang out with the ground forces." *Air Force Magazine* (September 2003).

Campbell-Kelly, Martin, and William Aspray. *Computer: A History of the Information Machine*. New York: Basic Books, 1996.

Campen, Alan D. *The First Information War: The Story of Communications, Computers and Intelligence Systems in the Persian Gulf War*. Fairfax, Va.: AFCEA International Press, 1992.

Carr, Nicholas G. *Does IT Matter? Information Technology and the Corrosion of Competitive Advantage*. Boston: Harvard Business School Press, 2004.

Carter, Phillip. "The Crucible: How the Iraq disaster is making the U.S. Army stronger." *Washington Monthly* (July/August 2004).

———. "Tomorrow's Soldier Today." *Slate* (March 11, 2004).

Chamberlain, Jeffrey. "FY2005 Defense Budget: Frequently Asked Questions." *CRS Report for Congress*. Washington, D.C.: Congressional Research Service, July 12, 2004.

Chambers, John Whiteclay II, ed. *The Oxford Companion to American Military History*. New York: Oxford University Press, 1999.

Chandler, Alfred, Jr. *Inventing the Electronic Century: The Epic Story of the Consumer Electronics and Computer Industries*. New York: Free Press, 2001.

———. *The Visible Hand: The Managerial Revolution in American Business*. Cambridge: Harvard University Press, 1977.

Chandler, David G., ed. *The Oxford History of the British Army*. Oxford: Oxford University Press, 1994.

Chang, Alicia. "Stanford Volkswagen Wins $2M Robot Race." Associated Press (October 9, 2004).

Chapman, Anne W. *The Origins and Development of the National Training Center, 1976–1984*. Fort Monroe, Virginia: Office of the Command Historian, U.S. Army Training and Doctrine Command, 1992.

Chase, Kenneth. *Firearms: A Global History to 1700*. Cambridge: Cambridge University Press, 2003.

Christensen, Clayton M. *The Innovator's Dilemma*. New York: HarperBusiness, 2000 [1997].

Churchill, Winston. *My Early Life: 1874–1904*. New York: Touchstone, 1996 [1930].

———. *The River War: An Historical Account of the Reconquest of the Sudan*, 2 vols. London: Longmans, Green and Co., 1900.

———. *The Second World War, Vol. I: The Gathering Storm*. Boston: Houghton Mifflin, 1985 [1948].

———. *The Second World War, Vol. II: Their Finest Hour*. Boston: Houghton Mifflin, 1985 [1949].

Cipolla, Carlo M. *Guns, Sails and Empires: Technological Innovation and the Early Phases of European Expansion, 1400–1700*. Manhattan, Kans.: Sunflower University Press, 1965.

Citino, Robert M. *Blitzkrieg to Desert Storm: The Evolution of Operational Warfare*. Lawrence: University Press of Kansas, 2004.

———. *Quest for Decisive Victory: From Stalemate to Blitzkrieg in Europe, 1899–1940*. Lawrence: University Press of Kansas, 2002.

Clancy, Tom, with General Chuck Horner (Ret.). *Every Man a Tiger*. New York: G. P. Putnam's Sons, 1999.

———, with General Fred Franks, Jr. (Ret.). *Into the Storm: A Study in Command*. New York: Berkley Books, 2004 [1997].

Clark, Asa A. IV, Peter W. Chiarelli, Jeffrey S. McKitrick, James W. Reed (eds.). *The Defense Reform Debate: Issues and Analysis*. Baltimore: Johns Hopkins University Press, 1984.

Clarke, Richard A. *Against All Enemies: Inside America's War on Terror*. New York: Free Press, 2004.

Clodfelter, Michael. *Warfare and Armed Conflicts: A Statistical Reference to Casualty and Other Figures, 1500–2000*. 2nd ed. Jefferson, N.C.: McFarland, 2002.

Cloud, David S. "Insurgents Using Bigger, More Lethal Bombs, U.S. Officers Say." *New York Times* (August 4, 2005).

———. "Iraqi Rebels Refine Bomb Skills, Pushing Toll of G.I.'s Higher." *New York Times* (June 22, 2005).

Coffey, Thomas M. *Iron Eagle: The Turbulent Life of General Curtis LeMay*. New York: Crown Publishers, 1986.

Cohen, Eliot A. "Defending America in the Twenty-First Century." *Foreign Affairs* (November–December 2000).

———. "A Revolution in Warfare." *Foreign Affairs* (March–April 1996).

———, Michael J. Eisenstadt, and Andrew J. Bacevich. *Knives, Tanks & Missiles: Israel's Security Revolution*. Washington, D.C.: Washington Institute for Near East Policy, 1998.

Colebrooke, T. E. *Life of the Honorable Monstuart Elphinstone*, 2 vols. London: John Murray, 1884.

Coll, Steve. *Ghost Wars: The Secret History of the CIA, Afghanistan, and Bin Laden, From the Soviet Invasion to September 10, 2001*. New York: Penguin Press, 2004.

———. "In the Gulf, Dissidence Goes Digital: Text Messaging Is New Tool of Political Underground." *The Washington Post* (March 29, 2005).

———. "What Bin Laden Sees in Hiroshima." *The Washington Post* (February 6, 2005).

———, and Susan B. Glasser. "Attacks Bear Earmarks of Evolving Al Qaeda." *The Washington Post* (July 8, 2005).
"Comment." *Harper's Weekly* (June 10, 1905).
Connaughton, Richard. *Rising Sun and Tumbling Bear: Russia's War with Japan*. London: Cassell, 2003.
Conway, James T. " 'Farther and Faster' in Iraq." *Naval Institute Proceedings* (January 2005).
Cook, Haruko Taya, and Theodore F. Cook. *Japan at War: An Oral History*. New York: The New Press, 1992.
Cooper, J. P., ed. *The New Cambridge Modern History. Volume IV: The Decline of Spain and the Thirty Years War, 1609–48/59*. Cambridge: Cambridge University Press, 1970.
Cooper, Randolf G. S. The *Anglo-Maratha Campaigns and the Contest for India: The Struggle for Control of the South Asian Military Economy*. Cambridge: Cambridge University Press, 2003.
Corbett, Julian S., ed. *Maritime Operations in the Russo-Japanese War, 1904–1905*. 2 vols. Annapolis: Naval Institute Press, 1994.
———. *Papers Relating to the Navy During the Spanish War, 1585–1587*. New York: Burt Franklin, 1970 [1898].
Cordesman, Anthony H., with Patrick Baetjer. *The Ongoing Lessons of Afghanistan: Warfighting, Intelligence, Force Transformation, and Nation Building*. Washington, D.C.: Center for Strategic and International Studies, May 2004.
Cordesman, Anthony H. "War against Iraq might not be a cake walk." *The San Diego Union-Tribune* (August 11, 2002).
Cornwell, Bernard. *Sharpe's Triumph: Richard Sharpe and the Battle of Assaye, September 1803*. New York: HarperPerennial, 2000.
Correa, Gaspar. *The Three Voyages of Vasco de Gama and His Viceroyalty from Lendas da India of Caspar Correa*. New York: Burt Franklin, 1963 [1869].
Correll, John T. "Casualties: Until recently, large numbers of killed and wounded were an inevitable part of warfare." *Air Force Magazine* (June 2003).
Corrigan, Gordon. *Wellington: A Military Life*. London: Hambledon and London, 2001.
Corum, James S. *The Roots of Blitzkrieg: Hans von Seeckt and German Military Reform*. Lawrence: University Press of Kansas, 1992.
Coyne, James P. "A Strike by Stealth." *Air Force Magazine* (March 1992).
Craig, Gordon A. *The Battle of Königgrätz: Prussia's Victory over Austria, 1866*. Philadelphia: University of Pennsylvania Press, 2003 [1964].
———. *The Politics of the Prussian Army, 1640–1945*. New York: Oxford University Press, 1964.
Crane, Conrad C. "Avoiding Vietnam: The U.S. Army's Response to Defeat in Southeast Asia." Carlisle, Penn.: Strategic Studies Institute, U.S. Army War College, September 2002.
———. *Bombs, Cities, and Civilians: American Airpower Strategy in World War II*. Lawrence: University Press of Kansas, 1993.
Craven, Frank Wesley, and James Lea Cate, eds. *The Army Air Forces in World War II. Volume Five: The Pacific: Matterhorn to Nagasaki, June 1944 to August 1945*. Chicago: University of Chicago Press, 1953.
———. *The Army Air Forces in World War II. Volume Six: Men and Planes*. Chicago: University of Chicago Press, 1955.
Creasy, Edward. *Fifteen Decisive Battles of the World*. Harrisburg, Penn.: Stackpole, 1960 [1851].
Creveld, Martin van. *Fighting Power: German and U.S. Army Performance, 1939–1945*. Westport, Conn.: Greenwood Press, 1982.
———. *Supplying War: Logistics from Wallenstein to Patton*. Cambridge: Cambridge University Press, 1977.
———. *Technology and War from 2000 B.C. to the Present*. New York: Free Press, 1989.
———. *The Transformation of War*. New York: Free Press, 1991.

Cringely, Robert X. *Accidental Empires: How the Boys of Silicon Valley Make Their Millions, Battle Foreign Competition, and Still Can't Get a Date*. New York: HarperBusiness, 1996 [1992].

Crosby, Alfred W. *Throwing Fire: Projectile Technology Through History*. Cambridge: Cambridge University Press, 2002.

Crouch, Tom. *The Bishop's Boys: A Life of Wilbur and Orville Wright*. New York: W.W. Norton, 1989.

———. *Wings: A History of Aviation from Kites to the Space Age*. New York: W.W. Norton, 2003.

Cullison, Alan. "Inside Al-Qaeda's Hard Drive: A fortuitous discovery reveals budget squabbles, baby pictures, office rivalries—and the path to 9/11." *The Atlantic Monthly* (September 2004).

Daggett, Stephen, Larry Nowels, Curt Tarnoff, and Rhoda Margesson. "FY2004 Supplemental Appropriations for Iraq, Afghanistan, and the Global War on Terrorism." Washington, D.C.: Congressional Research Service, October 15, 2003.

Danchev, Alex. *Alchemist of War: The Life of Basil Liddell Hart*. London: Weidenfeld & Nicolson, 1998.

Daso, Dik Alan. *Hap Arnold and the Evolution of American Airpower*. Washington: Smithsonian Institution Press, 2000.

Davis, Anthony. "How the Afghan war was won." *Jane's Intelligence Review*, February 2002.

Davis, Burke. *The Billy Mitchell Affair*. New York: Random House, 1967.

Davis, John R. *The Great Exhibition*. Phoenix Mill, England: Sutton, 1999.

DeBloise, Bruce M., Richard L. Garwin, R. Scott Kemp, and Jeremy C. Marwell. "Space Weapons: Crossing the U.S. Rubicon." *International Security* (Fall 2004).

Defense Science Board. "Study on Unmanned Aerial Vehicles and Uninhabited Combat Aerial Vehicles." Washington, D.C.: February 2004.

———. "Task Force on High Energy Laser Weapons Systems Applications—Final Report." Washington, D.C.: June 2001.

———. "Task Force on Sea Basing—Final Report." Washington, D.C.: August 2003.

De Gaulle, Charles. *The Army of the Future*. Philadelphia: J. B. Lippincott, 1941 [1934].

———. *The Complete War Memoirs of Charles de Gaulle*. Trans. Jonathan Griffin and Richard Howard. New York: Carroll & Graf, 1998 [1955–1960].

De Gheyn, Jacob. *The Exercise of Arms: A Seventeenth Century Military Manual*. Edited by David J. Blackmore. London: Greenhill Books, 1986 [1607].

Deighton, Len. *Blitzkrieg: From the Rise of Hitler to the Fall of Dunkirk*. Edison, N.J.: Castle Books, 2000 [1979].

Deitchman, S. J. "Military Force Transformation: Progress, Costs, Benefits and Tasks Remaining." Washington, D.C.: The Atlantic Council of the United States, December 2004.

Delbruck, Hans. *History of the Art of War within the Framework of Political History, Vol. IV: The Modern Era*. Trans. Walter J. Renfroe, Jr. Lincoln: University of Nebraska Press, 1990.

Michael DeLong with Noah Lukeman. *Inside Centcom: The Unvarnished Truth About the Wars in Afghanistan and Iraq*. Washington, D.C.: Regnery Publishing, 2004.

DeVore, Marc. *The Airborne Illusion: Institutions and the Evolution of Postwar Airborne Forces*. Cambridge: Massachusetts Institute of Technology, Security Studies Program Working Paper (June 2004).

Diamond, Jared. *Guns, Germs and Steel: The Fate of Human Societies*. New York: W.W. Norton, 1999.

Diamond, Larry. *Squandered Victory: The American Occupation and Bungled Effort to Bring Democracy to Iraq*. New York: Times Books, 2005.

Dicey, Edward. *The Battle-Fields of 1866*. London: Tinsley Brothers, 1866.

Divine, David. *The Blunted Sword*. London: Hutchinson, 1964.

Djerejian, Edward P., and Frank G. Wisner. "Guiding Principles for U.S. Post-Conflict Policy in Iraq: Report of the Independent Working Group Cosponsored by the

Council on Foreign Relations and the James A. Baker III Institute for Public Policy of Rice University." New York: Council on Foreign Relations, 2003.

Dobbins, James, et al. *America's Role in Nation-Building from Germany to Iraq*. Santa Monica, Calif.: Rand Corp., 2003.

Dodge, Theodore Ayrault. *Gustavus Adolphus*. New York: Da Capo Press, 1998 [1895].

Dombey, Daniel. "US Nato chief chides Europeans over budgets." *Financial Times* (June 9, 2005).

Donnelly, John M. "Rumsfeld Uses Pentagon's Emergency Purchase Power for Jamming Devices." *CQ Today* (May 5, 2005).

Donnelly, Thomas. *The Military We Need: The Defense Requirements of the Bush Doctrine*. Washington: AEI Press, 2005.

———. *Operation Iraqi Freedom: A Strategic Assessment*. Washington, D.C.: AEI Press, 2004.

Dorr, Robert F. *B-29 Superfortress Units of World War 2*. Oxford, England: Osprey, 2002.

Doughty, Robert Allan. *The Breaking Point: Sedan and the Fall of France, 1940*. Hamden, Conn.: Archon Books, 1990.

———. *The Seeds of Disaster: The Development of French Army Doctrine, 1919–1939*. Hamden, Conn.: Archon Books, 1985.

Downing, Brian M. *The Military Revolution and Political Change: Origins of Democracy and Autocracy in Early Modern Europe*. Princeton: Princeton University Press, 1992.

Dressler, Friedrich August. *Moltke in His Home*. Trans. Mrs. Charles Edward Barrett-Lennard. New York: E. P. Dutton, 1906.

Dudney, Robert S. "Long-Range Strike in Two Jumps." *Air Force Magazine* (June 2004).

Duelfer, Charles. *Comprehensive Report of the Special Advisor to the DCI on Iraq's WMD* (September 30, 2004).

Duff, James Grant. *History of the Mahrattas*, 2 vols. New Delhi: Associated Publishing House, 1971. Reprint of 1878 edition.

Duffy, Christopher. *The Military Experience in the Age of Reason*. London: Wordsworth Editions, 1998.

———. *Siege Warfare: The Fortress in the Early Modern World, 1494–1660*. London: Routledge & Kegan Paul, 1979.

Dugan, Michael. "The Air War." *U.S. News & World Report* (February 11, 1991).

Dull, Paul S. *A Battle History of the Imperial Japanese Navy (1941–1945)*. Annapolis: Naval Institute Press, 1978.

Dunn, Richard S. *The Age of Religious Wars, 1559–1689*. New York: W.W. Norton, 1970.

Dunnigan, James F., and Raymond M. Macedonia. *Getting It Right: American Military Reforms After Vietnam to the Persian Gulf and Beyond*. New York: William Morrow, 1993.

Dunstan, Simon. *Flak Jackets: 20th Century Military Body Armour*. London: Osprey, 1984.

Dupuy, Trevor N. *Attrition: Forecasting Battle Casualties and Equipment Losses in Modern War*. Fairfax, Va.: Hero Books, 1990.

———. *The Evolution of Weapons and Warfare*. Indianapolis: Bobbs-Merrill, 1980.

———. *A Genius for War: The German Army and General Staff, 1807–1945*. Garden City, N.Y.: Military Book Club, 2002 [1977].

———. *Understanding War: History and Theory of Combat*. New York: Paragon House, 1987.

Durant, Will. *The Story of Civilization, Part V: The Renaissance—A History of Civilization in Italy from 1304–1576 A.D.* New York: Simon & Schuster, 1953.

Dynes, Robert J. *The Lee: British Service Rifle from 1888 to 1950*. Bloomfield, Ontario: Museum Restoration Service, 1979.

Easterbrook, Gregg. "American Power Moves Beyond the Mere Super." *New York Times* (April 27, 2003).

———. "The End of War?" *New Republic* (May 30, 2005).

Echevarria, Anutlio II. "Globalization and the Nature of War." Carlisle, Penn.: Strategic Studies Institute, U.S. Army War College, March 2003.

Edoin, Hoito. *The Night Tokyo Burned: The Incendiary Campaign Against Japan, March–August, 1945*. New York: St. Martin's Press, 1987.

Elias, Paul. "Pentagon Invests in Unmanned 'Trauma Pod'." Associated Press (March 29, 2005).

Ellis, John. *The Social History of the Machine Gun*. Baltimore: Johns Hopkins University Press, 1986 [1975].

Engels, Friedrich. "Notes on the War in Germany, No. I." *Manchester Guardian* (June 20, 1866).

———. "Notes on the War in Germany, No. IV." *Manchester Guardian* (July 3, 1866).

———. "Notes on the War in Germany, No. V." *Manchester Guardian* (July 6, 1866).

Erasmus, Desiderius. *Pilgrimages to Saint Mary of Walsingham and Saint Thomas of Canterbury, With the Colloquy of Rash Vows, and the Characters of Archbishop Warham and Dean Colet*. Trans. by John Gough Nichols. London: John Murray, 1875.

Evans, David C., and Peattie, Mark R. *Kaigun: Strategy, Tactics and Technology in the Imperial Japanese Navy, 1887–1941*. Annapolis: Naval Institute Press, 1997.

Evans, Michael. "From Kadesh to Kandahar: Military Theory and the Future of War." *Naval War College Review* (Summer 2003).

———. "US-Style Military Reforms Will Cut British Firepower." *The* [London] *Times* (June 27, 2003).

Exum, Andrew. *This Man's Army: A Soldier's Story from the Front Lines of the War on Terrorism*. New York: Gotham Books, 2004.

Fainaru, Steve. "Opinions on Attire Not Quite Uniform." *The Washington Post* (April 4, 2005).

———. "Soldiers Defend Faulted Strykers." *The Washington Post* (April 3, 2005).

Fairbank, Katie. "Unmanned Aircraft Are Wowing Defense Industry." *The Dallas Morning News* (June 14, 2005).

Fairbanks, Charles H., Jr. "Afghanistan Reborn: The epic achievement of the Bush administration—and of America." *The Weekly Standard* (November 1–8, 2004).

Faith, Nicholas. *The World the Railways Made*. New York: Carroll & Graf, 1990.

Falk, Edwin A. *Togo and the Rise of Japanese Sea Power*. New York: Longmans, Green and Co., 1936.

Fallows, James. "Blind into Baghdad." *The Atlantic Monthly* (January/February 2004).

———. *National Defense*. New York: Vintage Books, 1982 [1981].

Featherstone, Donald. *Weapons and Equipment of the Victorian Soldier*. Poole, England: Blandford Press, 1978.

Feder, Barnaby J. "Technology Strains to Find Menace in the Crowd." *The New York Times* (May 31, 2004).

Ferguson, Niall. *The Cash Nexus: Money and Power in the Modern World, 1700–2000*. New York: Basic Books, 2001.

———. *Empire: The Rise and Demise of the British World Order and the Lessons for Global Power*. New York: Basic Books, 2003.

Ferguson, Roger W. "Productivity: Past, Present and Future." Remarks at the New York Association for Business Economics, July 7, 2004. http://www.federalreserve.gov/boarddocs/speeches/2004.

Filkins, Dexter. "In Fallujah, Young Marines Saw the Savagery of an Urban War." *The New York Times* (November 21, 2004).

Fineman, Mark, Robin Wright, and Doyle McManus, "Preparing for War, Stumbling to Peace: U.S. is paying the price for missteps made in Iraq." *Los Angeles Times* (July 18, 2003).

Finn, Peter. ". . . And His U.S. Partners; Wounded Army Captain Details Offensive Against Taliban." *The Washington Post* (December 11, 2001).

FitzGerald, Mary C. "The Impact of the Military-Technical Revolution on Russian Military Affairs," 2 vols. Washington and Indianapolis: Hudson Institute, August 20, 1993.

———. "Marshal Ogarkov on Modern War: 1977–1985." Alexandria, Va.: Center for Naval Analyses, November 1986.
———. "The New Revolution in Russian Military Affairs." London: Royal United Services Institute for Defense Studies, 1994.
———. "Soviet Views on SDI." Pittsburgh: University of Pittsburgh Center for Russian and East European Studies, May 1987.
Folcher, Gustave. *Marching to Captivity: The War Diaries of a French Peasant, 1939–45*. Trans. Christopher Hill. London: Brassey's, 1996.
Fontenot, Gregory, E. J. Degen, and David Tohn. *On Point: The United States Army in Operation Iraqi Freedom*. Annapolis: Naval Institute Press, 2005.
Fontane, Theodor. *Der deutsche Krieg von 1866*. Berlin: R.V. Decker, 1871.
Fortescue, J. W. *A History of the British Army, Vol. IV: 1803–1807*. London: Macmillan, 1921.
Fosten, D. S. V., and B. K. Fosten. *The Thin Red Line: Uniforms of the British Army between 1751 and 1914*. London: Windrow & Greene, 1989.
Fountain, Augustus W. III. "Transforming Defense Basic Research Strategy." *Parameters* (Winter 2004–5).
Franks, Tommy, with Malcolm McConnell. *American Soldier*. New York: ReganBooks, 2004.
———. "War of Words." *The New York Times* (October 19, 2004).
Freedman, Lawrence. *The Revolution in Strategic Affairs*, Adelphi Paper No. 318. London: International Institute for Strategic Studies, May 1988.
———. "War Evolves into the Fourth Generation: A Comment on Thomas X. Hammes." *Contemporary Security Policy* (August 2005), 1–10.
Freeman, Chris, and Louçã, Francisco. *As Time Goes By: From the Industrial Revolutions to the Information Revolution*. Oxford: Oxford University Press, 2002.
Freeman, Kathleen. *Ancilla to the Pre-Socratic Philosophers: A Complete Translation of the Fragments in Diels, Fragmente der Vorsokratiker*. Cambridge: Harvard University Press, 1971.
Freiberger, Paul, and Michael Swaine. *Fire in the Valley: The Making of the Personal Computer*, 2nd ed. New York: McGraw-Hill, 2000.
Friedel, Robert, and Paul Israel, with Bernard S. Finn. *Edison's Electric Light: Biography of an Invention*. New Brunswick, N.J.: Rutgers University Press, 1986.
Friedman, George, and Meredith Friedman. *The Future of War: Power, Technology and American Dominance in the 21st Century*. New York: St. Martin's Griffin, 1996.
Friedman, Norman. *Carrier Air Power*. New York: Rutledge Press, 1981.
———. *Terrorism, Afghanistan, and America's New Way of War*. Annapolis: Naval Institute Press, 2003.
Friedman, Thomas L. *The World Is Flat: A Brief History of the Twenty-First Century*. New York: Farrar, Straus and Giroux, 2005.
Friedrich, Otto. *Blood and Iron: From Bismarck to Hitler, the von Moltke Family's Impact on German History*. New York: HarperPerennial, 1995.
Frieser, Karl-Heinz, with John T. Greenwood. *The Blitzkrieg Legend: The 1940 Campaign in the West*. Annapolis: Naval Institute Press, 2005 [1996].
Fulghum, David A. "Disappearing Act: Daytime stealth is being designed into low-level, long-endurance combat aircraft." *Aviation Week & Space Technology* (October 25, 2004).
———. "New Threat, New Defense," "Cruise Missile Battle," "Testing Ground," "Net Changes." *Aviation Week & Space Technology* (May 31, 2004).
———. "A Night Over Iraq." *Aviation Week & Space Technology* (May 15, 2005).
———, and Douglas Barrie. "Su-30MK Beats F-15C 'Every Time.'" *Aviation Week & Space Technology* (May 24, 2002).
Fuller, J. F. C. *Armament and History: The Influence of Armament on History from the Dawn of Classical Warfare to the End of the Second World War*. New York: De Capo, 1998 [1945].

———. *The Decisive Battles of the Western World, and Their Influence Upon History*, 3 vols. London: Eyre & Spottiswoode, 1954.

———. *Memoirs of an Unconventional Soldier*. London: Ivor Nicholson and Watson Limited, 1936.

———. *The Reformation of War*. London: Hutchinson & Co., 1923.

Gabriel, Richard A., and Karen S. Metz. *A History of Military Medicine, Vol. II: From the Renaissance Through Modern Times*. Westport, Conn.: Greenwood Press, 1992.

Galbraith, Peter W. "Eyewitness to a failure in Iraq." *Boston Globe* (October 27, 2004).

Gallagher, P., and Cruickshank, D. W. *God's Obvious Design: Papers for the Spanish Armada Symposium, Sligo, 1988*. London: Tamesis Books Ltd., 1990.

Gander, Terry J. *Jane's Infantry Weapons, Twenty-Second Edition (1996–1997)*. Alexandria, Va.: Jane's Information Group, 1996.

Gardiner, Robert, ed. *Steam, Steel and Shellfire: The Steam Warship, 1815–1905*. Edison, N.J.: Chartwell Books, 2001.

Garreau, Joel. *Radical Evolution: The Promise and Peril of Enhancing Our Minds, Our Bodies—and What It Means to Be Human*. New York: Doubleday, 2005.

Gash, Norman. *Aristocracy and People: Britain, 1815–1865*. London: Edward Arnold, 1988.

Gat, Azar. *British Armour Theory and the Rise of the Panzer Arm: Revising the Revisionists*. New York: St. Martin's Press, 2000.

Gawande, Atul. "Casualties of War—Military Care for the Wounded from Iraq and Afghanistan." *New England Journal of Medicine* (December 9, 2004).

Gellman, Barton. "One Year Later: War's Faded Triumph." *The Washington Post* (January 16, 1992).

———, and Dafna Linzer. "Afghanistan, Iraq: Two Wars Collide." *The Washington Post* (October 22, 2004).

General Accounting Office. *Operation Desert Storm: Evaluation of the Air Campaign*. Gaithersburg, Md.: GAO Document Distribution Facility, 1997.

Gertz, Bill. "U.S. deploys warfare unit to jam enemy satellites." *The Washington Times* (September 22, 2005).

Glanz, James, and William J. Broad. "Looting at Weapons Plants Was Systematic, Iraqi Says." *The New York Times* (March 13, 2005).

Gleig, G. R. *The Life of Major-General Sir Thomas Munro, Bart. and K.C.B., Late Governor of Madras, With Extracts from his Correspondence and Private Papers*, 2 vols. London: Henry Colburn and Richard Bentley, 1830.

Goddard, C. H., and C. B. Marks. "DD(X) Navigates Uncharted Waters." *U.S. Naval Institute Proceedings* (January 2005).

Goerlitz, Walter. *History of the German General Staff, 1657–1945*. Trans. Brian Battershaw. New York: Praeger, 1953.

Goldman, Emily O., and Leslie C. Eliason, eds. *The Diffusion of Military Technology and Ideas*. Stanford: Stanford University Press, 2003.

Goldstein, Donald M., and Katherine V. Dillon, eds. *The Pearl Harbor Papers: Inside the Japanese Plans*. Dulles, Va.: Brassey's, 2000.

Goldstein, Joshua S. "The Worldwide Lull in War." *The Christian Science Monitor* (May 14, 2002).

Goldstein, Lyle, and William Murray. "Undersea Dragons: China's Maturing Submarine Force." *International Security* (Spring 2004).

Gordon, Andrew. *A Modern History of Japan from Tokugawa Times to the Present*. New York: Oxford University Press, 2003.

Gordon, Craig. "Troops seen vulnerable in Humvees." *The Boston Globe* (December 18, 2003).

Gordon, Greg. "Beam Burns into the Future." *Minneapolis Star Tribune* (May 30, 2004).

Gordon, John Steele. *A Thread Across the Ocean: The Heroic Story of the Transatlantic Cable*. New York: Walker, 2002.

Gordon, Michael R., and Bernard E. Trainor. *Cobra II: The Inside Story of the Invasion and Occupation of Iraq*. New York: Pantheon, 2006.

———. *The Generals' War: The Inside Story of the Conflict in the Gulf.* Boston: Back Bay Books, 1995.

Gordon, Michael R. "Catastrophic Success: The Strategy to Secure Iraq Did Not Foresee a 2nd War." *The New York Times* (October 19, 2004).

———. "Catastrophic Success: Poor Intelligence Misled Troops About Risks of Drawn-Out War." *The New York Times* (October 20, 2004).

———. "Catastrophic Success: Debate Lingering on Decision to Dissolve the Iraqi Military." *The New York Times* (October 21, 2004).

Gordon, Robert J. "Does the 'New Economy' Measure Up to the Great Inventions of the Past?" *Journal of Economic Perspectives* (Fall 2000), Vol. 14, No. 4, 49–74.

Gordon, Stewart. *The New Cambridge History of India, II:4: The Marathas, 1600–1818.* Cambridge: Cambridge University Press, 1993.

Gorman, Siobhan, and Sydney J. Freedberg, Jr., "Early Warning." *National Journal* (June 10, 2005).

Gothenberg, Gunther E. "Maurice of Nassau, Gustavus Adolphus, Raimondo Montecucoli, and the 'Military Revolution' of the Seventeenth Century." In *Makers of Modern Strategy from Machiavelli to the Nuclear Age,* edited by Peter Paret, with Gordon A. Craig and Felix Gilbert, 32–64. Princeton: Princeton University Press, 1986.

———. "Moltke, Schlieffen and the Doctrine of Strategic Envelopment," ibid., 296–325.

Gouré, Daniel. "The U.S. Army Meets Star Wars." Washington, D.C.: Lexington Institute, February 13, 2003. www.lexingtoninstitute.org/defense.asp?aid=90

Gourley, Scott R. "Zeus-Humvee Laser Ordnance Neutralization System." *Army* (December 2004).

Graduate Institute of International Studies. *Small Arms Survey 2004: Rights at Risk.* www.smallarmssurvey.org/publications/yb_2004.htm

Graham, Bradley. "Bravery and Breakdowns in a Ridgetop Battle: 7 Americans Died in Rescue Effort That Revealed Mistakes and Determination." *The Washington Post,* May 24, 2002.

———. "Commanders Plan Eventual Consolidation of U.S. Bases in Iraq." *The Washington Post* (May 22, 2005).

———. "General Says Army Reserve Is Becoming a 'Broken' Force." *The Washington Post* (January 6, 2005).

———. "Pentagon to Stress Foreign Languages." *The Washington Post* (April 8, 2005).

———. "A Wintry Ordeal at 10,000 Feet: Rangers Battled Weather and Enemy to Rescue Stranded Unit." *The Washington Post,* May 25, 2002.

———, and Vernon Loeb. "An Air War of Might, Coordination and Risks." *The Washington Post* (April 27, 2003).

———, and Dana Priest. "Insurgents Using U.S. Techniques." *The Washington Post* (May 3, 2005).

———, and Thomas E. Ricks. "Pentagon Blamed for Lack of Postwar Planning in Iraq." *The Washington Post* (April 1, 2005).

Grant, Rebecca. "The Bekaa Valley War." *Air Force Magazine* (June 2002).

———. "Eyes Wide Open." *Air Force Magazine* (November 2003).

———. "The Missing Aces." *Air Force Magazine* (September 2004).

———. "The War Nobody Expected." *Air Force Magazine* (April 2002).

Gray, Colin S. *Strategy for Chaos: Revolutions in Military Affairs and the Evidence of History.* London: Frank Cass, 2002.

———. "How Has War Changed Since the End of the Cold War?" *Parameters* (Spring 2005).

Gray, Edwyn. *The Devil's Device: Robert Whitehead and the History of the Torpedo.* Revised, updated edition. Annapolis: Naval Institute Press, 1991.

Grenz, Chris. "Legislature Honors Kansas Warrior." *The Topeka* [Kansas] *Capital-Journal* (April 11, 2002).

Grey, Stephen. "How the US Killed Al-Qaeda leaders by remote control." *The Sunday Times* (London) (November 18, 2001).

Grier, Peter. "Follow the Money." *Air Force Magazine* (August 2004).
———. "A Quarter Century of AWACS." *Air Force Magazine* (March 2002).
———. "Science Projects." *Air Force Magazine* (December 2003).
———. "The Sensational Signal." *Air Force Magazine* (February 2003).
Grossman, Elaine M. "U.S Forces in Iraq Face Obstacles in Getting Intelligence They Need." *Inside the Pentagon* (May 5, 2005).
Guderian, Heinz. *Panzer Leader*. Trans. Constantine Fitzgibbon. New York: Da Capo Press, 1996 [1952].
———. *Achtung—Panzer! The Development of Tank Warfare*. Trans. Christopher Duffy. London: Cassell, 1992 [1937].
Gudmundsson, Bruce I. *Stormtroop Tractics: Innovation in the German Army, 1914–1918*. Westport, Conn.: Praeger. 1989.
Guicciardini, Francesco. *The History of Florence*. Trans. Mario Domandi. New York: Harper & Row, 1970.
———. *The History of Italy*. Trans. Sidney Alexander. New York: Macmillan, 1969 [1561].
Guillain, Robert. *I Saw Tokyo Burning: An Eyewitness Narrative from Pearl Harbor to Hiroshima*. Trans. William Byron. New York: Playboy Paperbacks, 1981.
Guilmartin, John F., Jr. *Galleons and Galleys*. London: Cassell & Co., 2002.
Gurwood, John, ed. *The Dispatches of Field Marshal the Duke of Wellington During His Various Campaigns in India, Denmark, Portugal, Spain, the Low Countries, and France, from 1799 to 1818*, Vols. I–II. London: John Murray, 1834.
Gutman, Stephanie. *The Kinder, Gentler Military: Can America's Gender-Neutral Fighting Force Still Win Wars?* New York: Scribner, 2000.
Habeck, Mary R. *Storm of Steel: The Development of Armor Doctrine in Germany and the Soviet Union, 1919–1939*. Ithaca and London: Cornell University Press, 2003.
Hafner, Katie, and Matthew Lyon. *Where Wizards Stay Up Late: The Origins of the Internet*. New York: Touchstone, 1998 [1996].
Hagan, Kenneth J. *This People's Navy: The Making of American Sea Power*. New York: Free Press, 1992 [1991].
Hagerman, Edward. *The American Civil War and the Origins of Modern Warfare: Ideas, Organization and Field Command*. Bloomington: Indiana University Press, 1988.
Halberstam, David. *War in a Time of Peace: Bush, Clinton and the Generals*. New York: Scribner, 2001.
Hale, J. R. *The Civilization of Europe in the Renaissance*. New York: Atheneum, 1994.
———. *War and Society in Renaissance Europe, 1450–1620*. Baltimore: Johns Hopkins University Press, 1985.
Hall, Bert S. *Weapons and Warfare in Renaissance Europe: Gunpowder, Technology, and Tactics*. Baltimore: The Johns Hopkins University Press, 1997.
Hallahan, William H. *Misfire: The History of How America's Small Arms Have Failed Our Military*. New York: Charles Scribner's Sons, 1994.
Hallion, Richard P. *Storm Over Iraq: Air Power and the Gulf War*. Washington, D.C.: Smithsonian Institution Press, 1992.
———. *Taking Flight: Inventing the Aerial Age from Antiquity Through the First World War*. New York: Oxford University Press, 2003.
Hamilton, Jack, and Tom Walker. "Dane named as umbrella killer." *The Sunday Times* (London) (June 5, 2005).
Hammes, Thomas X. *The Sling and the Stone: On War in the 21st Century*. St. Paul, Minn.: Zenith Press, 2004.
Hammond, Grant T. "Myths of the Gulf War: Some 'Lessons' Not Learned." *Aerospace Power Journal* (Fall 1998).
Hansell, Haywood S., Jr. *The Strategic Air War Against Germany and Japan*. Washington, D.C.: Office of Air Force History, 1986.
Hanson, Neil. *The Confident Hope of a Miracle: The True Story of the Spanish Armada*. New York: Alfred A. Knopf, 2005.

Hanson, Victor Davis. *Carnage and Culture: Landmark Battles in the Rise of Western Power*. New York: Doubleday, 2001.

Harrington, Peter, and Frederic A. Sharf, eds. *Omdurman 1898: The Eye-Witnesses Speak*. London: Greenhill Books, 1998.

Harris, Robert, and Jeremy Paxman. *A Higher Form of Killing: The Secret History of Chemical and Biological Warfare*. New York: Random House, 2002.

Harrison, James P. *Mastering the Sky: A History of Aviation from Ancient Times to the Present*. New York: Sarpedon, 1996.

Hasim, Ahmed S. "The Revolution in Military Affairs Outside the West." *Journal of International Affairs* (Winter 1998).

Headrick, Daniel R. *The Invisible Weapon: Telecommunications and International Politics, 1851–1945*. New York: Oxford University Press, 1991.

———. *The Tentacles of Progress: Technology Transfer in the Age of Imperialism, 1850–1940*. New York: Oxford University Press, 1988.

———. *The Tools of Empire: Technology and European Imperialism in the Nineteenth Century*. New York: Oxford University Press, 1981.

Hebert, Adam J. "It Means 'We Didn't Buy Enough.'" *Air Force Magazine* (July 2003).

———. "New Horizons for Combat UAVs." *Air Force Magazine* (December 2003).

———. "Operation Reachback: In Gulf War II, US-based units sent forward 30,000 intelligence reports and spotted some 1,000 targets." *Air Force Magazine* (April 2004).

Hecht, Jeff. *City of Light: The Story of Fiber Optics*. New York: Oxford University Press, 1999.

Hendren, John, and Maura Reynolds. "Bombs Spur Drive for Vehicle Hardier Than a Humvee." *Los Angeles Times* (May 6, 2005).

———. "The Untold War: The U.S. Bomb That Nearly Killed Karzai." *Los Angeles Times* (March 27, 2002).

Heppenheimer, T. A. *Turbulent Skies: The History of Commercial Aviation*. New York: John Wiley & Sons, 1995.

Herman, Arthur. *To Rule the Waves: How the British Navy Shaped the Modern World*. New York: HarperCollins, 2004.

Hersh, Seymour M. "Escape and Evasion: What Happened When the Special Forces Landed in Afghanistan?" *The New Yorker* (November 12, 2001).

———. "King's Ransom: How Vulnerable Are the Saudi Royals?" *The New Yorker* (October 22, 2001).

Hibbert, Christopher. *Wellington: A Personal History*. Reading, Mass.: Addison-Wesley, 1997.

Higueras Rodriguez, D., and Aladren, M. P. San Pio. "Irish Wrecks of the Great Armada: The Testimony of the Survivors." In *God's Obvious Design: Papers for the Spanish Armada Symposium, Sligo, 1988*, edited by P. Gallagher and D. W. Cruickshank, 143–167. London: Tamesis Books Ltd., 1990.

Hirsh, Michael, John Barry, and Babak Dehghanpisheh, "Hillbilly Armor." *Newsweek* (December 20, 2004).

The History and Proceedings of the House of Commons from the Restoration to the Present Time, 14 vols. London: Richard Chandler, 1741–44.

Hobhouse, Hermione. *The Crystal Palace and the Great Exhibition: Art, Science and Productive Industry—A History of the Royal Commission for the Exhibition of 1851*. London: Athlone Press, 2002.

Hodges, Andrew. *Alan Turing: The Enigma*. New York: Simon and Schuster, 1983.

Hogg, Ian V. *The Illustrated History of Firearms*. London: New Burlington Books, 1983.

———. *The Story of the Gun*. New York: St. Martin's Press, 1996.

Holborn, Hajo. "The Prusso-German School: Moltke and the Rise of the General Staff." In *Makers of Modern Strategy from Machiavelli to the Nuclear Age*, edited by Peter Paret, Gordon A. Craig, and Felix Gilbert, 281–296. Princeton: Princeton University Press, 1986.

———. *A History of Modern Germany, 1648–1840*. New York: Alfred A. Knopf, 1964.

Holmes, Richard. *Redcoat: The British Soldier in the Age of Horse and Musket*. New York: W. W. Norton, 2001.

Holt, P. M. *The Mahdist State in the Sudan, 1881–1898: A Study of Its Origins, Development and Overthrow*. Oxford: Oxford University Press, 1958.

Honan, William H. *Visions of Infamy: The Untold Story of How Journalist Hector C. Bywater Devised the Plans That Led to Pearl Harbor*. New York: St. Martin's Press, 1991.

Hone, Thomas C., Norman Friedman, and Mark D. Mandeles. *American and British Carrier Development, 1919–1941*. Annapolis: Naval Institute Press, 1999.

Hooker, Gregory. *Shaping the Plan for Operation Iraqi Freedom: The Role of Military Intelligence Assessments*. Washington: Washington Institute for Near East Policy, 2005.

Horikoshi, Jiro. *Eagles of Mitsubishi: The Story of the Zero Fighter*. Trans. Shorjiro Shindo and Harold N. Wantiez. Seattle: University of Washington Press, 1981.

Horne, Alistair. *To Lose a Battle: France 1940*. Boston: Little, Brown, 1969.

Hough, Richard. *The Fleet That Had to Die*. Edinburgh: Birlinn Ltd., 2000 [1958].

Houlgate, Kelly. "Urban Warfare Transforms the Corps." *U.S. Naval Institute Proceedings* (November 2004).

Howarth, David. *The Voyage of the Armada: The Spanish Story*. Guilford, Conn.: Lyons Press, 2001 [1981].

Hoyt, Edwin. *Japan's War: The Great Pacific Conflict, 1853 to 1952*. New York: McGraw-Hill, 1986.

———. *Three Military Leaders: Heihachiro Togo, Isoroku Yamamoto, Tomoyuki Yamashita*. Tokyo: Kodansha International, 1993.

Hozier, H. M. *The Seven Weeks' War: Its Antecedents and Incidents*. Minneapolis, Minn.: Absinthe Press [1867].

Huber, Peter W., and Mark P. Mills. "How Technology Can Defeat Terrorism." *City Journal* (Winter 2002).

Hudson, Rex A. *Peru: A Country Study*. Washington: Federal Research Division, Library of Congress, 1992.

Hughes, B. P. *Firepower: Weapons Effectiveness on the Battlefield, 1630–1850*. New York: Sarpedon, 1974.

Hughes, Daniel J., ed. *Moltke on the Art of War: Selected Writings*. Trans. Daniel J. Hughes and Harry Bell. Novato, Calif.: Presidio, 1993.

Hume, Martin A. S. *Calendar of Letters and State Papers Relating to English Affairs, Preserved In, Or Originally Belonging to, the Archives of Simancas, Vol. IV: Elizabeth, 1587–1603*. London: Eyre and Spottiswoode, 1899.

Ignatieff, Michael. "Nation Building Lite." *The New York Times Magazine* (July 28, 2002).

Ikle, Fred C. "The Next Lenin: On the Cusp of Truly Revolutionary Warfare." *The National Interest* (Spring 1997).

Intel. "Expanding Moore's Law: Exponential Opportunity," Fall 2002 Update. ftp://download.intel.com/labs/eml/download/EML_opportunity.pdf [accessed July 23, 2004]

International Institute for Strategic Studies. *The Military Balance, 2003–4*. Oxford: Oxford University Press, 2003.

Ireland, Bernard. *War at Sea, 1914–45*. London: Cassell, 2002.

Ironside, Sir Edmund. *Time Unguarded: The Ironside Diaries, 1937–1940*. Edited by Roderick Macleod and Denis Kelly. New York: David McKay Company, 1962.

Isaacson, Walter. "The Winds of Reform: Runaway weapons costs prompt a new look at military planning." *Time* (March 7, 1983).

Jaffe, Greg. "As Chaos Mounts in Iraq, U.S. Army Rethinks Its Future." *The Wall Street Journal* (December 8, 2004).

———. "Pentagon Procurement Slows Supply Flow." *The Wall Street Journal* (May 24, 2005).

———. "Rumsfeld's Push for Speed Fuels Pentagon Dissent." *The Wall Street Journal* (May 16, 2005).

James, Lawrence. *Raj: The Making and Unmaking of British India.* New York: St. Martin's Griffin, 1997.

Jehl, Douglas. "Better at Langauges, U.S. Spy Agencies Still Lag." *The New York Times* (March 4, 2004).

Jehl, Douglas, and Eric Schmitt. "Errors Are Seen in Early Attacks on Iraqi Leaders." *The New York Times* (June 13, 2004).

Jennings, Christian. *Midnight in Some Burning Town: British Special Forces Operations from Belgrade to Baghdad.* London: Weidenfeld and Nicolson, 2004.

Johnson, David E. *Fast Tanks and Heavy Bombers: Innovation in the U.S. Army, 1917–1945.* Ithaca: Cornell University Press, 1998.

Johnson, Paul. *The Birth of the Modern: World Society 1815–1830.* New York: HarperCollins, 1991.

———. *Modern Times: The World From the Twenties to the Eighties.* New York: Perennial Library, 1985.

———. *The Renaissance: A Short History.* New York: Modern Library, 2002.

Jones, Archer. *The Art of War in the Western World.* Champaign: University of Illinois Press, 1987.

Jonnes, Jill. *Empires of Light: Edison, Tesla, Westinghouse, and the Race to Electrify the World.* New York: Random House, 2003.

Kagan, Frederick W., and Robin Higham, eds. *The Military History of Tsarist Russia.* New York: Palgrave, 2002.

Kagan, Frederick W. "The Art of War." *The New Criterion* (November 2003).

———. "War and Aftermath." *Policy Review* (August–September 2003).

Kahn, Gabriel. "Making Labels for Less." *The Wall Street Journal* (August 12, 2004).

Kaiser, David. *Politics and War: European Conflict from Philip II to Hitler.* Enlarged edition. Cambridge: Harvard University Press, 2000.

Kalathil, Shanthi, and Taylor C. Boas. *Open Networks, Closed Regimes: The Impact of the Internet on Authoritarian Rule.* Washington, D.C.: Carnegie Endowment for International Peace, 2003.

Kamen, Henry. *Empire: How Spain Became a World Power, 1492–1763.* New York: HarperCollins, 2003.

———. *Philip of Spain.* New Haven: Yale University Press, 1997.

Kanigel, Robert. *The One Best Way: Frederick Winslow Taylor and the Enigma of Efficiency.* New York: Viking, 1997.

Kann, Robert A. *A History of the Habsburg Empire, 1526–1918.* Berkeley: University of California Press, 1974.

Kaplan, Robert D. "Five Days in Fallujah." *The Atlantic Monthly* (July/August 2004).

———. *Imperial Grunts: The American Military on the Ground.* New York: Random House, 2004.

———. "The Airport Road." *The Wall Street Journal* (January 27, 2005).

Kaplan, Lawrence. "Survivor: Iraq—America's Never-Invisible Wounded." *The New Republic* (October 13, 2003).

Karp, Jonathan. "Smarter People, Weapons and Networks." *The Wall Street Journal* (January 31, 2005).

Keaney, Thomas A., and Eliot A. Cohen. *Revolution in Warfare? Air Power in the Persian Gulf.* Annapolis: Naval Institute Press, 1995.

Keegan, John. *The First World War.* New York: Alfred A. Knopf, 1999.

———. *A History of Warfare.* New York: Alfred A. Knopf, 1993.

———. *The Iraq War.* New York: Alfred A. Knopf, 2004.

———. *The Price of Admiralty: The Evolution of Naval Warfare.* New York: Viking, 1988.

Keels, Donn W., Jr. "A Different Kind of Pick-Me-Up." *Naval Institute Proceedings* (July 2005).

Keeton, Ann. "Sensors on Containers May Offer Safer Shipping." *The Wall Street Journal* (March 31, 2005).

Keller, Bill. "The Fighting Next Times." *The New York Times Magazine* (March 10, 2002).

Kelly, Jack. *Gunpowder: Alchemy, Bombards, and Pyrotechnics: The History of the Explosive That Changed the World.* New York: Basic Books, 2004.
Kelly, Michael. "Across the Euphrates." *The Washington Post* (April 3, 2003).
Kelsey, Harry. *Sir Francis Drake: The Queen's Pirate.* New Haven: Yale University Press, 1998.
Kennedy, David M. *Freedom from Fear: The American People in Depression and War, 1929–1945.* New York: Oxford University Press, 1999.
———. *Over Here: The First World War and American Society.* Oxford: Oxford University Press, 1980.
Kennedy, Paul. *The Rise and Fall of the Great Powers: Economic Change and Military Conflict from 1500 to 2000.* New York: Vintage, 1989.
———. *The Rise and Fall of British Naval Mastery.* Amherst, N.Y.: Humanity Books, 1998.
Kennett, Lee. *A History of Strategic Bombing.* New York: Charles Scribner's Sons, 1982.
Keown-Boyd, Henry. *A Good Dusting: A Centenary Review of the Sudan Campaigns, 1883–1899.* London: Leo Cooper 1986.
Kerkstra, Patrick. "Shadowy Insurgency in Iraq Is Built on Homemade Bombs." *Philadelphia Inquirer* (October 28, 2004).
Kessler, Andy. *How We Got Here: A Silicon Valley and Wall Street Primer: A Slightly Irreverent History of Technology and Markets.* Palo Alto, Calif.: Escape Velocity Press, 2004.
King, David C., and Zachary Karabell. *The Generation of Trust: How the U.S. Military Has Regained the Public's Confidence Since Vietnam.* Washington, D.C.: AEI Press, 2003.
King, Neil, Jr. "Power Struggle: Race to Get Lights on in Iraq Shows Perils of Reconstruction." *The Wall Street Journal* (April 2, 2004).
King, Steven, and Geoffrey Timmins. *Making Sense of the Industrial Revolution.* Manchester, U.K.: Manchester University Press, 2001.
Kiper, Richard L. "In the Dark: The 3/75th Ranger regiment." *Special Warfare* (September 2002).
Kirkpatrick, Melanie. "The Anti-Doomsday Machine." *Wall Street Journal* (April 25, 2005).
Kitchen, Martin. *The Cambridge Illustrated History of Germany.* Cambridge: Cambridge University Press, 1996.
Kitfield, James. "Attack Always." *National Journal* (April 26, 2003).
———. *Prodigal Soldiers: How the Generation of Officers Born of Vietnam Revolutionized the American Style of War.* Washington, D.C.: Brassey's, 1997 [1995].
———. *War & Destiny: How the Bush Revolution in Foreign and Military Affairs Redefined American Power.* Washington, D.C.: Potomac Books, 2005.
Klado, Nicolas. *The Battle of the Sea of Japan.* Trans. J. H. Dickinson and F. P. Marchant. London: Hodder and Stoughton, 1906.
Knox, MacGregor, and Williamson Murray. *The Dynamics of Military Revolution, 1300–2050.* Cambridge: Cambridge University Press, 2001.
Koblentz, Gregory. "Pathogens as Weapons: The International Security Implications of Biological Warfare." *International Security* (Winter 2003/04).
Koch, Andrew. "Electro-Magnetic Railguns: Fire Support Revolution." *Jane's Defense Weekly* (April 16, 2003).
Komarow, Steven. "Tanks Take a Beating in Iraq." *USA Today* (March 29, 2005).
———. "Troops Wary of 'RPG Alley'." *USA Today* (December 24, 2004).
Krepinevich, Andrew F. *The Army and Vietnam.* Baltimore: Johns Hopkins University Press, 1986.
———. "Cavalry to Computer: The Pattern of Military Revolutions," *National Interest* (Fall 1994), 30–42.
———. "The Military-Technical Revolution: A Preliminary Assessment." Washington, D.C.: Center for Strategic and Budgetary Assessments, 2002.
Kubicek, Robert V. "The Role of Shallow-Draft Steamboats in the Expansion of the British Empire, 1820–1914." *International Journal of Maritime History* (June 1994).

Kucera, Joshua. "Iraq Conflict Raises Doubts on FCS Survivability." *Jane's Defense Weekly* (May 19, 2004).

Kudyba, Stephen, and Romesh Diwan. *Information Technology, Corporate Productivity, and the New Economy*. Westport, Conn.: Quorum Books, 2002.

Kuhn, Thomas S. *The Structure of Scientific Revolutions*, 3rd ed. Chicago: University of Chicago Press, 1996.

Kurlantzick, Joshua. "Dictatorship.com." *The New Republic* (April 5, 2004).

Labbé, Theola. "Trained for War, U.S. Soldiers Learn to Keep the Peace." *The Washington Post* (October 26, 2003).

La Forte, Robert S., and Ronald E. Marcello, eds. *Remembering Pearl Harbor: Eyewitness Accounts by U.S. Military Men and Women*. New York: Ballantine Books, 1991.

Lamberson, Donald L., Edward Duff, Don Washburn, and Courtney Holmberg. "Whither High-Energy Lasers?" *Air & Space Power Journal* (Spring 2004).

Lambeth, Benjamin S. *Air Power Against Terror: America's Conduct of Operation Enduring Freedom*. Santa Monica, Calif.: RAND National Defense Research Institute, 2005.

———. *NATO's Air War for Kosovo: A Strategic and Operational Assessment*. Santa Monica, Calif.: RAND, Project Air Force, 2001.

———. *The Transformation of American Air Power*. Ithaca: Cornell University Press, 2000.

Landes, David S. *The Unbound Prometheus: Technological Change and Industrial Development in Western Europe from 1750 to the Present*, 2nd ed. Cambridge: Cambridge University Press, 2003.

———. *The Wealth and Poverty of Nations: Why Some Are So Rich and Some Are So Poor*. New York: W.W. Norton, 1998.

Langewiesche, William. "Welcome to the Green Zone: The American Bubble in Baghdad." *Atlantic Monthly* (November 2004).

Larrabee, Eric. *Commander in Chief: Franklin Delano Roosevelt, His Lieutenants and Their War*. New York: Harper & Row, 1987.

Lasseter, Tom. "U.S. may have won, but at a great personal cost." Knight-Ridder Newspapers (November 24, 2004).

Laughton, John Knox. *The Defeat of the Spanish Armada, Anno 1588*, 2 vols. New York: Burt Franklin, 1971 [1894].

Learmonth, Michael. "Military use boosts commercial satellite companies." Reuters (May 25, 2004).

LeMay, Curtis E., with MacKinlay Kantor. *Mission with LeMay: My Story*. Garden City, N.Y.: Doubleday & Co., 1965.

———. with Bill Yenne. *Superfortress: The Story of the B-29 and American Air Power*. New York: Berkley Books, 1988.

Lerman, David. "Submarine Numbers Shrinking Gradually." *Newport News Daily Press* (June 9, 2005).

Levy, Frank, and Richard J. Murnane. *The New Division of Labor: How Computers Are Creating the Next Job Market*. Princeton: Princeton University Press, 2004.

Lewis, Bernard. "Freedom and Justice in the Modern Middle East." *Foreign Affairs* (May/June 2005).

Lewis, William W. *The Power of Productivity: Wealth, Poverty, and the Threat to Global Stability*. Chicago: University of Chicago Press, 2004.

Liddell Hart, B. H., ed. *The Defence of Britain*. New York: Random House, 1939.

———. *The German Generals Talk*. New York: Quill, 1979.

———. *The Rommel Papers*. Trans. Paul Findlay. New York: Da Capo Press, N.D. [1953].

———. *Strategy*, 2nd ed. New York: Frederick A. Praeger, 1968.

Lind, William S. "Understanding Fourth Generation War." *Military Review* (September–October 2004).

Lipton, Eric. "3-D Maps from Commercial Satellites Guide G.I.'s in Iraq's Deadliest Urban Mazes." *The New York Times* (November 26, 2004).

———. "U.S. to Spend Billions More to Alter Security Systems." *The New York Times* (May 8, 2005).

Loades, David. *The Tudor Navy: An Administrative, Political and Military History.* Aldershot, U.K.: Scolar Press, 1992.

Loeb, Vernon. "Bursts of Brilliance: How a string of discoveries by unheralded engineers and airmen helped bring America to the pinnacle of modern military power." *Washington Post Magazine* (December 15, 2002).

Locher, James R. III. "Taking Stock of Goldwater-Nichols." *Joint Forces Quarterly* (Autumn 1996).

Loder, Natasha. "Small Wonders: A Survey of Nanotechnology." *The Economist* (January 1, 2005).

Lohr, Steve. "New Economy: Researchers seem confident that technology has made American workers more efficient. Now some think they even know why." *The New York Times* (February 2, 2004).

Lok, Corie. "Life Vest: Lester Shubin and Nicholas Montanarelli turned Kevlar into lifesaving armor." *MIT Technology Review* (February 2005).

London, Simon. "Intel quits race for faster chips." *The Financial Times* (October 15, 2004).

———. "Man who foresaw the rise of chips with everything." *The Financial Times* (March 17, 2005).

Longford, Elizabeth. *Wellington: The Years of the Sword.* New York: Harper & Row, 1969.

Love, Robert W., Jr. *History of the U.S. Navy, Vol. One: 1775–1941.* Harrisburg, Pa.: Stackpole Books, 1992.

Lowry, Richard. "What Went Wrong? The miscalculations and missteps that led to the current situation in Iraq." *National Review* (October 25, 2004).

———. "What Went Right: How the U.S. began to quell the insurgency in Iraq." *National Review* (May 9, 2005).

Lubold, Gordon. "Remote-Control Robot Debuts in War." *Marine Corps Times* (July 19, 2004).

Lupfer, Timothy J. *The Dynamics of Doctrine: The Change in German Tactical Doctrine During the First World War.* Ft. Leavenworth, Kans.: Combat Studies Institute, U.S. Army Command and General Staff College, July 1981.

Lynn, John, ed. *Acta of the XXVIIIth Congress of the International Commission of Military History.* Chicago: McCormack Foundation, 2003.

———. *Battle: A History of Combat and Culture.* Boulder, Colo.: Westview, 2003.

———. "The Evolution of Army Style in the Modern West, 800–2000." *International History Review* (August 1996).

———. *Giant of the Grand Siècle: The French Army, 1610–1715.* Cambridge: Cambridge University Press, 1997.

———, ed. *Tools of War: Instruments, Ideas, and Institutions of Warfare, 1445–1871.* Champaign: University of Illinois Press, 1990.

Lyons, Eugene. *David Sarnoff: A Biography.* New York: Harper & Row, 1966.

Macfarlane, Allison. "Assessing the Threat: To Predict Bioweapons' Effects, We Need More Data." *Technology Review* (March/April, 2006).

Macgregor, Douglas A. *Transformation Under Fire: Revolutionizing How America Fights.* Westport, Conn.: Praeger, 2003.

———. "Army Transformation: Implications for the Future." Testimony before the House Armed Services Committee (July 15, 2004).

Machalaba, Daniel, and Andy Pasztor. "Thinking Inside the Box: Shipping Containers Get 'Smart.'" *The Wall Street Journal* (January 15, 2004).

Mackenzie, Richard. "Apache Attack." *Air Force Magazine* (October 1991).

Macksey, Kenneth. *Guderian: Panzer General.* London: Greenhill Books, 2003.

MacLeod, Christine. *Inventing the Industrial Revolution: The English Patent System, 1660–1800.* Cambridge: Cambridge University Press, 1988.

MacRae, Catherine. "The Promise and Problem of Laser Weapons." *Air Force Magazine* (December 2001).

Mahan, A. T. *The Influence of Sea Power Upon History, 1660–1783*. New York: Dover, 1987 [1894].
Mahbubani, Kishore. *Beyond the Age of Innocence: Rebuilding Trust Between America and the World*. New York: PublicAffairs, 2005.
Malone, Thomas W. "Pioneers that cultivate a new model of work." *Financial Times* (August 12, 2004).
Manchester, William. *A World Lit Only by Fire: The Medieval Mind and the Renaissance: Portrait of an Age*. Boston: Little, Brown, 1992.
———. *The Arms of Krupp: The Rise and Fall of the Industrial Dynasty That Armed Germany at War*. Boston: Back Bay Books, 2003 [1968].
Mann, Golo. *Wallenstein: His Life Narrated by Golo Mann*. Trans. Charles Kessler. New York: Holt, Rinehart, and Winston, 1976.
Mann, James. *Rise of the Vulcans: The History of Bush's War Cabinet*. New York: Penguin Books, 2004.
Mansoor, Peter R. *The GI Offensive in Europe: The Triumph of American Infantry Divisions, 1941–1945*. Lawrence, Kans.: University Press of Kansas, 1999.
Marconi, Degna. *My Father, Marconi*. New York: McGraw-Hill, 1962.
Marder, Arthur J. *Old Friends, New Enemies: The Royal Navy and the Imperial Japanese Navy. Vol. I: Strategic Illusions, 1936–1941*. Oxford: Clarendon Press, 1981.
Marshall, Chester. *Sky Giants Over Japan: A Diary of a B-29 Combat Crew in WWII*. Winona, Minn.: Apollo Books, 1984.
Marshall, P. J., ed. *The Oxford History of the British Empire, Volume II: The Eighteenth Century*. Oxford: Oxford University Press, 1998.
Marshall, Tyler. "Operation Limited Freedom." *Los Angeles Times* (January 22, 2005).
Martin, Colin, and Geoffrey Parker. *The Spanish Armada*, rev. ed. Manchester, U.K.: Mandolin, 1999.
Martin, Paula. *Spanish Armada Prisoners: The Story of the Nuestra Senora del Rosario and her crew, and of other prisoners in England, 1587–97*. Exeter, U.K.: University of Exeter, 1988.
Massie, Robert. *Castles of Steel: Britain, Germany, and the Winning of the Great War at Sea*. New York: Random House, 2003.
———. *Dreadnought: Britain, Germany and the Coming of the Great War*. New York: Ballantine Books, 1992.
Matthews, William. "Rise of the Military Robot." *Defense News* (May 31, 2004).
Mattingly, Garrett. *The Defeat of the Spanish Armada*. London: Jonathan Cape, 1959.
Matus, Victorino. "Metal Storm: Rise of the Machines." *The Weekly Standard* (July 16, 2003).
May, Ernest R. *Strange Victory: Hitler's Conquest of France*. New York: Hill and Wang, 2000.
McCartney, Scott. *ENIAC: The Triumphs and Tragedies of the World's First Computer*. New York: Walker, 1999.
McClellan, James E. III, and Harold Dorn. *Science and Technology in World History*. Baltimore: The Johns Hopkins University Press, 1999.
McDonnell, Patrick J. "No Shortages of Fighters in Iraq's Wild West." *Los Angeles Times* (July 25, 2004).
McDougall, Walter A. *The Heavens and the Earth: A Political History of the Space Age*. New York: Basic Books, 1985.
McKelway, St. Clair. "A Reporter with the B-29s: I—Possum, Rosy, and the Thousand Kids." *The New Yorker*, June 9, 1945.
———. "A Reporter with the B-29s: II—The Doldrums, Guam, and Something Coming Up." *New Yorker*, June 16, 1945.
———. "A Reporter with the B-29s: III—The Cigar, the Three Wings, and the Low-Level Attacks." *New Yorker*, June 23, 1945.
———. "A Reporter with the B-29s: IV—The People." *New Yorker*, June 30, 1945.
McLaughlin, Abraham. "The quest to create a futuristic battle suite, one micron at a time." *The Christian Science Monitor* (June 10, 2003).

McLeay, Alison. *The Tobermory Treasure: The True Story of a Fabulous Armada Galleon*. London: Conway Maritime Press, 1986.

McMartin, Pete. "Laser's creator will be beaming at party." *Vancouver Sun* (May 13, 2000).

McMaster, H. R. "Crack in the Foundation: Defense Transformation and the Underlying Assumption of Dominant Knowledge in Future War." U.S. Army War College, Student Issue Paper (November 2003). www.comw.org/rma/fulltext/0311mcmaster.pdf.

McNeill, William H. *The Pursuit of Power: Technology, Armed Force and Society Since A.D. 1000*. Chicago: University of Chicago Press, 1982.

———. *Keeping Together in Time: Dance and Drill in Human History*. Cambridge: Harvard University Press, 1995.

Mearsheimer, John R. *Liddell Hart and the Weight of History*. Ithaca: Cornell University Press, 1988.

Menon, Rajan. "Terrorism Inc: Amid Globalization, Al Qaeda Looks a Lot Like GM." *Los Angeles Times* (August 22, 2004).

Merle, Renae. "Low-Tech Grenades a Danger to Helicopters." *The Washington Post* (November 18, 2003).

———. "Running Low on Ammo: Military Turns to Overseas Suppliers to Cover Shortages." *The Washington Post* (July 22, 2004).

Messimer, Dwight R. *Find and Destroy: Antisubmarine Warfare in World War I*. Annapolis: Naval Institute Press, 2001.

Micheletti, Eric. *Special Forces: War on Terrorism in Afghanistan*. Trans. Cyril Lombardini. Paris: Histoire & Collections, 2003.

Middleton, Drew. *Our Share of Night: A Personal Narrative of the War Years*. New York: Viking, 1946.

Mihm, Stephen. "The Quest for the Nonkiller App." *The New York Times Magazine* (July 25, 2004).

Miller, Judith, Stephen Engelberg, William Broad. *Germs: Biological Weapons and America's Secret War*. New York: Touchstone, 2002 [2001].

Miller, T. Christian. "Violence Trumps Rebuilding in Iraq." *Los Angeles Times* (February 21, 2005).

———. "Projects in Iraq to Be Reevaluated." *Los Angeles Times* (April 9, 2005).

Millett, Allan R. *Semper Fidelis: The History of the United States Marine Corps*. Revised and expanded. New York: The Free Press, 1991.

———, and Peter Maslowski. *For the Common Defense: A Military History of the United States of America*. Revised and expanded. New York: Free Press, 1994.

Mokyr, Joel. *The Gifts of Athena: Historical Origins of the Knowledge Economy*. Princeton: Princeton University Press, 2002.

———. *The Lever of Riches: Technological Creativity and Economic Progress*. New York: Oxford University Press, 1990.

Moltke, Helmuth Karl Bernhard von. *Essays, Speeches, and Memoirs of Field Marshal Count Helmuth von Moltke*, 2 vols. Trans. Flint McClumpha, C. Barter, and Mary Herms. New York: Harper & Brothers, 1893.

———. *Letters of Field-Marshal Count Helmuth von Moltke to His Mother and His Brothers*. Trans. Clara Bell and Henry W. Fischer. New York: Harper & Brothers, 1892.

———. *Moltke: His Life and Character Sketched in Journals, Letters, Memoirs, A Novel and Autobiographical Notes*. Trans. Mary Herms. New York: Harper & Brothers, 1892.

———. *Moltke's Letters to His Wife and Other Relatives*, 2 vols. Trans. J. R. McIlraith. London: Kegan, Paul, Trench, Trubner & Co., 1896.

———. *Strategy, Its Theory and Application: The Wars for German Unification, 1866–1871*. Westport, Conn.: Greenwood Press, 1971.

Monro, Robert. *Monro, His Expedition with the Worthy Scots Regiment Called Mac-Keys*. Edited by William S. Brockington, Jr. Westport, Conn.: Praeger, 1999 [1637].

Montgomery, Bernard Law. *The Memoirs of Field Marshal the Viscount Montgomery of Alamein, K.G.* Cleveland: Word Publishing Company, 1958.
Moore, Matt. "World Military Spent $956 billion." *Toronto Star* (June 10, 2004).
Moore, Robin. *The Hunt for Bin Laden: Task Force Dagger: On the Ground with the Special Forces in Afghanistan.* New York: Random House, 2003.
Morgan, Robert, with Ron Powers. *The Man Who Flew the Memphis Belle: Memoir of a WWII Bomber Pilot.* New York: Dutton, 2001.
Morison, Samuel Eliot. *History of United States Naval Operations in World War II. Vol. 3: The Rising Sun in the Pacific, 1931–April 1942.* Champaign: University of Illinois Press, 2001 [1948].
Morris, David J. *Storm on the Horizon: Khafji—The Battle That Changed the Course of the Gulf War.* New York: Free Press, 2004.
Morris, Donald R. *The Washing of the Spears: A History of the Rise of the Zulu Nation under Shaka and Its Fall in the Zulu War of 1879.* New York: Simon & Schuster, 1965.
Morris, Edmund. *Theodore Rex.* New York: Random House, 2001.
Morris, James. *Heaven's Command: An Imperial Progress.* San Diego: Harcourt Brace Jovanovich, 1973.
Morris-Suzuki, Tessa. *The Technological Transformation of Japan from the Seventeenth to the Twenty-first Century.* Cambridge: Cambridge University Press, 1994.
Moss, Michael. "Many Missteps Tied to Delay of Armor to Protect Soldiers." *The New York Times* (March 7, 2005).
"The Most Underrated CEO Ever." *Fortune* (April 5, 2004): 242–248.
Mueller, John. *The Remnants of War.* Ithaca: Cornell University Press, 2004.
Muravchik, Joshua. "The End of the Vietnam Paradigm?" *Commentary* (May 1991).
Murphy, Caryle, and Bassam Sebti. "Power Grid in Iraq Far from Fixed." *The Washington Post* (May 1, 2005).
Murray, James B., Jr. *Wireless Nation: The Frenzied Launch of the Cellular Revolution in America.* Cambridge, Mass.: Perseus Publishing, 2001.
Murray, Williamson, and Allan R. Millett, eds. *Calculations: Net Assessment and the Coming of World War II.* New York: Free Press, 1992.
———. *Military Innovation in the Interwar Period.* Cambridge: Cambridge University Press, 1996.
———. *A War to Be Won: Fighting the Second World War.* Cambridge: Harvard University Press, 2000.
Murray, Williamson. *The Change in the European Balance of Power, 1938–1939: The Path to Ruin.* Princeton: Princeton University Press, 1984.
Murray, Williamson, and Robert H. Scales, Jr. *The Iraq War: A Military History.* Cambridge: Harvard University Press, 2003.
National Science Board. *Science and Engineering Indicators 2004.* www.nsf.gov/sbe/srs/seind04/c0/c0s1.htm
Nairn, Alasdair. *Engines That Move Markets: Technology Investing from Railroads to the Internet and Beyond.* New York: John Wiley & Sons, 2002.
Nakamoto, Michiyo. "Caught in its own trap: Sony battles to make headway in the networked world." *Financial Times* (January 27, 2005).
Nalty, Bernard C., ed. *Winged Shield, Winged Sword: A History of the United States Air Force, Volume II: 1950–1997.* Washington, D.C.: Air Force History and Museums Program, 1997.
Napoleon I. *Correspondance de Napoleon 1er.* Paris: H. Plon, J. Dumaine, 1858–1870.
National Intelligence Council. "Mapping the Global Future." Washington, D.C.: December 2004.
———. "Foreign Missile Developments and the Ballistic Missile Threat Through 2015." Washington, D.C.: December 2001.
National Commission on Terrorist Attacks Upon the United States. *The 9/11 Commission Report.* New York: W.W. Norton, 2004.

Naylor, Sean. *Not a Good Day to Die: The Untold Story of Operation Anaconda.* New York: Berkley Books, 2005.

———. "Ready or Not: An armored unit grapples with broken gear and equipment shortages as it races toward its next war deployment." *Army Times* (November 22, 2004).

Nelson, James L. *Reign of Iron: The Story of the First Battling Ironclads, the Monitor and the Merrimack.* New York: William Morrow, 2004.

Newman, Richard J. "The Little Predator That Could: It is not yet officially operational, but it proved itself in Afghanistan." *Air Force Magazine* (March 2002).

———. "The Joystick War: Run from afar, Predators and other spy gear signal a new era in remote-control warfare." *U.S. News & World Report* (May 19, 2003).

———. "Masters of Invisibility: In Afghanistan, the work of USAF Special Operations Forces was not seen but most assuredly felt." *Air Force Magazine* (June 2002).

Nicolle, David. *Fornovo 1495: France's Bloody Fighting Retreat.* Oxford: Osprey, 1996.

Nordland, Rod, Tom Masland, and Christopher Dickey. "Unmasking the Insurgents." *Newsweek* (February 7, 2005).

Novikoff-Priboy, A. *Tsushima.* Trans. Eden and Cedar Paul. New York: Alfred A. Knopf, 1937.

Nye, Russel B. *George Bancroft: Brahmin Rebel.* New York: Alfred A. Knopf, 1945.

O'Connell, Robert L. "The Great War Torpedoed: The weapon that could have won the war for Germany in 1915." In *What If? 2: Eminent Historians Imagine What Might Have Been* edited by Robert Cowley, 195–210. New York: G. P. Putnam's Sons, 2001.

———. *Of Arms and Men: A History of War, Weapons and Aggression.* New York: Oxford University Press, 1989.

———. *Soul of the Sword: An Illustrated History of Weaponry and Warfare from Prehistory to the Present.* New York: Free Press, 2002.

Odom, William E. *America's Military Revolution: Strategy and Structure After the Cold War.* Washington, D.C.: American University Press, 1993.

Office of Technology Assessment, U.S. Congress. "Technologies Underlying Weapons of Mass Destruction." Washington, D.C.: Government Printing Office, December 1993.

'An Officer.' *Sudan Campaign, 1896–1899.* London: Chapman & Hall, 1899.

O'Hanlon, Michael E. *Defense Strategy for the Post-Saddam Era.* Washington, D.C.: Brookings Institution Press, 2005.

———. "A Flawed Masterpiece." *Foreign Affairs* (May/June 2002).

———. *Neither Star Wars Nor Sanctuary: Constraining the Military Uses of Space.* Washington, D.C.: Brookings Institution Press, 2004.

———. *Technological Change and the Future of Warfare.* Washington: Brookings Institution Press, 2000.

———, and Adriana Lins de Albuquerque. "Iraq Index: Tracking Variables of Reconstruction and Security in Post-Saddam Iraq." www.brookings.edu/iraqindex.

Okumiya, Masatake, and Jiro Horikoshi. *Zero! The Air War in the Pacific During World War II from the Japanese Viewpoint.* Washington, D.C.: Zenger, 1956.

Oman, Charles. *A History of the Art of War in the Sixteenth Century.* New York: E. P. Dutton, 1937.

Orwell, Sonia, and Ian Angus, eds. *The Collected Essays, Journalism and Letters of George Orwell, Vol. 2: My Country Right or Left, 1940–1943.* Boston: Nonpareil Books, 2000 [1968].

Overy, Richard. *The Air War, 1939–1945.* New York: Stein and Day, 1981.

———. *Why the Allies Won.* New York: W.W. Norton, 1995.

Owens, Bill, with Ed Offley. *Lifting the Fog of War.* New York: Farrar, Straus and Giroux, 2000.

Paarlberg, Robert L. "Knowledge as Power: Science, Military Dominance, and U.S. Security." *International Security* (Summer 2004): 122–51.

Packer, George. *The Assassins' Gate: America in Iraq.* New York: Farrar, Straus and Giroux, 2005.

Padfield, Peter. *Battleship*. Edinburgh: Birlinn Ltd., 2000.
———. *Maritime Supremacy and the Opening of the Western Mind*. Woodstock: Overlook Press, 2000.
———. *Tide of Empires: Decisive Naval Campaigns in the Rise of the West*, 2 vols. London: Routledge & Kegan Paul, 1979.
Pae, Peter. "Not Such an Inflated Notion." *Los Angeles Times* (November 11, 2002).
———. "Pentagon's Robot Race Stalls in Gate." *Los Angeles Times* (March 11, 2004).
Pagonis, William G., with Jeffrey L. Cruikshank. *Moving Mountains: Lessons in Leadership and Logistics from the Gulf War*. Boston: Harvard Business School Press, 1992.
Pakenham, Thomas. *The Scramble for Africa: White Man's Conquest of the Dark Continent from 1876 to 1912*. New York: Avon Books, 1991.
Palmer, Michael E. *Command at Sea: Naval Command and Control Since the Sixteenth Century*. Cambridge: Harvard University Press, 2005.
Palmer, R. R., Joel Colton, and Lloyd Kramer. *A History of the Modern World*, 6th ed. New York: Alfred A. Knopf, 2002.
Paltrow, Scot J. "Many Antiterror Recommendations Wither." *The Wall Street Journal* (April 26, 2005).
Paret, Peter, ed., with Gordon A. Craig and Felix Gilbert. *Makers of Modern Strategy: From Machiavelli to the Nuclear Age*. Princeton: Princeton University Press, 1986.
Parker, Geoffrey. *The Cambridge Illustrated History of Warfare*. Cambridge: Cambridge University Press, 1995.
———. "From the House of Orange to the House of Bush: 400 Years of 'Revolutions in Military Affairs,'" in John Lynn, ed., *Acta of the XXVIIIth Congress of the International Commission of Military History*. Chicago: McCormack Foundation, 2003.
———. *The Grand Strategy of Philip II*. New Haven: Yale University Press, 1998.
———. *The Military Revolution: Military Innovation and the Rise of the West, 1500–1800*, 2nd ed. Cambridge: Cambridge University Press, 1996.
———. "Military Revolutions, Past and Present" and "Random Thoughts of a Hedgehog." *Historically Speaking* (April 2003), 2–7, 13–14.
———. *Success Is Never Final: Empire, War, and Faith in Early Modern Europe*. New York: Basic Books, 2002.
———, ed., *The Thirty Years' War*. London: Routledge, 1987.
Parrish, Thomas. *The Submarine: A History*. New York: Viking, 2004.
Peattie, Mark.R. *Sunburst: The Rise of Japanese Naval Airpower, 1909–1941*. Annapolis: Naval Institute Press, 2001.
Perrin, Noel. *Giving up the Gun: Japan's Reversion to the Sword, 1543–1879*. Boston: David R. Godine, 1979.
Perry, William J. "Desert Storm and Deterrence." *Foreign Affairs* (Fall 1991).
———. "Military technology: an historical perspective." *Technology in Society* (April-August, 2004).
Pesola, Maija. "Asian hackers bombard UK government with attacks on vital financial networks." *The Financial Times* (June 16, 2005).
Peters, Ralph. *Beyond Terror*. Mechanicsburg, Pa.: Stackpole Books, 2002.
———. *Fighting for the Future: Will America Triumph?* Mechanicsburg, Pa.: Stackpole Books, 1999.
———. "A Grave New World: 10 lessons from the war in Iraq." *Armed Forces Journal* (April 2005).
———. "In Praise of Attrition." *Parameters* (Summer 2004).
Phillips, David L. *Losing Iraq: Inside the Postwar Recontruction Fiasco*. New York: Westview, 2005.
Pierce, Terry C. *Warfighting and Disruptive Technologies: Disguising Innovation*. London: Frank Cass, 2004.
Pierson, Peter. *Commander of the Armada: The Seventh Duke of Medina Sidonia*. New Haven: Yale University Press, 1989.
Pincus, Walter. "Pentagon Has Far-Reaching Defense Spacecraft in Works." *The Washington Post* (March 16, 2005).

———. "Predator to See More Combat." *The Washington Post* (March 22, 2005).
Pitre, K. G. *The Second Anglo-Maratha War, 1802–1805: A Study in Military History.* Poona, India: Dastane Ramchandra & Co., 1990.
Pleshakov, Constantine. *The Tsar's Last Armada: The Epic Voyage to the Battle of Tsushima.* New York: Basic Books, 2002.
Poletti, Therese. "Defense Dept. Hope to Enlist AI in War Against Terrorism." *San Jose Mercury News* (August 2, 2004).
Politovsky, Eugene S. *From Libau to Tsushima.* Trans. F. R. Godfrey. London: John Murray, 1906.
Pollack, Kenneth M. *Arabs at War: Military Effectiveness, 1948–1991.* Lincoln: University of Nebraska Press, 2002.
———. *The Threatening Storm: The Case for Invading Iraq.* New York: Random House, 2002.
Pollock, John. *Kitchener: Architect of Victory, Artisan of Peace.* New York: Carroll & Graf, 1998.
Polmar, Norman. "Composing Command and Control." *U.S. Naval Institute Proceedings* (February 2005).
———. "Antiship Ballistic Missiles . . . Again." *Naval Institute Proceedings* (July 2005).
Porter, Bruce D. *War and the Rise of the State: The Military Foundations of Modern Politics.* New York: Free Press, 1994.
Posen, Barry R. "Command of the Commons: The Military Foundation of U.S. Hegemony." *International Security.*
———. *The Sources of Military Doctrine: France, Britain, and Germany Between the World Wars.* Ithaca: Cornell University Press, 1984.
Potter, G. R. (ed.). *The New Cambridge Modern History, Volume I: The Renaissance, 1493–1520.* Cambridge: Cambridge University Press, 1957.
Pound, Edward T. "Seeds of Chaos." *U.S. News & World Report* (December 20, 2004).
Powell, Bill, and Aparisim Ghosh. "Paul Bremer's Rough Ride." *Time* (June 28, 2004).
Powell, Colin L., with Joseph E. Persico. *My American Journey.* New York: Ballantine Books, 1996 [1995].
Prange, Gordon W., with Donald M. Goldstein and Katherine V. Dillon. *At Dawn We Slept: The Untold Story of Pearl Harbor.* New York: Penguin Books, 1991.
———. *December 7, 1941: The Day the Japanese Attacked Pearl Harbor.* New York: McGraw-Hill, 1988.
———. *God's Samurai: Lead Pilot at Pearl Harbor.* Washington, D.C.: Brassey's, 2004.
———. *Pearl Harbor: The Verdict of History.* New York: McGraw-Hill, 1986.
Prescott, William H. *History of the Conquest of Mexico, and History of the Conquest of Peru.* New York: Modern Library, 2001 [1856].
Preston, Richard A., Alex Roland, and Sydney F. Wise. *Men in Arms: A History of Warfare and Its Interrelationships with Western Society*, 5th ed. Fort Worth, Tex.: Holt, Rinehart and Winston, 1991.
Priest, Dana. "In War, Mud Huts and Hard Calls: As U.S. Teams Guided Pilots' Attacks, Civilian Presence Made Task Tougher." *The Washington Post* (February 20, 2002).
———. "New Spy Satellite Debated on Hill." *The Washington Post* (December 11, 2004).
———. "'Team 555' Shaped a New Way of War: Special Forces and Smart Bombs Turned Tide and Routed Taliban." *The Washington Post* (April 3, 2002).
Pritchard, Tim. *Ambush Alley: The Most Extraordinary Battle of Operation Iraqi Freedom.* New York: Ballantine Books, 2005.
Prussian General Staff, Department of Military History. *The Campaign of 1866 in Germany.* Trans. Colonel Von Wright and Captain Henry M. Hozier. London: His Majesty's Stationery Office, 1872.
Qiao, Liang, and Wang Ziangsui. *Unrestricted Warfare: China's Master Plan to Destroy America.* Panama City, Panama: Pan American Publishing Co., 2002 [1999].
"Quadrennial Defense Review Report." Washington, D.C.: Department of Defense, February 6, 2006.

Ralston, David B. *Importing the European Army: The Introduction of European Military Techniques and Institutions into the Extra-European World, 1600–1914.* Chicago: University of Chicago Press, 1990.

Ratner, Daniel, and Mark A. Ratner. *Nanotechnology and Homeland Security: New Weapons for New Wars.* Upper Saddle River, N.J.: Prentice Hall, 2004.

Ratnesar, Romesh, and Michael Weisskopf Weisskopf. "Portrait of a Platoon." *Time* (December 29, 2003).

Raudzens, George. "Military Revolution or Maritime Evolution? Military Superiorities or Transportation Advantages as Main Causes of European Colonial Conquests to 1788." *The Journal of Military History* 63, no. 3 (July 1999), 631–641.

Reed, Fred. "Robots Eyed for Urban Warfare." *Washington Times* (April 15, 2004).

Reed, Thomas C. *At the Abyss: An Insider's History of the Cold War.* New York: Ballantine Books, 2004.

Regan, Michael P. "Army Prepares 'Robo Soldier' for Iraq." Associated Press (January 22, 2005).

———. "Pentagon Looks to Directed-Energy Weapons." Associated Press (October 28, 2004).

Reid, T. R. *The Chip: How Two Americans Invented the Microchip and Launched a Revolution.* New York: Random House, 2001 [1984].

Reiter, Dan, and Allan C. Stam III. *Democracies at War.* Princeton: Princeton University Press, 2002.

Reitman, Janet. "Apocalypse Now." *Rolling Stone* (November 25, 2004).

Reynolds, David. "1940: Fulcrum of the Twentieth Century?" *International Affairs* (April 1990), 325–350.

———. *In Command of History: Churchill Fighting and Writing the Second World War.* New York: Random House, 2005.

Reynolds, Nicholas E. *Basrah, Baghdad, and Beyond: The U.S. Marine Corps in the Second Iraq War.* Annapolis: Naval Institute Press, 2005.

Riasanovsky, Nicholas V. *A History of Russia*, 4th ed. New York: Oxford University Press, 1984.

Rice, Eugene F., with Anthony Grafton. *The Foundations of Early Modern Europe, 1460–1559*, 2nd ed. New York: W. W. Norton, 1994.

Rich, Ben R., and Leo Janos. *Skunk Works: A Personal Memoir of My Years at Lockheed.* Boston: Back Bay Books, 1994.

Richburg, Keith B., and Colum Lynch. "Afghan Victors Agree to Talks in Berlin." *The Washington Post* (November 21, 2001).

Richelson, Jeffrey T. *America's Space Sentinels: DSP Satellites and National Security.* Lawrence, Kans.: University Press of Kansas, 1999.

Ricks, Thomas E. "Soldiers Record Lessons from Iraq." *The Washington Post* (February 8, 2004).

———. "U.S. Adopts Aggressive Tactics on Iraqi Fighters." *The Washington Post* (July 28, 2003).

Rip, Michael Russell, and James M. Hasik. *The Precision Revolution: GPS and the Future of Aerial Warfare.* Annapolis: Naval Institute Press, 2002.

Ritter, Gerhard. *The Sword and the Scepter: The Problem of Militarism in Germany, Vol. I: The Prussian Tradition, 1780–1890.* Princeton Junction, N.J.: The Scholar's Bookshelf, 1988 [1964].

Roan, Shari: "Science Quickens Its Steps." *Los Angeles Times* (March 9, 2006).

Roberts, Andrew. *Napoleon and Wellington: The Battle of Waterloo—and the Great Commanders Who Fought It.* New York: Simon & Schuster, 2001.

Roberts, J. M. *The Penguin History of the World.* 3rd ed. London: Penguin Books, 1995.

Roberts, Michael. *Gustavus Adolphus: A History of Sweden, 1611–1632*, 2 vols. London: Longmans, Green and Co., 1958.

Robinson, Linda. *Masters of Chaos: The Secret History of the Special Forces.* New York: PublicAffairs, 2004.

Rodger, N. A. M. *The Command of the Ocean: A Naval History of Britain, 1649–1815*. New York: W.W. Norton, 2004.
———. *The Safeguard of the Sea: A Naval History of Britain, 660–1649*. New York: W.W. Norton, 1997.
Rodriguez-Salgado, M. J., and Simon Adams, eds. *England, Spain and the Gran Armada, 1585–1604: Essays from the Anglo-Spanish Conferences London and Madrid 1988*. Edinburgh, Scotland: John Donald Publishers Ltd., 1991.
Rogers, Clifford, ed. *The Military Revolution Debate: Readings on the Military Transformation of Early Modern Europe*. Boulder, Colo.: Westview Press, 1995.
Rohde, David. "Portrait of a U.S. Vigilante in Afghanistan." *The New York Times* (July 11, 2004).
Romjue, John L. *The Army of Excellence: The Development of the 1980s Army*. Fort Monroe, Va.: Office of the Command Historian, U.S. Army Training and Doctrine Command, 1997.
———. *Prepare the Army for War: A Historical Overview of the Army Training and Doctrine Command, 1973–1993*. Fort Monroe, Va.: Office of the Command Historian, U.S. Army Training and Doctrine Command, 1993.
Roosevelt, Ann. "Checkpoints in Iraq, Afghanistan Get New Inspection Technology." *Defense Daily* (August 4, 2004).
Rosen, Stephen Peter. *Winning the Next War: Innovation and the Modern Military*. Ithaca: Cornell University Press, 1991.
———. *Societies and Military Power: India and Its Armies*. Ithaca: Cornell University Press, 1996.
———. "The Future of War and the American Military: Demography, technology, and the politics of modern empire." *Harvard Magazine* (May–June 2002).
Rostow, W. W. *The World Economy: History and Prospect*. Austin: University of Texas Press, 1978.
Rotte, Ralph, and Christoph M. Schmidt, "On the Production of Victory: Empirical Determinants of Battlefield Success in Modern War." Bonn, Germany: Institute for the Study of Labor, May 2002. http://ssrn.com/abstract=314204
Roy, Kaushik. "Military Synthesis in South Asia: Armies, Warfare, and Indian Society, c. 1740–1849." *The Journal of Military History* (July 2005), 651–690.
Royle, Trevor. *Death Before Dishonour: The True Story of Fighting Mac*. Edinburgh, Scotland: Mainstream, 1982.
———. *The Kitchener Enigma*. London: Michael Joseph, 1985.
Rubin, Allissa J., and Doyle McManus. "Why America Has Waged a Losing Battle on Fallouja." *Los Angeles Times* (October 24, 2004).
Rumsfeld, Donald H., et al. *Report of the Commission to Assess United States National Security Space Management and Organization*. Washington, D.C., January 11, 2001.
———. "Why Defense Must Change." *The Washington Post* (July 18, 2000).
Rush, Robert Sterling. *Hell in Hürtgen Forest: The Ordeal and Triumph of an American Infantry Regiment*. Lawrence, Kans.: University Press of Kansas, 2001.
Russell, W. H. "The Battle of Sadowa." *The Times* (London) (July 11, 1866).
Sageman, Marc. *Understanding Terror Networks*. Philadelphia: University of Pennsylvania Press, 2004.
Saint-Exupéry, Antoine de. *Airman's Odyssey*. New York: Reynal & Hitchcock, 1942.
Sale, Tony. *The Colossus Computer, 1943–1996, and How It Helped to Break the German Lorenz Cipher in WWII*. Cleobury Mortimer, Shropshire, U.K.: M&M Baldwin, 2000.
Samuels, Richard J. *"Rich Nation, Strong Army": National Security and the Technological Transformation of Japan*. Ithaca: Cornell University Press, 1994.
Sanderson, Ward. "Raptors for Langley." *Newport News Daily Press* (May 10, 2005).
Santayana, George. *Soliloquies in England and Later Soliloquies*. Ann Arbor: University of Michigan Press, 1967.
Scales, Robert H., Jr. *Certain Victory: The U.S. Army in the Gulf War*. Washington, D.C.: Brassey's, 1994.
———. "Urban Warfare: A Soldier's View." *Military Review* (January–February 2005).

———. *Yellow Smoke: The Future of Land Warfare for America's Military*. Lanham, Md.: Rowan and Littlefield, 2003.

Scammell, G. V. *The First Imperial Age: European Overseas Expansion c. 1400–1715*. London: Unwin Hyman, 1989.

Scarborough, Rowan. "Enemy Force uses 'Guerrilla Tactics' in Iraq, Abizaid Says." *Washington Times* (July 17, 2003).

———. *Rumsfeld's War: The Untold Story of America's Anti-Terrorist Commander*. Washington, D.C.: Regnery, 2004.

Schama, Simon. *A History of Britain, Vol. III: The Fate of Empire, 1776–2000*. New York: Miramax Books, 2003.

Schiller, Friedrich von. *The History of the Thirty Years' War in Germany*. New York: John D. Williams, N.D. [Composed 1791–93.]

Schlesinger, James R., Harold Brown, Tillie K. Fowler, and Charles A. Horner. *Final Report of the Independent Panel to Review DoD Detention Operations* (August 2004).

Schmitt, Eric. "Medal of Honor to Be Awarded to Soldier Killed in Iraq, a First." *The New York Times* (March 30, 2005).

———. "Remotely Controlled Aircraft Crowd Dangerous Iraqi and Afghan Skies." *The New York Times* (April 5, 2005).

———, and Thom Shanker. "Estimates by U.S. See More Rebels with More Funds." *The New York Times* (October 22, 2004).

Schroen, Gary C. *First In: An Insider's Account of How the CIA Spearheaded the War on Terror in Afghanistan*. New York: Ballantine Books, 2005.

Schulze, Hagen. *Germany: A New History*. Trans. Deborah Lucas Schneider. Cambridge: Harvard University Press, 1998.

Schwarzkopf, H. Norman, with Peter Petre. *It Doesn't Take a Hero: The Autobiography: General H. Norman Schwarzkopf*. New York: Bantam Books, 1992.

Scott, William B. " 'Space' at War." *Aviation Week & Space Technology* (October 25, 2004).

Segal, Adam. "Is America Losing Its Edge? Innovation in Globalized World." *Foreign Affairs* (November–December 2004).

Semenoff, Vladimir. *The Battle of Tsu-shima Between the Russian and Japanese Fleets, Fought on 27th May 1905*. Trans. A. B. Lindsay. London: John Murray, 1906.

———. *Rasplata (The Reckoning): His Diary During the Blockade of Port Arthur and the Voyage of Admiral Rojestvensky's Fleet*. London: John Murray, 1909.

Sen, Surendra Nath. *The Military System of the Marathas*. Bombay: Orient Longmans, 1958.

"Sen. Warner Plans to Renew Push for Unmanned Vehicles." *Aerospace Daily* (November 8, 2002).

Seppings Wright, H. C. *With Togo: The Story of Seven Months' Active Service Under His Command*. London: Hurst and Blackett Ltd., 1905.

Sevastopulo, Demetri. "Games software and plywood harden US troops for Iraq battle." *Financial Times* (November 26, 2004).

Shachtman, Noah. "When a Gun Is More Than a Gun." *Wired* (March 20, 2003).

———. "More Robot Grunts Ready for Duty." *Wired* (December 1, 2004).

Shanker, Thom. "Hussein's Agents Are Behind Attacks in Iraq, Pentagon Finds." *The New York Times* (April 29, 2004).

———. "Rewarding Skill in a Mission with Few Thrills." *The New York Times* (October 17, 2002).

———. "Rumsfeld Faults Turkey for Barring Use of Its Land in '03 to Open Northern Front in Iraq." *The New York Times* (March 21, 2005).

———, and Eric Schmitt. "Past Battles Won and Lost Helped in Falluja Assault." *The New York Times* (November 22, 2004).

Shapiro, Samantha M. "The War Inside the Arab Newsroom." *The New York Times Magazine* (January 2, 2005).

Shea, Dana A. "Terrorism: Background on Chemical, Biological, and Toxin Weapons and Options for Lessening Their Impact." Washington, D.C: Congressional Research Service, December 1, 2004.

Sheridan, P. H. *Personal Memoirs of P. H. Sheridan, General United States Army*. 2 vols. Wilmington, N.C.: Broadfoot, 1992 [1888].

Sherry, Michael S. *The Rise of American Air Power: The Creation of Armageddon*. New Haven: Yale University Press, 1987.

Shirer, William L. *Berlin Diary: The Journal of a Foreign Correspondent, 1934–1941*. Boston: Little, Brown, 1941.

Showalter, Dennis E. "Information Capabilities and Military Revolutions: The Nineteenth-Century Experience." *The Journal of Strategic Studies* (June 2004).

———. *Railroads and Rifles: Soldiers, Technology, and the Unification of Germany*. Hamden, Conn.: Archon Books, 1975.

———. *The Wars of German Unification*. London: Hodder Arnold, 2004.

Shubik, Martin. "Terrorism, Technology, and the Socioeconomics of Death." *Comparative Strategy* (1997), 399–414.

Shulman, Mark R. *Navalism and the Emergence of American Sea Power, 1882–1893*. Annapolis: Naval Institute Press, 1995.

Shultz, Richard H., Jr. "Showstoppers: Nine reasons why we never sent our Special Operations Forces after al Qaeda before 9/11." *The Weekly Standard* (January 26, 2004).

Sichel, Daniel E. *The Computer Revolution: An Economic Perspective*. Washington, D.C.: Brookings Institution Press, 1997.

Singer, P. W. "The Ultimate Military Entrepreneur." *MHQ: The Quarterly Journal of Military History* (Spring 2003).

Singman, Jeffrey L. *Daily Life in Elizabethan England*. Westport, Conn.: Greenwood Press, 1995.ß

Sipress, Alan. "Hunt for Hussein Led U.S. to Insurgent Hub." *The Washington Post* (December 26, 2003).

Smith, R. Jeffrey. "Study Faults Army Vehicle." *The Washington Post* (March 31, 2005).

Smith, Lewis Ferdinand. *A Sketch of the Rise, Progress and Termination of the Regular Corps, Formed and Commanded by Europeans in the Service of the Native Princes of India, with Details of the Principal Events and Actions of the Late Marhatta War*. Calcutta: J. Greenway, 1805.

Smith, Vincent. *The Oxford History of India*, 4th ed. New Delhi: Oxford University Press, 1981.

Solberg, Carl. *Conquest of the Skies: A History of Commercial Aviation in America*. Boston: Little, Brown, 1979.

Sondhaus, Lawrence. *Naval Warfare, 1815–1914*. London: Routledge, 2001.

Sowell, Thomas. *Conquests and Cultures: An International History*. New York: Basic Books, 1998.

Spears, Edward L. *Assignment to Catastrophe, Vol. I: Prelude to Dunkirk, July 1939–May 1940*. New York: A. A. Wyn, 1954.

———. *Assignment to Catastrophe, Vol. II: The Fall of France, June 1940*. New York: A. A. Wyn, 1955.

Spector, Ronald H. *At War at Sea: Sailors and Naval Combat in the Twentieth Century*. New York: Viking, 2001.

———. *Eagle Against the Sun: The American War with Japan*. New York: Vintage Books, 1985.

Spencer, Alfred, ed. *Memoirs of William Hickey, Vol. IV: 1790–1809*. New York: Alfred A. Knopf, 1925.

Spencer, Jack, and James Jay Carafano. "The Use of Directed-Energy Weapons to Protect Critical Infrastructure." Washington, D.C.: Heritage Foundation, August 2, 2004. www.heritage.org/Research/NationalSecurity/bg1783.cfm.

Spiegel, Peter. "Aerial Combat: Why there are doubts for the US and its allies over this $200bn jet." *Financial Times* (January 31, 2005).

Spinner, Jackie. "Medics Testify to Fallujah's Horrors." *The Washington Post* (November 24, 2004).

Stanhope, Philip Henry, 5th Earl. *Notes of Conversations with the Duke of Wellington, 1831–1851*. New York: Da Capo Press, 1973 [1888].

Starr, S. Frederick. "Silk Road to Success." *The National Interest* (Winter 2004/05).
Steevens, G. W. *With Kitchener to Khartum*. New York: Dodd, Mead, 1898.
Stross, Randall. "How the iPod Ran Circles Around the Walkman." *New York Times* (March 13, 2005).
Studemann, Frederick. "UK shakes up military in shift toward rapid reaction." *Financial Times* (July 22, 2004).
Sumida, Jon Tetsuro. *In Defense of Naval Supremacy: Finance, Technology and British Naval Policy, 1889–1914*. Boston: Unwin Hyman, 1989.
Swainson, Bill. *Encarta Book of Quotations*. New York: St. Martin's Press, 2000.
Swartz, Jon, and Edward Iwata. "Invented to save gas, Kevlar now saves lives." *USA Today* (April 15, 2003).
———. "New Breed of Robots Takes War to Next Level." *USA Today* (May 12, 2003).
Talbot, David. "How Technology Failed in Iraq." *Technology Review* (November 2004).
———. "Terror's Server: Fraud, gruesome propaganda, terror planning: the Net enables it all. The online industry can help fix it." *Technology Review* (February 2005).
Tallett, Frank. *War and Society in Early-Modern Europe, 1495–1715*. London: Routledge, 1992.
Taubman, Philip. *Secret Empire: Eisenhower, the CIA, and the Hidden Story of America's Space Espionage*. New York: Simon & Schuster, 2003.
Taylor, Frederick. *Dresden: Tuesday, February 13, 1945*. New York: HarperCollins, 2004.
Tether, Tony. "Statement Submitted to the Committee on Science, U.S. House of Representatives," May 14, 2003.
Thibo, Charles A. "U-Boat!" *Naval Institute Proceedings* (June 2005).
Third Battalion, 5th Special Forces Group. "The Liberation of Mazar-e Sharif: 5th SF Group Conducts UW in Afghanistan." *Special Warfare* (June 2002).
Third Infantry Division (Mechanized). "After Action Report: Operation Iraqi Freedom."
Thomas, Evan. "A Street Fight." *Newsweek* (April 29, 2002).
———, Rod Nordland, and Christian Caryl. "Operation Hearts & Minds." *Newsweek* (December 29, 2003).
———, John Barry, and Christian Caryl. "A War in the Dark." *Newsweek* (November 10, 2003).
Thomas, Hugh. *Conquest: Montezuma, Cortes, and the Fall of Old Mexico*. New York: Simon & Schuster, 1993.
Thompson, Warren E. *Bandits Over Baghdad: Personal Stories of Flying the F-117 Over Iraq*. North Branch, Minn.: Specialty Press, 2000.
Thorn, William. *Memoir of the War in India, Conducted by General Lord Lake, Commander-in-Chief, and Major-General Sir Arthur Wellesley, Duke of Wellington, from Its Commencement in 1803, to its Termination in 1806, on the Banks of the Hyphasis*. London: T. Egerton, 1818.
Thucydides. *History of the Peloponnesian War*. Trans. Rex Warner. New York: Penguin Books, 1986 [1954].
Till, Geoffrey. *Air Power and the Royal Navy, 1914–1945: A Historical Survey*. London: Jane's, 1979.
Tilly, Charles. *Coercion, Capital and European States, AD 990–1992*. Cambridge, Mass.: Blackwell, 1990.
———. *The Formation of National States in Western Europe*. Princeton: Princeton University Press, 1975.
Tirpak, John A. "Enduring Freedom: USAF's heavy bombers dominated events in Afghanistan, but the success story was much broader than that." *Air Force Magazine* (February 2002).
———. "In Search of Spaceplanes." *Air Force Magazine* (December 2003).
———. "Precision: The Next Generation." *Air Force Magazine* (November 2003).

———. "Setting a Course for the Airborne Laser." *Air Force Magazine* (September 2003).

Toffler, Alvin, and Heidi Toffler. *War and Anti-War: Making Sense of Today's Global Chaos*. New York: Warner Books, 1993.

Toland, John. *The Rising Sun: The Decline and Fall of the Japanese Empire, 1936–1945*. New York: Modern Library, 2003 [1970].

Tolstoy, Leo. *War and Peace*. New York: Modern Library, 1994.

Tomayko, James E. *Computers in Spaceflight: The NASA Experience*. Washington, D.C.: NASA History Office, March 1988. www.hq.nasa.gov/office/pao/History/computers/Compspace.html

Toner, Mark. "Robots Far from Leading the Fight." *Atlanta Journal-Constitution* (March 14, 2004).

Trimble, William F. *Admiral William A. Moffett: Architect of Naval Aviation*. Washington: Smithsonian Institution Press, 1994.

Trippi, Joe. *The Revolution Will Not Be Televised: Democracy, the Internet, and the Overthrow of Everything*. New York: ReganBooks, 2004.

Trulock, Notra III, Kerry L. Hines, and Anne D. Herr. "Soviet Military Thought in Transition: Implications for the Long-Term Military Competition." A report prepared by Pacific-Sierra Research Corp. for the Office of Secretary of Defense/Net Assessment, May 1988.

Trythall, Anthony John. *'Boney' Fuller: Soldier, Strategist, and Writer, 1876–1966*. New Brunswick, N.J.: Rutgers University Press, 1977.

Tucker, Jonathan. *War of Nerves: Chemical Warfare from World War I to Al Qaeda*. New York: Pantheon Books, 2006.

Tyson, Ann Scott. "Two Years Later, Iraq War Drains Military." *The Washington Post* (March 19, 2005).

The United States Army in Afghanistan: Operation Enduring Freedom, October 2001–March 2002. Fort McNair, Wash.: U.S. Army Center for Military History, N.D.

Usher, Abbott Payson. *A History of Mechanical Inventions*. Revised edition. New York: Dover, 1988 [1954].

Usherwood, Stephen, ed. *The Great Enterprise: The History of the Spanish Armada, as Revealed in Contemporary Documents*. London: Folio Society, 1978.

U.S. Air Force. "Counterspace Operations." Air Force Doctrine Document 2–2.1 (August 2, 2004).

"U.S. Air Force's New Chief Eyes Speeding Up Acquisition." *Aerospace Daily & Defense Report* (September 13, 2005).

U.S. Central Command Air Forces. "Operation Iraqi Freedom by the Numbers." (April 30, 2003).

The United States Strategic Bombing Survey, Summary Report (European War). Washington, D.C., September 30, 1945.

The United States Strategic Bombing Survey, Summary Report (Pacific War). Washington, D.C., July 1, 1946.

U.S. News & World Report. *Triumph Without Victory: The Unreported History of the Persian Gulf War*. New York: Times Books, 1992.

Vandergriff, Donald. *The Path to Victory: America's Army and the Revolution in Human Affairs*. Novato, Calif.: Presidio, 2002.

Van Der Vat, Dan. *The Atlantic Campaign: World War II's Great Struggle at Sea*. New York: Harper & Row, 1988.

Vandervort, Bruce. *Wars of Imperial Conquest in Africa, 1830–1914*. Bloomington: Indiana University Press, 1998.

Van Natta, Don, Jr., and Desmond Butler. "How Tiny Swiss Cellphone Chips Helped Track Global Terror Web." *New York Times* (March 4, 2004).

Vernon, Alex, with Neil Creighton, Jr., Greg Downey, Rob Holmes, and Dave Trybula. *Eyes of Orion: Five Tank Lieutenants in the Persian Gulf War*. Kent, Ohio: Kent State University Press, 1999.

Vickers, Michael G., and Robert C. Martinage. *The Revolution in War.* Washington, D.C.: Center for Strategic and Budgetary Assessments, December 2004.
Vogel, Steve. "Over Afghanistan, Gantlets in the Sky; U.S. Pilots Are Tested by Complex and Sometimes Perilous Missions." *The Washington Post* (October 29, 2001).
Wade, Nicholas. "Exotic Military Arts: On the Scent of Terrorists." *The New York Times* (January 5, 2003).
Waller, Douglas. *A Question of Loyalty: Gen. Billy Mitchell and the Court-Martial That Gripped the Nation.* New York: HarperCollins, 2004.
———. "Secret Warriors." *Newsweek* (June 17, 1991).
Walsh, Elsa. "Learning to Spy: Can Maureen Baginski Save the FBI?" *The New Yorker* (November 8, 2004).
Ware, Michael. "Into the Hot Zone." *Time* (November 22, 2004).
Warner, Denis, and Peggy Warner. *The Tide at Sunrise: A History of the Russo-Japanese War, 1904–1905.* New York: Charterhouse, 1974.
Warner, Philip. *Kitchener: The Man Behind the Legend.* London: Hamish Hamilton, 1985.
Washburn, Mark. "Advances Make War More Survivable." *Miami Herald* (June 13, 2005).
Watkins, Owen Spencer. *With Kitchener's Army: Being a Chaplain's Experiences with the Nile Expedition, 1898.* London: S. W. Partridge, 1899.
Watt, Donald Cameron. *Too Serious a Business: European Armed Forces and the Approach to the Second World War.* Berkeley: University of California Press, 1975.
Wawro, Geoffrey. *The Austro-Prussian War: Austria's War with Prussia and Italy in 1866.* Cambridge: Cambridge University Press, 1996.
Weaver, Mary Anne. "Lost at Tora Bora." *The New York Times Magazine* (September 11, 2005).
Webb, Al. "Briton Arrested in Military Hacking." *The Washington Times* (June 9, 2005).
Wedgwood, C. V. *The Thirty Years War.* London: Methuen, 1981 [1938].
Weigley, Russell F. *The Age of Battles: The Quest for Decisive Warfare from Breitenfeld to Waterloo.* Bloomington: Indiana University Press, 1991.
Weiner, Tim. "Air Force Seeks Bush's Approval for Space Weapons Programs." *The New York Times* (May 18, 2005).
———. "Arms Fiascos Lead to Alarm Inside Pentagon." *The New York Times* (June 8, 2005).
———. "Navy's Fleet of Tomorrow Is Mired in Politics of Yesterday." *The New York Times* (April 19, 2005).
———. "Pentagon Envisioning a Costly Internet for War." *The New York Times* (November 13, 2004).
———. "A Vast Arms Buildup, Yet Not Enough for Wars." *The New York Times* (October 1, 2004).
Weintraub, Stanley. *Long Day's Journey Into War: December 7, 1941.* New York: Truman Talley Books, 1991.
Weiss, Rick. "For Science, Nanotech Poses Big Unknowns," and "Applications Abound for Unique Physical, Chemical Properties." *The Washington Post* (February 1, 2004).
———. "Nanotech Is Booming Biggest in U.S., Report Says." *The Washington Post* (March 28, 2005).
Weiss, Stanley A., and William A. Owen. "An Indefensible Military Budget." *The New York Times* (February 7, 2002).
Weller, Jac. *Wellington in India.* London: Greenhill Books, 2000 [1972].
Wellington, 7th Duke of, ed. *The Conversations of the First Duke of Wellington with George William Chad.* Cambridge, U.K.: The Saint Nicolas Press, 1956.
———. *Supplementary Dispatches and Memoranda of Field Marshal Arthur Duke of Wellington, Vol. IV: India, 1797–1805.* London: John Murray, 1859.
Welsh, James. *Military Reminiscences Extracted from a Journal of Nearly Forty Years' Active Service in the East Indies.* London: Smith, Elder, 1830.
Werrell, Kenneth P. *Blankets of Fire: U.S. Bombers over Japan During World War II.* Washington, D.C.: Smithsonian Institution Press, 1996.

———. *Chasing the Silver Bullet: U.S. Air Force Weapons Development from Vietnam to Desert Storm*. Washington, D.C.: Smithsonian Books, 2003.

Werth, Alexander. *The Last Days of Paris: A Journalist's Diary*. London: Hamish Hamilton, 1940.

West, Bing. "Maneuver Warfare: It Worked in Iraq." Naval Institute Proceedings (February 2004).

———. *No True Glory: A Frontline Account of the Battle for Fallujah.* New York: Bantam Books, 2005.

———, and Ray L. Smith. *The March Up: Taking Baghdad with the 1st Marine Division.* New York: Bantam Books, 2003.

Westin, David. "Don't Blame the Networks." *The Washington Post* (July 20, 2004).

Westwood, J. N. *Witnesses of Tsushima*. Tokyo: Sophia University, 1970.

Whalen, David J. *The Origins of Satellite Communications, 1945–1965.* Washington, D.C.: Smithsonian Institution Press, 2002.

Wheeler, Keith. *Bombers over Japan*. Alexandria, Va.: Time-Life Books, 1982.

White, Lynn, Jr. *Medieval Technology and Social Change.* London: Oxford University Press, 1962.

Williams, Carol J. "Suicide Attacks Rising Rapidly." *Los Angeles Times* (June 2, 2005).

Williams, Cindy. "Draft Lessons from Europe." *The Washington Post* (October 5, 2004).

———. "From Conscripts to Volunteers: NATO's Transition to All-Volunteer Forces." *Naval War College Review* (Autumn 2004).

Willis, Clint, ed. *Boots on the Ground: Stories of American Soldiers from Iraq and Afghanistan*. New York: Thunder's Mouth Press, 2004.

Willmott, H. P., with Tohmatsu Horuo and W. Spencer Johnson. *Pearl Harbor*. London: Cassell, 2001.

Wilson, Isaiah (Ike). "Thinking Beyond War: Civil-Military Operational Planning for Northern Iraq." Unpublished paper prepared for delivery at Cornell University, October 14, 2004.

Wilson, James Q. *Bureaucracy: What Government Agencies Do and Why They Do It.* New York: Basic Books, 1989.

Wilson, Jim. "Beyond Bullets." *Popular Mechanics* (April 2003).

———. "Stealth Strike Force." *Popular Mechanics* (November 2003).

Wingate, Ronald. *Wingate of the Sudan: The Life and Times of General Sir Reginald Wingate, Maker of the Anglo-Egyptian Sudan*. London: John Murray, 1955.

Wirls, Daniel. *Buildup: The Politics of Defense in the Reagan Era*. Ithaca: Cornell University Press, 1992.

Witzel, Morgan. "When a company's success breeds failure." *Financial Times* (March 14, 2005).

Wohlstetter, Roberta. *Pearl Harbor: Warning and Decision*. Stanford: Stanford University Press, 1962.

Wolfe, Tom. *Hooking Up*. New York: Farrar, Straus and Giroux, 2000.

Wong, Leonard. *Developing Adaptive Leaders: The Crucible Experience of Operation Iraqi Freedom.* Carlisle, Pa.: Strategic Studies Institute, U.S. Army War College, July 2004.

———. *Stifled Innovation? Developing Tomorrow's Leaders Today.* Carlisle, Pa.: Strategic Studies Institute, U.S. Army War College, April 2002.

Woods, Kevin, James Lacey, and Williamson Murray. "Saddam's Delusions: The View from the Inside." *Foreign Affairs* (May/June 2006).

Woodward, Bob. *Bush at War*. New York: Simon & Schuster, 2002.

———. *Plan of Attack*. New York: Simon & Schuster, 2004.

Wright, Evan. *Generation Kill: Devil Dogs, Iceman, Captain America and the New Face of American War*. New York: G. P. Putnam's Sons, 2004.

Wright, Lawrence. "The Terror Web: Were the Madrid bombings part of a new, far-reaching jihad being planned on the Internet?" *New Yorker* (August 2, 2004).

Wright, Patrick. *Tank: The Progress of a Monstrous War Machine*. New York: Viking, 2002.

Wright, Robin, and Thomas E. Ricks. "Bremer Criticizes Troop Levels." *The Washington Post* (October 5, 2004).

Wriston, Walter B. "Bits, Bytes, and Diplomacy." *Foreign Affairs* (September/October 1997).

———. *The Twilight of Sovereignty: How the Information Revolution Is Transforming Our World*. New York: Charles Scribner's Sons, 1992.

Yardley, Jim. "A Hundred Cellphones Bloom, and Chinese Take to the Streets." *The New York Times* (April 25, 2005).

Yergin, Daniel. *The Prize: The Epic Quest for Oil, Money & Power*. New York: Simon & Schuster, 1991.

Yusopov, Felix. *Lost Splendor*. Trans. Ann Green and Nicolas Katkoff. New York: G. P. Putnam's Sons, 1953.

Zakaria, Fareed. "Rejecting the Next Bill Gates." *The Washington Post* (November 23, 2004).

Ziegler, Philip. *Omdurman*. New York: Dorset Press, 1987 [1973].

Zinsmeister, Karl. *Boots on the Ground: A Month with the 82nd Airborne in the Battle for Iraq*. New York: St. Martin's Press, 2003.

———. *Dawn over Baghdad: How the U.S. Military Is Using Bullets and Ballots to Remake Iraq*. San Francisco: Encounter Books, 2004.

Zucchino, David. "At the Front: In Iraq, where danger is constant, bases offer troops a taste of home." *Los Angeles Times* (March 27, 2005).

———. "Iraq's Swift Defeat Blamed on Leaders." *Los Angeles Times* (August 11, 2003).

———. *Thunder Run: The Armored Strike to Capture Baghdad*. New York: Atlantic Monthly Press, 2004.

Zulfo, 'Ismat Hasan. *Karari: The Sudanese Account of the Battle of Omdurman*. Trans. Peter Clark. London: Frederick Warne, 1980.

NOTES

For full citations, see the Bibliography.

AUTHOR'S NOTE

Army organization www.army.mil/organization/unitdiagram.html; www.globalsecurity.org/military/intro/org.htm.

PROLOGUE

xv "*War is both king of all*" Freeman, *Ancilla to the Pre-Socratic Philosophers*, 28.

"*Moral considerations account for three-quarters*" Napoleon, *Correspondance de Napoleon*, XVII:472.

"*99 percent of victory*" From an official British government memorandum that Fuller wrote in 1919. Cited in Fuller, *Armament and History*, 30.

1 *Physical description of Charles VIII* David Abulafia, "Introduction," in Abulafia, *The French Descent into Renaissance Italy*, 16.

Twitched and stuttered Bradford, *Cesare Borgia*, 37.

"*No great value*" Quoted in Potter, *The New Cambridge Modern History*, I:292–93.

Charles's lack of literacy Potter, *The New Cambridge Modern History*, I:293.

2 *Size of Charles's army* Hale, *War and Society in the Renaissance*, 62, gives the figure as 28,000. Of these, he writes, 13,000 were cavalry, 15,000 infantry. Other sources give slightly different numbers. The French scholar Ferdinand Lot estimated Charles VIII's army at 22,000 to 27,200. See Lynn, *Giant of the Grand Siècle*, 39.

Only one man died Felix Gilbert, "Machiavelli," in Paret, *Makers of Modern Strategy*, 21.

First standing army Porter, *War and the Rise of the State*, 31; John A. Lynn, "The Pattern of Army Growth, 1445–1945," in Lynn, *Tools of War*, 1.

4 *Three rounds in a day* Kelly, *Gunpowder*, 60.

Corned powder Brodie, *From Crossbow to H-Bomb*, 52, claims that corned powder was not discovered until 1520, but a more recent and more definitive investigation (Hall, *Weapons and Warfare in Renaissance Europe*, 88) found that "corned powders are mentioned often in fifteenth-century accounts and inventories." For more on corned powder, see Chase, *Firearms*, 61.

The shape of artillery for 350 years John A. Lynn, "Preface," in Lynn, *Tools of War*, viii, writes that "if a gunner from the artillery train of Charles VIII in 1494 had been transported to the battlefield of Gettysburg in 1863, he could have served a muzzle-loading bronze Napoleon cannon with much the same technique as that he employed to fire his primitive culverin." McNeill, *The Pursuit of Power*, 88, writes, "In essence the siege gun design developed in France and Burgundy between 1465 and 1477 lasted until the 1840s, with only marginal improvement." Likewise, the Brodies, in *From Crossbow to H-Bomb*, 42–43, write, "There was no radical change in the design of cannon from the time of the first cast-bronze tube of the late fifteenth century to about the middle of the nineteenth century."

Improvements in artillery Simon Pepper, "Castles and cannon in the Naples campaign," in Abulafia, *The French Descent into Renaissance Italy*, 263–64; Duffy, *Siege Warfare*, 8–11; Nicolle, *Fornovo 1495*, 18–19.

Fall of Mordano Cecil H. Clough, "The Romagna campaign of 1494," in Abulafia, *The French Descent into Renaissance Italy*, 211.

5 *Fall of Monte San Giovanni* Simon Pepper, "Castles and cannon in the Naples campaign," in Abulafia, *The French Descent into Renaissance Italy*, 271–72.

The French brought a much handier engine Francesco Guicciardini cited in Simon Pepper, "Castles and cannon in the Naples campaign," in Abulafia, *The French Descent into Renaissance Italy*, 263. A slightly different translation is offered in Guicciardini, *The History of Italy*, 50–51.

Population of Naples Manchester, *World Lit Only by Fire*, 47.

"Beneath a canopy of gold cloth" Durant, *The Story of Civilization*, V:612.

Cracked his head David Abulafia, "Introduction," in Abulafia, *The French Descent into Renaissance Italy*, 24.

Syphilis Arnaldi, *Italy and Its Invaders*, 136–37. The term "syphilis" was not coined until 1530.

French suffered for not having adopted muskets McNeill, *The Pursuit of Power*, 94.

6 *Impact of invasion on Italy* McNeill, *The Pursuit of Power*, 89, notes, "In Europe the major effect of the new weaponry was to dwarf the Italian city-states and to reduce other small sovereignties to triviality."

Italians' reputation for being poor soldiers Arnaldi, *Italy and Its Invaders*, 137–38.

"When war broke out" and "The French came upon this like a sudden tempest" Parker, *The Military Revolution*, 159–60. For a slightly different translation see Guicciardini, *The History of Florence*, 89.

INTRODUCTION

8 *Revolution* The *Oxford English Dictionary*, 2nd ed. (1989), defines it as "an instance of great change or alteration in affairs or in some particular thing."

Defining revolutions For instance, the futurists Alvin and Heidi Toffler argue that there have been only three revolutions (or waves, as they call them): agricultural, industrial, and information. See Toffler, *War and Anti-War*. The economists Chris Freeman and Francisco Louca are convinced there have been five waves just since 1800: "The Age of Cotton, Iron and Water Power"; "The Age of

Iron Railways, Steam Power and Mechanization"; "The Age of Steel, Heavy Engineering, and Electrification"; "The Great Depression and the Age of Oil, Automobiles, Motorization, and Mass Production"; and the emerging "Age of Information and Communication Technology." See Freeman and Louca, *As Time Goes By*. The military strategist William Lind thinks there have been four generations of war since 1648: The first generation was line and column tactics; the second was mass firepower; the third was maneuver warfare; and the fourth is guerrilla warfare and terrorism. See Lind, "Understanding Fourth Generation War." Another military strategist, Andrew Krepinevich, posits ten revolutions since the fourteenth century (not counting the current RMA): Revolutions in artillery; sail and shot; fortresses; gunpowder; Napoleonic tactics and organization; land warfare in the mid-nineteenth century; naval warfare with the coming of steel; mechanization, aviation, and information in the interwar years; and nuclear weapons. See Krepinevich, "Cavalry to Computer." Other writers have different ideas. Much of this is a semantic and fairly arbitrary dispute over how to classify trends that are generally agreed upon.

Some economists and historians deny altogether that any revolutions have occurred, arguing that it is inaccurate to refer to the "Industrial Revolution" because all change is the result of slow, gradual improvement. For a defense of the use of "revolutionary" terminology, see Landes, *The Wealth and Poverty of Nations*, 186–99, and Mokyr, *The Lever of Riches*, 12–13, 81–82, 289–92. Mokyr argues that while there have been many "small, incremental" microinventions throughout history, the less common development of radically new macroinventions has been far more important "than their small number would suggest" (291): "There have, and probably always will be, large and discrete changes in technology that sweep the world off its feet. . . . [W]ithout novel and radical departures, the continuous process of improving and refining existing techniques would run into diminishing returns and eventually peter out" (12–13).

8 *Soviet origins of "military technical revolution"* See FitzGerald, "The Impact of the Military-Technical Revolution on Russian Military Affairs"; Trulock, Hines, and Herr, "Soviet Military Thought in Transition."

RMA Stephen Peter Rosen, a Harvard political scientist, credits Andrew Marshall, longtime head of the Pentagon's Office of Net Assessment, with coining the term. Rosen, "The Future of War and the American Military."

Origins of "RMA" See Owens and Offley, *Lifting the Fog of War*, 82–85; Krepinevich, "The Military-Technical Revolution"; Williamson Murray and MacGregor Knox, "Thinking about revolutions in warfare," in Knox and Murray, *The Dynamics of Military Revolutions*, 1–14.

Current RMA There is an ever-growing literature on the subject. Among books and monographs, see Freedman, *The Revolution in Strategic Affairs;* Friedman, *The Future of War*; Owens and Offley, *Lifting the Fog of War*; Odom, *America's Military Revolution*; Qiao and Wang, *Unrestricted Warfare*; O'Hanlon, *Technological Change and the Future of Warfare*; Rip and Hasik, *The Precision Revolution*; Vickers and Martinage, *The Revolution in War.*

Histories of military revolutions The seminal work is Roberts's *The Military Revolution* (1956). More recent volumes include Parker, *The Military Revolution,* and Rogers, *The Military Revolution Debate.* Revolutions are also covered in more general military histories such as Keegan, *A History of Warfare*; Van Creveld, *Technology and War;* Brodie, *From Crossbow to H-Bomb*; and McNeill, *The Pursuit of Power.*

Comparative studies Knox and Murray, *The Dynamics of Military Revolutions*; Goldman and Eliason, *The Diffusion of Military Technology and Ideas*; Pierce,

Warfighting and Disruptive Technologies; Rosen, *Winning the Next War.* For a short, accessible overview see Krepinevich, "Cavalry to Computer."

9 *Nontechnological factors* Lynn, "The Evolution of Army Style in the Modern West," argues that nontechnological factors were actually the most important. He divides Western military history since the year 800 into stages according to how armies were raised—"feudal, medieval-stipendiary, aggregate-contract, state-commission, popular-conscript, mass-reserve, and volunteer-technical." These changes were obviously real and important but they were also closely linked to technical developments, as Lynn concedes (p. 509): "Weapons such as the flintlock musket/bayonet combination changed the composition and structure of infantry battalions at the end of the seventeenth century, while contemporary weapons systems demand better-educated soldiers, and therefore influence recruitment and training. Advances in transport sometimes affected armies even more profoundly; the railroad made the deployment and maintenance of mass armies feasible in the nineteenth century and went a long way towards compelling armies to adopt general staffs after the model pioneered by the Prussians in the early 1800s. Improvements in communications exerted a similarly significant impact in the late nineteenth and twentieth centuries."

"Two dangerous determinisms" Hedrick, *Tools of Empire*, 10.

10 *Paradigm shift* The seminal work is Kuhn, *The Structure of Scientific Revolutions.*

Not mutually contradictory For instance, Diamond's work may help to explain why Western culture took shape in ancient Greece, which Hanson and Landes argue accounts for Europe's domination of the world.

PART I: THE GUNPOWDER REVOLUTION

THE RISE OF THE GUNPOWDER AGE

17 *"We all know"* From a statement made by three members of Parliament, including the prime minister, Sir Robert Walpole, in *History and Proceedings of the House of Commons*, IX:44.

19 *The dividing line* See, e.g., Oman, *A History of the Art of War in the Sixteenth Century*, 30: ". . . in the history of the Art of War, the break between the Middle Ages and modern times can be fixed very definitely at the start of Charles VIII of France on his great expedition of 1494."

"Do these brutes imagine . . . ?" Erasmus, *Pilgrimages*, 53–54.

20 Europeans usually lost: Lynn, *Battle*, 23–27.

Landmass controlled in 1450 Geoffrey Parker, "From the House of Orange to the House of Bush: 400 Years of 'Revolutions in Military Affairs,'" in Lynn (ed.), *Acta of the XXVIIIth Congress of the International Commission of Military History.*

Landmass controlled in 1800 and 1914 Headrick, *Tools of Empire*, 3.

Introduction of paper Mokyr, *The Lever of Riches*, 41.

21 *Number of books* Rice and Grafton, *Foundations of Early Modern Europe*, 7, is the source for the 6 million figure. Estimates vary. Johnson in *The Renaissance*, 21, gives the figure as 9 million. The figure of 15 million to 20 million comes from Roberts, *Penguin History of the World*, 523.

Stirrups White, *Medieval Technology and Social Change*, 1–39.

Infantry revolution Clifford J. Rogers, "Military Revolutions in the Hundred Years War," in Rogers (ed.), *The Military Revolution Debate*, 55–93.

Longbow This represented a considerable advance in range and penetrating power over the shorter compound bow that had been in use since antiquity.

22 *Origins of guns* This summary is based on recent archaeological evidence indicating that the Chinese developed firearms earlier than previously thought—hundreds of years before Europeans did. For the older view (which held that the Chinese developed gunpowder first but created firearms at about the same time as Europeans did), see McNeill, *Pursuit of Power*, 38–39. For the latest evidence, see Chase, *Firearms*, 1, 30–33, 58–61 (my sentence beginning "If one of the essential characteristics of modernity . . ." is a paraphrase of ibid, 31); Parker, *The Cambridge Illustrated History of Warfare*, 106; Hall, *Weapons and Warfare*, ch. 2.

"So many brave and valiant men" Quoted in Rice and Grafton, *Foundations of Early Modern Europe*, 15.

"Instrument sent from hell" Quoted in Chase, *Firearms*, 59.

Cut off hands Delbruck, *History of the Art of War*, IV:31.

Why Europe pulled ahead In addition to the arguments cited here, a new theory has emerged. According to Chase, *Firearms*, western Europe and Japan were able to concentrate on firearms development because, unlike most of Asia and eastern Europe, they were not forced to combat marauding nomadic hordes. Early firearms were too cumbersome to fight the Mongols' fast-moving light cavalry and hence languished in areas where they were the dominant threat, but they proved more useful in the siege warfare that dominated western Europe and Japan in the early modern period.

Hungarian gunmaker Cipolla, *Guns, Sails, and Empires*, 93–94.

23 *"No walls exist"* Quoted in Parker, *The Military Revolution*, 10.

Trace italienne John A. Lynn, "The Trace Italienne and the Growth of Armies: The French Case," in Rogers (ed.), *The Military Revolution Debate*, 169–99. McNeill, *The Pursuit of Power*, 91, notes that the trace italienne "put a very effective obstacle in the way of the political consolidation of Europe into a single imperial entity at almost the same time that such a possibility become conceivable, thanks to the extraordinary collection of territories that the Hapsburg heir, Charles V of Ghent, acquired between 1516 and 1521."

"Money, more money . . ." Quoted in Rice and Grafton, *Foundations of Early Modern Europe*, 118; Hale, *War and Society in Renaissance Europe*, 232.

24 *"War made the state"* Tilly, *Formation of National States*, 42. McNeill, *The Pursuit of Power*, 117, makes a similar point about a "self-sustaining feedback loop."

Size of Cortés's force Thomas, *Conquest*, 150–152. Aztec population: ibid, 609–14. Size of Pizarro's force and population of Inca empire: Hudson, *Peru: A Country Study*, 8.

25 *Habsburg bid for mastery* Kennedy, in *The Rise and Fall of the Great Powers*, ch. 2, offers a superb summary. For a fuller account, see Kamen, *Empire*.

"If my own son was a heretic" Quoted in Hanson, *The Confident Hope of a Miracle*, 22.

Between 1480 and 1700 Tallett, *War and Society in Early Modern Europe*, 13. The next sentence, about the protracted conflicts, is a paraphrase of ibid.

State entered French vernacular Porter, *War and the Rise of the State*, 44.

CHAPTER 1: SAIL AND SHOT

26 *July 29* In 1582 Pope Gregory XIII discarded the old Julian calendar named after Julius Caesar. Spain immediately accepted the revised, Gregorian calendar; England did not. Its calendar remained ten days behind the rest of Europe until 1752. Thus, what the Spanish called July 29 was known to the English as July 19. All dates here are given in the New Style.

Drake at bowls The earliest reference to the English commanders at bowls came in a 1624 pamphlet, and it made no mention of Drake. See F. Fernandez Armesto, "Armada Myths: The Formative Phase," in Gallagher and Cruickshank, *God's Obvious Design*, 20; Kelsey, *Sir Francis Drake*, 321–22.

27 "*Ocean groaning under their weight*" From a contemporary chronicle, quoted in Padfield, *Maritime Supremacy*, 38.

28 *2,000 cannonballs fired* Hanson, *The Confident Hope of a Miracle*, 251.

"*We durst not adventure*" Howard to Walsingham, July 31, 1588, in Usherwood, *The Great Enterprise*, 100, and Laughton, *Defeat of the Spanish Armada*, I:288.

First major battle There had been smaller battles between sailing ships armed with artillery during the inconclusive Baltic War of 1563–1570, pitting Sweden against Denmark and the free city of Lübeck, as well as a clash in 1582 off the Azores in which a Spanish-Portuguese fleet defeated the Portuguese pretender Dom Antonio. At Lepanto in 1571, the Holy League of Spain, the Papal States, and Venice used ship-borne artillery to defeat the Ottoman navy. But both fleets were composed of oared galleys and some galleases, not pure sailing ships. Guilmartin, *Galleons and Galleys*, 99–102, 126–51; Martin and Parker, *The Spanish Armada*, 72.

Events of July 29–31 For Medina Sidonia's account, see Duke of Medina Sidonia to Philip II, August 21, 1588, in Usherwood, *The Great Enterprise*, 124–125. Howard's description may be found in Howard to Privy Council, August 1588, in ibid, 113–14. For another contemporary English account see "A Relation of Proceedings," in Laughton, *The Defeat of the Spanish Armada*, I:6–8. This description is also based on Howarth, *The Voyage of the Armada*, 119–25; Mattingly, *The Defeat of the Spanish Armada*, 231–43; Martin and Parker, *The Spanish Armada*, 10–30, 139–49.

30 "*The growth of Athenian power*" Thucydides, *History of the Peloponnesian War*, 49.

"*Twenty times greater*" Pedro Salazar de Mendoza, quoted in Kamen, *Empire*, 305.

31 *More than forty years* Philip II did not become king until the abdication of his father, Charles V, in 1555–56. But he had been regent of Spain since 1543, when he was just sixteen.

Six months of peace Parker, *Grand Strategy of Philip II*, 2.

Philip's reading "Of the 42 books in the king's bedside bookcase all but one (a serious historical tome) concerned the Christian faith," according to Parker and Martin, *Spanish Armada*, 84

"*Ten whole bodies*" Kamen, *Philip of Spain*, 189.

"*The same thing*" Quoted in Hanson, *The Confident Hope of a Miracle*, 22–23.

Bureaucrat King Hanson, *The Confident Hope of a Miracle*, 22.

Philip II The definitive biography is Kamen, *Philip of Spain*. This description also draws on excellent character sketches in Parker, *Philip II*; Parker, *Grand*

Strategy of Philip II; Parker and Martin, *Spanish Armada*, 82–88; Dunn, *The Age of Religious Wars*, 16–18.

Ducats At the time there were approximately four ducats to the pound sterling. Parker, *Philip II*, 2.

32 *Drake* There are numerous biographies. The best recent one is Kelsey, *Sir Francis Drake*. The physical description of Drake is from 226–27; his harshness is noted at 271. The emerald-studded crown and diamond cross are mentioned at 217.

33 *"Great terror of our enemies"* Drake to John Foxe, April 27, 1587, in Usherwood, *The Great Enterprise*, 40.

"Singed the King of Spain's beard" Quoted in Hanson, *The Confident Hope of a Miracle*, 79.

Medina Sidonia The only modern English-language biography is Pierson, *Commander of the Armada*.

"I should not give a good account" Medina Sidonia to Juan d'Idiaquez, secretary to Philip II, February 16, 1588, in Usherwood, *The Great Enterprise*, 65.

34 *Artillery* I. A. A. Thompson, "Spanish Armada Gun Procurement and Policy," in Gallagher and Cruickshank, *God's Obvious Design*, 83.

Arquebusiers and musketeers Martin and Parker, *Spanish Armada*, 19.

"If we come to close quarters" Quoted in Mattingly, *Defeat of the Armada*, 191–92, and Rodger, *The Safeguard of the Sea*, 259. Philip II was also aware of this problem. He told Medina Sidonia that his aim must be to bring the enemy "to close quarters and grapple with him," but he offered no suggestions on how to achieve this elusive goal. From Phillip's instructions to Medina Sidonia, April 1, 1588, in Hume, *Calendar of Letters and State Papers*, 247.

35 *Spanish ships* C. Martin, "The Ships of the Spanish Armada," in Gallagher and Cruickshank, *God's Obvious Design*, 41–69.

"Scurviest ship" and "dreadfully slow" Medina Sidonia to Philip II, July 30, 1588, in Hume, *Calendar of Letters and State Papers*, 356.

"Honourable terms" Medina Sidonia to Philip II, June 24, 1588, in Hume, *Calendar of Letters and State Papers*, 318.

36 *A tenth of Philip's revenues* Rodger, *Safeguard of the Sea*, 248.

Growth of English naval administration The best account is Rodger, *Safeguard of the Sea*, chs. 17, 23.

Emergence of caravels and carracks Mokyr, *The Lever of Riches*, 46.

37 *Hawkins coat of arms* Hanson, *The Confident Hope of a Miracle*, 24.

Blueprints Hanson, *The Confident Hope of a Miracle*, 177.

Galleon design Loades, *Tudor Navy*; Rodger, *Safeguard of the Sea*, ch. 16; Cipolla, *Guns, Sails and Empires*, 75–89; Guilmartin, *Galleons and Galleys*, 158–64; Block, *To Harness the Wind*, ch. 4.

Size of Navy Royal Martin and Parker, *The Spanish Armada*, 40.

38 *Truck carriages* Martin and Parker, *The Spanish Armada*, 32–33, 198–200; Guilmartin, *Galleons and Galleys*, 68–69.

English cannons Cipolla, *Guns, Sails and Empires*, 36–46; Martin and Parker, *Spanish Armada*, 277.

Two hundred to three hundred yards Guilmartin, *Galleons and Galleys*, 70.

"*Your own judgment*" Quoted in Rodger, *Safeguard of the Sea*, 263.

39 *Sailors per ton* Martin and Parker, *The Spanish Armada*, 38.

Speaking a dozen different languages Hanson, *The Confident Hope of a Miracle*, 108. For two hundred Englishmen, see ibid, 115.

"*Worthier ships than these*" Howard to Burghley, February 29, 1588, in Laughton, *The Defeat of the Spanish Armada*, I:85.

40 "*Hot fight*" Ibid, 116.

41 "*Had not embarked*" Quoted in Martin and Parker, *Spanish Armada*, 168.

42 "*So tremendous was the fire*" Statement by the Purser Pedro Coco Calderon, September 24, 1588, in Hume (ed.), *Calendar of Letters and State Papers*, 444.

"*Within speech of one another*" Quoted in Padfield, *Maritime Supremacy*, 51.

Account of the San Felipe Statement of the Purser Pedro Coco Calderon, September 24, 1588, in Hume (ed.), *Calendar of Letters and State Papers*, 444.

"*Their force is wonderful great*" Howard to Walsingham, August 8, 1588, in Usherwood, *The Great Enterprise*, 105.

43 "*Winds and tide were against us*" Medina Sidonia to King Philip II, August 21, 1588, in Usherwood, *The Great Enterprise*, 133.

"*Just enough being served*" Medina Sidonia to King Philip II, August 21, 1588, in Hume (ed.), *Calendar of Letters and State Papers*, 394.

Blasphemy forbidden In his initial instructions to Medina Sidonia, issued on April 1, 1588, Philip II wrote: "In the first place, as all victories are the gifts of God Almighty, and the cause we champion is so exclusively His, we may fairly look for His aid and favour, unless by our sins we render ourselves unworthy. You will therefore have to exercise special care that such cause of offence shall be avoided on the Armada, and especially that there shall be no sort of blasphemy." In Hume (ed.), *Calendar of Letters and State Papers*, 246.

44 "*To apprehend and execute*" Quoted in Martin and Parker, *Spanish Armada*, 216.

La Trinidad Valencera The story of its crew is related in the "Statement of Juan de Nova and of Francisco de Borja," in Hume (ed.), *Calendar of Letters and State Papers*, 507–508.

Six thousand shipwrecked, 750 survived Figures from Higueras and San Pio, "Irish Shipwrecks of the Great Armada," in Gallagher and Cruickshank, *God's Obvious Design*, 158.

"*Terrible hardships*" *and* "*cruel deaths*" From the letter of Francisco de Cuellar in Gallagher and Cruickshank, *God's Obvious Design*, 223–47.

"*Foul and beastly*" "Their ships are kept foul and beastly, like hog-sties and sheep-cots in comparison with ours," Sir William Monson wrote. Quoted in Martin and Parker, *Spanish Armada*, 39.

"*The troubles and miseries*" Medina Sidonia to King Philip II, September 23, 1588, in Hume (ed.), *Calendar of Letters and State Papers*, 432.

45 "*I sent my ships*" Quoted in Mattingly, *Defeat of the Spanish Armada*, 325.

One-third more firepower Martin and Parker, *Spanish Armada*, 185.

Spanish crews left their guns Martin and Parker, *Spanish Armada*, 191–92.

Three or four times more proficient Hanson, *The Confident Hope of a Miracle*, 182.

"At ranges" Mattingly, *Defeat of the Spanish Armada*, 267.

46 *"To go within musket shot"* Lord Howard to Privy Council, August 1588, in Usherwood, *The Great Enterprise*, 116.

Why the Armada lost This analysis draws heavily on the groundbreaking work of Martin and Parker, *Spanish Armada*, ch. 11.

Victorian historians Quoted in Fernandez-Armesto, "Armada Myths," 24.

"Not really an English victory" See ibid.

47 *Population figures* Jan de Vries, "Population," in Brady, Oberman, and Tracy (eds.), *Handbook of European History*, 13.

48 *Geographical advantages* The seminal work is Mahan, *Influence of Sea Power*, 25–44. Kennedy's *The Rise and Fall of the Great Powers*, 89, notes that during the wars of 1689–97 and 1702–14, France allocated less than 10 percent of its total spending on the navy and more than 57 percent on the army. Britain, by contrast, spent 35 percent of its budget on the navy and 40 percent on the army.

"Seafaring and trade begets merchants" Padfield, *Maritime Supremacy*, 3.

Naval power and states For the role of the Royal Navy in the development of English democracy, see Porter, *War and the Rise of the State*, 48–49, 103, and Tallett, *War and Society in Early-Modern Europe*, 210–11, 213–14.

49 *"No amount of money"* Rodger, *Safeguard of the Sea*, 431.

CHAPTER 2: MISSILE AND MUSCLE

52 *"A scene of horrors"* Schiller, *History of Thirty Years' War*, 158.

Numbers of soldiers Sources vary, with the range for the Imperial army being 31,000 to 40,000, for the Swedes 20,000 to 25,000, and for the Saxons 15,000 to 18,000. It is doubtful that even the commanders had an exact count.

Sweetmeats E. A. Beller, "The Thirty Years War," in Cooper (ed.), *The New Cambridge Modern History*, IV:314.

54 *One round every two minutes* Parker, *The Military Revolution*, 18.

56 *"Something so majestic"* Quoted in Ahlund, *Gustav Adolf*, 112. Weigley, in *Age of Battles*, 32, makes the point that Gustavus did not fit the classic image of the fighting man.

Breitenfeld Quotes from Monro are from *Monro, His Expedition*, 189–95. This account of the battle is also based on the following sources: Weigley, *The Age of Battles*, 3–23; Roberts, *Gustavus Adolphus*, Vol. 2, 250–53, 262–64; Wedgwood, *Thirty Years War*, 296–301; Schiller, *Thirty Years' War*, 169–75; Fuller, *A Military History of the Western World*, 49–64; Parker, *Thirty Years' War*, 125–26; Jones, *The Art of War*, 232–37; Dodge, *Gustavus Adolphus*, 257–71; Delbruck, *History of the Art of War*, IV: 202–7; Parker, *Cambridge Illustrated History of Warfare*, 156–58.

57 *Infantry* An alternative explanation is that "infantry" comes from "infant," referring to small foot soldiers. See Barzun, *From Dawn to Decadence*, 104.

Spanish tercios For the background, see Oman, *History of the Art of War*, 51–61; McNeill, *Pursuit of Power*, 94–95.

58 *"A slug of some weight"* O'Connell, *Soul of the Sword*, 120.

"Service of a matchlock" McNeill, *Keeping Together in Time*, 187.

Wheel lock O'Connell, *Soul of the Sword*, 121.

59 *Manual of arms* See De Gheyn, *Exercise of Arms*, for an English translation. Parker, "From the House of Orange to the House of Bush," makes the point that this was the first illustrated how-to book.

Schola Militaris Parker, *Cambridge Illustrated History of Warfare*, 156.

60 *Muscular bonding and drill* The seminal work is McNeill, *Keeping Together in Time*.

Japanese volley fire Parker, *The Military Revolution*, 140–42; Parker, "From the House of Orange to the House of Bush." According to Parker, there was no evidence of Japanese influence on European warfare.

Dutch drill sergeants in demand Parker, "From the House of Orange to the House of Bush."

Dutch reforms This summary is based on Parker, *The Military Revolution*, 16–23; Parker, "Military Revolutions, Past and Present"; McNeill, *Pursuit of Power*, 125–43; McNeill, *Keeping Together in Time*; Gunter Rothenberg, "Maurice of Nassau, Gustavus Adolphus, Raimondo Montecuccoli, and the 'Military Revolution' of the Seventeenth Century," in Paret (ed.), *Makers of Modern Strategy*, 32–45; Weigley, *Age of Battles*, 5–14; Delbruck, *History of the Art of War*, IV: 155–71; Oman, *History of the Art of War*, 568–69.

Gustavus His Swedish name was Gustav Adolf but he was more popularly known by his Latinized name, Gustavus Adolphus.

61 *"Half-isolated"* Roberts, *Gustavus Adolphus*, 24.

"If we were all as cold" Wedgwood, *Thirty Years War*, 273. Dodge, *Gustavus Adolphus*, 400, and Ahlund, *Gustav Adolf*, 97, give slightly different versions of this anecdote.

62 *"At least two generations ahead"* Ralston, *Importing the European Army*, 10.

"We serve our master honestly" Sir James Turner quoted in Parker, *The Military Revolution*, 52.

"No gallows" Roberts, *Gustavus Adolphus*, II:244.

63 *Gustavus's innovations* Brzezinski, *The Army of Gustavus Adolphus*, argues that the Swedish king was not the originator of shortened pikes, lighter muskets, and some other innovations attributed to him by other historians. Early seventeenth-century documents are sufficiently murky and incomplete that it is impossible to render a definitive judgment. Even if Gustavus did not create all these things, however, he was the first to put them together in an effective way.

"He thought nothing well done" Monro, *Monro, His Expedition*, 147. Another soldier said of him: "He was not content to be commander-in-chief, he must needs be captain, subaltern, engineer, gunner and private—in short, everything." Quoted in Ahlund, *Gustav Adolf*, 121.

Armies growing tenfold Parker, *The Military Revolution*, 1, and Tallett, *War and Society*, 9. Roberts's seminal article, "The Military Revolution" (see Rogers [ed.], *The Military Revolution Debate*, 13–35), suggested that the growth in army size was caused by Dutch/Swedish tactical reforms, but Parker offers the likelier explanations cited here. See Parker, "The Military Revolution—a Myth?" in *The Military Revolution Debate*, 44–48. In *Firearms*, 75, Chase emphasizes the growing manpower pool created by the use of firearms.

64 *Gustavus Adolphus and his reforms* This section relies primarily upon Michael Roberts' superb two-volume biography, *Gustavus Adolphus*, and his one-volume revision, *Gustavus Adolphus and the Rise of Sweden* ("war must pay for itself" may be found at 122). Two other biographies are helpful: Ahlund, *Gustav Adolf*, and Dodge, *Gustavus Adolphus*. In addition, there is an excellent character sketch in Wedgwood, *Thirty Years War*, 270–72.

Close to a major river Van Creveld, *Supplying War*, 10, notes that "the strategic mobility of seventeeth-century armies was severely limited by the course of the rivers."

66 *"The custom of the time"* Schiller, *Thirty Years' War*, 245. In *Monro, His Expedition*, 131, Monro relates the tale of a sergeant's wife aboard a troop transport, "who without the help of any women was delivered of a boy, which all the time of the tempest she carefully did preserve, and being come ashore, the next day, she marched near four English miles, with that in her arms, which was in her belly the night before."

Logistics Van Creveld, *Supplying War*, 5–17. He concludes, at 7, that "the armies of this period were probably the worst supplied in history."

"Where's your king?" Wedgwood, *Thirty Years War*, 315. Dodge, in *Gustavus Adolphus*, 315, gives a different account of this conversation which, in any case, may be apocryphal.

67 *"Tall, thin, forbidding"* Wedgwood, *Thirty Years War*, 171.

Wallenstein The most comprehensive modern biography is Mann, *Wallenstein*. See also the character sketches in Parker, *Thirty Years' War*, 138; Schiller, *Thirty Years' War*, 130–31, 223–37; Wedgwood, *Thirty Years' War*, 170–74, 347; Singer, "The Ultimate Military Entrepreneur."

68 *"Oftimes an army"* Monro, *Monro, His Expedition*, 185.

"Now in very truth" Quoted in Roberts, *Gustavus Adolphus*, II:748.

69 *Wallenstein's ditch* The ditch is still visible today, but the windmills are long gone. Based on the author's visit.

72 *Wallenstein's decision to retreat* Parker, *The Thirty Years' War*, 131.

"With such fury" Cited in Mann, *Wallenstein*, 655.

Lützen This account is based on the following sources: Monro, *Monro, His Expedition*, 294–97; Weigley, *The Age of Battles*, 32–36; Roberts, *Gustavus Adolphus*, II: 253, 763–72; Wedgwood, *Thirty Years War*, 324–27; Schiller, *Thirty Years' War*, 256–65; Mann, *Wallenstein*, 648–65; Fuller, *A Military History of the Western World*, 67–71; Parker, *Thirty Years' War*, 130–32; Jones, *The Art of War*, 241–43; Dodge, *Gustavus Adolphus*, 386–97; Delbruck, *History of the Art of War*, IV: 207–9.

"While the world stands" Monro, *Monro, His Expedition*, 301.

Napoleon's tribute Roberts, *Gustavus Adolphus and the Rise of Sweden*, 114.

Skeptics See, e.g., David A. Parrott, "Strategy and Tactics in the Thirty Years' War," in Rogers (ed.), *The Military Revolution Debate*, 227–51. Parrott claims, at 235, that Gustavus's "tactical development was virtually irrelevant to the battles after the Swedish invasion of Germany." He explains away the outcome of Breitenfeld as "a typical case of a massively confident, 'professional' army pitted against a force that was demoralized and inexperienced." Leaving aside the question of whether the confidence and professionalism of the Swedish army should be attributed to Gustavus's reforms (what other cause could there be?), this attempt to demean the quality of the losing Imperial army is not terribly convincing. In fact, on the preceding page (234), Parrott himself concedes that Tilly's veterans had the same "military spirit" as the Swedes, which grew out of "a long series of wars and victories." Actually, the Imperialists had won far more significant victories by 1631 than had the Swedes, so, if anything, they should have been more spirited than their enemies. In fact, at Breitenfeld, Tilly's army routed the hapless Saxons from the field. It is hard to avoid the conclusion that Gustavus's decisive triumph against such an experienced force, even after he was abandoned by his allies, was due in large measure to his tactical innovations.

74 *Russia imitated the Swedish model* Richard Hellie, "Warfare, Changing Military Technology, and the Evolution of Muscovite Society," in Lynn, *Tools of War*, 94–97.

Military revolution Parker uses "field warfare revolution" in *The Military Revolution*, 16. John F. Guilmartin prefers "combined arms" revolution. See his "The Military Revolution: Origins," in Rogers (ed.), *The Military Revolution Debate*, 307. This summary of the different revolutions is inspired by (though not identical to) Krepinevich's "Cavalry to Computer" and the work of Clifford Rogers cited below.

The term *military revolution* comes from a seminal 1955 lecture by Michael Roberts, a historian of Sweden, titled, "The Military Revolution, 1560–1660." It sparked decades of scholarly debate. A good summary may be found in Rogers, *The Military Revolution Debate*, which includes Roberts's original essay as well as numerous articles critiquing his arguments. Few critics deny that the introduction of gunpowder weapons constituted a revolution. Most critiques concern two smaller matters: (1) The timing of the revolution, with many historians suggesting that the "Roberts century" was less critical than developments that preceded it and followed it. (2) The location of the revolution, with some historians suggesting that Sweden and the Dutch Republic were no more important as incubators of reform than Italy, France, Spain, Switzerland, or other countries.

The most convincing synthesis was made by Clifford Rogers in "Military Revolutions of the Hundred Years War," in *The Military Revolution Debate*, 55–93. Rogers argues for a "punctuated equilibrium model" borrowed from evolutionary biology in which "a whole series of revolutions"—from the infantry revolution of the fourteenth century to the Dutch/Swedish revolution of the seventeenth century—"combined to create the Western military superiority of the eighteenth century." This model subsequently was endorsed by the dean of early modern military historians, Geoffrey Parker, in the second edition of *The Military Revolution*, 158. (Parker also extended the idea of revolutions to naval and colonial warfare, which Roberts had not discussed.)

A few historians such as Jeremy Black argue, "it is unclear that the concept of a revolution lasting a century or longer is really helpful" (Black, *European Warfare, 1494–1660*, 45). But most experts agree with Geoffrey Parker that the use of "revolution" is justified. See Parker's *The Military Revolution*, 157–76, for a convincing refutation of the naysayers.

"*The Swedish pattern*" Rothenberg, "Maurice of Nassau," 48. On the Dutch-style reforms being brought to North America, Parker writes in "Military Revolutions, Past and Present," 4–5: "every governor of Virginia between 1610 and 1621 had served as an officer under Maurice. Indeed the Virginia Company in London actively recruited Englishmen serving in the Dutch army. Many leaders of other English colonies had also served in the Dutch army, including Thomas Dudley in the Caribbean and Miles Standish at Plymouth (where men began drilling in the Dutch fashion as soon as they disembarked from the *Mayflower* in 1620). A decade later, in Massachusetts Bay, John Winthrop entrusted each of the colony's four militia companies to the veterans of the Dutch army whom he had persuaded to join him."

Spending on war These are all wartime figures; the percentages were slightly lower in peacetime. Parker, *The Military Revolution*, 62, 148. Porter gives a slightly lower figure for Frederick in *War and the Rise of the State*, 116.

Definition of bureaucracy www.m-w.com. For its first use, see the Oxford English Dictionary.

75 *"The success of this type"* Porter, *War and the Rise of the State*, 66. Max Weber made the essential point earlier: "The discipline of the army gives birth to all discipline . . . military discipline is the ideal model for the modern capitalist factory, as it was for the ancient plantation." Max Weber, "The Meaning of Discipline," in *From Max Weber: Essays in Sociology*, 261, cited in ibid, 21.

Prussia and Sweden For this comparison I am indebted to Ralston, *Importing the European Army*, 10. McNeill, *The Pursuit of Power*, 154, notes that the Great Elector's buildup was a reaction to Swedish depredations in the Thirty Years' War.

Political implications of the military revolution Kennedy offers a nuanced and succinct account in *The Rise and Fall of the Great Powers*, 31–72. See also Porter, *War and the Rise of the State*, 23–104.

76 *French reforms* See Lynn, *Giant of the Grand Siècle*. While Lynn argues that the French independently evolved some of the innovations associated with Maurice of Nassau and Gustavus Adolphus, "[n]onetheless," he writes (474), "there is no question that the French turned to the Dutch, and later to the Swedes, to improve their own battalion formations and tactics."

CHAPTER 3: FLINTLOCKS AND FORBEARANCE

78 *Kailna River* It was incorrectly rendered in Wellington's dispatches—and in most subsequent histories—as the Kaitna. See Cooper, *The Anglo-Maratha Campaigns*, 372.

79 *A former East India Company sergeant* This was Anthony Pohlman. Most accounts of Assaye report that he was leading his Maratha regiment in person, but according to Cooper, *The Anglo-Maratha Campaigns*, 276–78, by this time Pohlman had resigned from the Maratha service because, as a British subject, he refused to fight British troops. Pohlman had been imprisoned by the Marathas prior to the battle and not released until February 1804. He then rejoined the East India Company as a lieutenant colonel.

British and Maratha order of battle Thorn, *Memoir of the War in India*, 274. One modern account, Roy, "Military Synthesis in South Asia," gives a much higher figure for Wellesley's army—10,500 men—but this is at odds with contemporary accounts (such as Thorn's memoir) as well as other secondary accounts, e.g., Cooper, *The Anglo-Maratha Campaigns*, 101.

"No time to be lost" Wellesley to Lieutenant General Stuart, September 24, 1803, in Gurwood (ed.), *The Dispatches of Field Marshal the Duke of Wellington*, II: 327.

March to Assaye Wellesley's correspondence contains numerous details, especially about provisions and preparations. See Gurwood (ed.), *Dispatches of Field Marshall the Duke of Wellington*, I:357–II:345. In addition, there are detailed accounts in Weller, *Wellington in India*, 170–75; Blackiston, *Twelve Years of Military Adventure*; Thorn, *Memoir of the War in India*, 270–73; Longford, *Wellington: The Years of the Sword*, 89–90; and a lightly fictionalized but largely accurate account in Cornwell, *Sharpe's Triumph*, 171–97. The details on British soldiers are also drawn from Brett-James, *Life in Wellington's Army*, especially chs. 1, 2, 4, and 5; Holmes, *Redcoat*; and Fosten, *The Thin Red Line* (which contains handsome full-color plates of uniforms). For details on Maratha encampments, see Sen, *Military System of the Marathas*, 127–39.

80 *Advances in shipbuilding and navigation* For an argument emphasizing their importance in European colonial conquests, see Raudzens, "Military Revolution or Maritime Evolution?"

"*Curry made to eat*" Correa, *Three Voyages of Vasco da Gama*, 331. For more details on da Gama's victory, see Guilmartin, *Galleons and Galleys*, 77–83; Padfield, *Tide of Empires*, 42–53; John F. Guilmartin, Jr., "The Military Revolution: Origins and First Tests Abroad," in Rogers (ed.), *The Military Revolution Debate*, 314–15. In a footnote to the latter article, Guilmartin explains (331–32) that "Dhow is a generic European term for the large, lateen rigged Arab sailing vessels of the Indian Ocean. . . . Prau, or prahu, from Malay for boat or vessel, is a similarly generic term for relatively small local Indian and Indonesian sailing craft."

Chinese navy A good overview may be found in McNeill, *Pursuit of Power*, 44–47; Chase, *Firearms*, 49–52; and Landes, *Wealth and Poverty of Nations*, 93–98. The reasons usually cited for China's isolationist turn are the Confucian court's scorn for merchants, their jealousy of Cheng Ho, and their desire to concentrate their limited resources on defending their northern frontier against nomads.

82 "*Barely sufficient force*" Cornelis Nieuwenroode, quoted in Cipolla, *Guns, Sails and Empire*, 139.

More than four million warriors Rosen, *Societies and Military Power*, 130–31.

"*This matter should be dropped*" Cipolla, *Guns, Sails and Empire*, 141.

83 *Percentage of landmass conquered by Europeans* Headrick, *Tools of Empire*, 3.

What happened? I am indebted to Parker's *The Military Revolution*, 117, for the formulation of this problem. This section relies heavily on his invaluable work.

"*Our business is trade*" Parker, *The Military Revolution*, 133.

"*Trade cannot be maintained*" The famous remark by Jan Pieterszoon Coen, proconsul of Batavia (today's Jakarta), is quoted in numerous places, e.g., Parker, *The Military Revolution*, 132, and Landes, *The Wealth and Poverty of Nations*, 143.

European expansion This is a highly abbreviated and necessarily simplified version of a long and complex tale. Debate still rages about many matters, such as why the conquistadors prevailed. I have relied on the astute summaries in Landes, *The Wealth and Poverty of Nations*, 60–185; Kennedy, *The Rise and Fall of the Great Powers*, 16–30; Scammell, *The First Imperial Age*; Roberts, *Penguin History of the World*, 607–32; Parker, *The Military Revolution*, 92–145; Cipolla, *Guns, Sails and Empire*, 90–148; John F. Guilmartin, Jr., "The Military Revolution: Origins and First Tests Abroad," in Rogers (ed.), *The Military Revolution Debate*, 299–333.

84 *The division* McNeill, *Pursuit of Power*, 162–63; Weigley, *Age of Battles*, 264.

85 *Artillery* Hughes, *Firepower*, 35–38.

"*Accuracy of measurement*" Hughes, *Firepower*, 10. The author was a retired major general in the British army.

86 *Accuracy* In 1814 a British army colonel wrote that the Brown Bess "will strike a figure of a man at 80 yards. . . . It may even be at 100, but a soldier must be

very unfortunate indeed who shall be wounded by a common musket at 150 yards provided his antagonist aims at him: and as to firing at a man at 200 yards with a common musket you may as well fire at the moon and have the same hope of hitting your object." Quoted in Hughes, *Firepower*, 26.

Range of M-16A2 Chambers, T*he Oxford Companion to American Military History*, 405. Note that 600 yards is the effective range for hitting a "point" target (e.g., one enemy soldier); the range for an "area" target (e.g., a line of enemy soldiers) is 870 yards. See: www.armystudyguide.com/m16/studyguide.htm; www.fas.org/man/dod-101/sys/land/m16.htm.

"Ferocious aggressiveness" and "passive disdain" O'Connell, *Of Arms and Men*, 119.

87 *U.S. casualties in Gulf War* Based on Pentagon statistics: http://web1.whs.osd.mil/mmid/casualty/GWSUM.pdf. Note that in addition to 147 battle deaths there were also 235 "non-hostile" U.S. deaths from accident, disease, and suicide.

Casualties The figure for Zorndorf comes from Weigley, *Age of Battles*, 189. The figure of 1 in 15 comes from Parker, *The Military Revolution*, 148.

"To the right and left" Lieutenant William Grattan commenting on a visit to a temporary hospital in 1812. Quoted in Brett-James, *Life in Wellington's Army*, 264–65.

Treatment of sick and wounded In addition to ibid, see Gabriel and Metz, *A History of Military Medicine*, II:97–141; Holmes, *Redcoat*, 95–98, 249–62.

88 *"Scum of the earth"* This famous quote may be found in, among other places, David Gates, "The Transformation of the Army 1783–1815," in Chandler (ed.), *Oxford History of the British Army*, 145.

Eighteenth-century warfare There are good summaries in Weigley, *Age of Battles*; O'Connell, *Of Arms and Men*, 148–66; O'Connell, *Soul of the Sword*, 157–80; Holmes, *Redcoat*; Archer, *The Art of War in the Western World*, 252–319; McNeill, *Pursuit of Power*, 144–84.

"Battle culture of forbearance" John A. Lynn, "Forging the Western army in seventeenth-century France," in Knox and Murray, *Dynamics of Military Revolutions*, 45; Lynn, *Battle*, 147–49.

Brutality of Western warfare This is a theme developed by Hanson in *The Western Way of War* and *Carnage and Culture*.

Ripping hearts out of chests Prescott, *History of the Conquest of Mexico*, 46–49.

"The most gallant Apaches" Hanson, *Carnage and Culture*, 9.

Warriors' mind-set See Keegan, *History of Warfare*, and Davis, *Carnage and Culture*, both of which describe Aztec warfare.

89 *"It is not valour"* Quoted in Parker, *The Military Revolution*, 128.

Controlled 75 percent of India Cooper, *The Anglo-Maratha Campaigns*, 8.

Rabble loyal to jagirdars Roy, "Military Synthesis in South Asia," 657.

"Collections of skilled individuals" Lynn, *Battle*, 175.

Origin of sepoy www.merriam-webster.com, or *Merriam-Webster's Collegiate Dictionary*, 11th ed.

90 *Marathas* See Stewart, *The Marathas*; Sen, *The Military System of the Marathas*; Smith, *Oxford History of India*, 410–15, 434–42. For a newer account, which challenges some of the assumptions of these earlier works, see Cooper, *The Anglo-Maratha Campaigns*. Cooper is intent on disproving the conventional

wisdom that Indian armies were not as effective as those of the West, which he dismisses as "cultural imperialism" (285) propagated by "ethnocentric twentieth-century Western military analysts" (310). He does not think the British were superior in "technology, discipline, or drill" (284), which raises the question of why, if the Marathas were so advanced, they felt compelled to hire Europeans to run their best military units. No European armies, by contrast, were run by Indians. Even Cooper, for all his revisionism, has to concede a basic difference between the Western and Maratha ways of war: While Europeans valued victory above all and were willing to suffer heavy losses to achieve it, the Marathas were more focused on grabbing loot and were not as willing to die "for a transitory political cause" (58). The practical result was that British soldiers would slug it out in close-quarters infantry combat while their Maratha counterparts preferred to rely on cavalry raids that were much less likely to achieve decisive results.

"Veterans who had ever been victorious" Smith, *A Sketch of the Rise, Progress and Termination of the Regular Corps*, 45.

"To be born in a stable" Corrigan, *Wellington: A Military Life*, 3.

91 *"Fit only for powder"* Corrigan, *Wellington: A Military Life*, 6. Hibbert, in *Wellington: A Personal History*, 3, renders this quote as "My ugly boy Arthur is food for powder and nothing more," which makes less sense.

"Whores and lice" Quoted in Holmes, *Redcoat*, 144.

History of Woolwich and Sandhurst For a brief overview, see the official website: http://www.atra.mod.uk/atra/rmas/history/index.htm.

92 *"What one ought not to do"* The Duke of Wellington, October 12, 1839, quoted in Stanhope, *Notes of Conversations with the Duke of Wellington*, 182.

Wellesley There are many biographies. The one I found most helpful was Corrigan, *Wellington: A Military Life*. See also Langford, *Wellington: Years of the Sword*; Hibbert, *Wellington: A Personal History*; Weller, *Wellington in India*; Roberts, *Napoleon & Wellington*.

Bengal population James, *Raj*, 30.

93 *Tiger mauling officer* James, *Raj*, 68.

94 *Background to the 1803 Anglo-Maratha War* Cooper, *The Anglo-Maratha Campaigns*, 70–81.

"Having never troubled" Welsh, *Military Reminiscences*, 156.

"These English are a strange people" Quoted in Welsh, *Military Reminiscences*, 164. For a modern account of the siege, see Cooper, *The Anglo-Maratha Campaigns*, 87–92.

96 *"Dash at the first fellows"* Wellesley to Colonel Stevenson, August 20, 1803, in Gurwood (ed.), *Dispatches of Field Marshal the Duke of Wellington*, II:219.

"Common sense" The Duke of Wellington, January 23, 1834, quoted in Stanhope, *Notes of Conversations with the Duke of Wellington*, 49.

"Rather an ugly beginning" Blackiston, *Twelve Years of Military Adventure*, I:161.

"Most steady manner" Blackiston, *Twelve Years of Military Adventure*, I:161.

"Vomited forth death" Blackiston, *Twelve Years of Military Adventure*, I:164.

97 *"No sooner"* Thorn, *Memoir of the War in India*, 277.

"Most terrible cannonade" Wellesley, "Memorandum on the Battle of Assye," in Gurwood (ed.), *Dispatches of Field Marshal the Duke of Wellington*, II:331.

98 *"With great slaughter"* Thorn, *Memoir of the War in India*, 276.

"Had their numerous cavalry" Letter from an unknown cavalry officer, September 24, 1803, in Bennell (ed.), *The Maratha War Papers*, 290.

"I had lost all my friends" Brett-James, *Wellington at War*, 82. Colebrooke, *Life of the Honorable Monstuart Elphinstone*, I:68, noted the dead officer and the wounded officer next to Wellesley.

Gerald Lake For a summary of his campaign, see Lynn, *Battle*, 167–70.

99 *Of which he was proudest* In 1844, George William Chad recorded the following conversation with the Duke of Wellington: "'Pray Duke what is the best thing you ever did in the fighting line?' The Duke was silent for about 10 seconds & then answered 'Assaye.' He did not add a word." Wellington, *Conversations of the First Duke of Wellington*, 20.

Assaye Wellington's own accounts may be found in Gurwood (ed.), *Dispatches of Field Marshal the Duke of Wellington*, II:323–54; Wellington (ed.), *Supplementary Dispatches*, 179–90; and Gleig (ed.), *Life of Major General Sir Thomas Munro*, I:347–52. Three of Wellesley's officers left detailed accounts: Blackiston, *Twelve Years of Military Adventure*, I:154–82; Letter from Colin Campbell in Wellington (ed.), *Supplementary Dispatches*, IV:184–87; and Colebrooke, *Life of the Honorable Monstuart Elphinstone*, I:63–73. Another British officer wrote a semiofficial account: Thorn, *Memoir of the War in India*, 273–87. The best secondary source is Weller's *Wellington in India*, 170–94, which has a detailed reconstruction. Also helpful are: Cooper, *The Anglo-Maratha Campaigns*, 96–116; Corrigan, *Wellington: A Military Life*, 73–77; Pitre, *Second Anglo-Maratha War*, 64–78; Fortescue, *History of the British Army*, IV:21–35; and Duff, *History of the Mahrattas*, II:276–79.

Ten times the revenue Landes, *Wealth and Poverty of Nations*, 156.

100 *"So good"* Arthur Wellesley to Henry Wellesley, October 3, 1802, in Gurwood (ed.), *Dispatches of Field Marshal the Duke of Wellington*, II:371.

Ganimi Kava Roy, "Military Synthesis in South Asia," 659.

Wellesley's provisions Roy, "Military Synthesis in South Asia," 687–88.

Europeans per battalion Weller, *Wellington in India*, 291, 298; Pitre, *Second Anglo-Maratha War*, 25; Cooper, *The Anglo-Maratha Campaigns*, 226. One officer who worked for Sindia wrote: "had they acquired more European officers, and been furnished with European arms, they could have been as good soldiers as any that have been formed from the natives of the country, by European skill and example." Smith, *A Sketch of the Rise, Progress and Termination of the Regular Corps*, 10.

Defections of European officers Cooper, *The Anglo-Maratha Campaigns*, 213–83, argues this was the key factor explaining the Maratha defeat, without examining why the Marathas had almost no native officers. He also asserts that the defection of the Europeans represented an "insurmountable" intelligence windfall for the British (306), ignoring the fact that Wellesley did not have good intelligence before the Battle of Assaye. He was surprised by the location, size, and firepower of the Maratha forces. Whatever accounted for the British triumph, it was not knowledge of the enemy.

101 *"Rather than providing a defense"* Ralston, *Importing the European Army*, 179.

102 *The cost of military modernization* This argument is based primarily on Ralston, *Importing the European Army*, cf. 173–80, as well as Rosen, *Societies and Military Power*; Sen, *Military System of the Marathas*; Hanson, *Carnage and Culture*; Kennedy, *Rise and Fall of the Great Powers*, 30.

THE CONSEQUENCES OF THE GUNPOWDER REVOLUTION

103 *Differing opinions* Michael Roberts, who first applied the term "military revolution" to early modern Europe, argued that it lasted from 1550 to 1660. Geoffrey Parker, the foremost expert on the subject today, suggests that the sixteenth century was of "central importance." See Parker, "In Defense of *The Military Revolution*," in Rogers, *The Military Revolution Debate*, 341. Other historians have suggested other periods; see Rogers, *The Military Revolution Debate*.

104 *Manufacturing output* Kennedy, *Rise and Fall of the Great Powers*, 149.

105 *Innovation stopped* These dates are from Mokyr, *The Lever of Riches*, 81. Other historians give slightly different dates.

Why southern Europe fell behind Many reasons have been cited for Spain's decline, in particular, including imperial "overstretch," too many wars, and poorly developed state finances and administrative apparatus. Landes stresses intellectual stagnation growing out of the Inquisition. In *Wealth and Poverty of Nations*, 133–36, he points out that by 1600 the Iberians had lost their onetime leadership in navigational theory and practice: "The crypto-Jewish scientists, mathematicians and physicians of yesteryear fled; no dissenters appeared to take their place."

Cultures and innovation Mokyr, *The Lever of Riches*, does a good job of analyzing and synthesizing various theories. But his conclusion is a concession of defeat: "It is hard to think of conditions that would be either necessary or sufficient for a high level of technological creativity. A variety of social, economic, and political factors enter into the equation to create a favorable climate for technological progress. At the same time, a favorable climate may itself be insufficient if the new technological ideas fail to arise" (299).

PART II: THE FIRST INDUSTRIAL REVOLUTION

THE RISE OF THE INDUSTRIAL AGE

107 *"The gun can be discharged"* Ellis, *Social History of the Machine Gun*, 29.

109 *Great Exhibition* The most helpful book I found was Beaver's *The Crystal Palace*, which is amply illustrated (for the queen's reaction, see 40). I also consulted Hobhouse, *The Crystal Palace and the Great Exhibition*; Davis, *The Great Exhibition*; Morris, *Heaven's Command*, 198; Schama, *A History of Britain*, III:142–50.

110 *"Metropolis of the commercial system"* James Kay, 1832, quoted in King and Timmins, *Making Sense of the Industrial Revolution*, 15.

British population Figures are from Palmer, Colton and Kramer, *A History of the Modern World*, 432; Johnson, *The Birth of the Modern*, 203; Gash, *Aristocracy and People*, 15.

"Nothing can be conceived more grand" Quoted in Gash, *Aristocracy and People*, 10.

Peterloo massacre On August 16, 1819, more than sixty thousand people gathered at St. Peter's Field in Manchester to protest for radical reform. The militia tried to arrest one of the speakers. In the ensuing melee eleven people were killed and four hundred injured. The incident became known satirically as "Peterloo" in reference to the great military victory at Waterloo four years before. Gash, *Aristocracy and People*, 94–95.

Average wages Kennedy, *Rise and Fall of the Great Powers*, 147.

"Swifter than a bird flies" Quoted in Faith, *The World the Railways Made*, 34, and Johnson, *The Birth of the Modern*, 191.

"Rail travel at high speed" Dr. Dionysus Lardner, professor of natural philosophy at University College, London, quoted in Nairn, *Engines That Move Markets*, 1.

Five thousand miles Gash, *Aristocracy and People*, 320.

111 *Submarine cables* The story of the first one laid across the Atlantic is deftly told in Gordon's *A Thread Across the Ocean.*

Life changed E.g., Mokyr writes in *The Lever of Riches*, 81, "In two centuries [meaning apparently 1750–1950] daily life changed more than it had in the 7,000 years before."

Why the industrial revolution developed in Britain Mokyr, *The Lever of Riches*, addresses this question in ch. 10.

112 *Steam engine* For a brief overview see Mokyr, *The Lever of Riches*, 84–90.

Power looms Figures are from Ashton, *The Industrial Revolution*, 61, and McClellan and Dorn, *Science and Technology in World History*, 284.

Iron production Figures are from McClellan and Dorn, *Science and Technology in World History*, 287; Headrick, *Tools of Empire*, 146; Kennedy, *Rise and Fall of the Great Powers*, 130; Landes, *The Unbound Prometheus*, 86.

Global track mileage Faith, *The World the Railways Made*, 28.

"Routinization of discovery" Landes, *Wealth and Poverty of Nations*, 204.

Patents issued Ashton, *The Industrial Revolution*, 72; MacLeod, *Inventing the Industrial Revolution*, 150.

Science after 1850 Mokyr, *The Lever of Riches*, 113.

113 *British share* Roberts, *Penguin History of the World*, 685; Landes, *The Unbound Prometheus*, 124.

World manufacturing output Kennedy, *Rise and Fall of the Great Powers*, 148–49.

114 *"The hardware of war"* Weigley, *Age of Battles*, xvi.

Mobilization Frederick the Great had 4 percent of his population under arms (two hundred thousand men), leading Porter to conclude in *War and the Rise of the State*, 115, "In purely quantitative terms, the mass army had arrived in Europe even before the French Revolution." For conscription under Peter the Great, see Lynn, "Army Style in the Modern West," 528, who notes that "the Russians had adopted a key element of the popular-conscript army long before the French Revolution."

Army size Kennedy's *The Rise and Fall of the Great Powers*, 99, contains a useful chart on army size from 1690 to 1814. According to John A. Lynn, "The Pattern of Army Growth, 1445–1945," in Lynn, *Tools of War*, 4, Louis XIV mobilized around 400,000 men for his wars, while the French contingent of Napoleon's army stood at no more than 600,000 to 650,000 at one time. He notes (at 6) that in percentage terms the biggest jump in French army size occurred in the mid-seventeenth century (700 to 1,000 percent), not in the late eighteenth century (150 to 250 percent). Duffy, in *The Military Experience in the Age of Reason*, 18, writes that in the eighteenth century it was generally estimated that no general could effectively control more than 50,000 men in the field; this figure did not rise much by the early nineteenth century. For the size of the *Grand Armée*, see McNeill, *The Pursuit of Power*, 200. Van Creveld, *Supplying War*, ch. 2, examines the logistical challenges of Napoleonic warfare.

Napoleonic innovations For a summary, see Geoffrey L. Herrera and Thomas G. Mahnken, "Military Diffusion in Nineteenth-Century Europe," in Goldman and Eliason, *The Diffusion of Military Technology and Ideas*, 241.

Napoleon A modern historian writes: "He added nothing particularly new here; he just handled his forces more effectively, more often, than any of his enemies. . . . The *corps d'armée*, for example, was not Napoleon's idea. Nor was the *ordre mixtre* that he used so successfully in battle. There was no real advance in weaponry during the Napoleonic period." Citino, *Quest for Decisive Victory*, 5, 8. A soldier-scholar agrees: "Napoleon introduced neither a new weapon nor a new tactical system." Dupuy, *The Evolution of Weapons and Warfare*, 317. McNeill, The *Pursuit of Power*, 146, further notes that "Old Regime military institutions continued to regulate the French *levée en masse* of 1793. As a result, Napoleon's defeat in 1815 allowed the victorious powers to restore a plausible simulacrum of the Old Regime. The traditional military order did not begin to break up irretrievably until the 1840s, when new industrial techniques began to affect naval and military weaponry and organization in radical and fundamental ways." Even Peter Paret, who labels the Napoleonic period a "Revolution in War," suggests that the French Revolution simply "expanded the scope" of "innovations" that had already been under way: "Profound changes in military institutions and practice, some already firmly established under the Old Regime, others still tentative and experimental, were adopted by the Revolution, and developed further." Paret, "Napoleon and the Revolution in War," in Paret (ed.), *Makers of Modern Strategy*, 124.

CHAPTER 4: RIFLES AND RAILROADS

116 *Königgrätz* Also known as the Battle of Sadowa, after the name of a nearby village. The chapter title is inspired by Showalter's *Railroads and Rifles.*

Austria was bigger In 1865 Prussia had 19.2 million people, 216,000 active duty soldiers, and a defense budget of 5.9 million British pounds. Austria had 34.3 million people, 300,000 soldiers, and a defense budget of 9.1 million pounds. Statistics come from the University of Michigan's Correlates of War Project: www.umich.edu/~cowproj/dataset.html#Capabilities.

117 "*Odds are against the Prussians*" Engels, "Notes on the War in Germany, No. I." Another English writer wrote: "At the time I left England not only was public sympathy very strongly in favour of the Austrians, but the almost universal conviction was, that if France did not interfere with her, Prussia would inevitably be defeated." Dicey, *Battle-Fields of 1866*, 3.

"*An officer proposing such a plan*" Engels, "Notes on the War in Germany, No. IV."

"*A downpour*" and "*gray and cheerless*" Russell, "The Battle of Sadowa."

"*Extremely strong position*" Hughes (ed.), *Moltke on the Art of War*, 59.

Army strengths Different sources give slightly different figures. These numbers are from the Prussian General Staff's official history, *The Campaign of 1866*, 186–87.

119 "*The Austrian artillery shot exceedingly well*" Moltke to wife, July 4, 1866, in Moltke, *Moltke's Letters to His Wife*, 186.

"*Terrible . . . cannonade*" Unsigned editorial, *The Times*, July 11, 1866.

"*Felt we were in God's hands*" Quoted in Craig, *The Battle of Königgrätz*, 102.

"*Every minute more critical*" Prussian staff, *Campaign of 1866*, 233.

120 *"I know General Fransecky" and "Stand or die"* Craig, *The Battle of Koniggratz*, 109.

"Let's see you fight" Craig, *The Battle of Königgrätz*, 103.

"We are losing" Wawro, *The Austro-Prussian War*, 229.

"Your majesty will win today" Friedrich, *Blood and Iron*, 131; Craig, *The Battle of Königgrätz*, 110; Goerlitz, *History of the German General Staff*, 88.

121 *"Heartless, wooden, half-educated"* Quoted in Friedrich, *Blood and Iron*, 43.

Prussian reforms Kitchen, *The Cambridge Illustrated History of Germany*, 152–58; Schulze, *Germany: A New History*, 101–23; Holborn, *A History of Modern Germany*, 393–421, 434–61.

122 *"Support incompetent generals"* Dupuy, *A Genius for War*, 25.

"Plans conceived at leisure" Keegan, *The First World War*, 24.

Prussian General Staff The classic accounts are Craig, *The Politics of the Prussian Army*, 37–76; Goerlitz, *History of the German General Staff*; and Ritter, *The Sword and the Scepter*, I. Holborn, "The Prusso-German School," provides a good summary. Also helpful are: Bucholz, *Moltke and the German Wars*, 12–36; Dupuy, *A Genius for War*, 17–69; and Delbruck, *History of the Art of War*, IV:449–56.

123 *"Strict, even harsh"* Moltke, *Moltke: His Life and Character*, 19.

124 *Moltke's habits* Dressler, *Moltke in His Home*, 27–55.

"A mixture of reflection and shyness" Memoir of Marie Ballhorn, daughter of a cousin, in Moltke, *Essays, Speeches and Memoirs*, II: 143.

Moltke Aside from the memoirs cited previously, the best account of his life that I found was Friedrich, *Blood and Iron*. Also useful was Bucholz, *Moltke and the German Wars*, and Goerlitz, *History of the German General Staff*, 69–102.

125 *First railway* Schulze, *Germany: A New History*, 130.

Uses of railways See, e.g., Moltke's 1843 report, "Considerations in the Choice of Railway Routes," in Moltke, *Essays, Speeches and Memoirs*, 227–35. Railways were so new at that point that Moltke felt compelled to begin with a definition ("the railway is a way with tracks of strong cast-iron rails").

The American Civil War Showalter, "Information Capabilities and Military Revolutions," 233, writes: "In the summer of 1863, for example, a Confederacy reeling from the double disasters of Gettysburg and Vicksburg responded by transferring an entire army corps by rail from Virginia to Tennessee. The reinforcements enabled the near-destruction of the Union Army of the Cumberland at Chickamaugua and its subsequent blockading in Chattanooga. The situation was saved when the North shifted troops, again by rail, to raise the siege." In *The Wars of German Unification*, 165, Showalter notes that, contrary to popular misconception, Moltke and the Prussian General Staff carefully studied the American Civil War.

126 *Prussian use of railways* Van Creveld, *Supplying War*, 78–79.

Standard time Showalter, "Information Capabilities and Military Revolutions," 235.

Railroads and telegraphs The best source remains Showalter's *Railroads and Rifles*, 19–70.

"No plan of operations" Quoted in Holborn, "The Prusso-German School," 289.

"An order shall contain" Quoted in Holborn, "The Prusso-German School," 291.

127 *"I want to sleep in Schleswig"* Bucholz, *Moltke and the German Wars*, 83.

Rate of misfires Percussion muskets misfired once in 166 shots; flintlocks misfired once in 6.5 shots. Featherstone, *Weapons and Equipment*, 15.

Most wounds caused by small arms Gabriel and Metz, *A History of Military Medicine*, II:181.

128 *Breech-loaders* Dreyse did not invent the modern breech-loading rifle. Two of the leading contenders for this achievement are Johannes Pauly, a Swiss gunmaker who in 1812 developed a breech-loading sporting gun, and the American gunsmith John Hall, who patented his rifle in 1811 while employed at the Harper's Ferry Armory. Pauly's invention was never adopted for military use. Hall's was slightly more successful: Although spurned by the U.S. Army infantry, American cavalrymen did carry Hall's carbine. See Hogg, *The Story of the Gun*, 50–51, for Pauly's claim, and Hallahan, *Misfire*, 52–83, for Hall's.

"A ten-thousand-man unit" Hallahan, *Misfire*, 128.

129 *Spencer and Sharps rifles* For details see Bruce, *Lincoln and the Tools of War*; Hallahan, *Misfire*; Perret, *Lincoln's War*, 144–55.

Needle gun Showalter, *Railroads and Rifles*, 77–125; O'Connell, *Soul of the Sword*, 194–200.

Artillery Figures on Austrian vs. Prussian cannons come from Craig, *Battle of Königgrätz*, 8. This also draws on Showalter, *Railroads and Rifles*, 138–217; Showalter, *The Wars of German Unification*, 101–2; O'Connell, *Soul of the Sword*, 200–6.

130 *86 percent* Dupuy, *The Evolution of Weapons and Warfare*, 171.

"Iron and blood" Friedrich, *Blood and Iron*, 108.

131 *"Germany is too small"* Friedrich, *Blood and Iron*, 105.

"The struggle will be terrific" Moltke to his brother Adolf, May 26, 1866, in Moltke, *Letters of Moltke to His Mother and Brothers*, 177.

Planning For Moltke's first memorandum on fighting Austria, written in 1860, see Moltke, *Strategy*, 4–19.

132 *"Great decisive battles"* Moltke to the king, April 14, 1866, in Moltke, *Strategy*, 58.

"Incalculable loss" Moltke to Stosch, April 24, 1866, in Moltke, *Strategy*, 73.

133 *"His Majesty commands"* Hughes (ed.), *Moltke on the Art of War*, 245.

Moltke issuing orders directly Order of the King in Cabinet of June 2, 1866, in Moltke, *Strategy*, 25.

"Who is General Moltke?" Wawro, *The Austro-Prussian War*, 232. Goerlitz, in *History of the Prussian General Staff*, 85–86, gives a slightly different version of this famous anecdote.

"Muddy boots" and "Professors in shoulder straps" Showalter, *The Wars of German Unification*, 98.

"Strong stomach" Wawro, *The Austro-Prussian War*, 26.

Baedeker Craig, *The Battle of Königgrätz*, 11.

134 *Austrian weakness* Kann, *A History of the Habsburg Empire*, 235–366; Kennedy, *Rise and Fall of the Great Powers*, 162–66.

"The wood and the road were plastered" Quoted in Wawro, *The Austro-Prussian War*, 129.

Opening battles They are described in great detail in the official Prussian General Staff history, *The Campaign of 1866*, 68–159.

135 *"Peace at any price"* Craig, *Battle of Königgrätz*, 80.

"Physically and morally broken" Wawro, *Austro-Prussian War*, 203. Wawro believes that Benedek was not planning a battle in front of Königgrätz, whereas Craig, in *Battle of Königgrätz*, credits the Austrian commander with more premeditation. Showalter, in *The Wars of German Unification*, 178–79, plausibly argues that Benedek wasn't thinking very hard but instead reacting based on emotions.

"If my old luck does not desert me" Quoted in Showalter, *The Wars of German Unification*, 179.

"Extremely strong position" Cited in Hughes (ed.), *Moltke on the Art of War*, 59.

136 *"War cannot be conducted"* Quoted in Hughes (ed.), *Moltke on the Art of War*, 77.

Inedible bread Showalter, *Railroads and Rifles*, 70.

"Surviving on potatoes" Craig, *Battle of Königgrätz*, 72.

"Two chocolate bonbons" Moltke to his wife, July 4, 1866, in Moltke, *Moltke's Letters to His Wife*, 188.

137 *"Your Royal Highness . . ."* Moltke to Second Army, July 2, 1866, in Moltke, *Strategy*, 55, and in Prussian General Staff, *Campaign of 1866*, 166.

"Good God!" "The Austrian Defeat."

138 *"Phantoms" and "yellow dogs"* Craig, *Battle of Königgrätz*, 133.

"The air was literally filled" Craig, *Battle of Königgrätz*, 145.

139 *"With loud cheers"* Hozier, *Seven Weeks' War*, 192.

"19 hours on the march" Prussian staff, *Campaign of 1866*, 294.

"Literally covered" Dicey, *Battle-fields of 1866*, 99.

140 *"Every cottage"* Hozier, *The Seven Weeks' War*, 194.

"My Emperor no longer has an army" Craig, *Battle of Königgrätz*, 165.

141 *"An army with a state"* Dupuy, *A Genius for War*, 16, attributes this wisecrack to an eighteenth-century Prussian official, Baron Friedrich Leopold von Schrotter. Other authors attribute it to the Marquis Honoré de Mirabeau, author of a 1788 book, *The Prussian Monarchy under Frederick the Great*.

"No one, of course, dreamed" Quoted in Hughes (ed.), *Moltke on the Art of War*, 61.

"It may be doubted" Engels, "Notes on the War in Germany, No. V."

142 *197,000 men and 55,000 horses* Prussian staff, *Campaign of 1866*, 24.

"Men of education" Dicey, *Battle-fields of 1866*, 238.

"More in earnest" Ibid, 17.

American Civil War For its impact on modern warfare, see Hagerman, *The American Civil War and the Origins of Modern Warfare*.

"Great successes" Cited in Hughes (ed.), *Moltke on the Art of War*, 263.

143 *Imitated spiked helmets* Lynn, "The Evolution of Army Style in the Modern West," 510.

Franco-Prussian War For the latest study, see Wawro, *The Franco-Prussian War.*

Rearming McNeill emphasizes the revolutionary impact of new production techniques in *Pursuit of Power*, 235.

144 *475 miles of trenches* Keegan, *The First World War*, 136.

145 *Offensive technology* Keegan, in *The First World War*, 22–23, stresses the lack of radio communications as being the biggest hindrance for attackers. For the German development of offensive doctrine near the end of World War I, see Lupfer, *The Dynamics of Doctrine.*

CHAPTER 5: MAXIM GUNS AND DUM DUMS

146 *"A mighty rumbling"* Burleigh, *Khartoum Campaign*, 148.

"Gigantic banners" Bennett, *Downfall of the Dervishes*, 162. Bennett called the Dervishes' garments *gibbehs*; the word has been changed here to the more common spelling (*jibbahs*) to avoid confusion.

Dervish A corruption of the Persian *darvish*, used to describe adherents to Muslim Sufi religious orders. It was not a designation used by the Mahdists, who called themselves the "ansar," or helpers. Holt, *Mahdist State*, 17.

147 *"Some burst high in the air"* Churchill, *The River War*, II:116.

"There is no God" Ziegler, *Omdurman*, 119.

Al Khalil struck Zulfo, *Karari*, 172.

"Wilder courage" Burleigh, *Khartoum Campaign*, 157.

148 *"The fire discipline"* "An officer," *Sudan Campaign*, 192.

"Cease fire" Ziegler, *Omdurman*, 133.

Dervishes wounded There is no exact count. Churchill estimated 2,000 dead and 2,000 wounded in the morning action. See *River War*, II:128. A modern Sudanese author puts the figure at 2,800 dead and 4,200 wounded. See Zulfo, *Karari*, 179.

"A fearful sight" Harrington and Sharf, *Omdurman 1898*, 144.

"Regular inferno" Ibid, 127.

"It was not a battle" Steevens, *With Kitchener to Khartum*, 264. Steevens was a *Daily Mail* correspondent.

149 *Cordite* The British government refused to pay Nobel royalties and he lost a patent-infringement lawsuit. Credit for cordite has gone to two British chemists, Sir Frederick Abel and Sir James Dewar; the latter is better known as the inventor of the thermos bottle and liquid hydrogen.

150 *Used in the Iraq War* Keegan, *The Iraq War*, 179.

Repeating rifles See Hogg, *Story of the Gun*, 57–81; O'Connell, *Soul of the Sword*, 183–200; Bruce, *Lincoln and the Tools of War;* Hallahan, *Misfire*; Featherstone, *Weapons and Equipment*, 11–34; Dynes, *The Lee.* I also found the following web sites helpful: www.geocities.com/lee_enfield_rifles/history.html; www.webpages.uidaho.edu/~stratton/history.htm; www.militaryrifles.com/britain/metford.htm.

"In 1882 I was in Vienna" Quoted in Ellis, *Social History of the Machine Gun*, 33–34.

151 *Machine guns* The best account of their development, upon which this section draws heavily, remains Ellis's brilliant book, *The Social History of the Machine Gun*. A more recent work is Smith, *Machine Gun*. For more on Maxim, see McCallum, *Blood Brothers*.

Military-industrial complex Manchester provides a vivid description of how the process worked in *Arms of Krupp*.

152 *"Who were the naked Matabele?"* Ellis, *Social History of the Machine Gun*, 91.

"Whatever happens" Belloc, *The Modern Traveler*, 41.

Portuguese repelled Tonga warriors Vandervoort, *Wars of Imperial Conquest in Africa*, 154.

Rorke's Drift For the best account see Morris, *Washing of the Spears*.

154 *Adowa* For a short summary see Vandervoort, *Wars of Imperial Conquest*, 156–66.

Abd el-Kader Vandervoort, *Wars of Imperial Conquest*, 60–62. Samori Touré: Ibid, 126–36.

155 *London to Cape Town* Ferguson, *Empire*, 168.

"Every river is laid open" Headrick, *Tools of Empire*, 17.

156 *Gunboats* Headrick, *Tools of Empire*, 17–58, 129–57; Kubicek, "The Role of Shallow-Draft Steamboats"; Vandervoort, *Wars of Imperial Conquest*, 51–53, 114.

Railroads Faith, *The World the Railways Made*, 144–74; Headrick, *Tools of Empire*, 180–204.

157 *"The telegraph saved India"* Quoted in Headrick, *The Invisible Weapon*, 52.

Cables Headrick, *Tools of Empire*, 157–64; Headrick, *The Invisible Weapon*, 28–115. In the latter work, at 94, Headrick notes that in 1908 Britain controlled 56.2 percent of the worldwide cable network, followed by the U.S. with 19.5 percent and France with 9.4 percent.

Medicine Headrick, *Tools of Empire*, 58–83. "77 percent . . . perished" is at 63; "first-year death rates . . . dropped" is at 70.

"Lower the cost" Headrick, *Tools of Empire*, 206.

158 *"World domination on the cheap"* Ferguson, *Empire*, 245. This is also the source for the figures on British troops and defense spending.

159 *"Formidable mixture"* Quoted in David Steel, "Lord Salisbury, the 'False Religion' of Islam, and the Reconquest of the Sudan," in Spiers (ed.), *Sudan: The Reconquest Reconsidered*, 15.

Sudan This account is based on Holt, *Mahdist State*; Zulfo, *Karari*; Pakenham, *Scramble for Africa*; Spiers (ed.), *Sudan: The Reconquest Reappraised*.

160 *"He stands"* Steevens, *With Kitchener to Khartum*, 45.

"Permanent encumbrance" Pollock, *Kitchener*, 213.

Kitchener The most recent and most complete biography is Pollock, *Kitchener*. For slightly older biographies, see Warner, *Kitchener*, and Royle, *Kitchener Enigma*.

"His precision is so inhumanly unerring" Steevens, *With Kitchener to Khartum*, 46.

161 *"The deadliest weapon"* Steevens, *With Kitchener to Khartum*, 22.

From four months to eleven days Churchill, *River War*, II:3.

"We aren't particular to a man or two" Pollock, *Kitchener*, 107.

163 *Worst possible strategy* For this point I am grateful to Van Creveld, *Technology and War*, 229–30. Zulfo explains why the Khalifa adopted this strategy in *Karari*, 108.

"Yellow brown pointed dome" Churchill, *River War*, II: 82.

Wingate No modern biography exists of this extraordinary soldier-administrator, who ran Sudan from 1899 to 1916. There is only his son's hagiographic work, *Wingate of the Sudan*.

"What God chooses" Zulfo, *Karari*, 153.

165 *Two thousand warriors* Zulfo, *Karari*, 196; Edward M. Spiers, "Campaigning Under Kitchener," in Spiers, *Sudan: The Reconquest Reappraised*, 70.

Charge of the 21st Lancers All quotations are from Churchill, *River War*, II:130–42.

166 *"Bravest of the brave"* Burleigh, *Khartoum Campaign*, 193. For Macdonald's life, see Royle, *Death Before Dishonour*.

167 *"A good dusting"* Ziegler, *Omdurman*, 178.

"Three men and a machine gun" Quoted in Porter, *War and the Rise of the State*, 149.

"Arms of science" Churchill, *River War*, II:164.

"Boers are not like the Soudanese" Pollock, *Kitchener*, 181.

168 *"To the verge of annihilation"* Quoted in Michael Howard, "Men Against Fire: The Doctrine of the Offensive in 1914," in Paret (ed.), *Makers of Modern Strategy*, 521.

Increase in firepower Figures are from Biddle, *Military Power*, 29.

Balkan Wars In these wars pitting Turkey against a coalition of Greece, Serbia, Bulgaria, and Montenegro, the Ottomans lost most of their Balkan possessions. For a succinct and intelligent overview of warfare between the Crimean War and World War I, see Citino, *Quest for Decisive Victory*, 1–141. Citino notes that the lessons of those years did not point exclusively toward the dominance of the defense: they also showed that armies on the offensive could prevail, as long as they were willing to endure high casualties.

CHAPTER 6: STEEL AND STEAM

170 *"Slept in a corner" and "reveille at dawn"* Novikoff-Priboy, *Tsushima*, 141.

Eighteen thousand miles and eight months Hough, *The Fleet That Had to Die*, 24.

Battle of the Yellow Sea Connaughton, *Rising Sun and Tumbling Bear*, 210–13.

"Yellow monkeys" Hough, *The Fleet That Had to Die*, xiii.

171 *Fleet set off October 15* Corbett, *Maritime Operations*, II:31.

Dogger Bank incident Pleshakov, *The Tsar's Last Armada*, 96–112.

Hamburg-America Line Pleshakov, *The Tsar's Last Armada*, 59–60.

Twelve thousand officers and crew Connaughton, *Rising Sun and Tumbling Bear*, 297.

"It was like a cemetery" Novikoff-Priboy, *Tsushima*, 58.

Forty-two ships Connaughton, *Rising Sun and Tumbling Bear*, 297.

"Only shame and dishonour" The surgeon of the *Izumrud*, quoted in Westwood, *Witnesses of Tsushima*, 71.

Rozhdestvensky His name is sometimes rendered in English as "Rozhestvensky" or "Rozhdestvenski."

172 *"Outwardly"* Novikoff-Priboy, *Tsushima*, 126.

Appearance of Russian ships Connaughton, *Rising Sun and Tumbling Bear*, 291.

Time when Japanese fleet spotted Corbett, *Maritime Operations*, II:223.

173 *Celebration aboard the* Suvorov Westwood, *Witnesses of Tsushima*, 164–65; Semenoff, *Battle of Tsu-shima*, 45–46; Hough, *The Fleet That Had to Die*, 160–61.

Voyage of the Second Pacific Squadron The best account remains Hough's *The Fleet That Would Not Die*. Pleshakov's more recent book, *The Tsar's Last Armada*, makes for challenging reading but does contain some new information from Russian archives. In addition to the eyewitness accounts already cited, I have also referred to the diaries kept by two Russian officers: Politovsky, *From Libau to Tsushima*, and Semenoff, *Rasplata*.

Demologos Sondhaus, *Naval Warfare*, 18–19.

174 *Paixhans* O'Connell, *Sacred Vessels*, 41.

Battle of Sinop Sondhaus, *Naval Warfare*, 58.

La Gloire Sondhaus, *Naval Warfare*, 73–74.

"Their Lordships feel" Quoted in McNeill, *Pursuit of Power*, 226.

HMS Warrior Sondhaus, *Naval Warfare*, 74–75.

Monitor and Merrimack These ships have been the subject of a voluminous literature. For the latest account (as of this writing) see Nelson, *Reign of Iron*.

175 *"Damn the torpedoes"* There is some question as to whether this is exactly what was said. See Love, *History of the U.S. Navy*, I:310.

Hunley Gray, *The Devil's Device*, 189.

Whitehead For his story, see Gray, *The Devil's Device*.

Torpedo ranges These figures for Whitehead torpedoes come from Gray, *Devil's Device*, 253. Other torpedoes had slightly different capabilities.

Torpedo boats and destroyers O'Connell, *Sacred Vessels*, 142–43.

Britain built sailing ships until the 1880s Gardiner, *Steam, Steel, and Shellfire*, 102–3.

176 *"Noiseless"* Capt. Reginald Bacon of the *Dreadnought*, quoted in Massie, *Dreadnought*, 475.

Development of steam engines For a summary see Gardiner, *Steam, Steel, and Shellfire*, 170–78.

Armor Padfield, *Battleship*, 148–49.

Target practice at fourteen thousand yards Sumida, *In Defense of Naval Supremacy*, 250.

Marksmanship Sumida, *In Defense of Naval Supremacy*, discusses the Royal Navy's failure to buy an accurate fire-control mechanism created by an inventor named Arthur Pollen.

177 *Majestic* vs. *Victory* Figures for *Victory* come from www.hms-victory.com/factsandfigures.htm, for *Majestic* from www.navalhistory.flixco.info/H/198098/8330/a0.htm; Gardiner (ed.), *Steam, Steel and Shellfire*, 117, 169; and Padfield, *Battleship*, 147–50.

Three and a half times as much metal The *Majestic*'s four main guns together could hurl 3,400 pounds of shells at a time; *Victory*'s biggest guns were thirty 32-pounders that collectively could hurl 960 pounds of metal at a time, though only half of that total could be directed against one target.

"A period of transition" Padfield, *Battleship*, 110.

Battle of Lissa For a summary, see Palmer, *Command at Sea*, 218–23.

178 *"The ram is fast supplanting the gun"* Quoted in Padfield, *Battleship*, 68.

Jeune École Sondhaus, *Naval Warfare*, 139–59. For more on submarines, see Parrish, *The Submarine: A History*.

Mahan's impact See, Shulman, *Navalism and the Emergence of American Sea Power*.

Nineteenth-century naval developments Unless otherwise specified, this summary is based on Sondhaus, *Naval Warfare, 1815–1914*; Padfield, *Battleship*; O'Connell, *Sacred Vessels*; Gardiner, *Steam, Steel and Shellfire*.

179 *Japan remained traditional* For the state of Japanese technology before the Meiji Restoration, see Morris Suzuki, *Technological Transformation of Japan*.

180 *Bombardment of Kagoshima* See Blond, *Admiral Togo*, 36–46; Bodley, *Admiral Togo*, 47–55; Hoyt, *Three Military Leaders*, 24.

Togo The account of Togo's life is based on Blond, *Admiral Togo*; Bodley, *Admiral Togo*; Falk, *Togo and the Rise of Japanese Sea Power*; Hoyt, *Three Military Leaders*, 17–76.

181 *"Rich Country, Strong Army"* For the origins of this slogan, see Samuels, *"Rich Nation, Strong Army,"* 34–42.

Japan's modernization This account is based on Buruma, *Inventing Japan*; Gordon, *A Modern History of Japan*; Morris Suzuki, *The Technological Transformation of Japan*; Ralston, *Importing the European Army*, ch. 6.

182 *"Nothing to be afraid of"* Quoted in Blond, *Admiral Togo*, 87. See also Bodley, *Admiral Togo*, 75, and Hoyt, *Three Military Leaders*, 38.

Sino-Japanese War Sondhaus, *Naval Warfare*, 169–73; Evans and Peattie, *Kaigun*, 41–51.

Russia vs. Japan Russia's population was 142 million; Japan's, 45.5 million. Russia produced 2.4 million tons annually of steel and iron; Japan, 40,000 tons. Russia spent 48 million British pounds on defense; Japan, 15 million pounds. Russia had 1.1 million men under arms; Japan, 214,000. Figures are from the University of Michigan's Correlates of War Project: www.umich.edu/~cowproj/dataset.html#Capabilities

Japanese naval strength Evans and Peattie, *Kaigun*, 64.

183 *Togo* A British correspondent who met him in 1904 described him as follows: "He is a short, well-built man with rather a slight stoop. . . . His is a kindly face, but it is was marked by lines of care, the result of the anxious watching and thought of the last six months. Although it might be the face of an ordinary, studious man, it indubitably impresses one. The eyes are brilliant and black, like those of all Japanese, and a slight pucker at the corners suggests humour. A small drooping nose shades a pursed-up mouth with the under lip slightly protruding. He has a large head, which is a good shape and shows strongly defined bumps, and the hair is thin and worn very short. A slight beard fringes the face and it is whitening on the chin, and the moustache is thin and black." Seppings-Wright, *With Togo*, 57.

"*Togo is lucky*" Quoted in Evans and Peattie, *Kaigun*, 82.

Russian reform Riasanovsky, *A History of Russia*, 391–403, 422–34; Jacob W. Kipp, "The Imperial Russian Navy, 1696–1900: The Ambiguous Legacy of Peter's 'Second Arm,'" in Kagan and Higham, *Tsarist Military History*, 151–81.

Three to four knots advantage Spector, *At War at Sea*, 14, says it was four knots; Connaughton, *Rising Sun and Tumbling Bear*, 328, says it was three knots.

184 *Four times the explosive power* Spector, *At War at Sea*, 18.

Fire twice as fast Spector, *At War at Sea*, 12.

"*Officers fraternise with the men*" Seppings Wright, *With Togo*, 104.

"*Rotten biscuits*" Novikoff-Priboy, quoted in Westwood, *Witnesses of Tsushima*, 60.

"*These noblemen's sons*" Quoted in Westwood, *Witnesses of Tsushima*, 59.

"*Convict-like appearance*" Quoted in Westwood, *Witnesses of Tsushima*, 66.

185 *Port Arthur attack* Connaughton, *Rising Sun and Tumbling Bear*, 44–51.

Two Russian battleships sank Connaughton, *Rising Sun and Tumbling Bear*, 60–61.

Two Japanese battleships sunk Connaughton, *Rising Sun and Tumbling Bear*, 62.

Battle of the Yellow Sea Evans and Peattie, *Kaigun*, 102–7.

Fall of Port Arthur Warner, *The Tide at Sunrise*, 472–79.

Vladivostok squadron Evans and Peattie, *Kaigun*, 107–10.

186 "*Piercing black eyes*" *and* "*pepper-and-salt beard*" Novikoff-Priboy, *Tsushima*, 20.

"*Tall, grave, virile*" Novikoff-Priboy, *Tsushima*, 131.

Abuse of officers Novikoff-Priboy, *Tsushima*, 132–34.

"*I wish I had such splendid admirals*" Hough, *The Fleet That Had to Die*, 19.

"*Fall to pieces*" Novikoff-Priboy, *Tsushima*, 23.

Rozhdestvensky This sketch is based on Pleshakov, *The Tsar's Last Armada*, 37–56; Westwood, *Witnesses of Tsushima*, 53–59; Novikoff-Priboy, *Tsushima*, 129–36; Connaughton, *Rising Sun and Tumbling Bear*, 293–95.

187 *Eight battleships* Since "battleship" has no strict definition, some authors count heavy cruisers and coastal defense ships as battleships and thus write that the Russians had twelve battleships, not eight. The classification employed here follows Corbett's *Maritime Operations*, Appendix E.1, an official study written for the British Admiralty.

127 guns vs. 92 Figures are from Evans and Peattie, *Kaigun*, 115.

Improving gunnery British naval attachés reported that prior to the Battle of Tsushima the Japanese fleet expended a year's peacetime allowance of ammunition in practice *every week*. See Spector, *At War at Sea*, 12.

Russian fleet sighted at 1:39 P.M. Corbett, *Maritime Operations*, II:238.

"*The fate of the empire*" Translations of this message vary slightly in their wording. Compare Spector, *At War at Sea*, 14, with Connaughton, *Rising Sun and Tumbling Bear*, 328–29, and Corbett, *Maritime Operations*, II:239. Togo was inspired by the message Nelson sent before the Battle of Trafalgar: "England expects that every man will do his duty."

188 *Turn ordered at 2:02 P.M.* Corbett, *Maritime Operations*, II:245.

189 "*How rash*" Quoted in Semenoff, *Battle of Tsu-shima*, 53–54.

Seven thousand yards Corbett, *Maritime Operations*, II:246.

Single line vs. double line As Captain Nicolas Klado, one of Rozhdestvensky's officers, explained, "A single column can wriggle like a snake, and change direction again and again. If there are two columns, when such movements are suddenly made, and in the heat of battle, collisions must result." Klado, *The Battle in the Sea of Japan*, 72.

Opened fire at 2:10 P.M. Corbett, *Maritime Operations*, II:246.

"*I had not only never witnessed*" Semenoff, *Battle of Tsu-shima*, 62–65.

Fell out of line at 2:40 P.M. Corbett, *Maritime Operations*, II:251.

Suvorov *sank at 7:30 P.M.* Corbett, *Maritime Operations*, II:291.

190 "*Making a hole*" Novikoff-Priboy, *Tsushima*, 193.

Sinking of the Oslyabya Corbett, *Maritime Operations*, II:253.

A few minutes after 6 P.M. According to Corbett, *Maritime Operations*, II:286, by 6:12 P.M. the *Mikasa* was within 6,500 yards of the *Borodino*, the maximum effective range.

Alexander III *sank at 6:55* Corbett, *Maritime Operations*, II:290.

"*The loss of the* Borodino" Quoted in Corbett, *Maritime Operations*, II:292. Corbett's account suggests that the sinking occurred around 7:30 P.M.

twenty-one destroyers, thirty-two torpedo boats Corbett, *Maritime Operations*, II:309.

191 *Japanese losses* According to Corbett, *Maritime Operations*, II:310, two Japanese torpedo boats were sunk by shell fire and one by collision, while three destroyers were disabled by collision and four by shell fire.

Loss of the Navarin Corbett, *Maritime Operations*, II:304.

Loss of the Sisoi Veliky Corbett, *Maritime Operations*, II:308.

Remnants of Russian fleet Corbett, *Maritime Operations*, II:315.

twenty-seven warships Corbett, *Maritime Operations*, II:320.

"*Both officers and men lost hope*" Quoted in Corbett, *Maritime Operations*, II:320.

Casualty figures Corbett, *Maritime Operations*, II:333.

"*The success attained by Admiral Togo*" "Comment," *Harper's Weekly*. The other headlines are from the newspapers cited.

192 *Treaty of Portsmouth* For an account of how it was negotiated, see Morris, *Theodore Rex*, 386–414.

"*The war with Japan*" Yusopov, *Lost Splendor*, 51.

Nehru realized that Asians could defeat Europeans Mahbubani, *Beyond the Age of Innocence*, 13.

193 "*The long day*" Mattingly, *Defeat of the Spanish Armada*, 238.

Dreadnought For its construction process and specifications, see Massie, *Dreadnought*, 468–97.

Convoying Winston Churchill explained why this strategy worked: "The size of the sea is so vast that the difference between the size of a convoy and the size of a single ship shrinks in comparison to insignificance. There was in fact very nearly as good a chance of a convoy of forty ships in close order slipping unperceived between the patrolling U-boats as there was for a single ship; and each time this happened, forty ships escaped instead of one." Quoted in Massie, *Castles of Steel*, 733.

194 *Battleship critics* See, e.g., O'Connell's *Sacred Vessels* and "The Great War Torpedoed."

"*The submarine was not an effective weapons system*" Holger H. Herwig, "The Battlefleet Revolution, 1885–1914," in Knox and Murray, *The Dynamics of Military Revolution*, 127.

Battle of Jutland See Massie, *Castles of Steel*, 579–685; Bennett, *Naval Battles of the First World War*, 155–246.

THE CONSEQUENCES OF THE INDUSTRIAL REVOLUTION

197 "*We showed ourselves*" Quoted in Porter, *War and the Rise of the State*, 147.

198 *Mobilized and killed* Figures come from Porter, *War and the Rise of the State*, 170–71; McNeill, *Pursuit of Power*, 314; Kennedy, *Rise and Fall of the Great Powers*, 274, 278. Kennedy notes, at 278, that even excluding Russia at least 5 million civilians were also killed by "war-induced causes," such as disease and famine.

199 *State spending* The figure for 1910 comes from Rostow, *The World Economy*, 60. For 2003, see www.oecd.org/dataoecd/8/4/1874420.pdf.

State and military expenditures Porter, *War and the Rise of the State*, 163. Porter notes, at 162, that Britain's income tax rose during World War I from 6.25 percent to more than 30 percent.

200 *100 percent more* Porter, *War and the Rise of the State*, 162.

"*The National-Socialist ethos*" Quoted in Porter, *War and the Rise of the State*, 217.

201 *National Recovery Administration* Kennedy, *Over There*, 140–41.

Impact of World War I This section draws heavily on Porter's *War and the Rise of the State*, 149–297, and, to a lesser extent, on McNeill's *Pursuit of Power*, ch. 9, and Johnson, *Modern Times*, 12–22. McNeill concludes, at 345, "In light of the global propagation of managed economies in the second half of the twentieth century, this is likely to seem the major historical significance of World War I in time to come."

PART III: THE SECOND INDUSTRIAL REVOLUTION

THE RISE OF THE SECOND INDUSTRIAL AGE

203 *"The whole future of warfare"* Quoted in Corum, *The Roots of Blitzkrieg*, 31.

205 *Edison* For a recent biography see Baldwin, *Edison: Inventing the Century*.

Electrical power For an overview of its development see Jonnes, *Empires of Light*. For a brief summary of its impact, see Gordon, "Does the 'New Economy' Measure Up to the Great Revolutions of the Past?", 59.

206 *Automobile* Daimler and Benz each claimed to be the motor vehicle's originator. Ironically, these bitter rivals would have their names inextricably linked when, following their deaths, their companies merged in 1926.

Automobile history The best summaries I have found are in Curcio's *Chrysler*, 127–215; Nairn's *Engines That Move Markets*, ch. 6; and Mokyr, *The Lever of Riches*, 131–35.

"An age of oil" Quoted in Yergin, *The Prize*, 254.

Two-thirds of world petroleum Murray and Millett, *A War to Be Won*, 528.

207 *Ford* See Brinkley, *Wheels for the World*.

Ninety percent of world output Freeman and Louca, *As Time Goes By*, 260. For half of all American families owning a car, see ibid, 289. Rostow, *The World Economy*, 210–211, notes that in 1938 there were 29.4 million registered motor vehicles in the U.S. and only 8.3 million in Europe—72 percent fewer. Overy, *Why the Allies Won*, 224, writes that in 1937 the U.S. produced over 4.8 million vehicles, Germany produced 331,000, Italy 71,000, Japan 26,000.

The Wright brothers The best biography is Crouch, *The Bishop's Boys*.

208 *Aviation in World War I* The figures for increases in speed come from Harrison, *Mastering the Sky*, 67. The rest of this paragraph is based on Hallion, *Taking Flight*, 352–77.

Commercial aviation See Heppenheimer, *Turbulent Skies*; Harrison, *Mastering the Sky*; and Crouch, *Wings*.

Aircraft production In 1938, for example, the USSR produced 7,500 aircraft; Germany 5,235; Japan 3,201; Britain 2,827; Italy 1,850; the United States 1,800, and France 1,382. The next year the U.S. fell behind France too. Overy, *The Air War*, 26.

209 *Bell system* Brooks's *Telephone* offers a succinct history. For figures on stockholding, see 176.

Taylor See Kanigel, *The One Best Way*.

Development of management Chandler's *The Visible Hand* is the classic account.

Radio See Lyons, *Sarnoff*; Headrick, *The Invisible Weapon*, 116–213; Marconi, *My Father, Marconi*.

210 *Development of radar* Buderi, *The Invention That Changed the World*, offers the definitive account.

British decline Various explanations have been advanced for this phenomenon, which was a decline only in relative terms: not enough technical schools; a lot of antiquated industrial and transportation infrastructure (the cost of being an early mover in the First Industrial Revolution); and a lack of its own oil, at least until the discovery of North Sea fields in the 1970s, whereas it had rich stocks of coal, the fuel of the First Industrial Revolution. For a summary of the debate about why (and whether) Britain declined, see Landes, *Wealth and Poverty of Nations*, 450–64.

Figures on manufacturing output Kennedy, *Rise and Fall of the Great Powers*, 202, 330. For figures on defense spending, see ibid, 332.

211 *Efficient war machine* American war correspondent William Shirer wrote of the German army in 1940: "It is a gigantic, impersonal war machine, run as coolly and efficiently, say, as our automobile industry in Detroit." Shirer, *Berlin Diary*, 379.

CHAPTER 7: TANKS AND TERROR

212 *"Most glorious spring day"* Alanbrooke, *War Diaries*, 59.

"I was sleeping so soundly" Boothe, *Europe in the Spring*, 223.

"So much for security" Ironside, *Time Unguarded*, 301.

"We sat down" Boothe, *Europe in the Spring*, 233.

213 *"With a pleased and martial air"* Beaufre, *1940: The Fall of France*, 180.

"This is the moment" General André Corap, Ninth Army commander, quoted in May, *Strange Victory*, 386.

214 *"Nobody who has ever heard"* Bloch, *Strange Defeat*, 54. "It left me profoundly shaken": Ibid, 55. "This dropping of bombs from the sky": Ibid, 57.

"Look out—they're Germans" Bloch, *Strange Defeat*, 47–48.

"Somewhere in the north of France" Saint-Exupéry, *Airman's Odyssey*, 350.

"The most pathetic sight" Alanbrooke, *War Diaries*, 67.

215 *"Half mad"* Folcher, *Marching to Captivity*, 99.

"Fatal leisureliness" Horne, *To Lose a Battle*, 265.

"Our own rate of progress" Bloch, *Strange Defeat*, 45.

"All our doctrine" Beaufre, *1940: The Fall of France*, 185. "We were facing an adversary": Ibid, 181.

216 *Churchill's role* Wright, *Tank*, 24.

Water containers Wright, *Tank*, 35.

"A little man" and "stood out at once" Evan Charteris, *HQ Tanks*, quoted in Wright, *Tank*, 69.

"Heretic" Trythall, *'Boney' Fuller*, 94.

Fuller and Crowley Wright, *Tank*, ch. 10; Trythall, *'Boney' Fuller*, 20–26.

217 *"Shot through the brain"* Quoted in Wright, *Tank*, 136. "To attack the nerves": Ibid, 146.

"It smells like a garage" Danchev, *Alchemist of War*, 130.

"Pluto-mobocracy" and "Cancer of Europe" Wright, *Tank*, 181. For more on Fuller and Mosley see Trythall, *'Boney' Fuller*, ch. 8.

"The intention of annoying someone" Citino, *Quest for Decisive Warfare*, 185.

"The cartoonist's idea of a learned professor" Bernard Newman, *Spy*, quoted in Danchev, *Alchemist of War*, 2.

218 *Liddell Hart* The best biography is Danchev, *Alchemist of War*.

"*Totally unfit*" Montgomery, *Memoirs*, 46.

Colonel Blimp A figure created in the 1930s by cartoonist David Low, he was meant to signify the reactionary viewpoint of Britain's ruling class.

"*The well-bred horse*" Quoted in Divine, *The Blunted Sword*, 147.

219 *British strategy and rearmament* A leading historian of the subject stresses "the strategic and political environment"—not simply the stupidity of the army high command—as the explanation for British failures to utilize the tank in the 1930s. See Williamson Murray, "Armored Warfare: The British, French, and German Experiences," in Murray and Millett (eds.), *Military Innovation in the Interwar Period*, 12. For a short summary of British military affairs in the interwar period, see Bond, *British Military Policy*, and Brian Bond, "The Army Between the Two World Wars 1918–1939," in Chandler (ed.), *The Oxford History of the British Army*, 256–72.

"*Autonomous operation*" De Gaulle, *The Complete War Memoirs*, 14.

"*Army of shock troops*" De Gaulle, *The Army of the Future*, 104. "Entirely on caterpillar wheels": Ibid, 99. "Rapidly move round": Ibid, 150. This was the English language edition of *Toward a Professional Army*.

"*One must not exaggerate*" Quoted in Jackson, *Fall of France*, 24.

220 *French war planning* Doughty, *Seeds of Disaster*, offers the best overview.

"*Handed the initiative over*" De Gaulle, *The Complete War Memoirs*, 8.

221 *Description of Seeckt* Fraser, *Knight's Cross*, 89.

Reichsheer The German army and navy (Reichsmarine) together were known as the Reichswehr.

Storm troopers The seminal study is Gudmundsson's *Stormtroop Tactics*.

"*The whole future of warfare*" Quoted in Corum, *The Roots of Blitzkrieg*, 31.

"*To hell with combat*" Quoted in Citino, *Quest for Decisive Victory*, 183.

222 *The British influence* This has been a matter of scholarly debate for many years. Liddell Hart and Fuller both worked assiduously after World War II to claim credit for the German blitzkrieg. Mearsheimer's *Liddell Hart and the Weight of History* pointed out that Liddell Hart even added a paragraph to the English-language edition of Guderian's memoir fulsomely thanking Liddell Hart for his contributions to German tank development. This deception has led some writers to deprecate the overall contribution of Fuller and Liddell Hart. See, e.g., Corum, *Roots of Blitzkrieg*, 136–43. The pendulum has now swung back, with at least one recent scholar arguing that "Liddell Hart was on the whole working to get the credit he deserved" (Gat, *British Armour Theory*, 95). The argument about Liddell Hart is largely a sideshow, however, because no one denies Fuller's much larger influence on tank theory around the world. See, e.g., Habeck, *Storm of Steel*, xi–xii.

Hitler's fear of horses May, *Strange Victory*, 88.

"*Hitler was much impressed*" Guderian, *Panzer Leader*, 29. Although Guderian gives 1933 at the date of this meeting, most historians suggest it occurred in 1934 or 1935. See Deighton, *Blitzkrieg*, 121.

"*Far-sighted soldiers*" Guderian, *Panzer Leader*, 20. Guderian's *Achtung—Panzer!* Includes an extensive discussion of British and French tank developments. For a biography of Guderian, see Macksey, *Guderian*.

223 *"I became convinced"* Guderian, *Panzer Leader*, 24.

"Tactical" bombing An important caveat must be added: In 1939–40 cooperation between the German air and ground forces was not good enough for the Luftwaffe to provide much close air support to mechanized forces on the move. Not until 1941 did the Luftwaffe master this difficult skill. But even in 1939–40 the Luftwaffe was more adept than its Allied counterparts at hitting static targets in the way of a ground offensive. It was also better at providing air cover so that German troops would not be attacked by enemy airplanes. See Murray and Millett, *A War to Be Won*, 30–35, 46.

"Prostitution of the air force" May, *Strange Victory*, 391.

224 *German weapons and methods* Deighton offers a superb summary in *Blitzkrieg*, 99–176. He notes German individualism at 152.

"I hope you were pleased with your children?" This story, which Fuller often repeated, is recounted in Wright, *Tank*, 224, and Trythall, *'Boney' Fuller*, 205.

225 *Polish cavalry vs. German tanks* Wright disentangles this tale in *Tank*, chs. 12–13. According to his research, the incident which gave rise to the myth of gallant cavaliers foolishly attacking panzers occurred after some Polish cavalry had successfully charged a German infantry battalion. The horsemen were then surprised by German tanks or armored cars that emerged from some nearby woods. The cavalry suffered terrible losses before they could get away. Wright, *Tank*, 258–59.

Bewegungskrieg Citrino, *Quest for Decisive Victory*, 213. For a revisionist account, which nevertheless defends the use of "blitzkrieg," see Frieser, *The Blitzkrieg Legend*.

A small portion motorized Of 54 divisions thrown against Poland, six were heavy tank divisions, four were light tank divisions, and four were motorized infantry divisions. Murray and Millett, *A War to Be Won*, 46.

"Strongest in Europe" This comment was made by General Ludwig Beck in May 1938, shortly before Hitler fired him. May, *Strange Victory*, 68.

Germans outnumbered These figures come from May, *Strange Victory*, 477. Other sources give slightly different numbers but do not dispute the general conclusion. As May observes in an endnote, "no numbers are certain. They should all be allowed a range of error or at least plus or minus 5 percent." May, *Strange Victory*, 520.

226 *112 divisions* This figure comes from May, *Strange Victory*, 352.

Ten to one against Jackson, *Fall of France*, 32.

"The systematic French" May, *Strange Victory*, 236.

229 *Rommel* Fraser, *Knight's Cross*, is the best biography. Horne, *To Lose a Battle*, 269–72, and Deighton, *Blitzkrieg*, 53–54, have good character sketches.

"Our boats were being destroyed" Rommel, *Rommel Papers*, 8.

"Looking decidedly unhealthy" and "alarmingly small number" Rommel, *Rommel Papers*, 11.

230 *"The gunners stopped firing"* Horne, *To Lose a Battle*, 291.

"No co-ordination" Montgomery, *Memoirs*, 53. For more on conflicting headquarters, see Deighton, *Blitzkrieg*, 94–95.

231 *"A perfect road"* Quoted in Horne, *To Lose a Battle*, 358. For the most complete reconstruction of the Meuse breakthrough, see Doughty, *The Breaking Point*.

"*Aucune*" Churchill, *The Second World War*, II:42.

"*Venerable officials pushing wheel-barrows*" Ibid. Subsequent research suggests that Churchill probably embellished this account for literary effect. See Reynolds, *In Command of History*, 166–67.

232 "*We met a body of fully armed French motorcyclists*" Rommel, *Rommel Papers*, 16.

"*Reconnaissance in force*" Guderian, *Panzer Leader*, 110.

"*We must be very careful*" Churchill, *The Second World War*, II:103.

233 "*Not much more than a third*" Horne, *To Lose a Battle*, 584. The casualty figures cited here are from ibid; other sources give slightly different figures.

"*Trahi*" Boothe, *Europe in the Spring*, 250.

Lynching enemy agents Gen. Brooke wrote on May 27 that he had seen a body lying in the gutter opposite his temporary headquarters in France. "They have just shot that chap!" his orderly explained. "When I said: 'Who shot him?' he replied, 'Oh! Some of those retiring French soldiers, they said he was a spy, but I think the real reason was that he refused to give them any cognac!' " Alanbrooke, *War Diaries*, 71. For paranoia about Fifth Columnists, see Barlone, *A French Officer's Diary*, 66–68.

Carnage of the Great War France lost four and a half million killed or wounded from 1914 to 1918. May, *Strange Victory*, 6.

"*Resigned but resolute*" Quoted in May, *Strange Victory*, 187.

234 *French losses compared to the Korean and Vietnam Wars* May, *Strange Victory*, 7.

Loss of will not being a major factor See, e.g., May's *Strange Victory* and Jackson's *The Fall of France*. For a different perspective, which emphasizes low French morale, see Horne, *To Lose a Battle*.

Total number of bombers and fighters May, *Strange Victory*, 479.

Missing French aircraft Deighton, *Blitzkrieg*, 269–70.

Ten tank divisions and ten motorized Citino, *Blitzkrieg to Desert Storm*, 3.

235 "*A triumph of intellect*" Bloch, *Strange Defeat*, 36. "The ruling idea of the Germans": Ibid, 37.

"*Almost a miracle*" Guderian, *Panzer Leader*, 106.

236 *Return of decisiveness* I am indebted for this argument to Citino, *Quest for Decisive Victory*.

A minor sideshow Rommel had only four divisions in 1942, while 178 were deployed in Russia. Overy, *Why the Allies Won*, 19.

Size of German forces Overy, *Why the Allies Won*, 5. Overy, at 211, estimates that the Soviets had fifteen thousand tanks in 1941. Figures for aircraft come from ibid, 212, 215.

A few hundred miles Murray and Millett, *A War to Be Won*, 119, citing the testimony of Major General Eduard Wagner, the Wehrmacht's chief logistics officer.

Logistical difficulties in Russia Van Creveld, *Supplying War*, ch. 5.

237 *Deep battle* For a summary of its development, see Habeck, *Storm of Steel*, which notes, at x, "the adoption by the Red Army and the Wehrmacht of the very same doctrine." Elsewhere (at 296) Habeck refers to the "eerie similarity" between deep battle and blitzkrieg.

U.S. army and tanks For a recent survey, see Johnson, *Fast Tanks and Heavy Bombers.*

Development of armored warfare by Germany's enemies For a summary see Thomas G. Mahnken, "Beyond Blitzkrieg: Allied Responses to Combined-Arms Armored Warfare During World War II," in Goldman and Eaton, *The Diffusion of Military Technology and Ideas*, 243–66.

238 *"Combat effectiveness"* Dupuy, *A Genius for War*, 4.

One hundred Germans equivalent to two hundred Russians Dupuy, *Understanding War*, 224.

Dupuy's findings They have been seconded and amplified by Van Creveld in *Fighting Power*. Both analysts attribute the German army's proficiency primarily to a lean, decentralized organizational structure that effectively promoted and rewarded fighting excellence and punished failure. Their conclusions are amplified in works by such historians as John Keegan, Russell Weigley, and Max Hastings. For a different perspective, which emphasizes the decreasing effectiveness of German soldiers and the increasing effectiveness of Allied soldiers by war's end, see Mansoor, *The GI Offensive in Europe*, and Rush, *Hell in Hürtgen Forest.* Comparing the German and American armies at their zenith—June 1941 for the Germans, April 1945 for the Americans—Mansoor concludes (at 14) that "one would be hard-pressed to choose between the two forces on the basis of technical or tactical proficiency at the division level." He explains (at 4) that "German units could maneuver well and generally had excellent leadership, discipline (however ruthlessly applied), unit cohesion and training," but fell short in such areas as "logistics, fire support, intelligence, and interservice cooperation." The American army, by contrast, was distinguished by "its ability to adapt to changing conditions on battlefields across the globe, its use of intelligence, outstanding fire support, the ability to execute joint operations, and, most important, its endurance." But not even Mansoor claims that American units were on average superior to German divisions at the height of their power—only after severe attrition had caused the quality of German formations to deteriorate.

"47,000 aircraft" Overy, *Why the Allies Won*, 192.

"The Ford company alone" Overy, *Why the Allies Won*, 195.

90 percent of world oil production Overy, *Why the Allies Won*, 228. Germany and Japan tried to make up for this shortfall with synthetic oil manufactured from coal.

If the Allies had fought incompetently Overy makes the point in *Why the Allies Won* that the outcome of World War II cannot be attributed solely to the Allies' having bigger and better-equipped forces. "God," he notes, at 3, "does not always march with the big battalions. In World War I Britain, France, and Russia mustered 520 divisions in the middle of 1917, but could not prevail over 230 German divisions and 80 Austrian."

239 *"Helped reshape international politics"* Reynolds, "1940: Fulcrum of the Twentieth Century?", 328. What follows is based on Reynolds's article.

"It was at that time" Lewis, "Freedom and Justice in the Modern Middle East."

240 *"Dazzled by the early German victories"* Guillain, *I Saw Tokyo Burning*, 14–15.

"Half Judaized, half Negrified" Quoted in Toland, *Rising Sun*, 243.

CHAPTER 8: FLATTOPS AND TORPEDOES

242 *"Ten times more"* The average U.S. pilot had 300 hours of flying time. Prange, *At Dawn We Slept*, 189. Belote, *Titans of the Seas*, 3, gives the figure as 305.

Fuchida This biographical sketch is based on Prange, *God's Samurai*. 1–31; Prange, *At Dawn We Slept*, 195–96.

"Glistening in the sun" Quoted in Prange, *December 7, 1941*, 109.

"What a majestic sight" Quoted in Prange, *December 7, 1941*, 109.

The first reaction of many For instance, Fireman Dan Wentrcek, of the USS *Nevada*, recalled he was showering at the time: "There were two or three other guys in the shower who said: 'Aw, of all the damned times to have a drill!'" La Forte and Marcello, *Remembering Pearl Harbor*, 72.

243 *"Chuck potatoes at them"* Machinist's Mate Second Class Leon Bennett, USS *Neosho*, in La Forte and Marcello, *Remembering Pearl Harbor*, 80–81.

"The best goddamn drill" Prange, *At Dawn We Slept*, 510.

Warned of the danger The most prescient warning came from the Martin-Bellinger report issued on March 31, 1941. Rear Admiral Patrick Bellinger and Major General Frederick Martin, the senior air officers on Oahu, warned of the possibility of a "dawn air attack" that "would most likely be launched from one or more carriers which would probably approach inside of three hundred miles." Prange, *At Dawn We Slept*, 93–95.

"Strongest fortress in the world" Prange, *At Dawn We Slept*, 122. This quote comes from an aide-mémoire Marshall presented to President Roosevelt in late April 1941.

Japanese fliers were not capable Prange, *At Dawn We Slept*, 461.

Why intelligence failed The classic account is Wohlstetter, *Pearl Harbor: Warning and Decision*, which pointed out the difficulty of separating useful information from background noise.

Entertainment in 1941 I am indebted for this list to Berry, "*This Is No Drill*," 1–11.

245 *Broomsticks and eggs* Weintraub, *Long Day's Journey Into War*, 14.

"This is no drill" Morison, *The Rising Sun in the Pacific*, 101.

"No shit" Musician First Class Warren G. Harding in La Forte and Marcello, *Remembering Pearl Harbor*, 86. Numerous other sailors from other ships recalled similar messages. This book is based on 350 interviews conducted by the University of North Texas Oral History Program with Pearl Harbor survivors.

"Adrenaline pumping" Seaman Martin Matthews in La Forte and Marcello, *Remembering Pearl Harbor*, 29.

Second wave at 9 A.M. Prange, *At Dawn We Slept*, 532. Other sources give slightly different timing.

Welch and Taylor shot down seven aircraft Prange, *At Dawn We Slept*, 538.

"It was Whoosh" La Forte and Marcello, *Remembering Pearl Harbor*, 17.

"A rending of metal" La Forte and Marcello, *Remembering Pearl Harbor*, 74.

Casualties aboard the Arizona The exact toll was 1,103 killed or missing, plus 44 wounded, out of a total of 1,511 men. Morison, *The Rising Sun in the Pacific*, 109.

"Steel fragments in the air" Seaman Matthews in La Forte and Marcello, *Remembering Pearl Harbor*, 30.

"Hair was burned off" La Forte and Marcello, *Remembering Pearl Harbor*, 18.

246 *"Sea of fire"* Seaman Jack Kelley, USS *Tennessee*, in La Forte and Marcello, *Remembering Pearl Harbor*, 56.

Contributed nothing to the attack Willmott, in *Pearl Harbor*, 115, challenges this traditional account, claiming that one midget submarine hit the *California* and *West Virginia* with its torpedoes. He does not provide any proof, however.

Fuchida wanted to take another crack There is some ambiguity about what exactly was said on the bridge of the *Agaki*. See Willmott, *Pearl Harbor*, 154–57.

Attack on Pearl Harbor The definitive sources are Prange's *At Dawn We Slept* and *December 7, 1941*. Also extremely useful are Lord, *Day of Infamy;* Morison, *The Rising Sun in the Pacific*, 80–127; Toland, *The Rising Sun*, 211–29; Belote, *Titans of the Seas*, 1–15; Weintraub, *Long Day's Journey into War*.

247 *Pioneering flights* For a brief summary see Hone et al., *American and British Carrier Development*, 15, and Love, *History of the U.S. Navy*, I: 450–51, though Love gives the wrong date for the landing on the cruiser *Birmingham* (it was in 1910, not 1909).

248 *Cuxhaven Raid* Massie, *Castles of Steel*, 361–74.

British aviation in World War I This summary is based on Spector, *At War at Sea*, 122–25; Ireland, *War at Sea*, 79–88.

249 *German and Italian navies* For an examination of why they never built carriers, see Emily Goldman, "Receptivity to Revolution: Carrier Air Power in Peace and War," in Goldman and Eliason (eds.), *The Diffusion of Military Technology and Ideas*, 288–97. A lot of the answer has to do with geography: Both the German and Italian navies, expecting to operate around the coasts of Europe, planned to rely on land-based aircraft. Both Germany and Italy, like Britain but unlike Japan or the U.S., also set up independent air forces (the Luftwaffe and the Regia Aeronautica) that siphoned off aviation assets from their navies.

250 *British naval aviation after World War I* Till, *Air Power and the Royal Navy*; Ireland, *War at Sea*; Herman, *To Rule the Waves*, 522–24; www.naval-history.net; Belote, *Titans of the Seas*, 34–37; Budiansky, *Air Power*, 256–57.

251 *"Nonsensical and impossible"* Quoted in O'Connell, *Sacred Vessels*, 255.

"Immense, round, helpless sea animal" Austin Parker, reporter for *The World*, quoted in Davis, *Billy Mitchell Affair*, 108. Waller, *A Question of Loyalty*, 154, writes: "Some admirals sobbed like babies."

"We showed old Admiral Tubaguts" Quoted in Davis, *Billy Mitchell Affair*, 111.

"Epoch-making" Davis, *Billy Mitchell Affair*, 113.

"Most competent and intrepid pilot" Davis, *Billy Mitchell Affair*, 150.

"Plumed fellow" Davis, *Billy Mitchell Affair*, 3.

252 *"Incompetency, criminal negligence"* Waller, *A Question of Loyalty*, 20; Davis, *Billy Mitchell Affair*, 218.

"The most charitable way" Trimble, *Admiral William A. Moffett*, 162.

Royal Navy suffered without a Moffett I am grateful for this point to Murray and Millett, *A War to Be Won*, 36.

Moffett deceived the "gun club" See Pierce, *Warfighting and Disruptive Tech-*

nologies, 127–30. The author, a U.S. Navy captain, argues that innovators like Moffett often have to disguise their intentions from a conservative military hierarchy.

253 *Moffett* See Trimble, *Admiral William A. Moffett*, for a workmanlike biography.

Carriers Unless otherwise stated, "carrier" refers to fleet carriers, designed, as the name implies, to operate with the fleet. There were also smaller "escort" carriers designed to protect merchant convoys. They would prove vital in the Battle of the Atlantic. The first one, the USS *Long Island*, was commissioned in June 1941. It is sometimes added to the total of U.S. aircraft carriers on the eve of war, to produce a total of 8, not 7. See Belote, *Titans of the Seas*, 33. It should also be noted that by 1941 the original U.S. aircraft carrier, *Langley*, had been converted into a seaplane tender, so it no longer counted in the carrier total.

254 *U.S. Navy up to 1941* Love, *History of the U.S. Navy*, I: 584–617, 655–67; Belote, *Titans of the Seas*, 16–24; Hone et al., *American and British Aircraft Carrier Development*, 48–81. For a listing of all U.S. aircraft carriers, see: http://www.history.navy.mil/avh-1910/APP03.PDF.

255 *"The slightest chance"* Quoted in Reynolds, *In Command of History*, 103.

Aircraft production For the figures, see Kennedy, *Rise and Fall of the Great Powers*, 324, 354. For the number of aircraft in the Japanese navy, see Peattie, *Sunburst*, 162.

256 *Zero* For the story of its development, written by its developer, see Horikoshi, *Eagles of Mitsubishi*.

Japanese aircraft The definitive source is Peattie, *Sunburst*, 86–101.

"War is so easy" Quoted in Marder, *Old Friends, New Enemies*, I:292.

257 *Need for fighter escorts* Peattie, *Sunburst*, 122–28, discusses the lessons the Japanese aviators learned in China.

Japanese training Peattie, *Sunburst*, 31–33, 131–34; Marder, *Old Friends, New Enemies*, I:265–95; Evans and Peattie, *Kaigun*, 583.

Yamamoto admired Lincoln Agawa, *The Reluctant Admiral*, 84–85.

U.S. advantages In 1941 the U.S. had a total population of 133 million vs. 72 million for Japan. Total U.S. steel and iron output was 75 million tons vs. 6.8 million for Japan. However, Japan had over 3 million men in the military vs. 1.8 million for the U.S. Statistics from: www.umich.edu/~cowproj/dataset.html#Capabilities. Kennedy, *The Rise and Fall of the Great Powers*, 332, notes that in 1937 total U.S. income was $68 billion (1.5 percent spent on defense) while Japan's was $4 billion (28.2 percent spent on defense). Ibid, 330, notes that in 1937 the U.S. had 35.1 percent of world manufacturing output vs. 3.5 percent for Japan.

"If I am told to fight" Prange, *At Dawn We Slept*, 10.

"A major calamity" Quoted in Agawa, *The Reluctant Admiral*, 186.

258 *"Secure raw materials"* Prange, *At Dawn We Slept*, 174.

Perhaps wrongly Spector argues in *Eagle Against the Sun*, 84, that, absent an attack against the U.S., FDR would have had a hard time rallying the nation to war in defense of European colonies.

"People think I'm a gangster" Agawa, *The Reluctant Admiral*, 64.

"Eighty Sen" Agawa, *The Reluctant Admiral*, 2.

259 *"The three great follies"* Agawa, *The Reluctant Admiral*, 93.

Genesis of Pearl Harbor plan Yamamoto noted that he was inspired by the Russo-Japanese War in a letter to Navy Minister Koshiro Oikawa, January 7, 1941, contained in Goldstein and Dillon, *The Pearl Harbor Papers*, 116. For an argument that Yamamoto was "powerfully influenced" by the British journalist and spy Hector Bywater, see Honan, *Visions of Infamy*.

6,788 miles Prange, *At Dawn We Slept*, 215.

260 *10 percent hits* Minoru Genda and Masataka Chihaya, "How the Japanese Task Force Idea Materialized," in Goldstein and Dillon (eds.), *The Pearl Harbor Papers*, 11.

Refueling Sadao Chigusa, "Conquer the Pacific Ocean Aboard Destroyer Akigumo: War Diary of the Hawaiian Battle," in Goldstein and Dillon (eds.), *The Pearl Harbor Papers*, 208–9.

Hit rates Budiansky, *Air Power*, 267.

Torpedoes Peattie, *Sunburst*, 36; Evans and Peattie, *Kaigun*, 583.

Bombs Peattie, *Sunburst*, 140.

First Air Fleet Peattie, *Sunburst*, 152.

"Epoch-making progress" and "revolutionary change" Minoru Genda and Masataka Chihaya, "How the Japanese Task Force Idea Materialized," in Goldstein and Dillon (eds.), *The Pearl Harbor Papers*, 10.

261 *War game* Toland, *Rising Sun*, 161.

"This operation is a gamble" Prange, *At Dawn We Slept*, 263; Toland, *The Rising Sun*, 155.

262 *"Date which shall live in infamy"* The quotation, of course, is from FDR's famous war message to a Joint Session of Congress delivered on December 8. He began: "Yesterday, December 7, 1941—a date which shall live in infamy—the United States of America was suddenly and deliberately attacked by naval and air forces of the Empire of Japan."

Geographical reach unequaled Okumiya and Horikoshi, *Zero!*, 65, note this point with pride.

263 *Sinking of Task Force Z* See Marder, *Old Friends, New Enemies*, Vol. 1; Toland, *Rising Sun*, 238–43; Okumiya and Horikoshi, *Zero!*, 93–120; Marder, *To Rule the Waves*, 537–42.

"Against the Zero fighters" Okumiya and Horikoshi, *Zero!*, 64. It should be noted that no Zeros were present that day, but even against Nell and Betty bombers the Buffaloes would not have been very effective.

"Tremendous and terrible vindication" Quoted in Marder, *OId Friends, New Enemies*, I:495.

"We have lost command of the sea" Alanbrooke, *War Diaries*, 210.

264 *"The Japanese were giddy"* Guillain, *I Saw Tokyo Burning*, 48.

265 *Warship and aircraft production* Murray and Millett, *A War to Be Won*, 535.

U.S. aircraft carriers Willmott, *Pearl Harbor*, 140, gives the breakdown on different kinds of carriers. For a listing of individual carriers, see: www.history.navy.mil/avh-1910/APP03.PDF.

Frontline aircraft in 1945 Overy, *The Air War*, 120.

"*At peak production*" Willmott, *Pearl Harbor*, 140.

266 *Pacific War* This summary is based on Spector, *Eagle Against the Sun;* Toland, *Rising Sun*; Murray and Millett, *A War to Be Won.*

"*Greatest single victory*" Willmott, *Pearl Harbor*, 180.

"*Prestige has been completely shattered*" Quoted in Reynolds, "1940," 349.

267 "*Without the sinking*" Herman, *To Rule the Waves*, 542.

CHAPTER 9: SUPERFORTRESSES AND FIREBOMBS

268 "*Bees*" For the use of this nickname, see Toland, *The Rising Sun*, 670.

"*Long, glinting wings*" Guilllain, *I Saw Tokyo Burning*, 182.

"*Stark crazy*" Marshall, *Sky Giants Over Japan*, 144.

269 "*Guys cursed General LeMay*" Marshall, *Sky Giants Over Japan*, 145.

"*We'll get the holy hell shot out*" Wheeler, *Bombers Over Japan*, 168.

"*Blown out of the sky*" Morgan, *The Man Who Flew the Memphis Belle,* 309.

"*Unprecedented, daring*" McKelway, "A Reporter with the B-29s—III."

Lose three-quarters of aircraft Coffey, *Iron Eagle*, 156.

Visions of dead airmen LeMay, *Mission with LeMay*, 350–52.

"*I can't sleep*" McKelway, "A Reporter with the B-29s—III."

270 "*We'd flown more than 2,000 missions*" Morgan, *The Man Who Flew the Memphis Belle,* 302.

Bombing in World War I For a summary, see Hallion, *Taking Flight*, 350–65; Biddle, *Rhetoric and Reality in Air Warfare*, ch. 1.

271 "*Swept away by an avalanche of terror*" Sherry, *The Rise of American Air Power*, 24.

"*The bomber will always get through*" Quoted in Werrell, *Blankets of Fire*, 9.

"*We thought of air warfare in 1938*" Quoted in Budiansky, *Air Power*, 139.

272 *Development of precision bombing* Crane, *Bombs, Cities and Civilians*, ch. 2.

"*Loss of any of these systems*" Hansell, *The Strategic Air War*, 13.

B-17 characteristics Craven and Cate, *The Army Air Forces in World War II*, VI:205. These figures are for the B-17B; the 600-mile figure is the tactical radius. All performance figures for the B-17, as for every other major U.S. aircraft, improved slightly with wartime modifications. The B-17 went through A,B,C,D,E,F, and G versions. Craven and Cate, VI:196–97, offer the following useful definitions: "The *range* of a plane is the total distance it can fly without refueling. . . . The *combat range* is necessarily somewhat less than the maximum range because the plane is combat-loaded—that is, it carries bombs and other items required for combat in lieu of additional fuel. The *tactical radius* of a plane is the maximum distance it can fly away from its base with a normal combat load and return without refueling, allowing for all safety and operating factors. For practical purposes during World War II, tactical radius was considered to be three-eighths to two-fifths of the combat range."

Bomber production Between January 1940 and August 31, 1945, 12,692 B-17s were produced vs. 18,190 B-24s. Craven and Cate, *The Army Air Forces in World War II*, VI:206.

Norden bombsight Budiansky, *Air Power*, 173–75; Werrell, *Blankets of Fire*, 6. The cost of the Manhattan Project was $1.9 billion, according to a Brookings Institution study: www.brook.edu/FP/PROJECTS/NUCWCOST/MANHATTN.HTM.

273 *Accompanied by bodyguards* Friedman, *The Future of War*, 219.

Thirteen B-17s Werrell, *Blankets of Fire*, 21.

536 light bombers Biddle, *Rhetoric and Reality in Air Warfare*, 183.

"When the war finally broke out" Overy, *The Air War*, 33.

British vs. German air strength Budiansky, *Air Power*, 233.

274 *Jamming of German signals* Buderi, *The Invention That Changed the World*, 192–96.

Battle of Britain This summary is based on Budiansky, *Air Power*, 230–54; Murray and Millett, *A War to Be Won*, 83–89.

"We can take it" Budiansky, *Air Power*, 243.

Berlin bombed Biddle, *Rhetoric and Reality in Air Warfare*, 188.

Fewer than 25 percent Biddle, *Rhetoric and Reality in Air Warfare*, 1; Overy, *Why the Allies Won*, 104.

Shift to area bombing Overy, *Why the Allies Won*, 112–13.

275 *Hamburg raid* *U.S. Strategic Bombing Survey, Summary Report (European War)*, 10.

"Elimination of German industrial cities" Budiansky, *Air Power*, 329.

Half of urban area Budiansky, *Air Power*, 317.

"Panacea" targets Murray and Millett, *A War to Be Won*, 307.

"The devil shall get no rest" Budiansky, *Air Power*, 316.

276 *88 mm specifications* Budiansky, *Air Power*, 224.

55,000 guns Overy, *Why the Allies Won*, 125.

Ploesti raid *U.S. Strategic Bombing Survey, Summary Report (European War)*, 20–22; Larrabee, *Commander in Chief*, 240.

Attacks on ball-bearing and aircraft plants *U.S. Strategic Bombing Survey, Summary Report (European War)*, 14–18.

Speer tripled production *U.S. Strategic Bombing Survey, Summary Report (European War)*, 8.

Casualties Budiansky, *Air Power*, 312; Sherry, *Rise of American Air Power*, 204.

"Nobody expected to live through it" Budiansky, *Air Power*, 314.

"The fighter experts asserted" Hansell, *The Strategic Air War*, 14.

277 *P-51 range* Craven and Cate, *The Army Air Forces in World War II*, VI:219.

Seven times fewer losses Larrabee, *Commander in Chief*, 603.

Fighter losses *U.S. Strategic Bombing Survey, Summary Report (European War)*, 19.

Luftwaffe pilot losses Murray and Millett, *A War to Be Won*, 325.

80 hours vs. 225 Millett and Murray, *A War to Be Won*, 325.

10 percent of bombers didn't make it home For instance, on March 30, 1944, in a raid on Nuremberg, British Bomber Command lost 95 of 795 bombers. Overy, *Why the Allies Won*, 121.

8,722 sorties vs. 250 Larrabee, *Commander in Chief*, 603.

278 *8th and 9th Luftwaffe* Budiansky, *Air Power*, 306.

McNair killed Murray and Millett, *A War to Be Won*, 429.

"*Bomb in a pickle barrel*" Budiansky, *Air Power*, 175.

Within two miles of target Larrabee, *Commander in Chief*, 604.

Only 20 percent of bombs within 1,000 feet *The U.S. Strategic Bombing Survey, Summary Report (European War)*, 13.

Swiss town hit Werrell, *Blankets of Fire*, 30.

Luftwaffe bombing Freiburg-im-Breisgau Fussell, *Wartime*, 14.

USAAF and area bombing Crane, *Bombs, Cities, and Civilians*, 6–7, explores differing views among U.S. generals over the efficacy and morality of bombing civilians. Some, like General Carl Spaatz, were troubled by such attacks; many others, like General Curtis LeMay, were not. Because individual commanders had considerable autonomy, precision bombing remained the practice in some commands but not in others. In *Rhetoric and Reality in Air Warfare*, 292, Biddle notes, "While claiming adherence to 'precision' bombing of industrial targets, [the American air force in Europe] often engaged in area attacks on cities."

Dresden For a revisionist study which argues that the raid had reasonable justification, see Taylor, *Dresden*.

"*Morale is a military target*" Budiansky, *Air Power*, 284.

300,000 deaths *U.S. Strategic Bombing Survey, Summary Report (European War)*, 6.

German morale *The U.S. Strategic Bombing Survey, Summary Report (European War)*, 11–12, found that while "the morale of the German people deteriorated under aerial attack," they "lacked either the will or the means to make their dissatisfaction evident."

279 *Impact of attacks on oil facilities* *U.S. Strategic Bombing Survey, Summary Report (European War)*, 23.

Attacks on transportation infrastructure *U.S. Strategic Bombing Survey, Summary Report (European War)*, 30–33.

83 percent of bomb tonnage Overy, *Why the Allies Won*, 125. 1.4 million tons: *U.S. Strategic Bombing Survey, Summary Report (European War)*, 84.

Output 30 percent below what was planned Overy, *Why the Allies Won*, 131.

Two million Germans Werrell, *Blankets of Fire*, 32.

Ten thousand 88s Larrabee, *Commander in Chief*, 599; Murray and Millett, *A War to Be Won*, 332.

Electronics and optical instruments Budiansky, *Air Power*, 330.

20 percent of all ammunition Overy, *Why the Allies Won*, 131.

24,000 fighters Murray and Millett, *A War to Be Won*, 333.

Civilians killed, wounded and made homeless U.S. Strategic Bombing Survey, *Summary Report (European War)*, 6, 36.

"Essential to the defeat" Murray and Millett, *A War to Be Won*, 335.

280 *Losses over Europe* Budiansky, *Air Power*, 330. The finding that U.S. air crews suffered a 40 percent higher fatality rate than ground forces in Europe is derived from statistics in Beebe and De Bakey, *Battle Casualties*, 44. According to the authors, out of every thousand men serving in American combat air crews in the European theater from 1941 to 1945, an average of 0.532 were killed every day. The comparable figure for U.S. ground forces in Europe was 0.38—a difference of 40 percent. The air crews also had a much higher percentage of missing personnel. The percentage of ground forces who were wounded was, however, higher.

"Back into the Stone Age" LeMay, *Mission with LeMay*, 565. According to Coffey, *Iron Eagle*, 3–4, this statement was penned by LeMay's ghostwriter, novelist MacKinlay Kantor.

Dr. Strangelove Buck Turgidson, the trigger-happy general played by George C. Scott, who advocates a preemptive attack on the Soviet Union, was said to be based on LeMay.

Youngest major general Coffey, *Iron Eagle*, 107.

"I had to spend my extracurricular hours" LeMay, *Mission with LeMay*, 32.

281 *Navigational feats* Coffey, *Iron Eagle*, 240–42.

"Iron Ass" Coffey, *Iron Eagle*, 21.

"We were still a sorry outfit" LeMay, *Mission with LeMay*, 218.

282 *Combat Box* Coffey, *Iron Eagle*, 32–38.

Regensburg raid Coffey, *Iron Eagle*, 80–92.

LeMay's move to B-29 command Coffey, *Iron Eagle*, 107–9. For LeMay as an "operator"—the description comes from General Lauris Norstad, Arnold's chief of staff—see ibid, 132.

"The greatest U.S. gamble of the war" Larrabee, *Commander in Chief*, 580.

1,664 ordered Craven and Cate, *Army Air Forces in World War II*, V:7.

B-29 performance characteristics Craven and Cate, *Army Air Forces in World War II*, VI:210; Wheeler, *Bombers over Japan*, 31–32; Werrell, *Blankets of Fire*, 58–59.

Wings, propellers, tail fin Wheeler, *Bombers over Japan*, 22, 26.

Nose art Wheeler, *Bombers over Japan*, 172–73.

283 *150 electric motors, seven generators, eleven miles of wiring* Werrell, *Blankets of Fire*, 61.

A year to train Wheeler, *Bombers Over Japan*, 35.

B-29 testing and crash LeMay, *Superfortress*, 60–63; Wheeler, *Bombers Over Japan*, 27.

"B-29s had as many bugs" LeMay, *Mission with LeMay*, 321.

Four times the rate of fire Werrell, *Blankets of Fire*, 71.

1.5 million parts vs. 15,000 Overy, *Why the Allies Won*, 196.

27,000 pounds of aluminum, 1,000 pounds of copper Werrell, *Blankets of Fire*, 74.

284 *More than 40 percent women* Werrell, *Blankets of Fire*, 79.

Productivity figures Werrell, *Blankets of Fire*, 80.

B-29 production Wheeler, *Bombers over Japan*, 38–39; Larrabee, *Commander in Chief*, 606–9; Werrell, *Blankets of Fire*, 74–80.

3,432 produced Werrell, *Blankets of Fire*, 81. LeMay, *Superfortress*, 49, gives a figure of 3,628.

Six logistics flights for every combat flight Wheeler, *Bombers over Japan*, 52.

285 *The one commander whose company FDR enjoyed* Larrabee, *Commander in Chief*, 209.

18 of 23 airmen died Larrabee, *Commander in Chief*, 206.

USAAF expansion LeMay, *Superfortress*, 55.

"*Destroy the capacity and the will*" Quoted in Larrabee, *Commander in Chief*, 220.

Hap Arnold For a biography see Daso, *Hap Arnold*. Larrabee provides an excellent character sketch in *Commander in Chief*, ch. 5. For Arnold's own story see Arnold, *Global Mission*.

Not even LeMay Coffey, *Iron Eagle*, 111.

Four missions a month LeMay, *Mission with LeMay*, 332.

Northern Mariana Islands Murray and Millett, *A War to Be Won*, 353–62; Spector, *Eagle Against the Sun*, 301–21.

287 "*No other facilities*" Hansell, *The Strategic Air War*, 174.

"*They had built tennis courts*" LeMay, *Mission with LeMay*, 340.

"*Simply awful*" Hansell, *The Strategic Air War*, 185.

"*My greatest worry*" Quoted in Wheeler, *Bombers over Japan*, 118.

288 *November 24, 1944, raid* For the lead pilot's account, see Morgan, *The Man Who Flew the Memphis Belle*, 284–88. For an overview of the mission, see Craven and Cate, *Army Air Forces in World War II*, V:558–60.

"*Exercise in futility*" Morgan, *The Man Who Flew the Memphis Belle*, 291.

Get results fast The message was delivered by Arnold's aide, General Lauris Norstad, who said in effect, "You go ahead and get results with B-29. If you don't get results you'll be fired." LeMay, *Mission with LeMay*, 347.

Arrived January 19 Coffey, *Iron Eagle*, 129.

"*Hardheaded bastard*" Cecil Combs, quoted in Werrell, *Blankets of Fire*, 139.

"*Big, husky, healthy*" McKelway, "A Reporter with the B-29s—II."

"*Chillingly soft voice*" Morgan, *The Man Who Flew the Memphis Belle*, 304.

"*Couldn't make himself heard*" McKelway, "A Reporter with the B-29s—II."

"*Aura as a borderline sociopath*" Morgan, *The Man Who Flew the Memphis Belle*, 298.

Bell's palsy Coffey, *Iron Eagle*, 23–24.

289 "*Grimace is a smile*" McKelway, "A Reporter with the B-29s—III."

"*Stinko*" LeMay, *Mission with LeMay*, 343.

Napalm Sherry, *The Rise of American Air Power*, 226.

More than 1,600 tons of bombs According to *The U.S. Strategic Bombing Survey, Summary Report (Pacific War)*, 85, the exact figure was 1,657 tons.

"Bursts of light" and "barely a quarter of an hour" Guillain, *I Saw Tokyo Burning*, 182.

290 *"We were in hell"* Funato Kazuyo, "The Burning Skies," in Cook and Cook, *Japan at War*, 346.

"Baked as if in a casserole" Edoin, *The Night Tokyo Burned*, 83.

"Bombing with damn near impunity" Morgan, *The Man Who Flew the Memphis Belle*, 311.

291 *Two-thirds the area of Manhattan* McKelway, "A Reporter with the B-29s—III."

"Single greatest disaster" Quoted in LeMay, *Mission with LeMay*, 10. In addition to the sources previously listed, this account draws on the descriptions in Craven and Cate, *Army Air Forces in World War II*, V:614–18; Toland, *The Rising Sun*, 670–77; Caidin, *A Torch to the Enemy*.

"The turning-point in our aerial campaign" Morgan, *The Man Who Flew the Memphis Belle*, 317.

12,054 mines Craven and Cate, *Army Air Forces in World War II*, V:751.

three hundred vessels Budiansky, *Air Power*, 340.

292 *"The destruction of Japan's ability to wage war"* LeMay to Lauris Norstad, April 1945, quoted in LeMay, *Mission with LeMay*, 373.

Iwo Jima This account is based on a recent revisionist study: Burrell, "Breaking the Cycle of Iwo Jima Mythology." Most histories cite claims that 2,251 Superforts landed at Iwo Jima to suggest that as many as 24,000 crewmen would have been lost without the availability of this airfield. But, Burrell writes, "Of the 2,251 touchdowns popularized in most histories, the vast majority did not result from crucial or unavoidable crises."

Japanese balloons Kennedy, *Freedom from Fear*, 848–49.

293 *Toll from atomic bombs* See Sherry, *The Rise of American Air Power*, 406, for a survey of various estimates. Subsequent research suggests that the estimates of the *U.S. Strategic Bombing Survey, Summary Report (Pacific War)*, 100–1, were low. The survey reported 60,000 to 70,000 killed in Hiroshima and 40,000 in Nagasaki.

Only three planes needed over Hiroshima Friedman, *The Future of War*, 74.

Percentage of damage caused by A-bombs Werrell, *Blankets of Fire*, 220.

Japanese casualties and refugees *U.S. Strategic Bombing Survey, Summary Report (Pacific War)*, 92, 95.

3,415 killed and missing Sherry, *Rise of American Air Power*, 209. 359 B-29s lost: Burrell, "Breaking the Cycle of Iwo Jima Mythology," 1178.

Statements of Prince Konoye and Admiral Suzuki Craven and Cate, *Army Air Forces in World War II*, V:756.

294 *"We just weren't bothered about the morality"* LeMay, *Mission with LeMay*, 381.

"More immoral to use less *force"* LeMay, *Mission with LeMay*, 382.

"Cut off the dog's tail" LeMay, *Mission with LeMay*, 384.

"An odor" and "something awful" Morgan, *The Man Who Flew the Memphis Belle*, 311, 314.

THE CONSEQUENCES OF THE SECOND INDUSTRIAL REVOLUTION

295 *What changed* This section relies in part on the excellent summary in Murray and Millett, *A War to Be Won*, Appendix 3 ("Weapons").

Paratroopers For the general failure of this concept, see DeVore, *The Airborne Illusion.*

296 *Development of amphibious warfare* Millett, *Semper Fidelis*, ch. 12.

Main rifle dated from 1905 Overy, *Why the Allies Won*, 222.

297 *"55 percent of Japan's losses"* Spector, *Eagle Against the Sun*, 487.

298 *"Had Japan developed such bombers"* Okumiya and Horikoshi, *Zero!*, 227.

Room 40 For the Royal Navy's failure to take advantage of its intelligence in the Battle of Jutland, see Massie, *Castles of Steel*, 640–42.

299 *Code breaking* For a valuable summary see Budiansky, *Battle of Wits.*

Battle of the Atlantic For a summary, see Murray and Millett, *A War to Be Won*, 234–61.

Window Buderi, *The Invention That Changed the World*, 208–9.

U.S. casualties in World War II and Civil War Boot, "Reality Check—This Is War." These figures were derived from Chambers, *The Oxford Companion to American Military History*, 849, and official Defense Department data: web1.whs.osd.mil/mmid/casualty/WCPRINCIPAL.pdf.

300 *"Dispersion has actually increased more rapidly"* Dupuy, *The Evolution of Weapons and Warfare*, 310. All figures on lethality and dispersion up to 1973 are from ibid, 312. The figure for the Gulf War comes from Rip and Hasik, *The Precision Revolution*, 126.

301 *Military medicine* For a summary of developments see Gabriel and Metz, *A History of Military Medicine*, II: 147–257. For figures on deaths in World War II and Korea, see ibid, 255–57, and Beebe and De Bakey, *Battle Casualties*, 77 (which also contains data on nineteenth-century wars). According to *Battle Casualties*, 21, out of every thousand Union soldiers in the Civil War, 71.2 died of disease every year; the comparable figure for World War II was just 0.6.

"Does not explain the outcome of wars" Overy, *Why the Allies Won*, 316.

302 *425 different kinds of aircraft* Overy, *Why the Allies Won*, 201.

303 *Federal spending and employment* Porter, *War and the Rise of the State*, 279, 283.

PART IV: THE INFORMATION REVOLUTION

THE RISE OF THE INFORMATION AGE

305 *"It's a constant struggle of one-upsmanship"* Cited in Wong, *Developing Adaptive Leaders*, 11.

307 *U.S. atomic weapons* For a short summary see Millett and Maslowski, *For the Common Defense*, 534–44. For 12,305 warheads, see 541; for the number of SAC bombers, see 538. For the Pentomic army, see Bacevich, *The Pentomic Era*.

308 *First digital computers* These included the Model 1 Relay Computer completed at Bell Labs in 1939; the Atansasoff-Berry Computer completed at Iowa State University in 1939; and the Harvard Mark I completed by IBM at Harvard in 1943. According to Campbell-Kelly and Aspray, *Computer*, 69, at least 10 such machines were built by 1945.

309 *120 degrees* Wolfe, *Hooking Up*, 33.

Colossus Hodges, *Alan Turing*, 289–305; Sale, *The Colossus Computer;* Kessler, *How We Got Here*, 112–15; Nairn, *Engines That Move Markets*, 313; Budiansky, *Battle of Wits,* 314–15.

ENIAC The definitive account is McCartney, *ENIAC*. See also Campbell-Kelly and Aspray, *Computer*, 79–104; Freiberger and Swaine, *Fire in the Valley*, 8–10 (see 9 for "whirring tape drives" and "clanking teletype machines").

310 *Microchip* See Reid, *The Chip*. Among the first buyers were NASA and the Air Force, which needed smaller computers to control their spacecraft and ballistic missiles. The Apollo program to land a man on the moon probably would not have been possible without the integrated circuit. Tomayko, *Computers in Spaceflight.*

Transistors, integrated circuits, microprocessors For a summary see Kessler, *How We Got Here*, 119–32; Wolfe, *Hooking Up*, 17–65. For Hoff 's own story, see the oral history interview at www.sul.stanford.edu/depts/hasrg/histsci/silicongenesis/hoff-ntb.html [accessed July 26, 2004].

311 *Not built to survive nuclear war* Hafner and Lyon, *Where Wizards Stay Up Late*, 10; Abbate, *Inventing the Internet*, 37, 43–47.

Internet users These statistics come from the International Telecommunications Union: www.itu.int/ITU-D/ict/statistics/at_glance/KeyTelecom99.html.

Forty million Web sites Carr, *Does IT Matter?*, 66. For the Internet's origins, see Hafner and Lyon, *Where Wizards Stay Up Late*; Abbate, *Inventing the Internet.*

312 *Cell phone users* www.itu.int/ITU-D/ict/statistics/at_glance/KeyTelecom99.html. For the history of cell phones, see Murray, *Wireless Nation*. For only twenty-four people being able to use mobile phones in New York in 1981, see ibid, 19.

Number of AT&T calls "A Brief History: The Bell System." www.att.com/history/history4.html.

80 percent on fiber-optic lines www.ciscopress.com/articles/article.asp?p=170740. For a summary of fiber-optic technology, see Hecht, *City of Light*; Kessler, *How We Got Here*, 137–43.

A billion transistors www.intel.com/technology/magazine/silicon/mooreslaw0405.pdf; www.wired.com/wired/archive/10.01/gilder_pr.html.

Unprecedented improvement Garreau, *Radical Evolution*, 50–51.

313 *Moore's Law* Originally Moore held that the number of transistors would double every 12 months, a figure that was later increased to 24 months, then decreased to 18 months. See Kessler, *How We Got Here*, 127; Gordon, "Does the 'New Economy' Measure Up to the Great Inventions of the Past?", 51; London, "Man who foresaw the rise of chips with everything."

280 million miles and circle the earth 11,320 times *Business Week*, cited in Carr, *Does IT Matter?*, 66.

Cost dropping by 50 percent every eighteen months Garreau, *Radical Evolution*, 51.

Costs of transistors Intel, "Expanding Moore's Law," 2. According to Sichel, *The Computer Revolution*, 3, the cost of computers and peripheral equipment declined at an average annual rate of 15.1 percent between 1970 and 1994.

Information technology Carr, *Does IT Matter?*, xii, offers a handy definition: "all the technology, both hardware and software, used to store, process, and transport information in digital form."

Boeing Levy and Murnane, *The New Division of Labor*, 32.

Innovate organizationally For the importance of "organization capital," as it is dubbed by economists, see Lohr, "New Economy."

"*The early commitment to technology*" Quoted in "The Most Underrated CEO Ever," 244. For more on Wal-Mart's success, see Lewis, *The Power of Productivity*, 94; Carr, *Does IT Matter?*, 93–94.

Owed their rise to railroads Carr, *Does IT Matter?*, 24, points out that "the great speed, capacity, and coverage of the railroads" made it "economical to ship finished products, rather than just raw materials and industrial components, over long distances."

314 *Globalization* For an example of how this phenomenon works in the apparel industry, see Kahn, "Making Labels for Less." This article recounts how Liz Claiborne Inc. designs clothes in the U.S. that are produced by "250 suppliers in 35 countries as diverse as Saipan, Mexico and Cambodia." The company is planning to focus its production in southern China in order to reduce the time involved in getting new styles into stores from ninety days to sixty days.

Israel vs. Saudi Arabia According to the World Bank, Israel's gross national income per capita was $17,380 in 2004, while Saudi Arabia's was $10,430. http://devdata.worldbank.org/data-query [accessed January 10, 2006].

"*Competition for the best information*" Wriston, "Bits, Bytes, and Diplomacy," 176.

70 to 75 percent service workers Lewis, *The Power of Productivity*, xv; Kudyba and Diwan, *Information Technology*, 41. According to Lewis, manufacturing accounts for 20 to 25 percent of the economy, agriculture for less than 5 percent.

Driving world economic growth According to the National Science Board, *Science and Engineering Indicators 2004*: "During the 22-year period examined (1980–2001), high-technology production grew at an inflation-adjusted average annual rate of nearly 6.5 percent compared with 2.4 percent for other manufactured goods."

Average annual productivity growth Ferguson, "Productivity: Past, Present, and Future."

315 *Not as important as electricity or plumbing* This point is taken from Gordon, "Does the 'New Economy' Measure Up to the Great Inventions of the Past?", 72.

"*The Goliath of totalitarianism*" From Reagan's speech at London's Guildhall, June 14, 1989, cited in Kalathil and Boas, *Open Networks, Closed Regimes*, 1.

Cell phones with text messaging For the use of this technology by Arab dissidents, see Coll, "In the Gulf, Dissidence Goes Digital."

Twisting it for their own purposes Kurlantzick, in "Dictatorship.com," writes: "the Web may actually be helping to keep some dictatorships in power. Asian dissidents have told me that the Web has made it easier for authoritarian regimes to monitor citizens." This conclusion is supported by Kalathil and Boas, *Open*

Networks, Closed Regimes. Based on a study of China, Cuba, Singapore, Vietnam, Burma, the United Arab Emirates, Saudi Arabia, and Egypt, the authors conclude, at 136, "the authoritarian state is hardly obsolete in the era of the Internet." However, see also reports of how text messaging and cell phones are undermining the control of Beijing: Yardley, "A Hundred Cellphones Bloom, and Chinese Take to the Streets."

"We've moved from a media oligarchy" Westin, "Don't Blame the Networks."

"Giant vote-counting machine" Wriston, *The Twilight of Sovereignty*, 9.

316 *"Opt out of the Information Standard"* Wriston, *The Twilight of Sovereignty*, 70.

Mobilizing with the Internet An example was Howard Dean's campaign for the Democratic presidential nomination in 2004. His campaign manager explains how he utilized the Internet: Trippi, *The Revolution Will Not Be Televised.*

The market state See Bobbitt, *The Shield of Achilles.*

U.S. economy vs. Japan and Germany See Lewis, *The Power of Productivity*, for an explanation of differing growth rates.

Leading producer National Science Board, *Science and Engineering Indicators 2004.* Japan was at 12.9 percent, France (the EU leader) at 5.5 percent. The EU collectively was at 22.8 percent. Using another yardstick, of the twenty biggest information technology companies in the world in 1996, twelve were based in the U.S., seven in Japan, one in Taiwan, and one in Europe. See table 7.3 in Chandler, *Inventing the Electronic Century*, 232.

650 million computers www.itu.int/ITUD/ict/statistics/at_glance/KeyTelecom99.html.

317 *Patents* "The United States. accounts for 42 percent of patents filed at the World Intellectual Property Organization, well ahead of Germany's 13 percent and Japan's 10 percent." Odom and Dujarric, *America's Inadvertent Empire*, 191.

Two-thirds of the most influential scientists work in the U.S. Paarlberg, "Knowledge as Power," 128. This is based on the Institute for Scientific Information's survey of the world's 1,222 most "highly cited scientists."

Almost 50 percent of global R&D National Science Board, *Science and Engineering Indicators 2004.* For more than $250 billion annually in R&D spending, see Paarlberg, "Knowledge as Power," 129.

More Web sites than any other country Nye, *Soft Power*, 34.

CHAPTER 10: PRECISION AND PROFESSIONALISM

318 *Iraqi air defenses* Hallion, *Storm over Iraq*, 153.

319 *Task Force Normandy* Scales, *Certain Victory*, 157–60; Mackenzie, "Apache Attack"; Waller, "Secret Warriors"; Atkinson, *Crusade*, 13–50; Hallion, *Storm over Iraq*, 166–167; Gordon and Trainor, *The Generals' War*, 209–10.

Cody He went on to become a four-star general and, starting in 2004, vice chief of staff of the army.

F-117 attack on Nukhayb Coyne, "A Strike by Stealth."

320 *Most heavily defended city ever attacked* Baghdad's defenses were seven times as dense as Hanoi's during Operation Linebacker II in 1972. Hallion, *Storm over Iraq*, 169.

CNN off the air and "blackslapping" Clancy with Horner, *Every Man a Tiger*, 541.

"Shit hot" Atkinson, *Crusade*, 40.

321 *Thirty-eight lost* Keaney and Cohen, *Revolution in Warfare?*, 52.

Almost seven hundred friendly airplanes In the first 24 hours, 668 aircraft few 1,300 sorties. Hallion, *Storm over Iraq*, 166.

"Most awesome" Dugan, "The Air War."

322 *"This war didn't take 100 hours to win"* Quoted in Scales, *Certain Victory*, 35.

Drug use Surveys showed that 60 percent of soldiers in Vietnam had used marijuana and 23 percent had used heroin. King and Karabell, *The Generation of Trust*, 34.

Eight hundred fraggings King and Karabell, *Generation of Trust*, 22; Scales, *Certain Victory*, 6. The term "fragging" comes from a weapon frequently used in such attacks: the fragmentation grenade.

32 percent had a great deal of confidence King and Karabell, *Generation of Trust*, 21.

Half high-school dropouts King and Karabell, *Generation of Trust*, 23; Kitfield, *Prodigal Soldiers*, 207; Scales, *Certain Victory*, 16.

323 *"Hollow army"* Quoted in King and Karabell, *Generation of Trust*, 23.

Defense spending The defense budget doubled between 1980 ($143.9 billion) and 1985 ($294.7 billion). In inflation-adjusted terms this was an increase of 54.7 percent. Wirls, *Buildup*, 36.

Trappist monk Kitfield, *Prodigal Soldiers*, 209.

Second-best jingle of the century www.adage.com/century/jingles.html.

"Be All You Can Be" King and Karabell, *Generation of Trust*, 70–74; Kitfield, *Prodigal Soldiers*, 209–14, 228–29.

97 percent high school graduates Kitfield, *Prodigal Soldiers*, 326.

Drug use fell from 27.6 percent to 3.4 percent King and Karabell, *Generation of Trust*, 35.

324 *"More equal opportunity"* Powell, *My American Journey*, 487. The number of black officers and NCOs in the army increased from 14 percent in 1970 to 31 percent in 1990. By the time of the Gulf War there were twenty-four black army generals (more than 7 percent of the total). King and Karabell, *Generation of Trust*, 43.

153,000 British soldiers in 1992 John Strawson, "Conflict in Peace, 1963–1992," in Chandler, *The Oxford History of the British Army*, 343.

Soviet force size From the University of Michigan's Correlates of War Project: www.umich.edu/~cowproj/capabilities.html.

Soviet advantage For instance in 1980 the U.S. had 10,700 tanks vs. 48,000 for the USSR. Dunigan and Macedonia, *Getting It Right*, 159.

325 *Fritz X* Rip and Hasik, *The Precision Revolution*, 203–4.

Inaccuracy of V-2 Rip and Hasik, *The Precision Revolution*, 54–55.

Smart bombs between World War II and Vietnam Craven and Cate, *The Army Air Forces in World War II*, VI:253–62; Werrell, *Chasing the Silver Bullet*, 139–42; Budiansky, *Air Power*, 406–7.

326 *Barely able to hold their own in air-to-air combat* "In comparison to an 8 to 1 victory-loss ratio in World War II and a 10 to 1 ratio in Korea, the United States [air forces] mustered no better than a 2.4 to 1 ratio over North Vietnam, and for a while it was roughly 1 to 1." Hallion, *Storm over Iraq*, 20. See also Lambeth, *The Transformation of American Air Power*, 45.

More than 1,500 aircraft downed, 95 percent from ground fire Budiansky, *Air Power*, 395.

327 *Paveway vs. dumb bombs* Figures are from Werrell, *Chasing the Silver Bullet*, 152.

Primarily by laser guidance Some other "smart" bombs and missiles were also employed, such as the Walleye, which was guided by a nose camera linked to a small TV screen in the cockpit, but they were less reliable.

Paveway development and Thanh Hoa bridge Loeb, "Bursts of Brilliance"; Werrell, *Chasing the Silver Bullet*, 147–53; Anderegg, *Sierra Hotel*, 123; Reed, *At the Abyss*, 196–97.

28,000 Paveways Budiansky, *Air Power*, 408; Loeb, "Bursts of Brilliance." By contrast, only 17,000 precision munitions were used in the Gulf War. Keaney and Cohen, *Revolution in Warfare*, 191–93.

0.2 percent were precision guided Rip and Hasik, *The Precision Revolution*, 224.

Relatively few in number In the mid-1980s the U.S. Air Force, Navy, and Marine Corps were spending $167 billion annually on new fighters while the entire laser-guided bomb program was allocated just $1.8 billion—an indication of skewed priorities. Budiansky, *Air Power*, 409.

328 "*During World War II*" This is a paraphrase of a point John A. Warden frequently made. See Halberstam, *War in a Time of Peace*, 52. For Warden's own (less pithy) rendition, see Warden, *The Air Campaign*, 147. According to Werrell, *Chasing the Silver Bullet*, 259, the circular error probable (CEP) of bombs—defined as the radius of a circle in which a bomb or missile will land at least half the time—was 3,300 feet (not 2,300 feet) in World War II. In Vietnam it was down to 400 feet. By the Gulf War it had fallen to 200 feet.

Brown and Perry backed precision weaponry In 1978, Perry said: "Precision guided weapons, I believe, have the potential of revolutionizing warfare." Vickers and Martinage, *The Revolution in War*, 8–9. For Perry's own account see his essay, "Military Technology."

B-2 For its history and specifications, see the websites of the Federation of American Scientists (www.fas.org/nuke/guide/usa/bomber/b-2.htm) and Global security.org (www.globalsecurity.org/wmd/systems/b-2.htm).

F-117 Ben Rich, head of the Skunk Works when the F-117 was designed, tells his story in Rich and Janos, *Skunk Works*. Thomas Reed, Air Force secretary in the Ford administration and in the early days of Carter, provides his perspective in Reed, *At the Abyss*, 202–3. For the account of F-117 pilots see Warren, *Bandits over Baghdad*. I have also relied on Werrell, *Chasing the Silver Bullet*, 120–37; Budiansky, *Air Power*, 410–12. Note that although the F-117 was designated a fighter and is about the same size as the F-15 it is really a light bomber; it does not carry weaponry for air-to-air combat.

Fighter-bombers Both the F-16 and F-15 were designed initially for air-to-air combat, the designers' slogan being "Not a pound for air-to-ground." But both were eventually utilized extensively in a ground-attack role. For the story of their development, see Werrell, *Chasing the Silver Bullet*, 77–99. Hallion, *Storm over Iraq*, 33–46, provides an overview of all the "super fighters" built in the 1970s.

329 *A-10* Its official designation is the Thunderbolt, but it is universally known as the Warthog. For its development, see Werrell, *Chasing the Silver Bullet*, 99–120.

A few were canceled The Sergeant York antiaircraft gun and the Roland surface-to-air missile never made it into mass production. The B-1 was canceled, then revived. See Hallion, *Storm over Iraq*, 71–72.

"Magic weapons" and "more and more complex" Fallows, *National Defense*, 35.

"Slavish devotion" Isaacson, "The Winds of Reform."

Hart For his 1986 critique of the military see Hart and Lind, *America Can Win*. For an overview of the military reform debate, see Wirls, *Buildup*, 79–101.

What the reformers did not realize I am indebted for this point to Budiansky, *Air Power*, 401–2. Perry made the same point earlier, in "Defense Reform and the Quantity-Quality Quandary," in Clark et al. (eds.), *The Defense Reform Debate*, 191–92: "Military reformers often confuse technology with complexity, but they are not synonymous." See also Lambeth, *The Transformation of American Air Power*, 80–81.

330 *Depleted uranium* www.globalsecurity.org/military/systems/ground/m1a1.htm.

M1 Almost all of the Abrams tanks in the Gulf War were M1A1s, an upgrade from the original M1, which has since been superseded by the M1A2. For a summary of its capabilities see www.globalsecurity.org/military/systems/ground/m1intro.htm and www.fas.org/man/dod-101/sys/land/m1.htm.

Rounds per kill The Abrams was estimated to need 1.2 rounds per kill. Scales, *Certain Victory*, 81.

Night-vision devices This paragraphs draws on www.globalsecurity.org/military/systems/ground/nvg.htm.

AWACS Grier, "A Quarter Century of AWACS"; Werrell, *Chasing the Silver Bullet*, 194–99; Hallion, *Storm over Iraq*, 309–10. Grier, "Science Projects," notes that the AWACS was the successor to the EC-121 Warning Star, the first airborne radar system, which was deployed beginning in 1953 to protect North American airspace and was later sent to Vietnam.

JSTARS Werrell, *Chasing the Silver Bullet*, 199–205; Rip and Hasik, *The Precision Revolution*, 147–51; Scales, *Certain Victory*, 167–68; Hallion, *Storm over Iraq*, 310–11.

331 *U.S. surpassed USSR in space* The first U.S. satellite, *Explorer I*, was launched in 1958. By the end of the year, the U.S. had successfully launched seven satellites to only one for the USSR. Whalen, *Origins of Satellite Communications*, 39. For an engaging overview of early U.S. satellite efforts based on declassified documents, see Taubman, *Secret Empire*.

The role of satellites See Sir Peter Anson and Dennis Cummings, "The First Space War: The Contribution of Satellites to the Gulf War," in Campen (ed.), *The First Information War*, 128–31; Rip and Hasik, *The Precision Revolution*, 140; Friedman, *The Future of War*, 313–24; Werrell, *Chasing the Silver Bullet*, 211, 265; Scales, *Certain Victory*, 171–72.

GPS Rip and Hasik, *The Precision Revolution*, is the best source. See also Werrell, *Chasing the Silver Bullet*, 262–65. Grier, "Science Projects," notes that the navy had developed its own satellite navigation system, known as Transit, in the 1960s but it was accurate to just 83 feet, provided positions in only two dimensions, and required a navy vessel to go very slowly while using it.

332 *Top Gun improved performance* "Between 1970 and 1973 . . . the Navy had improved its kill-loss ratio by a factor of five, notching twelve enemy kills for every loss." Kitffield, *Prodigal Soldiers*, 162. See also Chapman, *The Origins and Development of the National Training Center*, 15.

Red Flag Kitfield, *Prodigal Soldiers*, 165–74; Lambeth, *The Transformation of American Air Power*, 60–69. In 1984, the navy set up its own training range for air strikes, as opposed to air-to-air fighting: the Strike Warfare Center at Fallon Naval Air Station, Nevada. Hallion, *Storm over Iraq*, 103.

National Training Center Chapman, *The Origins and Development of the National Training Center*; Kitfield, *Prodigal Soldiers*, 306–18; Scales, *Certain Victory*, 20–23.

"Everyone had been there before" Clancy and Horner, *Every Man a Tiger*, 131.

Training and Doctrine Command For its history see Romjue, *Prepare the Army for War*.

333 *"Weight of materiel" and "Prepare to win the first battle"* Quoted in Citino, *Blitzkrieg to Desert Storm*, 256.

"Rapid, unpredictable, violent, and disorienting" Ibid, 264.

AirLand Battle In addition to the sources previously cited, see the Tofflers, *War and Anti-War*, 48–63, for the story of its development.

334 *Other commands* Not all of them were established by Goldwater-Nichols. The Special Operations Command, for instance, was created in 1987 by the Nunn-Cohen Act. Other changes were made subsequently, such as the creation of the Northern Command, the Joint Forces Command, and Strategic Command, and the elimination (or, more properly, downgrading) of Atlantic Command and Space Command.

Goldwater-Nichols See Kitfield, *Prodigal Soldiers*, 277–99; Locher, "Taking Stock of Goldwater-Nichols."

335 *"Luckiest man alive"* Clancy and Horner, *Every Man a Tiger*, 434.

"This is dumb! This is stupid!" Atkinson, *Crusade*, 220.

"Screaming" and "thrived on confrontation" Clancy and Horner, *Every Man a Tiger*, 11.

"Not an easy commander" Schwarzkopf, *It Doesn't Take a Hero*, 299.

Worked 18–20 hours a day Atkinson, *Crusade*, 70.

336 *"The worst battalion"* Schwarzkopf, *It Doesn't Take a Hero*, 158.

"I had to be a complete son of a bitch" Ibid, 159.

"Bad shape" Ibid, 178.

Vuono's sponsorship Gordon and Trainor, *The Generals' War*, 41–42.

Reorienting Centcom's war plan Gordon and Trainor, *The Generals' War*, 42–46.

337 *Force size in November* Atkinson, *Crusade*, 54; Scales, *Certain Victory*, 97.

Final coalition force size Figures from Pagonis, *Moving Mountains*, 11–12; Keaney and Cohen, *Revolution in Warfare?*, 7; Hallion, *Storm over Iraq*, 157.

338 *Iraqi troop strength* Scales, *Certain Victory*, 161. Estimates vary and no exact figure exists. For a slightly lower estimate, see Pollack, *The Threatening Storm*, 41.

Computers, cell phones, 3×5 cards Scales, *Certain Victory*, 60–61.

Logistical feats All figures are from Pagonis, *Moving Mountains*. For figures on tracked vehicles and wheeled vehicles, see 6. For figures on meals and fuel, see 1. For figures on mail, see 128. The analogy to Atlanta is from Scales, *Certain Victory*, 40–41.

250,000 troops, 64,000 vehicles Scales, *Certain Victory*, 145. Atkinson, *Crusade*, 304, gives a slightly higher figure for vehicles.

Division logistical requirements These are for the 24th Infantry Division. See Scales, *Certain Victory*, 255. A large part of the logistics load was due to the fact that each M1A1 consumed six gallons per mile (not miles per gallon). Atkinson, *Crusade*, 251–52.

"Logisticians will not let you down" Pagonis, *Moving Mountains*, 140.

339 *Description of Horner* This draws on Atkinson, *Crusade*, 39.

No four-star bomber pilots Budiansky, *Air Power*, 397.

"Airheaded airmen" Clancy and Horner, *Every Man a Tiger*, 15.

Warden For his theories, see Warden, *The Air Campaign*.

Warden vs. Horner Accounts of this confrontation may be found in Clancy and Horner, *Every Man a Tiger*, 255–65; Gordon and Trainor, *The Generals' War*, 91–96; Budiansky, *Air Power*, 413–18; Keaney and Cohen, *Revolution in Warfare?*, 30–31. For a contrasting account sympathetic to Warden and disdainful of Horner, see Halberstam, *War in a Time of Peace*, 47–56.

340 *Bulk of strikes against ground forces* According to the Gulf War Air Power Survey, only 15 percent of strikes were against strategic targets while at least 56 percent were against Iraqi ground forces. Keaney and Cohen, *Revolution in Warfare?*, 55.

Warden as author of air campaign For instance, Halberstam wrote in *War in a Time of Peace*, 47: "If one of the newsmagazines had wanted to run on its cover the photograph of the man who had played the most critical role in achieving victory, it might well have chosen Warden instead of Powell or Schwarzkopf."

Modeled on Israel in 1982 During the invasion of Lebanon, Israel destroyed 19 Syrian SAM batteries and 84 Syrian aircraft with no losses. Grant, "The Bekaa Valley War"; Hallion, *Storm over Iraq*, 97–98; Lambeth, *The Transformation of American Air Power*, 92–96.

Scud hunt Hallion, *Storm over Iraq*, 177–88; Gordon and Trainor, *The Generals' War*, 228–48.

Suffered more in thirty minutes Gordon and Trainor, *Generals' War*, 286; Pollack, *Arabs at War*, 244.

Khafji Morris, *Storm on the Horizon*; Atkinson, *Crusade*, 208–13; Gordon and Trainor, *The Generals' War*, 267–88; Scales, *Certain Victory*, 190–91; Keaney and Cohen, *Revolution in Warfare?*, 16–17.

341 *Interdiction strikes* For an analysis of their impact, see Keaney and Cohen, *Revolution in Warfare?*, 80–87.

Destroying tanks Gordon and Trainor, *The Generals' War*, 322–23; Atkinson, *Crusade*, 262–65; Scales, *Certain Victory*, 188; Hallion, *Storm over Iraq*, 203; Lambeth, *The Transformation of American Air Power*, 124–25 (for 1,300 vehicles destroyed).

28 million leaflets Scales, *Certain Victory*, 197.

Seduced by Bart Simpson Scales, *Certain Victory*, 196.

342 *Casualties on the first day* U.S. losses were eight dead, 27 wounded. Powell, *My American Journey*, 503.

343 *More than 11,000 shells and rockets* Atkinson, *Crusade*, 394, and Scales, *Certain Victory*, 226. Clancy and Franks, *Into the Storm*, 342, and Gordon and Trainor, *The Generals' War*, 379, give a figure of about 6,500.

Eddie Ray As a lieutenant-colonel in command of a Light Armored Reconnaissance battalion, he would also distinguish himself during the invasion of Iraq in 2003. See Gordon and Trainor, *Cobra II*, 243.

344 *Battle of Al Burqan* The most dramatic narrative is Kitfield, *Prodigal Soldiers*, 406–10. This account also draws on Atkinson, *Crusade*, 411–15; Gordon and Trainor, *The Generals' War*, 363–69; U.S. News & World Report, *Triumph Without Victory*, 318–19.

As fast as any tank unit in history For a list of the fastest combat movements of the twentieth century see Dunigan and Macedonia, *Getting It Right*, 211. Their list suggests the U.S. advance in 1991 was three times as fast as the Wehrmacht's in 1940 and almost twice as fast as Israel's in the Sinai in 1967.

Advance of the 24th Infantry Division Scales, *Certain Victory*, 254–60. For an account by five tank lieutenants who served in the 24th see Vernon et al., *The Eyes of Orion*.

"Slow, ponderous pachyderm mentality" Schwarzkopf, *It Doesn't Take a Hero*, 433.

75 percent of armor intact Keaney and Cohen, *Revolution in Warfare?*, 92. According to Biddle, *Military Power*, 142–43, the Republican Guard lost less than 24 percent of its prewar strength in the bombing campaign, as compared with ordinary Iraqi units, which lost 48 percent of their tanks, 30 percent of their armored personnel carriers, and 60 percent of their artillery.

345 *Battle of 73 Easting* The best description of McMaster's actions (on which I have drawn heavily) is Scales, *Certain Victory*, 1–4. See also Atkinson, *Crusade*, 441–48; Gordon and Trainor, *The Generals' War*, 387–92; Biddle, *Military Power*, 134; U.S News & World Report, *Triumph Without Victory*, 332–42; Clancy and Franks, *Into the Storm*, 443–45. McMaster later served with distinction as a colonel commanding the 3rd Armored Cavalry Regiment in Iraq in 2005–6.

"One poor soul" Kelly, *Martyrs' Day*, 238.

346 *"Cooked to the point of carbonization"* Ibid, 240.

Senior policy makers worried National Security Adviser Brent Scowcroft wrote: "We had all become increasingly concerned over impressions being created in the press about the 'highway of death' from Kuwait City to Basra." Bush and Scowcroft, *A World Transformed*, 485.

"Gut reaction" and "I don't have any problem" Schwarzkopf, *It Doesn't Take a Hero*, 469–70. For Powell's version, see Powell, *My American Journey*, 504–8.

"The gates are closed" Gordon and Trainor, *The Generals' War*, 417.

Iraqis managed to get out of Kuwait Keaney and Cohen, *Revolution in Warfare?*, 92–93.

No one was willing to challenge Powell or Schwarzkopf Gordon and Trainor, *The Generals' War*, 419, 423.

"Destroy[ing] the Republican Guard" Schwarzkopf, *It Doesn't Take a Hero*, 382.

347 *FM radios didn't work well* Gordon and Trainor, *The Generals' War*, 387.

Grease pencils Scales, *Certain Victory*, 375.

92 percent dumb bombs GAO, *Operation Desert Storm*. According to Keaney and Cohen, *Revolution in Warfare?*, 191, a total of 227,000 munitions were dropped, of which 17,000 were precision-guided. Other studies give slightly different figures for the percentage of unguided bombs. According to Rip and Hasik, *The Precision Revolution*, 212, and Atkinson, *Crusade*, 227, it was 93 percent. According to Hallion, *Storm over Iraq*, 188, it was 91 percent.

Smart bombs did not find their targets all the time A postwar Government Accounting Office report found, for instance, that the hit rate for F-117 bombs was no more than 60 percent—still high by historic standards but less than the 80 percent rate reported by the Air Force. GAO, *Operation Desert Storm*.

"*Did not have a clear picture*" Bush and Scowcroft, *A World Transformed*, 484.

Friendly fire 35 out of 147 U.S. battle deaths (or almost 24 percent) were attributed to friendly fire. Hammond, "Myths of the Gulf War."

Experts predicted tens of thousands of casualties For a summary of faulty predictions, see Muravchik, "The End of the Vietnam Paradigm?" and Friedman, *The Future of War*, 253–54.

16,000 body bags Atkinson, *Crusade*, 183.

Total casualties Biddle, *Military Power*, 133. Other sources give slightly different figures. See, for example, Correll, "Casualties," which lists 247 battle deaths and 901 wounded.

U.S. casualties Based on Pentagon statistics: http://web1.whs.osd.mil/mmid/casualty/GWSUM.pdf. Note that in addition to 147 battle deaths there were also 235 "non-hostile" deaths from accident, disease, and suicide.

U.S. aircraft losses Hallion, *Storm over Iraq*, 196.

Young men safer in Gulf War than at home "The average death rate for those personnel deployed in the Gulf was 69 per one hundred thousand. For males 20 to 30 years of age living in the United States during the same period, the death rate was 104 per one hundred thousand." Hammond, "Myths of the Gulf War."

348 "*Lowest cost in human life ever recorded*" Scales, *Certain Victory*, 5.

Allied vs. Iraqi strength at beginning of war Scales, *Certain Victory*, 216. No more than 350,000 Iraqis at end of air campaign: Pollack, *Arabs at War*, 239.

Three-to-one advantage needed The late U.S. Army Colonel Trevor Dupuy, in *Understanding War*, 34, called this "a rule of thumb so widely accepted that it has become virtually a military principle."

349 *Twelve-year gap* Biddle, *Military Power*, 102.

2 percent of combat sorties but hit 40 percent of "strategic" targets Keaney and Cohen, *Revolution in Warfare?*, 190; Hallion, *Storm over Iraq*, 174.

Soviet collapse For the impact of Western military advances on the Soviet political and military leadership, see Trulock, Hines, and Herr, "Soviet Military Thought in Transition." For Russian perceptions of the Gulf War as a turning point, see FitzGerald, "The Impact of the Military-Technical Revolution on Russian Military Affairs."

350 "*Skyrocketed*" Bush and Scowcroft, *A World Transformed*, 492.

Falling defense spending The U.S. military budget fell in real terms from $299 billion in 1990 to $272 billion in 1995—a 9 percent decrease. The active-duty force went from 2 million to 1.5 million personnel in the same period—a 25 percent decrease. Based on statistics from the Center for Strategic and Budgetary Assessments (www.csbaonline.org).

CHAPTER 11: SPECIAL FORCES AND HORSES

352 *Insertion of ODA 595* In addition to the *Frontline* interviews ("Campaign Against Terror"), see Briscoe et al., *Weapon of Choice*, 118–22. There is also an account in Moore, *The Hunt for bin Laden*, 62–69, but the book is riddled with inaccuracies. The author relied heavily on "my longtime Green Beret friend [Jonathan] Keith Idema" (xi) who, unfortunately, turned out to be a fraud and a felon. See Blake, "Tin Soldier," for the story of this con job.

353 *Special Forces* For background, see Robinson, *Masters of Chaos*. The term *Special Forces* (SF) properly refers only to the Green Berets. All other members of the U.S. Special Operations Command, which is composed of Navy, Air Force, and Army personnel ranging from the Night Stalkers air regiment to the SEALs, are known as Special Operations Forces (SOF). The distinction often gets lost in casual use.

Mixup of Mohammed Attah and Attah Mohammed Author's interview with Major Dean (Last Name Withheld), leader of ODA 534. This book will not provide the last names of Special Forces personnel without their consent unless they have already been publicly identified in the news media. Dean, for instance, does not want his last name used. But other special operators, such as Mark Nutsch and Jason Amerine, have already been publicly identified. And others have told the author they do not mind if their full names are used.

"It was like the sand people" Chief Warrant Officer Bob, in Frontline, "Campaign Against Terror: Interview: U.S. Special Forces ODA 595."

354 *Anti-Taliban movement* Dostum was not technically part of the Northern Alliance, but rather of the National Islamic Movement.

"Irritable bear" Ignatieff, "Nation Building Lite."

15,000 for Northern Alliance vs. 40,000 Taliban Briscoe et al., *Weapon of Choice*, 95. O'Hanlon, "A Flawed Masterpiece," gives a figure of 50,000 to 60,000 for the Taliban. Lambeth, *Air Power Against Terror*, 77, estimates Taliban strength at 45,000.

More than eighty years General John J. Pershing led cavalrymen in pursuit of Pancho Villa in northern Mexico in 1916–17. See Boot, *The Savage Wars of Peace*, ch. 8.

"It was pretty painful" Robert Young Pelton, "The Legend of Heavy D and the Boys," *National Geographic Adventure* (March 2002), in Willis, *Boots on the Ground*, 43.

355 *"This is General Dostum speaking"* Briscoe et al., *Weapon of Choice*, 127.

"The real fight" Author's interview with a navy officer who requested anonymity, January 23, 2006.

Dispatching more soldiers and the specter of Vietnam Woodward, *Bush at War*, 256–57, 291. For an alarmist article written at the time see Apple, "A Military Quagmire Remembered."

Match their triumph over the Soviet Union According to a journalist who examined one of al Qaeda's computers, "The jihadis expected the United States, like the Soviet Union, to be a clumsy opponent. Afghanistan would again become a slowly filling graveyard for the imperial ambitions of a superpower." Cullison, "Inside Al-Qaeda's Hard Drive."

356 *Maintained the computer's files* Cullison, "Inside Al-Qaeda's Hard Drive." For more on al Qaeda's reliance on the Internet, see Wright, "The Terror Web," and Talbot, "Terror's Server." Both articles quote one expert as saying that, as of 2004, there were over 4,000 terrorist web sites.

"*Virtual Caliphate*" From a briefing at which the author was present at Centcom's headquarters, McDill Air Force Base, Tampa, Florida, on April 11, 2005.

Netwar doesn't require advanced technology Arquilla and Ronfeldt, "The Advent of Netwar (Revisited)," in Arquilla and Ronfeldt, *Networks and Netwar*, 11.

"*Rely on high-speed transportation*" Menon, "Terrorism Inc." For more on al Qaeda as a multinational entity, akin to a high-tech company, see Friedman, *The World Is Flat*, 429–36, 441–52.

Al Qaeda spent $30 million a year National Commission on Terrorist Attacks Upon the United States, *The 9/11 Commission Report*, 171.

357 "*Eleven keystrokes to complete a search*" Walsh, "Learning to Spy."

NSA's problems Bamford, *Body of Secrets*, 456–70.

Tracking cell phones Van Natta and Butler, "How Tiny Swiss Cellphone Chips Helped Track Global Terror Web."

"*Some of the best operational security*" Berkowitz, *The New Face of War*, 10.

History of al Qaeda This section is based on National Commission on Terrorist Attacks Upon the United States, *The 9/11 Commission Report.*

358 *U.S. response* For a short explanation of why the U.S. did not take stronger steps to stop al Qaeda, see Shultz, "Showstoppers," which stresses military as well as civilian reluctance to risk casualties or take chances. For the longer version see Coll, *Ghost Wars*. For the account of the NSC's counterterrorism director, see Clarke, *Against All Enemies*.

359 "*Fell into the void*" National Commission on Terrorist Attacks Upon the United States, *The 9/11 Commission Report*, 263.

360 *Fourth airplane might have hit its target* "NORAD officials have maintained that they would have intercepted and shot down United 93. We are not so sure." National Commission on Terrorist Attacks Upon the United States, *The 9/11 Commission Report*, 45.

Economic toll On September 4, 2002, New York City Comptroller William Thompson, Jr., issued a report that estimated the total economic damage of 9/11 at $82 billion to $94 billion. www.comptroller.nyc.gov/bureaus/bud/reports/impact-9-11-year-later.pdf.

Less than $500,000 National Commission on Terrorist Attacks Upon the United States, *The 9/11 Commission Report*, 169.

Description of al Qaeda and planning for 9/11 This section is based primarily on *The 9/11 Commission Report*, supplemented by Benjamin and Simon, *The Age of Sacred Terror*, and Coll, *Ghost Wars*.

361 "*We must change*" www.defenselink.mil/speeches/2001/s20010910-secdef.html.

Size of the military shrank 30 percent Chamberlain, "FY2005 Defense Budget," 22. Military spending fell 15 percent in constant dollars during that period.

Cost of cruise missiles www.globalsecurity.org/military/systems/munitions/bgm-109-specs.htm.

Cost of laser-guided bombs www.globalsecurity.org/military/systems/munitions/gbu-24.htm.

JDAM accuracy Rip and Hasik, *The Precision Revolution*, 235–38.

29 percent precision munitions in Kosovo Lambeth, *NATO's Air War for Kosovo*, 88.

362 *Predator and Global Hawk specifications* www.globalsecurity.org/intell/systems/predator.htm; www.globalsecurity.org/intell/systems/global_hawk.htm; Bone and Bolkcom, *Unmanned Aerial Vehicles: Background and Issues for Congress.*

Monitor an area the size of Illinois Brasher, "Unmanned Aerial Vehicles and the Future of Air Combat."

"It's like having a low-earth orbit satellite" Newman, "The Little Predator That Could."

363 *Description of Joint Operations Center* Based on the author's visit, April 11, 2005, and on Franks, *American Soldier*, 287.

100 T-1 lines Grant, "The War Nobody Expected." For the capacity of a T-1 line, see http://computer.howstuffworks.com/question372.htm.

Special Forces communications equipment Ackerman, "Special Operations Forces Become Network-Centric."

"As the Information Age matured" Franks, *American Soldier*, 174–75.

364 *"One of the last bastions of central planning"* www.defenselink.mil/speeches/2001/s20010910-secdef.html.

"Transformed organizations" Ibid.

"Pentagon bureaucracy" as "adversary" Ibid.

"The sweepstakes have already begun" Quoted in Mann, *Rise of the Vulcans*, 291. For more on Rumsfeld's pre-9/11 conflicts, see ibid, 288–93; Scarborough, *Rumsfeld's War*, 118–27; Kitfield, *War & Destiny*, 28–32.

365 *Tenet first presented plan* National Commission on Terrorist Attacks Upon the United States, *The 9/11 Commission Report*, 332; Woodward, *Bush at War*, 74–92.

"Franks is not a whiner" Retired general Crosbie Saint, quoted in Boyer, "The New War Machine."

Not a deep thinker Author's interview with a U.S. Army colonel who would prefer to remain anonymous, January 18, 2005. Boyer, "The New War Machine," also notes that Franks "is not one of the military's deep thinkers—he spends his free time watching movies on his portable DVD player, his taste running to Eddie Murphy comedies."

"Spoiled, unfocused, immature" Franks, *American Soldier*, 33.

"Near-mutinous ill-discipline" Ibid, 119.

366 *"Harsh taskmaster"* Ibid, 313.

"Genetically impatient" Ibid, 287.

"I had no tolerance for this parochial bullshit" Ibid, 276.

Franks and Rumsfeld Newt Gingrich, who knew both men, is quoted as saying: "I think that one of the things that made Tommy Franks so successful was that Franks figured out in the second or third meeting that we just got to go toe-to-toe and disagree bluntly or I'll never get anywhere with this guy. And as a result they developed a very good relationship." Scarborough, *Rumsfeld's War*, 138.

"He not only thought outside the box" Franks, *American Soldier*, 232.

367 *Two million rations* Lambeth, *Air Power Against Terror*, 268.

"Coming up with a target list" DeLong, *Inside Centcom*, 35.

Mullah Omar's convoy For two different versions of the story see DeLong, *Inside Centcom*, 37–38, and Franks, *American Soldier*, 289–96. A third version comes from Hersh, "King's Ransom," who first brought this incident to light but whose information was garbled.

Death of Atef Grey, "How the US Killed Al-Qaeda Leaders by Remote Control"; Richburg and Lynch, "Afghan Victors Agree to Talks in Berlin."

Predators fired 115 missiles and designated 525 targets Vickers and Martinage, *The Revolution in War*, 35.

Collateral damage in Afghanistan For a list of significant incidents see Cordesman, *The Ongoing Lessons of Afghanistan*, 63–64; Lambeth, *Air Power Against Terror*, 100–1.

"The technology itself" Budiansky, *Air Power*, 429.

368 *Commanders' caution* Lambeth notes in *Air Power Against Terror*, xxvi, a study commissioned by Centcom, that "the exceptional stringency of the rules of engagement caused by collateral damage concerns led to a target-approval bottleneck at CENTCOM that allowed many fleeting attack opportunities to disappear."

Confusion at SOCOM Schroen, *First In*, 148, complains about "the vacillation and indecision on putting Special Operations troops into the [Panjshir] valley with us."

"I don't want bin Laden and his thugs captured" Schroen, *First In*, 38. Woodward, *Bush at War*, 141, offers a similar account of this conversation.

"That amount is bulky" Schroen, *First In*, 28.

Task Force Dagger Its formal name was Joint Special Operations Task Force North (JSOTF-N). Southern Afghanistan was given to Joint Special Operations Task Force South, known as Task Force K-Bar. For command relationships see Bonin, U.S. *Army Forces Central Command in Afghanistan*, 8.

Setting up K2 Briscoe et al., *Weapon of Choice*, 64–82.

369 *"When is something going to happen?"* Franks, *American Soldier*, 296.

Raids on airfield and Omar's compound Franks, *American Soldier*, 302–6; Briscoe et al., *Weapon of Choice*, 109–15; Kiper, "Into the Dark." For a fanciful account of the raids, which are characterized as a "near disaster," see Hersh, "Escape and Evasion." As Franks notes, Hersh claimed that 16 AC-130s participated in the raid whereas no more than 3–4 were available for all of Afghanistan on any given night. Hersh's article also contains this odd quotation, which appears humorous in hindsight: "'This is no war for Special Operations,' one officer said—at least, not as orchestrated by CENTCOM and its commander, General Tommy R. Franks, of the Army, on October 20th."

ODA 534 landed on November 2 The date is given as November 5 in Briscoe et al., *Weapon of Choice*, but according to an e-mail to the author from Captain Dean, the team leader, the date was actually November 2.

371 *Plastic shower sandals* Captain Dean, in Frontline, "Campaign Against Terror: Interview: U.S. Special Forces ODA 534."

Two-thirds did not have guns Author's interview with Captain Dean.

Fourteen-hour round-trip Briscoe et al., *Weapon of Choice,* 100.

Sony PlayStation Author's interview with Major Cory Peterson.

Free medical care ODA 534's medic recalled how he performed an amputation on one Northern Alliance soldier whose leg had been mangled by a landmine. He laid the man down on a blanket spread over a mud courtyard at night and, by the light of a headlamp, sawed off the remains of his leg using the foldout blade from a Leatherman pocket tool, while another sergeant warded off a stray dog that tried to grab the bone. Staff Sergeant Jason, in *Frontline,* "Campaign Against Terror: Interview: U.S. Special Forces ODA 534"; Donatella Lorch, "The Green Berets up Close," *Newsweek* (January 14, 2002), in Willis, *Boots on the Ground,* 29.

"*I am surprised that we have not been slaughtered*" Grenz, "Legislature Honors Kansas Warrior"; Moore, *The Hunt for bin Laden,* 72.

"*He actually watched the round go in*" Author's interview with Captain Dean.

372 "*Three or four bombs hit*" Robert Young Pelton, "The Legend of Heavy D and the Boys," in Willis, *Boots on the Ground,* 46. For other accounts of the Battle of Bai Beche, see Biddle, *Afghanistan and the Future of Warfare,* 38–40; 3rd Battalion, 5th Group, "The Liberation of Mazar-e Sharif."

"*Like a scene out of a World War II movie*" Sgt. First Class Bobby in *Frontline,* "Campaign Against Terror: Interview: U.S. Special Forces ODA 534."

Demolished the girls' school 3rd Battalion, 5th Special Forces Group, "The Liberation of Mazar-e Sharif." The information about the deaths of the hostages and Attah's negotiators comes from the author's interview with Dean.

373 *AC-130* www.globalsecurity.org/military/systems/aircraft/ac-130-specs.htm.

"*It was pretty bad*" Author's interview with Sonntag. This account of the uprising is also based on Briscoe et al., *Weapon of Choice,* 104, 158–65; *The United States Army in Afghanistan,* 16–18; Robert Young Pelton, "The Legend of Heavy D and the Boys," in Willis, *Boots on the Ground,* 33–56.

Prison uprising planned in advance Author's interviews with Major Sonntag and Captain Dean.

374 "*We have the angel of death*" Mark Nutsch, in *Frontline,* "Campaign Against Terror: Interview: U.S. Special Forces ODA 595."

Lengths of fighter and bomber missions Grant, "The War Nobody Expected"; Lambeth, *Air Power Against Terror,* 293.

375 *Description of F-15 pilots' living conditions* Bowden, "The Kabul-Ki Dance." For the navy aviators' perspective, see Vogel, "Over Afghanistan, Gantlets in the Sky."

Calling in air strikes According to Colonel Robert Holmes, commander of the Air Force Special Operations Command's 720th Special Tactics Group, air force combat controllers called in about 85 percent of all air strikes in Enduring Freedom. See Newman, "Masters of Invisibility." For background on controllers, see Callandar, "Controllers." Combat controllers are specially trained to work with Special Operations forces; terminal attack controllers receive less commando training and generally work with regular army units, although some were employed with Special Forces in Afghanistan.

Control tower For a description, see Schroen, *First In*, 240, 290.

376 *Frustration over not enough aircraft* Schroen, *First In*, 290–92.

ODA 555 All quotes from Zastrow are from the author's interview with him. For more on ODA 555, see Priest, "In War, Mud Huts and Hard Calls"; Priest, " 'Team 555' Shaped a New Way of War"; Bacon, "To Hell and Back"; *Frontline* interviews; Moore, *The Hunt for Bin Laden*, 89–103 (with the usual caveats about this book's unreliability). For the account of the CIA officer who succeeded Schroen as leader of the Jawbreaker team, see Berntsen, *Jawbreaker*, ch. 10.

"*Karzai spoke English 'better than 50 percent of [my] guys'* " Lt. Col. David Fox, *Frontline*, "Campaign Against Terror: Interviews."

377 *Defense of Tarin Kowt* This account is based on the author's interview with Amerine; *Frontline*, "Campaign Against Terror: Interview: U.S. Special Forces ODA 574"; Briscoe et al., *Weapon of Choice*, 155–58; Finn, ". . . And His U.S. Partners."

"*Turning point*" *Frontline*, interview with Karzai, "The Campaign Against Terror."

Amerine coordinating air strikes Author's interview with Amerine.

Marines at Rhino See Reynolds, *Basrah, Baghdad, and Beyond*, 1–7, for the official account.

378 "*Mad Max*" The comparison comes from Lt. Col. David Fox (identified by the nom de guerre "Forsythe"), who was part of the caravan. Briscoe et al., *Weapon of Choice*, 173.

Bombing of ODA 574 Based on author's interview with Amerine; *Frontline*, interview with Amerine, "Campaign Against Terror: Interview: U.S. Special Forces ODA 574"; *Frontline*, interview with Lt. Col. David Fox; *Frontline*, interview with Hamid Karzai; Briscoe et al., *Weapon of Choice*, 179–82; Hendren and Reynolds, "The Untold War: The U.S. Bomb That Nearly Killed Karzai." For the PJs' viewpoint see Newman, "Masters of Invisibility."

"*Squat, bushy-haired man*" Anderson, "Letter from Afghanistan: After the Revolution."

379 *Sherzai's operation* Briscoe et al., *Weapon of Choice*, 165–71, 174–75, 178–79, 182–83.

Bin Laden at Tora Bora Franks wrote in an op-ed article, "War of Words," published during the presidential campaign in the fall of 2004, "We don't know to this day whether Mr. bin Laden was at Tora Bora in December 2001. Some intelligence sources said he was; others indicated he was in Pakistan at the time; still others suggested he was in Kashmir. Tora Bora was teeming with Taliban and Qaeda operatives, many of whom were killed or captured, but Mr. bin Laden was never within our grasp." However, Franks's deputy, General "Rifle" DeLong, wrote in his own memoir, *Inside Centcom*, 56, that bin Laden "was definitely there [at Tora Bora] when we hit the caves." A subsequent Pentagon investigation apparently confirmed DeLong's conclusion. See Weaver, "Lost at Tora Bora."

"*Failing to capture*" Fontenot et al., *On Point*, 25.

380 *Tora Bora* For interviews with local Afghans who took part in the operations see Bergen, "The Long Hunt for Osama"; Brown, "U.S. Lost Its Best Chance to Decimate Al-Qaida in Tora Bora"; Weaver, "Lost at Tora Bora." For a British viewpoint see Jennings, *Midnight in Some Burning Town*, 162–64. For the American viewpoint see Robinson, *Masters of Chaos*; Naylor, *Not a Good Day to Die*, 17–21; interview with ODA 572 in *Frontline*'s "Campaign Against Terror: Inter-

views"; Briscoe et al., *Weapon of Choice*, 213–16; DeLong, *Inside Centcom*, 55–56; Berntsen, *Jawbreaker*, chs. 16–20. While Briscoe et al. write that only two A-teams were at Tora Bora (ODAs 572 and 561), Robinson says that ODA 563 led by Chief Warrant Officer Randy Wurst also participated. The army's official history, *The United States Army in Afghanistan*, refers to "several SF teams" (26).

Failed to locate at least half of their hiding holes Biddle, *Afghanistan and the Future of Warfare*, 29.

Anaconda For a gripping narrative, see Naylor, *Not a Good Day to Die*. Although the official U.S. estimate of enemy dead was at least 800, Naylor estimates (p. 375) that the actual figure was between 150 and 300. For U.S. Special Operations Command's official version see Briscoe, et al., *Weapon of Choice*, 279–328.

Afghanistan political developments For summaries see Starr, "Silk Road to Success," and Fairbanks, "Afghanistan Reborn."

al-Harithi and Gray Fox Scarborough, *Rumsfeld's War*, 25.

381 "*Are we capturing, killing, or deterring*" www-hoover.stanford.edu/publications/books/fulltext/practical/xxv.pdf.

382 *Watch list* Paltrow, "Many Antiterror Recommendations Wither."

"*We had accomplished in eight weeks*" DeLong, *Inside Centcom*, 55.

Manpower Lambeth, *Air Power Against Terror*, 258, gives a figure of 316 Special Forces troops divided into eighteen A-Teams, four company-level teams, and three battalion-level teams. Berntsen, *Jawbreaker*, 308, writes that there were 110 CIA officers.

U.S. casualties All figures are from the Department of Defense: http://web1.whs.osd.mil/mmid/casualty/OEFDEATHS.pdf, http://web1.whs.osd.mil/mmid/casualty/oef_date_of_death_list.pdf.

100,000 Red Army troops and 15,000 fatalities Deitchman, "Military Force Transformation," 10.

Percentage of PGMs 57 percent of the total munitions used in Afghanistan were precision-guided, according to Cordesman, *The Ongoing Lessons of Afghanistan*, 28. Tirpak, "Enduring Freedom," and Lambeth, *Air Power Against Terror*, 249, give the figure as 60 percent.

A-Teams coordinated Northern Alliance This point was stressed in an e-mail to the author from Captain Dean, July 5, 2005.

383 "*If I was a conventional army guy*" Interview with Master Sergeant Paul in *Frontline*, "Campaign Against Terror: Interviews: U.S. Special Forces ODA 595."

"*Flintstones meet the Jetsons*" Captain Mark Nutsch, in *Frontline*, "Campaign Against Terror: Interview: U.S. Special Forces ODA 595."

"*Compliance instead of creativity*" Wong, *Stifling Innovation?* The author of this study was a retired lieutenant colonel employed by the Army War College's Strategic Studies Institute.

"*An extraordinarily wonderful amount of authority*" Author's interview with Dean.

CHAPTER 12: HUMVEES AND IEDs

385 *Cordon off the capital* For the original plan, see Gordon and Trainor, *Cobra II*, 79–80, 374–75.

Origins of "thunder run" Zucchino, *Thunder Run*, 13.

386 *"Are you fucking crazy"* Zucchino, *Thunder Run*, 6; Gordon and Trainor, *Cobra II*, 378.

First thunder run This account is based mainly on Zucchino's vivid narrative, *Thunder Run*, 6–65. See also Fontenot et al., *On Point*, 340–46; Gordon and Trainor, *Cobra II*, 374–89.

387 *"I don't think any of us had a dry eye"* Quoted in Fontenot et al., *On Point*, 346.

"We butchered the forces present at the airport" Zucchino, *Thunder Run*, 72; Fontenot et al., *On Point*, 347.

"In no uncertain terms" Interview with Perkins, Frontline, "The Invasion of Iraq: Interviews."

"We own Baghdad" Zucchino, *Thunder Run*, 131. For the comparison to seizing Washington see Fontenot et al., *On Point*, 354.

388 *"There was not a soldier on Curley"* Corporal Warren Hall, quoted in Fontenot et al., *On Point*, 366.

Second thunder run This account draws heavily on Zucchino, *Thunder Run*, 86–260; Fontenot et al., *On Point*, ch. 6; and Gordon and Trainor, *Cobra II*, 391–410.

Marines For their experiences see West and Smith, *The March Up*, by two veterans traveling with the 1st Marine Division, and Wright, *Generation Kill*, by a reporter embedded with a Force Reconnaissance platoon.

"Getting killed by idiotic technology" Quoted in Scott, " 'Space' at War."

389 *"America's interests in security"* www.whitehouse.gov/news/releases/2003/02/print/20030226-11.html. For the intellectual and political process that led to the invasion, see Packer, *The Assassins' Gate*, ch. 2.

"Out of date" Franks, *American Soldier*, 315. "It didn't account for": Ibid, 329.

"Iraq is still the most effective military power" Cordesman, "War against Iraq might not be a cake walk."

390 *Franks did not request a larger force* Although many military planners wanted a bigger force, in the end Franks declared himself satisfied with a ground force of 150,000, with more on the way if the invasion ran into trouble. Undersecretary of Defense Douglas Feith told the author, "No one ever came to Rumsfeld and asked for more troops and got turned down." Of course, this may have been because senior military officers were too intimidated by the defense secretary to make a request that they knew he would frown upon. Gordon and Trainor detail in *Cobra II* (see, e.g., at 4) how Rumsfeld pressured the generals to go with the minimal force possible. He also "micromanag[ed] the deployment process" (101).

"Several hundred thousand" and "off the mark" Packer, *The Assassins' Gate*, 114.

"Running start" Technically the plan was dubbed the Hybrid, a combination of two competing schemes called Generated Start and Running Start, but as Gordon and Trainor note in *Cobra II*, 89, "The Hybrid was essentially a larger version of the Running Start."

Fifteen-hour gap between air and ground operations Fontenot et al., *On Point*, 93; Reynolds, *Basrah, Baghdad, and Beyond*, 26.

"A revolutionary concept" Franks, *American Soldier*, 367. For more on the development of Oplan 1003V, see Hooker, *Shaping the Plan for Operation Iraqi Freedom*, and Gordon and Trainor, *Cobra II*, 3–117.

391 "*Speed and momentum*" Franks, *American Soldier*, 395.

Fifty air strikes Jehl and Schmitt, "Errors Are Seen in Early Attacks on Iraqi Leaders."

393 *Comparison with "left hook"* This point was made by Colonel Steven Rotkoff, deputy intelligence officer of the Combined Forces Land Component Command, who wrote in August 2003: " 'G before A' [ground before air] was this war's equivalent of the 'left hook' of DESERT STORM." Fontenot et al., *On Point*, 94.

Patriots Among their victims were a British Tornado fighter and a Navy FA-18C fighter, both shot down by accident. Cahlink, "Better 'Blue Force' Tracking."

Basra See Keegan, *The Iraq War*, 165–83; Murray and Scales, *The Iraq War*, 129–54.

394 *Fedayeen* For their origins and organization, see Woods, Lacey, and Murray, "Saddam's Delusions," and Gordon and Trainor, *Cobra II*, 61–62. The Fedayeen were so determined to resist in part because the penalty for failure was execution.

Battle of Nasiriyah Pritchard, *Ambush Alley*; Reynolds, *Basrah, Baghdad and Beyond*, 73–86.

Medina raid Gordon and Trainor, *Cobra II*, ch. 14; Fontenot, et al., *On Point*, 179–90.

"*The color of bile*" Wright, *Generation Kill*, 126.

"*The weather really sucked*" William Scott Wallace in *Frontline*, "The Invasion of Iraq."

"*We'd had no warning*" Franks, *American Soldier*, 486. Franks's chief Iraq intelligence analyst, Gregory Hooker, writes in *Shaping the Plan for Operation Iraqi Freedom*, 78, "Iraq's use of the security services to ambush coalition forces and interdict lines of communications was a surprising development."

395 *Commander of the coalition ground force* McKiernan was head of 3rd Army, which was cross-designated as Combined Forces Land Component Command. In Desert Storm, Schwarzkopf had not appointed a land forces commander; he had performed the job himself, for which he was subsequently criticized. Franks decided to appoint a land-component commander because he had to divide his attention between Afghanistan and Iraq.

82nd and 101st Divisions For their experiences see Atkinson, *In the Company of Soldiers*, by a reporter embedded with the 101st, and Zinsmeister, *Boots on the Ground*, by a reporter embedded with the 82nd.

Most important decision Fontenot et al., *On Point*, 245.

"*One of the fiercest, and most effective*" Franks, *American Soldier*, 503.

396 *79 percent in support of ground forces* U.S. Central Command Air Forces, "Operation Iraqi Freedom by the Numbers." This figure includes both killbox interdiction and close air support.

68 percent PGMs Ibid.

"*We were able to make decisions quite quickly*" Author's interview with Col. Chris Haave.

397 *Less than ten minutes* For instance, on April 9, 2003, a Predator found two tanks in a tree line. The imagery was instantly analyzed by intelligence experts back in the U.S. who were in contact with the CAOC in Saudi Arabia. Seven minutes later the tanks had been destroyed. See Hebert, "Operation Reachback."

Predator controllers several thousand miles away Newman, "The Joystick War," notes that roughly half of the fifteen Predators in the Middle East were controlled by operators in the U.S. However, maintenance crews still had to be in the region to keep them flying.

Red and blue The use of blue to designate friendly forces and red for the enemy went back to Prussian war games in the nineteenth century.

More bandwidth used Communications satellites alone provided 30 times more bandwidth in 2003 than in 1991. Vickers and Martinage, *The Revolution in War*, 115–16.

Communications problems at Nasiriyah West and Smith, *The March* Up, 41.

398 "*Every echelon found it nearly impossible*" Fontenot et al., *On Point*, 257.

Problems with digital communications Talbot, "How Technology Failed in Iraq"; Fontenot et al., *On Point*, 60–63, 174–76, 394–96; West, "Maneuver Warfare"; 3rd Infantry Division, "After Action Report," 2–5.

399 "*For about a mile and a half*" Interview with Lt. Col. Rock Marcone, from *Frontline*, "The Invasion of Iraq: Interviews." This account also draws on Fontenot et al., *On Point*, 283–99; Talbot, "How Technology Failed in Iraq"; Kelly, "Across the Euphrates" (Michael Kelly was a journalist embedded with TF 3-69; he died in a Humvee accident on the evening of April 3); Kitfield, *War & Destiny*, 200–3; Gordon and Trainor, *Cobra II*, 348–53.

Marines across the Tigris West and Smith, The March Up, 133–52.

"*Had Saddam by the balls*" Quoted in Kitfield, "Attack Always."

Fight for the airport Interview with Lt. Col. Rock Marcone, from Frontline, "The Invasion of Iraq: Interview"; Fontenot et al., *On Point*, 299–310; Kitfield, *War & Destiny*, 212–14. For his actions during this battle, Sergeant First Class Paul R. Smith became the first American since 1993 to be awarded the Medal of Honor, the nation's highest military decoration. It was a posthumous award. See Schmitt, "Medal of Honor to be Awarded to Soldier Killed in Iraq"; www.army.mil/medalofhonor/citation.

400 *Coalition achievement* "The coalition loss rate of fewer than one in 2,300 troops killed in action was among the lowest ever for major mechanized campaigns." Biddle, *Toppling Saddam*, 1.

Saddam's delusions For an overview, based on interviews with Iraqi officials and captured Iraqi documents, see Woods, Lacey and Murray, "Saddam's Delusions," and Gordon and Trainor, *Cobra II*, ch. 4.

Saddam got information from the Internet Duelfer, *Comprehensive Report*, 8.

The southern attack was a feint Interview with Lt. Gen. Raad Al-Hamdani, a Republican Guard commander, in *Frontline*, "The Invasion of Iraq: Interviews."

Not a serious threat Woods, Lacey, and Murray, "Saddam's Delusions."

"*Culture of lies*" Duelfer, *Comprehensive Report*, 11.

"*People were lying to Saddam*" Quoted in Branigan, "A Brief, Bitter War for Iraq's Military Officers."

Hinder coordination Woods, Lacey and Murray, "Saddam's Delusions."

"*Service interactions were broad*" Biddle, *Toppling Saddam*, 35.

401 *Looting* "Driving around Baghdad during the first week in April, I see handcarts, flatbed trucks, cars, mules, even baby strollers laden with choice items—gigantic vases, photocopies, TVs, silk carpets—from Saddam's palaces and the various luxurious outposts of the Baath Party ministries." Paul William Roberts, "Beyond Baghdad," *Harper's* (July 2003), in Willis (ed.), *Boots on the Ground*, 207.

"*We paid a big price for not stopping it*" Quoted in Wright and Ricks, "Bremer Criticizes Troop Levels."

Special Operations forces For their experiences, see Robinson, *Masters of Chaos*, and Veritas, the Journal of Army Special Operations History (Winter 2005).

"*By the time Baghdad was taken*" Shanker, "Rumsfeld Faults Turkey."

"*Never understood at a visceral level*" Author's interview with Todd Megill. In a similar vein, Ralph Peters writes in "In Praise of Attrition," "A number of the problems we have faced in the aftermath of Operation Iraqi Freedom arose because we tried to moderate the amount of destruction we inflicted on the Iraqi military. The only result was the rise of an Iraqi *Dolchstosslegende* [stab-in-the-back myth], the notion that they weren't really defeated, but betrayed."

Ratio of troops to peacekeepers Dobbins, *America's Role in Nation-Building*, 198.

20,000 to 25,000 mercenaries Bergner, "The Other Army"; Boot, "The Iraq War's Outsourcing Snafu."

402 "*While the U.S. could take Iraq with three divisions*" Quoted in Fineman, Wright and McManus, "Preparing for War, Stumbling to Peace." Bremer endorsed this analysis. According to *My Year in Iraq*, 9–10, he sent a copy of Dobbins's report to Rumsfeld in April 2003 but never heard back from the defense secretary. Bremer further says (*My Year in Iraq*, 106) that in July he told National Security Adviser Condoleezza Rice that "the Coalition's got about half the number of soldiers we need here." Bremer's predecessor, Jay Garner, also said, "The gut problem is the force is too small." See Bergner, "The Other Army."

"*It was never really fleshed out*" Author's interview in the summer of 2005 with an official who requested anonymity. This tallies with the comments of military planners, such as Lt. Col. Michael F. Morris of the 1st Marine Expeditionary Force, who told the author, "No one spent a lot of time focusing on Phase IV." Likewise, the official Marine Corps history—Reynolds, *Basrah, Baghdad and Beyond*, 42—states that Phase IV "received very little attention." Franks, for his part, concedes (*American Soldier*, 525): "We had neither the money nor a comprehensive set of policy decisions that would provide for every aspect of reconstruction, civic action, and governance." Gregory Hooker, Franks's chief Iraq intelligence analyst, writes in *Shaping the Plan for Operation Iraqi Freedom*, 36–37, that in Centcom's view "the sustained effort for relief and reconstruction fell outside the sphere of CENTCOM." For more on Phase IV planning—or the lack thereof—see Packer, *The Assassins' Gate*, ch. 4, and Gordon and Trainor, *Cobra II*, ch. 8. It was subsequently claimed in some media accounts that the State Department's Future of Iraq Project had produced a postwar plan that was ignored by the Defense Department, but as Gordon and Trainor note in *Cobra II*, 159, the State Department study was "of uneven quality" and "far short of a viable plan."

Assumed there would be a functioning government Author's interview with Douglas Feith. See also Gordon and Trainor, *Cobra II*, 145.

403 *Disbanding the Iraqi military* Gordon, "Debate Lingering on Decision to Dissolve the Iraqi Military"; Gordon and Trainor, *Cobra II*, 479–485. For Bremer's justification of his directives, see Bremer, *My Year in Iraq*, 39–60.

No idea how run-down everything was Author's interview with Douglas Feith. See also Gordon and Trainor, *Cobra II*, 150, which quotes Maj. Gen. Carl Strock, who served on Jay Garner's team: "I had a dramatic underestimation of the condition of the Iraqi infrastructure, which turned out to be one of our biggest problems." For private, prewar estimates of Iraq reconstruction needs, see Djerejian and Wisner, "Guiding Principles for U.S. Post-Conflict Policy in Iraq," 12.

Only $2 billion spent by December 2004 Boot, "The Struggle to Transform the Military." According to Miller, "Violence Trumps Rebuilding in Iraq," security costs were estimated to eat up 43 percent of reconstruction funds. For more on contracting woes, see Miller, "Projects in Iraq to Be Reevaluated."

Phase IV This has been the subject of a voluminous literature. For contrasting views, see Lowry, "What Went Wrong" (written from a conservative, pro-administration perspective), and Fallows, "Blind into Baghdad" (written from a liberal, anti-administration perspective).

404 "*The terrorists took all the explosives*" Brigadier Abdul Kadir Moniem Said, quoted in Kerkstra, "Shadowy Insurgency in Iraq Is Built on Homemade Bombs."

M-14 making plans for guerrilla war Duelfer, "Comprehensive Report"; Shanker, "Hussein's Agents Are Behind Attacks in Iraq, Pentagon Finds"; Nordland et al., "Unmasking the Insurgents"; Pound, "Seeds of Chaos"; Bremer, *My Year in Iraq*, 126–27.

Failure to anticipate the insurgency Hooker, Centcom's senior intelligence analyst for Iraq, writes in *Shaping the Plan for Operation Iraqi Freedom*, 38, that "the military intelligence estimates for Phase IV were lacking in one important respect: their underestimation of the insurgency that targeted the coalition." The overall U.S. intelligence community made the same error. According to Gordon, "Catastrophic Success: Poor Intelligence Misled Troops About Risk of Drawn-Out War": "The National Intelligence Council, senior experts from the intelligence community, prepared an analysis in January 2003 on postwar Iraq that discussed the risk of an insurgency in the last paragraph of its 38-page assessment. 'There was never a build-up of intelligence that says: "It's coming. It's coming. It's coming. This is the end you should prepare for," ' said Gen. Tommy R. Franks, the former head of the United States Central Command and now retired, referring to the insurgency." One of the few experts to warn about the danger of "protracted guerrilla war against the occupation (or liberation)" was retired Marine Colonel Gary Anderson, who published an article in the *Washington Post* on April 4, 2003, "Baghdad's Fall May Not be the End."

Changing guerrilla tactics Ricks, "U.S. Adopts Aggressive Tactics on Iraqi Fighters."

"*Headed off*" Quoted in Gordon and Trainor, *Cobra II*, 494–495.

405 "*Classic, guerrilla-type campaign*" Scarborough, "Enemy force uses 'guerrilla tactics' in Iraq, Abizaid says."

Massagetae ambush Asprey, *War in the Shadows*, 6.

Insurgents don't show up very well Schmitt and Shanker, "The Conflict in Iraq: Estimates by U.S. See More Rebels with More Funds."

406 "*Can't Provide Anything*" Boot, "Reconstructing Iraq." Reynolds, *Basrah, Baghdad and Beyond*, 257, notes: "Criticism of ORHA and CPA was almost universal among Marines interviewed by [Marine] field historians."

"*Little potentates*" Conway, " 'Farther and Faster' in Iraq."

Arabic speakers In 2000 the U.S. military had 2,581 Arabic speakers, a figure that had gone up to just 2,864 by the end of 2004. Jaffe, "Rumsfeld's Push for Speed Fuels Pentagon Dissent." In March 2004, a congressional study found that the CIA and other spy agencies had only 30 percent of the capacity they needed in Arabic, Persian, Pashto, Urdu, and other critical languages. Jehl, "Better at Languages, U.S. Spy Agencies Still Lag."

Peacekeeping Institute to close Carter, "The Crucible."

Counterinsurgency doctrine The Army's interim Field Manual 3-07.22 ("Counterinsurgency Operations") came out in October 2004, eighteen months after the fall of Baghdad. Field Manual 3.0 ("Operations"), issued in June 2001 and in effect in 2003, devoted only one page to counterinsurgency, and that focused primarily on the support that U.S. forces could provide to foreign governments. Crane, "Avoiding Vietnam," details the army's resistance to thinking about counterinsurgency doctrine in the wake of Vietnam.

West Point introduced counterinsurgency class in 2005 Boot, "The Struggle to Transform the Military." West Point did address counterinsurgency in parts of previous classes.

Most soldiers did not train in peacekeeping The 3rd Infantry Division's "After Action Report" acknowledges (at 17–18) that the division "did not have a fully developed plan for the transition to SASO [stability and support operations] and civil military operations in Baghdad prior to entering the city. . . . The division as a whole did not focus on CMO [civil military operations] training prior to the beginning of combat operations, instead focusing on mid-intensity operations."

Disdain for peacekeeping and nation-building In a 2000 debate with Vice President Al Gore, candidate Bush said, "I don't think our troops ought to be used for what's called nation building" ("2nd Presidential Debate Between Gov. Bush and Vice President Gore"). In a speech at The Citadel in 1999, Bush said, "We will not be permanent peacekeepers, dividing warring parties. This is not our strength or our calling" (http://citadel.edu/r3/pao/addresses/pres_bush.html).

"*We're doing a lot of things that we didn't know we were getting into*" Lt. Col. William Rabena, commander of the 2nd Battalion, 3rd Field Artillery Regiment, cited in Labbé, "Trained for War, U.S. Soldiers Learn to Keep the Peace." In a similar vein, Captain Michael Kirkpatrick, who commanded a company in the 2nd Armored Cavalry Regiment in Iraq in 2003, told the author, "We had trained strictly on high-intensity conflict."

"*OIF requires junior leaders*" Wong, *Developing Adaptive Leaders*, 4.

407 *Peninsula Strike, Desert Scorpion, etc.* Donnelly, *Operation Iraqi Freedom*, 85–98; Thomas, Nordland, Caryl, "Operation Hearts & Minds."

"*We made things worse for ourselves*" Author's interview with Mike Iverson. In a similar vein, Marine Colonel T. X. Hammes writes in *The Sling and the Stone*, 187, "Repeated humiliation of Iraqis in their own homes and at checkpoints is turning many neutrals to the ACF [anti-coalition forces]."

"*If I were treated like this*" Quoted in Aylwin-Foster, "Changing the Army for Counterinsurgency Operations," 3. The author, a British brigadier who served with U.S. forces in Iraq in 2004, concluded that the U.S. Army "was too 'kinetic.'"

Hearts and minds For an account of such programs in northern and southern Iraq in August 2003, see Boot, "Reconstructing Iraq."

Lack of electricity According to O'Hanlon and de Albuquerque, "Iraq Index," even by February 2005 Iraq still had not reached the same level of electricity production as before the war. See also Murphy and Sebti, "Power Grid in Iraq Far from Fixed." For an account of early struggles to generate electricity, see King, "Race to Get Lights on in Iraq Shows Perils of Reconstruction."

408 *Danger at U.S. bases* In addition to the danger of the mortar and rocket rounds, a suicide bomber managed to sneak into a U.S. base in Mosul. Twenty-four people, including nineteen U.S. soldiers, died in the explosion on December 21, 2004.

"*Little oases*" Brigadier General Karl Horst, quoted in Zucchino, "At the Front."

More than one hundred bases Graham, "Commanders Plan Eventual Consolidation of U.S. Bases in Iraq."

Soldiers in between raids Boot, "Hurry Up and Wait." Based on the author's visits to Iraq.

"*Little America embedded in the heart of Baghdad*" Langewiesche, "Welcome to the Green Zone." For more on the Green Zone, see Marshall, "Operation Limited Freedom"; Diamond, *Squandered Victory*, 74–76, 93–98.

IED casualties Graham and Priest, "Iraqis Using U.S. Techniques," says IEDs accounted for roughly 50 percent of U.S. casualties. Hendren, "Bombs Spur Drive for Vehicle Hardier Than a Humvee," says the figure is 70 percent.

409 *Eighty Abramses disabled* Komarow, "Tanks Take a Beating in Iraq."

One of eight armored Gordon, "Troops seen vulnerable in Humvees."

"*We're kind of sitting ducks*" Lieutenant Colonel Vincent Montera, cited in Gordon, "Troops seen vulnerable in Humvees."

"*Hillbilly armor*" Hirsh, Barry, Dehghanpisheh, "Hillbilly Armor."

Even armored vehicles could be blown up Cloud, "Insurgents Using Bigger, More Lethal Bombs, U.S. Officers Say," reports on the appearance of powerful "shaped" explosive charges capable of blowing up a 25-ton Marine armored amphibious assault vehicle.

Shortages of rifles, night-vision goggles, etc. Naylor, "Ready or Not."

Shortage of small-caliber ammunition Merle, "Running Low on Ammo."

410 *History of body armor* Dunstan, *Flak Jackets*.

Development of Kevlar Lok, "Life Vest"; Swartz and Iwata, "Invented to save gas, Kevlar now saves lives."

Interceptor body armor www.globalsecurity.org/military/systems/ground/interceptor.htm.

Medical care Gawande, "Casualties of War"; Kaplan, "Survivor: Iraq—America's Near-Invisible Wounded"; Washburn, "Advances Make War More Survivable."

Advances in prosthetics Roon, "Science Quickens Its Steps."

Delays in fielding body armor Moss, "Many Missteps Tied to Delay of Armor to Protect Soldiers," offers a comprehensive account.

411 *Toy-car controller to jam IEDs* Ricks, "Soldiers Record Lessons from Iraq."

Robotic bomb inspection and disposal Lubold, "Remote-Control Robot Debuts in War"; Roosevelt, "Checkpoints in Iraq, Afghanistan Get New Inspection Technology."

"*No technology silver bullet*" Lt. Col. Ernie Benner, quoted in Baum, "Battle Lessons."

Insurgents responded to jammers Hendren, "Bombs Spur Drive for Vehicle Hardier Than a Humvee"; Cloud, "Iraqi Rebels Refine Bomb Skills, Pushing Toll of G.I.'s Higher."

Pentagon had trouble keeping up Jaffe, "Pentagon Procurement Slows Supply Flow."

30 to 40 percent of IEDs defused Graham and Priest, "Iraqis Using U.S. Techniques."

"It's a constant struggle of one-upsmanship" Cited in Wong, *Developing Adaptive Leaders*, 11.

Sixty- to ninety-day rotations for CIA officers Gellman and Linzer, "Afghanistan, Iraq: Two Wars Collide."

Convoy Skills Trainer Based on the author's visit to the trainer at Fort Hood, Texas, on April 14, 2005. For more see Sevastopulo, "Games software and plywood harden US troops for Iraq battle."

412 *Training for the 1st Cavalry Division* Author's interview with Colonel Keith Walker.

Joint Readiness Training Center Based on the author's visit to Fort Polk, April 15–16, 2005.

Cavnet Author's interview with Keith Walker.

Companycommand.com, platoonleader.org Baum, "Battle Lessons."

Diagram Sipress, "Hunt for Hussein Led U.S. to Insurgent Hub."

413 *Bureaucratic hindrance* In "Changing the Army for Counterinsurgency Operations," British Brigadier Aylwin-Foster, who worked with U.S. forces in Iraq in 2004, noted a trend toward "micro-management, with many hours devoted to daily briefings and updates."

Dozens of different databases Grossman, "U.S Forces in Iraq Face Obstacles in Getting Intelligence They Need."

"We are a hierarchy" Schoomaker did, however, add, "But I also believe we are getting better." See Jaffe, "As Chaos Mounts in Iraq, U.S. Army Rethinks Its Future."

Killing civilians at checkpoints "Daily reports compiled by Western security companies chronicle many incidents in which Iraqis with no apparent connection to the insurgency are killed or wounded by American troops who have opened fire on suspicion that the Iraqis were engaged in a terrorist attack." Burns, "U.S. Checkpoints Raise Ire in Iraq."

"We're really good at teaching privates" Author's interview with Ken Tovo.

"No evidence" Schlesinger et al., *Final Report of the Independent Panel*, 5.

"A failure to plan for a major insurgency" Ibid, 11.

414 *Decision to invade Fallujah* West, *No True Glory*, 5–7.

Offensive called off Bremer, *My Year in Iraq*, 334–35; West, *No True Glory*, 119–23; Diamond, *Squandered Victory*, 233–36.

Information war Nabil Khatib of the Al-Arabiya satellite television channel said, "To my surprise, the opposition is doing better, P.R.-wise, than the official Americans and Iraqis, who are not as readily available for comment to give their side as the opposition. The militants are ready with a video of masked men and a person available for comment a half-hour after the story breaks." Shapiro, "The War Inside the Arab Newsroom."

"Al Jazeera kicked our butts" Quoted in West, *No True Glory*, 322.

415 *Fall of Fallujah* For a compelling account see West, *No True Glory*, 53–221.

Battle against Al Mahdi militia Berenson, "Fighting the Old-Fashioned Way in Najaf"; Berenson, "After the Siege, a City in Ruins"; Reitman, "Apocalypse Now."

Fallujah retaken West, *No True Glory*, 253–317; Ware, "Into the Hot Zone"; Lasseter, "U.S. may have won, but at a great personal cost"; Shanker and Schmitt, "Past Battles Won and Lost Helped in Falluja Assault"; Spinner, "Medics Testify to Fallujah's Horrors"; Filkins, "In Falluja, Young Marines Saw the Savagery of an Urban Fight."

"*Ghost fighters*" Marine Corporal Glenn Hamby, quoted in McDonnell, "No Shortages of Fighters in Iraq's Wild West."

Speeding up the process to self-rule Bremer, *My Year in Iraq*, 186–90; Lowry, "What Went Right."

416 *$148 billion* The figure comes from Belasco, "The Cost of Operations in Iraq, Afghanistan, and Enhanced Security." By fiscal year 2006, the cost had risen to about $192 billion. This was more than the cost of the Revolutionary War, Mexican War, Civil War, and Spanish American War combined, but only one-third the cost, in real terms, of the Vietnam War. See Daggett, Nowels, et al., "FY2004 Supplemental Appropriations for Iraq, Afghanistan, and the Global War on Terrorism."

Strain on all-volunteer force For an overview see Tyson, "Two Years Later, Iraq War Drains Military."

"*Broken force*" Graham, "General Says Army Reserve Is Becoming a 'Broken' Force."

Casualties http://icasualties.org/oif/.

417 *D-Day casualties* Clodfelter, *Warfare and Armed Conflicts*, 524. During World War II the U.S. lost an average of three hundred soldiers a day. During the occupation of Iraq, the figure was roughly two soldiers a day. Boot, "History Can Offer Bush Hope."

U.S. personnel in Iraq The Defense Department does not provide separate figures for Afghanistan and Iraq. But as of November 2004, according to information provided to the author by the Pentagon, 955,609 personnel had served in Afghanistan or the Persian Gulf region since September 11, 2001—the vast majority in Iraq.

Urban Tactical Planner Lipton, "3-D Maps from Commercial Satellites Guide G.I.'s in Iraq's Deadliest Urban Mazes."

JSTARS Fulghum, "A Night Over Iraq," offers an account of a typical JSTARS mission in the spring of 2005.

Over seven hundred UAVs Schmitt, "Remotely Controlled Aircraft Crowd Dangerous Iraqi and Afghan Skies."

"*95 percent of our intelligence*" Author's interview with Mattis.

THE CONSEQUENCES OF THE INFORMATION REVOLUTION

419 "*Basic propulsion systems*" O'Hanlon, *Technological Change and the Future of Warfare*, 71.

Ship speed hasn't increased *Arleigh Burke*–class destroyers, which first went operational in 1989, have a top speed of 30 knots. The top speed of the USS *Bainbridge*, a destroyer commissioned in 1902, was 29 knots. O'Hanlon, *Technological Change and the Future of Warfare*, 80.

420 *Nine thousand M1s* Paarlberg, "Knowledge as Power."

Armor advances Composite armor, also known as Chobham armor after the facility in Britain where it was first made, combines steel and ceramics to provide more protection than steel alone. Reactive armor, pioneered by Israeli researchers, explodes when struck by a high-explosive projectile, destroying it before it can penetrate the tank's armor. Freidman, *The Future of War*, 132–35.

Advantage shifted against armor O'Hanlon, *Technological Change and the Future of Warfare*, 81–82, notes, "Even as tanks keep getting better, overall trends appear to favor the weapons shooting at them, which have been proving lethal at increasingly long ranges for decades."

FCS This description is based on a briefing provided to the author on May 11, 2004, by FCS's civilian and military project managers, Dennis Muilenberg of Boeing and Brigadier General Donald F. Schenk of the U.S. Army, and on the author's attendance at an FCS "technology demonstration" held at Fort Aberdeen, Maryland, on September 21, 2005.

"*Perfect situational awareness*" Retired Army Colonel Douglas Macgregor told Congress: "Perfect situational awareness, the key underlying assumption of the Army's future combat system, is an illusion, or perhaps a delusion. . . . Timely and useful information is important, but it cannot substitute for firepower, mobility, and armored protection." For other critiques of FCS, see McMaster, "Crack in the Foundation"; Kucera, "Iraq Conflict Raises Doubts on FCS Survivability"; Kagan, "War and Aftermath."

421 *XM29* Shachtman, "When a Gun Is More Than a Gun"; Mihm, "The Search for the Nonkiller App."; www.globalsecurity.org/military/systems/ground/m29-oicw.htm.

Electronic guns This "revolutionary new weapons technology"—the words of a DARPA assessment—has been developed by Metal Storm, an Australian company that has received Pentagon funding. See Matus, "Metal Storm: Rise of the Machines"; Adams, "Is This What War Will Come To?"

250 million small arms Posen, "Command of the Commons," 31.

1,249 suppliers in ninety countries Graduate Institute of International Studies, *Small Arms Yearbook.*

The U.S. edge decreases "With the possible exceptions of night-vision devices, Global Positioning Systems, and shoulder-fired missiles," writes retired Major General Robert H. Scales, Jr., a former commander of the Army War College, in "Urban Warfare: A Soldier's View," "there is no appreciable technological advantage for an American infantryman when fighting the close battle against even the poorest, most primitive enemy." And night vision devices, GPS, and shoulder-fired missiles are rapidly proliferating around the world.

U.S. aircraft carriers The fleet is likely to shrink momentarily to eleven with the retirement of the oil-fired *John F. Kennedy*.

422 *Aircraft carriers* Baker, "World Navies in Review"; International Institute for Strategic Studies, *The Military Balance, 2004–2005.*

Aegis can track nine hundred targets Briefing by Captain James M. Carr aboard the Aegis cruiser *Anzio*, April 12, 2005.

Supercavitating torpedo By using a gas ejector to create an air bubble around the torpedo to reduce water drag, it can go 200 mph as opposed to traditional torpedoes, which go 30–40 mph. The Soviets invented the concept and the U.S. Navy is now in the process of developing its own, improved version. Adams, "Is This What War Will Come To?"

Submarine forces International Institute for Strategic Studies, *The Military Balance, 2004–2005*, 24–25, 105; Baker, "World Navies in Review." Of France's nuclear subs, four are ballistic boats and six are attack boats. Of Britain's fifteen, four are ballistic boats and eleven are attack subs, though the latter figure is projected to shrink to six to eight by 2012. It should be noted that because the U.S. is not building enough submarines to replace aging *Los Angeles*–class boats the number of attack submarines in its fleet is expected to decline to 40 by 2028. Lerman, "Submarine Numbers Shrinking Gradually."

"Fastest, quietest, and most heavily armed" Paarlberg, "Knowledge as Power."

FORCEnet Polmar, "Composing Command and Control."

423 *75,000 antiship missiles* O'Hanlon, *Technological Change and the Future of Warfare*, 102.

Ballistic antiship missiles For their resurgence, see Polmar, "Antiship Ballistic Missiles . . . Again."

Antiship cruise missiles Vickers and Martinage, *The Revolution in War*, 88–92.

Danger of sinking fuel tankers Vickers and Martinage, *The Revolution in War*, 87–88.

Diesel submarines The newest German model, sold for export, can stay submerged for at least two weeks and costs only a tenth of a nuclear submarine. See Thibo, "U-Boat!"

Diesel submarines Vickers and Martinage, *The Revolution in War*, 92–96; Baker, "World Navies in Review"; Goldstein and Murray, "Undersea Dragons."

Mines Vickers and Martinage, *The Revolution in War*, 96–98; Posen, "Command of the Commons," 38.

424 *Width of Persian Gulf and Taiwan Strait* From *Encyclopaedia Britannica* (Britannica.com).

DD(X) For a promotional overview, see Goddard and Marks, "DD(X) Navigates Uncharted Waters."

Littoral Combat Ship Wilson, "Stealth Strike Force."

Sea bases For an overview, see Defense Science Board Task Force on Sea Basing, "Final Report."

425 *No aces since 1972* The number of American aces has been in steady decline, from 708 in World War II, to 39 in the Korean War, 3 in the Vietnam War, and none since (although in the Gulf War 2 F-15C pilots scored three kills apiece). It is interesting to note that the last ace, Captain Jeffrey Feinstein, was not a pilot but a weapons officer aboard an F-4; he achieved his kills by firing AIM-7 missiles. Grant, "The Missing Aces."

Visual stealth Fulghum, "Disappearing Act."

Sensors can detect stealth planes Vickers and Martinage, *The Future of War*, 112.

Serbs shot down an F-117 Posen, "Command of the Commons," 26.

Range of 23,000 feet *Small Arms Survey*, 82.

Shoulder-fired missiles Vickers and Martinage, *The Future of War*, 190–92; www.globalsecurity.org/military/intro/manpads.htm.

426 *50 pounds of explosives* This is the Small Diameter Bomb which weighs 250 pounds in all. See Tirpak, "Precision: The Next Generation."

Sensor-Fuzed Weapon and Tactical Tomahawk Vickers and Martinage, *The Revolution in War*, 20–21.

U.S. long-range bombers Dudney, "Long-Range Strike in Two Jumps."

Eighty aim points per sortie Dudney, "Long-Range Strike in Two Jumps"; "Quadrennial Defense Review Report," 46.

Transport and cargo aircraft Russia has 20 tanker and 318 cargo aircraft; China has 10 tanker and 513 cargo aircraft. Figures are from International Institute for Strategic Studies, *The Military Balance, 2004–2005*: 27, 29 (for U.S. figures which combine the Navy, Marine Corps, and Air Force), 108 (Russia), 172 (China).

Increase in JSTARS Posen, "Command of the Commons," 15.

Not enough to go around Hebert, "It Means 'We Didn't Buy Enough.'"

U.S. satellites in orbit Figures from Posen, "Command of the Commons," 12. O'Hanlon, *Neither Star Wars Nor Sanctuary*, 42, gives a lower figure—60 U.S. military satellites operational in 2004.

U.S. spends 90 percent of world total O'Hanlon, *Neither Star Wars Nor Sanctuary*, 5. But only about a third of all satellite launches recently have been from the U.S. One-third have been from Russia and one-third from the rest of the world. Ibid, 35.

Pick out a six-inch object O'Hanlon, *Technological Change and the Future of Warfare*, 34. This is an estimate for Keyhole imaging satellites, which can work at day or night but cannot penetrate cloud cover. Lacrosse or Onyx systems that utilize radar imaging can work in all kinds of weather. They can reportedly distinguish objects three to nine feet across. Satellite capabilities are strictly classified; these are only informed guesses. See O'Hanlon, *Neither Star Wars Nor Sanctuary*, 43.

Stealth satellite Priest, "New Spy Satellite Debated on Hill."

427 *$30-billion-a-year industry* Grier, "The Sensational Signal."

At least forty countries with satellites O'Hanlon, *Neither Star Wars Nor Sanctuary*, 37. In addition, various multinational organizations such as the Asia Satellite Corp., Arab Satellite Communications Organization, International Telecom Satellite Organization, and European Space Agency have launched their own satellites.

1.5 feet wide Rumsfeld et al., *Report of the Commission to Assess United States National Security Space Management and Organization*, 35.

"*Their own reconnaissance satellites*" Quoted in Vickers and Martinage, *The Revolution in War*, 117–18. For more on ImageSat, see www.imagesatintl.com.

U.S. reliant on private satellites O'Hanlon, *Neither Star Wars Nor Sanctuary*, 4. Learmonth, "Military use boosts commercial satellite companies," reported that, as of May 25, 2004, an estimated 84 percent of U.S. military traffic in the Middle East was carried on commercial satellites.

Missile proliferation For a summary of nations with existing programs, see www.fas.org/irp/threat/missile/summary.htm. In "Foreign Missile Developments and the Ballistic Missile Threat Through 2015," the National Intelligence Council estimated that at least a dozen nations would have land-attack cruise missiles by 2015.

Ballistic missile defense For an overview from the U.S. Missile Defense Agency, see www.mda.mil/mdalink/pdf/bmdsbook.pdf.

Cruise missile defenses The U.S. Air Force is working on networked sensors that will pick out cruise missiles and relay their coordinates to stealthy, supersonic F-22 or F-35 fighters which could knock them down with AIM-120 air-to-air missiles. See Fulghum, "New Threat, New Defense," "Cruise Missile Battle," "Testing Ground," "Net Changes."

428 *Limitations of sensors* For a discussion of this problem see Biddle, *Military Power*, 52–77.

Global Information Grid "Quadrennial Defense Review Report," 58–59; Weiner, "Pentagon Envisioning a Costly Internet for War," which quotes Robert J. Stevens, CEO of Lockheed Martin Corp., speaking of a "God's-eye view."

Radar that sees through walls or foliage Vickers and Martinage, *The Revolution in War,* 110–12; O'Hanlon, *Technological Change and the Future of Warfare*, 40.

Better communications The U.S. Advanced Extremely High Frequency Satellite Communications System, due to be launched into geosynchronous orbit (i.e., 22,300 miles above the earth) starting in 2007, has five times the communications capacity of the existing Milstar (Military Strategic and Tactical Relay) system. O'Hanlon, *Neither Star Wars Nor Sanctuary,* 44. Also under developments are high-capacity laser relays that can transfer over one gigabit per second—more than the entire bandwidth used in the Afghanistan War—or even more. O'Hanlon, *Neither Star Wars Nor Sanctuary*, 81–82.

"In the palm of your hand" Quoted in Ratner and Ratner, *Nanotechnology and Homeland Security*, 81.

"Lifting the fog of war" This is the title of a book by retired Admiral Bill Owens, a former vice chairman of the Joint Chiefs of Staff (1994–96) and a leading proponent of information technology.

429 *Masters of the commons* I am indebted for this point to Posen, "Command of the Commons."

"The strongest the world has ever known" Easterbrook, "American Power Moves Beyond the Mere Super."

Size of military forces All figures from the International Institute for Strategic Studies, *The Military Balance, 2004–2005*, 353–58.

430 *Israel's security revolution* See Cohen, Eisenstadt, and Bacevich, *Knives, Tanks, & Missiles.*

Soviets unable to keep up with U.S. A 1988 U.S. Defense Department report (Trulock, Hines, Herr, "Soviet Military Thought in Transition," 118) concluded: "It seems evident that by the time General Secretary Gorbachev assumed leadership of the Soviet Union, the Soviet military was concerned that their control over the long-term military competition with the West was seriously in danger. Western perceptions of Soviet military programs had motivated particularly the United States to attempt to exploit its vast scientific-technical potential to overturn Soviet military advantages accumulated during the late 1960s and throughout the 1970s." For more on the effect of the U.S.-led military technical revolution on the Soviet Union, see various writings by Mary FitzGerald listed in the Bibliography.

"An off season" Gray, "How Has War Changed Since the End of the Cold War?" In a similar vein, political scientist Joshua Goldstein writes in "The Worldwide Lull in War": "We probably owe this lull [in wars] to the end of the cold war, and to a unipolar world order with a single superpower to impose its will in places like Kuwait, Serbia, and Afghanistan. The emerging world order is

not exactly benign—Sept. 11 comes to mind—and Pax Americana delivers neither justice nor harmony to the corners of the earth. But a unipolar world is inherently more peaceful than the bipolar one where two superpowers fueled rival armies around the world." For a general examination of why the incidence of conventional war has declined, see Mueller, *The Remnants of War*, ch. 9, which stresses growing "aversion to war" across the world.

431 *Costs of submarines, carriers, fighters* See Weiner, "A Vast Arms Buildup, Yet Not Enough for Wars"; Weiner, "Navy's Fleet of Tomorrow Is Mired in Politics of Yesterday"; Spiegel, "Aerial combat."

U.S. spends almost as much as the rest of the world U.S. defense spending in 2003 accounted for 41 percent of the global total, but the U.S. defense budget has increased since then, according to data derived from the International Institute for Strategic Studies, *The Military Balance 2004–2005*, 353, 358. The Stockholm International Peace Research Institute, which conducts a competing survey of global military spending using slightly different methodology, reports that the U.S. spent 47 percent of the global total in 2004, followed by Japan at 5 percent and, at 4 percent each, Britain, France, and China. http://yearbook2005.sipri.org/ch8/ch8.

U.S. R&D spending Paarlberg, "Knowledge as Power," 132; O'Hanlon, *Defense Strategy for the Post-Saddam Era*, 18.

Defense budget These figures are from the International Institute for Strategic Studies, *The Military Balance, 2004–2005*, 353–55. Estimates for China's military spending vary from $30 billion to $90 billion. Beijing does not publish reliable budget figures.

Relying less on draftees Williams, "From Conscripts to Volunteers" and "Draft Lessons from Europe."

"*Neither copy nor steal*" Paarlberg, "Knowledge as Power."

Challenging U.S. power The CIA's National Intelligence Council concluded in "Mapping the Global Future" (at 17): "While no single country looks within striking distance of rivaling US military power by 2020, more countries will be in a position to make the United States pay a heavy price for any military action they oppose."

"*The survivability of aircraft carriers*" Vickers and Martinage, *The Revolution in War*, 75. In a similar vein, Friedman, *The Future of War*, 375, contends, "The ability of conventional weapons platforms—tanks and aircraft carriers—to survive in a world of precision-guided munitions is dubious."

432 "*Render ineffective*" Quoted in Vickers and Martinage, *The Revolution in War*, 72.

"*Swarming*" John Arquilla and David Ronfeldt (inventors of the term "netwar"), posit, in *Swarming & the Future of Conflict*, 45, that future wars will be fought by small military units known as "pods" that operate in "clusters" and use "swarming" tactics—"a seemingly amorphous, but deliberately structured, coordinated, and strategic way to strike from all directions."

"*They no longer will exist*" Bevin, *How Wars Are Won*, 13–14.

433 *Irregular tactics* In *A History of Warfare*, 5, Keegan writes of "the endemic warfare of non-state, even pre-state peoples, in which there was no distinction between lawful and unlawful bearers of arms, since all males were warriors; a form of warfare which had prevailed during long periods of human history."

100 bullets a minute The AK-47's theoretical capacity is 600 rounds a minute, but 100 rounds a minute is the practical maximum, because a typical magazine holds only 30 rounds. Gander, *Jane's Infantry Weapons*, 136.

Suitcase nukes In 1997 Russian General Alexander Lebed claimed that the Soviet Union had suitcase nukes, and that a hundred of them had gone missing. The U.S. has also worked on such projects in the past. Allison, *Nuclear Terrorism*, 43–46.

"American Hiroshima" Coll, "What Bin Laden Sees in Hiroshima"; Allison, *Nuclear Terrorism*, 3.

"It is not a matter of if" Eugene Habiger, quoted in Allison, *Nuclear Terrorism*, 6. For an overview of highly porous U.S. defenses, see Gorman and Freedberg, "Early Warning."

Using IT to fight terrorism For a summary see Huber and Mills, "How Technology Can Defeat Terrorism."

Microchips in cargo containers Machalaba and Pazstor, "Thinking Inside the Box: Shipping Containers Get 'Smart'"; Keeton, "Sensors on Containers May Offer Safer Shipping."

Picking out terrorists in a crowd Feder, "Technology Strains to Find Menace in the Crowd."

434 *Odor recognition* Wade, "Exotic Military Arts."

Computer programs Poletti, "Defense Dept. Hopes to Enlist AI in War Against Terrorism."

High-tech sensors For instance, Lawrence Livermore National Laboratory developed the Autonomous Pathogen Detection System, a $200,000 machine that scans the air for 95 separate agents ranging from anthrax to plague. Kirkpatrick, "The Anti-Doomsday Machine." But security systems don't always deliver what their makers promise. See Lipton, "U.S. to Spend Billions More to Alter Security Systems."

Phraselator Carter, "Tomorrow's Soldier Today."

Remedies For an overview of some proposals, see Boot, "The Struggle to Transform the Military." For other ideas see Scales, *Yellow Smoke;* Scales, "Urban Warfare: A Soldier's View"; Graham, "Military to Stress Foreign Languages"; Vandergriff, *The Path to Victory.*

Divisional structure In 2004, army chief of staff General Peter Schoomaker announced that henceforth, "modular," self-sustaining brigades, rather than divisions, would be the "basic units of action." Macgregor argues in "Army Transformation" that the new modular brigades will not have sufficient resources for independent action.

Organizational structure There have been some changes. The air force, for instance, has organized Aerospace Expeditionary Forces to make it easier to rotate units abroad. And the navy has transformed Carrier Battle Groups and Amphibious Ready Groups into smaller Carrier Strike Groups and Expeditionary Strike Groups that allow for more deployments.

Eighteen months per duty station Vandergriff, *The Path to Victory*, 120.

435 *Time to produce new weapons systems increased* Rumsfeld, "Why Defense Must Change." For instance, the F-15 took six years to develop in the 1970s, whereas its successor, the F/A-22, took fourteen years, according to "U.S. Air Force's New Chief Eyes Speeding Up Acquisition."

"Shorten the time frame" Fountain, "Transforming Defense Basic Research Strategy." For more on procurement reform, see Owens and Offley, *Lifting the Fog of War*, 23, 203.

"The Defense Department gets a D" Quoted in Weiner, "Arms Fiascos Lead to Alarm Inside Pentagon."

Tooth-to-tail ratio Figures from Owens and Offley, *Lifting the Fog of War*, 45.

"An astounding 70 percent" Weiss and Owens, "An Indefensible Military Budget."

"Up the chain of command" Hammes, *The Sling and the Stone*, 192–93.

"Dinosauric, vertical bureaucracy" Kaplan, *Imperial Grunts*, 227.

PART V: REVOLUTIONS PAST, PRESENT, FUTURE

CHAPTER 13: REVOLUTIONS TO COME

437 *"Victors don't learn"* Quoted in Keller, "The Fighting Next Time."

440 *DD(X) controlled by remote sensors* Goddard and Marks, "DD(X) Navigates Uncharted Waters." The DD(X) will carry fewer than half the number of crew members of older *Spruance*-class destroyers.

At least forty countries Fairbank, "Unmanned Aircraft Are Wowing Defense Industry."

UAVs For a survey see Bone and Bolkcom, "Unmanned Aerial Vehicles," and Defense Science Board, "Study on Unmanned Aerial Vehicles."

Predator Initially the RQ-1 Predator was jury-rigged with two Hellfire missiles. A more advanced model, the MQ-9A Predator B, can fly higher and faster and carry more ordnance. Pincus, "Predator to See More Combat."

"Flat as a pancake" and "kite-shaped" Brzezinski, "The Unmanned Army."

Combat UAVs Hebert, "New Horizons for Combat UAVs."

Fire Scout Keels, "A Different Kind of Pick-Me-Up."

Futuristic UAVs Vickers and Martinage, *The Revolution in War*, 157–64.

441 *LOCAAS* www.globalsecurity.org/military/systems/munitions/locaas.htm; Tirpak, "Precision: The Next Generation."

Small UAVs O'Hanlon, *Technological Change and the Future of Warfare*, 37, 76; Vickers and Martinage, *The Revolution in War*, 41, 161.

"Some are as big as a backhoe" Matthews, "Rise of the Military Robot."

Armed Talon Regan, "Army Prepares 'Robo Soldier' for Iraq"; Shachtman, "More Robot Grunts Ready for Duty."

DARPA race For the 2005 results, see Chang, "Stanford Volkswagen Wins $2M Robot Race." For 2004, see Pae, "Pentagon's Robot Race Stalls in Gate."

442 *Unmanned Ground Vehicles* Reed, "Robots Eyed for Urban Warfare"; Swartz, "New Breed of Robots Takes War to Next Level"; Regan, "Army Prepares 'Robo Soldier' for Iraq." Also based on the author's visit to the Future Combat System Technology Demonstration at Fort Aberdeen, Maryland, September 21, 2005.

Exoskeleton Garreau, *Radical Evolution*, 5; Vickers and Martinage, *The Revolution in War*, 167–68; Carter, "Tomorrow's Soldier Today"; www.berkeley.edu/news/media/releases/2004/03/03_exo.shtml.

Naval robots Vickers and Martinage, *The Revolution in War*, 43–45.

Much cheaper Each X-45 is expected to cost $15 million to $20 million, about one-third the cost of an F-22 or F-35, according to Brzezinski, "The Unmanned Army."

Ratcheting up spending Elias, "Pentagon Invests in Unmanned 'Trauma Pod'," reports that the Defense Department is expected to spend over $10 billion on UAVs between 2000 and 2010, up from $3 billion in the previous decade. For more on UAV spending, see Karp, "Smarter People, Weapons and Networks."

One-third must be unmanned "Sen. Warner Plans to Renew Push for Unmanned Vehicles."

Dull, dirty, dangerous Toner, "Robots Far from Leading the Fight."

443 *"Space Pearl Harbor"* Rumsfeld et al., *Report of the Commission to Assess United States National Security Space Management and Organization*, xv.

Electromagnetic pulse This could cause even more damage if exploded within the earth's atmosphere. A single nuclear bomb exploded three hundred miles above North America would disrupt and destroy electronic and electrical systems across the U.S., Canada, and Mexico. See the report of the Commission to Assess the Threat to the United States from Electromagnetic Pulse Attack: http://empcreport.ida.org.

Pellet cloud DeBloise et al., "Space Weapons: Crossing the U.S. Rubicon," 60–61.

Jamming Vickers and Martinage, *The Revolution in War*, 126–27.

Mobile satellite-jamming teams Gertz, "U.S. Deploys Warfare Unit to Jam Enemy Satellites."

Mini-satellites O'Hanlon, *Neither Star Wars nor Sanctuary*, 86.

Space debris from blowing up a satellite DeBloise et al., "Space Weapons: Crossing the U.S. Rubicon," 64.

444 *Small satellites* In 2003 the U.S. Air Force launched the 62-pound XSS-10 (Experimental Satellite System); in 2005 it followed up with a 305-pound XSS-11. www.designation-systems.net/dusrm/app3/ss-10.html; www.designation-systems.net/dusrm/app3/xss-11.html. According to O'Hanlon, *Neither Star Wars Nor Sanctuary*, 86, the U.S. is also said to have secretly launched a miniature satellite weighing less than two pounds.

British university launched a satellite O'Hanlon, *Neither Star Wars Nor Sanctuary*, 86–87.

Metal shield DeBloise et al., "Space Weapons: Crossing the U.S. Rubicon," 59.

"Rods from God" DeBloise et al., "Space Weapons: Crossing the U.S. Rubicon," 70; Weiner, "Air Force Seeks Bush's Approval for Space Weapons Programs"; Adams, "Is This What War Will Come To?"

Common Aero Vehicle and space plane Pincus, "Pentagon Has Far-Reaching Defense Spacecraft in Works"; Vickers and Martinage, *The Revolution in War*, 179; Tirpak, "In Search of Spaceplanes."

"Space superiority" U.S. Air Force, "Counterspace Operations," i.

445 *Space Force* The Rumsfeld Commission discussed the possibility of creating a Space Corps within the Air Force as a way station toward the creation of a Space Department led by its own Assistant Secretary of Defense for Space.

Zeus laser Gourley, "Zeus-Humvee Laser Ordnance Neutralization System."

Solid-state lasers Figures on their capabilities, and the need for at least 50 to 100 kilowatts, come from Lamberson et al., "Whither High-Energy Lasers?"

Airborne Laser Tirpak, "Setting a Course for the Airborne Laser"; O'Hanlon, *Neither Star Wars nor Sanctuary*, 73–77; Beason, *The E-Bomb*, 127–60.

Tactical High-Energy Laser Wilson, "Beyond Bullets"; Goure, "The U.S. Army Meets Star Wars"; Spencer and Carafano, "The Use of Directed-Energy Weapons to Protect Critical Infrastructure"; Beason, *The E-Bomb*, 11–12, 171–72.

446 *160,000 times faster than a bullet* Beason, *The E-Bomb*, 17.

1/1000th of a second Ibid.

Cost of Tactical High Energy Laser vs. Patriot 3 Spencer and Carafano, "The Use of Directed-Energy Weapons to Protect Critical Infrastructure."

Lasers Defense Science Board, "Task Force on High-Energy Laser Weapons Systems Applications"; Lamberson, et al., "Whither High-Energy Lasers?"; DeBloise et al., "Space Weapons: Crossing the U.S. Rubicon"; MacRae, "The Promise and Problem of Laser Weapons."

Active Denial System Mihm, "The Search for the Nonkiller App."; www.globalsecurity.org/military/systems/ground/v-mads.htm; Goure, "The U.S. Army Meets Star Wars"; Regan, "Pentagon Looks to Directed-Energy Weapons"; Gordon, "Beam Burns into the Future"; Beason, *The E-Bomb*, 113–25.

Long-Range Acoustic Device www.cnn.com/2004/TECH/ptech/03/03/sonic.weapon.ap; Braiker, "Master Blaster"; Arkin, "The Pentagon's Secret Scream."

447 *Above 150 decibels* Alexander, *Future War*, 97.

Electromagnetic guns Adams, "Is This What War Will Come To?"; Koch, "Electro-Magnetic Railguns."

"Reset pump speeds" Reed, *At the Abyss*, 265.

448 *Fictional scenarios* These are all from Alexander, *Future War*, 103, except for the mention of *24*, which comes from Bank and Richmond, "Where the Dangers Are."

"Start a war in a computer room" Qiao and Wang, *Unrestricted Warfare*, 32.

Thirty countries developing cyberattack Echevarria, "Globalization and the Nature of War," 4.

DARPA developed firewalls Tether, "Statement Submitted."

U.S. military computers attacked According to Brookes, "The Art of (Cyber) War," in 2004 U.S. Defense Department computer networks were attacked 79,000 times.

Some hackers succeed in disrupting networks For instance, a disgruntled British computer engineer hacked into almost one hundred U.S. military computers and managed to render inoperable for three days the computer system of the Military District of Washington. Webb, "Briton Arrested in Military Hacking." Meanwhile, "extremely well organized and well structured" Asian hackers have attacked critical financial and government infrastructure in Britain and the U.S., according to Pesola, "Asian hackers bombard UK government with attacks on vital financial networks."

Infrastructure computers under attack Bank and Richmond, "Where the Dangers Are."

Helpless against a customized Trojan horse This is a paraphrase of a sentence in Bank and Richmond, "Where the Dangers Are."

Electromagnetic pulse warheads in the Gulf War Tofflers, *War and Anti-War*, 175.

E-bomb Beason, *The E-Bomb*; Vickers and Martinage, *The Revolution in War*, 137; www.globalsecurity.org/military/systems/munitions/hpm.htm.

Radio-frequency cannon Vickers and Martinage, *The Revolution in War*, 102.

A $500 radio-frequency weapon Vickers and Martinage, *The Revolution in War*, 193.

"The smarter the weapon" Quoted in Beason, *The E-Bomb*, 48.

449 *Nanotech battlesuit* Based on the author's visit to the Institute for Soldier Nanotechnologies, November 12, 2003, as well as Ratner and Ratner, *Nanotechnology and Homeland Security*, 49–55; McLaughlin, "The quest to create a futuristic battle suit, one micron at a time"; web.mit.edu/isn/aboutisn/index.html.

Superheroes I am grateful for the analogy to Garreau, *Radical Evolution*, 20–21.

Definition and applications of nanotech Ratner and Ratner, *Nanotechnology and Homeland Security*, 13–14; Loder, "Small Wonders"; Weiss, "For Science, Nanotech Poses Big Unknowns" and "Applications Abound for Unique Physical, Chemical Properties."

Nanotech's importance For instance, retired admiral David Jeremiah says, "Military applications of molecular manufacturing have even greater potential than nuclear weapons to radically change the balance of power." Quoted in Ratner and Ratner, *Nanotechnology and Homeland Security*, 29.

450 *Nanotech spending* Weiss, "Nanotech Is Booming Biggest in U.S., Report Says."

Monkey controlling robot arms Garreau, *Radical Evolution*, 19–20.

"More than two millennia ago" Miller, Engelberg, Broad, *Germs*, 38–39.

451 *Heydrich assassination* Harris and Paxman, *A Higher Form of Killing*, 90–96.

Killing thousands of Chinese civilians Miller, Engelberg, Broad, *Germs*, 40.

Building on Japanese research Ibid, 40, 49.

"Twice as toxic as cobra venom" Ratner and Ratner, *Nanotechnology and Homeland Defense*, 35.

Markov's assassination Hamilton and Walker, "Dane named as umbrella killer"; Alibek, *Biohazard*, 173–74.

"Yellow rain" Miller, Engelberg, Broad, *Germs*, 78, 93.

Chemical weapons in World War I Tucker, *War of Nerves*, 20. Harris and Paxman, *A Higher Form of Killing*, 34, gives slightly different figures. No exact accounting exists.

Interwar uses of chemicals Harris and Paxman, *A Higher Form of Killing*, 46, 50–52.

Discovery of nerve gas Tucker, *War of Nerves*, ch. 2; Harris and Paxman, *A Higher Form of Killing*, 55.

452 *"To abstain from"* Harris and Paxman, *A Higher Form of Killing*, 7.

Chemical weapons in Vietnam Harris and Paxman, *A Higher Form of Killing*, 193–99.

Iraqi use of chemicals Harris and Paxman, *A Higher Form of Killing*, 241–42; Tucker, *War of Nerves*, 255–59, 269–73, 279–86.

Aum Shinrikyo attacks Tucker, *War of Nerves*, 326–50; Harris and Paxman, *A Higher Form of Killing*, 250–51.

The use of germs Millions of Indians were killed in the New World by the arrival of Old World germs, but the bulk of the evidence suggests that almost all of these plagues were accidental—not deliberate biological warfare.

Seventy nanograms Office of Technology Assessment, "Technologies Underlying Weapons of Mass Destruction," 80. Keep in mind, however, that such estimates are based on limited knowledge. MIT's Macfarlone cautions, in "Assessing the Threat," that "the data are too thin to make accurate projections of the effect of bioweapons attacks."

A pound could kill a billion Miller, Engelberg, Broad, *Germs*, 39.

Ricin Office of Technology Assessment, "Technologies Underlying Weapons of Mass Destruction," 181.

220 pounds of anthrax Koblentz, "Pathogens as Weapons," 88.

Pentagon experiment to manufacture anthrax Miller, Engelberg, Broad, *Germs*, 297–99.

Dissemination of biological weapons Koblentz, "Pathogens as Weapons."

Al Qaeda and biological weapons Vickers and Martinage, *The Revolution in War*, 196.

Seventeen countries Echevarria, "Globalization and the Nature of War," 3.

Soviet superbugs Miller, Engelberg, Broad, *Germs*, 94–95, 136–37, 300–5; Alibek, *Biohazard*; Vickers and Martinage, *The Revolution in War*, 140–41; interview with former Soviet scientist Sergei Popov in *Frontline*, "Bioterror."

453 *"Brainpox"* Miller, Engelberg, Broad, *Germs*, 225.

Crops and livestock targeted An outbreak of foot and mouth disease in Taiwan in 1997 took $4 billion to eradicate and cost an estimated $15 billion in lost exports. The cost of bovine spongiform encephalopathy (mad cow disease), which broke out in Britain in the 1990s, is estimated at more than $10 billion. Vickers and Martinage, *The Revolution in War*, 146.

Antimaterial uses of biological and chemical weapons Alexander, *Future War*, 71–76, 119–23.

Gene vaccine Miller, Engelberg, Broad, *Germs*, 306; Garreau, *Radical Evolution*, 30–31.

"*Energizer bunny in fatigues*" Quoted in Garreau, *Radical Evolution*, 32.

454 "*Key to survival and operational dominance*" Quoted in Garreau, *Radical Evolution*, 22. This section on super-soldiers is based on Garreau's ch. 2, "Be All You Can Be."

"*Be all you can be and a lot more*" Quoted in Garreau, *Radical Evolution*, 32.

EPILOGUE

456 *Yom Kippur War* For a short summary of its lessons, see Bolia, "Overreliance on Technology in Warfare."

Ottoman success as an obstacle to change See Ralston, *Importing the European Army*, ch. 3.

Failure of minicomputer makers Christensen, *The Innovator's Dilemma*, xiii. For the failure of sailing ship makers, see ibid, 86. For more on business complacency, see Witzel, "When a company's success breeds failure."

Sony's struggles Stross, "How the iPod Ran Circles Around the Walkman"; Nakamoto, "Caught in its own trap: Sony battles to make headway in the networked world."

457 *"Innovations differ"* Wilson, *Bureaucracy*, 227. Rosen, *Winning the Next War*, 243, agrees that "uncertainties about the enemy and about the costs and benefits of new technologies make it impossible to identify the single best route to innovation."

Maiman and laser McMartin, "Laser's creator will be beaming at party," quotes Maiman as saying, "It was almost a bootleg project for me. They [Hughes] tried to pull funding from me twice."

458 *"We can no more explain"* Mokyr, *The Lever of Riches*, viii.

Discount the importance of technology Economists Rotte and Schmidt argue in "On the Production of Victory: Empirical Determinants of Battlefield Success in Modern War" that "technology is . . . a negligible factor" in determining the outcome of battles. But they base this conclusion on an analysis of a database that seems to consist mainly of battles between industrialized nations. Certainly technology was not "negligible" in determining the outcome at Omdurman or lots of other imperial clashes.

German soldiers killed Parker, "From the House of Orange to the House of Bush."

459 *"Sustaining" vs. "disruptive" inventions* See Christensen, *The Innovator's Dilemma*. For an application of this concept to military affairs, see Pierce, *Warfighting and Disruptive Technologies*.

460 *Influence of civilians* Rosen, *Winning the Next War*, 255, finds that the "role civilian political leaders and scientists have had in the initiation and management of military innovation" has been "relatively minor."

Kennedy's lack of success Krepinevich, *The Army and Vietnam*.

"To the extent that civilians can control" Cohen, "Defending America in the Twenty-First Century." In a similar vein, Christensen notes that senior decision-makers in a business are often under the illusion that they are calling the shots whereas, whether they realize it or not, "many of the really critical resource allocation decisions have actually been made long before senior management gets involved: Middle managers have made their decisions about which projects they'll back and carry to senior management—and which they will allow to languish." Christensen, *The Innovator's Dilemma*, 95.

462 *Democracies spend less on the military* Ferguson, *Cash Nexus*, 405.

Democratic soldiers don't fight better Reiter and Stam, *Democracies at War*, 4, argue that "democracies' emphasis on individuals and their concomitant rights and privileges produces better leaders and soldiers more willing to take the initiative on the battlefield." But Brooks, "Making Military Might," points out many flaws in Reiter and Stam's argument.

German prowess See Van Creveld, *Fighting Power*, and Dupuy, *A Genius for War*. For a contrasting perspective, which emphasizes American combat effectiveness and German deficiencies, see Mansoor, *The GI Offensive in Europe*.

Similarities between armies George Orwell made this point: "Discipline, for instance, is ultimately the same in all armies. Orders have to be obeyed and enforced by punishment if necessary, the relationship of officer and man has to be the relationship of superior and inferior." Orwell and Angus, *The Collected Essays, Journalism and Letters of George Orwell*, II:250. Of course discipline and punishment is usually far harsher in the armies of authoritarian states than in those of democracies.

463 *"The will of the people"* Reiter and Stams, *Democracies at War*, 4.

DARPA For an internal history highlighting its achievements see www.darpa.mil/body/pdf/transition.pdf.

Realizing a military revolution This is a point confirmed by numerous other analysts. Retired U.S. Army Colonel Douglas Macgregor writes in *Transformation Under Fire*, 12: "Revolutionary change exists only at the intersection of organizational, doctrinal, personnel, cultural and technological change." Rand analysts John Arquilla and John Ronfeldt write in *Swarming & the Future of Conflict*, 4–5: "Technology matters, yes, but so does the form of organization that is adopted or developed to embrace it. A proper organizational form can empower the technology, as, for example, the creation of the panzer division allowed blitzkrieg to flourish. Today, the key form of organization on the rise is the network, especially the all-channel network."

464 *Libya-Chad war* See Pollack, *Arabs at War*, 375–412.

"The incompetence of Arab tactical leadership" Pollack, *Arabs at War*, 581.

465 *"In no profession"* Colonel John Mitchell, quoted in Rosen, *Winning the Next War*, 2.

Discrimination in the Imperial Germany Navy Holger W. Herwig, "The Battlefleet Revolution, 1885–1914," in Knox and Murray, *The Dynamics of Military Revolution*, 123–24.

Line versus staff officers Hagan, *The People's Navy*, 117–18.

UAV operators and the air force Shanker, "Rewarding Skill in a Mission with Few Thrills."

"A pasty-faced scholar" Qiao and Wang, *Unrestricted Warfare*, 32.

"The cultural challenge" Cohen, "A Revolution in Warfare."

466 *Militaries too eager to change* For more on this point see Biddle, *Military Power*, 198–99, 206.

467 *European strategy won't work* General James Jones, the NATO commander, warned in June 2005, "Even in countries where they are trying to call it [defense] transformation, some of them are not just reducing their forces but also their budgets. . . . That is not transformation in my mind." Quoted in Dombey, "US Nato chief chides Europeans over budgets." For more on how Britain, the leading military power in Europe, is trying to justify downsizing as "transformation," see Evans, "US-Style Military Reforms Will Cut British Firepower," and Studemann, "UK shakes up military in shift toward rapid reaction."

Revolutionary change through evolutionary increments Williamson Murray and MacGregor Knox, "The Future Behind Us," in Knox and Murray, *The Dynamics of Military Revolution*, 185, note: "The most successful organizations avoided wild leaps into the future; their innovations remained tied to past experience, derived from conceptually sophisticated and honestly assessed experiments, and depended on the ability to learn from both success and failure."

A little cutting-edge technology can go a long way This is a point Andy Marshall made in an interview with the author. Bill Keller in "The Fighting Next Time" paraphrases Marshall as follows: "He talks in terms of changing 10 to 15 percent of the force from old to new."

"How to make a transition" Author's interview with Marshall.

468 *War games* Retired Marine Colonel T. X. Hammes notes in *The Sling and the Stone*, 236, "Unfortunately, today, instead of realistic, free-play exercises, we use mostly scripted exercises with no chance of losing."

U.S. hegemony endangered The Hart-Rudman Commission on U.S. National Security in the 21st Century, which issued its report on February 15, 2001, made "recapitalizing" U.S. scientific and engineering expertise its second most urgent priority. www.au.af.mil/au/awc/awcgate/nssg/ phaseIIIfr.pdf.

38 percent of Ph.D.'s are foreign born Segal, "Is America Losing Its Edge?"

Post-9/11 visa restrictions Between 2001 and 2003 the number of successful visa applicants to the U.S. fell from 10 million to 6.5 million. The number of temporary worker visas for science and technology jobs fell even faster. Paarlberg, "Knowledge as Power," 146.

"Silent Sputnik" For instance, Zakaria, "Rejecting the Next Bill Gates," warns that "the foreign visa crisis" if "left unattended" is "going to have deep and lasting effects on American security and competitiveness." See also Segal, "Is America Losing Its Edge?"

It doesn't take a lot of money to innovate On this point see Rosen, *Winning the Next War*, 252.

469 *Cost per person doubled* Deitchman, "Military Force Transformation," 6. Britain, France and Germany continue to spend no more per person than the U.S. spent in 1970.

"Money, more money" King Louis XII's adviser, quoted in Rice and Grafton, *Foundations of Early Modern Europe*, 118; Hale, *War and Society in Renaissance Europe*, 232.

"Burkina Faso or Paraguay" Hasim, "The Revolution in Military Affairs Outside the West."

"God is on the side of the big battalions" Quoted in Biddle, *Military Power*, 14.

40 percent greater GNP and 170 percent greater population Ferguson, *Cash Nexus*, 402.

470 *Seven Years' War* Kennedy, *The Rise and Fall of the Great Powers*, 113, notes "that the Anglo-Prussian combination remained superior in three vital aspects: leadership, financial staying power, and military/naval expertise."

Bigger side won only half the time Biddle, *Military Power*, 21.

"Superior numbers can be decisive or almost irrelevant" Ibid, 3. "How forces are used": Ibid, 5.

Making war "pay for itself" According to Gellman, "One Year Later: War's Faded Triumph," allies paid $52.4 billion of the $61 billion it cost the U.S. to fight the Gulf War. In *Cash Nexus*, 397–98, Niall Ferguson presents a litany of states that have succeeded in making war pay for itself. For instance, Britain managed to cover 40 percent of its defense budget for 1842 with the £5.8 million indemnity it extracted from China in the First Opium War; Prussia covered its defense budget four times over with the 5 billion francs it extracted from France in 1871; and Japan made China pay a sum amounting to double the cost of defeating it in 1895.

471 Obsolescence of war: For the most cogent forms of this argument see Mueller, *The Remnants of War*, and Easterbrook, "The End of War?"

"Only the dead have seen the end of war" Santayana, *Soliloquies in England*, 102.

Military analysts who stress guerrilla warfare See, e.g., Lind, "Understanding Fourth Generation War"; Peters, *Fighting for the Future* and *Beyond Terror*; Hammes, *The Sling and the Stone*; Van Creveld, *The Transformation of War*. Many of these analysts, led by Lind, maintain that guerrilla warfare represents the "fourth generation" of warfare, even though such tactics are as old as warfare itself. It is true, however, that modern technology has made terrorism more destructive. For a trenchant critique of the "fourth generation" approach, see Freedman, "War Evolves into the Fourth Generation."

Stress on high-tech weapons See, e.g., Owens, *Lifting the Fog of War*; Freidman, *The Future of War*; and Toffler, *War and Anti-War*.

472 *"Tame the big wildcats"* Evans, "From Kadesh to Kandahar."

473 *Unconventional strategies* Qiao and Wang, *Unrestricted Warfare*.

INDEX